Stored-Grain Ecosystems

BOOKS IN SOILS, PLANTS, AND THE ENVIRONMENT

Soil Biochemistry, Volume 1, edited by A. D. McLaren and G. H. Peterson
Soil Biochemistry, Volume 2, edited by A. D. McLaren and J. Skujiņš
Soil Biochemistry, Volume 3, edited by E. A. Paul and A. D. McLaren
Soil Biochemistry, Volume 4, edited by E. A. Paul and A. D. McLaren
Soil Biochemistry, Volume 5, edited by E. A. Paul and J. N. Ladd
Soil Biochemistry, Volume 6, edited by Jean-Marc Bollag and G. Stotzky
Soil Biochemistry, Volume 7, edited by G. Stotzky and Jean-Marc Bollag
Soil Biochemistry, Volume 8, edited by Jean-Marc Bollag and G. Stotzky

Organic Chemicals in the Soil Environment, Volumes 1 and 2, edited by C. A. I. Goring and J. W. Hamaker
Humic Substances in the Environment, M. Schnitzer and S. U. Khan
Microbial Life in the Soil: An Introduction, T. Hattori
Principles of Soil Chemistry, Kim H. Tan
Soil Analysis: Instrumental Techniques and Related Procedures, edited by Keith A. Smith
Soil Reclamation Processes: Microbiological Analyses and Applications, edited by Robert L. Tate III and Donald A. Klein
Symbiotic Nitrogen Fixation Technology, edited by Gerald H. Elkan
Soil-Water Interactions: Mechanisms and Applications, edited by Shingo Iwata, Toshio Tabuchi, and Benno P. Warkentin
Soil Analysis: Modern Instrumental Techniques, Second Edition, edited by Keith A. Smith
Soil Analysis: Physical Methods, edited by Keith A. Smith and Chris E. Mullins
Growth and Mineral Nutrition of Field Crops, N. K. Fageria, V. C. Baligar, and Charles Allan Jones
Semiarid Lands and Deserts: Soil Resource and Reclamation, edited by J. Skujiņš
Plant Roots: The Hidden Half, edited by Yoav Waisel, Amram Eshel, and Uzi Kafkafi
Plant Biochemical Regulators, edited by Harold W. Gausman
Maximizing Crop Yields, N. K. Fageria
Transgenic Plants: Fundamentals and Applications, edited by Andrew Hiatt
Soil Microbial Ecology: Applications in Agricultural and Environmental Management, edited by F. Blaine Metting, Jr.
Principles of Soil Chemistry: Second Edition, Kim H. Tan

Water Flow in Soils, edited by Tsuyoshi Miyazaki
Handbook of Plant and Crop Stress, edited by Mohammad Pessarakli
Genetic Improvement of Field Crops, edited by Gustavo A. Slafer
Agricultural Field Experiments: Design and Analysis, Roger G. Petersen
Environmental Soil Science, Kim H. Tan
Mechanisms of Plant Growth and Improved Productivity: Modern Approaches, edited by Amarjit S. Basra
Selenium in the Environment, edited by W. T. Frankenberger, Jr., and Sally Benson
Plant–Environment Interactions, edited by Robert E. Wilkinson
Handbook of Plant and Crop Physiology, edited by Mohammad Pessarakli
Handbook of Phytoalexin Metabolism and Action, edited by M. Daniel and R. P. Purkayastha
Soil–Water Interactions: Mechanisms and Applications, Second Edition, Revised and Expanded, Shingo Iwata, Toshio Tabuchi, and Benno P. Warkentin
Stored-Grain Ecosystems, edited by Digvir S. Jayas, Noel D. G. White, and William E. Muir
Agrochemicals from Natural Products, edited by C. R. A. Godfrey

Additional Volumes in Preparation

Seed Development and Germination, edited by Jaime Kigel and Gad Galili

Nitrogen Fertilization in the Environment, edited by Peter Edward Bacon

Phytohormones in Soils: Microbial Production and Function, W. T. Frankenberger, Jr., and Muhammad Arshad

Stored-Grain Ecosystems

edited by

Digvir S. Jayas
University of Manitoba
Winnipeg, Manitoba, Canada

Noel D. G. White
Agriculture and Agri-Food Canada
Winnipeg, Manitoba, Canada

William E. Muir
University of Manitoba
Winnipeg, Manitoba, Canada

Marcel Dekker, Inc. New York • Basel • Hong Kong

Library of Congress Cataloging-in-Publication Data

Stored grain ecosystems / edited by Digvir S. Jayas, Noel D. G. White, William E. Muir.
p. cm. — (Books in soils, plants, and the environment)
Papers presented at an International Symposium on Stored Grain Ecosystems, held in Winnipeg, Manitoba, Canada, from June 7–10, 1992.
Includes bibliographical references and index.
ISBN 0-8247-8983-0 (acid-free paper)
1. Grain—Storage—Congresses. 2. Grain—Storage—Diseases and injuries—Congresses. I. Jayas, Digvir S. II. White, Noel D. G. III. Muir, William E.
IV. International Symposium on Stored Grain Ecosystems (1992 : Winnipeg, Man.) V. Series.
SB190.S87 1995
633.1'0468—dc20

94-22852
CIP

The publisher offers discounts on this book when ordered in bulk quantities. For more information, write to Special Sales/Professional Marketing at the address below.

This book is printed on acid-free paper.

Marcel Dekker, Inc.
270 Madison Avenue, New York, New York 10016

Current printing (last digit):
10 9 8 7 6 5 4 3 2 1

PRINTED IN THE UNITED STATES OF AMERICA

Preface

The problem of feeding the increasing world population has been treated by policy makers and researchers in two dimensions: world population and production of food materials. A third dimension to this problem is the protection of already produced grain. About 900 million tonnes of grain are currently in storage throughout the world at any given time, about one half of annual world production. Protection of this commodity is critical to human survival. Properly engineered and managed systems for preserving harvested grain provide wholesome food, free of insect damage, live insects and mites, insect fragments, microflora, mycotoxins, and pesticides; reduce the pressure on the environment by requiring less farmland; and reduce food-related illnesses in domestic animals and humans.

The International Symposium on Stored Grain Ecosystems was held in Winnepeg, Manitoba, Canada, June 7–10, 1992, to synthesize and integrate knowledge from several areas of grain storage research and to provide directions of future research for optimal management of stored grain in the 21st century. The symposium was sponsored by the Department of Agricultural Engineering, University of Manitoba, and the Winnipeg Research Centre of Agriculture and Agri-Food Canada. More than 190 participants from 23 countries attended the symposium, and the period during which it was held was proclaimed "Grain Storage Week" by both the Province of Manitoba and the City of Winnipeg.

The Winnipeg research group of biologists and engineers has led the world in studying stored grain bulks as ecosystems in which the living grain interacts with its living (e.g., insects, fungi) and nonliving (e.g., temperature) environment. The driving force of this group was Dr. Ranendra N. Sinha, an entomologist/ecologist with Agriculture and Agri-Food Canada. He prepared the first chapter of this book which sets the context for following chapters. A wide range of topics dealing with the application of biology, cereal chemistry, economics, engineering, mathematical modeling, and toxicology to stored-grain ecosystems is covered.

The 20 chapters of this book were written by the invited speakers at the symposium who have reviewed the scientific material in the context of stored grain ecosystems and who have set directions for future research. To obtain a high standard of presentation each chapter was sent for critical review to at least two competent people in the discipline of the author. The three book editors also reviewed each chapter and a final editing was done after each author had considered the reviewers' suggestions and revised the manuscript.

In the planning and preparation of this book we received help from many sources—from people who provided encouragement, helpful advice, critical reviews, and illustrative material, and from organizations that provided funds and secretarial assistance. We are grateful to them all. We especially want to thank the Symposium Organizing Committee; faculty, staff, and graduate students of the Department of Agricultural Engineering, University of Manitoba; and the researchers and technicians of the Agriculture and Agri-Food Canada Winnipeg Research Centre for offering their time and talents to make this project a success. Also, we thank our dedicated authors for committing their time and resources to this project. We thank Ms. Debby Watson for providing secretarial assistance during the symposium including the typing and formatting of many chapters of the book.

Finally, the editors wish to thank the following individuals for their thoughtful reviews of the material in this book: Dr. D. Abramson, Agriculture and Agri-Food Canada; Dr. P. C. Annis, CSIRO Division of Entomolgy; Dr. R. T. Arbogast, USDA-Agricultural Research Service; Mr. P. Bailey, Scottish Institute of Agricultural Engineering; Dr. F. W. Bakker-Arkema, Michigan State University; Dr. M. Calderon, Institute for Technology and Storage of Agricultural Products; Dr. B. Champ, CSIRO Division of Entomology; Dr. R. R. Cogburn, USDA-Agricultural Research Service; Dr. R. Corrigan, Purdue University; Dr. C. P. F. DeLima, Western Australian Department of Agriculture; Dr. F. Dunkel, Montana State University; Dr. M. Evans, University of Manitoba; Mr. O. H. Friesen, Manitoba Department of Highways; Dr. M. C. Gough, Natural Resources Institute; Dr. D. Hagstrum, U.S. Grain Marketing Research Laboratory; Dr. E. G. Jay, USDA-Agricultural Research Service; Dr. J. C. Jofriet, University of Guelph; Dr. H. Kawamoto, Noguchi Thyroid Clinic & Hospital Foundation; Ms. E. Larmond, Canadian Grain Commission; Dr. J. LeBars, Laboratorie de Pharmacologie-Toxicologie INRA; Mr. J. van Loon, Canadian Grain Commission; Dr. O. Lukow, Agriculture and Agri-Food Canada; Dr. N. Magan, Cranfield Institute of Technology; Dr. R. V. Morey, University of Minnesota; Professor E. B. Moysey, University of Saskatchewan; Dr. J. Mumford, Silwood Centre for Pest

Management; Dr. R. M. Peart, University of Florida; Dr. W. O. Pruitt, University of Manitoba; Dr. K. Richer, Institute for Tropical Agriculture; Mr. B. E. Ripp, CAFCO; Dr. I. J. Ross, University of Kentucky; Dr. D. B. Sauer, USDA-Agricultural Research Service; Dr. G. C. Shove, University of Illinois; Dr. R. N. Sinha, Agriculture and Agri-Food Canada; Dr. F. W. Sosulski, University of Saskatchewan; Dr. C. L. Storey, USDA-Agricultural Research Service; Dr. R. Stroshine, Purdue University; Dr. G. Thorpe, Victoria University of Technology; Dr. E. Tyrchniewicz, University of Alberta; Dr. P. Williams, Institute of Plant Sciences; Dr. V. F. Wright, International Center for Seed and Grain Science; and Dr. T. D. Yadav, Indian Agricultural Research Institute.

The Symposium was financially sponsored by two major contributors, the Canadian Agri-Food Development Initiative of Agriculture Canada and Prairie Pools Inc.; and by 13 other federal, provincial, and industrial organizations, including the faculty of Agricultural and Food Sciences and the faculty of Engineering at the University of Manitoba.

The Symposium on Stored Grain Ecosystems offered a unique opportunity for research, regulatory, and industry personnel to discuss grain storage practices, including prevention and control of pests and contaminants in stored grain, within an international forum. This focused book is a result of the symposium and will directly benefit a worldwide audience concerned with the practical and safe storage of an important food commodity.

Digvir S. Jayas
Noel D. G. White
William E. Muir

Contents

Contributors

Jonathan Banks, Ph.D. Head, Stored Grain Research Laboratory, CSIRO Division of Entomology, Canberra, Australia

Myron G. Britton, Ph.D. Professor, Agricultural Engineering Department, University of Manitoba, Winnipeg, Manitoba, Canada

David M. Bruce, B.Sc. Head, Biomaterials Group, Bio-Engineering Division, Silsoe Research Institute, Silsoe, Bedford, United Kingdom

Stefan Cenkowski, D.Sc., Ph.D., P.Eng. Associate Professor, Department of Agricultural Engineering, University of Manitoba, Winnipeg, Manitoba, Canada

Steven A. Clarke, M.S. Project Coordinator, International Agricultural Programs, University of Minnesota, St. Paul, Minnesota

Paul Fields, Ph.D. Research Scientist, Research Centre, Agriculture and Agri-Food Canada, Winnipeg, Manitoba, Canada

Paul W. Flinn, Ph.D. Research Biologist, United States Grain Marketing Research Laboratory, Agricultural Research Service, United States Department of Agriculture, Manhattan, Kansas

Jens C. Frisvad, Ph.D. Associate Professor, Department of Biotechnology, Technical University of Denmark, Lyngby, Denmark

Christopher P. Haines, Ph.D. Head, Food Storage Group, Natural Resources Institute, Chatham Maritime, Kent, United Kingdom

Phillip K. Harein, Ph.D. Consultant, Stored Product Pest Management, Inc., Crosslake, Minnesota

Digvir S. Jayas, Ph.D., P.Eng. Department of Agricultural Engineering, University of Manitoba, Winnipeg, Manitoba, Canada

Harvey B. Manbeck, Ph.D., P.Eng. Department of Agricultural Engineering, The Pennsylvania State University, University Park, Pennsylvania

F. Mazaud Senior Officer, Special Action Programme for the Prevention of Food Losses, Food and Agricultural Organization of the United Nations, Rome, Italy

J. T. Mills, D.I.C., Ph.D. Research Scientist, Research Centre, Agriculture and Agri-Food Canada, Winnipeg, Manitoba, Canada

Gerald W. Moore, B.S.A., M.S., P.Ag. Senior Project Manager, Economic Development Board of Cabinet, Government of Manitoba, Winnipeg, Manitoba, Canada

William E. Muir, Ph.D., P.Eng. Professor, Department of Agricultural Engineering, University of Manitoba, Winnipeg, Manitoba, Canada

Martin E. Nellist, Ph.D., C.Eng., F.I.Agr.E. Research Leader, Heat and Mass Flow Group, Process Engineering Division, Silsoe Research Institute, Silsoe, Bedford, United Kingdom

Anand Prakash, Ph.D. Senior Scientist, Department of Entomology, Central Rice Research Institute, Cuttack, Orissa, India

Virendra M. Puri, Ph.D. Professor, Department of Agricultural Engineering, The Pennsylvania State University, University Park, Pennsylvania

Jagadiswari Rao, Ph.D. Senior Scientist, Department of Entomology, Central Rice Research Institute, Cuttack, Orissa, India

Ranendra N. Sinha, Ph.D. Principal Research Scientist (Ecology), Research Centre, Agriculture and Agri-Food Canada, Winnipeg, Manitoba, Canada

Edward A. Smith, Ph.D. Senior Lecturer, Department of Mathematics and Statistics, University of Paisley, Paisley, United Kingdom

Robert H. Smith, Ph.D. Professor of Environmental Biology, School of Biological Sciences, University of Leicester, Leicester, United Kingdom

O. J. Sode Consultant and Storage Specialist, Food and Agricultural Organization of the United Nations, Rome, Italy

James E. Throne, Ph.D. Research Entomologist, Stored-Product Insects Research and Development Laboratory, Agricultural Research Service, United States Department of Agriculture, Savannah, Georgia

Keith H. Tipples, Ph.D. Director, Grain Research Laboratory, Canadian Grain Commission, Winnipeg, Manitoba, Canada

Francis Troude Agricultural Engineering Service, Agricultural Services Division, Food and Agricultural Organization of the United Nations, Rome, Italy

Noel D. G. White, Ph.D. Research Scientist and Section Head, Ecology of Field and Stored Product Pests Section, Research Centre, Agriculture and Agri-Food Canada, Winnipeg, Manitoba, Canada

Donald T. Wicklow, Ph.D. Lead Scientist, Mycotoxin Research Unit, National Center for Agricultural Utilization Research, Agricultural Research Service, United States Department of Agriculture, Peoria, Illinois

Qiang Zhang, Ph.D., P.Eng. Assistant Professor, Department of Agricultural Engineering, University of Manitoba, Winnipeg, Manitoba, Canada

Stored-Grain Ecosystems

1

The Stored-Grain Ecosystem

Ranendra N. Sinha

Agriculture and Agri-Food Canada, Winnipeg, Manitoba, Canada

SUMMARY

The stored-grain ecosystem is viewed as an agroecosystem that is part of the broader human-subsidized solar-powered ecosystem. Farm households in both the developing and developed countries of the world create and manage three interlinked agroecosystems: field crop, stored grain, and livestock ecosystems within the backdrop of socio-economic and environmental realities. The pattern of distribution, abundance, consumption, and deterioration of stored grain, and ways to protect it by proper management practices in developing and developed countries are described within a broad ecological context with specific examples from Ghana, Togo, Ethiopia, and Canada. Economic realities within which stored-grain ecosystems are developed are discussed. Methods of scientific study for rational use of stored-grain ecosystems and future directions in stored-grain ecosystem research and management are outlined.

1. INTRODUCTION

For thousands of years, humankind has thrived on this planet feeding its members primarily on the fruits of cultivated grass crops of the family Gramineae. The fruits are those of eight principal cereal crops—wheat, barley, rice, corn, oats, rye, sorghum, and millet. Production, storage, and distribution of these principal crops are crucial to the survival and continued progress of human

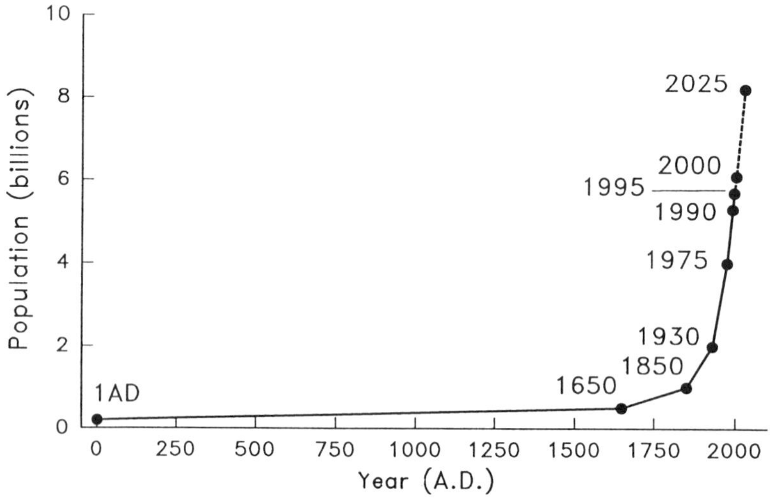

Figure 1 Growth of the world's human population since 1 AD. Broken line represents estimated projected increase. *Source*: World Almanac and Book of Facts 1991.

civilization. Grown with the aid of modern science and technology on approximately 5.5% (719 Mha) of the land surface of the earth, cereal plants have yielded increasing quantities of grain in the later part of the 20th century from 1.8 t/ha in 1969–71 to 2.3 t/ha in 1981 (Kent 1983). Through the centuries, cereal carbohydrate has been the main fuel energy for running the human machine—worldwide averages of energy yield are 460 kJ/ha from wheat, 737 kJ/ha from rice, and 837 kJ/ha from corn (Kent 1983).

On one side of the ledger, growing cereals and protecting them until used for seeding or consumption have been major preoccupations for humans since the inception of agriculture about 10,000 years ago. On the other side, the human population which consumes cereal food was estimated to be about 200 million in 1 A.D., as compared with 5.3 billion in 1990, and projections of 6.1 billion in 2000 and 8.1 billion in 2025 (Fig. 1, World Almanac and Book of Facts 1991). In the last 25 years, technological advancement and the need to feed rapidly increasing populations have prompted many countries to produce large quantities of cereal grains, although in developed countries the dependence on cereals for food energy has decreased. Nevertheless, populations of 53 developing countries still derive 40% of all food energy from cereals; the total production of cereal crops, rose from 13.9 Gt in 1974–76 to 18.6 Gt in 1989 (World Almanac and Book of Facts 1991) (Fig. 2). Global food production has scarcely kept pace with the galloping increase in population which seems to have exceeded the biological and environmental carrying capacity of earth's

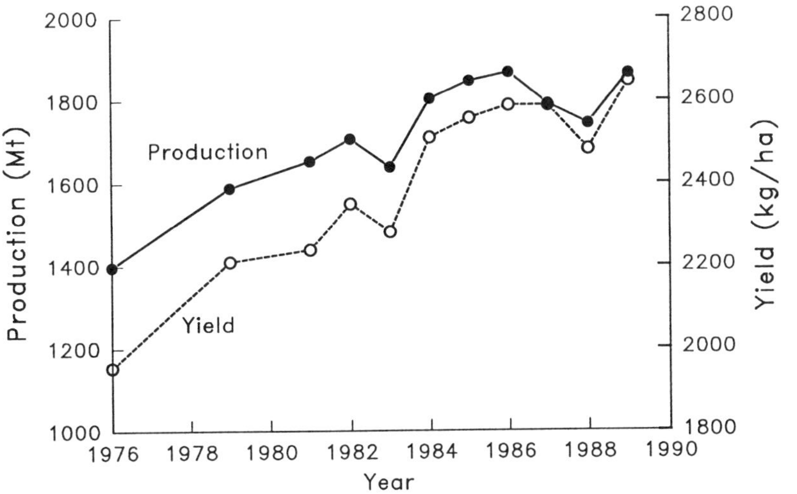

Figure 2 Total cereal production and average yield for the world during 1974–1989. *Source*: World Almanac and Book of Facts 1991.

land, water, and energy resources (Pimentel 1991). The impact of diminishing food supply and the spectre of starvation in many parts of the world can be greatly minimized if all cereals produced could be equally distributed to all consumers regardless of their regional location and ability to pay and with adequate compensation to producers. Given the past record of behavior of nation states and present political and socio-economic conditions, however, such a Utopian solution is unlikely to be found. Consequently, those who produce less grain than they need and those who produce more than they can consume must both try their best to preserve grain that has been domestically produced, imported, earmarked for long-term storage, or export.

Storage is a repeated interim phase in the complex logistics of moving grain from producer to processor and grain products from processor to consumer (Anderson 1973). Its purpose is to equilibriate fluctuations between two major factors, supply and demand. These factors could be seasonal, annual, demand-induced, weather-induced, due to logistic and transport delays or to other causes; storage need is a function of variability of these factors and the level of protection one desires from the effects of fluctuation (Hindmarsh and Trotter 1990). Managing stored grain wisely with minimal loss while maintaining its nutritional quality is a major concern at all stages. How much, where, how, and when to exercise this control depends largely on political and socio-economic conditions within and outside a country. Improper management of harvested grain causes significant quantitative and qualitative losses. Postharvest food losses are estimated to range from 9% in the United States (Pimentel 1991) up to 50% in some parts of the developing nations. Much of the loss results from invasion of the grain mass by microfloral, arthropod, and vertebrate pests.

At present, the world is divided into two main groups: several surplus-grain countries or blocks of countries such as Australia, Canada, the European Community, and the United States; and many food-short, developing countries in Africa, Asia, and South America. Ironically, the climate and storage conditions are generally more favorable for pests in the food-short countries, as compared with food-surplus countries. In fact, most stored-grain pests have originated in the tropics where many food-short countries lie. Although the destructive potential of a pest organism varies with the environment in which it lives and multiplies, its distinct life history and physiological mechanism for interaction with other biotic and abiotic variables during the process of grain destruction, are usually similar regardless of the country in which the pests are active. Sound management of stored grain requires an understanding of political, socio-economic, and meteorological factors along with those related to the physical storage system and individual pest organisms.

Grain spoilage and quality loss seldom result from the action of one organism or one factor. Organisms that become associated with the grain kernel interact with the grain itself or amongst themselves in response to changing weather and environment within the framework of an ecological system. In this chapter, I shall attempt to describe the stored-grain ecosystem and how it is managed for human welfare both at the farm and global levels and also to provide some guidelines about investigating such ecosystems.

2. ECOSYSTEMS OF THE WORLD

2.1. Ecosystem Definition, Classification, and Description

Tansley (1935) first used the term ecosystem for biotic and abiotic components considered as a whole. Odum (1989) has defined the term ecosystem for general usage as: the *biotic community* and its nonliving environment treated together as a functional system of complimentary relationships including transfer and circulation of energy and material. The biotic community is comprised of a group of populations of plants and animals in a designated area (*habitat*) and the term *population* denotes a group of individuals of a single species living together in a designated area. Ecosystems are open systems in which energy and materials constantly enter and leave. Ecological succession or replacement of one kind of animal or plant community by another occur within an ecosystem. The size and boundary for an ecosystem can be arbitrary. Thus, a bag of grain, a silo from which the grain is drawn, or the region where many silos and grain-handling equipment are located can all be considered ecosystems.

Based on their source and level of energy, Odum (1989) has grouped the world's ecosystems into the following four classes. Their estimated rates of energy flow per year kJ/m^2 are given in parentheses.

1. Unsubsidized, natural, solar-powered ecosystems (4.2–42; e.g., open oceans, upland forests).
2. Naturally-subsidized, solar-powered ecosystems (42–168; e.g., tidal estuaries).
3. Human-subsidized, solar-powered ecosystems (42–168; e.g., pastures and field crop agroecosystems, granaries).
4. Urban-industrial, fuel-powered ecosystems (420–12,600; e.g., large modern cities, London, Tokyo).

The stored-grain agroecosystem—exemplified by farm granaries, silos, warehouses, and large terminal elevators—is classified as a human-subsidized, solar-powered ecosystem.

The following four properties distinguish an agroecosystem from a natural ecosystem (Odum 1984).

1. In addition to solar energy, a part of the energy that contributes to productivity is fossil fuel, and human and animal labor.
2. Diversity of animal and plant species is minimized by human management to maximize the yield of specific food or fiber.
3. The dominant plants and animals are under artificial selection instead of natural selection.
4. Control is goal-oriented rather than an internal feedback system (as in natural ecosystems).

2.2. Agroecosystems in a Farm Household

A modern stored-grain ecosystem on a farm may be best understood when it is considered within the context of a farm household. A farmer is the creator and manager of the basic stored-grain ecosystem which could be as small as an earthen pot or as large as a concrete silo. Being the owner and nerve center of a farming unit, a farmer not only provides physical labor but also manages two or more distinct agroecosystems which are affected indirectly by political and socio-economic forces originating from within and outside his or her country, and directly by meteorological forces, existing resources, and other ecosystems (Fig. 3). A farm household, when it is managed with specific objectives, is responsible for the creating, nurturing, and destroying of three agroecosystems:

1. Field-crop ecosystem: the objective is to obtain food energy as edible seeds from growing plants.

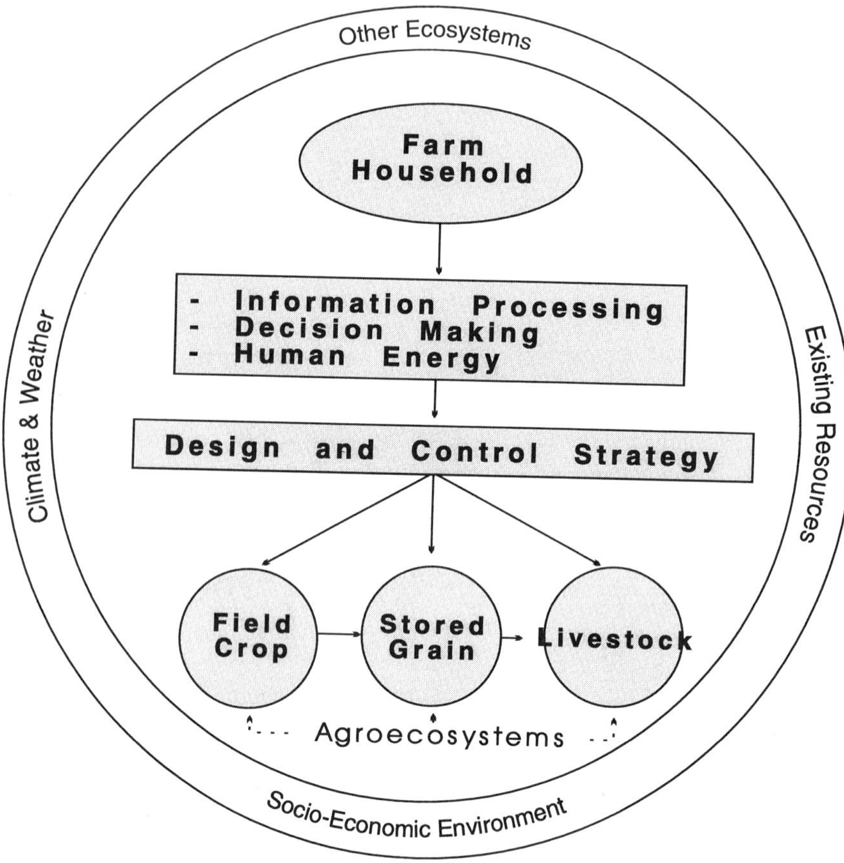

Figure 3 A schema of interrelations of three agroecosystems and their management strategy; these agroecosystems are created and managed by a farm household.

2. Stored-grain ecosystem: the objective is to preserve the harvested crop with minimum loss in quantity and quality.
3. Livestock ecosystem: the objective is to transform cereal food energy to animal products for human consumption.

The role of the farm household has a special significance, particularly for farmers in developing countries where the stored crop is primarily used to feed the farmer's family, thereby converting cereal food energy to produce and maintain human energy. Sound management of field-crop and stored-grain ecosystems is crucial for producing and preserving the quantity and quality of food grain. How is this done? Having gathered political and socio-economic information (outer circle, Figure 3), the farmer synthesizes and processes these data to make decisions about when, how many, how large, and what type of agroecosystems should be created and how they should be managed. For each operation, decision-making precedes the specific design, control, and impact of manual labor strategies. Then, the farmer faces a challenge: how to minimize the adverse effect of weather and manipulate the ecological processes that are invariably involved in the destructive action of undesirable plants (weeds, etc.), pathogens (fungi), and arthropod or vertebrate pests. The overall goal is to maximize the yield of high-quality grain through the field-crop ecosystem, preserve the harvested grain with minimum loss of quality and quantity through the stored-grain ecosystem, and convert the grain energy most efficiently into animal products for human food through the livestock ecosystem. All these farming operations affect the availability of resources and contributions from other agroecosystems or natural ecosystems. The disposal of end-products from these ecosystems which would have maximum advantage for the farm household depends primarily on political and socio-economic factors, the timing, and the quantity of grain stock released would be crucial factors in farmers' decision making.

Throughout prehistory and historical times when human science and technology were not well-developed, natural evolutionary and ecological processes dominated the agroecosystems. Although the human goal was to gather food from naturally growing crops in the wilderness, ancient hunter-gatherers did not manipulate natural grassland ecosystems to suit their needs. North American Indians still preserve this ancient practice today by gathering wild rice. Socio-economic factors were often not as effective as ecological ones in determining the success of grain production and protection. Modified versions of natural agriculture are still common in the developing world particularly in parts of Africa, Asia, and South America where many farmers still practice subsistence agriculture (i.e., they are neither net buyers nor net sellers of crops, only the direct consumers).

The unprecedented progress and application of modern science and technology in the developed world, however, has now made political and socio-economic factors more important in the production, preservation, processing, and consumption of the grain than natural ecological factors.

2.3. Stored-Grain Ecosystem

All harvested grain is seldom immediately consumed by people or domestic animals. Generally, grain is stored in weather- and pest-proof containers or specially designed storage structures so that its viability, food energy, nutritional quality, and marketability at a future date can be assured. Like any other living organism, grain is subject to decay in time; any prolongation of its life, quality, or retention of its quantity requires an understanding of the deterioration process and its multiple causal factors. Only when a mass of stored grain is considered an ecosystem (Sinha 1973, Calderon 1981, Dunkel 1992), can it be managed with the goal of preserving the maximum amount of grain in the best possible condition for the longest period. A management strategy should begin with an understanding that in stored-grain ecosystems, the abiotic physico-chemical environment interacts with the grain itself, and an array of associated flora and fauna become involved as consumers and decomposers of the grain.

Unlike a natural ecosystem (such as an undisturbed tropical forest) in which many of the processes are controlled by herbivores, predators, parasites, and pathogens, a stored-grain ecosystem is created and controlled by human beings. Management decisions that directly influence the structure and function of the ecosystem are made in two parts: design decisions and control decisions (Hart 1984).

It seems that basic principles of management of stored-grain ecosystems—storing grain in proper structures and distributing it when it is needed—were well-understood by human societies in ancient times; fears of poor harvests and famine must have prompted such actions. Thus, a biblical reference in Genesis 41:47 cites (Thompson Chain-Reference Bible 1983):

> During the seven years of abundance the land produced plentifully. Joseph collected all the food [wheat and barley] produced in those seven years of abundance in Egypt and stored it in the cities ... Joseph stored up huge quantities of grain, like the sand of the sea; it was so much that he stopped keeping records because it was beyond measure.... When the famine spread over the whole country, Joseph opened the storehouses and sold grain to the Egyptians for the famine was severe throughout Egypt.

Archaeological research has revealed that large reed baskets (diameter 0.3–1.8 m, height 0.3–0.6 m) or clay jars (height ca. 1 m) embedded in soil were archetypes of granaries used by neolithic people of the Nile Delta in Lower Egypt (Levinson and Levinson 1989). Granaries during the first dynasty (ca. 2920–2770 BC) comprised cylindrical earthen silos with roof openings and two

Figure 4 A wooden model of a granary found in Saqqara, Egypt, originating from the Middle Kingdom (ca. 2040–1785 BC). *Source*: Levinson and Levinson (1989).

square apertures at different heights. During the Middle Kingdom (ca. 2040–1785 BC) and New Kingdom (ca. 1554–1080 BC) cylindrical chambers with a vaulted roof were common (Levinson and Levinson 1989). The Egyptian Museum in Cairo has a wooden model of a 4000-year-old granary with a flat roof discovered in Saqqara (Fig. 4). It had five adjoining storage chambers surrounded by a high wall with an entrance. The figure shows a worker carrying a grain bag to the roof, two workers pouring grain from bags into the openings of two storage bins; and a scribe and assistant recording the amount of grain stored. Two workers are filling the bags with grain while two supervisors are holding sticks in the courtyard (Levinson and Levinson 1989).

2.4. The Pattern of Production, Distribution, and Consumption of Stored Grain in Developed and Developing Countries

Stored-grain ecosystems in technology-dependent, developed countries are controlled primarily by the politics and economics of the marketplace and to a lesser extent by weather and other factors. Stored-grain ecosystems in the rural agriculture of a developing country, however, function like a natural ecosystem in which decisions are primarily affected by ecological factors, weather, and to a lesser extent by socio-economic factors. In the nearly self-sustaining villages of developing countries, production is directly regulated by local demands for cereals, feeds, and fuel for the consumer population (Mitchell 1984). Because such villages often lack chemical pesticides and grain drying and cooling technology, pest organisms compete directly with humans in consuming harvested grain. Facing little interference, pests follow their normal predator-prey relationships, ecological succession, and trophic-dynamic energy-transfer to remove energy destined for human use. Although these types of stored-grain ecosystems have a high level of biological activity with consequent losses in grain quantity and quality, grain is usually consumed within a short period by humans. An understanding of the interactions among energy, material, and information processes within the context of political, socio-economic, and ecological realities can contribute greatly to devising a sound strategy to solve grain-storage problems.

2.5. Decomposition in Stored-Grain Ecosystems

Energy flow in a terrestrial ecosystem, which could include a forest, an agricultural plot, or other land area, occurs through three major components: production or plant subsystem, consumption or herbivore subsystem, and decomposition or microorganism subsystem (Swift et al. 1979). Decomposition and consumption by nonhuman herbivores dominate the stored-grain ecosystem in many developing countries (Fig. 5). These two processes are not involved in the gathering of solar energy and soil nutrients through cereal crop plants. The consumption (herbivore) subsystem includes direct feeding on stored grain by rodents, birds, insects, and mites. In many developing countries, pest organisms such as the rice weevil, angoumois grain moth, or Norway rat dominate the ecosystem and reduce the value of cereals earmarked for human use. In this case, humans lose food and nutrients while in competition with granivorous pests. Pest organisms multiply according to biological and ecological principles regardless of the countries and socio-economic environments in which they are found.

In developed countries, the normal ecological processes that involve consumption and decomposition of grain are arrested by application of human technology to a point at which grain-storage managers have become complacent. But during the last two decades, the destructive potential of certain pest organisms (e.g., toxigenic stored-grain fungi which produce harmful mycotoxins) has been recognized by the grain industries and governments of both developing and developed countries.

In developing countries, vast populations of humankind have to contend with accelerated deterioration and decomposition of their inadequately protected cereal food. Here, the consumption subsystem contains major granivorous herbivores: rodents, birds, insects, and mites. Unfortunately, the climates in many developing countries are subtropical or tropical which encourage faster multiplication of cosmopolitan insects and fungal species than in largely temperate, developed countries.

When there is much biological activity in a stored-grain ecosystem, decomposition is the most important process and human ingenuity and initiative are directed to defer this process for as long as possible. The decomposition subsystem may have many players: mites, beetles, moths, psocids, fungi, yeast, actinomycetes, and bacteria (Fig. 6). Insects, mites and fungi break down grain

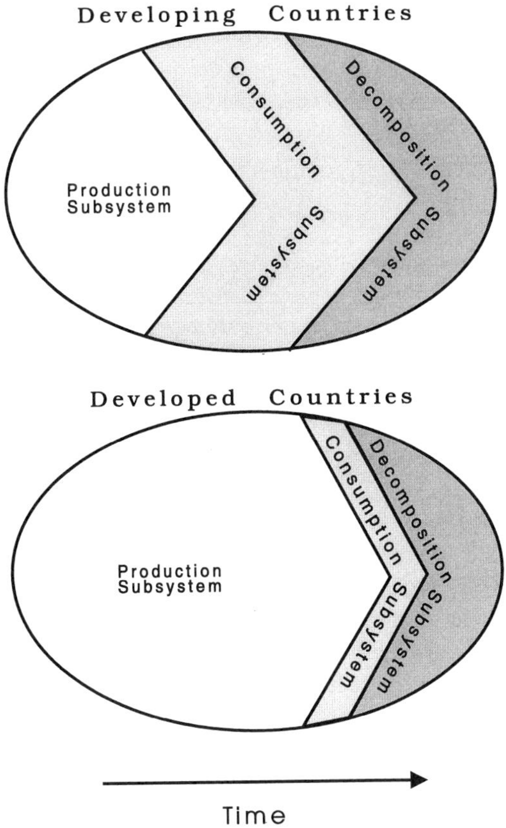

Figure 5 Three components (using biomass, energy use, and pest-infestation as criteria variables) of a stored-grain ecosystem. The ratio of the consumption (excluding human and domesticated animal use) and decomposition subsystems in relation to each other and the production subsystem fluctuates from year to year and region to region, the magnitude of fluctuation is greater in the developing countries than in the developed countries.

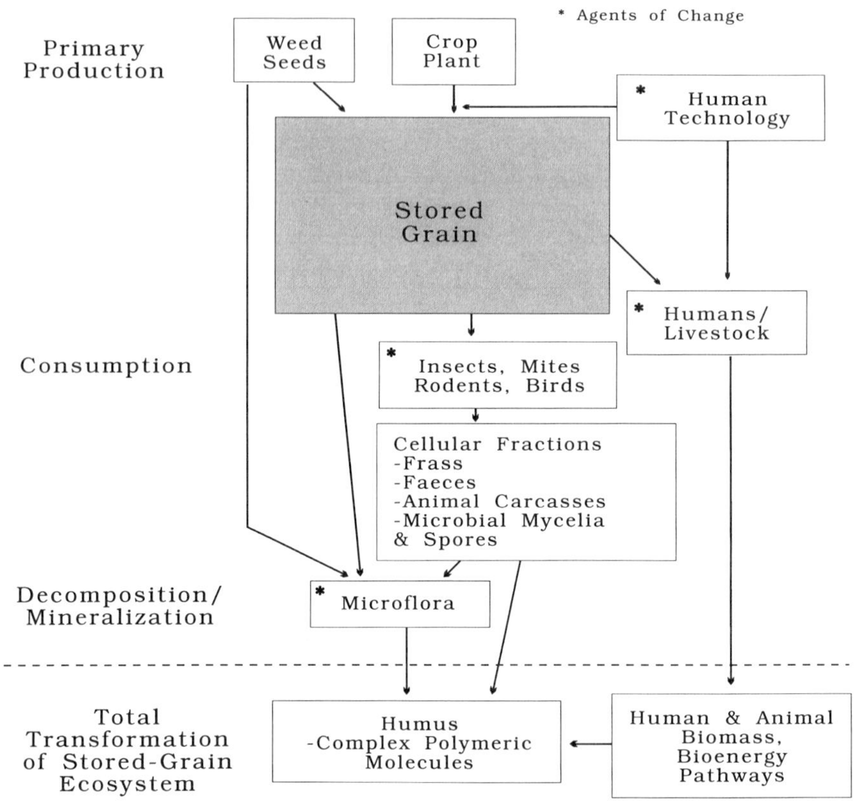

Figure 6 Stages of decomposition of grain with agents of change in a stored grain ecosystem.

particles, feed on and spread fungal spores and microflora. Microflora are the primary decomposing organisms which transform a grain mass and carcasses of animal pests into debris. The main storage microflora belong to the *Penicillium* and *Aspergillus* groups of fungi, and *Erwinia* bacteria. Because most cereals have a carbon to nitrogen ratio (wheat 15:1–26:1, barley 18:1, oat 20:1, and rice 25:1) within the optimum range of microbial activity (15:1 to 30:1), these microflora can effectively decompose grain provided that grain temperature and moisture content are favorable (Sinha 1973).

In developed countries, decomposition and consumption often play a minor role, and the production system dominates (Fig. 5). It is now possible in these countries to reduce their impact by drying and cooling the grain by using ventilation systems, constructing sound granary structures, storing the grain in controlled environments with artificially created gas ratios, or by using pesticides. Decomposition of grain in the stored-grain ecosystem can be arrested or delayed by implementing sound management practices. But whenever humans fail to arrest the natural ecological process of transformation and decay of organic matter that occurs with cereal grain, a sequence of events is set in motion: gradually, the grain germ, endosperm carbohydrate, and associated materials are converted to carbon dioxide (CO_2), free fatty acids, and numerous organic and inorganic compounds rendering the grain unfit or less suitable for human consumption.

Figure 7 describes, in schematic form, the decomposition of primary materials leading to secondary resources and finally to decomposed grain debris. If not completely eaten by humans or their livestock, the grain is eventually consumed and decomposed at varying rates by various pest organisms. Following either of the two pathways, the remainder of the grain along with animal products and byproducts will ultimately become nutrients and minerals which will eventually support future plant growth. Only the CO_2 released as a byproduct during activity of these largely aerobic organisms will go into the atmosphere.

2.6. Major Global Problems in Stored-Grain Ecosystems

Movement of grain from grain-surplus countries to grain-short countries of the world has become a common practice in modern times. This uncontrolled admixture of foreign grains with domestic grain has changed the nature and pathways of grain deterioration in stored-grain ecosystems in many countries. For example, introducing pesticide-resistant insect species, predators, parasites, and toxicogenic fungal strains has modified the ecological sequence of events in stored-grain ecosystems in grain-importing countries.

A global survey of pesticide susceptibility of stored-grain pests in 1976 showed that resistance to malathion and lindane in several major stored-grain beetles was widespread (Champ and Dyte 1976). Resistances to most residual insecticides have now been found in at least 31 stored-product pest species; resistance to fumigants including phosphine and methyl bromide has occurred in nine major storage species (Champ 1985). Resistance in *Cryptolestes ferrugineus* (Stephens), *Tribolium castaneum* (Herbst), *Sitophilus oryzae* (L.), *S. granarius* (L.), and *Rhyzopertha dominica* (F.) to the most common and effective fumigant, phosphine (Champ 1985, Zettler and Cuperus 1990), is of great concern because most of the world's grain-storage managers depend on fumigation by this insecticide for both a routine disinfestation treatment and a means of controlling insect pests that are resistant to other pesticides. To compound the problem, insect resistance to phosphine is on the increase worldwide because easily applied phosphine tablets are often improperly used to fumigate granaries

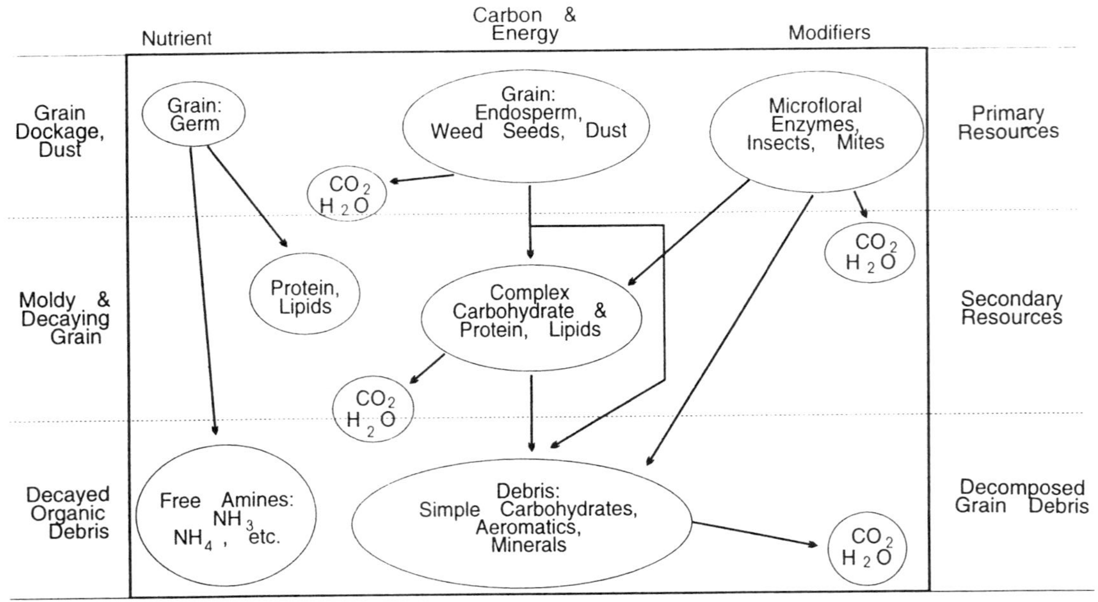

Figure 7 Pathways of decomposition of primary materials leading to secondary resources, and finally grain debris. (*Source*: after Swift et al. 1979)

that are not sufficiently gas-tight. As gas escapes, some insects survive and pass on their low-level inherent resistance to their progeny. Ironically, phosphine fumigation is humankind's main line of defence to combat stored-grain insects and thereby prevent enormous losses of stored grain (Taylor 1989). Our ability to store grain safely would be seriously challenged if the genetics of pest species continue to change in this direction. Research on the impact of large-scale introduction of resistant insect-pest species in stored-grain ecosystems is scarce.

3. RURAL STORED-GRAIN ECOSYSTEMS IN DEVELOPING COUNTRIES

3.1. Traditional Stored-Corn Ecosystems in Togo

The sharp contrast in the structure and function of stored-grain ecosystems between rural farms in developing and developed countries can be shown through examples of these ecosystems from a few selected countries.

A schema of a typical farm in eastern Ghana and Togo with three ecosystems is given in Figure 8. Corn grown intercropped with other crops (A, Fig. 8) is usually stored in the husk either near the field-crop ecosystem or is transported to the compound of the family home and stored in cribs or in bags (B, C, Fig. 8). The field-crop ecosystem could be several kilometers away from the home. The corn is then periodically dehusked and shelled for family consumption and sale to local villagers. Damp-harvested corn is dried in the sun by placing it on a mat or a concrete pad (D, Fig. 8). Shelled corn may also be stored in bags, baskets, and metal containers inside living quarters for day-to-day consumption (C, Fig. 8). Infestations of corn by insects and rodents are common. Infested and spoiled corn is normally fed to sheep, goats, and poultry in the livestock ecosystem within the compound of the home (E, Fig. 8).

Lepigre and Pointel (1971) gave an excellent account of traditional farm granaries in Togo. The main food crop, corn, is harvested and stored in the husk, either in July/August or in January/February, in traditional open granaries (outside the home of the farmer) built on wooden platforms (Fig. 8). The storage period lasts 8–10 mo when there is only one good harvest and 4–5 mo if there are two. The cobs are aligned and piled side-by-side with their tips facing inwards and their bases placed outwards along the periphery of the granaries. As this "wall" is raised, cobs are deposited inside the bulk. Lianas, a climbing and twining tropical-forest plant, are used as a belt around the cobs every five rows to prevent collapse of the corn-bulk cylinder by internal pressure. A conical straw roof is then placed on top extending over the side "walls". The wood floor may be raised 0.50–1.50 m above the ground on stakes. Such granaries are called "Ebli-Va" or "Kedelin" and contain 2–20 t of corn in the husk.

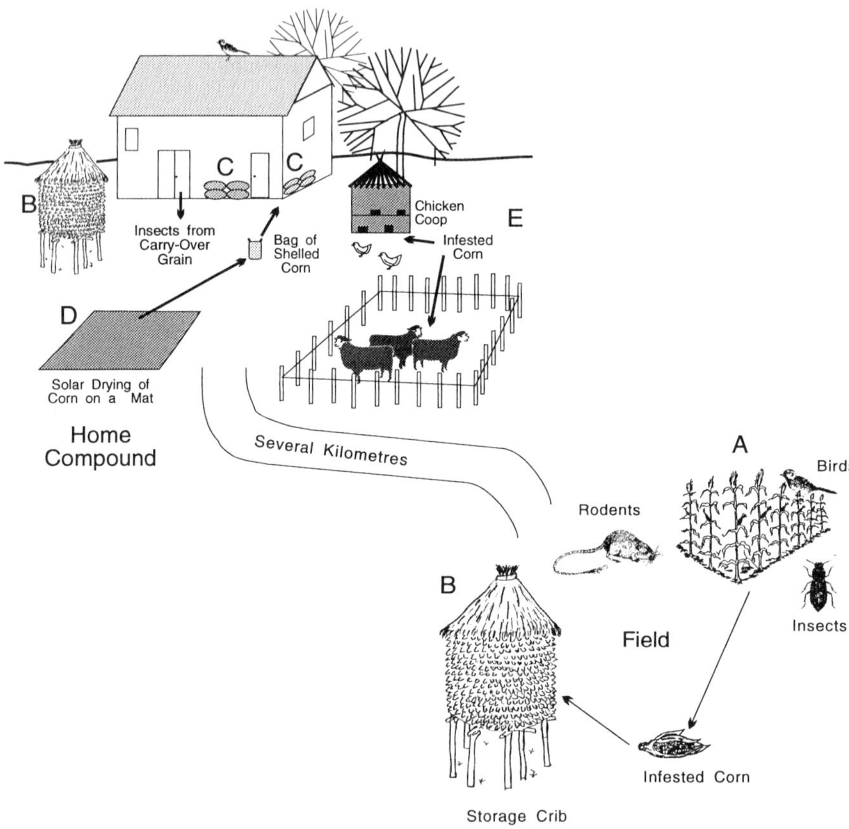

Figure 8 A schema of three agroecosystems created and managed by farm households in eastern Ghana and Togo. A. Field-crop ecosystem away from the home. B. Stored-grain ecosystem in which corn is stored in a crib. C. Stored-grain ecosystem in which cobs or shelled corn is stored in bags. D. Solar drying of damp shelled corn on a mat or a concrete pad. E. Livestock ecosystem in the home compound where sheep, goats, and chickens are fed with grain that is not normally suitable for human consumption.

The construction and use of these stored-corn ecosystems are greatly influenced by socio-economic factors in the village. Owning several of these visible stored-corn ecosystems conveys a high economic and social status to the owner. Most of the corn removed from a granary is directly consumed by the owner or local village buyers with little relation to the international grain market. When farmers remove corn cobs for their own consumption or for sale, they cut one or more lianas and the freed cobs fall on the edge of the wooden floor or onto the ground. The owner and buyer settle on the price of the volume or quantity purchased in relation to the diameter of the granary and the number of lianas cut.

Although corn is transferred from owner to the buyer who quickly consumes it as food, it is subject to severe insect infestations which usually occur in cobs with damaged sheaths. The sheath damage that facilitates insect invasion, however, occurs in the field-crop ecosystem long before the stored-corn ecosystems are created. *Sesimia* spp. and other moth larvae make small holes in the sheaths of maturing cobs on corn plants. Female maize weevils, *Sitophilus zeamais* Mots., lay eggs in corn kernels through the incompletely dried sheaths; generally, only one minute hole is pierced in each cob. The insect completes development inside the cob while it remains on the standing corn plant; the adult weevil then pierces the dry sheaths and emerges to mate. Its exit hole then becomes an entry point through which females can easily reach the kernels to lay more eggs. Vertebrate pests, such as various rodents and birds, also break open cobs on the plant in the field, allowing insects to enter the corn seeds either in the field-crop or stored-grain ecosystem.

In the stored-corn ecosystem, much of the seed on the cob is consumed in 4–5 mo by insect pests which thrive in the local climate; often only the epidermis and one or two kernels are all that remain (Lepigre and Pointel 1971). This example illustrates how ecological factors (corn, insect, climate) dominate the fate of such stored-grain ecosystems in the tropical environment of a developing country.

3.2. Underground Stored-Grain Ecosystem in Ethiopia

Buxall (1974) has described traditional, underground, stored-grain ecosystems which are used by 60% of the farmers in Harar province of Ethiopia. Each granary usually contains 0.5–20 t of sorghum or corn. Conical, flask-shaped storage pits from 75–800 cm high, with diameters of 50–125 cm at the mouth, and 50–400 cm at the base, are dug above the water table inside the farmer's compound (Fig. 9). The pits are lined with straw, cardboard, or concrete, closed at the top with timber strips, flat boards, or rocks, and then sealed with dung or a mud-dung mixture. Pits are filled with sorghum or corn usually with safe moisture

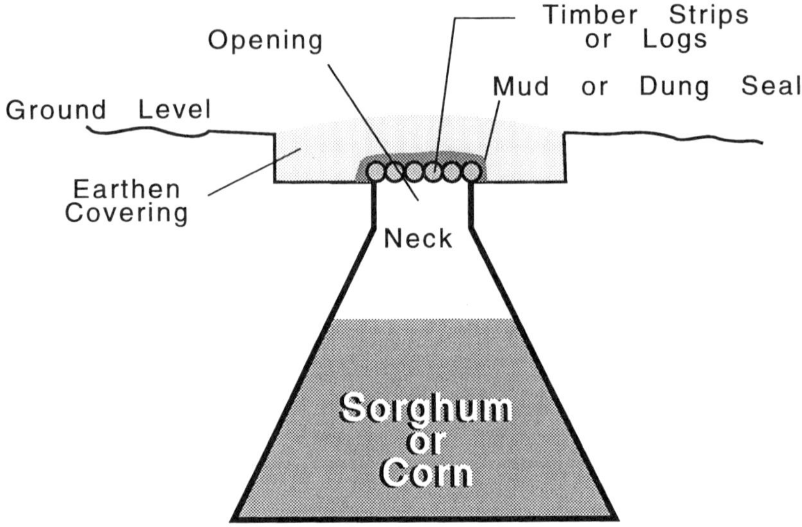

Figure 9 A schema of an underground stored-grain ecosystem, a grain storage pit, in the Harar province of Ethiopia.

contents (below 13%) and stored from 2 mo to 5 yr. In long-term storage, the grain is used during years of shortage or when grain prices are high.

In such an ecosystem, insects and fungi cause considerable loss of grain mass, market value, and nutritive value of the grain; occasionally rodents also consume and contaminate the grain. Infestation by major insect pests such as *Sitophilus oryzae*, *S. granarius*, and *S. zeamais* occur in the field-crop ecosystem about 5 wk before harvest, originating from endemic residual insect populations near the bases of the sorghum stalks. Other harmful insects include *Tribolium castaneum*, *Rhyzopertha dominica*, *Sitotroga cerealella* (Olivier), and *Ephestia cautella* (Walker). In damp pits, mycotoxin-producing fungi such as *Aspergillus flavus* Link, *A. ochraceus* Wilhelm, and *A. chevalieri* (Mangin) Thom and Church heavily infect a 2–5-cm-thick peripheral layer around the grain bulk causing grain spoilage.

When pits were opened and accessed once a month, grain losses caused by fungi and insects were 25 and 38% (Buxall 1974), respectively; these losses were reduced when pits were opened less frequently. Storage experts have suggested many improvements to reduce human food losses particularly by improving the pit roofing and lining. For example, in Morocco, Bartali et al. (1990) have shown that the use of a polyethylene lining instead of traditional straw layer in underground conical barley storage pits (Matmora) reduces dry mass loss, loss of germination, *A. flavus* infection as well as infestation levels of *R. dominica* and *S. granarius*. Nevertheless, this example again demonstrates how pest organisms can flourish and reduce the human food supply if population succession in a stored-grain ecosystem is allowed to progress without interference from humans.

4. STORED-GRAIN ECOSYSTEM IN A DEVELOPED COUNTRY

Modern grain-storage technology and expert knowledge from several disciplines have been successfully applied over 12 yr using a comprehensive ecosystem-management strategy on a 2000 ha, mixed farm in Manitoba, Canada. The main objective of this field project has been to determine whether cereal grain and animal health could be protected without the use of any chemical pesticides. The success of this project has resulted from extensive and regular use of systems to monitor grain infestation and quality, a natural air ventilation system, and alertness of the farm manager. Although portions of the cereals stored were contaminated with ochratoxin and other harmful mycotoxins in three crop years, and heavily infested by stored-grain insects such as *Tribolium castaneum*, *Cryptolestes ferrugineus*, and *Alphitobius* in one crop year, the grain was protected with negligible economic loss and health risk.

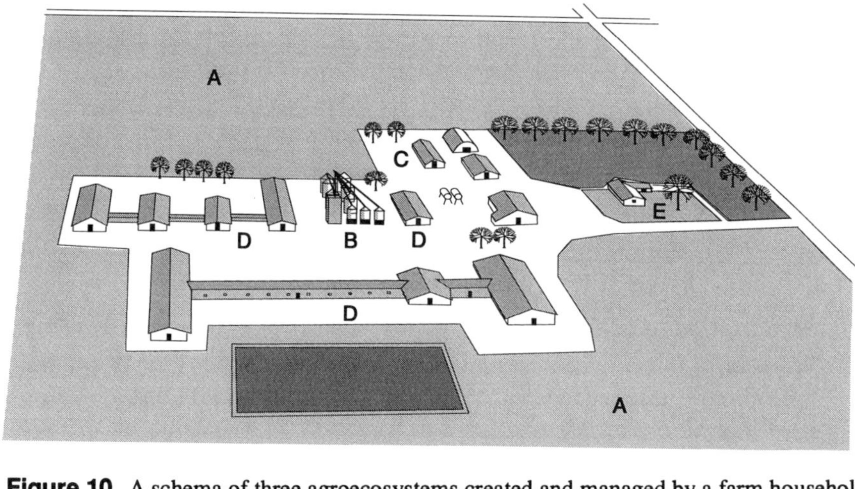

Figure 10 A schema of three agroecosystems created and managed by a farm household in Manitoba, Canada. A. Field-crop ecosystems near the farm in which wheat and barley crops are grown. B. Stored-grain ecosystems containing six ventilated steel bins. C. Stored-grain ecosystems containing nonventilated buildings. D. Livestock ecosystems containing hogs and chickens. E. Farmer's home.

This typical North American, well-planned farmstead (Fig. 10) used in the project contains all three typical ecosystems: field crop, stored cereal, and livestock. The emphasis here is on production of swine and poultry for sale to domestic and foreign markets (although excess grain is occasionally sold off the farm). The main crops grown in the field-crop ecosystems (A, Fig. 10) are barley and wheat which are subsequently stored in six bins. In average years, 3000 t of grain is stored in large, ventilated metal bins (B, Fig. 10). When there is a bumper crop, however, additional surplus grain is stored in nonventilated barns or machine sheds (C, Fig. 10). Daily requirements for approximately 900 breeding sows, and 4300 laying hens dictate year-round grinding and mixing of wheat and barley with feed supplements and nutrients in an on-farm feed mill. The swine and poultry ecosystems (D, Fig. 10) are operated with state-of-the-art technology. Insect infestation and fungal infection (including mycotoxin production) in most years are prevented by using sound pest-management strategies. The monitoring of grain-quality changes by measuring levels of temperature, moisture, CO_2, free fatty acids, seed germination, and seed-borne microflora has been conducted by Agriculture Canada and University of Manitoba scientists. Insect infestation is monitored by the Berlese funnel method and pitfall traps (in the machine sheds). When mycotoxins or harmful insects occasionally occur in the grain because of weather conditions at harvest, the farmer has appropriate information and training to deal with these problems before they cause health problems and economic loss.

5. MANAGEMENT OF STORED-GRAIN ECOSYSTEMS

5.1. Management of Stored-Grain Ecosystems in Developing Countries

Management of stored-grain ecosystems on small farms in developing countries is considerably more difficult than in developed countries. The main reasons for such difficulty are listed by Lindblad and Druben (1980):

1. Wide variety of structures used.
2. Presence of climatic conditions that are more suitable for growth and multiplication of insects, rodents, and fungi than those in developed countries.
3. Low income of the farmers.
4. Lack of knowledge and training about sound grain-storage management.

Although basic principles of sound grain storage by way of keeping grain cool and dry while protecting it from insects, rodents, and microflora are known, application of these principles often requires good knowledge of modern storage technology and well-planned design and control strategies.

The following basic strategy for sound grain storage on small farms can be used in developing countries (Lindblad and Druben 1980, Dunkel et al. 1986).

1. *Waterproofing* granaries and containers as much as possible and locating them where they will not be flooded by ground water or run-off during heavy rains.
2. *Rodent-proofing* the containers and granaries as much as possible.
3. *Cleaning* storage containers to remove all old grain, dust, straw, and insects, and storing only clean grain.
4. *Drying* grain to 12–13% moisture content before initial storage.
5. Keeping grain *cool* and protecting it from large changes in outdoor temperatures by storing grain away from direct sunlight and using appropriate building materials.
6. *Protecting* grain from insect infestation by drying, applying insecticides (chemical, botanical, and other alternatives), or by placing grain into airtight storages. Because regular use of one insecticide increases the potential for insect species to develop resistance, alternate insecticides and integrated pest control practices should be used.
7. *Checking* grain regularly for infestations and unsafe storage conditions and correcting any problems that arise.

5.2. Management of Stored-Grain Ecosystems in Developed Countries

The modern trend throughout the world has been to store grain in large granaries and silos, and to manage it with trained professionals whose training is updated regularly as grain-storage technology advances. Large terminal elevators, silos, and warehouses operated by cooperative farm groups, grain companies, or governments in Australia, Canada, France, United Kingdom, and the United States already use this strategy successfully to reduce losses in stored grain. Nevertheless, occasionally there are major problems, natural or human-made, such as above-normal rainfall during harvest, large carryover of the previous year's grain, or prolonged storage of grain caused by poor sales. When such problems occur farmers have difficulty in coping with pest infestations, mycotoxin contamination, and other associated quality loss problems; consequently, they may suffer economic losses.

The types and circumstances of farming vary widely among regions that have different climatic, political, and socio-economic conditions. Basic criteria used for monitoring grain spoilage and enhancing the keeping quality of the grain, however, are broadly similar (Fig. 11). The main difference between the countries in developed and developing regions lies in the extent to which knowledge and existing technology are used in response to consumer demand and market pressure. In developed countries that have sophisticated grain-storage

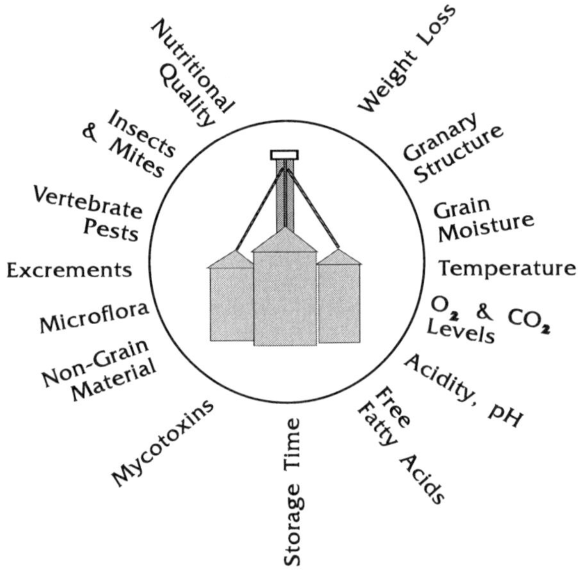

Figure 11 A schema of the main quality monitoring and control variables applied to a well-managed, stored-grain ecosystem in North America.

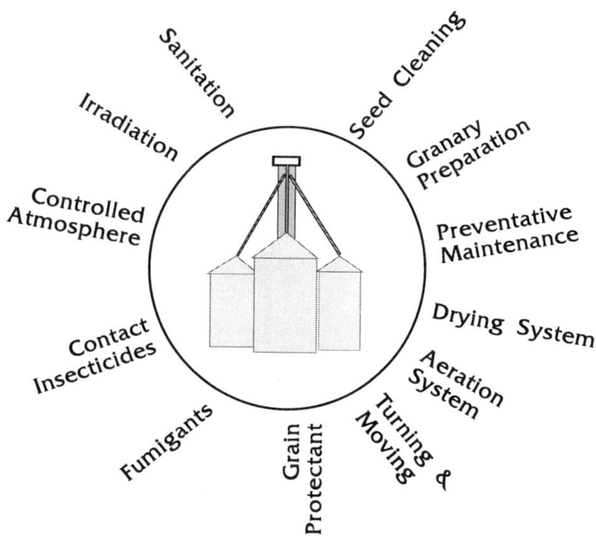

Figure 12 A schema of the main control tools used in a well-managed stored-grain ecosystem in North America.

systems, several variables are monitored (Fig. 11) used in conjunction with existing, well-tested control tools and strategies (Fig. 12). Some of the important monitoring variables are moisture, temperature, intergranular gas, mass loss, grain quality, insects, microflora, foreign materials, and mycotoxins.

The major prevention and control tools and strategies used by the farmer for farm-stored grain ecosystems in the midwestern United States are seed cleaning, granary preparation, preventive maintenance, drying and aeration systems, pest detection, and chemical control of pests with contact insecticides or fumigation (Cink and Harein 1989).

Despite chronic low prices of grain through the world because of subsidies to growers, many farmers have been able to use advanced technology and sound grain-storage practices to their advantage. Consequently, natural growth and reproduction of various pests of stored grain have been kept to a low level in most years. The result is availability of high-quality, nutritious cereal grain for human and livestock consumption at an incredibly low cost.

6. ECONOMICS AND STORED-GRAIN ECOSYSTEMS

6.1. Overview

Of all the factors affecting the agroecosystems in our times, economics is probably the most important. Because of the great disparity in peoples' income within a country or among countries of the world, rational solutions for the creation and management of agroecosystems on the basis of legitimate human needs and expectations have often eluded humankind. Agricultural scientists, cereal chemists, biologists, and engineers have made great strides in solving technological problems such as developing good quality and high-yielding cereal varieties, systems of aeration and drying, integrated pest management strategies, and pesticide compounds and formulations for preserving stored grain. Considerably less progress has been made towards solving sociological, economic, and educational problems underlying production, storage, distribution, and consumption of cereal food. Thus, different people still prefer one kind of cereal over another because of tradition or social convention, regardless of their nutritional need. People starve while there is an ample supply of grain within the country, or farmers go bankrupt by producing grain for which no market can be found at home or abroad. These dislocations in grain marketing occur because producers, consumers, and managers are all trapped within inadequate socioeconomic systems.

Political and economic policies of a country determine the level of food security, storage, import, and export of cereals leading to the well-being of its citizens. Through its grain production, domestic consumption, imports, and exports economic policy makers of a country attempt to reduce fluctuations between supply and demand. These fluctuations are caused by factors such as varying production levels from year to year, weather changes, market demand, transport problems, or political events.

6.2. Ecosystems and Food Chains in Developing Countries

The human food chain or "conveyer belt", containing whole and processed grain connects producers and consumers at two opposite ends. The food chain is generally shorter in developing countries than in developed countries. The service sector of the food chain includes assemblers, government warehouse-persons, millers, and retailers who play a major role in moving grain to consumers who in turn use grain products to meet the basic needs of feeding their families with simple and adequate diets. Raw grain materials and grain products are stored, bought, and sold in a relatively stable market. The stability of the market results from consumers' modest "needs" and the government's involvement in the food chain. For example, grain marketing in many African countries has usually involved obligatory farm sales to state marketing boards. Stored-grain ecosystems are often managed by trained personnel when grain is stored in government warehouses; however, with liberalized policies in many developing countries, the grain trade is being privatized and prices decontrolled. In the long run, the free market is likely to benefit both producers and consumers. Nevertheless, storage of grain by producers and increased control by small traders, most of whom are not adequately trained to protect stored grain, may lead to grain losses, mycotoxin contamination, logistic problems, and chaotic price fluctuations. These difficulties may gradually be overcome by improved understanding of the market systems, developing local infrastructure and transport, promoting proper storage and an active market presence by farmers and traders, and reforming financial systems (Coulter and Compton 1991). Modern training on stored-grain ecosystem management and marketing of grain, however, would be essential.

The need for management of a stored-grain ecosystem is a function of the variability of these factors and the level of protection from the effect of fluctuation one desires (Hindmarsh and Trotter 1990). Proper storage of grain, particularly for periods longer than a year, is expensive. Part of the cost comes from resolving physical and biological problems that cause qualitative and quantitative losses which inevitably lead to economic loss. Hindmarsh and Trotter (1990) identified six basic problems with economic consequences, particularly in African developing countries:

1. Unclear marketing and storage goals.
2. Undefined or poorly defined target stock levels, procurement, and release rules.

3. Storage facilities not synchronized with distribution needs in terms of locations, capacity, and design.
4. Inadequate relationship between transport and storage.
5. Inadequate stored stocks; underpaid and poorly motivated staff.
6. Lack of ability to dispose of surpluses.

The best way to resolve these logistic and socio-economic problems, which are not very different from those in developed countries, is to relate them with the ecological problems associated with a stored-grain ecosystem and make interventions that are scientifically sound and economically acceptable. Management of stored-grain ecosystems must have two interwoven strategies: an ecologically sound, grain-loss prevention strategy, and a socio-economically sound, control strategy for the intake and release of grain stocks. Grain stocks can be double-edged weapons, as grain traders and farmers in both developing and developed countries often discover. If released when supplies are low, they can help stabilize the price and supply. If released at an inappropriate time, place, or price, they may enhance food insecurity and income loss (Hindmarsh and Trotter 1990).

6.3. Ecosystems and Food Chains in Developed Countries

The task of storing food grain and its products and the creation of stored-grain ecosystems is interwoven in the global food chain. The raw grain produced by a farmer on a farm passes through various transformations and stages as it moves through the marketplace to the consumer; it is assembled, graded, packed, processed, and distributed. The grain, in one form or another in varying quantities, and for varying lengths of tenure, has to be stored after nearly every step. The players in these operations are storage managers, packers, food processors, manufacturers, and retailers, most of whom enjoy considerably more value-added opportunities for profit making than the producer. The entire food-chain system is in a state of flux, being constantly subjected singly or collectively to economic, market, institutional, and technical pressures. Although all players want wholesome and saleable food for their survival, efficient operation, and maximum profits, the consumer holds the most important card. Through his or her "needs" (i.e., for a basic diet) and "wants" (i.e., food requirement for novelty and diverse tastes) the consumer ultimately commands what grain crop will be produced, stored, or processed. In this respect, Adam Smith's two-century-old doctrine (Smith 1937) still holds true.

> Consumption is the sole and end purpose of all production and the interests of the producer ought to be attended to, only so far as it may be necessary for promoting that of the consumer.

In the developed countries, there has been an ongoing oversupply of food grain for several decades. Moreover, consumers in these countries have undergone radical changes in their lifestyle, income patterns, social affiliation, and also their awareness of environmental degradation and pollution. Their demand for specific types of food have intensified and they have exerted considerable pressure on the producer through the marketplace. Consumers are increasingly becoming more quality conscious but also are demanding more diverse, usually short-lived, food products.

The relatively strong service sector of the food chain has responded to this consumer demand, competition, and price pressure by using high-technology solutions. For example, many variations of food have been produced in the breakfast cereal market simply to satisfy the sugar-craving taste buds of the affluent children in developed countries. The consumer-driven market is now interested in pesticide-free food with zero insect tolerance; however, a potential alternative such as irradiated, insect-free food is viewed as unacceptable. The emphasis is to produce high-demand foods of high-quality and novel foods with short shelf-lives. But any fundamental change in grain-food quality has to be undertaken using advanced research tools such as biotechnology, biological pest control techniques, and computer simulation modeling only at the producer level, which is financially the weakest sector in the food chain. These original suppliers and initial preservers of the grain are debt-ridden because of high production costs and low grain prices and they do not benefit from the value-added opportunities of the processors and retailers.

Production of novel, high-quality grain in the field-crop ecosystem, and preservation of this grain in the stored-grain ecosystem without having it contaminated by insect pests, toxic pesticides, or mycotoxins, require imaginative long-term basic research. For consumer satisfaction, research is needed in all sectors of the food chain, but it becomes more developmental, short-term, and applied when undertaken progressively along the food chain towards the consumer end. At the producer end, it must become more basic rather than applied and more long-term rather than short-term. Once a batch of poor-quality grain is produced or a high-quality batch is degraded through mycotoxin or pesticide contamination during storage on farms, little can be done to make it acceptable to the consumer.

6.4. STUDY AND USE OF STORED-GRAIN ECOSYSTEMS

The scientific study and rational use of stored-grain ecosystems involve multidisciplinary teamwork and technical skills because stored-grain ecosystems are extremely variable. They may vary in size, shape, and type of storage enclosure (bag, bulk, granary, or none), the type of grain with its unique physical and biological properties, preharvest and harvest history, geographical and climatic location of where the grain is stored, and the potential and real nutritional and economic value of the grain from the viewpoint of the producer, consumer, or both. The following are some of the basic steps for undertaking such a study:

1. *Developing clear objectives.*
2. *Identifying variables* that are relevant and measurable within the limits of resources.
3. *Setting up controls*, either through replicate granaries or through elaborate measurement of all chosen variables associated with the grain during its initial storage. In a granary study, at least two similar granaries should be used.
4. *Developing a uniform sampling plan*, including sample size, location, frequency, and subsampling with random kernel distribution for each variable; the plan should be consistently followed throughout the study.
5. *Measuring each variable* by using the most reliable techniques available in the discipline involved.
6. *Conducting several small-scale experiments* in the laboratory with chosen sets of variables to determine cause-effect relationships.
7. *Consulting regularly* with the multidisciplinary research team throughout the study about the progress of the study; direction of the study may be changed, if needed, and objectives may be modified; new variables may also be added or old variables discarded if their relative importance changes.
8. *Analyzing data* by using descriptive, univariate, and multivariate statistical techniques.
9. *Interpreting results* from both unidisciplinary and multidisciplinary viewpoints.
10. *Developing* (if the results justify and resources are available) both explanatory and predictive mathematical models.
11. *Developing expert systems* that will include all available knowledge on a problem for use by farmers, grain storage managers, processors, retailers, and policy makers. An expert system is a computer-based consultation program that integrates information from historical data, laboratory and field studies, and expert knowledge to help classify, diagnose, or plan; with its knowledge base and a logical control structure an expert system could be a valuable tool for the grain storage manager in the future.
12. *Setting up data banks* and expert systems on typical stored-grain ecosystems (e.g., crop-based, structure-based, etc.) for various bioclimatic, agronomic, and economic regions and storing the information in computers for future use.

7. FUTURE DIRECTIONS IN STORED-GRAIN ECOSYSTEM RESEARCH AND MANAGEMENT

An enormous amount of change is taking place in the political and economic spheres of today's world. Changeover from communist-type political systems to democratic and free-enterprise systems in the grain-producing countries of eastern Europe and elsewhere is likely to have a major impact on stored-grain ecosystems in these countries. It is possible that an increasing number of farmers will grow and store grain on their own farms, adopting a capitalist system of operation in which the market demand will regulate the storage time for the farmers' grain.

Generally, large grain-storage premises, if properly managed, provide good protection for stored grain. But, on a visit to Poland, Sinha et al. (1991) asked many small farmers whether cooperatively owned and managed storage and drying facilities would be acceptable; most farmers replied negatively, expressing their unwillingness to become involved with farmer-owned cooperative grain storages because of their bad experiences with so-called "cooperatives" forced on them by their past Communist government. Because a glut of unsold grain exists in grain-surplus countries and vulnerability of grain to spoilage increases with the duration of storage period, grain-spoilage problems are likely to increase in the 1990s and well into the 21st century. The following suggestions are for the future improvement of the management of stored-grain ecosystems.

1. Exploratory surveys followed by 2- to 5-year-long research projects to determine the interrelations in stored-grain ecosystems on a regional basis should be undertaken. Socio-economic factors should be considered in such projects.
2. Several types of small granaries, incorporating modern knowledge of grain-storage engineering, mycotoxicology, and pest-management practices, etc., should be developed for use in a country or region that has a characteristic climate and a tradition of growing certain types of crop and farming practices.
3. Expert systems based on meteorological, agronomic, economic, and pest-management data should be developed for use by grain-storage managers of stored-grain ecosystems. Because the use of computers and software is rapidly increasing, expert knowledge and statistical data can be synthesized to enhance management of stored-grain ecosystems.
4. Research emphasis should be given to developing optimum multidisciplinary ecosystem-management strategies (including economics and rural sociology) with the aid of computer simulation modeling techniques which could include grain supply and demand models, grain ventilation, controlled-atmosphere storage including heat sterilization of intergranular air

(both above and below ground), and pheromone-based monitoring and control methods, without relying heavily on chemical pesticides.

5. Research should be directed at developing effective new low-cost, environmentally sound pesticides.
6. Pest-management programs using chemicals should attempt to kill all pest insects during a treatment to reduce the chance of development of resistance; ongoing monitoring of pesticide resistance in stored-product insects on a global basis is needed.
7. Research on biological and other nonchemical control strategies for stored-product pests should be intensified because of the possibile loss of effective pesticides, such as malathion, chlorpyriphos-methyl and fumigants such as phosphine and methyl bromide within the next decade.
8. Extensive training on principles of sound grain-storage practice and the simple concept of grain-storage ecosystems should be given to grain-storage workers. Television and other modern telecommunication tools should be used for training with the latest scientific and economic information. Such knowledge should be provided, particularly to female populations of village farms in developing countries. Because women take a special responsibility in family planning, food preservation, and feeding the family, their training is crucial for proper management of stored-grain ecosystems.
9. Training of researchers both in developing and developed countries who would be capable of solving grain-storage problems particularly in rural areas in an overpopulated world in the 21st century is urgently needed; the number of qualified researchers has greatly diminished over the last few decades.

ACKNOWLEDGEMENTS

I thank R. Cofie-Agblor for providing information on rural storage conditions in Ghana, C.J. Demianyk and R. Sims for preparing the illustrations, and J.T. Mills, C.J. Demianyk, and J.B. Bole for reviewing the manuscript.

REFERENCES

Anderson, J. A. 1973. Problems of controlling quality of grain. pp. 1-14 in: Grain Storage: Part of a System. R. N. Sinha and W. E. Muir (eds.). Avi Publishing Co., Westport, CT.

Bartali, H., F. V. Dunkel, A. Said, and R. L. Sterling. 1990. Performance of plastic lining for storage of barley in traditional underground structures (Matmora) in Morocco. J. Agric. Eng. Res. 47: 297-314.

Buxall, R. A. 1974. Underground storage of grain in Harar Province, Ethiopia. Trop. Stored Prod. Inf. 28:39-48.

Calderon, M. 1981. The ecosystem approach for apprehending the extent of postharvest grain losses. Phytoparasitica 9: 157-167.

Champ, B. R. 1985. Occurrence of resistance to pesticides in grain storage pests. pp. 229-255 in: Pesticides of Humid Tropical Grain Storage Systems. B. R. Champ and E. Hagley (eds.). ACIAR Proceedings No. 14. Australian Centre for International Agricultural Research, Canberra.

Champ, B. R. and D. E. Dyte. 1976. Report of the FAO global survey of pesticide susceptibility of stored grain pests. FAO Plant Production and Protection Series No. 5. Food and Agriculture Organization, Rome. 297 pp.

Cink, J. H. and P. K. Harein (eds.). 1989. Stored grain pest management. Minnesota Extension Service, Univ. of Minn., St. Paul, MN. 280 pp.

Coulter, J. and J. A. F. Compton. 1991. Liberalization of cereals marketing in sub-Saharan Africa: Implementation issues. Marketing Ser. Vol. 1. Natural Resources Institute, Overseas Development Administration, London. 26 pp.

Dunkel, F. V. 1992. The stored grain ecosystem. J. Stored Prod. Res. 28: 73-87.

Dunkel, F. V., T. P. Wittenberger, N. R. Reid, and E. Munyarushoka. 1986. National storage survey of beans and sorghum in Rwanda. Misc. Publ. 46-1986. Minn. Agr. Res. Stn., Univ. of Minnesota, St. Paul, MN. 205 pp.

Hart, R. D. 1984. Agroecosystem determinants. pp. 105-120 in: Agricultural Ecosystems Unifying Concepts. R. Lawrence, B. R. Stinner, and G. H. House (eds.). John Wiley & Sons. New York.

Hindmarsh, P. and P. Trotter. 1990. Developments in grain storage for food security. pp. 126-132 in: Developing World Agriculture. Grosvenor Press International, London.

Kent, N. L. 1983. Technology of cereals. 3rd ed. Pergamon Press, Oxford. 306 pp.

Lepigre, A. L. and J. G. Pointel. 1971. Protection of maize stored in traditional Togolese granaries. Trop. Stored Prod. Inf. 21:7-12.

Levinson, H. Z. and A. R. Levinson. 1989. Food storage protection in ancient Egypt. Boletin de Saridad Vegetal Fuera de Serie 17:475-482.

Lindblad, C. and L. Druben. 1980. Small farm grain storage. Vol. 3. Storage Methods. Action/Peace Corps. Program and Training Journal Manual Series No. 2. Volunteers in Technical Assistance. VITA Publications Manual Series 35 E. Washington.

Mitchell, R. 1984. The ecological basis for comparative primary production. pp. 13-53 in: Agricultural Ecosystems Unifying Concepts. R. Lawrence, B. R. Stinner, and G. H. House (eds.). John Wiley & Sons. New York.

Odum, E. P. 1984. Properties of agroecosystems. pp. 5-11 in: Agricultural Ecosystems Unifying Concepts. R. Lowrance, B. R. Stinner, and G. J. House (eds.). John Wiley & Sons. New York.

Odum, E. P. 1989. Ecology and our endangered life-support systems. Sinauer Associates, Inc. Sunderland, MA. 283 pp.

Pimentel, D. 1991. World resources and food losses to pests. pp. 5-11 in: Ecology and Management of Food Industry Pests. J. R. Gorham (ed.). FDA Tech. Bull. 4. Assoc. Off. Anal. Chem., Arlington, VA.

Sinha, R. N. 1973. Ecology of storage. Ann. Technol. Agric. 22:351-369.

Sinha, R. N., W. E. Muir, and R. A. Sellen. 1991. Cereal storage loss study Poland mission—summer 1990. Int. Agr. Dev. Div., Int. Prog. Br., Agriculture Canada., Ottawa. Tech. Rept. 24 pp.

Smith, A. 1937. The Wealth of nations. Modern Library, New York, NY. 976 pp.

Street, P. R. 1990. A systems view of commercial supply and marketing links. pp. 159-203 in: Systems Theory Applied to Agriculture and the Food Chain. J. G. W. Jones and P. R. Street (eds.). Elsevier Applied Science, New York.

Swift, M. J., O. W. Heal, and J. M. Anderson. 1979. Decomposition in terrestrial ecosystems. Univ. Calif. Press., Berkeley. 372 pp.

Tansley, A. G. 1935. The use and abuse of vegetational concepts and terms. Ecology 16:284-307.

Taylor, R. W. 1989. Phosphine: a major grain fumigant at risk. Intern. Pest Control 31:10-14.

Thompson Chain-Reference Bible. 1983. New International Version. The Old and New Testament. F. C. Thompson (compl. and ed.). B. B. Kirkbridge Bible Co., Inc., Indianapolis, IN. 1877 pp.

World Almanac and Book of Facts. 1991. Scripts Howard Co., New York. 960 pp.

Zettler, J. L. and G. W. Cuperus. 1990. Pesticide resistance in *Tribolium castaneum* (Coleoptera: Tenebrionidae) and *Rhyzopertha dominica* (Coleoptera: Bostrichidae) in wheat. J. Econ. Entomol. 83:1677-1681.

2

Expert System Concept

Paul W. Flinn

Agricultural Research Service, United States Department of Agriculture, Manhattan, Kansas

William E. Muir

University of Manitoba, Winnipeg, Manitoba, Canada

SUMMARY

Expert systems are computer programs that solve complex problems within a given area. Unlike traditional programming languages, they can store both qualitative and quantitative information; they also act as storehouses of information that can be continually added to and improved over time. Expert systems can use models to predict when insects or molds will become a problem in stored grain, and suggest alternative strategies to prevent problems from developing. The inference engine gets the necessary information from the user, runs the models if needed, and interprets the output for the user. Expert systems can bridge gaps in knowledge by supplying heuristic information to the models when necessary. The predictive ability of models coupled with the user-friendliness of expert systems results in a very powerful system.

Stored Grain Advisor is an expert system that provides information on how to sample grain for insects, sets insect thresholds for determining when fumigation is required, and has an illustrated taxonomic key that helps the user identify the major insect pests of stored grain. Rules for the expert system came from two sources: 1) interviews from acknowledged experts in stored-grain management and, 2) computer simulations with stored-grain insect models.

In the future, it may be possible to automate insect sampling by using acoustical detection or carbon dioxide detection. An expert system could constantly monitor sensors in a grain bin to detect developing insect infestations at a much earlier date than is currently possible. The expert system could predict, using computer models of grain temperature and insect populations, when infestations will reach damaging levels based on estimates of existing populations, and recommend optimum management strategies to the grain manager.

Expert systems are a practical approach for developing decision-support systems for better grain management. Their importance will probably increase in the future as we come to rely more on biological and physical control methods and less on traditional insecticides, because these methods require more knowledge about the grain storage system to be used effectively.

1. INTRODUCTION

1.1. Farming is Risky Business

Farmers face many problems today in managing their farms as there are many unpredictable factors that can make farming a high risk proposition. These factors, such as insects and adverse weather, can cause high yield losses. The costs of agricultural chemicals are increasing and they are no longer as effective as they used to be due to increased insect and weed resistance. There is also increased pressure on the farmer to produce a product with little or no chemical residues. Market prices remain low due to increased domestic and foreign competition. In the future, successful farmers will need to become experts at integrating the biological, meteorological, economic, chemical, engineering, and legal aspects of farm management. Good management will be crucial to farm survival in the future.

1.2. Expert Systems Can Improve Farm Management

Expert systems are computer programs that can help farmers make better farm-management decisions. They can help the farmer predict what will happen in the future. They can assist the farmer in making complicated decisions by integrating diverse types of information, and by predicting the probability of future events based on his management decisions. Often expertise is scarce. Extension specialists are not always available, and even when they are available, they may not have the most up-to-date information. Because expert systems are computer software, they can be easily distributed and updated. Like books, they can make the knowledge of a single expert available to virtually everyone, and they can act as a storehouse of knowledge (Coulson and Saunders 1987). Under ideal conditions, this storehouse can be continually added to and improved over time. In this chapter, we will discuss the different types of expert systems that are available and their advantages and disadvantages. We will discuss how an expert system is developed, using as an example an expert system for the management of stored grain insects. In the last part of the chapter we will discuss future uses of expert systems in stored grain management.

2. DEFINITION AND TYPES OF EXPERT SYSTEMS

2.1. What are Expert Systems?

Expert systems are computer programs that solve complex problems within some defined domain (Plant and Stone 1991). They are different from traditional algorithmic-based approaches because they offer an alternative way to store and process information. They can store both qualitative and quantitative information. Expert systems store information using heuristics (logical rules-of-thumb). For example, the following is a typical rule for storage of wheat in Kansas:

IF initial grain moisture is less than 12%
    AND storage date is greater than June 15
    AND aerate date is less than Oct. 15
THEN probability of having an insect problem in this bin is low.

In agriculture, the arrival of integrated pest management (IPM) in the 1970s was heralded as the golden age of agricultural systems science. Models of crops and insects were viewed as a panacea. These models used mathematical equations to predict future pest outbreaks and yield losses. These models have been invaluable in some cases as a research tool for finding optimal IPM strategies (Onstad 1984). Few of these ambitious models, however, are being used by farm managers today. There are several reasons for their lack of use. First, lack of confidence in their predictions. Agricultural systems are very complex and it is doubtful that one could ever completely model these systems; there will always be some unknown factors that can affect the system. Many factors that affect the system are inherently unpredictable, such as weather and insect migration, which add to the uncertainty of the prediction. Second, these models are typically difficult to use. The data must be carefully selected and entered in the correct format before simulations can be run. Even if the user is able to run the model, the user still requires skill in interpreting the results, which are often no more than large tables of numbers.

Mathematical models of agricultural systems can be made more accessible to farm managers by including the models or their results in an expert system. Expert systems offer an additional method of representing farm management knowledge. Often gaps in knowledge exist that prevent models from being implemented. Expert systems can bridge this gap by supplying heuristic information to the models when necessary. This hybrid system allows more rapid

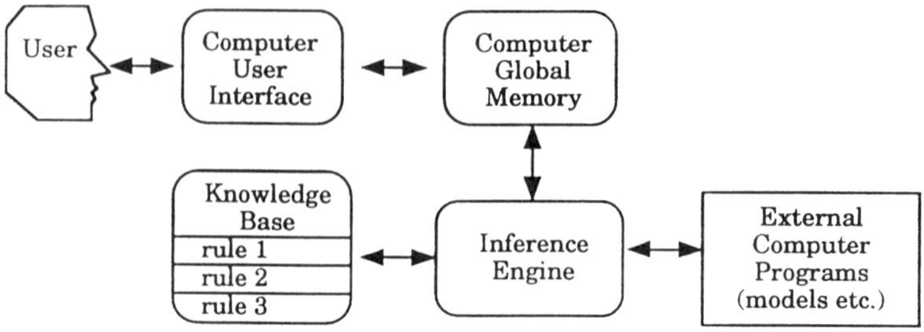

Figure 1 Major components of a simple expert system.

implementation. For the farm manager, making decisions based on a qualitative understanding is still better than making decisions without any understanding (Plant and Stone 1991). Expert systems are a practical approach for developing decision-support systems for better agricultural management.

2.2. Essential Parts of an Expert System

Expert systems typically consist of two main parts: an inference engine and a knowledge base (Fig. 1). The user interacts with the expert system through the user interface, which consists of the keyboard, monitor, and mouse. The inference engine has the "reasoning" capabilities in the expert system. It attempts to come to a conclusion by using information in the knowledge base and by gathering information from the user. The knowledge base contains the logical rules obtained from a person that is an expert in a particular area. Some expert systems are connected to external computer programs (often called demons) which can provide additional information to the inference engine. In most expert systems, the knowledge base is separate from the inference system. Expert system shells are available commercially that include the user interface, inference engine, and an empty knowledge base. Using the knowledge editor provided in the shell, the specific rules must be entered by the person developing the expert system.

People that collect the information needed to build the knowledge base are often called "knowledge engineers." This does not mean one has to be an engineer to build expert systems; some programming experience, however, is usually helpful.

2.3. Types of Expert Systems

Rule-based expert systems use heuristic (rules-of-thumb) information obtained from human experts. The information incorporated in the knowledge-base may be either subjective or objective. Rule-based systems are usually simple and relatively easy to develop and the most successful rule-based systems have focused on narrow problems. SOYBUG is a typical rule-based expert system that makes recommendations about managing four insect pests of soybeans (Beck et al. 1989). This system is based completely on heuristic information from two extension entomologists. It took approximately 3 yr to develop. The system was validated using written scenarios to compare its recommendations with an expert and with an extension bulletin. SOYBUG made the same recommendations as the expert more consistently than a person using the extension bulletin.

Other problems that have higher levels of complexity may require a *frame-based* expert system which uses an object-oriented approach. For example, a rule-based system may be adequate for a system that makes recommendations about economic thresholds for a single insect pest of a given crop. A frame-based approach may be necessary if insects, weeds, pathogens, and weather are included in the system. Frame-based expert systems are more suitable for complex interactions because of their ability to organize information (Minsky 1975). In a completely rule-based system the rules are unstructured except for their order. This leads to problems when one attempts to modify or add new rules to the system. Frame-based systems attempt to overcome this problem by organizing the information as frames. A frame can inherit values from other frames. Information is stored in "slots" within a frame. The value of a slot in a frame is called the "filler". The filler of a slot can also be a subframe of the "parent" frame. An example of possible slots within a "grain frame" are: type of grain, grain moisture, grain temperature, and how long it has been stored. An important feature of frame-based systems is object inheritance. "Child" frames can inherit, by default, the rules and attributes from "parent" frames, which describe the frame in general. Exceptions can be easily added with any child frame. This aids in the revision of the expert system because the knowledge is more organized and is not duplicated throughout the knowledge base.

Some of the more powerful expert system shells incorporate a "blackboard" decision-making scheme. Erman et al. (1980) originated the idea of a blackboard to develop a speech recognition system. The blackboard acts as an administrator, giving out assignments to different knowledge bases within the expert system. Each knowledge base could be viewed as a single expert that the administrator consults to solve a problem. The blackboard allows knowledge bases to communicate with each other, and decides which assignments will be given to a specific knowledge base. The goal of the blackboard is to choose procedures which are most likely to move the problem-state closer to a solution.

Model-based expert systems are coupled with one or more computer simulation models. The inference engine gets the necessary information from the user, runs the models if needed, and interprets the output for the user. Model-based

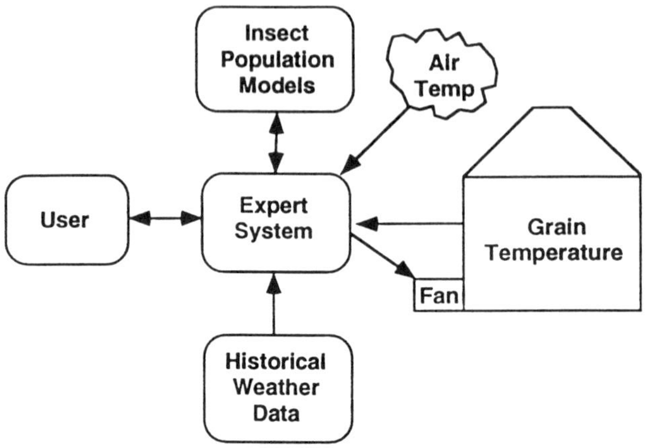

Figure 2 An expert system connected to temperature sensors and fan controls.

expert systems can be programmed to search for an optimum solution to a problem by running a model hundreds of times. The predictive ability of models coupled with the user-friendliness of expert systems results in a very powerful system. It can use the predictive ability of models to augment its heuristic knowledge.

GOSYM/COMAX, a model-based expert system that advises cotton farmers on irrigation and timing of fertilizer applications, is an example of how successful this combination can be (McKinion et al. 1989). COMAX obtains the necessary information from the user to run the simulation model GOSYM and makes recommendations to the user by interpreting the simulated results. Based on a survey of users, the estimated benefits of using COMAX software were between \$100 and \$350/ha (McKinion et al. 1989).

Many agricultural models have been developed that predict complex plant-insect interactions and economic loss. Few of these models, however, are being used by farmers. Coupling computer models with expert systems technology could encourage farmers and extension agents to use the powerful predictive ability of these models.

Expert systems can be connected to sensors to collect real-time data and to switches that automatically control equipment (Fig. 2). The expert system could monitor important variables (e.g., external air temperature and grain temperature). The expert system could use historical weather data and insect models to conduct cost/benefit analyses of different aeration strategies. It could use this information to directly control aeration fans.

Expert systems developed for real-time control of equipment have several advantages over traditional programming of microprocessor controls. Expert systems are more flexible as rules are written in English and can be easily modified. Powerful control strategies can be developed using simulation models and historical data. For example, a decision of whether to turn on the aeration fan might depend on: 1) current temperatures of grain and outside air, 2) estimated insect density in the grain, 3) forecasted air temperatures, 4) time until the grain may be sold, and 5) predicted insect population growth under nonaerated and aerated conditions.

3. ADVANTAGES AND LIMITATIONS OF EXPERT SYSTEMS

3.1. Advantages of Expert Systems

One of the primary advantages of expert systems is that they can act as a repository of knowledge (Coulson and Saunders 1987). Human experts eventually retire. Unless their expertise is passed on to others or is carefully written down their knowledge is lost. If the expert's knowledge is successfully captured by the knowledge engineer then the expert system becomes a permanent working model of the expert's knowledge. Over time the knowledge base can be continually improved by comparing it's diagnoses or predictions with actual case histories.

Some expert systems can "self-learn" by creating new rules based on results fed back to the computer during actual use. Such expert systems can potentially become smarter over time. While this sounds like a good idea, in practice it could be disastrous. Agricultural systems are by nature unpredictable and an expert system could easily learn a set of rules during one unusual year that would lead to poor decision making for several future years. The advantage of a nonlearning rule-based system is that you have complete control over what rules are in the knowledge base.

Expert systems can utilize both objective and subjective information. Agricultural systems are complicated and our information about the entire system is usually incomplete. Nevertheless, human experts often successfully diagnose problems in agricultural settings where objective scientific data are lacking. This type of problem is ideally suited for expert systems. Management systems can be designed that use a combination of objective information—scientific results and predictive models if they are available—and subjective knowledge "rules-of-thumb" from experts. This allows a management system to be quickly developed and put into use. One of the biggest obstacles to implementation of conventional models is that they are never quite completed. There is always some other part of the system that cannot be modeled due to a lack of research data. Expert systems can complement modeling by filling the gaps with experiential knowledge from an expert until more scientific information becomes available.

Expert systems usually have explanatory capabilities that enable the user to ask the program why it gave a certain diagnosis or prediction. Because the system remembers its chain of reasoning, it has a record of the rules that led to a conclusion. For example, the system might respond: "The expert system determined that the probability of insect infestation is high because duration of grain storage is greater than 140 d, initial grain temperature is high, and initial grain moisture is high." This capability offers a way to educate users. Teaching systems can be developed that use the explanatory capability of the expert system to educate users. In effect, they are using heuristics in the expert system to turn inexperienced managers into experienced decision makers. In addition, the expert system can adjust its subject coverage based on the student's answers. For example, if a student is having difficulty in a particular area, the expert system can provide additional information.

Expert systems can integrate information from several disciplines, such as entomology, engineering, and economics. Traditionally, farmers obtain pieces of information from several different disciplines. In most cases, it is up to them to integrate this information so as to best manage their farm operations. Several integrated agricultural expert systems have been developed (Saunders et al. 1987, Plant 1989) that help farmers make crop management decisions based on weeds, insects, soil fertility, disease, and economics. It takes a great deal of cooperation from a team of experts to develop these systems. Getting experts to agree on what happens when various factors interact is not always easy. Nevertheless, the potential benefits are high. These systems not only provide information from diverse disciplines but, more importantly, they consider the effects of multiple stresses on a crop, such as disease, weeds, and insects.

One of the first things a user notices when using an expert system is how easy they are to use. Most expert system shells include a graphical user-interface. They have mouse support, pull-down menus, on-line help, and easy access from one part of the expert system to another. Most expert systems have the ability to display graphics. Some examples are diagrams that could be used to identify weeds or insects, or to diagnose equipment problems.

3.2. Limitations of Expert Systems

Expert systems do not handle broad domains well. It is almost impossible to develop a useful expert system for an area that requires a very broad range of knowledge. For example, designing an expert system that would diagnose problems for all electrical appliances would be very difficult because of the diverse types of electrical appliances that exist. Unlike a human expert, the expert system would fail the first time the user wanted information on an unknown type; whereas the expert might be able to figure out how to fix the appliance based on his past experiences with similar appliances.

Because of their inability to handle broad problems, they appear to have a built-in rigidity or inflexibility compared with a human expert. Expert systems cannot easily jump to a new line of reasoning because they do not have human intuition. Some of the more powerful expert system shells have attempted to overcome this deficiency by incorporating a "blackboard" decision-making scheme (Erman et al. 1980), which was previously described.

Another short-coming of expert systems is that their recommendations are only as good as the information provided by the user. If the user provides incorrect information, the system may arrive at an erroneous conclusion. This is also a problem that human experts frequently encounter. Humans, however, may be more adept at determining the accuracy of the information and may ask additional questions to verify key information. The expert system can be designed to reduce such errors by including practical limits for input data and by including default values.

4. ORDER OF EVENTS IN DEVELOPING AN EXPERT SYSTEM

4.1. Problem Identification

The first step in developing an expert system is to carefully define the problem that the system will solve (Prerau 1985). While this may seem to be obvious, in many cases it is overlooked. By carefully defining the problem domain, knowledge engineers can avoid creating irrelevant expert systems. At this stage developers also should ask themselves whether the problem is suitable for an expert system. The problem should be solved using mostly declarative rather than procedural methods. For example, a problem that could be solved using mathematical equations would be more easily solved using a procedural language such as FORTRAN, whereas a problem in identification or diagnosis would be appropriate for an expert system. This is not a rigid rule because many problems require both types of knowledge. The system developers must determine whether the problem is important enough to develop an expert system: how many people would use the expert system and benefit from its use? The knowledge engineers should also determine whether the problem is too small. A simple test is to make a flow diagram of the problem. If the diagram is too difficult to be used as a decision aid in an extension manual, then it is probably well-suited for an expert system (Plant and Stone 1991).

4.2. Creating the Knowledge Base

Knowledge acquisition is one of the most difficult steps. The knowledge engineer extracts information from an expert, or several experts, written reports, simulation models, etc. When obtaining knowledge from an expert, a series of interviews is conducted by the knowledge engineer. During the interview, the expert is asked to explain some aspect of their problem-solving behavior. In the early stages of rule formation the rule will take the form of a hypothesis (Pederson 1989). The expert will then self-test proposed "theories" by mentally comparing them with previous case histories. During this process, the expert will probably find some exceptions to the first version of their rule. They will then reformulate the rule so that it includes the exceptions. The knowledge engineer's task is to help the expert put nonverbal thinking processes into language.

Developing a rule-base is fairly straightforward, and does not require a degree in computer science. In many cases the expert can act as the knowledge engineer. Coding the knowledge in a rule-based system is done using either forward chaining (data driven) or, more commonly, backward chaining (goal driven). Some expert systems allow a combination of both methods. Forward chaining is similar to traditional procedural programming. It attempts to gather all known facts before testing which hypotheses are true.

Backward chaining starts with a hypothesis and "chains" together rules to attempt to verify the hypothesis. For example, an expert system may have the following hypotheses:

1. D is true.
2. E is true.

In attempting to verify the first hypothesis, the system looks for rules that have the hypothesis as a consequent (the "then ..." part of a rule). The inference engine finds the following rule in the knowledge base:

If C is true **Then** D is true.

It next searches for rules with C as a consequent and finds:

If B is true **Then** C is true.

It then searches for rules with B as a consequent and finds:

If A is true **Then** B is true.

Finally, it searches for rules with A as a consequent, if it finds none, it may then ask the user if A is true. Statements A, B, and C must all be true if the first hypothesis is to be verified. If at any point, one of the antecedents (the "if..." part of a rule) is false, the system will jump to the next hypothesis. This type of reasoning is called backward chaining because it starts with the hypothesis as a goal and then attempts to verify the hypothesis by linking together rules until the hypothesis is either verified or rejected.

The type of reasoning method the knowledge engineer selects depends on the problem. Backward chaining is appropriate for problems that have a limited number of hypotheses. Forward chaining is appropriate for problems that have many possible solutions, as, for example, selecting all possible aeration fans for a certain size of grain bin. Backward chaining is more similar to human reasoning and is generally more efficient; the questions follow each other in a logical sequence. In a forward chaining system the questions don't follow a logical order, which can often be confusing to the end user.

There are several good expert system shells that run on many hardware platforms. Some common commercial shells that run on IBM-compatible personal computers are: Nexpert Object, Level 5 Object, ExSys, Knowledge Pro, and PC plus. Artificial intelligence (AI) languages such as LISP or PROLOG can also be used but are more difficult to program. Up to the early 1980s, most expert systems were developed on expensive LISP workstations. Today, most expert system shells are written for less expensive personal computers as well as for workstations. This facilitates delivery of the system to the end-user because the developer is assured that in most cases the expert system will run on hardware that is available to the end user.

While it is easier to develop an expert system when only one expert is involved, it is often desirable to obtain information from several experts. This often increases the reliability of the system because it is based on a broader range of experiences. The drawback of using several experts is that it is frequently difficult for them to agree on a common solution to a problem. This is usually because agriculture is not an exact science and there may also be more than one solution to a problem. Frequently, experiments in agricultural research have not been conducted to determine the best solution or there are so many variables that it is impossible for an expert to know the outcome for every combination. Often it is up to the knowledge engineer to try and reach an agreement by determining which explanations are common, and which are most consistent with available test results.

4.3. Testing the Expert System

The expert system must be thoroughly "debugged" before it is put into use. This involves both verification and validation (O'Keefe et al. 1987). Verification entails making sure that there are no errors in the logic of the programming code (i.e., that rules do not contradict each other and that there are no misspellings). Validation involves determining whether the expert system agrees with human experts and case histories, and is usually more difficult than verification. This is probably why many published reports of expert systems do not provide any procedures for evaluating the quality of the recommendations made by the system (Geissman and Shultz 1988).

To validate SOYBUG, Beck et al. (1989) used a series of validation and calibration cycles. The current version of the expert system was checked in a validation step by testing the rules calibrated in the last cycle with the expert's recommendations for new scenarios. Rules in SOYBUG were revised in response to any disagreements with the expert. The process continued until there was good agreement between the expert system and the expert, which took four or five cycles.

The knowledge engineer also needs to verify that all possible outcomes are represented in the rulebase. If they are not, the system will not be able to conclude anything if that combination of rules is not in the rulebase. As with simulation models, it is extremely difficult to validate expert systems for every possible combination of variables. It is easier to validate a simple model than a complex one; however, if the model or expert system is too simple it becomes trivial. The important point is to develop a rulebase that provides the best advice to the user, which may not necessarily be the one that most experts agree on.

4.4. Implementing Expert Systems

Implementing expert systems is a difficult step, especially with agricultural expert systems, because it often involves transfer of the system from a research to an "extension service" mode. Although extension personnel are often involved in development, most expert systems have been developed by researchers. Ideally, the developers of the expert system would be involved in the implementation stage. Realistically, researchers receive little credit for extension activities. Researchers need to be given more credit by administrators for transferring their products into practical use. The involvement of extension personnel in the project can also ease the transition. Extension personnel can often indicate how well the system will be accepted by farmers, because they are much more in touch with farmer's needs than are researchers.

It is also important that the personnel responsible for introducing the software to users are adept at using it themselves, so that they can quickly handle problems that end users encounter. In addition, the expert system manual needs to be concisely and clearly written. Although expert systems are touted for their user friendliness, they are more difficult to support than extension bulletins.

4.5. Maintaining Expert Systems

Expert system maintenance is also a critical part of the development cycle. The expert system needs to be viewed as an evolving work, one that will never be completed but will be continually improved as it is being used in the field. Providing the necessary support for the software is critical to its success. This is especially true for agricultural expert systems in which the users often have low levels of computer skills. Personnel must be available to answer users' questions and to provide training in the use of the system. The expert system will need to be continually updated to include new information, correct problems, and add improvements to the program. Thus, researchers should be involved in the maintenance program. Developers of expert systems and administrators need to be aware of the long-term financial and manpower commitments that are required before projects are initiated.

5. EXAMPLE EXPERT SYSTEM STORED GRAIN ADVISOR

5.1. How Stored Grain Advisor was developed

At least two expert systems have been developed for managing insect pests of stored grain. Grain Pest Advisor was developed by several British researchers (Denne 1988, Wilkin et al. 1991), and Stored Grain Advisor was developed by US scientists (Flinn and Hagstrum 1991). We will discuss Stored Grain Advisor, an expert system that was developed for US grain storage conditions.

The first step was to decide which problems were to be solved by the system. Flinn and Hagstrum (1990b) initially selected the following two problem areas:

1. Predicting the probability that the grain will become infested by insects.
2. Determining if insect populations in the grain have reached densities that require fumigation.

The first problem involves predicting whether or not insects would reach economically important densities in stored grain under a given set of grain storage conditions. Population growth of grain insects is a function of time, grain temperature, and grain moisture. Other management factors also affect insect population growth, such as insecticide concentration in the grain and aeration. These relationships are dynamic and it is thus very difficult for a stored-grain expert to accurately forecast the quantitative effects that these variables would have on insects in stored grain. Validated computer models are available that accurately describe population growth for five of the most important insect pests of stored wheat, as a function of time, insecticide, grain temperature, and grain moisture (Flinn and Hagstrum 1990a, Hagstrum and Flinn 1990, Kawamoto et al. 1989). While many expert systems have been developed that directly include simulation models, there are drawbacks to this approach, the foremost of which is the time required to run a simulation. It takes approximately 6 min to run a 1-yr simulation for one species and one set of environmental conditions on a 33 MHz IBM PC-compatible 80386 with a math coprocessor. Because of the

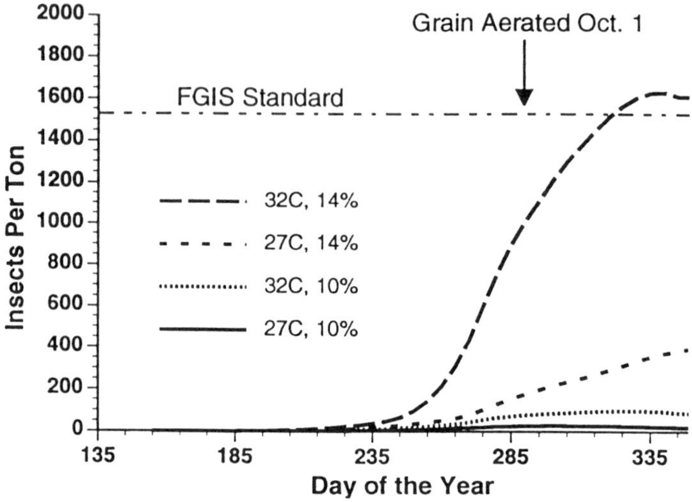

Figure 3 An example of simulation outputs from the model in which initial grain temperature and moisture were varied. The horizontal dashed line indicates an insect density that is likely to be detected with current marketing practices in the United States.

rather long time needed to run the model, Flinn and Hagstrum (1990b) used the model as a source of expertise to generate rules. A majority of the rules in the management submodule were derived from over 500 simulations conducted with the computer models. Variables that were changed in the simulations included: insect species, bin sanitation, storage date, initial grain temperature, initial grain moisture, aeration, and insecticidal protectant. The length of time that the grain is stored above 20°C, before it is either aerated in the fall or cools naturally, is a critical variable that affects population growth potential. Three storage dates: June 1, July 1, and August 1 were used to simulate wheat storage in the southern, middle, and northern wheat growing regions of the United States. The results of these simulations were then translated into rules for the knowledge base. The rules are predicated on the present US marketing regulations that allow up to two live insects/kg sample of wheat.

The rules were generated by examining graphs of predicted insect density versus time (Fig. 3). Low, moderate, and high potentials for insect infestation were assigned if the predicted number of insects, on the expected date of sale, was less than 27 adults/t, 27–2700 adults/t, or greater than 2,700 adults/t, respectively. If a high probability of infestation is predicted, the user is advised of two options: cool the grain using aeration in September or consider selling the grain earlier than planned. In addition, the system notifies the user that they should sample the grain for insects, starting about 1 mo after storage. If a moderate probability of infestation is diagnosed, the manager is advised that he should sample the grain for insects about 60 d after storage to see if treatment is needed. If a low probability of infestation is diagnosed, the manager is advised that the grain will probably be safe and that it is unnecessary to sample the grain for insects unless he decides to sell the grain at a later date than planned.

The second problem (determining if fumigation is necessary) required a combination of heuristic and algorithmic knowledge. Fumigation thresholds for probe-trap catch were based on consultations with a stored-grain entomologist with extensive field experience with both probe traps and grain triers. Rules are based on the number and kinds of insects caught in grain triers or probe traps, length of time the traps are left in the grain, time until sale, and time until fall cooling. The expert system uses statistical equations that convert probe-trap catch to a 5-d standard (Cuperus et al. 1990), based on duration of sampling and species differences in trap catch susceptibility. Because probe-traps tend to catch more insects than traditional sampling equipment (grain trier), there is a greater chance that grain may be unnecessarily fumigated. Stored Grain Advisor should reduce this problem because it automatically interprets probe-trap catch for the user.

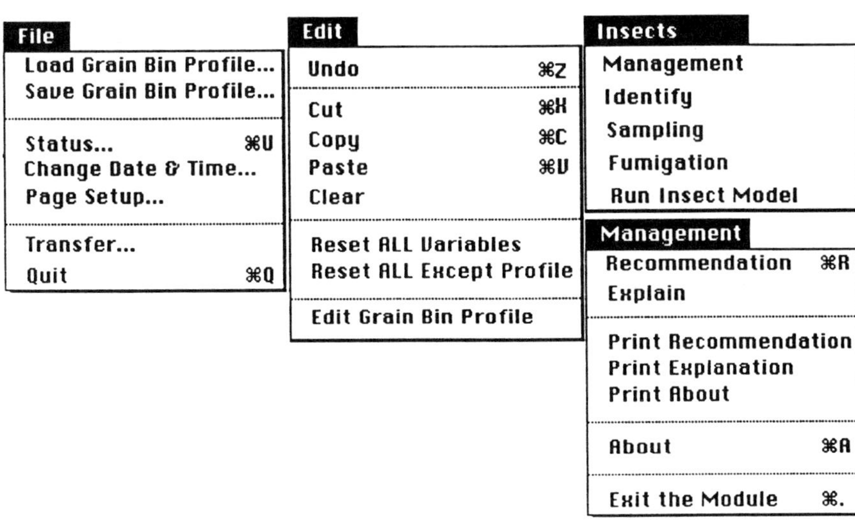

Figure 4 Menu headings available in the main start-up screen and headings within the "management" submodule of the expert system, Stored Grain Advisor.

5.2. How the User Interacts with Stored Grain Advisor

The expert system has typical Macintosh pull-down menus (Fig. 4). The main menus are: File, Edit, and Insects. Under the "File" menu, the user can load or save a bin profile, change the current date and time, or quit the program. The bin profile contains stored information about the management facility and individual bins (one profile is used for each bin). Information stored in the profile is relatively static. Some examples are: bin size, storage date, protectant, initial grain temperature, and moisture content.

Upon starting the expert system, the manager (user) is greeted with the start-up screen and a dialog box asking whether the manager wishes to open a bin profile. If the manager selects yes, then he is asked to select from previously stored profiles. If he selects no, then he is asked for the information needed to complete a new bin profile. A grain manager would have one bin profile for each bin. Upon exiting the expert system, the manager is asked if he or she wants to save the current bin profile. If yes is selected, then the profile is saved under the same file name, or it can be renamed.

Submodules under the Insect menu are: "Management," "Identify," "Sampling," and "Run Insect Model." For simplicity, we will only describe how the "management" submodule works. After selecting "management" from the menu, the user answers the following questions (ANSWERS).

1. Was the bin sampled for insects? NO
2. What type of bin sanitation was used? GOOD SANITATION
3. What is the storage date and the expected date of sale? JUNE 15 AND DECEMBER 30
4. What was the initial storage temperature and moisture content (wb) of the grain? >29°C AND 11–12%
5. Will the grain be aerated in the fall? NO
6. Was a grain protectant applied at the time of storage? NO

After answering the questions as indicated, the user is presented with the following recommendation:

This bin has been diagnosed as having a high probability for insect infestation. We advise:

1. Sampling the grain using probe traps or a grain trier about 60 days after the grain was harvested to be sure you do not have an insect problem.
2. Cool the grain using aeration in September or early October (if you plan to hold the grain past the end of October). This may eliminate the need to fumigate the grain.

At the bottom of the screen the user is asked whether he or she wants to calculate a safe storage period. If the user selects the yes button, the expert system calculates the length of time the user should be able to store his grain before insects reach detectable levels. In this case the expert system displays:

> Insect populations in your grain should remain below economically damaging levels for about 93 d, which is approximately until the 22 of September.

With this expert system it is easy to quickly investigate many different management scenarios. This is accomplished by selecting "reset all variables" from the menu and then selecting "recommendation." This erases any previous investigations from memory and starts a new consultation. By playing with the expert system the user can determine which management practices work best under their conditions.

6. FUTURE EXPERT SYSTEMS FOR STORED-GRAIN MANAGEMENT

6.1. Future Expert Systems for Farm-Stored Grain

Anything that can be done to reduce labor will be appreciated by most grain managers (assuming reasonable costs and high reliability). Automation of aeration equipment, as well as sampling of grain temperature, grain moisture, and insect density in the grain should be goals for future expert systems. Most farmers today do not sample their grain for insects for several reasons: lack of sampling equipment, unpleasant and often hazardous conditions in the grain bin, and lack of economic incentives. In the future, it may be possible to automate insect sampling by using acoustical detection (Hagstrum et al. 1991) or carbon dioxide detection (Muir et al. 1985).

An expert system could constantly monitor microphones in a grain bin (Fig. 5) to detect developing insect infestations at a much earlier date than is currently possible. The expert system could predict, using computer models of grain temperature and insect populations, when infestations will reach damaging levels based on estimates of existing populations (Flinn and Hagstrum 1990a).

In the design and management of many stored-grain ecosystems the growth of microorganisms and mites must be considered along with the risk of insects. Several agricultural engineering researchers (e.g., Bloome and Shove 1971, Metzger and Muir 1983, Huminicki et al. 1986) developed design and management guidelines for drying and aerating stored grain for several sets of condi-

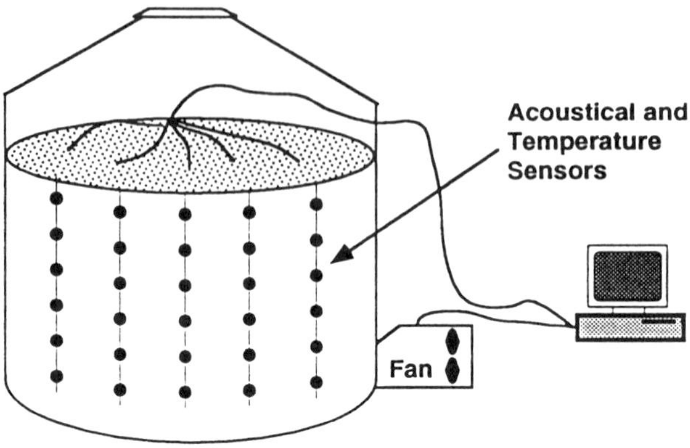

Figure 5 An expert system connected to temperature and acoustic sensors in a grain bin.

tions. The guidelines were developed using mathematical models to simulate heat and moisture transfer in the grain bulk, and statistical relationships to predict the effects of fungal deterioration and time on the stored grain. These computer programs can be difficult to use and require long computer running times. In addition, output results can be difficult to interpret. Expert systems can use these models to design grain ventilation systems and can simplify the inputting of data and operation of the computer models. The developer of the expert system can include heuristic rules based on expert knowledge that reduces the computer simulations and running times. The principal benefits of the expert system is the interpretation of the simulation results. The expert knowledge in the system can assist the user in designing and assessing the design alternatives (e.g., adding heat during drying). Peart et al. (1985) developed an expert system to control a fan to dry corn with air at near-ambient conditions in Indiana. Comparison of computer simulation results indicates that in-bin drying controlled by the expert system removed moisture more efficiently with less spoilage risk than continuous fan operation. Advanced expert systems could receive data directly from weather stations (past, present, and forecasted conditions) and from sensors located in the grain bulk and ventilating air. The expert system could use these data in conjunction with mathematical models, rules in the knowledge base, and the manager's objectives, to directly control bin-ventilation equipment. Expert systems could be developed that learn from the operating experience with a specific bin. For example, using sensor data collected during the operation, the expert system could correct initial estimates of airflow rates for a particular fan and bin combination.

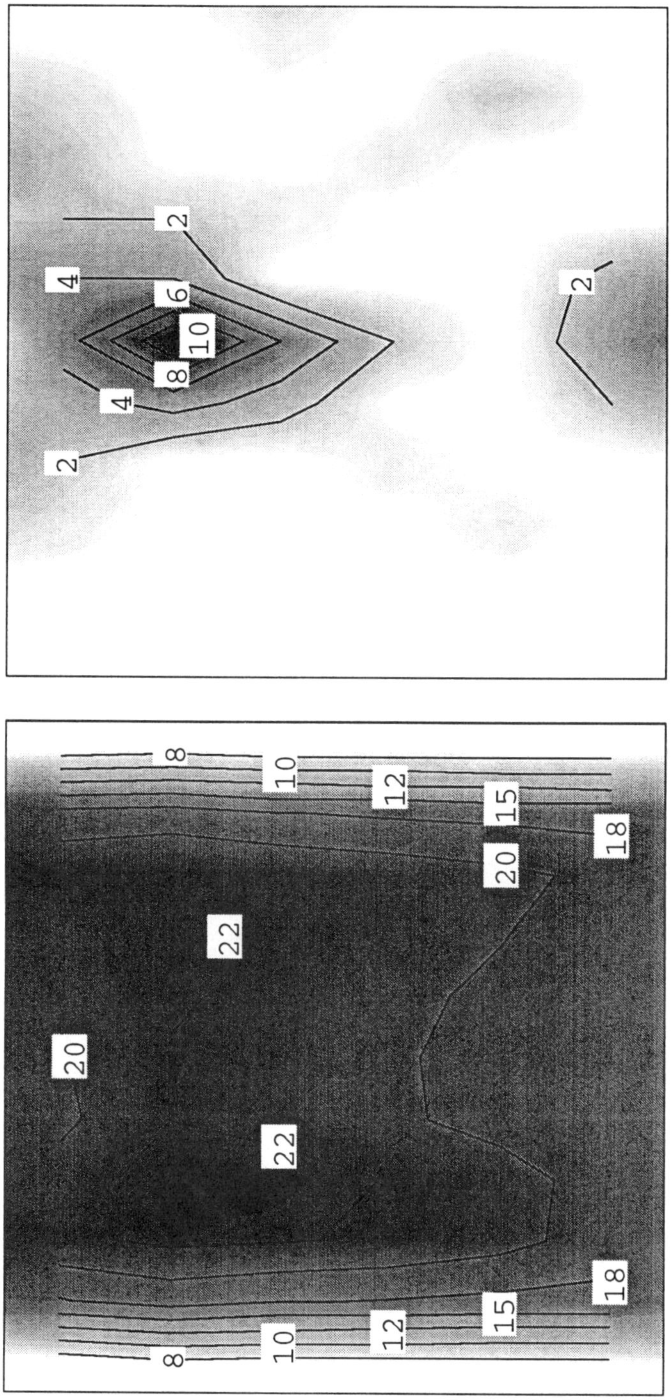

Figure 6 Hypothetical examples of grain temperatures (left) and insect densities (right) that an expert system could display to the user of an expert system.

6.2. Stored-Grain Expert Systems for Large Commercial Facilities

As with farm-stored grain, emphasis should be placed on providing the grain manager with accurate information on current and predicted grain conditions. This is even more important in large commercial facilities because of the huge amounts of grain involved. Many commercial facilities already have cables installed in grain bins that monitor grain temperature. Managers often use this information to detect insect hot spots in the grain. Acoustical sensors, however, have the advantage of detecting insect densities that are two or three orders of magnitude lower than those that temperature sensors can detect. By detecting infestations at a much earlier date (1–2 mo), managers would have more options available to them; cool the grain using aeration, reduce moisture content of grain by drying or mixing with drier grain, or ship it at an earlier date. As with management systems for farm-stored grain, expert systems could automatically control aeration to minimize shrink and losses due to insects. Acoustical sensors could provide accurate estimates of initial insect populations in the grain. Models in the expert system then could predict population trends for different management scenarios and recommend optimum management strategies to the grain manager. In addition, the system could provide graphical displays of insect densities and grain temperatures within a bin (Fig. 6). This could help in spot treatments and provide guidance in preventing future infestations.

Expert systems are a practical approach for developing decision-support systems for better grain management. Their importance will probably increase in the future as we lose the traditional insecticides due to environmental concerns (e.g., malathion, phosphine, and methyl bromide). In the future, we will be forced to rely more on biological and physical control methods to manage insect pests of stored grain and less on traditional chemical controls. Because these methods will require much more knowledge about the grain storage system to be used effectively, grain managers could benefit from a decision support system. These expert systems will probably include models that predict the effects of physical and biological controls on the pest organisms, and will aid the manager in selecting environmentally and economically sound control strategies.

REFERENCES

Beck, H. W., J. Pierce, and J. W. Jones. 1989. SOYBUG: an expert system for soybean insect pest management. Agric. Sys. 30:269-286.

Bloome, P. D. and G. C. Shove. 1971. Near equilibrium simulation of shelled corn drying. Trans. ASAE (Am. Soc. Agric. Eng.) 14:709-712.

Coulson, R. N. and M. C. Saunders. 1987. Computer-assisted decision-making as applied to entomology. Ann. Rev. Entomol. 32:415-437.

Cuperus, G. W., W. S. Fargo, P. W. Flinn, and D. W. Hagstrum. 1990. Variables affecting capture of stored-grain insects in probe traps. J. Kansas Entomol. Soc. 63:486-489.

Denne, T. 1988. An expert system for stored grain pest management. Unpub. Ph.D. thesis. University of London, London.

Erman, L. D., F. Hayes-Roth, V. R. Lesser, and D. R. Reddy. 1980. The Hearsay-II speech-understanding system: Integrating knowledge to resolve uncertainty. Comput. Surv. 12:213-253.

Flinn, P. W. and D. W. Hagstrum. 1990a. Simulations comparing the effectiveness of various stored-grain management practices used to control *Rhyzopertha dominica* (Coleoptera: Bostrichidae). Environ. Entomol. 19:725-729.

Flinn, P. W. and D. W. Hagstrum. 1990b. Stored Grain Advisor: a knowledge-based system for management of insect pests of stored grain. AI Appl. 4:44-52.

Flinn, P. W. and D. W. Hagstrum. 1991. An expert system for stored grain management. pp. 2011-2018 *In:* Proceedings 5th Intern. Working Conf. Stored-Prod. Prot. Vol. 3. Impremerie du Médoc at Bordeaux-Blanquefort, France.

Geissman, J. R. and R. D. Shultz. 1988. Verification and validation of expert systems. AI Expert, February 26-33.

Hagstrum, D. W. and P. W. Flinn. 1990. Simulations comparing insect species differences in response to wheat storage conditions and management practices. J. Econ. Entomol. 83:2469-2475.

Hagstrum, D. W., K. W. Vick, and P. W. Flinn. 1991. Automatic monitoring of *Tribolium castaneum* populations in stored wheat with computerized acoustical detection system. J. Econ. Entomol. 84:1604-1608.

Huminicki, D. N., C. I. Kitson, O. H. Friesen, and W. E. Muir. 1986. Computerized design of ventilation systems for stored grain. Paper NCR86-604. Am. Soc. Agric. Eng., St. Joseph, MI. 17 pp.

Kawamoto, H., S.M. Woods, R.N. Sinha, and W.E. Muir. 1989. A simulation model of population dynamics of the rusty grain beetle, *Cryptolestes ferrugineus* in stored wheat. Ecol. Model. 48:137-157.

McKinion, J. M., D. N. Baker, F. D.Whisler, and J. R. Lambert. 1989. Application of the GOSSYM/COMAX system to cotton crop management. Agric. Syst. 31:55-65.

Metzger, J. F. and W. E. Muir. 1983. Aeration of stored wheat in the Canadian Prairies. Can. Agric. Eng. 25:127-137.

Minsky, M. 1975. A framework for representing knowledge, pp. 211-271. *In* P. H. Winston (ed.) The psychology of computer vision, McGraw-Hill, New York. 282 pp.

Muir, W. E., D. Water, and R. N. Sinha. 1985. Carbon dioxide as an early indicator of stored cereal and oilseed spoilage. Trans. ASAE (Am Soc. Agric. Eng.) 28:1673-1675.

O'keefe, R. M., O. Balci, and E. P. Smith. 1987. Validating expert system performance. IEEE Expert 2:81-90.

Onstad, D. W. 1984. Management of potato leafhopper, *Empoasca fabae* (Homoptera: Cicadellidae), on alfalfa with the aid of systems analysis. Environ. Entomol. 13:1046-1058.

Peart, R. M., Y. C. Li, and J. R. Barrett. 1985. Simulation of a computerized grain drying decision support system. Paper 85-3014. Am. Soc. Agric. Eng., St. Joseph, MI.

Pederson, K. 1989. Expert systems programming: practical techniques for rule-based systems. John Wiley & Sons, Inc. New York. 298 pp.

Plant, R. E. 1989. An artificial intelligence based method for scheduling crop management actions. Agric. Syst. 31:127-155.

Plant, R. E. and N. D. Stone. 1991. Knowledge-based systems in agriculture. McGraw-Hill, Inc., New York. 364 pp.

Prerau, D. S. 1985. Selection of an appropriate domain for an expert system. AI magazine. Summer. 26-30.

Saunders, M. C., C. W. Haeseler, J. W. Travis, B. J. Miller, R. N. Coulson, K. D. Loh, and N. D. Stone. 1987. GRAPES: an expert system for viticulture in Pennsylvania. A.I. Appl. Nat. Res. Manag. 1:13-19.

Wilkin, D. R., J. D. Mumford, and G. A. Norton. 1991. An expert system for stored grain management. pp. 2039-2046 *In:* Proc. 5th Intern. Working Conf. Stored-Prod. Prot. Vol. 3. Impremerie du Médoc at Bordeaux-Blanquefort, France.

3

Grain Storage in the Tropics

Christopher P. Haines

Natural Resources Institute,
Chatham Maritime, Kent, United Kingdom

SUMMARY

An overview is presented of some of the main features of grain storage ecosystems in the tropics and our understanding of them. The diversity of scale, complexity, and character of these systems, together with the high rate of biological activity under tropical conditions, make them more difficult to analyze than those in temperate climatic zones. Scarce research resources in developing countries are more often devoted to short-term, pragmatic investigations. Pest communities are complex and, in most cases, their ecology is not well understood, though some notable advances have been made. Approaches to pest management are changing, especially because of failures of conventional techniques due to insecticide resistance and also because of increasing global concerns about the use of synthetic pesticides. Other influences on the storage ecosystem, including macro-economic policy, are also discussed. Suggestions for future research priorities and needs are presented.

1. INTRODUCTION

Grain storage systems in the tropics are extremely diverse in scale, complexity, and character. An ecosystem approach (as described by Sinha and Muir 1973, Multon 1988, and Sinha 1991) has been advocated for the analysis of grain losses in these warm-climate systems (Calderon 1979, Dunkel 1992). However, although many authors take a holistic view of tropical postharvest grain systems (e.g., the proposals for sorghum presented by Chung and Deyoe 1982), there are very few examples of truly ecosystematic analyses in tropical storage. A literature search of agricultural databases on the subject of storage ecosystems reveals many references to analyses of temperate systems (e.g., Campbell and Sinha 1990, Cuff and Hardman 1980, Hagstrum and Heid 1988, Longstaff and Cuff 1984, Sinha et al. 1986, Wallace et al. 1983, White and Sinha 1979). By contrast, there are fewer specific references to storage ecosystems in the tropics and some of these (e.g., Prakash and Rao 1987) use the term to define the area of interest but do not include any ecological analysis.

Thus, although the value of a holistic approach is promoted or acknowledged by many authors, there have been few major ecological studies of storage systems in the tropics. The systems themselves are less amenable to study than their equivalents in temperate climates because:

1. there is a much greater range of types of storage structures;
2. the marketing systems are often much more complex and commonly include informal market sectors;
3. the physical conditions (high temperature and, in some zones, high water activity) promote greater biological activity and rates of change;
4. the enhanced biological activity leads to greater dynamic variability in the ecological systems; and
5. the biological complexity and diversity is correspondingly greater.

Therefore, an ecosystem approach to tropical grain storage is much more challenging than in temperate climates because the higher complexity, diversity, and rate of change create much greater data needs. These data needs for ecosystem analysis in tropical developing countries have to compete with the need for narrowly-focused studies to address immediate problems affecting current food security. It is not surprising that the more pragmatic short-term needs are usually given a higher priority for the limited research and development resources.

Where resources are available for such research, a holistic ecosystem approach is often hindered by the lack of realistic synecological studies of tropical pest species (Haines 1982). Indeed, even autecological understanding is limited for some important tropical pests, as in the following examples.

1. A literature survey by Hodges (1979) revealed less than 70 references to the biology and control of *Corcyra cephalonica* (Stainton), only a few of which added substantially to our knowledge of this species, in spite of its importance as a secondary pest of cereal grains and their products in many tropical countries. It is possibly sometimes misidentified as *Ephestia cautella* (Walker), which—with its wider distribution, including the southern United States—has been much better studied (Abreu et al. 1982 lists 605 references for this species).

2. Very little is known about the ecology of *Liposcelis* spp. (except *L. bostrychophilus* Badonnel, which is a pest in temperate climates) and even the pantropical *L. entomophilus* (Enderlein), which reaches population densities of over 4000 insects/kg, has only recently been investigated (Rees and Walker 1990, Pike et al. 1991, Natural Resources Institute 1991b).
3. More is known about the cosmopolitan *Sitophilus oryzae* (L.) than the pantropical *S. zeamais* Motschulsky, and the association between these two important pest species and, respectively, paddy and white rice is a relatively recent observation (Anonymous 1981).

Without these autecological "building blocks" (as summarized briefly by Haines 1991), synecological research on pest communities is very difficult.

Because the external ambient conditions throughout the year in the tropics are often similar to the conditions in storage, there is a greater interaction between storage ecosystems and external habitats than in temperate systems. Although some authors (e.g., Chakravorty et al. 1983) emphasize the influence of climatic seasonality, it is more likely that tropical storage pest populations are influenced by the temporal patterns of storage and consumption, and by residual populations of the pests and their natural enemies, than by ambient climatic fluctuations.

In the literature on tropical grain storage systems, there is moderate (though often laboratory-based) information on the autecology and population ecology of major pests (especially those that are also of economic importance in developed countries). But with only a few notable exceptions in the fields of community ecology and ecosystem analysis, there is then a large gap between studies of selected species and the overviews of storage system interactions. Major synecological studies are rare.

2. GRAIN MANAGEMENT IN THE TROPICS

2.1. Conceptual Approaches

Grain production in most areas is generally confined, by climatic seasonality, to one or two (exceptionally three) harvests per year. Storage can be viewed as a mechanism for minimizing fluctuations between supply and demand due to seasonal availability of grain, inter-annual production levels, weather, changing consumer demand, transport delays, etc. The need for storage thus depends on the variation in these factors and the level of protection required from the undesirable effects of the fluctuations (Hindmarsh and Trotter 1990). This view of the grain storage system is, of course, highly amenable to socio-economic analysis, not only of the options for storage technologies but also of the costs and benefits of storage itself (as opposed to other market strategies for stabilizing the fluctuations of availability and price).

An alternative view of storage systems is that they are a component of the energy flow system from production of grain biomass to human consumption, with accompanying energy transfers involved in the process of spoilage. Energy budgets have been calculated by Stepien et al. (1973) for the common tropical storage mite *Tyrophagus putrescentiae* (Schrank) and Sinha (1982) presented energy budgets for six storage beetles, five of which are common in the tropics: *Cryptolestes ferrugineus* (Stephens), *Oryzaephilus surinamensis* (L.), *Tribolium castaneum* (Herbst), *Rhyzopertha dominica* (Fabricius), and *Sitophilus oryzae* (L.). The evaluation of such detailed budgets is time-consuming and there are probably insufficient data to attempt construction of a total energy budget based on the biotic and abiotic components of even the simplest tropical-storage ecosystems. A less detailed approach may yield a worthwhile analysis, however. For example, Nisanka and Misra (1990), using standard energy conversions for food items and external fuel sources to supplement their observations on biomass energy, have produced a total energy budget for an Indian village ecosystem.

Socio-economic and energetics approaches to grain storage provide quantifiable and objective analyses of the systems. However, Weymes (1980) points out that "a grain-storage system forms a part of the whole web of social and cultural networks within which a society operates" and that the social structures and functions related to grain storage should be understood before any changes in storage techniques are contemplated. Such social development issues are especially important in village-level storage and a lack of consideration of them can lead to the transfer of inappropriate technologies and thus to extension failures. One such concern is the role of women in on-farm or communal postharvest operations, and thus the importance of gender issues in relation to: targeting of extension messages, possible shifts in balance of economic status, physical workloads, and social status generally.

2.2. Food Security

The various definitions of food security have been discussed by Maxwell (1990). Although some of the more complex definitions address particular social and political issues, the definition given by the World Bank (1986) is widely accepted, namely: "Access by all people at all times to enough food for an active, healthy life. Its essential elements are the availability of food and the ability to acquire it. Food insecurity, in turn, is the lack of access to enough food." This and similar definitions of food security are described by Hindle (1990), Kennes (1990), Huddleston (1990), and Dearden and Cassidy (1990).

Food security is a complex issue; it involves interactions between production and marketing on the supply side, and availability and poverty on the demand side. The importance of both physical and economic access to food is the key point. As Hindmarsh and Trotter (1990) have emphasized, improved food supply does not necessarily lead to improved food security (e.g., if domestic production is stimulated by higher prices this may decrease food security for the urban or landless rural poor).

The approach to food security adopted in any given situation, which is determined by social, macro-economic, and political forces, determines the applicability of technological options for grain storage and thus influences the storage ecosystems. The linkages between policy objectives and causes of loss in the storage system are illustrated in Figure 1 (based on Tyler 1981). In this typical model, agricultural policy and macro-economic (marketing and pricing) policy are determined at the political level but are often not integrated. By contrast, storage policy is usually determined at an administrative level. As a result, grain storage often has to act as the buffer between the conflicting demands of the production and marketing policies (Tyler, personal communication).

There is much current debate on food security issues, particularly in sub-Saharan Africa, concerning the extent to which governments should retain an operational involvement in managing strategic reserves. Structural adjustment policies are being implemented and many are having a profound effect on food security and the grain ecosystem. Direct government control over grain procurement, storage, and marketing is being relaxed and private enterprise is being enabled to participate. However, if liberalization is poorly planned or executed, there is a risk that standards of storage management and grain quality may be sacrificed. More grain may remain for a longer period in inferior storage systems on-farm or with small traders and the implications for the quality of food grain are obvious. These issues have been discussed by Coulter and Compton (1991).

2.3. Marketing Systems

Although the postharvest system can be quite simple, especially where the majority of production is kept on the farm for family use or barter, it is more often very complex, with grain stocks held on the farm, in cooperatives, by private millers and traders, in large commercial warehouses, and in government-owned (parastatal) central depots. Thus, even though a spatial ecological community may be relatively simple at a single point in the marketing system, the temporal community may be complex, with the species successions being dependent on the specific history of a particular batch of grain.

Marketing systems are variable and are subject to major fluctuations dependent on macro-economic and agricultural policies (Fig. 1). Wright et al. (1981) have shown that network analysis can evaluate the benefits of alternative investments in grain storage and identify bottlenecks in the system.

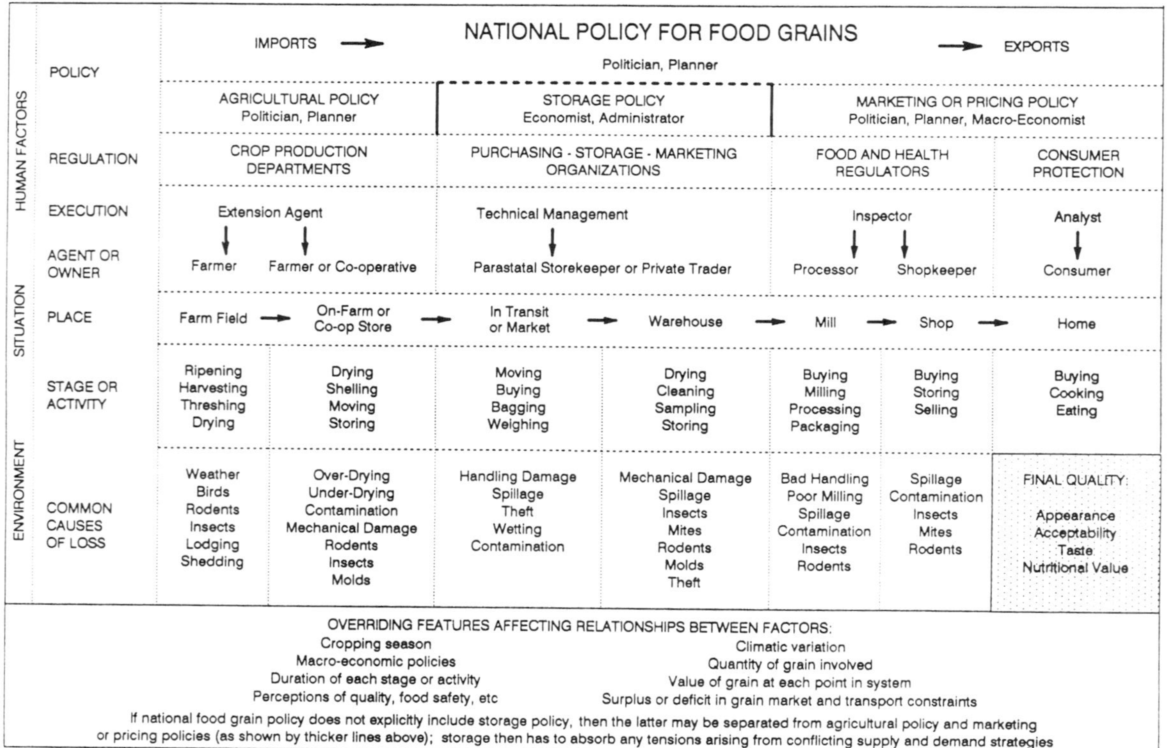

Figure 1 Relationships of factors affecting food grains in postharvest systems, and thus determining the final quality of grain or grain products for the consumer. *Source*: based on a diagram from Tyler (1981), with modifications.

2.4. Types of Storage

The types of storage in the tropics are extremely varied (Anonymous 1988, Mabbett 1986), from the vast silos of large-scale importers and exporters to the small quantities handled by individual subsistence farmers in traditional storage structures (Boxall 1987, Girish et al. 1990). Differences in technical management depending on ownership, extension, and accessible technology add further variety. There are thus considerable variations in the ecosystems that develop in these different storage facilities in response to the physical conditions and the management practices applied. For example, in a survey of pests of stores in Java, Indonesia, Haines and Pranata (1982) found that ownership and type of store had a considerable effect on: the average number of tenebrionid beetle species present per store (e.g., 1.3 species in government godowns, but 4.5 species in cooperatives and retail stores); the frequency of *Tribolium castaneum* in samples (e.g., 94% of samples from commercial godowns, 57% from government godowns, but only 17% from retail stores); and the frequency of occurrence of *Liposcelis entomophilus* (46% of samples from government godowns, 17% from commercial godowns, 7% from private traders' stores, and none in other stores).

On-farm storage is typically small-scale bulk storage, with the structural material and design being influenced by availability of materials, the form of the grain (whether threshed or shelled), and the requirement for in-store drying. Each type of store also has an associated pattern of management and use. Attempts to introduce improved structures have often failed because of a lack of understanding of the local management patterns and the balance between the storage ecosystem and the external environment (Boxall 1987).

Cooperatives and private traders normally store and handle grain in bags. In larger-scale storage, there is a choice between bag or bulk handling, each of which has advantages and disadvantages (Müller 1983). This choice is a complex one, and the factors to be considered are not only those of economics and engineering (Acasio and Maxon 1991) but also the management, logistics, and social issues. These and other factors have been fully analyzed by Friendship and Compton (1991), who have formulated two decision trees (Figs. 2 and 3) to assist in the analysis of the suitability of a storage system for bag or bulk handling. These are intended to be used in prefeasibility studies of whether to change from bag to bulk handling, and particularly to answer the two successive questions: "Is bulk handling possible?" and "Does bulk handling have potential advantages?" Friendship and Compton (1991) emphasize that, if the analysis indicates that bulk handling is possible and has potential advantages, this should be followed by a multidisciplinary feasibility study involving engineering, socio-economics, storage management, transportation, and nutrition.

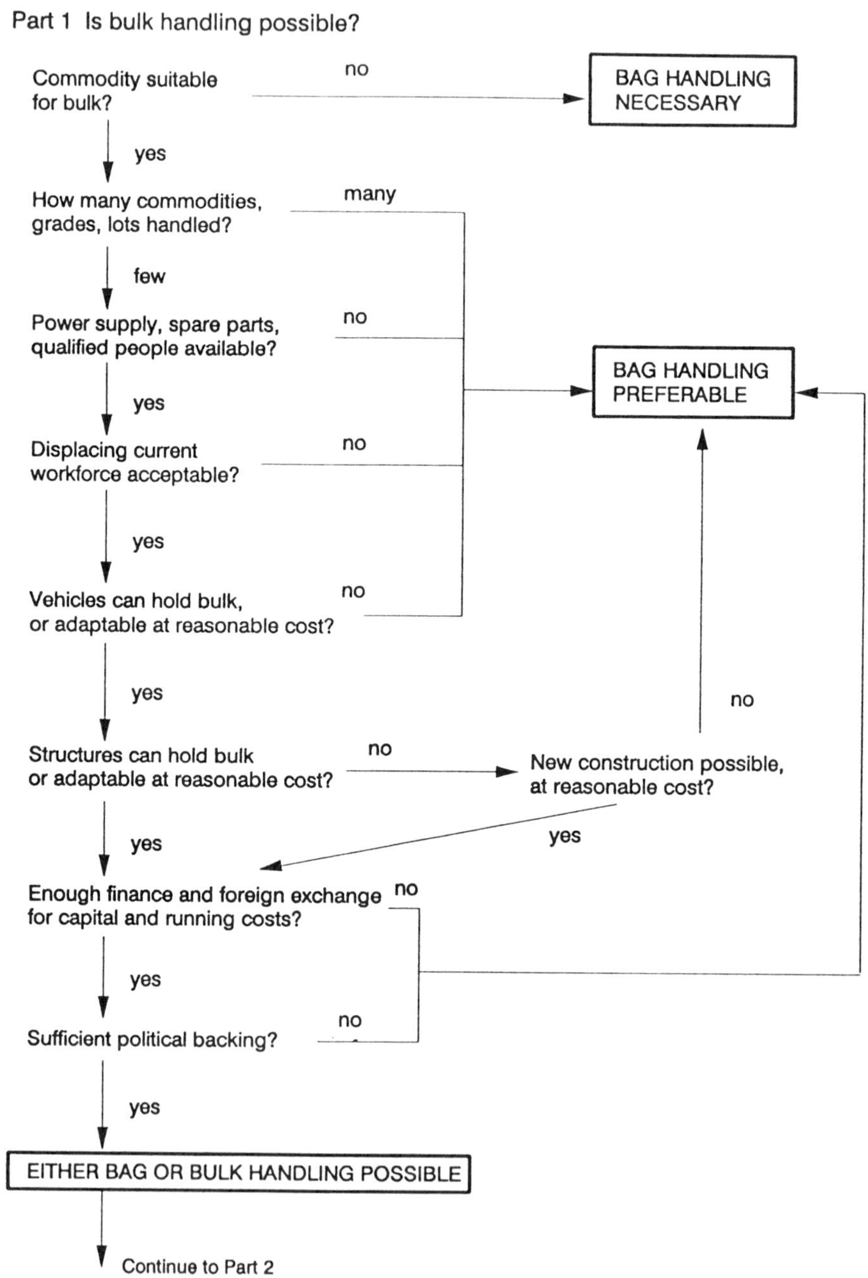

Figure 2 Decision tree for initial study of bag and bulk handling options. Part 1: Is bulk handling possible? *Source*: Friendship and Compton (1991). UK Crown copyright.

Part 2 Does bulk handling have potential advantages?

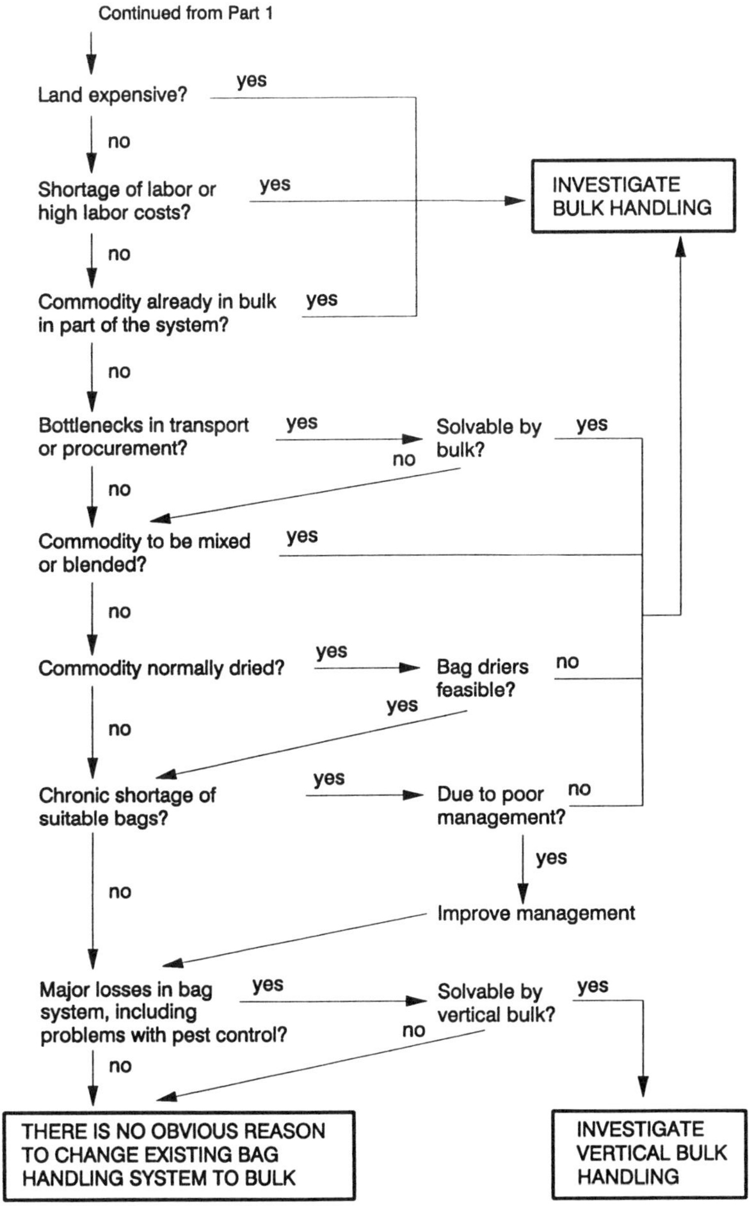

Figure 3 Decision tree for initial study of bag and bulk handling options. Part 2: Does bulk handling have potential advantages? *Source*: Friendship and Compton (1991). UK Crown copyright.

2.5. Causes of Loss

The main causes of loss in storage systems have been discussed by many authors (e.g., Hall 1970, TDRI 1983, Rowley 1984, Okereke and Nwosu 1987, Gwinner et al. 1990, Haines 1991). Many of the agents of loss are interrelated (e.g., under-drying or wetting leads to molds and mites, many mites and some insects feed on molds, insects spread fungal spores, and insect activity leads to increased moisture content). The common causes of loss in a typical postharvest grain system are illustrated in Figure 1, which shows how loss agents are also related to other factors in the system. The loss agents generally fall into three classes: unavoidable (e.g., weather), human-induced (e.g., contamination, spillage, theft), and pest-induced (e.g., insects, mites, rodents).

As grain moves through the postharvest system (Fig. 1), losses of both quantity and quality occur. At the end of the system, irrespective of quantity losses, the final value of the commodity to the consumer is determined by its quality (appearance, acceptability, taste, and nutritional value). Therefore, although the energy-budget approach (Sinha 1982) to losses in the system is more closely linked to an ecological view, it is less effective as an analytical tool for decision-making than the economic approach (e.g., Hindmarsh and Trotter 1990).

Following the review by Adams (1977) of grain storage losses and the manual of loss assessment methodology by Harris and Lindblad (1978), there was a decade of loss assessment activity in the tropics. In some cases, the results showed that earlier estimates had been unduly pessimistic: in others, the reliability of the methodologies was questioned. A critical review of the loss assessment methodology was presented by Boxall (1986) and many authors (e.g., Tyagi and Girish 1978, de Lima 1979, Giga and Katerere 1986, and Giga et al. 1988) have evaluated losses in grain stored in particular areas (see Chapter 4).

3. TROPICAL STORAGE PEST PROBLEMS

3.1. Ecological Linkages

The main biological agents of loss in stored grains are molds, insects, mites, and rodents. The incidence of fungal attack (and mycotoxins) is closely linked to the water activity of the stored grain (see Chapter 8). Rodents are a common component of the storage fauna but their populations in stores are usually an extension of a wider population, either the domestic commensal species of urban areas (with their additional public health implications) or the field species of rural areas (see Chapter 9). Mites are apparently less of a problem in the tropics than in temperate climates, but in the humid tropics they are much more common in grain stores than is generally appreciated (Haines 1991). Insects (especially beetles, moths, and also psocopterans) are ubiquitous in normal grain storage under tropical conditions. Approaches to the control of molds and mites generally depend on drying and monitoring. Rodent control is best achieved by exclusion using barriers, supplemented where necessary by poison-baiting. Insects can neither be excluded (except by special packaging for high-value foods) nor controlled by drying, and thus require complex pest management strategies (see Chapters 5 and 11).

Ecological linkages between the pest groups have not been well studied. The interaction that has attracted the most interest is the one between insects and fungi (Casas 1977, Sinha 1979b). Dunkel (1988) reviewed this interaction and concluded that "some storage insect species are disseminators of storage fungi and some are exterminators", whereas "some storage fungi attract storage insects and promote their population increase and others repel and secrete toxins harmful to insects". Insects as disseminators of fungi in cereal grains, and thus as promoters of mycotoxins, have been studied by several authors in India and Pakistan (Hashmi et al. 1983, Pande et al. 1989, Sinha and Sinha 1990, 1991), whose field and laboratory studies have shown that the presence of storage insects and their damage leads to increased incidence of fungi, especially *Aspergillus* and *Penicillium* species. Studies of interactions between mites and insects are mainly limited to the effects of predatory mites (e.g., Haines 1981b) on insect pests rather than linkages between the respective pest groups.

The wider linkages between one or more of these pest groups and abiotic factors in the storage ecosystem have been considered by several authors (see Chapter 14). Calderon (1979) has emphasized the importance of temperature and humidity in storage ecosystems. Grain temperature and moisture (Pixton 1982) are in turn influenced by climate (Smith and Gough 1990), and Sinha (1975) related the bionomics of many storage insects to their climatic plasticity. Lacey (1988, 1989) has reviewed the effects on storage fungi of many abiotic and biotic factors, and Prakash and Goel (1983) have reviewed insect losses in relation to such factors. Barney et al. (1991) have studied the interactions between *Sitophilus zeamais* and several factors in a stored corn (maize) ecosystem, and Surtees (1964) studied the three-dimensional dispersion patterns of five beetle species in relation to population density, temperature, and moisture content. Prevett and Halliday (1961) analyzed the distribution of temperature inside stacks of bagged groundnuts in Nigeria prior to a study of the effects of temperature and humidity on the dispersion patterns of storage pests in such stacks (Prevett 1964). Also in Nigeria, Giles (1964a) studied the effects of store type, natural control methods, temperature, and moisture content on insects infesting sorghum, millet, and corn in farmers' stores.

Some of the most extreme changes in the abiotic factors affecting grain are caused when it is moved between climatic zones. Studies of grain shipped from temperate climates to tropical ones, either as normal commercial transactions or as emergency food aid, have monitored changes in water activity, fungal contamination, insect infestation, and grain quality (Natural Resources Institute 1991a). In general, normal grain standards for moisture content and dockage in temperate climates are inappropriate for grain that is to be exported to the tropics. The increase in temperature during transit and after import frequently causes the water activity of such grain to exceed 0.7, with consequent problems of mold growth and mycotoxin contamination, and increased insect development.

3.2. Pest Communities

The first essential information for an ecological analysis of biotic factors in the storage ecosystem is an understanding of which principal species are involved. The major insect pests of stored grains in the tropics are quite well known. However, by comparison with the understanding of grain pests in temperate climates, there are notable data gaps. For example, the status of *Sitophilus zeamais* as a dominant pest of milled rice as well as corn, and the dominance of *Sitophilus oryzae* on paddy, was not noted (McFarlane 1978, Osman 1980, Anonymous 1981, Haines 1983) until many years after attention was drawn to the differentiation of the two species, which are probably still misidentified in routine inspections. Similarly, the identity of *Liposcelis entomophilus* as the psocopteran that often occurs in vast numbers on fumigated rice in the humid tropics was first noted by Haines and Pranata (1982), who suggested that it might be tolerant of phosphine. Since then, *Liposcelis* spp. have been reported on stored grain in several tropical countries and were recorded as the most frequent storage insects in a survey of milled rice stores in Guyana (Permual 1991). Research at the Natural Resources Institute (NRI) has shown that the eggs of *L. entomophilus* are, indeed, tolerant of phosphine (Pike, personal communication). The status of *Callosobruchus* species in Indonesia is a further example of misidentification of major pests: Haines (1989) found that *C. analis* (Fabricius) was the dominant pest of pulses in the humid western areas of the country, whereas most research in these areas has been based on the assumption that the main pest was *C. chinensis* (L.).

Reliable and wide-ranging check-lists of insects found in tropical grain stores are uncommon. Examples come from only a few countries, such as: Honduras (Hoppe 1986); Indonesia (Haines 1986); Jamaica (McFarlane 1963); Nepal and Bhutan (Conway 1986); and Nigeria (Giles 1965, Cornes 1973). Without such authoritative check-lists of the main species, the results of any ecological analysis are questionable.

A simplified view of the succession and generalized food web of insects in a typical stored grain ecosystem is given in Figure 4. Although the general principles of succession are well understood, specific studies are uncommon and there are certainly few tropical equivalents of the detailed long-term analysis by Arbogast and Mullen (1988) of succession in stored corn in a humid temper-

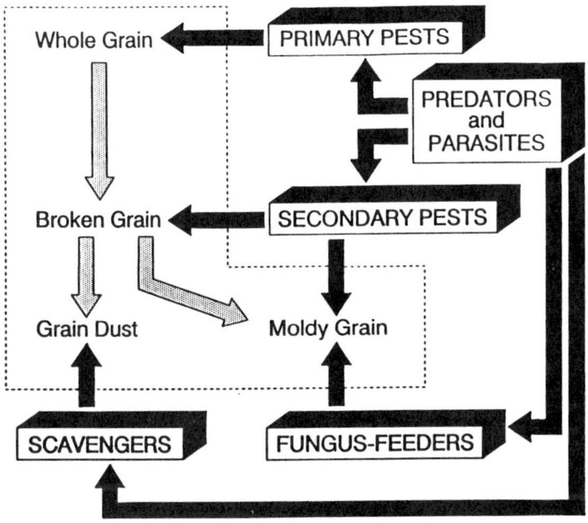

Figure 4 Generalized food web of pests and other species on stored cereal grains. *Source*: Haines (1991) with minor amendments. UK Crown copyright.

ate location (though a similar study of unprotected corn in the humid tropics would probably reach its end-point in one year rather than eight). Chakravorty et al. (1983) in India and Giles (1964b) in Nigeria have reported general patterns of species succession in cereal grains through the storage season, Cornes and Riley (1961) observed rapid successions of pests on stored corn in cribs in Nigeria, and de Lima (1978) studied aspects of insect succession and primary pest competition in farm-stored corn in Kenya. Prevett (1964) made a large-scale study of the distribution and succession of insects infesting stacks of groundnuts in Nigeria and found that instead of the usual increase in complexity of communities during storage, several of the initial pests died out due to changes in physical conditions, leaving *Tribolium castaneum* as the only significant pest at the end of storage. Similarly, in a laboratory study of the succession of the four commonest pests of stored sorghum in Sudan, Shazali and Smith (1990) found that *T. castaneum* was the most likely species to persist after 7 mo.

Detailed ecological studies of on-farm grain storage in the tropics are rare. A notable exception is the major study by Markham (1981) of the ecology of insects in corn cribs in Nigeria. The incidence of individual species was observed from preharvest infestation through to 10 mo of storage and followed a consistent pattern, dominated by *Sitophilus zeamais* but including a great diversity of other pest species and natural enemies. Colonization of the stored corn was shown to be mainly due to active migration of insects to the newly-loaded crib.

Table 1 Most Frequent Insect and Mite Species Found on Stored Products (Mainly Food Grains) in Indonesia.* (Pests Arranged in Order of Frequency of Occurrence.)

Genus and species	Order[+] and family	Habit[x]	% of samples
Tribolium castaneum	Col Tenebrionidae	SPest	39.2
Liposcelis entomophilus	Pso Liposcelidae	SPest	26.8
Sitophilus zeamais	Col Curculionidae	PPest	20.0
Cheyletus malaccensis	Pro Cheyletidae	Pred.	15.4
Rhyzopertha dominica	Col Bostrichidae	PPest	13.4
Oryzaephilus surinamensis	Col Silvanidae	SPest	12.2
Cryptolestes ferrugineus	Col Cucujidae	SPest	11.7
Cryptolestes pusillus	Col Cucujidae	SPest	11.3
Ahasverus advena	Col Silvanidae	MPest	9.7
Tyrophagus putrescentiae	Ast Acaridae	MPest	9.6
Tenebroides mauritanicus	Col Trogossitidae	SPest°	9.2
Sitophilus oryzae	Col Curculionidae	PPest	8.8
Latheticus oryzae	Col Tenebrionidae	SPest	8.8
Corcyra cephalonica	Lep Pyralidae	SPest	8.6
Xylocoris flavipes	Hem Anthocoridae	Pred.	8.6
Carpophilus pilosellus	Col Nitidulidae	SPest	7.3
Ephestia cautella	Lep Pyralidae	SPest	6.2
Araecerus fasciculatus	Col Anthribidae	SPest	5.5
Lophocateres pusillus	Col Lophocateridae	SPest	5.5
Suidasia pontifica	Ast Acaridae	SPest	5.4
Typhaea stercorea	Col Mycetophagidae	MPest	5.4
Choetospila elegans	Hym Pteromalidae	Para.	5.1
Alphitobius laevigatus	Col Tenebrionidae	MPest	4.9
Alphitobius diaperinus	Col Tenebrionidae	MPest	4.8
Palorus subdepressus	Col Tenebrionidae	SPest	4.5
Sitotroga cerealella	Lep Gelechiidae	PPest	4.4

Markham's study is unusual in giving thorough attention to the ecology of the many minor pests and natural enemies. Such studies, the check-lists referred to earlier, and other detailed records (e.g., Haines 1974, 1981a), all reveal that the insect community on stored grain can be more complex than is generally acknowledged. Table 1, for example, lists the 43 most frequent species found in a survey of Indonesian food stores, including 5 primary pests, 21 secondary pests, 8 mycophagous pests, 4 predators, 4 parasites, and 1 scavenger. It is interesting that the fourth most common species was a predatory mite, *Cheyletus malaccensis* Oudemans, that preys on the second most frequent species, the little-studied *Liposcelis entomophilus*. Table 2 shows the taxonomic and habit distribution of the 4931 records of 265 species found in this survey. Although

Table 1 (cont.)

Genus and species	Order+ and family	Habit×	% of samples
Oryzaephilus mercator	Col Silvanidae	SPest	3.9
Aleuroglyphus ovatus	Ast Acaridae	MPest	3.6
Tydeus australensis	Pro Tydeidae	Pred.	3.6
Anisopteromalus calandrae	Hym Pteromalidae	Para.	3.5
Leiodinychus krameri	Mes Uropodidae	MPest	3.4
Palorus genalis	Col Tenebrionidae	SPest	3.1
Cephalonomia waterstoni	Hym Bethylidae	Para.	3.1
Palorus ratzeburgii	Col Tenebrionidae	SPest	3.0
Lasioderma serricorne	Col Anobiidae	SPest	3.0
Bracon hebetor	Hym Braconidae	Para.	2.9
Carpophilus dimidiatus	Col Nitidulidae	SPest	2.3
Blattisocius keegani	Mes Ascidae	Pred.	2.3
Tyroborus miripes	Ast Acaridae	MPest	2.2
Liposcelis paetus	Pso Liposcelidae	SPest	2.1
Pheidole megacephala	Hym Formicidae	Scav.	2.1
Callosobruchus analis	Col Bruchidae	PPest	2.0
Necrobia rufipes	Col Cleridae	SPest°	2.0

* Data are predominantly from collaborative surveys conducted by NRI (UK) and SEAMEO-BIOTROP (Indonesia) in Java, Sumatra and Sulawesi between 1979 and 1984, supplemented by other NRI records. The frequency of occurrence is shown as the percentage of all commodity samples (total 1235) in which each species was found; species found in less than 2% of samples are not listed.

+ Orders are abbreviated to the first three letters: Astigmata; Coleoptera; Hemiptera; Hymenoptera; Lepidoptera; Mesostigmata; Prostigmata; and Psocoptera.

× Habits are: PPest, primary pest of cereal or pulse grains; SPest, secondary pest; MPest, mycophagous pest; Pred., predator; Para., parasite; Scav., scavenger.

° Also a facultative predator.

the survey included some commodities (e.g., dried cassava) other than grain, which will have increased the diversity, the great majority of the species and records were from cereals or pulses.

There is an extreme paucity of observations on mites in tropical storage; even key texts such as Hughes (1976) contain very few records from warm climates and Sinha (1979a) and others have concluded that mites are unimportant outside temperate zones. Although usually not noted by the collectors, samples sent to NRI for identification of insects have often also been found to contain mites (Haines 1974, 1981a). Where mites are noticed they are often referred to as "*Acarus siro* L." on the basis of pictures of this species in pesticide companies' posters and leaflets, even though it is very rarely found in tropical

Table 2 Taxonomic and Habit Grouping of Insects and Mites Found on Stored Products in Indonesia, by Number of Species and Number of Records of Species in Samples Analyzed by NRI*. (Orders and Families Listed in Systematic Order.)

Taxonomic group		No. of species					No. of records				
		Pe	Pr	Pa	Sc	Un	Pe	Pr	Pa	Sc	Un
Acarina:	Mesostigmata	7	4	0	1	5	63	56	0	2	9
	Prostigmata	2	10	2	0	3	12	282	5	0	8
	Astigmata	13	0	0	0	7	321	0	0	0	34
	Cryptostigmata	0	0	0	0	5	0	0	0	0	7
Other Arachnida		0	8	0	1	0	0	101	0	1	0
Thysanura		0	0	0	1	0	0	0	0	6	0
Dermaptera		0	0	0	1	0	0	0	0	6	0
Dictyoptera		0	0	0	4	0	0	0	0	23	0
Psocoptera:	Liposcelidae	5	0	0	1	0	417	0	0	1	0
	Other families	0	0	0	9	0	0	0	0	23	0
Hemiptera:	Anthocoridae	0	3	0	0	0	0	111	0	0	0
	Other families	0	2	0	0	4	0	4	0	0	15
Thysanoptera		0	0	0	0	2	0	0	0	0	3
Lepidoptera:	Tineidae	3	0	0	0	0	6	0	0	0	0
	Gelechiidae	1	0	0	0	0	54	0	0	0	0
	Pyralidae	5	0	0	0	0	202	0	0	0	0
	Other families	0	0	0	0	2	0	0	0	0	2
Siphonaptera		0	0	1	0	0	0	0	3	0	0
Diptera		0	1	0	1	2	0	2	0	1	9
Hymenoptera:	Braconidae	0	0	2	0	0	0	0	45	0	0
	Pteromalidae	0	0	5	0	0	0	0	56	0	0
	Bethylidae	0	0	6	0	0	0	0	90	0	0
	Formicidae	0	0	0	20	0	0	0	0	134	0
	Other families	0	0	2	0	3	0	0	7	0	5

Table 2 (cont.)

Taxonomic group		No. of species					No. of records				
		Pe	Pr	Pa	Sc	Un	Pe	Pr	Pa	Sc	Un
Coleoptera:	Carabidae	0	17	0	0	3	0	19	0	0	4
	Histeridae	0	2	0	0	0	0	6	0	0	0
	Dermestidae	4	0	0	0	0	9	0	0	0	0
	Thorictidae	1	0	0	0	0	7	0	0	0	0
	Anobiidae	2	0	0	0	0	40	0	0	0	0
	Bostrichidae	6	0	0	0	0	204	0	0	0	0
	Trogossitidae	1	0	0	0	0	114	0	0	0	0
	Lophocateridae	1	0	0	0	0	68	0	0	0	0
	Cleridae	1	1	0	0	0	25	23	0	0	0
	Nitidulidae	7	0	0	0	0	142	0	0	0	0
	Cucujidae	3	0	0	0	0	286	0	0	0	0
	Silvanidae	7	0	0	0	0	326	0	0	0	0
	Cerylonidae	1	0	0	0	0	3	0	0	0	0
	Merophysiidae	1	0	0	0	0	10	0	0	0	0
	Lathridiidae	2	0	0	0	0	4	0	0	0	0
	Mycetophagidae	1	0	0	0	0	67	0	0	0	0
	Tenebrionidae	11	0	0	4	6	871	0	0	18	11
	Bruchidae	5	0	0	0	0	54	0	0	0	0
	Anthribidae	1	0	0	0	1	68	0	0	0	1
	Curculionidae	3	0	0	0	0	358	0	0	0	0
	Other families	1	3	0	5	10	1	10	0	25	31
Totals		95	51	18	48	53	3732	614	206	240	139

* Data are predominantly from collaborative surveys conducted by NRI (UK) and SEAMEO-BIOTROP (Indonesia) in Java, Sumatra, and Sulawesi between 1979 and 1984, supplemented by other NRI records. Habit abbreviations are: Pe, pest (including primary, secondary, and mycophagous pests); Pr, predator; Pa, parasite; Sc, scavenger; Un, uncertain or unknown.

Table 3 Mesostigmatan Mites Found on Stored Products (Mainly Food Grains) in Indonesia.* (Entries Arranged in Conventional Systematic Order.)

Taxonomic list	No. of samples
Sejina	
Ichthyostomatogasteridae	
Uropodella laciniata Berlese	2
Gamasina	
Macrochelidae	
Macrocheles sp (near *muscaedomesticae* (Scopoli))	2
Ameroseiidae	
Kleemannia sp (near *plumigera* Oudemans)	1
Kleemannia sp (near *plumosus* [Oudemans])	3
Digamasellidae	
Digamasellus sp	1
Phytoseiidae	
Typhlodromus sp	3
Ascidae	
Blattisocius dentriticus (Berlese)	4
Blattisocius keegani Fox	28
Blattisocius tarsalis (Berlese)	22
Dermanyssidae	
Androlaelaps casalis casalis (Berlese)	2
Uropodina	
Polyaspidae	
?Polyaspis sp	2
?Dinychidae	
Genus and species indeterminate	1
Uropodidae	
Genus and species indeterminate (2 spp, not *Leiodinychus* sp)	12
Leiodinychus sp (not *krameri* or *interstructura*)	1
Leiodinychus ?interstructura (Hirschmann and Zirngiebl-Nicol)	4
Leiodinychus krameri (G & R Canestrini)	42

* Data are from collaborative surveys conducted by NRI (UK) and SEAMEO-BIOTROP (Indonesia) in Java, Sumatra, and Sulawesi between 1979 and 1984.

Table 4 Prostigmatan Mites Found on Stored Products (Mainly Food Grains) in Indonesia.* (Entries Arranged in Conventional Systematic Order.)

Taxonomic list	No. of samples
Pyemotidae	
Acarophenax tribolii Newstead & Duvall	2
Pyemotes sp	3
Tarsonemidae	
Genus and species indeterminate (not *Tarsonemus* sp)	2
Tarsonemus sp	10
Tydeidae	
Pronematus bachewingi Baker	10
Pronematus davisi Baker	3
Tydeus australensis Baker	44
Cunaxidae	
Cunaxa setirostris (Hermann)	1
Stigmaeidae	
Storchia pacificus (Summers)	1
Caligonellidae	
Neognathus sp A (near *summersi* Gerson)	3
Neognathus sp B	4
Cheyletidae	
?Caudacheles sp	1
Chelacheles sp	18
?Cheletonata sp	1
Cheyletus fortis Oudemans	12
Cheyletus malaccensis Oudemans	190
Ker sp	2

* Data are from collaborative surveys conducted by NRI (UK) and SEAMEO-BIOTROP (Indonesia) in Java, Sumatra, and Sulawesi between 1979 and 1984.

stores (except in cool upland areas). Before the aforementioned Indonesian survey was undertaken, there were no modern records of mites in stored food there. The survey found mites in well over 90% of stores, with an average of over five species per store. The records are given in Table 3 (Mesostigmata), Table 4 (Prostigmata), and Table 5 (Astigmata). As already noted, the most frequent species was a predator; the next was the acarid pest *Tyrophagus putrescentiae* (Schrank), which was tenth in the list of frequent insects and mites (Table 1). Some of the predatory species of Cheyletidae, Tydeidae, and Ascidae may play a significant role in the ecosystem, especially the relationship between *Cheyletus malaccensis* and *Liposcelis entomophilus*. The

Table 5 Astigmatan Mites Found on Stored Products (Mainly Food Grains) in Indonesia.* (Entries Arranged in Conventional Systematic Order.)

Taxonomic list	No. of samples
Acaridae	
Genus and species indeterminate (2 spp)	16
Aleuroglyphus ovatus (Troupeau)	45
Caloglyphus hughesi (Samšiňák)	2
Caloglyphus oudemansi (Zachvatkin)	9
Lardoglyphus konoi (Sasa and Asanuma)	1
Madaglyphus javensis Haines and Lynch	2
Michaelopus gallegoi (Portus and Gomez)	3
Suidasia nesbitti Hughes	11
Suidasia pontifica Oudemans	67
Tyroborus lini Oudemans	19
Tyroborus miripes (Athias-Henriot)	27
Tyrophagus palmarum Oudemans	5
Tyrophagus ?perniciosus Zachvatkin	1
Tyrophagus putrescentiae (Schrank)	118
Tyrophagus tropicus Robertson	15
Glycyphagidae	
Genus and species indeterminate	5
Austroglycyphagus sp	1
Blomia tropicalis Bronswijk, Cock, and Oshima	7
Glycyphagus ?domesticus (DeGeer)	1

* Data are from collaborative surveys conducted by NRI (UK) and SEAMEO-BIOTROP (Indonesia) in Java, Sumatra, and Sulawesi between 1979 and 1984.

acarid mites were not often found in large numbers, but on one occasion a very heavy infestation of *Caloglyphus* spp. was found on paddy (rough rice) that was not fully dried; given the high frequency of mite presence there is clearly a potential for mite problems in association with poor drying.

3.3. Autecology of Tropical Storage Pests

Basic information is available on the bionomics (e.g., Howe 1963, 1965) and autecology of the acknowledged major pests. The extent and depth of such information, however, is often closely related to whether they are cosmopolitan species, which are also of importance in the palearctic and nearctic zones, or whether they are mainly pantropical. The examples of *Sitophilus zeamais, Corcyra cephalonica,* and *Liposcelis* spp. were referred to in the introduction and it is interesting to note that a study of *C. cephalonica* by Shazali and Smith (1986) concluded that its pest potential has been underestimated in the past. Similar bias exists for mite species; the temperate *Acarus siro* has been well studied under a range of conditions but there have been very few studies of common tropical species such as *Tyrophagus putrescentiae* and *Suidasia pontifica* Oudemans. The available information on recognition, life-history, ecology, commodity association, and distribution of most of the insects and mites found in tropical stored products is summarized by Haines (1991).

Similar data gaps are found in the literature on rodents. There is a wealth of information on the cosmopolitan urban species, *Rattus norvegicus* (Berkenhout), *R. rattus* (L.), and *Mus musculus* L., but much less on the tropical rural species such as *R. argentiventer* Robinson and Kloss, *R. tiomanicus* (Miller), and *R. exulans* (Peale).

3.4. Synecology of Tropical Storage Pests

As suggested earlier, synecological studies of tropical storage pests are rare. The study of pest populations in groundnut stacks in northern Nigeria by Prevett (1964) was complemented by a smaller study of a similar ecosystem in southern Nigeria (Smith 1963). Giles (1964b) made observations on abiotic and biotic factors affecting pests of stored sorghum, also in Nigeria. In Kenya, Graham (1970a,b,c) made ecological studies of pests infesting bagged corn in warehouses and particularly examined the various factors affecting populations of *Ephestia cautella* (Walker). Also in Kenya, de Lima (1978) conducted laboratory and field studies on the factors affecting the development of *Sitophilus zeamais* and *Sitotroga cerealella* (Olivier) in farm-stored corn. The most detailed study to date has been that of Markham (1981), which examined the interactions of the insect community in corn cribs with moisture content, temperature, grain damage, time of harvest, husk cover, and field damage.

It is pertinent that the last two studies referred to above were undertaken as PhD research and were thus relatively cheap. Field synecology investigations of such dynamic and variable communities are very time-consuming and can thus be very expensive. The only alternative is to make the best use of data arising from other applied research studies. For example, a calibration trial of bait-bags for monitoring pest populations in bag-stacks of milled rice in Indonesia (Haines et al. 1991), provided an opportunity for supplementary analysis of the data on the pest communities and population levels.

Figure 5 shows the population densities of seven beetle pests on the 57 stacks sampled during the calibration trial. Most stacks were dominated by *Tribolium*, often to the near-exclusion of the other pests, and this pest was clearly the commonest agent of cross-infestation in the store complex; it was probably also present on most stocks arriving at the site. *Rhyzopertha* infesta-

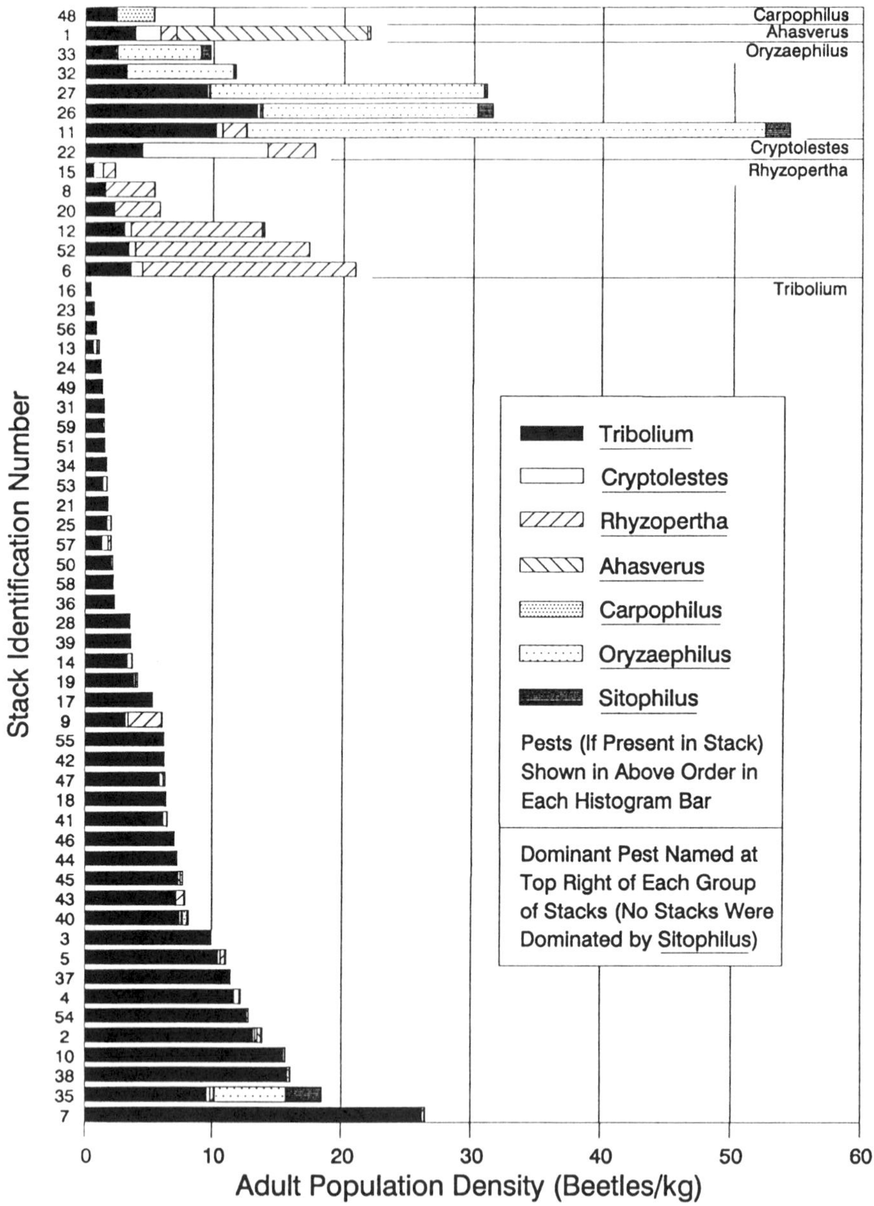

Figure 5 Densities of adult *Tribolium, Cryptolestes, Rhyzopertha, Ahasverus, Carpophilus, Oryzaephilus*, and *Sitophilus* in 57 bag stacks of milled rice at the main rice depot in Jakarta, Indonesia. Results are grouped by dominant pest (as shown at the right) and then arranged in order of total density of the seven beetle species.

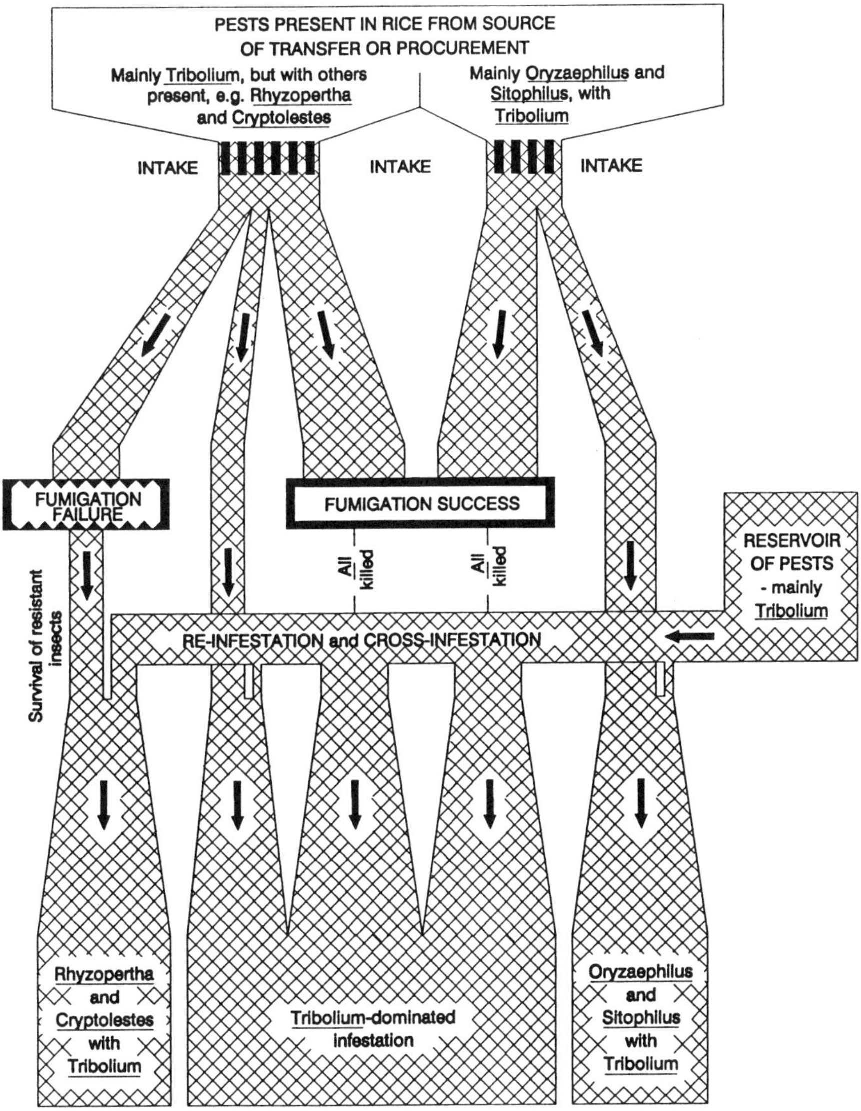

Figure 6 Development of the three main types of infestation found in bag stacks of milled rice at the main rice depot in Jakarta, Indonesia. Schematic diagram showing how they arise from the interaction of: pests present at intake, the use and success of fumigation, and the reservoir of pests in the storage complex.

tions were usually associated with *Cryptolestes* (which in one case was more numerous than *Rhyzopertha*); samples of both species from such stacks were found to be resistant to phosphine. Populations of *Oryzaephilus* were usually associated with *Sitophilus* infestations and these were usually on recently received grain that had not yet been fumigated. Only two stacks did not fit into one of these three patterns.

On the basis of these observations, when compared with records of the histories of the stacks, it was possible to construct a schematic diagram of the development of the three types of infestation (Fig. 6). The pest populations present at intake are dominated either by *Tribolium* or by *Oryzaephilus* with *Sitophilus*. Figure 6 shows how the use and effectiveness of fumigation, together with the reservoir of pests (mainly *Tribolium*) available for cross-infestation, can modify the balance of the different pests and give rise to the three types of infestation.

Multidisciplinary synecological investigations are currently being undertaken by NRI in collaboration with storage technologists in Zimbabwe and elsewhere to determine the cause and prevention of the recent problem of "stackburn" in corn. The symptoms of this problem are cycles of dry-grain heating to about 42°C, followed by thermal disinfestation and slow cooling; this leads to browning of the grain. Initial findings suggest that the heating is caused by insect activity and that the change to use of cheap polypropylene, rather than jute, bags may be implicated.

4. PEST MANAGEMENT IN TROPICAL STORAGE

4.1. Approaches to Management

The options for pest management are dependent on the scale and type of storage system. Integrated and structured approaches to pest management in tropical grain stores have been proposed by several authors (e.g., McFarlane 1981, 1989; Hindmarsh and McFarlane 1983; Haines 1982; Bengston 1984). Figure 7 shows a structured flow-chart for the development and implementation of appropriate pest management methods and Figure 8 shows the system interactions that influence options for pest management and loss reduction. Both diagrams emphasize the need for feedback and adjustment, and the importance of factors other than technical feasibility alone.

Approaches to management of storage pests are increasingly influenced by the development of resistance, not only to contact insecticides but to one of the key fumigants. They are also directly influenced by health and environment considerations related to residues of pesticides in grain, operator safety, disposal of containers, and (in the case of methyl bromide) ozone-depletion. Pest control

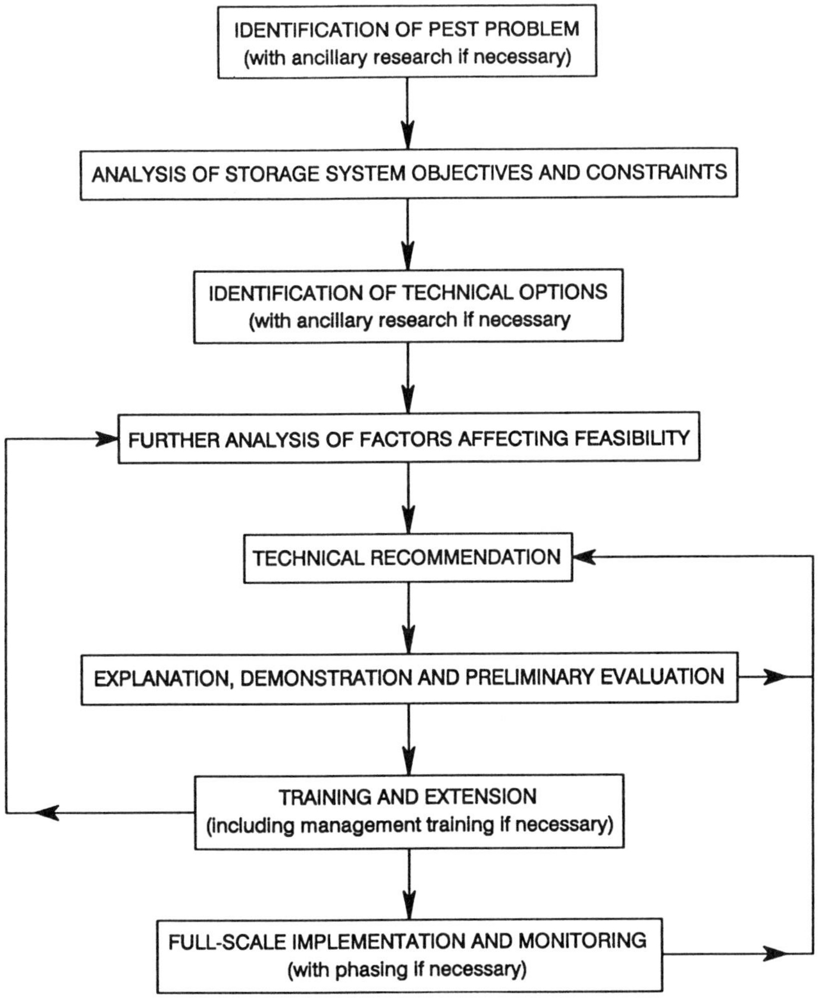

Figure 7 A structured approach to the development of pest management improvements. *Source*: McFarlane (1989), as adapted from Hindmarsh and McFarlane (1983). UK Crown copyright.

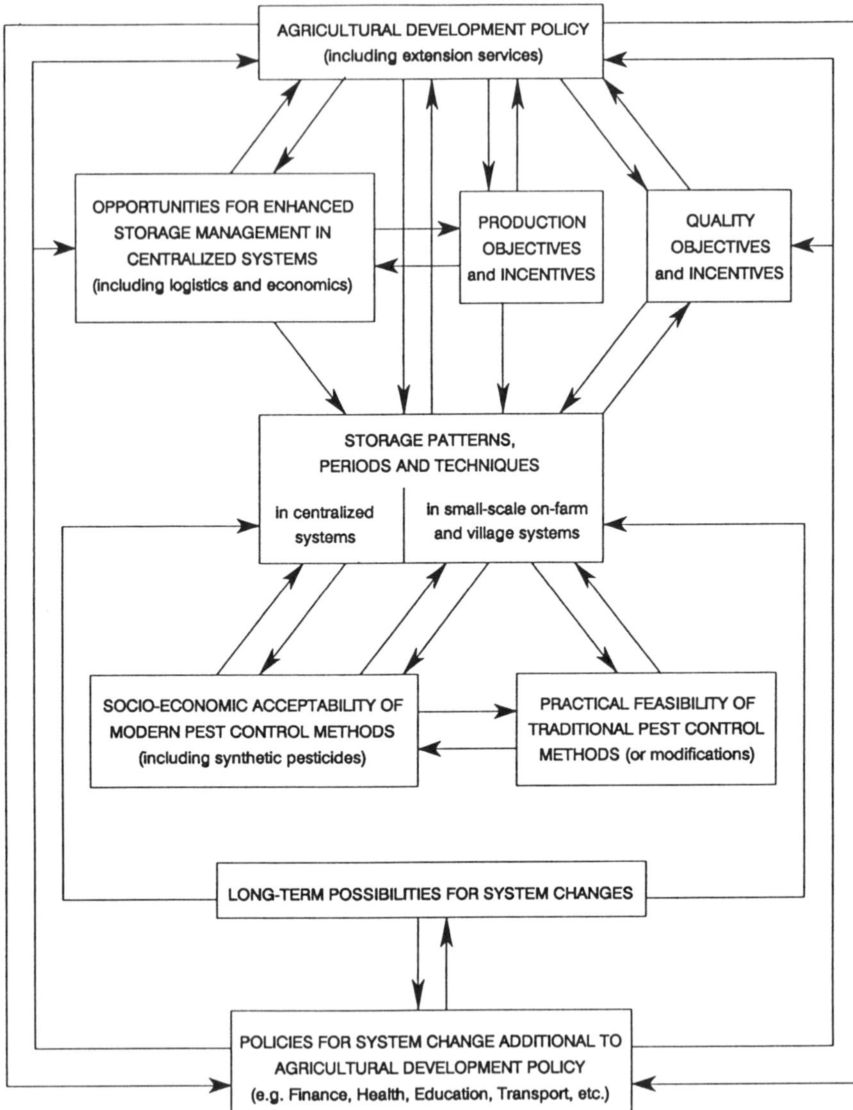

Figure 8 Factor interactions and key issues for pest management and loss reduction in grain storage. *Source*: McFarlane (1989) with minor amendments. UK Crown copyright.

options are also indirectly affected by macro-economic policies, which influence where the grain is stored and who is responsible for its management.

4.2. Options for Control Measures

The potential choice of pest control options in grain storage is great because it is a man-made ecosystem in which control of most of the environmental factors is possible. However, in central stores, with economies of scale (including lower training costs for a limited number of pest control operators), conventional control with synthetic contact insecticides and fumigants has for many years been the cheapest option. In on-farm storage, without economies of scale and with the hazards of untrained handling of toxic materials, cheap and effective options are scarce. Only when farmer or consumer perceptions of pest damage are unusually high (e.g., in areas affected by the larger grain borer, *Prostephanus truncatus* (Horn), in Africa (Dick 1988)) is there sufficient impetus for an effective and sustainable change in pest control practice in farm storage.

One barrier to the development of integrated approaches is the scarcity of studies (such as those of Prevett 1961 and Faroni et al. 1982) of interactions of different control measures. Dobie (1984) proposed modifications to conventional management-systems in grain warehouses (Figs. 9 and 10) and in on-farm corn stores (Figs. 11 and 12), which would incorporate biological approaches to monitoring and control as a step towards integrated pest management.

4.3. Efficacy of Contact Insecticides

Problems of resistance of storage pests to contact insecticides have been steadily increasing over the last three decades. Over the last decade or so, the rate of development of new insecticides for use in grain stores has declined greatly. At the same time, worldwide public concern about pesticide use has discouraged reliance on this method of control. Although admixture to the grain of those insecticides with low mammalian toxicity is still efficacious where resistance is not a problem, this is not a common practice in the tropics, except where special problems require it (e.g., infestations of *Prostephanus truncatus*).

In large-scale tropical storage, contact insecticides are more commonly used in regular residual spraying of store and stack surfaces, especially to reduce re-infestation following fumigation. There has been concern for some time about whether such spraying is cost-effective in typical stores in tropical climates. Hodges et al. (1990) investigated the effect of regular spraying regimes in godowns in Indonesia and found that spraying other than at the time of fumigation was not cost-effective in relation to the refumigation interval. Hodges and Dales (1991) have shown in field trials in Ghana that the residual efficacy of insecticides applied to typical absorbent store surfaces (e.g., brick, concrete, etc.) was very poor, with effective control being lost in less than a week and

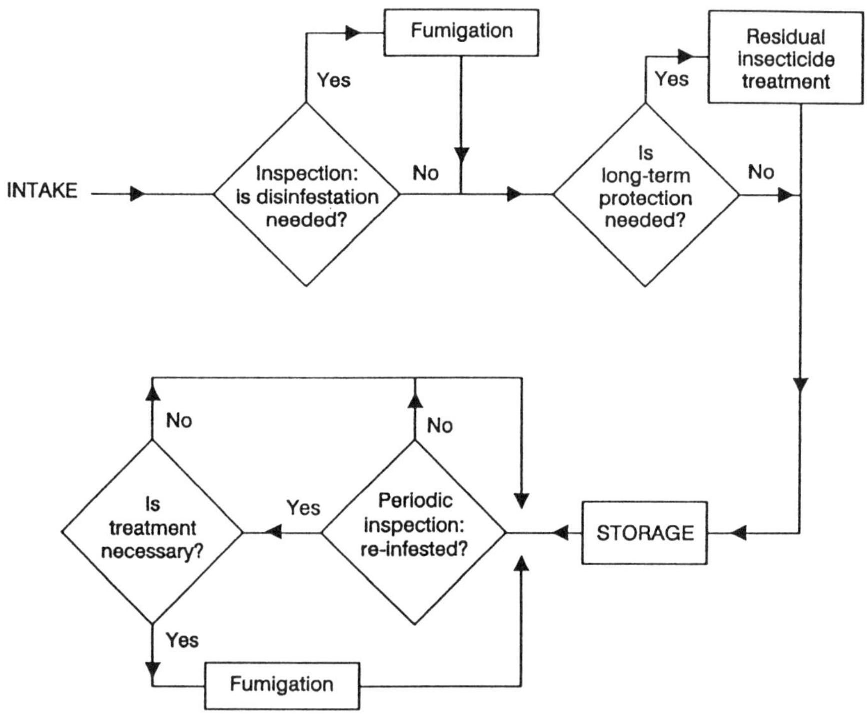

Figure 9 A common management pattern for the storage of bagged grain in a medium or large warehouse in the tropics, with dependence on chemical pest control and conventional inspection methods. *Source*: Dobie (1984) with minor amendments. UK Crown copyright.

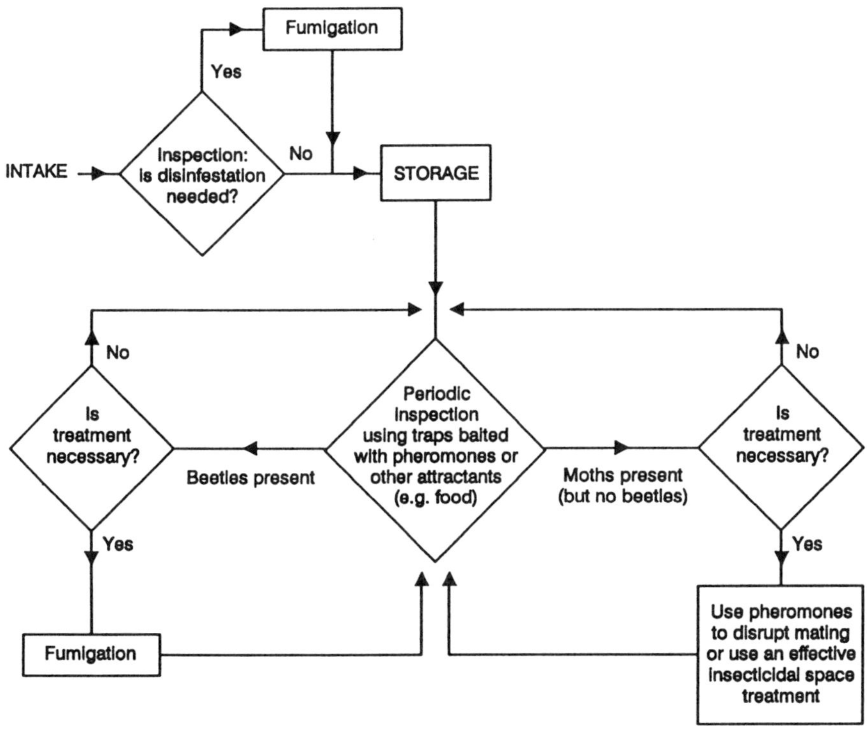

Figure 10 A modified management pattern for the storage of bagged grain in a medium or large warehouse in the tropics, with alternative control measures and improved pest monitoring methods. *Source*: Dobie (1984) with minor amendments. UK Crown copyright.

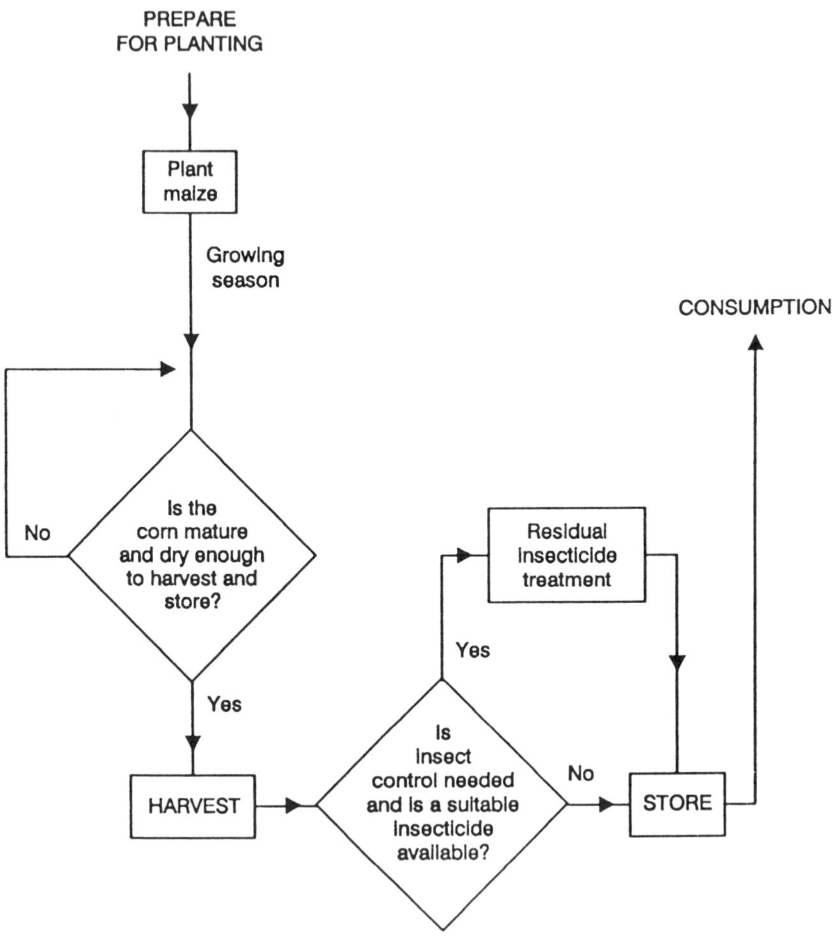

Figure 11 A common management pattern for the production and on-farm storage of corn in the tropics. *Source*: Dobie (1984) with minor amendments. UK Crown copyright.

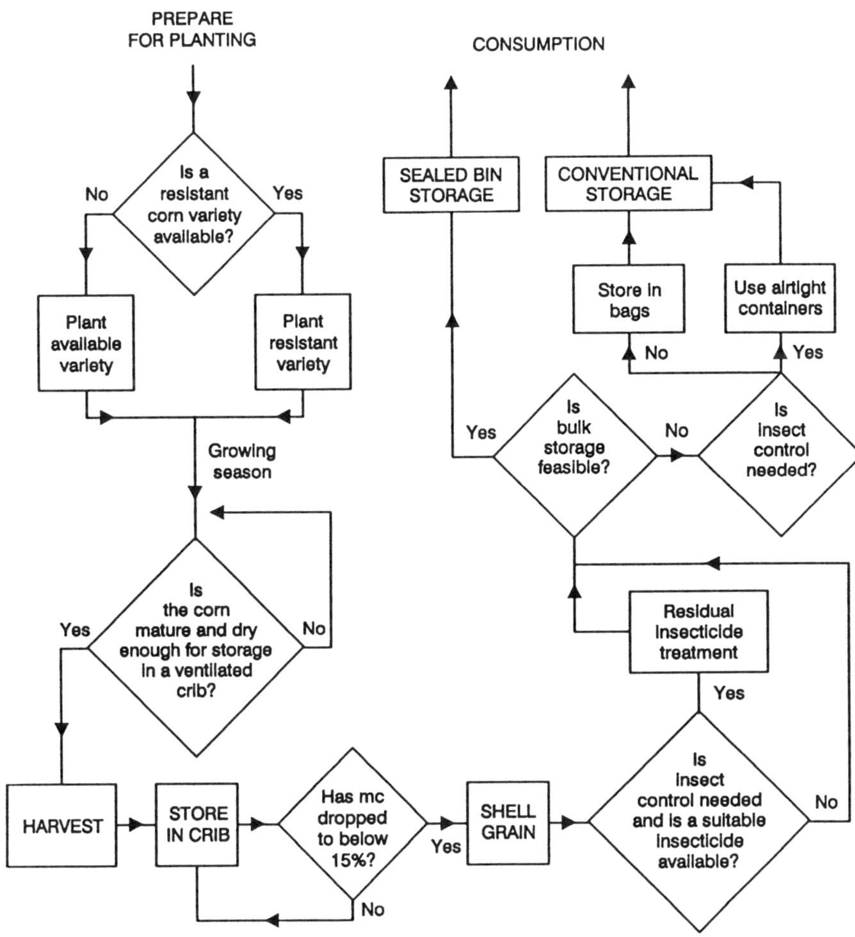

Figure 12 A modified management pattern for the production and on-farm storage of corn in the tropics, including optional use of resistant corn varieties, ventilated cribs, and types of storage for the shelled grain. *Source*: Dobie (1984) with minor amendments. UK Crown copyright.

often less than one day in most instances. Such treatments are clearly not cost-effective. The use of spraying at the time of fumigation to control insects outside the sheeted stack (Graham 1970d, Boon and Ho 1988) remains an effective measure.

4.4. Use of Fumigants

For many years, the use of phosphine as a grain fumigant has been gaining in popularity in tropical countries. Although fumigation periods of 5 d or more are required, this fumigant is easy to use and requires no specialized application equipment. Its ease of use by partly-trained or untrained operators, and the temptation to shorten the exposure period, however, have led to an increasing incidence and level of phosphine resistance among pest populations in several parts of the tropics (Taylor 1991). A considerable improvement in standards of dosing, application, sealing, and exposure periods is necessary if this fumigant is to remain useful in the long-term.

Methyl bromide remains effective as a grain fumigant, though it requires greater expertise and more capital investment than phosphine. It is the only widely-available practical control measure that can achieve a complete kill in as little as 24 h, with appropriate dosing, and it is thus of particular importance in the international grain trade at points of export and import. It is, however, implicated as an ozone depleter and may be banned by international agreement within a few years. It will be difficult to find a replacement control measure for rapid disinfestation.

4.5. Alternatives

The use of natural plant and mineral products with insecticidal or repellent properties is well established in traditional storage (Golob and Webley 1980). Although various plant products are widely promoted, any recommended use of extracts or synthetic mimics should be subject to the same mammalian toxicity testing and registration as conventional pesticides because naturally-occurring compounds are not by definition safe, as some authors claim. The use of edible vegetable oils has been studied by many authors (e.g., Ivbijaro et al. 1985) and avoids concerns about toxicity. The last decade or more of literature on these natural products and on other alternative methods, such as natural enemies and insect growth regulators, has been reviewed by Rees and Dales (1992).

Controlled-atmosphere (CA) storage is a technique that is receiving steadily increasing attention and long-term storage under carbon dioxide is now commonly used in countries such as Indonesia (see Chapter 11). To date, the emphasis has been on large-scale technology; the possible extension of CA techniques to small-scale, on-farm storage has not been addressed. The possibility of cooling grain by aeration for insect control has been studied by some authors (e.g., Thorpe et al. 1982) and the principles and benefits of aeration were reviewed by Thorpe (1986). Thorpe and Whitaker (1992a) have produced detailed mathematical models of heat and mass transfer in ventilated bulks of grain (see Chapters 14 and 15).

The potential for postharvest varietal resistance of grains to pest attack has been extensively studied. For example, a resistant variety of common bean has been developed and evaluated by Centro International de Agricultura Tropical (CIAT) and collaborators (Padgham et al. 1992, Cardona et al. 1992). Studies on cereals such as corn (e.g., Arnason et al. 1992) are still in the research stage of identification of possible biochemical resistance factors. The great advantage of varietal resistance is that the control technology is, in effect, applied centrally and extension is via established distribution systems for new varieties. The obvious application of modern biochemical and genetic techniques and gene transformation technology to this approach could circumvent previous constraints of bioassay delays and the difficulties of retaining recessive resistant genes in conventional plant breeding. The role of such genetic transforms in relation to legal, political, and economic issues for tropical developing countries is, however, still under international debate.

4.6. Monitoring of Pest Problems

Rates of biological development in tropical conditions, especially in the humid tropics, create a need for frequent monitoring and more rapid decisions on treatment than in temperate climates. There is a particular need for early warning of the presence of fungi and mycotoxins in stored grain before they reach levels that may cause spoilage or contamination. Coker (1991) has presented developments in sampling and analysis of mycotoxins, and Phillips et al. (1991) have reported the use of monoclonal antibodies as a highly specific and sensitive method for detecting fungi.

For insect pests, it has been shown by Hodges et al. (1985) that the conventional technique of routine spear-sampling cannot give useful estimates of population levels for the purposes of pest control decision-making but that certain trapping methods were sufficiently accurate for this purpose. On milled rice, the most appropriate trapping method is the use of brown-rice bait-bags and a field trial has calibrated the number of pests per bait-bag against population densities in the milled rice (Haines et al. 1991).

5. INFLUENCES ON TROPICAL STORAGE ECOSYSTEMS

Environmental issues are coming to the forefront of pest control in grain stores as elsewhere. Consumers, especially in urban areas, are demanding reduction of risks of residues in foods; operator safety is becoming an important concern in developing countries, as it has been for many years in developed ones; and there is a global view that the total use of conventional toxic pesticides has to be reduced across the board. As a result, research on alternative control technologies is now in the mainstream of many programs, where formerly it was in the province only of the enthusiast; this trend seems certain to continue, with consequent radical changes in practical pest control.

In general, most storage pests have long-established stable ranges, resulting from widespread distribution in the international trade in food commodities over the centuries. Major changes in the pest components of the ecosystem thus usually only occur when the storage environment itself is changed, leading to expansion or contraction of ranges. An important exception is the invasion and spread of the larger grain borer, *Prostephanus truncatus,* in Africa; this voracious pest from meso-America has had a profound effect on storage ecosystems in the affected areas.

As Locke and Boxall (1991) emphasize, even after narrowly-focused attention has been given to the reduction of particular losses, some losses persist due to failures of the system. To ensure that the maximum amount of food of the highest possible quality reaches the consumer, there is a need to integrate pest management and quality control with the financial and logistical systems of commodity management. Such developments are likely to involve computerized management information systems and expert systems (see Chapter 2).

Postharvest grain systems utilize considerable energy inputs for drying, processing, transport, and cooking. The grain ecosystem therefore incorporates not just the energy flow from the solar energy budget and agronomic inputs to produce the crop, but also these other postharvest inputs. As well as questions of efficiency of energy utilization, the postharvest requirements for energy raise issues of fuel resources (including biomass energy) and pollution.

Major changes to grain storage systems are caused by climatic extremes (drought and flood), macro-economic policy changes, and political crises, and these can have profound effects on the storage ecosystems and the options for manipulating them. These three over-riding factors may interact with each other, as in southern Africa in 1992; when they do, the future rate and direction of the evolution of storage systems becomes more difficult to predict.

6. FUTURE RESEARCH NEEDS

Tropical grain ecosystems are often complex, with high biological activity and energy transfers, especially in hot humid areas. In many countries, the organic components of these ecosystems have not been fully catalogued and analyzed, and there are very few examples of truly synecological studies. There is a continued need for more detailed analyses of such ecosystems and identification of the key factors influencing their behavior, especially in relation to grain deterioration and loss. This should not be seen as just a specialist academic pursuit for a few researchers. Rather, it should be viewed as an essential activity to which the wider community of storage technologists can contribute. This will, however, require not just an acceptance of the value of a holistic approach to the understanding of grain storage problems but also a more widespread knowledge in tropical countries of correct sampling procedures, techniques for measurement of physical factors, accurate recognition of pests and natural enemies, and the principles of ecology relevant to storage systems. Realization of the research needs is thus heavily dependent on improved training in the ecology and management of storage systems.

Where there is sufficient understanding of an ecosystem and the key factors influencing its behavior, there is then the opportunity to develop an expert system to assist operators to make optimal decisions on management of the system. There is considerable scope for the development of such decision-support systems, and the rapidly increasing access to personal computer systems is likely to expand the applicability of this approach in the future.

The search for alternatives to conventional pesticides will continue and intensify in the next decade. The availability of pesticides for use in grain storage systems has always been limited: this is due in part to the constraints on residues but it is also due to the limited volume of the market for storage pesticides. Against this background, the loss of formerly effective and acceptable compounds due to pest resistance and to environmental and consumer concerns has been a special concern of grain storage managers. The challenge for the development of alternative methods is to find ones that are truly cost-effective and socio-economically acceptable, as well as being demonstrably safe. The safety aspects are critical in relation to materials applied to grain because these must meet the same toxicity standards as conventional pesticides; it cannot be assumed that materials of natural origin are safe (since some of the most toxic compounds known to man are produced by flowering plants and fungi).

In large storages and in the international grain trade, the potential loss of the fumigant methyl bromide because of its ozone-depleting properties is a major threat to normal phytosanitary procedures. There is currently no acceptable alternative for rapid disinfestation of large quantities of grain, especially in transit at ports of export and import, in spite of many years of effort to identify one. When the time available for treatment is not short, phosphine provides an alternative. However, the steadily-growing problem of resistance to phosphine threatens its long-term use. There is also evidence (Haines and Pranata 1982; Pike, unpublished data) that the apparent increase in the incidence of heavy infestations of *Liposcelis entomophilus* in some tropical countries is due to increased dependence on the use of phosphine, to which the egg stage of this species is tolerant.

During the last decade, the larger grain borer, *Prostephanus truncatus,* has posed a major challenge to corn storage in Africa, and the threat increases as it steadily expands its range. The efforts to control this serious pest have to be integrated into strategies for managing other pests and maintaining quality in farm-stored corn. Phytosanitary systems have to be developed for the intra-African grain trade as well as for the import of corn from meso-America into Africa. The effects of the drought in southern Africa in 1991–1992 underline this problem, with large quantities of corn (some of it from meso-America) being imported along trade routes through infested zones to areas that are currently free of larger grain borer.

It has become clear that, more than any other important storage pest, the larger grain borer has major reservoir populations in nonstorage habitats from which it can infest stored corn and cassava. A research priority for this pest in Africa will thus be to understand its natural ecology and the relationship between this and its behavior in the storage ecosystem.

The problem of stackburn, especially in large-scale storages of corn in Africa, appears to be increasing. Preliminary studies suggest that it may be due to complex interactions of several environmental and management factors: initial grain quality and infestation, sack material and bulk or stack size, efficacy of fumigation, and grain respiration. If this problem continues to spread, especially in relation to large-scale storage of maize in international trade, more research effort on its apparently complex multidisciplinary cause will be necessary.

The quality of grain (especially corn, but also other cereals) in international trade has a significant effect on its potential utilization and deterioration in importing countries in the tropical developing world. This is particularly true when corn from global grain stocks is provided as humanitarian food aid in relief operations, as in those to southern Africa in 1992 following the 1991–92 drought. Corn that is at a safe moisture content for storage in a cool temperate climate can soon exceed the critical water-activity threshold when it is moved to a tropical climate, and rapid deterioration results. There is an urgent need for action to establish appropriate grain import-quality specifications for those tropical countries that do not have them and to match the grain quality standards of suppliers and donors with those of potential food-aid recipients.

New pests, different pesticides, altered physical conditions and structures, changing consumer preference for different commodities, and modified management practices all affect the grain storage ecosystem in tropical countries. Such factors can cause radical changes in previously stable ecosystems (e.g., the effects of the larger grain borer on small-scale corn storage in Africa [Dick 1988]). A better understanding is needed of the key factors which, by their interactions, can affect the stability of storage ecosystems and which thus need to be included in any predictive analysis of the effects of change.

A key factor affecting grain storage systems, especially in Africa at present, is the process of market liberalization as an element of structural adjustment. In many countries, the central marketing boards are being disbanded or their role is being greatly reduced. Although the aim is to enable the involvement of the private sector in grain marketing, the necessary conditions for this are not always in place, with the result that large amounts of grain, formerly stored in central warehouses, are now being stored on-farm or in village stores. The impact of liberalization on the storage ecosystem is thus considerable because of major differences in the constraints and options affecting grain management on-farm when compared with warehousing. Research is urgently needed on the options for greater enablement of private sector involvement, mechanisms for transfer of technology from the public to the private sector, and support for technical and financial services to farmers in a liberalized system.

Stored grain ecosystems are man-made and many of the key factors influencing the system are likewise anthropogenic. A major feature of the behavior of a storage ecosystem is thus the behavior of those who create and manage it. Many of the problems to be addressed in understanding and managing these systems arise, of course, from an imperfect understanding of their physical and biological interactions and the effects of key factors such as those noted previously. However, there are many examples of long-established storage systems that are adequately understood and for which the necessary management interventions are well known, but failure of the management of the system results in spoilage and loss of grain. Two examples are the application of fumigants to bag-stacks under torn fumigation sheets and without adequate sealing, and the shipment of grain at 14.5% moisture content and below 5°C to countries with ambient temperatures of over 27°C. In these cases the remedies are technically simple and widely disseminated but the managers of the systems have failed to implement them. Research on storage ecosystems should not be limited to the observable physical and biotic elements but should include the influence of human behavior, especially attitudes to improved management and perceptions of constraints to implementation of technologies. Ultimately, research that improves our comprehension of the grain storage ecosystem will be of little benefit unless the results are implemented through training programs and improved management systems.

ACKNOWLEDGEMENTS

I am grateful to my colleagues at the Natural Resources Institute for their help in the preparation of this chapter. In particular, I wish to thank Mr. P. S. Hindmarsh, Mr. P. S. Tyler, Mr. M. T. Locke, Ms. J. Compton, Mr. T. Donaldson,

Mr. J. A. McFarlane, Mr. D. P. Rees, and Dr. P. Wareing for their discussions, suggestions, or contributions.

REFERENCES

Abreu, J. M. de, R. N. Williams, and P. A. Rude. 1982. A revised bibliography of the almond moth (tropical warehouse moth) *Ephestia cautella* (Walker) (Lepidoptera: Phycitidae). Trop. Stored Prod. Inf. 44:15-36.

Acasio, U. A. and R. Maxon. 1991. Bulk handling and storage of grains, a Pakistan experience. ASEAN Grain Postharv. Quart. 3(1):7.

Adams, J. M. 1977. A review of the literature concerning losses in stored cereals and pulses, published since 1964. Trop. Sci. 19:1-28.

Anonymous. 1981. *Sitophilus* species on rice. Trop. Stored Prod. Inf. 42:2.

Anonymous. 1988. Tropical storage of cereal grains. Agric. Int. 40:166.

Arbogast, R. T. and M. A. Mullen. 1988. Insect succession in a stored-corn ecosystem in southeast Georgia. Ann. Entomol. Soc. Am. 81:899-912.

Arnason, J. T., J. Gale, B. Conilh de Beyssac, A. Sen, S. S. Miller, B. J. R. Philogene, J. D. H. Lambert, R. G. Fulcher, A. Serratos, and J. Mihm. 1992. Role of phenolics in resistance of maize grain to the stored grain insects, *Prostephanus truncatus* (Horn) and *Sitophilus zeamais* (Motsch.). J. Stored Prod. Res. 28:119-126.

Barney, R. J., J. D. Sedlacek, M. Siddiqui, and B. D. Price. 1991. Quality of stored corn (maize) as influenced by *Sitophilus zeamais* Motsch. and several management practices. J. Stored Prod. Res. 27:225-237.

Bengston, M. 1984. Integrated use of pesticides in grain storage in the humid tropics. pp. 9-10 in: ACIAR Grain Storage Newsletter No. 1. Aust. Centre Int. Agric. Res., Canberra, Australia.

Boon, K. H. and S. H. Ho. 1988. Factors influencing the post-fumigation reinfestation of *Tribolium castaneum* (Herbst) (Coleoptera: Tenebrionidae) in a rice warehouse. J. Stored Prod. Res. 24:87-90.

Boxall, R. A. 1986. A critical review of the methodology for assessing farm-level grain losses after harvest. Tropical Development and Research Institute Report No. G191, London. 139 pp.

Boxall, R. A. 1987. Storing grain on the farm. Approp. Technol. 14(2):14-16.

Calderon, M. 1979. The ecosystem approach for understanding the extent of postharvest grain losses in warm countries. pp. 263-280 in: Anales del Tercer Seminario Avanzado de Tecnologia de Alimentos (Proc. of the Third Advanced Seminar on Food Technology), COLCIENCIAS, Bogota, Colombia.

Campbell, A. and R. N. Sinha. 1990. Analysis and simulation modelling of population dynamics and bioenergetics of *Cryptolestes ferrugineus* (Coleoptera: Cucujidae) in stored wheat. Res. Popul. Ecol. 32:235-254.

Cardona, C., K. Dick, C. E. Posso, K. Ampofo, and S. M. Nadhy. 1992. Resistance of a common bean (*Phaseolus vulgaris* L.) cultivar to post-harvest infestation by *Zabrotes subfasciatus* (Boheman) (Coleoptera: Bruchidae). II. Storage tests. Trop. Pest Management 38:173-175.

Casas, E. de las. 1977. Stored product insects and microorganisms in grain ecosystems. pp. 31-34 in: Insect Ecology, Minnesota Agric. Exp. Station Tech. Bull. No. 310, St. Paul, MN.

Chakravorty, S., H. C. Das, and N. R. Das. 1983. Seasonal changes in densities of some insect pests in a population on stored grains. Environ. Ecol. 1:11-14.

Chung, D. S. and C. W. Deyoe. 1982. Grain postharvest technology for developing countries. pp. 527-532 in: Sorghum in the Eighties, Proc. Int. Sympo. Sorghum, Vol. 2, ICRISAT Center, Patancheru, India.

Coker, R. D. 1991. Recent developments in methods for sampling and analysis of mycotoxins. pp. 115-122 in: Fungi and mycotoxins in stored products. B. R. Champ, E. Highley, A. D. Hocking, and J. I. Pitt (eds.). ACIAR Proc. No. 36. Aust. Centre Int. Agric. Res., Canberra, Australia.

Conway, J. A. 1986. Insects and other arthropods recorded on stored food commodities in Nepal and Bhutan. Trop. Sci. 26:145-162.

Cornes, M. A. 1973. A check list of the insects associated with stored products in Nigeria. Technical Report No. 11. Report of the Nigerian Stored Products Research Institute 1971(1973):73-98.

Cornes, M. A. and J. Riley. 1961. Small scale storage of maize in Western Nigeria. pp. 85-96 in: West African Stored Products Research Unit Annual Report 1961, West African Stored Products Research Unit, Department of Marketing and Exports, Lagos, Nigeria.

Coulter, J. and J. A. F. Compton. 1991. Liberalization of cereals marketing in sub-Saharan Africa: implementation issues. NRI Marketing Series, Volume 1. Natural Resources Institute, Chatham, UK. iv + 26 pp.

Cuff, W. R. and J. M. Hardman. 1980. A development of the Leslie matrix formulation for restructuring and extending an ecosystem model: the infestation of stored wheat by *Sitophilus oryzae*. Ecol. Model. 9:281-305.

Dearden, P. J. and E. M. Cassidy. 1990. Food security: an ODA view. IDS Bull. 21(3):81-83.

de Lima, C. P. F. 1978. A study of the bionomics and control of *Sitophilus zeamais* Motschulsky and *Sitotroga cerealella* (Olivier) and associated fauna in stored maize under laboratory and field conditions in Kenya. PhD thesis, University of London. 241 pp.

de Lima, C. P. F. 1979. The assessment of losses due to insects and rodents in maize stored for subsistence in Kenya. Trop. Stored Prod. Inf. 38:21-25.

Dick, K. 1988. A review of insect infestation of maize in farm storage in Africa with special reference to the ecology and control of *Prostephanus truncatus*. Overseas Development Natural Resources Institute Bulletin No. 18, Chatham, UK. 42 pp.

Dobie, P. 1984. Biological methods for integrated control of insects and mites in tropical stored products. Trop. Stored Prod. Inf. 48:37-45.

Dunkel, F. V. 1988. The relationship of insects to the deterioration of stored grain by fungi. Int. J. Food Micro. 7:227-244.

Dunkel, F. V. 1992. The stored grain ecosystem: a global perspective. J. Stored Prod. Res. 28:73-87.

Faroni, L. R. d'A., R. Sinicio, and G. Roa. 1982. Comparaçao de três sistemas de secagem e armazenamento de milho em fazênda. Revista Brasileira de Armazenamento 7:45-51.

Friendship, C. A. R. and J. A. F. Compton. 1991. Bag or bulk? A decision-making checklist. Natural Resources Institute Bulletin No. 45, Chatham, UK. 19 pp.

Giga, D. P. and Y. Katerere. 1986. Rural grain storage in Zimbabwe: problems, loss assessment and prevention. University of Zimbabwe, Harare, Zimbabwe. 96 pp.

Giga, D. P., E. Tembo, I. Nehumai, and L. Matarirano. 1988. Grain storage in communal areas of Zimbabwe: results from a survey. Zimbabwe Sci. News 2:40-44.

Giles, P. H. 1964a. The storage of cereals by farmers in Northern Nigeria. Trop. Agric., Trinidad 41:197-212.

Giles, P. H. 1964b. The insect infestation of sorghum stored in granaries in Northern Nigeria. Bull. Entomol. Res. 55:573-588.

Giles, P. H. 1965. A record of stored product insects associated with northern Nigerian foodstuffs. Samaru Miscellaneous Paper No. 8. Inst. Agric. Res., Ahmadu Bello University, Zaria, Nigeria. 7 pp.

Girish, G. K., K. K. Arora, and S. K. Jain. 1990. Post harvest technology of food grains in India. Bull. Grain Technol. 28:66-88.

Golob, P. and D. J. Webley. 1980. The use of plants and minerals as traditional protectants of stored products. Tropical Products Institute Report No. G138, London. 32 pp.

Graham, W. M. 1970a. Warehouse ecology studies of bagged maize in Kenya—I. The distribution of adult *Ephestia (Cadra) cautella* (Walker) (Lepidoptera, Phycitidae). J. Stored Prod. Res. 6:147-155.

Graham, W. M. 1970b. Warehouse ecology studies of bagged maize in Kenya—II. Ecological observations of an infestation by *Ephestia (Cadra) cautella* (Walker) (Lepidoptera, Phycitidae). J. Stored Prod. Res. 6:157-167.

Graham, W. M. 1970c. Warehouse ecology studies of bagged maize in Kenya—III. Distribution of the immature stages of *Ephestia (Cadra) cautella* (Walker) (Lepidoptera, Phycitidae). J. Stored Prod. Res. 6:169-175.

Graham, W. M. 1970d. Warehouse ecology studies of bagged maize in Kenya—IV. Reinfestation following fumigation with methyl bromide gas. J. Stored Prod. Res. 6:177-180.

Gwinner, J., R. Harnisch, and O. Muck. 1990. Manual on the prevention of postharvest grain losses. GTZ, Hamburg, Germany. 294 pp.

Hagstrum, D. W. and W. G. Heid, Jr. 1988. U. S. wheat-marketing system: an insect ecosystem. Bull. Entomol. Soc. Am. 34:33-36.

Haines, C. P. 1974. Insects and arachnids from stored products: a report on specimens received by the Tropical Stored Products Centre 1972-73. Tropical Products Institute Report No. L39, London. 22 pp.

Haines, C. P. 1981a. Insects and arachnids from stored products: a report on specimens received by the Tropical Stored Products Centre 1973-77. Tropical Products Institute Report No. L54, London. 73 pp.

Haines, C. P. 1981b. Laboratory studies on the role of an egg predator, *Blattisocius tarsalis* (Berlese) (Acari: Ascidae), in relation to the natural control of *Ephestia cautella* (Walker) (Lepidoptera: Pyralidae) in warehouses. Bull. Entomol. Res. 71: 555-574.

Haines, C. P. 1982. Pest management in stored products. Prot. Ecol. 4:321-330.

Haines, C. P. 1983. The implications of some problems in the identification of stored-product arthropods. pp. 263-267 in: Proc. Sympo. Pest Ecol. Pest Management, C. P. Haines (ed.). BIOTROP Special Publication No. 18, Bogor, Indonesia.

Haines, C. P. 1986. A survey of stored-product insects and mites in Indonesia. Poster presented at the Royal Entomological Society of London Workshop on Recent Work on Stored-Product Insects, 14 October 1986, London.

Haines, C. P. 1989. Observations on *Callosobruchus analis* (F.) in Indonesia, including a key to storage *Callosobruchus* spp. (Col., Bruchidae). J. Stored Prod. Res. 25:9-16.

Haines, C. P. (ed.). 1991. Insects and arachnids of tropical stored products: their biology and identification. Natural Resources Institute, Chatham, UK. 246 pp.

Haines, C. P. and R. I. Pranata. 1982. Survey on insects and arachnids associated with stored products in some parts of Java. pp. 17-47 in: Progress in Grain Protection, N. C. Teter and A. S. Frio (eds.), Proc. 5th Annual Workshop on Grains Post-Harvest Technol. SEARCA, College, Laguna, Philippines.

Haines, C., D. Rees, K. Ryder, S. Sistyanto, and Y. Cahyana. 1991. Brown-rice bait-bags for monitoring insect pest populations in bag stacks of milled rice as an aid to pest control decision-making. pp. 1351-1358 in: Proc. Fifth Int. Working Conf. on Stored-Product Prot., Volume II, F. Fleurat-Lessard and P. Ducom (eds.), The Permanent Committee of International Conferences on Stored-Product Protection, Bordeaux, France.

Hall, D. W. 1970. Handling and storage of food grains in tropical and subtropical areas. Agricultural Development Paper 90. Food and Agriculture Organization, Rome, Italy. 350 pp.

Harris, K. L. and C. J. Lindblad (compilers). 1978. Post-harvest grain loss assessment methods. Am. Assoc. Cereal Chem., St. Paul, MN. 193 pp.

Hashmi, A. A., M. A. Nasir, and R. A. Javed. 1983. Fungi associated with stored grain insect pests at Faisalabad. Pakistan J. Agric. Res. 4:76-83.

Hindle, R. E. 1990. The World Bank approach to food security analysis. IDS Bull. 21(3):62-66.

Hindmarsh, P. S. and J. A. McFarlane. 1983. A programmed approach to food storage improvements. Trop. Stored Prod. Inf. 46:3-9.

Hindmarsh, P. and B. Trotter. 1990. Developments in grain storage for food security. pp. 126-132 in: Developing World Agriculture, A. Speedy (ed.). Grosvenor Press Int., London.

Hodges, R. J. 1979. A review of the biology and control of the rice moth *Corcyra cephalonica* Stainton (Lepidoptera: Galleriinae). Trop. Prod. Inst. Report No. G125, London. 20 pp.

Hodges, R. J. and M. J. Dales. 1991. Unpublished data on insecticide persistence on grain store surfaces in Ghana. Natural Resources Institute, Chatham, UK.

Hodges, R. J., H. Halid, D. P. Rees, J. Meik, and J. Sarjono. 1985. Insect traps tested as an aid to pest management in milled rice stores. J. Stored Prod. Res. 21:215-229.

Hodges, R. J., M. Sidik, H. Halid, and J. A. Conway. 1990. The efficacy of routine application of contact insecticides in protecting bagged milled rice under typical ASEAN storage conditions. Paper presented at the 13th ASEAN Seminar on Grain Postharvest Technology, 4-7 September 1990, Bandar Seri Begawan, Brunei Darussalam. 13 pp.

Hoppe, T. 1986. Storage insects of basic food grains in Honduras. Trop. Sci. 26:25-38.

Howe, R. W. 1963. The prediction of the status of a pest by means of laboratory experiments. World Rev. Pest Control 2:30-40.

Howe, R. W. 1965. A summary of estimates of optimal and minimal conditions for population increase of some stored products insects. J. Stored Prod. Res. 1:177-184.

Huddleston, B. 1990. FAO's overall approach and methodology for formulating national food security programmes in developing countries. IDS Bull. 21(3):72-80.

Hughes, A. M. 1976. The mites of stored food and houses. MAFF Tech. Bull. 9. Her Majesty's Stationery Office, London. 400 pp.

Ivbijaro, M. F., C. Ligan, and A. Youdeowei. 1985. Control of rice weevils, *Sitophilus oryzae* (L.), in stored maize with vegetable oils. Agric. Ecosys. Environ. 14:237-242.

Kennes, W. 1990. The European Community and food security. IDS Bull. 21(3):67-71.

Lacey, J. 1988. Grain storage: the management of ecological change. pp. 614-633 in: Biodeterioration 7. D. R. Houghton, R. H. Smith, and H. O. W. Eggins (eds.). Elsevier Applied Science, New York.

Lacey, J. 1989. Prevention of mould growth and mycotoxin production through control of environmental factors. pp. 161-168 in: Mycotoxins and phycotoxins '88 (A Collection of Invited Papers Presented at the Seventh Int. IUPAC Sympo. on Mycotoxins and Phycotoxins, Tokyo, Japan), S. Natori, K. Hashimoto, and Y. Ueno (eds.). Elsevier Science Pub., Amsterdam, The Netherlands.

Locke, M. T. and R. A. Boxall. 1991. Management strategies to minimise losses during grain storage. p. 1967 in: Proc. Fifth Int. Working Conf. Stored-Product Prot., Volume III, F. Fleurat-Lessard and P. Ducom (eds.), The Permanent Committee of International Conferences on Stored-Product Protection,Bordeaux, France.

Longstaff, B. C. and W. R. Cuff. 1984. An ecosystem model of the infestation of stored wheat by *Sitophilus oryzae*: a reappraisal. Ecol. Model. 25:97-119.

Mabbett, T. H. 1986. Crop storage in the tropics: problems and practice. Span 29:70-72.

Markham, R. H. 1981. The ecology of insect pest populations in maize storage cribs in Nigeria. PhD thesis, Imperial College, University of London, London. 326 pp.

Maxwell, S. 1990. Food security in developing countries: issues and options for the 1990s. IDS Bull. 21(3):2-13.

McFarlane, J. A. 1963. An annotated record of Coleoptera, Lepidoptera, Hemiptera and Hymenoptera associated with stored produce in Jamaica. Trop. Agric., Trinidad 40:211-216.

McFarlane, J. A. 1978. The significance and effects of insect infestation on milled rice. pp. 455-461 in: Proc. Workshop on Grain Post-Harvest Technol., Bangkok, Thailand. Southeast Asia Cooperative Post-Harvest Reseach and Development Programme, SEARCA, Laguna, Phillipines

McFarlane, J. A. 1981. The integrated approach to stored products pest management. pp. 12-15 in: Proc. BIOTROP Sympo. Pests Stored Prod., Bogor, Indonesia. BIOTROP, Bogar, Indonesia

McFarlane, J. A. 1989. Guidelines for pest management research to reduce stored food losses caused by insects and mites. Overseas Development Natural Resources Institute Bulletin No. 22, Chatham, UK. 62 pp.

Müller, P. 1983. Cereal post-harvest systems in the developing countries: problems and progress. The Courier 82:66-69.

Multon, J. L. 1988. Preservation and storage of grains, seeds and their by-products: cereals, oilseeds, pulses and animal feed. Lavoisier Pub., New York. 1095 pp.

Natural Resources Institute. 1991a. Strategies for preventing or reducing grain losses during national and international distribution of stocks. pp. 242-244 in: NRI Report on Operational Programmes: 1989-91. Natural Resources Institute, Chatham, UK.

Natural Resources Institute. 1991b. Pest status of liposcelids and options for their control. pp. 273-275 in: NRI Report on Operational Programmes: 1989-91. Natural Resources Institute, Chatham, UK.

Nisanka, S. K. and M. K. Misra. 1990. Ecological study of an Indian village ecosystem: energetics. Biomass 23:165-178.

Okereke, G. U. and V. C. Nwosu. 1987. Crop storage losses in southern Nigeria caused by the activities of micro-organisms. MIRCEN J. 3:201-210.

Osman, N. 1980. The distribution of *Sitophilus* spp. and *Tribolium* spp. and their resistance to insecticides in Indonesia. p. 40 in: BIOTROP Annual Report, July 1978-June 1979, Bogor, Indonesia. BIOTROP, Bogor, Indonesia.

Padgham, J., V. Pike, K. Dick, and C. Cardona. 1992. Resistance of a common bean (*Phaseolus vulgaris* L.) cultivar to post-harvest infestation by *Zabrotes subfasciatus* (Boheman) (Coleoptera: Bruchidae). I. Laboratory tests. Trop. Pest Management 38:167-172.

Pande, N., J. Saxena, and B. S. Mehrotra. 1989. Determination of fungal and insect infestation of stored cereal grains and their relationship. Bull. Grain Technol. 27:133-141.

Permual, D. 1991. Investigations of stored rice pest problems in Guyana. Unpublished PhD thesis, Imperial College, University of London. 244 pp.

Phillips, S., F. M. Dewey, D. R. Twiddy, and P. Wareing. 1991. Monoclonal antibodies for the detection of fungi in stored products. pp. 228-229 in: Fungi and mycotoxins in stored products. B. R. Champ, E. Highley, A. D. Hocking and J. I. Pitt (eds.). ACIAR Proc. No. 36. Aust. Centre Int. Agric. Res., Canberra, Australia.

Pike, V., D. Rees, and R. Hatch. 1991. Bionomics of *Liposcelis paetus* in stored grain (Psocoptera: Liposcelidae). pp. 175-179 in: Proc. Fifth Int. Working Conference Stored-Product Prot., Volume I, F. Fleurat-Lessard and P. Ducom (eds.), The Permanent Committee of International Conferences on Stored-Product Protection, Bordeaux, France.

Pixton, S. W. 1982. The importance of moisture and equilibrium relative humidity in stored products. Trop. Stored Prod. Inf. 43:16-29.

Prakash, A. and S. C. Goel. 1983. Factors affecting losses due to insects and their management in rice storage ecosystem. pp. 50-54 in: Insect Ecology and Resource Management. S. C. Goel (ed.). Sanatan Dharm College, Muzaffarnagar, India.

Prakash, A. and J. Rao. 1987. Use of chemicals as grain protectants in rice storage ecosystem and its consequence. Bull. Grain Technol. 25:65-69.

Prevett, P. F. 1961. An investigation into the bulk storage of sorghum in five cylindrical aluminium silos. pp. 73-82 in: West African Stored Products Research Unit Annual Report 1961, West African Stored Products Research Unit, Department of Marketing and Exports, Lagos, Nigeria.

Prevett, P. F. 1964. The distribution of insects in stacks of bagged groundnuts in Northern Nigeria. Bull. Entomol. Res. 54(1963):689-713 & Plate XII.

Prevett, P. F. and D. Halliday. 1961. An investigation into the distribution of temperature within a groundnut pyramid. pp. 44-52 in: West African Stored Products Research Unit Annual Report 1961, West African Stored Products Research Unit, Department of Marketing and Exports, Lagos, Nigeria.

Rees, D. P. and M. J. Dales. 1992. Unpublished bibliographic database on alternative methods for the control of stored-product pests. Natural Resources Institute, Chatham, UK. 151 pp.

Rees, D. P. and A. J. Walker. 1990. The effect of temperature and humidity on population growth of three *Liposcelis* species (Psocoptera: Liposcelidae) infesting stored products in tropical countries. Bull. Entomol. Res. 80:353-358.

Rowley, J. Q. 1984. An assessment of losses during handling and storage of millet in Mali. Trop. Stored Prod. Inf. 47:21-33.

Shazali, M. E. H. and R. H. Smith. 1986. Life history studies of externally feeding pests of stored sorghum: *Corcyra cephalonica* (Staint.) and *Tribolium castaneum* (Hbst). J. Stored Prod. Res. 22:55-61.

Shazali, M. E. H. and R. H. Smith. 1990. Ecological succession of four insect pests of stored sorghum in a laboratory ecosystem. Zeitschrift für Angewandte Zoologie 77:93-105.

Sinha, A. K. and K. K. Sinha. 1990. Insect pests, *Aspergillus flavus* and aflatoxin contamination in stored wheat. A survey at North Bihar (India). J. Stored Prod. Res. 26:223-226.

Sinha, K. K. and A. K. Sinha. 1991. Effect of *Sitophilus oryzae* infestation on *Aspergillus flavus* infection and aflatoxin contamination in stored wheat. J. Stored Prod. Res. 27:65-68.

Sinha, R. N. 1975. Climate and the infestation of stored cereals by insects. pp. 117-141 in: Proc. First Int. Working Conf. Stored-Product Entomol., Savannah, GA.

Sinha, R. N. 1979a. Role of Acarina in the stored grain ecosystem. pp. 263-272 in: Recent Advances in Acarology (Proc. Fifth Int. Cong. Acarol., East Lansing, MI). Vol. I. J. G. Rodriguez (ed.). Academic Press Inc., New York.

Sinha, R. N. 1979b. Ecology of microflora in stored grain. Annales de Technologie Agricole 28:191-209.

Sinha, R. N. 1982. Food losses through energy transfer from cereal grains to stored-product insects. Food and Nutri. Bull. 4(2):13-20.

Sinha, R. N. 1991. Storage ecosystems. pp. 17-30 in: Ecology and Management of Food-Industry Pests. FDA Technical Bulletin 4. J. R. Gorham (ed.). Assoc. Off. Anal. Chem., Arlington, VA.

Sinha, R. N. and W. E. Muir (eds.). 1973. Grain storage: part of a system. AVI Publishing Company Inc., Westport, CT. 481 pp.

Sinha, R. N., D. Abramson, and J. T. Mills. 1986. Interrelationships among ecological variables in stored cereals and associations with mycotoxin production in the climatic zones of western Canada. J. Food Prot. 49:608-614.

Smith, C. V. and M. C. Gough. 1990. Meteorology and grain storage. World Meteorological Organisation Technical Note No. 101, Geneva, Switzerland. 82 pp.

Smith, K. G. 1963. The study of an insect population living on bagged groundnuts stored in Southern Nigeria with particular reference to the behaviour of *Trogoderma granarium* Everts (Col., Dermestidae). J. West African Sci. Assoc. 8:44-57.

Stepien, Z. A., W. Goszcynski, and J. Boczek. 1973. The energy budget of *Tyrophagus putrescentiae* (Schr.) (Acaridae). pp. 373-378 in: Proc. Third Int. Cong. Acarol., Prague, Czechoslovakia.

Surtees, G. 1964. Laboratory studies on dispersion behaviour of adult beetles in grain. VI.—Three-dimensional analysis of dispersion of five species in uniform bulk. Bull. Entomol. Res. 55:161-171.

Taylor, R. W. D. 1991. Resistance to grain fumigants and future prospects for their use. Pesticide Outlook 2(2):22-24.

TDRI. 1983. Food storage manual. World Food Programme. Trop. Dev. Res. Inst., London. 263 pp.

Thorpe, G. R. 1986. Some fundamental principles and benefits of aeration of stored grain. pp. 31-34 in: Proc. Int. Seminar Preserving Grain Quality by Aeration and In-Store Drying, Kuala Lumpur, Malaysia. ACIAR Proc. No. 15, Aust. Centre Int. Agric. Res., Canberra, Australia.

Thorpe, G. R. and S. Whitaker. 1992a. Local mass and thermal equilibria in ventilated grain bulks. Part I: The development of heat and mass conservation equations. J. Stored Prod. Res. 28:15-27.

Thorpe, G. R. and S. Whitaker. 1992b. Local mass and thermal equilibria in ventilated grain bulks. Part II: The development of constraints. J. Stored Prod. Res. 28:27-54.

Thorpe, G. R., W. R. Cuff, and B. C. Longstaff. 1982. Control of *Sitophilus oryzae* infestation of stored wheat: an ecosystem model of the use of aeration. Ecol. Model. 15:331-351.

Tyagi, A. K. and G. K. Girish. 1978. Studies on the assessment of storage losses of foodgrains by insects. Part 2. Factor affecting extent of damage. Bull. Grain Technol. 15:46-66.

Tyler, P. S. 1981. Post harvest losses in durables. Paper presented at the First Consultative Meeting on Post Harvest Losses in the Caribbean, University of the West Indies, Trinidad and Tobago, 19-24 July 1981.

Wallace, H. A. H., P. L. Sholberg, R. N. Sinha, and W. E. Muir. 1983. Biological, physical and chemical changes in stored wheat. Mycopathologia 82:65-76.

Weymes, H. 1980. Psychological, educational, ethnic, religious constraints on crop conservation. Prog. Food and Nutri. Sci. 4(3/4):127-130.

White, N. D. G. and R. N. Sinha. 1979. Natural regulation of *Tarsonemus granarius* numbers in stored wheat ecosystems—a multivariate assessment. pp. 291-298 in: Recent Advances in Acarology (Proc. Fifth Int. Cong. Acarol., East Lansing, MI). Vol. I. J. G. Rodriguez (ed.). Academic Press, Inc., New York.

World Bank. 1986. Poverty and hunger: issues and options for food security in developing countries. A World Bank Policy Study. Washington, DC. 69 pp.

Wright, C. L., R. L. Meyer, and F. E. Walker. 1981. Analyzing bottlenecks in grain transportation and storage systems: a Brazilian case study. J. Dev. Studies 18:68-84.

4

Economics of Grain Storage

O. J. Sode, F. Mazaud, and Francis Troude

Food and Agricultural Organization of the United Nations, Rome, Italy

SUMMARY

Although world production of cereals has increased in the past 10 years and total production in developing countries is greater than in developed countries, the world population continues to grow rapidly at up to a 22% increase in 10 years in developing countries. Hunger threatens 800 million people in the developing world, 60% of them in Asia.

Postharvest losses in quantity and quality of cereals can be high in localized areas but on average tend to be under 10% even in developing countries. Postharvest loss reduction activities are outlined for several developing countries. Improvements of postproduction systems include creation of awareness of problems and institutional involvement, assessment of the postproduction system, effective pricing systems, improvement of storage techniques, pilot projects to test new storage techniques, and sustainability in production, storage, and marketing.

Future development and research should involve improvement of postharvest techniques at the farm level, determining the impact of poor quality grain on trade, maximum levels of mycotoxins in food, improving grain drying and aeration techniques, improving pest control in developing countries, and clarifying the link between storage techniques, quality of grain, and pricing policies.

1. INTRODUCTION

As background information, it is necessary to describe the general situation of grains in the world, their production, and their use. Table 1 shows the pattern of cereal production in the 1980s. Several observations need to be made:

1. World production increased 12% during the 10-yr period from 1979–1988.
2. Production of developing countries represents a higher amount than that of developed countries (969 Mt versus 774 Mt).
3. Average yields do not vary as much as could be expected. This is due to the relatively low yields in many developed countries. Although it is common for wheat or maize yields to exceed 10 t/ha in northern Europe, the long-term average yield in the developed countries is only 2.8 t/ha. This result is, to a certain extent, due to low yields in large producer countries which practise low-input cropping systems, such as Australia (1.7 t/ha), Canada (2.6 t/ha), and the former USSR (2.6 t/ha).
4. The population of developing countries continues to grow rapidly (up to 22% in 10 yr, against 7% in developed countries; Table 1). As a result, there is a tremendous gap in the food availability per head. Assuming no grain exports, each person in developed countries has about three times the cereals available than in developing countries (Table 1). Cereals and pulses form 75% of the basic staples produced in the developing world and more than 60% of the per capita energy intake.

The World Food Program, in its 1991 review, pointed out that food aid would be required on an increasing scale during the 1990s. Annual cereal imports to developing countries were projected to increase from about 100 Mt during the mid-1980s (of which only 12 Mt were met by food aid deliveries) to between 145–168 Mt by the year 2000.

Hunger today may be threatening the lives of as many as 800 million people in the developing world, with approximately 60% of the hungry living in Asia. The situation in sub-Saharan Africa, however, gives particular cause for concern. Total cereal supplies per person average 250 kg for developing countries but only 121 kg for the low-income food-deficit countries of sub-Saharan Africa. About one quarter of the population in sub-Saharan Africa—more than 100 million people—do not consume enough food to lead an active working life and this number is projected to increase to nearly 200 million by the year 2000.

Other differences related to storage economics between developed and developing countries include:

1. The economically active population in food production (which includes all economically active persons engaged principally in agriculture, forestry,

Table 1 Cereals: the World's food

	Developing countries			Developed countries			World total		
	1979	1984	1988	1979	1984	1988	1979	1984	1988
Production (Mt/yr)	752	992	969	804	883	774	1553	1805	1743
Area (Mha)	422	426	426	304	304	276	717	730	702
Yield (t/ha)	1.78	2.10	2.28	2.64	2.90	2.80	2.17	2.47	2.48
Population (Millions)	3180	3562	3879	1155	1202	1235	4335	4764	5115
	Data per person								
Production (kg/yr)	236	259	250	696	735	627	358	378	341
Energy Equiv*. (kJ/d)	10000	11000	10500	29000	31000	26000	15000	16000	14000
Protein** (g/d)	52	57	55	153	161	137	78	83	75

* Energy constant = 15.3 kJ/g (Smil 1991).
** Calculated assuming a protein level of 8% in cereals.
Source: extracted from FAO Yearbooks 1979, 1984, 1988

and fishing) represents only 4% in developed countries as opposed to 24% in developing countries.

2. A comparison of mechanization levels indicates that very few tractors are utilized in developing countries and combine-harvesters are even more scarce.

As the Food and Agriculture Organization (FAO) of the Unites Nations' activities are focused on developing countries, this chapter is mainly devoted to them.

2. THE NATURE OF LOSSES

2.1. Mass Loss

It is generally accepted that losses occur all along the line of grain handling. Figure 1 illustrates the flow of food grains as it might be in many cases. "Loss" in this connection means the disappearance of food of plant origin. Losses may be either quantitative or qualitative. It has been common to express the loss simply as a percentage because it is easy to convert into a monetary value.

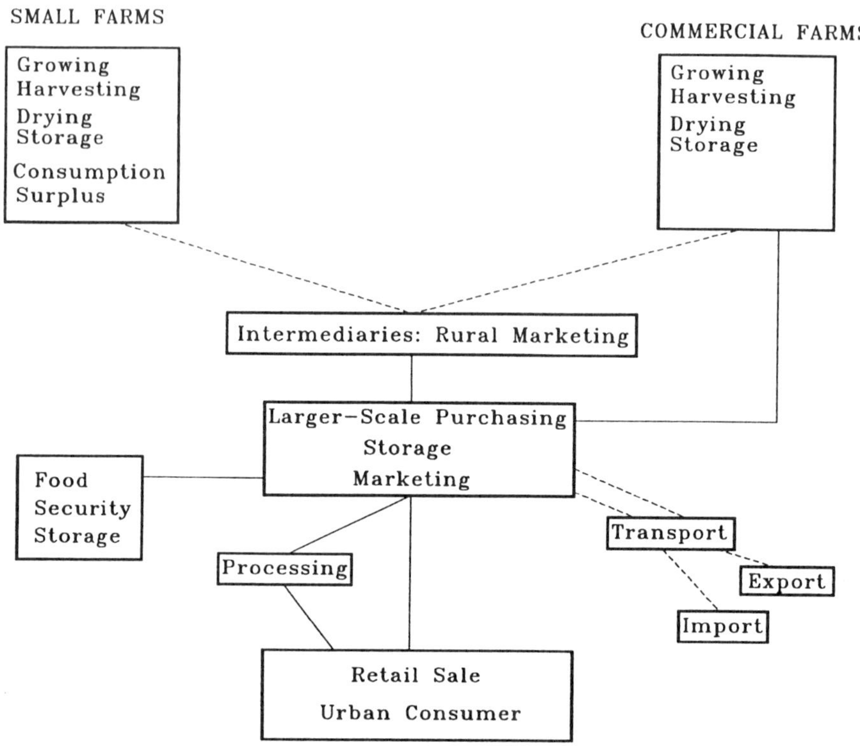

Figure 1 Flow of food grain from small and commercial farms as it is marketed, stored, transported, processed, and consumed.

2.2. Quality Loss

Qualitative losses are more difficult to value than quantitative losses (Anonymous 1986, Anthony and Arboleda 1987). Losses may consist, for example, of changes in the physical appearance, nutritional deterioration, or the development of mycotoxins. These factors can be measured in a wide variety of ways (De Lima 1982). Many countries have quality standards for trade and these are used as guidelines for the calculation of quality loses. In most developing countries, standards are only available for export crops such as coffee, cocoa, groundnuts, and cotton; few standards are available for rice, corn, and sorghum, and in these cases, loss is based on subjective judgments related to local preference.

Nutritional and germinative losses, which may be a combination of quantity and quality, are difficult to detect externally and are, therefore, scarcely measured.

The qualitative aspects of grain loss are probably of greater importance than the quantitative losses in developed countries. Deterioration of quality—for instance, the contamination by mold—might lead to the development of toxic agents, notably mycotoxins of different types (e.g., aflatoxins) which will downgrade the value of the crop.

Mycotoxins have been attracting global attention because of their adverse health effects and negative economic impact, especially related to trade. Aflatoxin as a contaminant was evaluated for the first time by the Joint FAO/World Health Organization Expert Committee on Food Additives (JEFCA) at its 31st meeting in 1987. JEFCA considered aflatoxin to be a potential human carcinogen but felt that insufficient information was available to establish a figure for a tolerable level of exposure. In the case of potent carcinogens, however, it believed that efforts should be made to limit their presence in food to irreducible levels. It defined an irreducible level as "that concentration of a substance which cannot be eliminated from a food without involving the discarding of the food altogether, severely compromising the ultimate availability of the major food supplies."

The significance of mycotoxins in international trade is being increasingly recognized by both the developed and developing countries. Often, especially in developing countries, the best quality products (i.e., free of mycotoxins) are exported, while the substandard products (i.e., not acceptable to foreign buyers) are distributed within the country with potential consequences to the health of the local population or on the productivity of animals.

3. THE CAUSES OF LOSS

The loss of food grains depends on a number of well known factors. The most critical are:

1. choice of varieties susceptible to insect infestation or shattering;
2. delays in harvesting after maturity (causing shattering, spillage, and damaged kernels); and
3. the care taken at each step from harvesting to milling.

The steps from harvesting to milling are numerous and must be clearly recognized in each particular situation:

1. shattering at harvest during cutting;
2. drying unthreshed in the field;
3. transportation from the field to the threshing area (on the head, with animals, bicycles, boats, etc);
4. threshing (manually, or with animals, threshers, or combines);
5. cleaning, winnowing;

6. drying (natural or artificial);
7. transportation to storage;
8. moisture content changes during storage (drying or rewetting);
9. insect attacks;
10. rodent attacks;
11. mold development;
12. handling, bagging, delivery to the mill or the market; and
13. milling recovery.

There are some specific aspects which we wish to characterize:

1. The greatest difference between developed and developing countries is that in developing countries the major part of grain and pulse crops, sometimes up to 80% of national production, is kept on the farms for home consumption while in developed countries grain is stored mainly for export in large storage facilities (e.g., United States) or on farms (e.g., Canada). In the so-called "traditional systems", there is normally an adequate and sufficient balance between production and postharvest practices. The losses which occur in the system are predictable and accepted as unavoidable with appropriate compensation being made in quantity as far as possible. Any changes affecting production will have postharvest consequences. The green revolution, for example, created an imbalance since the increased production was not matched by a corresponding increased capacity in the postharvest system. The introduction of high-yield hybrid varieties of corn in many developing countries in Africa was very attractive to farmers but these varieties are more susceptible to insect attack (partly because of less sheath cover) and increased storage losses have consequently been experienced. Similar problems arose in Asia with high-yield varieties of rice which are more susceptible to shattering and to insect attacks than the older varieties.

2. Government intervention in the marketing system of food grains, such as setting the annual price to the producers and subsidizing the consumer price, has been very common in many developing countries. This has often caused farmers to sell off their crops soon after harvest as there is no advantage in storing the grain on the farm in anticipation of an increase in the price. This causes temporary blockages at primary collection points of the national grain boards, which are (theoretically) forced to purchase whatever amount the farmers bring in. Often there is no control over quality (or the lack of facilities to enforce such a control), resulting in large quantities of the incoming grain being too wet, insect-infested, unclean and, in general, of substandard quality. If the facilities required for adequate treatment of the produce are not available or are of insufficient capacity, losses are bound to occur.

3. Incorrect pesticide applications and neglect of integrated pest management by store managers increase the cost of insect disinfestation and lead to an increase in resistance.

4. The transition from bag to bulk handling of grain, which is expected to take place in most developing countries in the future, will increase the risk of losses. Initial difficulties have been experienced in many countries where insufficient attention has been given to the unavoidable links among quality control, transport, management, and training of personnel in the operation of sophisticated monitoring equipment (Cutler 1987). As a result, some bulk storage programs have failed completely with high financial "losses."

4. THE MAGNITUDE OF LOSSES

The extent of postmaturity losses is highly variable. It is possible to find individual cases of losses ranging from almost 0% to 100%. Losses in connection with trade can be devastating if the grain does not meet the required quality standards. Shipments have been rejected without any alternative use for the grain.

In the developed countries, losses in food grains are generally at a level from which a further reduction in losses is not cost effective. Farmers in, for instance, northern Europe will accept about 1% spillage during combine harvesting. The spillage could probably be further reduced by applying a lower ground speed of the combine harvester and by operating only during optimum weather conditions. However, full utilization of labor, machinery, and time (i.e., 100% utilization of days with suitable weather for harvesting) has higher economic priority than avoiding a 1% spillage of grain.

The amount of food grains stored on farms will be decided on the basis of the expected price change up to the next harvesting season or on available markets (Creupelandt and Shepherd 1990). When good management is applied, quantitative losses are marginal. The losses over a period of 9–10 mo are in the order of 0.75–1%, mainly due to the wearing of grain through mechanical handling and oxidation.

The loss in the commercial trading sector is normally fixed at 0.75%, which probably reflects the actual loss fairly accurately. In some European countries, the loss during storage is declared nil from the financial point of view. This loss is, however, estimated at about 0.5% but there are no data available to support it. The situation is quite different for losses due to drying. The pricing scale introduces a penalty for overdrying equivalent to the mass of moisture removed below the marketable level.

From the qualitative point of view, reliable published information on the economic and trade aspects of contamination by mycotoxins is extremely limited and, when available, it is often not expressed in quantitative terms. The few studies and statistics available are mostly incomplete and are primarily restricted to aflatoxin. The only figure for losses related to mycotoxins was given at the

Second Joint FAO/WHO/UNEP International Conference on Mycotoxins, held in Bangkok in 1987. The estimated loss of 250,000 t of poultry meat (worth about US $143 million in 1983) was due to reduced body gains arising from mycotoxins in feeds. Mycotoxins cause significant, nontariff barriers to trade particularly for developing countries which depend on the export of aflatoxin susceptible crops used for food and animal feed (e.g., groundnut-cake).

In developing countries, estimates of losses have been made during the last two to three decades. Only in the late 1970s, however, were determined efforts started to stimulate a more realistic understanding and appreciation of postharvest losses. A significant event in this connection was the 1975 Resolution of an Interdepartmental Subcommittee of the VII Special Session of the UN General Assembly, which committed the member states to reducing postharvest losses by 50% by 1985. The available information on losses was reviewed and it was concluded that "there can be no agreed single figure for the percentage of postharvest losses on a global scale or even on a national basis" and there was a need "for more accurate assessment of these losses, to establish firm justification for the introduction of measures to reduce them."

Initially, the interest was mainly focused on identification of losses at the commercial level; however, it was realized that the main objective should be the small farms where most grain is stored. The difficulty encountered was initially the lack of reliable and acceptable methodologies for performing assessments. A first manual for loss assessment methodology was prepared by Harris and Lindblad (1978). The difficulties in developing the methodology were: 1) the previous experience originated mostly in developed countries, 2) a thorough knowledge of the wide variety of postharvest systems in developing countries was not available, and 3) the difficulties in applying the assessment methodology could not be foreseen.

FAO utilizes three types of loss assessment surveys:

1. Preliminary survey of specific problem points and on-site appraisals. This type of survey is normally conducted during project formulation and leads to an understanding of the postproduction system and identification of the causes of loss.
2. Nonrandomized survey in which a complete scientific-sampling-design is not followed. The expected result of such a survey is an estimate, with some reliability, of the losses taking place in the postproduction component studied.
3. Randomized survey aiming at obtaining statistically-reliable, quantitative data on losses. Only a few surveys can be qualified as such.

Tables 2–4 show results from different parts of the developing world where acceptable methodologies have been used in the assessments. They show how data are partial and variable even for a single operation of the postharvest system. They also demonstrate that, in most cases, losses must be tackled as a

Table 2 Examples of Studies to Measure Postmaturity Losses at the Farm and Village Level in Developing Countries

Country	Crop	Stage	Causes of loss	Estimated mass loss (%)	References
Dominican Republic	Paddy/rice	Harvesting/threshing	Incomplete threshing, scattering	18.1 ± 13.0	Boxall et al. (1981)
Philippines	Paddy/rice	Harvesting	Undefined	1–3	De Padua (1978)
		Handling	"	2–7	
		Threshing	"	2–6	
		Drying	"	1–5	
		Storage	"	2–6	
		Milling	"	2–10	
Bangladesh	Rice	Farm-level:			
		Harvesting		1.5	Greeley (1982)
		Field stooking		0.5	
		Transportation		0.5	
		Threshing		1.8	
		Storage		2.6	
Pakistan	Wheat	Harvesting	Handling-shattering	2.7–3.9	De Lima (1985a)
		Storing (field)	Rodents, weathering	1.5–1.6	
		Threshing	Machinery	0.1–1.0	
		Drying	Birds, rodents	0.01–0.02	
	Rice	Harvesting	Shattering	0.8–2.2	
		Threshing	Birds, rodents	0.6–1.7	
		Drying (field)	Rodents	0.2–0.6	
	Maize	Harvesting-drying	Rodents	0.4–0.7	

Table 3 Examples of Studies to Measure Losses During On-Farm Storage in Developing Countries

Country	Crop	Period of storage (mo)	Causes of loss	Estimated mass loss (%)	References
Zambia	Maize	7	Insects	1.7–5.6	Adams and Harman (1977, 1978)
India	Paddy	7	Insects, rodents, mold	4.26 ± 1.33	Boxall et al. (1979)
Kenya	Maize	Up to 9	Insects, rodents	3.53 ± 0.25	De Lima (1975)
Malawi	Maize	Up to 9	Insects		Golob (1981)
Lower Shire Valley				3.2–3.4	
Lilongwe Plans				1.6–3.5	
	Sorghum	Up to 9	Insects	1.7 ± 0.5	
Nepal	Maize	6	Insects, rodents	5.7 ± 3.2	Boxall and Gillett (1982)
	Wheat	3	Mold	2.4 ± 1.9	
	Paddy	8	Insects	3.4 ± 2.2	
Turkey	Wheat	8	Insects, molds	3.7 ± 1.9	Boxall (1983)
Swaziland	Maize	Unspecified	Insects	3.66	De Lima (1982)
			Molds	0.53	
			Rodents	0.16	
Bangladesh	Raw and parboiled paddy	3–4	Insects Rodents	2.4 (rice equivalent) (average for 3 crop seasons)	Huq (1980)
Honduras	Maize	7	Insects	5.5	De Breve et al. (1982)

Table 4 Examples of Studies to Measure Postharvest Losses at the Commercial Level in Developing Countries

Country	Commodity	Stage	Causes of loss	Estimated mass loss (%)	References
Cuba	Various processed	Storage (3 mo)	Rodents	1.0	Hernandez and Drummond (1984)
Mali	Millet	Storage (8 mo)	Insects	1.0	Rowley (1984)
Dominican Republic	Paddy/rice	Drying (rice mill)	Breakage, lower grade	8.1% reduction in whole grain	Boxall et al. (1981)
		Milling	Breakage	2.3% reduction in whole grain	
		Public sector storage	Spillage	0.35	
Pakistan	Paddy	Market storage	Various	1.8	Chaudhry (1980)
		Wholesale storage	Various	3.2	
		Public sector storage	Various	3.0–5.0	
		Processing (milling)	Various	1.1–1.4	De Lima (1985a)
	Wheat	Processing	Various	3.6–4.4	De Lima (1985b)
	Maize	Processing	Various	1.5–3.6	De Lima (1985a)

whole, each operation being correlated with the previous ones. Even if each particular loss is minimized, they are so numerous that collectively they represent significant amounts. It is helpful to have some estimates of total and particular losses in the system as the higher losses occur where the system is the weakest. Further, there has also been a tendency to use the "worst case" figures as justification for investment in loss-reducing activities or for obtaining funds for development projects.

Based on experience gained by FAO through its Special Action Programme for the Prevention of Food Losses, postharvest losses of food grain in developing countries have been conservatively estimated at 10–15%.

Many comments have been made in relation to the accuracy and the reliability of the different loss assessment methods. Due to natural variability of the grains themselves (sorghum or millet for instance), the statistical analysis shows that the confidence interval at 95% is often very large, reaching up to 5–8%, which is more than the level of loss to be assessed.

Looking back over more than a decade at FAO's activities of the Special Action Program for Prevention of Food Losses, there may seem to have been an over-investment in loss assessments and in developing loss assessment methodologies. There is no doubt, however, that the numerous assessment activities have created an awareness of postproduction problems by governments in the developing world, that an expertise in the subject has been created in many of the countries concerned, and that postharvest problems have been given more attention in research, training, and extension in many countries.

5. POSTHARVEST LOSS REDUCTION ACTIVITIES (TECHNICAL AND SOCIO-ECONOMIC ASPECTS)

5.1. Developed Countries

The present low value of cereal grains does not stimulate action to improve handling and processing techniques in the commercial sector. The cost of storage is relatively high. In France, for example, it represents 12% of the value of the grain (about US $200/t in 1991). Consequently neither farmers nor grain traders are prepared to invest in or experiment with new techniques (e.g., controlled atmosphere storage) due to the small margins on the commodity.

5.2. Developing Countries

It is often stated that losses of food grain in developing countries arise from deficiencies in farmers' handling and storage methods. It must, however, be borne in mind that the traditional system has evolved according to local conditions and has been developed over many years. The losses are normally kept to the lowest level that can be achieved with the available techniques. Excessive losses will occasionally occur when conditions change, be it adverse weather or an outbreak of pests. In modern times, however, the changes which might cause losses are often man-made. The introduction of new crops, the adjustment of marketing policies, and new production techniques are frequent factors which require the alteration of traditional practices to avoid losses. These changes have taken place and will take place with increasing speed, making the problem-solving process within the postharvest sector a continuing one.

The following examples of FAO-executed projects illustrate the variations in postharvest practices and problems and the complexity of developing methods to solve these problems.

5.2.1. *Corn in Africa*

Investigations made by several projects—notably in Nigeria, Swaziland, and Kenya—showed that considerable losses occurred between physical maturity and consumption or off-sale of corn. The major part of the loss occurred during field drying, from maturity at approximately 30% moisture content (mc) (wet mass basis) to 18–20% mc when threshing is possible without damaging the kernels. The mass loss varied between 10 and 15%, mostly due to attacks from rodents, birds, and termites, and to rotting and shattering. Losses in farmers' stores due to insects were in the order of 6–10%. Methods to drastically reduce these losses were developed. The methods included an earlier harvest at 25–30% mc instead of the traditional harvest at 18–20%, drying the husked cobs in naturally ventilated cribs down to about 13% mc (which normally could be achieved in 4–6 wk), shelling, chemical treatment against insect pests, and finally, storage in bags or solid-walled bins.

Although these methods have shown their technical and economic superiority to existing traditional methods, and farmers exposed to intensive introduction campaigns have adopted the techniques, widespread use has not been achieved for two main reasons:

1. An early harvest means that the peak labor requirements will occur some weeks earlier than in the traditional work calendar. This will then clash with other important work, such as the preparation for the second crop in southern Nigeria or Benin, or, as in the case of Swaziland, with a period when the children are in school and therefore cannot participate in the harvest and in the husking.
2. The lack of reasonable credit arrangements for the purchase of certain essential materials (e.g., pesticides and building materials) prevent many farmers from adopting improved techniques.

5.2.2. *Rice in Madagascar*

In Madagascar, labor constraints at harvest time result in delayed harvest and high shattering losses in rice. The introduction of small motor-powered reapers enabled farmers to harvest all their crops at the optimum time; this resulted in a better spread of the crop in the field so that drying could be completed in 2 days and the crop ready for threshing. An assessment of the potential profitability of small reaping machines—taking into account operating costs, loss reduction as against hand reaping, and the price of additional labor at harvesting time—showed that their introduction was a realistic and economically viable solution when spread over three seasons.

Additionally, a loss assessment survey found that losses in threshing by animal treading were not high. It is a low intensity method of threshing where losses, however, can occur because of unseparated grains, depending on the variety. Farmers, therefore, choose to grow a variety that is easy to thresh so as to avoid these losses. Consequently, they tend to refrain from growing a high-yielding, better quality short-grain, Japonica variety of rice, which has a higher milling output, better taste, and does not shatter easily during reaping. An attempt to introduce pedal-operated threshers failed because they yielded only 1000 kg of threshed paddy a day as compared to 3000 kg commonly obtained by the treading method.

5.2.3. *Rice in the Senegal River Valley*

In the Senegal River Valley region, the rice harvest is often delayed by 2–3 mo due to limited availability of harvesting equipment and labor shortages; consequently, the moisture content of the grain decreases to less than 10%, resulting in a high percentage of shattering and low head rice recovery. As new irrigation infrastructures enable double cropping, farmers harvest during the rainy season and now have to face problems connected with grain drying.

5.2.4. *Corn in Nicaragua*

In some cases, farmers tackle the postharvest problems in an original way, such as the "walking way" of Nicaragua. As marketing is impossible due to bad roads and insufficient drying and storage facilities, farmers convert their corn crops into pigs, which can "walk" to the markets!

5.2.5. *Larger Grain Borer* Prostephanus truncatus *in Tanzania*

Prostephanus truncatus is a storage pest from southern United States and Central America, and was accidentally introduced into Tanzania in the late 1970s. It can cause mass losses in farm-stored corn which are several times greater than those caused by indigenous insect pests. Mass losses as high as 34% have been observed in some corn cribs after only 3–6 mo of storage. Since 1980 the pest has spread into 17 of the country's 20 regions. An FAO-coordinated, multidonor funded project has developed recommendations to combat this pest; these include: shelling of the corn from the cob as soon as possible, treating the corn with an insecticide, and storing it in small silos to avoid the risk of reinfestation by insects and rodent attacks. These recommended practices are new to many small-scale Tanzanian farmers who normally field-dry corn and store it unthreshed on the cobs in traditional structures. To be adopted, these practices need a strong push through agricultural extension programs; however, food losses were less than 2% during a season when recommended practices were used.

5.2.6. Rice in the Philippines

Owing to the lack of a proper grading system for grain, farmers in the Philippines rarely obtain reasonable prices and thus have less incentive to produce a clean, dry, pure commodity. An FAO project (completed in 1986) established pilot grading stations equipped to make an objective assessment of the quality of paddy and milled rice. The characteristics chosen for grade determination and for which standards have been proposed for paddy procurement, are those of moisture, purity, maturity, and damaged and discolored grains. The work was directed towards simplified methods of measurement and four grades of milled rice were recommended. The implementation of the proposed system by the National Food Authority requires strong Government commitment and substantial capital expenditure. One year after the termination of the FAO project in 1986, it was reported that the project's recommendations were gradually being implemented by the National Food Authority's technical services.

6. IMPROVEMENT OF THE POSTPRODUCTION SYSTEM

6.1. Creation of Awareness and Institutionalization

To achieve the final goal of a fully-coordinated and integrated national effort to develop and support the postproduction sector it is necessary to take a step-by-step approach which has been implemented in the FAO Prevention of Food Losses Programme. Government, private institutions, and persons engaged in agricultural development need to be aware of what functions are expected to be performed by the postproduction system and to recognize where there is a problem. It is essential to ignite the political will to do something about development in this sector.

FAO, for example, is greatly concerned with the problem of mycotoxins in food, including their effect on food trade. FAO's assistance to member countries in the prevention and control of mycotoxins has been directed to initiate awareness actions such as an advisory role for the identification of mycotoxin prevention and control strategies at national, regional, and international levels. For example, the regional workshop for Asia, held in 1990, referred to the need for a bold Asian initiative for declaring the present decade as the "decade of prevention and control" with activities designed to minimize contamination in food. The Latin American workshop, held in 1991, resulted in the establishment of the Latin American Society on Mycotoxins. Nevertheless, a more precise norm than the "irreducible level" of aflatoxins in human food is still needed although it already exists for animal feeds (10 parts per billion).

6.2. Assessment of the Postproduction System

All existing practices and processes that take place from maturity of the crop in the farmers' field, through to the final consumer, need to be observed, assessed, and understood. Farmers, traders, middlemen, managers of cooperatives, other purchasers, millers, and market officials must be part of this investigation, which should focus on what actually takes place and why it is done in that way. Too many actions failed due to the lack of socio-economic figures and mainly because the role of women in the postharvest system had been underestimated.

6.3. Pricing Systems

The economic performance of the postharvest sector is linked primarily with the pricing system and its applicability. There must be correlation between improved techniques and market returns. If insufficient care is taken with the different economic margins, it might lead to complete failure. Two examples can illustrate this:

1. In many countries a subsidy for rice milling was established to increase producers' income ("floor price") and decrease consumers' expense ("ceiling price") but was applicable only to large public plants. As the output of those plants is insufficient, the pricing policy impedes the dissemination of small mills on the free market where paddy is bought at much lower than the floor price and where, without incentive for head rice, there is no investment in milling technology performance.
2. Another example was observed in some tropical Asian countries where the penalty for delivering high moisture grain was less than the cost of the loss in mass due to removal of moisture plus the cost of drying. It resulted in the delivery of tremendous quantities of wet paddy immediately after harvest, product deterioration, and excessive investment in drying facilities for the collector.

Although these problems seem to be easy to solve, they persist for years, even in developed countries where, for example, the market price is based on wet mass rather than on dry matter which results in undesirable practices such as the addition of water to dry-grain.

6.4. Improvement of Storage Techniques, Adoption of New Technology

Change for its own sake is unwelcome. B. Champ, the coordinator of the ACIAR Program, pointed out that appropriate technology for bulk handling in the tropics was available but the challenge was how to use it to provide benefits to all postharvest sectors from farmers to consumers. The choice of bulk handling depends on the quantities to be stored, the quality of roads and transport, and the amount of energy available to run the facilities, etc. (Cutler 1987). In many developing countries, bulk handling and storage is not really needed and, at the level of farmers and small merchants or cooperatives, it is more worthwhile to concentrate on general quality improvement in the different parts of the system which already exist. Even at the public-sector level there are examples of efficient, combined systems which use bags for country stores (up to several thousand tonnes) and bulk handling for large quantities near and in main towns.

An important issue is to establish the particular needs in terms of capacity and structure as the total investment required for storage facilities is a basic element in storage cost estimation. In the 1970s, marketing offices invested heavily in the construction of storage centers destined to regulate the flow of products, to set up stock reserves, or both. The determination of the optimal capacities which the centers should have is a complex problem due to the numerous factors (timing of inputs, timing of outputs, quantity of the harvest, and amount of stock left over) which must be taken into account as they are often interdependent and sometimes unpredictable.

In developed countries, all these factors are analyzed, using as a reference those years which are more indicative. Monthly estimates of the storage needs in a chosen area are established and capacities derived. These indicators are, unfortunately, very often unavailable in developing countries, where the related investment is high for bulk storage installations.

Typically, for a bulk storage facility:

1. The cost of bulk storage can vary from one to three times as much as for a bag storage system;
2. Depreciation costs alone represent more than 65% of the direct operating costs (apart from financial costs);
3. The depreciation of installations takes place over a long period, therefore, long-term strategies should be adopted;

4. Investment errors cannot be absorbed by a parallel increase in production or inflation; and finally,
5. The gross margin of storage activities is weak, approximately 15% of the total.

These factors explain why private and state investors are so reluctant to finance the construction of storage facilities.

The use of mechanical dryers has often not been adopted, due not only to the lack of incentives but also to the very low level of capacity utilization. This is applicable to Africa where some well-advised projects have shown the success of simple drying floors after years lost in trying to disseminate mechanical dryers. One additional advantage of the drying floor is that it is multipurpose (any type of product can be dried).

The postproduction sector will continuously be subject to demand for new technologies as a result of changes in the environment. Increasing urbanization in many developing countries has for instance, resulted in a scarcity of labor in agricultural peak periods. This is bound to lead to an increasing demand for the mechanization of certain postproduction operations, such as: harvesting, threshing, cleaning, drying, and processing. Grouping of small farmers is frequently the only way to run equipment economically and to allow access to credit.

6.5. Pilot Activities

Improved alternative techniques must be pilot tested and critically evaluated against existing practices. Evaluation will normally be carried out through pilot projects including a technical evaluation, an economic analysis, and an assessment of the acceptability by the target group.

The economic analysis of the planned intervention is essential but has nevertheless been neglected in many projects in the past. Only in specific cases can uneconomic storage be justified (e.g., food security buffer-stocks). As a principle, all interventions should be profitable (e.g., the benefit should be cost efficient).

6.6. Sustainability and Environment

An important aspect of sustainability is that the technical development to take place should be in accordance with expected future environmental requirements. For the postproduction sector this means that the focus should be put on the reduction of the use of fossil energy, health aspects in connection with the use of pesticides, and maintenance of healthy grains. The possibility of using alternative energy sources such as biomass, solar energy, and agricultural residues for heated-air drying and other energy consuming equipment must be exploited as far as possible. For example, rice husks which represent a tremendous source of energy (100 Mt), are frequently used for heating but no gasification technique is ready for extension.

The use of pesticides, in particular in tropical countries, is presently essential for maintaining food grains insect and rodent free (Booth et al. 1987). While their use in developed countries is governed by strict regulations, their introduction into the developing countries has been somewhat more hazardous. Strict measures for control of imported as well as locally manufactured chemicals should be developed and exact handling instructions prepared. Continuous monitoring of insects and the pesticide used should be maintained to detect the development of resistance amongst pests.

Education and training are priorities to achieve sustainability. There is a marked lack of manpower trained in postproduction principles and techniques at all levels in developing countries. Specialized courses must be organized to create an awareness at the decision making level.

Similar problems are found in the extension service, where postproduction techniques have been gravely neglected in the past. To develop postproduction systems for small farmers, however, extension workers are of crucial importance. A drawback of government extension services is that the services do not cover private entrepreneurs, such as traders, processors, artisans, and retailers. This group is supposed to handle most large quantities of grain in the future.

The lack of adequate marketing arrangements linked with quality standards in many developing countries have probably been the main reason for slow progress in technological development in the postproduction sector as a whole. The challenge is to develop ways to ensure that each participant in the system is reasonably rewarded for his or her input without rendering the produce too costly for the final consumer.

Improvements in the postproduction sector will not only increase the pressure on the marketing of produce but normally also increase the need for a better input supply. The inputs might be mechanical equipment and pesticides and these must be readily available at the right time. To save usually scarce foreign exchange and to ensure that spare parts and repair facilities are easily available, it is preferable that the required inputs be locally manufactured. This represents investment possibilities which could be taken up by the private sector. Unfortunately, larger companies are rarely interested in small-scale technology because the market may be small and the return on investment unattractive. Also, banks as well may be hesitant to finance such risky ventures, particularly as small enterprises are not regarded as credit-worthy. In many cases, attempts should be made to start local manufacture as a joint venture with firms from an industrialized country. Assistance is normally required from the public sector, not only with regard to financial support but also policy adjustments, such as taxation rules for raw materials (which often do not favor local manufacturing).

7. FUTURE RESEARCH

Technical development in the postproduction sector should be a dynamic process because changes occur continuously in and around the sector. It is therefore not possible to point out or give priority to specific subjects to be addressed, but areas which need attention can be defined:

1. Improvement of postharvest techniques at the farmer level. This is often more an economic than technical issue. Access to equipment is limited by:

    a. Lack of information; farmers in developing countries cannot visualize the equipment required unless a prototype is made available.
    b. Size of equipment adapted to cope with small-scale farms or farmers' associations; national research centers have to assess prospective machines and provide the design of prototypes to manufacturers when needed.
    c. Equipment cost; a great effort has to be made in many countries to use the national potential for manufacturing at a lower price, utilizing low cost manpower.
    d. Access to credit; in too many countries, especially in Africa, credit is difficult to obtain even though most equipment credit is duly reimbursed.

2. More attention should be given to reporting adverse trade effects of poor quality grains (low grade or contaminated grain). This is a difficult issue as most salesmen refuse to publish this type of information in view of the adverse effects it could induce.
3. Despite numerous recommendations, there is no agreement on a maximum level for mycotoxins in food.
4. Drying techniques need further improvement in the use of renewable energy and in the thermal performance of small dryers. Alternative techniques, such as storage under controlled atmosphere, have been developed to store moist grains but the techniques need further refinement, particularly on the design of economic structures for various volumes.
5. Aeration (i.e., ventilation of bulk grain with ambient air) is well developed for temperate and subtropical zones. There is a great need, however, for this technique in the humid tropics, in particular for temporary storage when bulk grain has to wait for drying or processing. Research work has already been carried out by ACIAR in Asian countries but no standards are available for the design and operation of forced aeration under various humid-tropic conditions.
6. In many tropical countries, too often pest control is based only on fumigation with phosphine under poor airtightness conditions which may lead to

resistance development in insects. The need for better pesticide resistance surveys and for new pesticides was addressed by FAO in 1990 (Schulten 1990); again, storage under modified atmosphere might be a realistic alternative for commercial traders handling grain either in bags or in bulk. For the small farmer, the introduction of low-cost, airtight structures should be of great interest to improve pesticide treatment efficiency. For the small-scale farmer in developing countries, the use of inert, locally produced, botanical insecticides as well as biological control might also be alternatives to chemical insecticides.

7. Last but most important, is the link between storage techniques, quality of grain, and pricing policies. Without incentives to improve quality the adoption of efficient technologies will remain an unfulfilled wish.

8. REFERENCES

Adams, J. M. and G. W. Harman. 1977. The evaluation of losses in maize stored in a selection of small farms in Zambia with particular reference to the development of methodology. Report of the Tropical Products Institute (Tropical Development and Research Institute), G. 109, 149 pp.

Adams, J. M. and G. W. Harman. 1978. Storage losses project. Trop. Stored Prod. Inf. 36:16-17.

Anonymous. 1986. Characteristics and limitations of methods to estimate losses in stored grain. Special Report No. 16, Kansas State University, Manhattan, KS. 23 pp.

Anthony, R. W. V. and J. R. Arboleda. 1987. Manual for loss assessment in rice postharvest systems, FAO/PFL, Rome, 58 pp.

Booth, R. H., A. Toet, and L. Bevan. 1987. Investigating in sustainable postharvest programs. FAO (AGS/PFL), Rome, 30 pp.

Boxall, R. A. 1983. Prevention of postharvest crop losses in Central Anatolia, Turkey. Storage loss assessment, Consultancy Report. Ag. GCPP/TUR/027/NET, Field Document 2, FAO, Rome, 50 pp.

Boxall, R. A. and R. Gillett. 1982. Farmer level storage losses in Eastern Nepal. Report of the Tropical Products Institute (Tropical Development and Research Institute), G. 157, 29 pp.

Boxall, R. A., M. Greeley, and D. S. Tyagi. 1979. Prevention of farm level food grain storage losses in India: a social cost benefit analysis (Save Grain Campaign, rice). Trop. Stored Prod. Inf. 37:11-17.

Boxall, R. A., J. La-Gra, E. Martinez, and J. Martinez. 1981. Post harvest losses of rice in Dominican Republic. Harvesting, milling and storage losses. Trop. Stored Prod. Inf. 42:5-10.

Chaudhry, M. A. 1980. Food grain losses at farm level in Pakistan. Dept. Agric. Marketing, Univ. Agric., Pakistan.

Creupelandt, H. and A. Shepherd. 1990. Developing food-grain production for the market. pp. 59-68 in: Structural Adjustment and Agricultural Marketing, Marketing and Credit Service, FAO, AGS, Rome.

Cutler, H. M. 1987. Key considerations in the adoption and application of bulk handling and storage systems for grain in developing countries. pp. 24-32 in: Bulk handling and storage of grain in the humid tropics B. R. Champ and E. Highley (eds.). ACIAR Proceedings No. 22, Aust. Center Int. Agric. Res., Canberra, Australia.

De Breve, M., G. Raboud, J. Sieber, J. A. Perdomo, and J. E. Velasquez. 1982. Proyecto postcosecha: informe sobre los primeros resultados. Ministerio de Recursos Naturales, Cooperaction Suiza al Desarrollo, Tegucigalpa, Honduras. 126 pp.

De Lima, C. P. F. 1975. The conduct of field infestation surveys and the economic use of their results. pp. 47-60 in: Proc. 1st Int. Work. Conf. Stored Prod. Entomol., Savannah, GA.

De Lima, C. P. F. 1982. An index of quantitative and qualitative food loss. Food Nutr. Bull. 4:10-12.

De Lima, C. P. F. 1985a. Survey of storage losses in major food grains in Pakistan—a socio-economic survey of the postharvest system. FAO/PFL/PAK002 Field document No. 4., FAO, Rome. 254 pp.

De Lima, C. P. F. 1985b. Survey of storage losses in major food grains in Pakistan—a survey of postharvest losses in wheat. FAO/PFL/PAK002 Field document No. 5. 379 pp.

De Padua. 1978. Research and development of postharvest losses. Trop. Stored Prod. Inf. 36:53.

FAO. 1979. Production yearbook. United Nations FAO, Rome. 309 pp.

FAO. 1984. Production yearbook. United Nations FAO, Rome. 309 pp.

FAO. 1988. Production year book. United Nations FAO, Rome. 309 pp.

Golob, P. 1981. A practical assessment of food losses sustained during storage by small holder farms in the Shire Valley Agricultural Development Project area of Malawi 1978/79. Report of the Tropical Products Institute (Tropical Development and Research Institute), G. 154, 47 pp.

Greeley, M. 1982. Farm-level postharvest food losses: a myth of the soft third option policies, Bangladesh. pp. 51-60 in: IDS Bull. Univ. Sussex Inst. Dev. Studies. Brighton, UK.

Harnandez, A. and D. C. Drummond. 1984. A study of rodent damage to food in some Cuban warehouses and the cost of preventing it . J. Stored Prod. Res. 20:83-86.

Harris, K. L. and C. J. Lindblad. 1978. Postharvest grain loss assessment methods. Am. Assoc. Cereal Chem., Minneapolis, MN. 193 pp.

Huq, A. K. F. 1980. Some estimates of farm-level storage losses in Bangladesh. Trop. Stored Prod. Inf. 39: 5-12.

Rowley, J. Q. 1984. As assessment of losses during handling and storage of millet in Mali. Trop. Stored Prod. Inf. 47:21-33.

Schulten, G. G. M. 1990. Stored products protection in warm climates. pp. 1557-1572 in: Proceedings 5th Int. Work. Conf. Stored Prod. Prot. Fleurat-Lessard and P. Ducom (eds.). Bordeaux, France.

Smil, V. 1991. General energetics—energy in biosphere and civilization. John Wiley and Sons, New York. 396 pp.

5

Insects, Mites, and Insecticides in Stored-Grain Ecosystems

Noel D. G. White

Agriculture and Agri-Food Canada, Winnipeg, Manitoba, Canada

SUMMARY

Insects have adapted to the storage environment because of several factors. These include: 1) their original and present natural reservoirs, 2) their range of tolerance for physical factors, 3) their range of food habits, 4) their reproductive rate and related population characteristics, 5) adult survival in the absence of food, and 6) morphological adaptations. Mites require relatively high moisture grain for survival and are classed as stored-product feeders, fungivores, predators, or parasites on rodents and birds.

The population dynamics of insects and mites in stored grain are regulated by various density-dependent (e.g., high mortality in immature stages) and density-independent mortality factors (e.g., temperature). Chemical ecology (e.g., grain and mold odors, insect and mite pheromones) can indicate pest infestations and potentially modify pest behavior. The bioenergetics of insects and mites give important information on efficiencies of energy use and accurate estimates of energy lost from stored food and can be used in pest-management decisions. Insecticides have a dramatic short-term impact on stored-grain ecosystems but insects and mites recover or re-invade the system as chemicals degrade. Heavy reliance on chemical control has led to widespread insecticide resistance and control failures and reduced interspecific competition in many communities.

The future directions for research related to stored-product insects and mites should include studying the detailed interactions in ecosystems in both the laboratory and the field. Computer modeling will permit more effective management of stored grain, based on physical variables, population dynamics, chemical ecology to maximize detection, and energy flow through the food web. We should broaden our boundaries of stored-product ecosystems for a more complete understanding of pest activity.

1. INTRODUCTION

Insects and mites are integral parts of stored-product ecosystems (Sinha 1973a,b; Calderon 1981; Lacey 1988). The ecosystem concept has received widespread usage since its introduction by Tansley (1935) to encompass the entire complex of physical and biological environmental factors operating in an arbitrary ecological unit. Ecosystems are not closed systems, isolated and independent, because energy and matter flow between adjacent ecosystems. The stored-product ecosystem is unique because this man-made system is not self-sustaining, having dormant seeds or processed material as "producers," and energy flow is unidirectional, out of the system as food biomass is converted to animal biomass and decomposed by microbes (Fig. 1). Energy input into the system is not directly from sunlight and photosynthesis but is indirectly supplied by sunlight and humans (e.g., petroleum products) during the growing of food (Odum 1989). In North America, labor, machinery, fuel, fertilizers, pesticides, drying, and

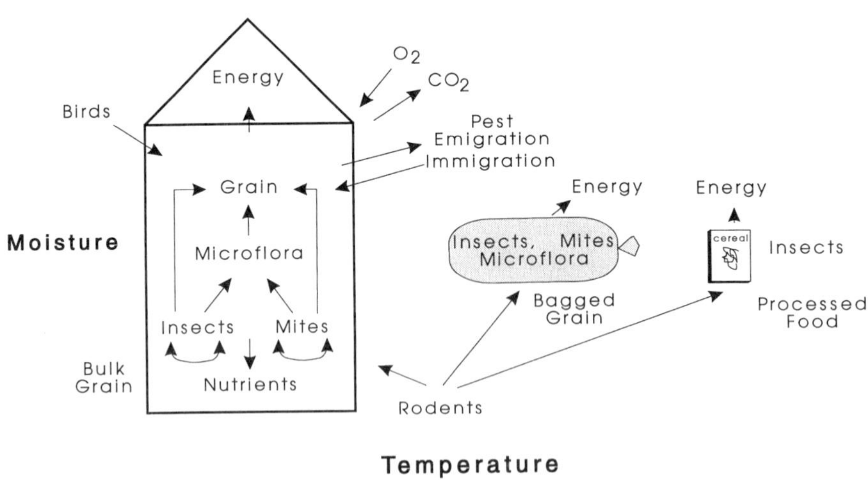

Figure 1 Schema of stored-grain ecosystems: a granary, a sack, and a box of processed cereal. Insects can attack the product regardless of the type of storage.

transport require about 3.5×10^{10} J/ha for the production of 7000 kg of corn/ha containing 1.0×10^{11} J (Pimentel 1984).

Humans have been aware of the interplay of physical factors, grain storage, and pest infestations for centuries and have developed local storage practices that have minimized food loss (Sigaut 1988). Many species of insects and mites have found stored products to be a favorable environment and have adapted to a situation where habitat and food are the same, and where entire communities of seed, microflora, and arthropods can be rapidly devastated by humans moving or consuming the product. The majority of research on long-term storage has been done with dry plant material such as seeds or their by-products.

It is estimated that about 900 Mt of cereal grains are in storage around the world at one time (half the annual production) and that annual losses to insects and rodents are about 10% in North America and 30% in Africa and Asia, but that higher losses and contamination often occur locally (Anonymous 1981, Hill 1990). In temperate climatic regions, low density insect and mite infestations tend to be perennial (Storey et al. 1982, 1983; Madrid et al. 1990) but outbreaks occur sporadically, while in tropical and subtropical regions infestations and outbreaks are common (Freeman and Heape 1968). Insects infested 11% of grain stores in a temperate region such as Germany in 1976 but declined to 1.2% in 1986 because of better storage management, indicating that effective techniques are available for pest control in developed countries (Stein 1991). Direct energy loss from grain is minimal in temperate regions but insects and mites cause significant qualitative losses such as increased levels of free fatty acids, seed damage, loss of seed germination, contamination with uric acid and exuviae, poor milling and baking quality of flour, and stimulation of microfloral growth (Van Bronswijk and Sinha 1971, White and Sinha 1980). Mycotoxins and "hotspots" are often associated with insect infestations (Sinha and Sinha 1990). Insects can be especially devastating to subsistence farmers who rely on their limited store of food for survival. Management of the stored-grain ecosystem with insecticides or physical manipulations is especially critical in tropical and subtropical regions where quantitative losses are most likely.

Over 100 species of pest insects have been found in stored grain and over 355 species of mites (Sinha 1991) have been recorded in stored products with 54 species from stored grain and flour (Hughes 1976) worldwide. The complex interactions among these animals in their roles as fungivore, granivore, predator, or parasite, and the environment, the microflora, and the stored grain or other food, result in ongoing population fluctuations, species succession, and degradation of human food (Sinha 1973a,b; Wallace and Sinha 1981).

This brief overview of the ecological basis for success of insects and mites in stored products is intended to put in perspective the role of these pests in storage ecosystems.

2. INSECTS

2.1. Successful Adaptations of Stored-Product Insects

Many insect pests of stored products are cosmopolitan. The reasons for their widespread presence range from evolutionary adaptations (morphological, physiological, and behavioral) to the actions of humans who transport them throughout the world and offer a protected habitat within stored foodstuffs (Freeman 1973).

The success of stored-product insect species in exploiting relatively dry human food is based on six major factors: 1) original and present reservoirs, 2) range of tolerance for physical factors in the environment, 3) range of food habits, 4) reproductive rate and other population parameters, 5) adult survival in absence of food, and 6) morphological adaptations. Stored-product insects and mites typically follow the adaptive strategies of colonizing species both genetically and ecologically (Parsons 1982).

2.2. Original and Present Natural Reservoirs of Stored-Product Insects

Humans began to cultivate plants and store plant foodstuffs around 8000 BC in the Middle East. It can be assumed that the storage pests of today evolved in natural habitats long before this time and moved into stores which provided conditions to which they were already adapted. *Tribolium castaneum* (Herbst) has been associated with stored-grains in Egyptian pyramids dating from 1345 BC and *Sitophilus granarius* (L.) has been found in grain from the 9th to 7th century BC in Israel (Buckland 1981). The following insects have been recorded from Iron Age or Roman period (0–400 A.D.) grain storages (Buckland 1981): *Stegobium paniceum* (L.) (Anobiidae); *Niptus hololeucus* (Fald.) (Ptinidae); *Tenebroides mauritanius* (L.) *(Trogositidae); Cryptolestes ferrugineus* (Steph.), *Cryptolestes* sp.; *Oryzaephilus surinamensis* (L.), *Oryzaephilus* sp. (Cucujidae); *Typhaea stercorea* (L.) (Mycetophagidae); *Aglenus brunneus* (Gyll.) (Colydiidae); *Blaps lethifera* Marsh; *T. castaneum*, *Tribolium* sp.; *Palorus ratzeburgi* (Wiss.), *Palorus subdepressus* (Woll.); *Tenebrio molitor* (L.), *Tenebrio obscurus* (F.) (Tenebrionidae); and *S. granarius*, *Sitophilus* sp. (Curculionidae).

Stored-products insects may be classified according to their sources of food (Linsley 1944):

1. *Seed-infesting species.* Some insects that naturally infest seeds will attack seeds stored by man. This behavior involves few adaptations with the exception of tolerance of low moisture content of the seeds. Surprisingly few modern stored-products pests are of this origin. Some examples can be found in the families Gelechiidae, Bruchidae, and Curculionidae. These insects can infest seeds in the field and continue development during storage (Yoshida 1983).

2. *Fungus-feeding species*. Many stored-product insect pests are primarily fungivorous and are associated with old, damp moldy products. Examples are in the Collembola, Lepidoptera (Tineidae, Pyralidae), and Coleoptera (Cryptophagidae, Lathridiidae, Mycetophagidae, Tenebrionidae families).
3. *Scavengers on dead plant materials*. The most notable examples are the phycitid moths.
4. *Scavengers on dead animal matter*. Principally the dermestid beetles.
5. *Scavengers or semipredators living under bark*. The bark habitat offers a favorable microclimate and a wide range of food types. Most insects in this group are tropical or semitropical in origin. Examples include the Trogositidae, Cucujidae, and Tenebrionidae (notably *Tribolium* spp.) families.
6. *Wood-borers and wood-scavengers*. Mainly the Anobiidae and Bostrichidae families.
7. *Scavengers in the nests of other insects*. The nests of bees, wasps, ants, and other insects offer favorable microenvironments and food supplies for many insects. Several of the Galleriidae, Phycitidae, Ptinidae, and Dermestidae insects may have originated in such habitats.
8. *Predators and parasites*. These include species from the Hemiptera, Diptera, and Hymenoptera. Predators and parasites of stored-product insects are probably attracted to a storage site by the product and habitat rather than their prey.

Current primary natural reservoirs for stored-product insect pests are bark and decomposing wood, nests and food caches of other insects, especially the social Hymenoptera, and nests of birds and rodents, all of which can be found near man-made storage structures (Linsley 1944) as well as infested grain residue and storage containers (Smith and Barker 1987, Cink and Harein 1989).

2.3. Range of Tolerance for Physical Factors in the Environment

Insects associated with stored products exhibit a wide range of tolerance for physical factors in the environment, especially moisture (Fig. 2). The response of even closely related species such as *C. ferrugineus* and *Cryptolestes pusillus* Schönherr can vary appreciably (Kawamoto et al. 1990) and there can be some variation within strains of a species (Jacob and Fleming 1989). These insects can probably tolerate more extreme conditions of temperature and humidity than many less-widely distributed species. Most of the important species can survive in foodstuffs with a moisture content considerably below the optimum for development. Buildings, warehouses, and storage structures tend to provide favorable environments.

A successful insect species must be able to multiply rapidly when conditions are favorable and be able to survive unfavorable conditions (Howe 1965). Certain very successful insect species are widely distributed but are usually

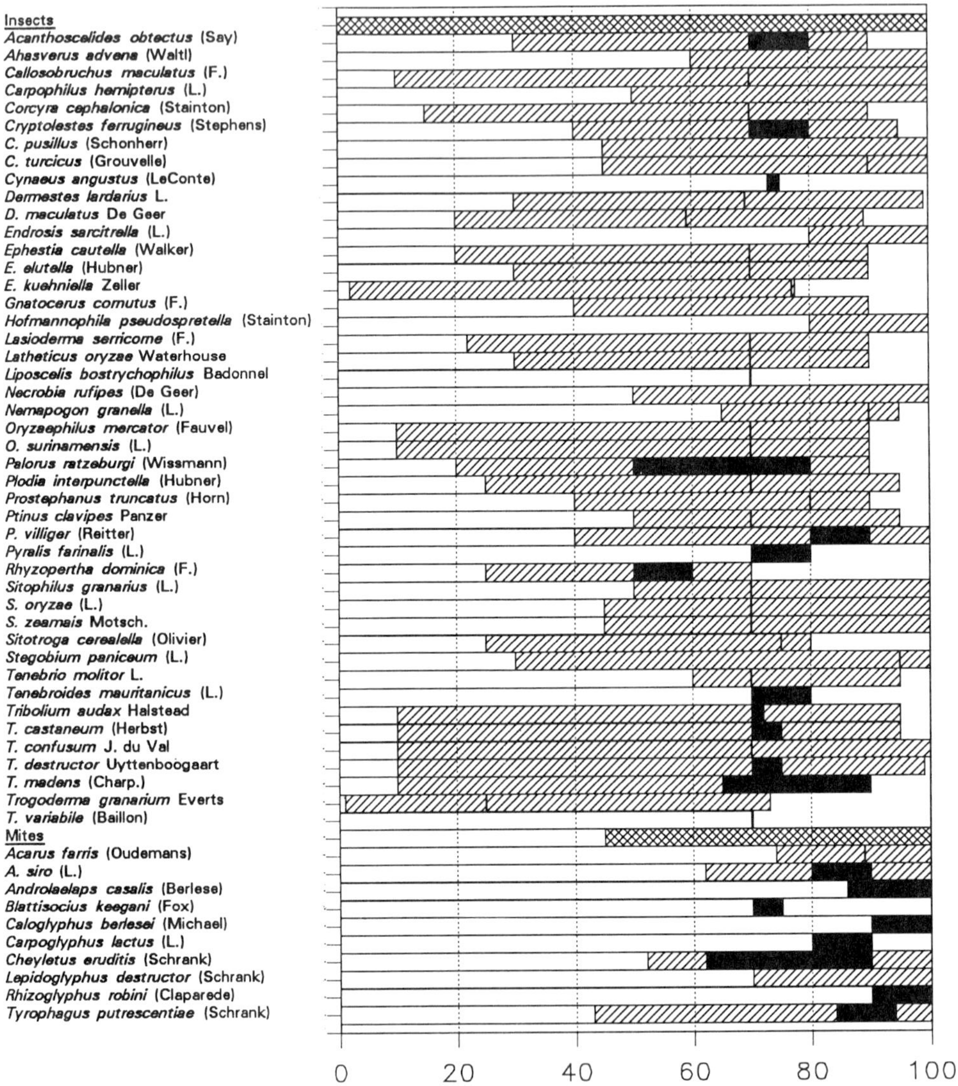

Figure 2 Range (cross-hatching) and optimal (solid) regions of relative humidity (%) for development of selected stored-product insects and mites. *Sources*: Sinha and Watters 1985, Sinha 1991.

thought of as tropical, temperate, or Mediterranean pests depending on their range of tolerances.

The range of temperatures in which stored-product insects can survive lies between 8–41°C (Sinha and Watters 1985) (Fig. 3) and usually development and multiplication is optimal near 30°C and 50–70% relative humidity (rh) (Howe 1965).

Pests which require high temperatures to develop may be classified as cold-hardy, moderately cold-hardy, and cold-susceptible. Insects which are cold hardy and tolerant of low relative humidities include *C. ferrugineus* and *O. surinamensis.* A species which is cold-hardy but requires a relative humidity above 25% is *Plodia interpunctella* (Hb.). Moderately cold-hardy species include *Tribolium confusum* J. du Val., *Rhyzopertha dominica* (F.), and *Lasioderma serricorne* (F.). *T. confusum* is tolerant of very low relative humidities, while *R. dominica* and *L. serricorne* require a moderate relative humidity. Cold-susceptible species which are tolerant of low relative humidities include *T. castaneum* and *Oryzaephilus mercator* (Fauvel).

Common insect pests which thrive at moderate temperatures may be classified as: cold-hardy needing a moderate relative humidity such as *Sitotroga cerealella* (Oliv.), cold-hardy needing a high relative humidity such as *S. granarius*, and moderately cold-hardy needing a high relative humidity such as *Sitophilus oryzae* (L.) (> 60%).

A more complete summary of estimates of optimal and minimal conditions for insect pests of stored-products can be found in Howe (1965) and Sinha and Watters (1985).

Diapause in stored-product insects is of adaptive significance in temperate climates. In most parts of the storage microenvironment, conditions are either relatively constant or vary irregularly. Few individuals of storage species encounter regular diurnal and seasonal variations although residual populations outside stored commodities may be exposed to variations. It is likely that diapause developed in a pre-storage environment, however diapause could be disadvantageous when food is continuously available. Pest species which demonstrate diapause include several Dermestidae, Ptinidae, *Ephestia elutella* (Hb.), *S. cerealella*, and *P. interpunctella* (Howe 1965).

Low temperature tolerance is of great importance to many stored-product insects (Fields 1992). The storage environment comprises two different habitats, the stored foodstuffs themselves in which conditions are usually stable and the surface of the food where conditions may fluctuate. Many species may live close to the surface and this could limit their distribution in various climatic regions of the world (Sinha 1974). Predictions of areas which are potential sites for specific pest infestations have been made for Japan by Sinha and Utida (1967) and a world-wide correlation between the physical requirements of stored-product insect pests and climate was used to develop a climatic plasticity

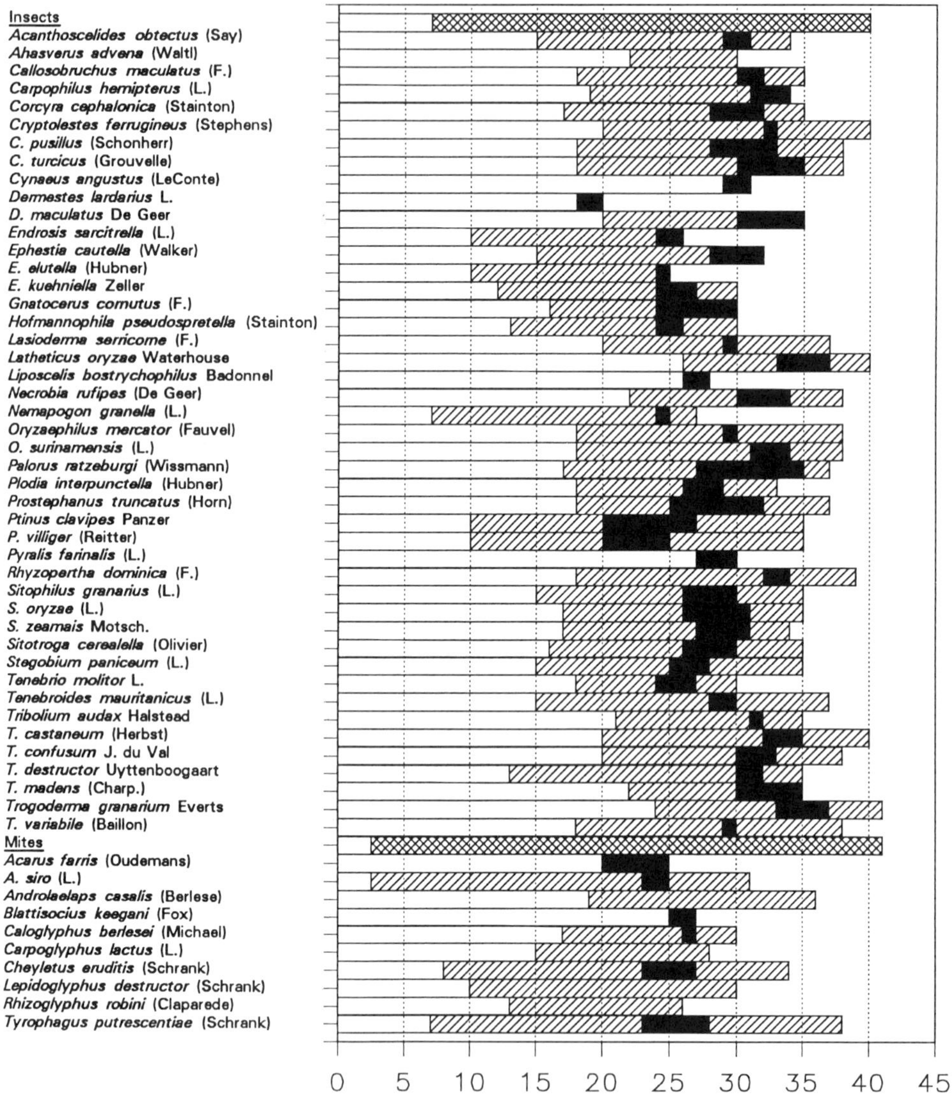

Figure 3 Range (cross-hatching) and optimal (solid) regions of temperature (°C) for development of selected stored-product insects and mites. *Sources*: Sinha and Watters 1985, Sinha 1991.

index by Sinha and Watters (1985). The most cosmopolitan pests, *T. castaneum*, *C. ferrugineus*, and *O. surinamensis*, have the highest plasticity indices (which incorporate the intrinsic rate of increase of a species and laboratory data on its physical limits for multiplication and survival).

Low temperatures in bulk grain in temperate regions are generally a limiting factor for insect pests in small granaries where ambient air temperature and grain temperature can equilibrate in a few months (Yaciuk et al. 1975) or where aeration is used to cool the grain.

2.4. Range of Food Habits

Most stored-product insects have a wide range of food habitats and may feed on a majority of the available dry food products of plant and animal origin. This food range is greater than that usually attributed to most other insects (Linsley 1944, Baker and Loschiavo 1987).

Stored grain is a principal diet for many storage pests but certain cultivars of grain are resistant to specific insects (Irshad et al. 1988). The quality of stored-grain is also important since many insect species are not whole-grain feeders and thrive only in the presence of dockage (Sinha 1975, Arthur and Redlinger 1988) or fungi associated with deterioration in grain quality (Sinha 1961, 1965, 1966, 1968, 1971; Sinha and Harasymek 1974). Some insects (e.g., *S. oryzae, T. castaneum, R. dominica*, and *Trogoderma granarium* Everts) have also been directly associated with seed damage, *Aspergillus flavus* growth, and the production of the mycotoxin Aflatoxin (Sinha and Sinha 1990). The wheat curl mite, *Eriophyes tulipae* (Keifer) is not an effective vector of *A. flavus* spores into corn kernels but the weevil *S. oryzae* is an effective vector in the United States (Barry et al. 1985).

The adaptive significance of fungus-feeding in several species of storage insects and mites has been studied by Sinha (1968), who pointed out that interspecific competition is often minimized by different preferences for specific fungi.

2.5. Reproductive Rate and Related Population Characteristics

Most stored-product insect species, with the exception of many moths, reproduce more or less continuously under favorable environmental conditions. Many species are long-lived as adults (Fig. 4) and some may live and reproduce over prolonged periods. Generally, species which can increase in numbers rapidly during favorable periods will be more abundant than those which increase in numbers slowly (Fig. 4). Dispersal is important to the success of insect species; stored-product insects usually walk in temperate regions but they may fly on warm days if they are morphologically able to do so, and in tropical regions they fly frequently (Taylor 1971). These insects have wide food preferences

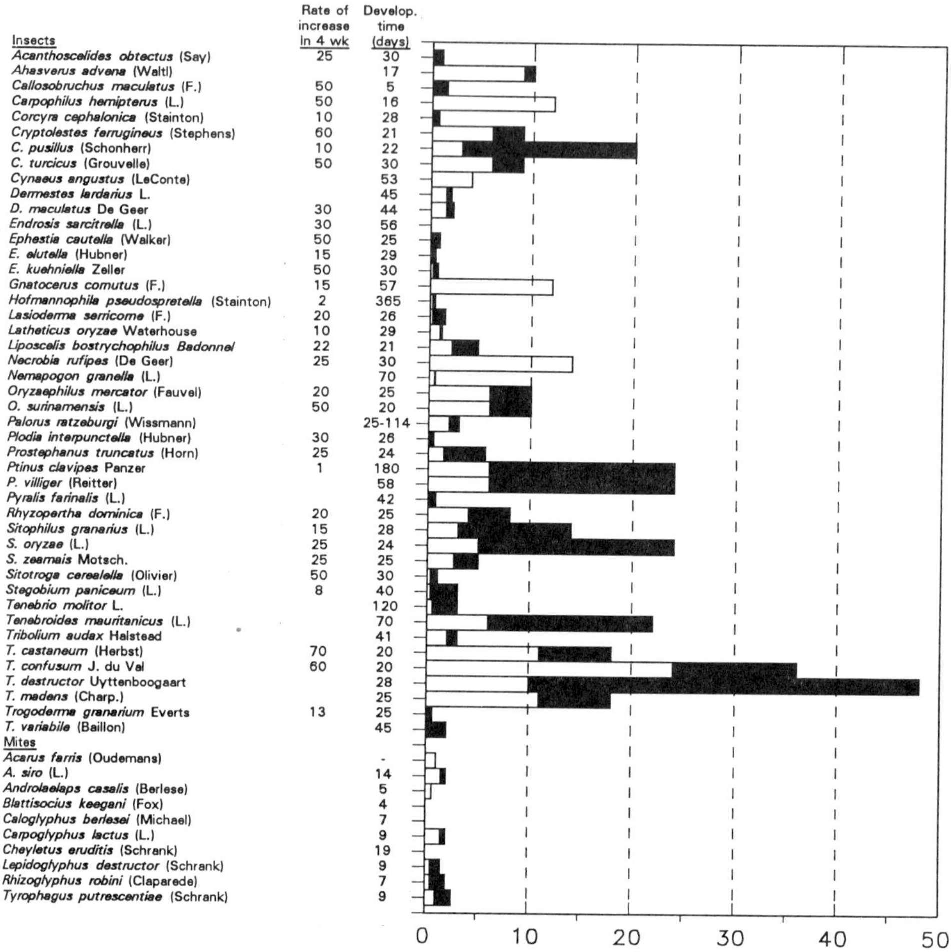

Figure 4 Range (white and black bars) and average (white bars) regions for adult longevity, known rates of increase in 4 wk, and optimal developmental times (d) for selected stored-product insects and mites. *Sources*: Sinha and Watters 1985, Sinha 1991.

and this allows them to move from one food product to another during storage and transportation (especially in warehouses, trucks, and the holds of ships) leading to cross-infestation and residual infestation (Freeman 1973).

2.6. Adult Survival in the Absence of Food

Some of the most important species can survive as adults for relatively long periods with little or no food, or in areas of poor sanitation (Coombs and Freeman 1955, Pinniger 1974). This fact favors their accidental transportation throughout the world especially in contaminated holds of ships. Many factors, such as sex, strain, and adult density affect the refuge-seeking behavior of *C. ferrugineus* (Cox et al. 1989, 1990) and other species.

2.7. Morphological Adaptations

The small size, dark coloration, and generally flattened shape of many beetle species may cause them to be overlooked and assist in their movement through the grain. Other species are more cylindrical and robust and can force their way between kernels.

3. MITES

3.1. General Characteristics

Mites are often found in association with stored products throughout the world (Sinha 1979) and can be classified into four groups according to their food (Hughes 1976): 1) stored-product feeders, 2) predators, 3) fungivores, and 4) parasites on rodents and birds which may be present in food stores.

A typical life-cycle consists of egg, larva, one to three nymphal stages with a quiescent period prior to each moult, and adult. Many Astigmatid mites (e.g., *Acarus* and *Lepidoglyphus* spp.) are capable of producing a relatively quiescent stage called a hypopus which takes the place of a deutonymphal stage. The hypopus is resistant to adverse conditions and is a stage during which passive dispersal occurs (wind, insects, etc.). *Acarus siro* L. is able to produce both active and inert hypopi while *Lepidoglyphus destructor* (Schrank) produces only inert forms (Sinha 1964). Hypopus formation is largely controlled by genetic and environmental interaction although dietary factors also play a role (Knulle 1987, 1991).

Respiration is either cutaneous or by means of trachea which open on the body surface by spiracles. The absence or presence and the number and position of these spiracles are the main diagnostic character on which the mites are subdivided. Stored-product mite pests belong to the suborder Astigmata (classification of Krantz 1971), while many predators and mycophagous species are represented in the suborders Mesostigmata, Prostigmata, and Cryptostigmata in that order of frequency of occurrence (Hall 1970). All stored-product mites, particularly the Astigmata, are sensitive to humidity (Fig. 2). In the Astigmata, all gas exchange occurs through the integument; other mites have respiratory organs but respiration through the body surface also occurs. Because water-loss cannot be controlled, low relative humidities lead to rapid death (Hughes 1976). Generally, stored-product mites can survive at temperatures ranges of 2.5–41°C (Fig. 3) although the optimum is around 30°C, and they require a high relative humidity (Sinha 1991). When temperature and moisture are favorable, mites typically have rapid developmental periods and short lifespans (Fig. 4).

3.2. Pest Status

Several mites associated with stored-products are important economic pests although they are often not noticed due to their small size (< 1 mm). When temperature conditions are favorable in materials with relatively high moisture contents these mites multiply rapidly to form dense populations. If grain is harvested dry, some of the kernels will be mechanically damaged providing a favorable substrate for mites when moisture increases occur in the grain-mass during storage. Also, if grain is moist at harvest, molds will develop which are alternate food sources for mites, and the high moisture content of the grain will favor mite development (Sinha 1964). Serious mite infestation occurs frequently in temperate climatic regions of the world but mites do not appear to be major pests of stored products in subtropical and tropical regions although many are cosmopolitan species (Hall 1970; Sinha 1973a, 1991).

The effects of mites in stored grain include: 1) the product may acquire an odor due to lipid secretions by mites, notably from *A. siro* infestations; 2) the grain germ may be consumed and, to some extent, the endosperm (Parkinson 1990), resulting in the loss of fat and vitamin content, notably by *A. siro*, *L. destructor*, and *Tyrophagus longior* (Gervais); 3) mites produce metabolic heat; however, "hot-spots" do not occur because mites do not penetrate bulk grain extensively, mites cannot tolerate high temperatures, and mites feed mainly on the seed germ and heavy infestations exhaust the available food supply at any given point before high temperatures can occur; 4) mites can cause intestinal disorders, dermatitis, and allergies in man and animals (Revsbeck and Anderson 1987); and 5) feeding on the germ kills the seed and affects milling quality (Solomon 1946). Mites also serve as vectors for several kinds of molds and bacteria (Zdarkova 1967, Lacey 1988) (Fig. 5) although heavy mite infestations can result in less visible mold on grain than on uninfested grain probably because of consumption of molds (Parkinson 1990).

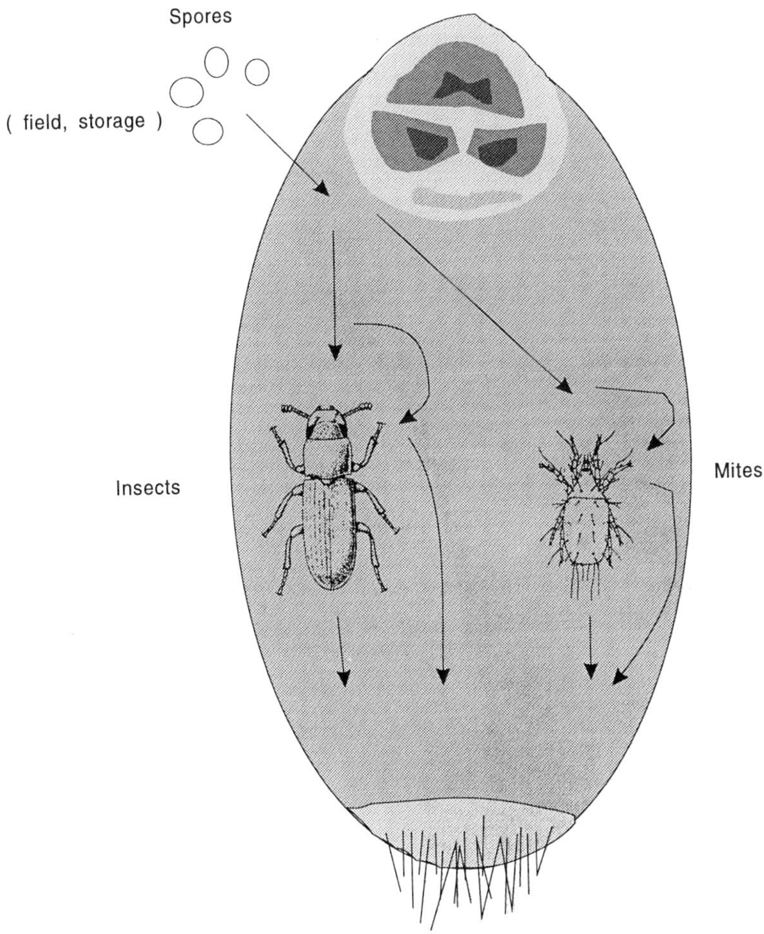

Figure 5 Schema of microfloral spore transmission to the stored product on or in insects and mites.

It has been hypothesized that mites found in stored products originally inhabited the nests and food caches of insects and the nests of rodents and birds (Linsley 1944, O'Connor 1979). Mite infestations of stored grain largely originate in the field and partially from debris on granary floors (Sinha 1968). Barker (1993) has shown that many stored-product mites can be carried on the bodies of flying stored-product beetles.

The most common stored-product mites in various countries in Europe and Asia are *L. destructor*, *A. siro*, *Cheyletus eruditus* (Shrank), *Acarus farris* Oudemans, *T. longior*, *Androlaelaps casalis* (Berlese), *Tyrophagus putrescentiae* (Schrank), Tydeidae, and *Tarsonemus granarius* Lindquist (Sinha 1979).

4. POPULATION DYNAMICS

Insect and mite populations in stored grain are difficult to quantify by sampling because of nonuniform distribution of the pests throughout a bulk (Zdarkova et al. 1983). Intense sampling using grain probes or probe traps, however, can be used to approximate the numbers of insects in a granary (Hagstrum et al. 1985, 1988; Lippert and Hagstrum 1987). A comprehensive review on the use of probe and pitfall traps in stored grain and problems associated with estimating insect density was given by White et al. (1990).

Arthropod populations in an optimal environment of temperature, relative humidity, and nutritionally suitable food increase exponentially (Lamb and Loschiavo 1981) until density-dependent factors decrease the population (Fig. 6)

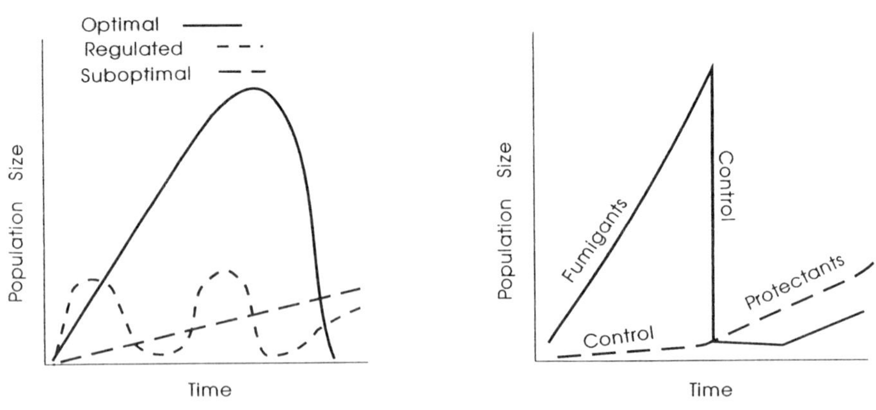

Figure 6 Patterns of insect and mite population growth in stored products under optimal, suboptimal or regulated (cycling) conditions or following the application of fumigants or protectant insecticides.

largely through mortality of immature stages (Kawamoto et al. 1989). Emigration of insects out of heavily infested grain is also extensive and is affected by species, temperature, insect density, grain age, and moisture content (mc). In an experiment of over 15 mo that used 27.3 kg bulks of wheat with an initial infestation of 100 adults, Sinclair and Adler (1984) found peak adult densities of 5086 *S. oryzae*/kg of wheat, 2405 for *R. dominica*, 332 for *T. castaneum*, 1721 for *C. pusillus*, 203 for *C. ferrugineus*, and 437 for *Cryptolestes pussilloides* (Steck and Howe). The average weekly adult emigration rate was estimated to be 24.9% for *S. oryzae*, 17.8% for *R. dominica*, 37.8% for *T. castaneum*, 28.3% for *C. pusillus*, 16.7% for *C. ferrugineus*, and 6.9% for *C. pusilloides*. Most flight activity occurred at temperatures above 27°C.

Another density-dependent factor that regulates insect populations in stored products is cannibalism, especially in *Tribolium* spp. (Stevens 1989) where adults and larvae feed on all life stages except for larvae feeding on larvae (Park et al. 1965). In species such as *C. ferrugineus* where up to three larvae may occasionally be present in the germ of a wheat kernel, only one eventually remains, forcing the others to leave the kernel to seek new food sources (White and Bell, unpublished data). Environmental conditions will often be suboptimal (low temperature or relative humidity) or will fluctuate causing slowed population growth. Natural enemies of pest insects and mites usually become abundant in stored grain only after heavy infestations and serious damage have occurred (Haines 1984), although Hagstrum (1987) showed that parasites could become abundant, and reduce the pest population by 50–70% before the pest population reached damaging levels. The lag in population growth between natural enemies and prey also occurs for the pteromalid wasp, *Anisopteromalus calandrae* (Howard) which is a parasitoid of the maize weevil, *Sitophilus zeamais* Motsch. (Arbogast and Mullen 1990). Artificial introduction of many *A. calandrae*, however, can suppress the initial increase in weevil populations. Biological control techniques need to be closely monitored but can be effective in raw agricultural products where insect bodies can be removed by cleaning prior to grain processing (Arbogast and Mullen 1990).

A long-term study of population fluctuations of insects in stored corn by Arbogast and Mullen (1988) revealed a succession of many species over 8 yr. They concluded that successional change results from: species-specific responses to grain degradation (insects and fungi) and interaction among species, the species that are present in the region, and potentially dominant species that are present at low densities. The type of food can strongly influence both rates of increase, through nutritional suitability (Baker and Loschiavo 1987) and, to a limited extent, sex ratios of insects (White and Jayas 1989) with type and variety of cereals affecting oviposition and development (White and Loschiavo 1988). Some oilseeds (e.g., rapeseed, flax, or sunflower) are an unsuitable diet for most insects (White and Jayas 1991) although *Tribolium* spp. and *Oryzaephilus* spp.

can survive on them. Other oilseeds, such as peanuts, are an excellent diet for many stored-product insects. Mites are common in bulk oilseeds and they often feed on associated fungi (Sinha 1979). Psocoptera also frequently occur in oilseeds (Sinha 1988).

Under favorable conditions (30°C and 70% rh) competition strongly affects the success of pest populations (Table 1). *Oryzaephilus surinamensis* is unable to compete successfully with the other insect species and an increase in most species is inhibited by the presence of other species competing for food and living space. The whole-seed feeders *S. oryzae* and *R. dominica* often initially benefit the broken-seed feeders *T. castaneum*, *C. ferrugineus*, and *O. surinamensis* by breaking wheat kernels open, but population pressure soon reverses this trend (White and Sinha 1980).

The distribution of insects in bulk grain is typically nonuniform and is determined by gradients of temperature and moisture, distribution of dockage and broken grain, and insect inter- and intra-species interactions. *Rhyzopertha dominica* and *S. granarius* are rarely found in grain bulks below 127 and 63 cm, respectively, (Keever 1983) while *C. ferrugineus* moves downward in a grain mass (White and Loschiavo 1986).

Sinha and Wallace (1973) monitored mite populations in Canada for 11 years in undisturbed granaries holding wheat and determined that four factors regulated mite numbers, especially *A. siro* and its mite predator *C. eruditus*: 1) basic determinants (temperature, moisture, rate of increase); 2) the main regulator (temperature); 3) influencers (dockage, seed cracks, microflora); and 4) seasonal and cyclic regulators (density-dependent factors). The population dynamics and distribution of nine species of mites in a bulk of undisturbed oats for 7 yr indicated clumped distributions at different depths and that *L. destructor* and *T. granarius* could be used as bio-indicators of spoilage in oats (Sinha and Kawamoto 1990).

5. TYPES OF ECOSYSTEMS

A distinction must be made between relatively stable stored-grain ecosystems (e.g., an undisturbed granary of wheat at 14% mc and moderate temperature stored for several years where pest density is low but diverse) and immature, unstable ecosystems with grain stored in either a hot-damp condition with rapidly increasing pest populations or when stored grain is managed intensely or held for only a short time.

In relatively stable stored-grain ecosystems, populations fluctuate in response to environmental (climatic) variables and predator-prey relationships. These systems are more like natural ecosystems where increased species diversity is related to system stability (Van Voris et al. 1980), a complex system of feedback regulation exists (Patten and Odum 1981), and population fluctuations may be "chaotic" and unpredictable (May and Oster 1976, May 1976). This could make computer simulation of arthropod numbers over a long-term difficult (i.e., over several generations). Various computer models have been developed for population growth of *C. ferrugineus* (Kawamoto et al. 1989), and *S. oryzae* (Longstaff 1988), and expert systems based on these models have been developed for grain management (Hagstrum and Flinn 1991).

It may be useful to consider a stored-grain ecosystem as a large geographical area (e.g., a municipality or a region) with the substrate (stored grain) clustered in granaries on farms. Insects and mites move freely among granaries from various reservoirs, often in a seasonal manner, and carry microorganisms with them (Fig. 7). In this model, pest migration and life histories outside of the grain bulks would need to be considered and current information on such larger (than a granary) ecosystems is very limited. The period of storage would be considered as one component in the production and utilization of seed (Fig. 7). A preliminary step in studying this type of system was taken by Wright et al. (1988) using four types of insect traps at different locations on farms throughout Kansas to compare traps and determine where stored-product insects were caught and how they moved during the year. Sinclair and Haddrell (1985) studied the movement of stored-product insects in a large grain-growing area in Queensland, Australia and Sinclair (1982) estimated the kinds and numbers of stored-product insects on farms in Australia; he estimated a total of 360,000 adults of *S. oryzae*, *R. dominica*, *Cryptolestes* spp., and *T. castaneum* per farm, 88% in bulk and bagged grain which is stored in relatively small quantities on Australian farms, mainly for seed or animal feed.

Another ecosystem is a discrete granary where insects and mites remain in grain residues and continually re-infested newly stored grain, the movement of the grain being only one phase in the ecosystem evolution.

An interesting commentary on the social and ecological context of food safety strategies, including storage, was given by Waltner-Toews (1991). He indicated that food safety strategies must be socially, geographically, and ecologically global and must reflect the diversity and complex interactions in natural ecosystems, and by implication, in man-made storage ecosystems.

6. CHEMICAL ECOLOGY

Grain, fungi, insects, and mites all produce an array of volatile odors (Sinha 1990). Grain has a distinct odor which becomes stronger as moisture content increases and enzymic reactions are stimulated in the seed. Grain odors are usually attractive to stored-product insects and mites. Some of the attractants from rice, corn, and wheat (Maga 1978) are C_5-C_7 fatty acids, 2-phenylethanol,

Table 1 Mean Number of Larvae and Adults of Five Species of Stored-Product Insects* in Bottles Holding 102 g Whole Wheat Initially Infested with 25 Adult Insects of the Indicated Species. Bottles were Incubated for 5, 10, and 15 wk at 30°C and 70% Relative Humidity.

		Week 5				Week 10				Week 15			
		Larvae		Adults		Larvae		Adults		Larvae		Adults	
	Insect	Mean**	SE**	Mean	SE	Mean	SE	Mean	SE	Mean	SE	Mean	SE
1	*Tribolium*	3.3	0.9	18.3	3.8	20.7	2.2	20.0	3.6	11.3	6.9	13.3	2.3
2	*Rhyzopertha*	27.3	5.6	394.0	4.6	185.3	13.1	1795.0	128.5	0	0	108.3	40.2
3	*Sitophilus*	0	0	37.7	16.8	0	0	446.3	107.2	0	0	204.0	132.4
4	*Oryzaephilus*	8.3	3.8	79.7	1.8	23.0	8.1	15.7	7.1	1.3	0.7	6.3	0.9
5	*Cryptolestes*	102.7	18.5	214.0	6.7	34.7	5.8	179.7	45.1	1.7	0.3	19.7	6.7
6	*Tribolium* +	5.3	0.9	46.0	11.9	0	0	8.1	3.2	0	0	5.3	0.7
	Rhyzopertha	2.0	2.0	157.3	79.0	222.3	27.2	1291.3	263.8	0	0	29.0	9.7
7	*Tribolium* +	15.3	1.8	54.3	1.2	11.3	8.5	11.0	1.5	8.0	8.0	10.0	5.3
	Sitophilus	0	0	93.7	27.5	0	0	305.7	67.3	0	0	483.7	243.4

8	*Tribolium* +	0	0	37.7	7.5	0.7	0.7	23.0	4.7	1.0	1.0	19.0	2.5
	Oryzaephilus	5.3	2.4	46.7	7.5	11.7	3.2	39.3	7.8	1.3	0.8	19.0	6.7
9	*Tribolium* +	3.3	1.5	21.7	4.3	0.3	0.3	17.7	2.4	0	0	21.0	3.0
	Cryptolestes	58.0	2.3	200.3	5.0	67.0	11.1	212.3	28.9	5.7	0.9	12.7	12.2
10	*Rhyzopertha*+	14.7	2.9	272.7	14.2	108.0	49.4	1516.3	42.3	0	0	12.7	12.2
	Sitophilus	0	0	36.3	11.6	0	0	59.3	11.9	0	0	0	0
11	*Oryzaephilus*+	0	0	8.3	0.9	0	0	0		0	0	0	0
	Cryptolestes	103.3	14.6	224.0	35.4	79.3	4.1	294.7	19.7	5.0	1.7	22.3	3.4
12	*Rhyzopertha* +	8.3	2.0	287.0	30.5	253.3	72.3	1030.7	128.7	0	0	10.7	5.9
	Sitophilus +	0	0	85.3	10.7	0	0	60.3	22.5	0	0	0	0
	Tribolium	1.7	1.2	33.3	4.3	0	0	9.0	2.1	0	0	4.7	1.2
13	*Cryptolestes* +	159.3	9.3	203.7	7.7	64.3	13.3	128.7	30.4	3.3	1.2	23.7	4.7
	Oryzaephilus +	0	0	6.7	2.0	0	0	1.7	0.7	0	0	0	0
	Tribolium	6.0	3.2	18.0	0.6	1.0	1.0	41.7	22.2	0	0	12.0	5.5

**Tribolium castaneum, Rhyzopertha dominica, Sitophilus oryzae, Oryzaephilus surinamensis, Cryptolestes ferrugineus.*

**Mean and standard error, three replicates per treatment per time.

Source: White, unpublished.

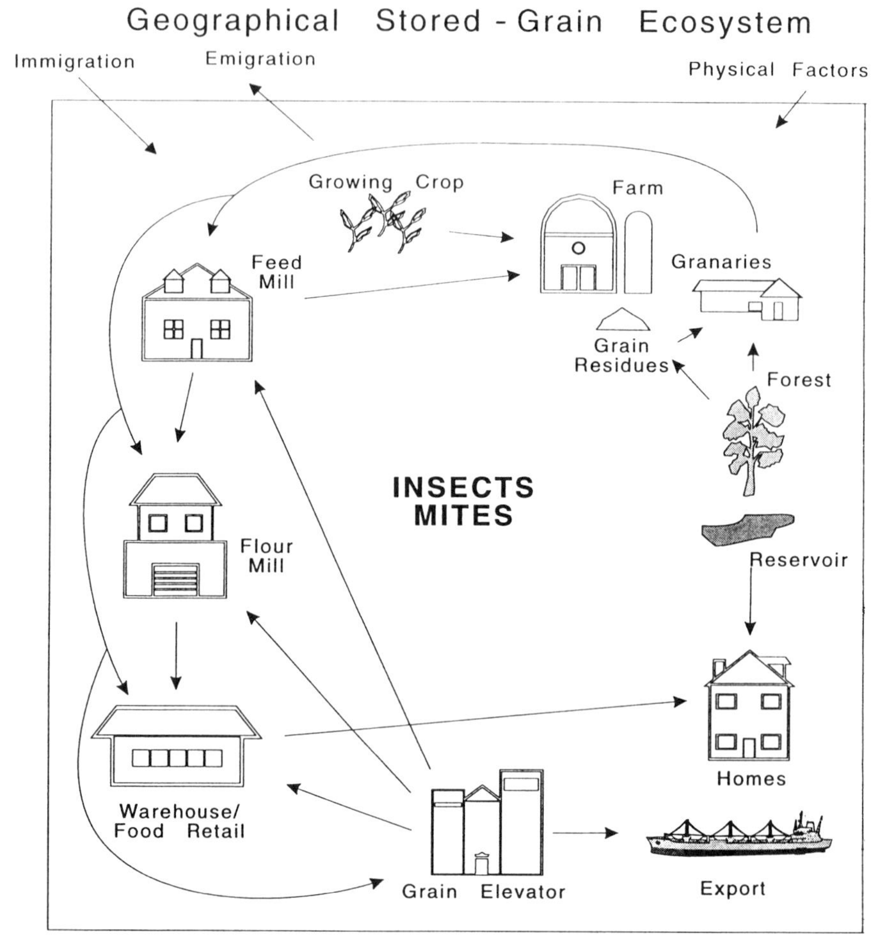

Figure 7 A regional ecosystem for the study of stored-product insects and mites indicating pest reservoirs and movement by flight or in contaminated goods.

2-nonen-4-olide, and nonan-4-olide (Honda and Ohsawa 1990). Growing fungi release volatile chemicals which can attract or repel insects and mites. *Cryptolestes ferrugineus* are attracted to the fungi *Penicillium corymbiferum* Westling, *Scopulariopes brevicaulis* (Sacc.) Bain, and *Fusarium* sp. (Dolinski and Loschiavo 1973). The fungal products 1-octen-3-ol and 3-methylbutanol are strongly attractive to cucujids with 1-octen-3-one, 3-octanol, and 3-octanone showing various degrees of attractiveness; *O. surinamensis* also responds positively to 2-phenylethanol and ethanol (Pierce et al. 1991a,b; Tuma et al. 1989).

Odors produced by insects and mites are usually pheromones for intraspecific communication, either sex pheromones produced by females to attract a mate in short-lived (< 1 month) adults (moths, dermestids, bruchids) or aggregation pheromones produced by males to attract other adults of the same species to a suitable habitat in long-lived adult insects (beetles) (Burkholder 1988). Aggregation pheromones also increase the likelihood of finding a mate and their effects are often synergistic with food odors. Extensive efforts to synthesize these pheromones have often been successful and many pheromones consist of various combinations of macro-cyclic lactones (Vanderwel and Oelschlager 1989). These compounds can increase capture of *C. ferrugineus* in pitfall traps placed in grain bins (Loschiavo et al. 1986).

Some adult insects such as *Tribolium* spp. produce benzoquinones in dense populations which contaminate the food substrate, provide protection from predators, inhibit fungal and bacterial growth (Burkholder 1982), and dissolve aggregation pheromone globules to cause the insects to disperse out of the crowded area (Faustini and Burkholder 1987). The mites *A. siro, Aeroglyphus robustus,* and *L. destructor* produce the alarm pheromone citral in wheat and all produce tridecane in varying amounts depending on their fungal diet (Tuma et al. 1990). Citral may inhibit fungal growth (Matsumoto et al. 1979); neral and geranial have also been reported as alarm pheromones in some mite species (Baker and Krantz 1984). Guanine, the nitrogenous waste product produced by many mites, and ammonia produced by fungi during nitrogen metabolism are both attractive to *A. siro* facilitating aggregation and food location (Levinson et al. 1991).

Odors produced by pests in grain may eventually be mechanically and remotely monitored and used as early warning detection methods by managers of stored grain and possibly as part of an integrated control plan using pheromones and food attractants to trap pests (Levinson and Levinson 1977).

7. BIOENERGETICS

Insects and mites in stored products are part of a food-web involving the "producer" (seed), "primary and secondary consumers" (granivores, fungivores, predators, parasitoids), and "decomposers" (fungi, bacteria, actinomycetes). Whenever a material is consumed, part of the energy it contains is assimilated and used for the production of biomass in the consumer while part is lost as by-products (feces, exuviae) and through respiration with the production of carbon dioxide and heat.

To determine the overall flow of energy in stored-grain ecosystems the bioenergetics of the seed (respiration) and all invading consumers need to be clarified and integrated (Sinha 1982). Insects, mites, and fungi cause grain nutrient loss (changes in vitamins, lipids, proteins, and carbohydrates), functional

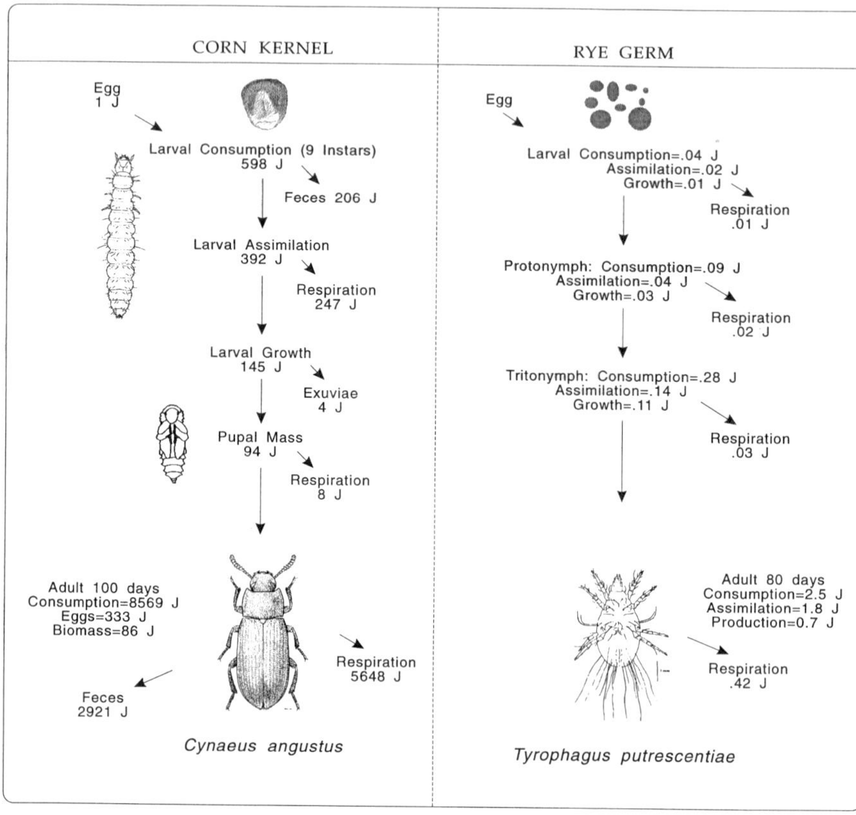

Figure 8 Approximate energy flow from a food source through the developmental stages and adult of *Cynaeus angustus* or *Tyrophagus putrescentiae*.

property losses (germinability), and aesthetic changes (discoloration, caking, odors) (Majumder 1975, White 1979). Also, mycotoxins that poison livestock and humans can be produced by the fungi (Pomeranz 1982).

The bioenergetics of the developmental stages and adults of the following mites and insects were studied: the mites *T. putrescentiae* (Stepien et al. 1971) and *Rhizoglyphus echinopus* (Stepien 1970, Klekowski and Uchmanski 1980); the bruchid beetle *Callosobruchus analis* (F.) (Wightman 1978a,b); the pyralid moths *Ephestia cautella* (Walker) (Sinha et al. 1986) and *P. interpunctella* (Imura and Sinha 1986); the cucujid beetles *C. ferrugineus* (Campbell and Sinha 1978) and *O. surinamensis* (White and Sinha 1981); the bostrichids *R. dominica* (Campbell and Sinha 1978) and *Prostephanus truncatus* Horn (Demianyk and Sinha 1988); the curculionids *S. granarius* (Campbell et al. 1976) and *S. oryzae*

(Singh et al. 1976); and the tenebrionids *T. castaneum* (Klekowski et al. 1967) and *Cynaeus angustus* (Le Conte) (White and Sinha 1987).

A comprehensive review of bioenergetic data on a wide range of insects, including those of stored products, was given by Wiegert and Petersen (1983). Bowker (1979) studied the energetics of interacting populations of *T. castaneum* and *T. confusum* at 25, 30, and 35°C and determined that *T. castaneum* had higher consumption and respiration levels. There was little interaction in populations at 25°C. Species interaction affected energy flows in the system with predation playing a significant role at 30°C and competition was significant only at 35°C.

The flow of energy from stored corn through the developmental stages and adults of the common beetle pest *C. angustus* (White and Sinha 1987) and from rye germ through the stages of the mite *T. putrescentiae* (Stepien et al. 1971) are given in Figure 8. The small mites consume very little energy on an individual basis. About 2448 adult *T. putrescentiae* consume as much energy as one adult *C. angustus* in 1 d. The efficiencies of energy usage are also important from an ecological viewpoint and are affected by the type of food (Campbell and Sinha 1978). The *assimilation efficiency* is (assimilation/ consumption) × 100%, the *gross production efficiency* is (production/consumption) × 100%, and the *net production efficiency* is (production/assimilation) × 100%. These are useful comparative relationships among species. The adult mite, *T. putrescentiae* has an assimilation efficiency of 72%, a gross production efficiency of 56%, and a net production efficiency of 77% (Stepien et al. 1971). These efficiencies are generally much higher than the cumulative efficiencies of the stored-product insects (White and Sinha 1987) (Table 2).

The consumption by various insects is given in Table 2 and the energy content of wheat and corn in Table 3 to indicate the amount of direct damage and energy loss from grain that occurs by the feeding of different pests during individual development. The respiration of wheat and associated microflora at different temperatures and moisture contents was determined by White et al. (1982) and the energy released under selected conditions is given in Table 4.

Campbell and Sinha (1990) have incorporated bioenergetics data for *C. ferrugineus* into a simulation model of population increase and accurately predicted insect numbers, consumption, assimilation, respiration levels, and the production of excreta and exuviae in small bulks of wheat. This type of analysis gives detailed knowledge of the destructive activity of pest populations.

Using published data on insect bioenergetics (Table 2) and on insect infestation levels for wheat stored in granaries in Kansas (Reed et al. 1991) as representative of conditions in the midwestern States and assuming a uniform insect distribution and an adult lifespan of 50 d (environmental effects shorten lifespan), about 2×10^8 J could be consumed by insects in 82 t of wheat (Table 5). Assuming an additional loss of 2×10^5 J for grain and microfloral respiration

Table 2 Energy Use, in Joules, by Several Species of Some Stored-Product Insects Reared on Major Cereal Crops Under Favorable Conditions (near 30°C and 70% rh)

Order	Family	Genus/species	Food	Cumulative energy consumption during development (J)
Coleoptera	Tenebrionidae	*Cynaeus*[1] *angustus*	Corn	598
		Tribolium[2] *castaneum*	Wheat	237
	Cucujidae	*Cryptolestes*[3] *ferrugineus*	Wheat	35
		Oryzaephilus[4] *surinamensis*	Rolled oats	38
	Curculionidae	*Sitophilus*[5] *granarius*	Wheat	345
		Sitophilus[6] *oryzae*	Wheat	130
	Bostrichidae	*Rhyzopertha*[3] *dominica*	Wheat	33(88)[10]
		Prostephanus[7] *truncatus*	Corn	252(547)[10]
Lepidoptera	Phycitidae	*Plodia*[8] *interpunctella*	Corn	657
		Ephestia[9] *cautella*	Wheat	605

Sources: [1]White and Sinha (1987); [2]Klekowski et al. (1967); [3]Campbell and Sinha (1978); [4]White and Sinha (1981); [5]Campbell et al. (1976); [6]Singh et al. (1976); [7]Demianyk and Sinha (1988); [8]Imura and Sinha (1986); [9]Sinha et al. (1986); [10]Numbers in parentheses indicate chewed plus ingested material.

Predicted cumulative energy consumption as adult (J)	Adult longevity (d)	Adult dry mass (mg)	Lifetime energy consumption (J)	Assimilation efficiency (%)	Net production efficiency (%)
8569	100	2.98	9167	56	27
3111	335	1.00	3348	46	18
311	180	0.11	336	66	25
272	180	0.30	310	90	35
1193	100	1.19	1538	76	14
300	150	0.69	430	79	15
414(2646)	150	0.35	447(2734)	29	30
4210(21383)	168	1.47	4462(21830)	47	34
—	—	1.99	657	50	23
—	—	3.45	605	26	46

Table 3 Energy Content of Wheat and Corn

Crop	Mass (wet)	No. kernels	Energy content (J)	Reference
Wheat	34 mg	1	752	(Sinha et al. 1986)
	1000 g	29,412	2.21×10^7	
	3.4 g	100	7.52×10^4	
	81.7*t	2.4×10^9	1.8×10^{12}	
Corn	225 mg	1	4,537	(Demianyk and Sinha 1988)
	1000 g	4,444	2.02×10^7	
	22.5 g	100	4.54×10^5	
	76.2* t	3.4×10^8	1.4×10^{12}	

* Standard 3000 bu granary holds 81.7 t wheat, 76.2 t corn.

Table 4 Energy Release in Stored Wheat by Respiration of Grain and Microflora

Temperature (°C)	Moisture content (%)	Storage time (d)[1]	Average CO_2 produced in 24 h (mg/kg wheat)	Energy released in 24 h (J/kg wheat)	Energy released in 24 h (J/81.7 t)
20	16	62	1	0.01	817
	18	23	5	0.05	4,087
	20	12	23	0.23	18,800
30	14	48	1	0.01	817
	16	18	5	0.05	4,087
	18	7	23	0.23	18,800
	20	2.5	80	0.80	65,392

[1]Maximum storage time at specified temperature and moisture level before 5–10% of seeds become nonviable.

1 L CO_2 = 1.98 g of CO_2; Production of 1 L CO_2 from carbohydrates results in release of 21.1 J.

Source: White et al. 1982.

Table 5 Insect Infestations in Bins of Wheat in Kansas and Estimated Energy Consumed by Insects During 50-d Lifespans

Insect	Mean no. adult insects per kg grain	No. insects per 82 t	Total energy consumption[2] (J)
		September	
Rhyzopertha dominica	0.3	24,600	2.38×10^7
Cryptolestes spp.	2.2	180,400	2.37×10^7
Tribolium castaneum	0.9	73,800	5.58×10^7
Oryzaephilus surinamensis	0.5	41,000	0.52×10^7
Others (fungus feeders, etc.)	0.4	32,800	—
	4.3	352,600	1.1×10^8
		November	
R. dominica	0.6	49,200	4.77×10^7
Cryptolestes spp.	0.7	57,400	0.74×10^7
T. castaneum	2.3	188,600	14.24×10^7
O. surinamensis	0	0	—
Others	0.5	41,000	—
	4.1	336,200	2.0×10^8

Total energy in a bin of 82 t = 1.8×10^{12} J.
[1]Assuming a uniform distribution for insects and [2]an adult lifespan of 50 d about 0.01% of energy in a bin would be consumed (2×10^5 J for grain and microfloral respiration added: 50 d, 16% MC, 30°C). For damage levels of 1% of kernels, each insect would need to damage an average of 7 kernels.

(50 d, 16% mc, 30°C), a total of 0.01% of energy in the bin could be consumed. If each insect damaged seven kernels there would be a 1% damage level. Insects such as *T. castaneum* damage a mean of 4.1 seed germs during development (White and Lambkin 1988) and extensive adult feeding makes the assumption of seven damaged kernels realistic.

In tropical regions of the world losses are considerably greater. Energy losses from 157 kg of wheat were calculated for 24 wk of storage (Table 6) in a tropical environment of 30°C, and two combinations of insect species (White and Sinha 1980). The total energy consumed by populations of *R. dominica, S. oryzae,* and *T. castaneum* was about 16% of the energy contained in the grain.

Table 6 Estimated Energy Loss from Undisturbed Wheat Infested with Three Species of Insects and Stored for 24 wk Under Simulated Tropical Conditions (30°C, 15.5% mc Increasing to 20% mc) in Small (157 kg) Bulks

	Mean no. (adults and larvae) per 126 g wheat on week					
Insects in grain bulk	6	12	18	24	Total no. insects per bulk in 24 wk	Cumulative energy loss over 24 wk (J)
Rhyzopertha dominica	125	250	80	50	6.3×10^5	1.0×10^8
Sitophilus oryzae	45	0	0	0	5.6×10^4	1.2×10^7
Tribolium castaneum	90	130	160	160	6.7×10^5	4.7×10^8
						5.8×10^8
Cryptolestes ferrugineus	70	150	70	100	4.9×10^5	5.4×10^7
Oryzaephilus surinamensis	40	0.3	0.3	0	5.1×10^4	5.6×10^6
Tribolium castaneum	70	125	95	90	4.7×10^5	3.3×10^8
						3.9×10^8

Energy in 157 kg wheat = 3.5×10^9 J.
Seed damage due to insect feeding combinations of pest species was > 25% by 24 wk. Assumes 50-d lifespan for adults and does not consider immatures that failed to reach adulthood.
Source: White and Sinha 1981.

The energy consumed by populations of *C. ferrugineus, O. surinamensis,* and *T. castaneum* was about 11%. Seed damage, however, due to insect feeding was > 25% in both systems although individual kernels were not extensively damaged by *C. ferrugineus*, *O. surinamensis*, and *T. castaneum.* Virtually 100% of the grain in the system with *R. dominica* and *S. oryzae* was spoiled because of whole-seed consumption and contamination by microflora, frass, and exuviae.

8. INSECTICIDES

Insecticides are used to prevent or control insect and mite populations in stored grain. Protectants are contact insecticides that are added to the grain as it enters storage and provide pest control throughout storage. Snelson (1987) presented a comprehensive review of the chemicals used throughout the world for this purpose and Longstaff (1988) has modeled the effect of temperature on toxicity of insecticides. Bond (1984) gave a comprehensive review of fumigants that

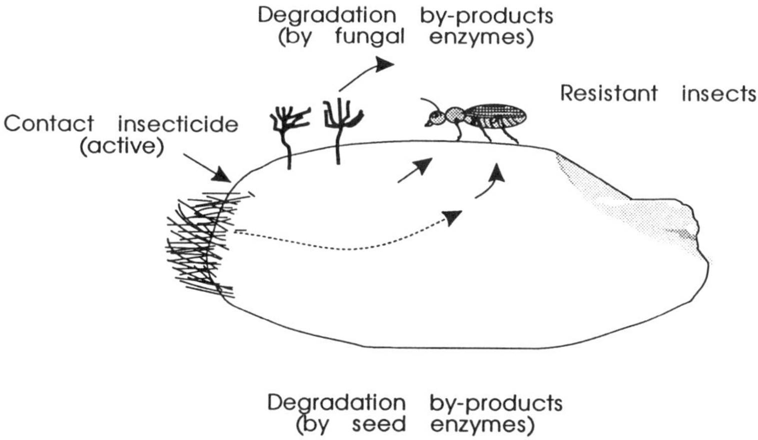

Figure 9 Schema of insecticide breakdown in stored grain ecosystems: purely chemical or by fungal or seed enzymes or by internal detoxification in resistant insects or mites.

are used throughout the world to disinfest stored products and Daft (1991) has discussed the cause and extent of fumigant residues in food. Price (1985) has reviewed the mode of action of the fumigants: hydrogen cyanide, phosphine, methyl bromide, carbon tetrachloride, ethylene dichloride, and ethylene dibromide. Currently, only phosphine and methyl bromide are widely used throughout most of the world.

An aspect of insecticide use in stored-grain ecosystems that has rarely been studied is the long-term impact on the flora and fauna present. Contact insecticides on grain are degraded by hydrolysis through physical processes or enzymatic action in the seed (Rowlands 1975) or microflora (Anderegg and Madisen 1983) (Fig. 9) and are selectively retained in parts of the seed which have a high lipid content such as the bran and germ of cereals (Mensah et al. 1979) and the oil of oilseeds (Dikshith et al. 1989). Ecosystems disturbed by the application of chemical agents such as insecticides will gradually recover as toxic levels of insecticide decrease. Some insecticides (e.g., the pyrethroid bioresmethrin) have repellent effects on insects such as *R. dominica* which will affect the distribution of insects in partially treated grain or grain with low levels of insecticide (Collins et al. 1988). It should be noted that some insecticides gradually become "bound" within seeds and are not seen with standard chemical analysis but the insecticide may remain biologically active in mammals eating the seed (Singh et al. 1993).

Sublethal exposures to insecticides usually have negative effects on an insect's longevity, fecundity, oviposition, and larval survival (Zettler and Le Cato 1974, Taher and Cutkomp 1983) but occasionally oviposition may increase as with *T. castaneum* exposed to chlorpyrifos-methyl or malathion (White and Sinha 1990a).

Malathion applied to 14 t of wheat in a granary in Manitoba, Canada was quantitatively monitored for residues monthly for 2 yr and the effects on the stored-grain ecosystem were determined (White et al. 1986). As the insecticide degraded, *C. ferrugineus* and *T. castaneum* reappeared in low numbers and populations of the mite *A. robustus* began to recover after 2 years. The fungus *Penicillium* was more abundant in the presence of malathion but this may have been an indirect effect because of less insect and mite feeding damage than in the control granary. The effect of chlorpyrifos-methyl on stored-oat ecosystems of 5.4 t in granaries in Manitoba, Canada was also studied for 2 yr (White and Sinha 1990b). These systems contained communities of eight species of mites as well as microflora. The insecticide had no effect on seed germination or microfloral infection. *Tarsonemus granarius* and *A. siro* were unaffected by the treatment and the original ecosystem had largely re-established itself in 2 years although predator diversity was reduced.

The impact of a phosphine fumigation on stored-wheat ecosystems containing 10 species of insects and mites was monitored by Sinha et al. (1967). The gas had no effect on seed germination or microflora and killed most insects and mites except *A. robustus*. Within 3 months *A. siro, L. destructor,* and *Spinibdella bifurcata* Atyeo reappeared in both bins and multiplied rapidly, partially because of elimination of the predator *C. eruditus.*

9. INSECTICIDE RESISTANCE

Resistance of insects to contact insecticides and fumigants is widespread globally (Champ and Dyte 1976, Taylor 1989) and is increasing in North America (Haliscak and Beeman 1983, White and Watters 1983, White and Loschiavo 1985, Subramanyam et al. 1989, Zettler and Cuperus 1990), Europe (Prickett et al. 1990), the Middle East (Navarro et al. 1986), Australia (Collins 1990), and elsewhere.

Malathion resistance is the most widespread form of resistance in insects because of the global use of this inexpensive chemical which has low mammalian toxicity, and because malathion-specific resistance involves one gene or a closely linked set of genes (Dyte and Rowlands 1968; Beeman 1983; Beeman and Nanis 1986; White and Bell 1988, 1990). Malathion resistance was first observed in

T. castaneum on peanuts in Nigeria in 1961 (Hayward 1962) and by 1974 it was regarded as common in this species throughout the world (Champ 1974).

Various contact insecticides have different toxicities to mite species found in grain (Wilkin and Hope 1973) and resistance to chemicals such as γ-HCH (lindane) (Stables 1980) and pirimiphos-methyl, used on maturing cheese, is also becomming common in *A. siro* and *L. destructor*. The mechanisms of insecticide resistance in stored-product pests have been discussed by Price (1991).

One mechanism for insecticide resistance in *T. castaneum* is a slowed rate of chemical penetration of the cuticle in cross-resistant insects (although enhanced carboxyesterase activity is the major mechanism in malathion-specific resistance; Walter and Price 1989). Resistance to other insecticides is increasing with various levels of cross-resistance (Beeman and Wright 1990). The development of resistance in insects is a pre-adaptive phenomenon with the required genes present in populations at low frequencies before exposure to the insecticide occurs (Oppennorth 1985). Resistance genes may have been selected in insect populations by natural toxicants such as the insecticidal organophosphates produced by microflora such as *Streptomyces antibioticus* (Neumann and Peter 1987) or plant-produced terpenoids which mimic the action of insecticides by inhibiting acetylcholinesterase (Ryan and Byrne 1988).

Resistance to phosphine was reported in 1982 (Mills 1983) and its impact on insect control has been discussed by Price and Mills (1988) and Mills et al. (1990). A twenty-fold increase in phosphine-specific resistance for five times the normal exposure period has been reported for some strains of stored-product insects (Price 1991). Low levels of phosphine resistance have been reported from the United States (Zettler et al. 1989). Even the bacterial insecticide *Bacillus thuringiensis* is of decreasing effectiveness as resistance is developing in strains of *P. interpunctella* (McGaughey 1985, 1990).

Insecticide resistance which is economically important is usually caused by one or two alleles which confer high levels of resistance (Roush and McKenzie 1987). Polygenic resistance has frequently been produced in the laboratory by subjecting small populations of insects to continuous finely regulated doses of insecticide but in the field where insects can move between populations, insecticide resistance requiring a combination of a number of genes is less likely to develop because of frequent interbreeding with susceptible insects. Single genes that allow insects to survive relatively high doses of insecticide can remain at low frequencies in a population which is not under selection and increase in frequency when repeated selection pressure occurs. Furthermore, these genes are often equal to the corresponding susceptible alleles in fitness characteristics such as reproductive rate and developmental period. In most studies where the *R* (resistance) and *S* (susceptible) alleles have been compared in isogenic backgrounds, the *R* alleles did not cause a reduction in fitness. Any apparent reduction in fitness in resistant strains is probably caused by inbreeding of populations

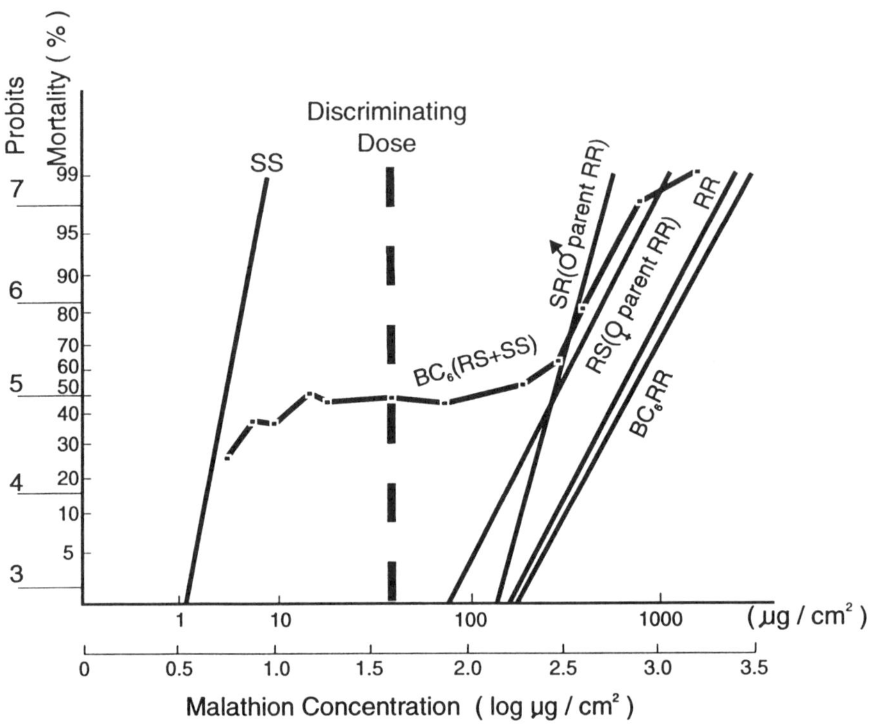

Figure 10 Dose-mortality response of adult *C. ferrugineus* exposed to malathion on filter paper: *SS* = susceptible strain; *SR* = heterozygous resistance (♂ parent *RR*); *RS* = heterozygous resistance (♀ parent *RR*); *RR* = homozygous resistance; BC_6 (*RS* + *SS*) = Sixth backcross, heterozygotes to susceptibles; BC_6*RR* = Sixth backcross reselected *RR* with the same genome as the *SS* strain except for the resistance genes (White and Bell 1990).

(Halliday 1990). In studies on *Culex quinquefasciatus* Say (Amin and White 1984), *T. castaneum* (Beeman and Nanis 1986), *Musca domestica* L. (Whitehead et al. 1985), and *C. ferrugineus* (White and Bell 1990) the fitness of the *SS* genome (homozygous susceptible) was not reduced when the *R* allele was substituted for its corresponding *S* allele. Once they are established in an insect population, major resistance genes will probably remain in high frequencies for a long time after use of the insecticide has been stopped. For a comprehensive review of the relative fitness of insecticide resistant genotypes see Roush and McKenzie (1987).

Homozygous resistant insects can often withstand much more insecticide than heterozygous insects and this effect is due entirely to the resistance gene in malathion-specific resistance (Fig. 10). Heterozygous resistant *T. castaneum* adults on malathion-treated wheat can produce more offspring than homozygotes, but homozygous larvae are much more resistant than heterozygotes (White and Bell 1988) which will affect population growth in a treated ecosystem.

Resistant and susceptible strains of *S. oryzae* exposed to sublethal concentrations of deltamethrin and pirimiphos-methyl have responses that can include increased fecundity and emergence (Longstaff 1991). Therefore, increasing the insecticide level will increase mortality, but population growth may remain high. Also insecticide resistance may decrease interspecific competition in stored-grain ecosystems by eliminating susceptible insect and mite species in a treated environment.

10. FUTURE RESEARCH

A great deal has been learned over the past 50 years about the insects and mites that infest stored grain and about their place in the stored-grain ecosystem but there are still many unanswered questions concerning the functioning of these systems that are important to effective stored-product management. The detailed interactions among variables within stored grain need to be studied in the laboratory and field to predict pest population dynamics and response to environmental and control factors (Prevett 1990, White 1992). The complex chemical ecology within bulk grain needs to be more clearly defined and techniques developed to use odors as warning of spoilage and, perhaps, for pest control. Knowledge of the bioenergetics of all organisms in stored-grain ecosystems will allow accurate prediction of stored-crop losses and how best to stop them. The process of developing simulation models and expert systems will identify research needs and assist in prediction of pest outbreaks. The impact of pesticides on stored-grain ecosystems and the impact of pesticide resistance on arthropod fitness and dispersal are also important.

The number of stored-product entomologists throughout the world is rapidly declining as research organizations place less emphasis on research related to the maintenance of stored food. This is ironic as food production and the world's population are increasing rapidly. Administrators assume that the technology of stored-product management has been largely perfected but insecticide resistance in insects is increasing rapidly, insecticides are being removed from the market because of threats to human health, crop-production patterns are changing worldwide, and increased globalization of trade transmits pests everywhere.

Insect and mite infestations in stored food are critical in trade and commerce: some countries have a zero tolerance for detectable insects in export grain and intense product management is required to prevent monetary or trade losses. Low numbers of insects in exports from countries in temperate climatic regions can rapidly become heavy infestations in tropical importing countries. A great deal of quality loss in grain and processed food still occurs in temperate countries and qualitative and quantitative losses in tropical countries can be devastating.

Research in developed countries should include the improvement of technology to prevent, detect, and control stored-product insects and mites. Research is needed on sources of infestation, improved storage structures, detection with pheromones, traps, and odors, storage characteristics of new crops, new or improved control practices including modified atmospheres, radiation, and aeration, and environmental synergists (Fields 1990) such as ice-nucleating bacteria. In developing countries, the technology of developed countries must be drastically modified before it can be applied to traditional storage systems. There is a need for more comprehensive studies of regional stored-product ecosystems which include pest and grain movement and the factors that affect grain quality loss.

Because the number of stored-product researchers is declining, it is critical that multidisciplinary training or a team approach be taken in stored-product research so that all relevant variables can be considered. A standardized, global approach to stored-product management and research is needed, and this may be possible as the systematic application of new knowledge-based computer programs reveals areas where research is needed.

Insects and mites that infest stored products will always threaten humans by spoiling hard-earned food supplies. These prolific and genetically adaptable pests must continually be studied if timely, reliable, and practical prevention and control techniques are to be available.

REFERENCES

Amin, A. M. and G. B. White. 1984. Relative fitness of organophosphate-resistant and susceptible strains of *Culex quinquefasciatus* Say (Diptera: Culicidae). Bull. Entomol. Res. 74:591-598.

Anderegg, B. N. and L. J. Madisen. 1983. Degradation of ^{14}C-malathion in stored corn and wheat inoculated with *Aspergillus glaucus*. J. Econ. Entomol. 76:733-736.

Anonymous. 1981. Postharvest food losses in developing countries. National Academy of Sciences. Washington. 206 pp.

Arbogast, R. T. and M. A. Mullen. 1988. Insect succession in a stored-corn ecosystem in southeast Georgia. Ann. Entomol. Soc. Am. 81:899-912.

Arbogast, R. T. and M. A. Mullen. 1990. Interaction of maize weevil (Coleoptera: Cuculionidae) and parasitoid *Anisopteromalus calandrae* (Hymeoptera: Pteromalidae) in a small bulk of stored corn. J. Econ. Entomol. 83:2462-2468.

Arthur, F. H. and L. M. Redlinger. 1988. Influence of loose-shell kernels and foreign material on insect damage in stored peanuts. J. Econ. Entomol. 81:387-390.

Baker, G. T. and G. W. Krantz. 1984. Alarm pheromone production in the bulb mite, *Rhizoglyphus robini* Claparede and its possible use as a control adjuvant in lily bulbs. pp. 686-692 in: Biological Control of Acari: Acarology VI, Vol. 2. D. A. Griffiths and C. E. Brown (eds.). Ellis Harwood Ltd., Chichester, UK.

Baker, J. E. and S. R. Loschiavo. 1987. Nutritional ecology of stored-product insects. pp. 321-344 in: Nutritional Ecology of Insects, Mites, and Spiders. F. Slansky Jr. and J. C. Rodriquez (eds.). John Wiley and Sons, New York.

Barker, P.S. 1993. Phoretic mites found on beetles associated with stored grain in Manitoba. Can. Entomol. 125:715-719.

Barry, D., M. S. Zuber, E. B. Lilkehoy, W. W. McMillian, N. J. Adams, W. F. Kwolek, and N. W. Widstrom 1985. Evaluation of two arthropod vectors as inoculators of developing maize ears with *Aspergillus flavus*. Environ. Entomol. 14:634-636.

Beeman, R. W. 1983. Inheritance and linkage of malathion resistance in the red flour beetle. J. Econ. Entomol. 76:737-740.

Beeman, R. W. and S. M. Nanis. 1986. Malathion resistance alleles and their fitness in the red flour beetle (Coleoptera: Tenebrionidae). J. Econ. Entomol. 79:580-587.

Beeman, R. W. and V. F. Wright. 1990. Monitoring for resistance to chlorpyrifos-methyl, pirimiphos-methyl, and malathion in Kansas populations of stored-product insects. J. Kansas Entomol. Soc. 63:385-392.

Bond, E. J. 1984. Manual of fumigation for insect control. FAO Plant Prod. and Prot. Paper 54, Rome, Italy. 432 pp.

Bowker, L. S. 1979. The energetics of interactions in populations of *Tribolium confusum* and *Tribolium castaneum*. Environ. Entomol. 8:840-849.

Buckland, P. C. 1981. The early dispersal of insect pests of stored products as indicated by archaeological records. J. Stored Prod. Res. 17:1-12.

Burkholder, W. E. 1982. Reproductive biology and communication among grain storage and warehouse beetles. J. Georgia Entomol. Soc. 17:1-10 (Second Suppl.).

Burkholder, W. E. 1988. Management of stored product insects with pheromones. pp. 77-85 in: Proceedings of the National Conference on Urban Entomology. University of Maryland, College Park, MD.

Calderon, M. 1981. The ecosystem approach for apprehending the extent of postharvest grain losses. Phytoparasitica 9:157-167.

Campbell, A. and R. N. Sinha. 1978. Bioenergetics of granivorous beetles, *Cryptolestes ferrugineus* and *Rhyzopertha dominica* (Coleoptera: Cucujidae and Bostrichidae). Can. J. Zool. 56:624-633.

Campbell, A. and R. N. Sinha. 1990. Analysis and simulation modeling of population dynamics and bioenergetics of *Cryptolestes ferrugineus* (Coleoptera: Cucujidae) in stored wheat. Res. Popul. Ecol. 32:235-254.

Campbell, A., N. B. Singh, and R. N. Sinha. 1976. Bioenergetics of the granary weevil *Sitophilus granarius* (L.) (Coleoptera: Curculionidae). Can. J. Zool. 54:786-798.

Champ, B. R. 1974. Pesticide resistance and its current significance in control of pests in stored products. pp. 159-181 in: Proc. 2nd Int. Working Conf. on Stored-Product Entomol., Ibadan, Nigeria.

Champ, B. R. and C. E. Dyte. 1976. Report of the FAO global survey of pesticide susceptibility of stored grain pests. Food and Agriculture Organization of the United Nations, Rome, Italy, Plant Prod. Prot. Series No. 5. 297 pp.

Cink, J. and P. Harein. 1989. Stored grain pest management. Burgess Printing Co., Minneapolis, MN. 282 pp.

Collins, P. J. 1990. A new resistance to pyrethroids in *Tribolium castaneum* Herbst. Pesticide Sci. 28:101-116.

Collins, P. J., E. R. Sinclair, C. J. Howitt, and R. L. Haddrell. 1988. Dispersion of grain beetles (Coleoptera) in grain partially treated with insecticides. J. Econ. Entomol. 81:1810-1815.

Coombs, C. W. and J. A. Freeman. 1955. The insect fauna of an empty granary. Bull. Entomol. Res. 46:399-417.

Cox, P. D., W. E. Parish, and M. A. Beirne. 1989. Variations in the refuge-seeking behavior of four strains of *Cryptolestes ferrugineus* (Stephens) (Coleoptera: Cucujidae) at different temperatures. J. Stored Prod. Res. 25:239-242.

Cox, P. D., W. E. Parish, and M. Ledson. 1990. Factors affecting the refuge-seeking behavior of *Cryptolestes ferrugineus* (Stephens) (Coleoptera: Cucujidae). J. Stored Prod. Res. 26:169-174.

Daft, J. L. 1991. Fumigants and related chemicals in foods:review of residue findings, contamination sources, and analytical methods. Sci. Total Environ. 100:501-518.

Demianyk, C. J. and R. N. Sinha. 1988. Bioenergetics of the larger grain borer *Prostephanus truncatus* (Horn) (Coleoptera: Bostrichidae), feeding on corn. Ann. Entomol. Soc. Am. 81:449-459.

Dikshith, T. S. S., S. N. Kimar, G. S. Tandon, R. B. Raizada, and P. K. Ray. 1989. Pesticide residues in edible oils and oilseeds. Bull. Environ. Contam. Toxicol. 42:50-56.

Dolinski, M. G. and S. R. Loschiavo. 1973. The effect of fungi and moisture on the locomotory behavior of the rusty grain beetle, *Cryptolestes ferrugineus* (Coleoptera: Cucujidae). Can. Entomol. 105:485-490.

Dyte, C. E. and D. G. Rowlands. 1968. The metabolism and synergism of malathion in resistant and susceptible strains of *Tribolium castaneum* (Herbst) (Coleoptera: Tenebrionidae). J. Stored Prod. Res. 4:157-173.

Faustini, D. L. and W. E. Burkholder. 1987. Quinone-aggregation pheromone interaction in the red flour beetle. Anim. Behav. 35:601-603.

Fields, P. G. 1990. The cold-hardiness of *Cryptolestes ferrugineus* and the use of ice nucleation-active bacteria as a cold synergist. pp. 1183-1191 in: Proc. 5th Int. Working Conf. Stored-Product Prot. F. Fleurat-Lessard and P. Ducom (eds.). Bordeaux, France.

Fields, P. G. 1992. The control of stored-product insects and mites with extreme temperatures. J. Stored Prod. Res. 28:89-118.

Freeman, J. A. 1973. Infestation and control of pests of stored grain in international trade. pp. 99-136 in: Grain Storage:Part of a System. R. N. Sinha and W. E. Muir (eds.). Avi Publ. Co., Inc. Westport, CT.

Freeman, J. A. and R. J. Heape. 1968. Infestation problems of imports. pp. 86-112 in: Infestation Control Report of the Director of the Infestation Laboratory, 1965-1967. H. M. Stationery Office, Harrow, UK.

Hagstrum, D. W. 1987. Seasonal variation of stored wheat environment and insect populations. Environ. Entomol. 16:77-83.

Hagstrum, D. W. and P. W. Flinn. 1991. Integrated pest management of stored grain insects. pp. 535-562 in: Storage of Cereal Grains and their Products. 4th ed. D. B. Sauer (ed.). Am. Assoc. Cereal Chem., St. Paul, MN.

Hagstrum, D. W., R. L. Meagher, and L. B. Smith. 1988. Sampling statistics and detection or estimation of diverse populations of stored-product insects. Environ. Entomol. 17:377-380.

Hagstrum, D. W., G. A. Milliken, and M. S. Waddell. 1985. Insect distribution in bulk-stored wheat in relation to detection or estimation of abundance. Environ. Entomol. 14:655-661.

Haines, C. P. 1984. Biological methods for integrated control of insects and mites in tropical stored products. III. Use of predators and parasites. Trop. Stored Prod. Info. 49:17-25.

Haliscak, J. P. and R. W. Beeman. 1983. Status of malathion resistance of five genera of beetles infesting farm-stored corn, wheat and oats in the United States. J. Econ. Entomol. 76:717-722.

Hall, D. W. 1970. Handling and storage of food grains in tropical and subtropical areas. FAO, Rome. 350 pp.

Halliday, W. R. 1990. Comparative fitness of malathion-resistant and susceptible Indianmeal moth (Lepidoptera: Pyralidae). J. Entomol. Sci. 25:239-245.

Hayward, L. A. W. 1962. Groundnuts. pp. 12-15 in: Report W. Africa Stored Prod. Res. Unit (1961). Nigeria.

Hill, D. S. 1990. Pests of stored products and their control. CRC Press, Boca Raton, FL. 274 pp.

Honda, H. and K. Ohsawa. 1990. Chemical ecology for stored product insects. J. Pesticide Sci. 15:263-270.

Howe, R. W. 1965. A summary of estimates of optimal and minimal conditions for population increase of some stored products insects. J. Stored Prod. Res. 1:177-184.

Hughes, A. M. 1976. The mites of stored foods and houses. Min. Agric. Fish. Food, Technol. Bull. 9, Her Majesty's Stationery Office, London. 400 pp.

Imura, O. and R. N. Sinha. 1986. Bioenergetics of the Indianmeal moth, *Plodia interpunctella* (Lepidoptera: Pyralidae). Ann. Entomol. Soc. Am. 79:96-103.

Irshad, M., W. A. Gillani, and A. Khan. 1988. Maize grain resistance to *Sitotroga cerealella* and *Sitophilus oryzae*. Pakistan J. Agric. Res. 9:539-542.

Jacob, T. A. and D. A. Fleming. 1989. The difference in the developmental period and mortality of some field strains of *Oryzaephilus surinamensis* (L.) at constant temperatures (Coleoptera: Silvanidae). J. Stored Prod. Res. 25:73-76.

Kawamoto, H., S. M. Woods, R. N. Sinha, and W. E. Muir. 1989. A simulation model of population dynamics of the rusty grain beetle, *Cryptolestes ferrugineus* in stored wheat. Ecol. Modeling 48:137-157.

Kawamoto, H., R. N. Sinha, and W. E. Muir. 1990. Effect of temperature and relative humidity on egg development of *Cryptolestes ferrugineus* compared with *C. pusillus* (Coleoptera: Cucujidae). Appl. Entomol. Zool. 25:35-41.

Keever, D. W. 1983. Distribution patterns of lesser grain borer (Coleoptera: Bostrichidae) in towers of wheat, and effects of the presence of the granary weevil (Coleoptera: Curculionidae). J. Econ. Entomol. 76:492-495.

Klekowski, R. Z., T. Prus, and H. Zyromska-Rudzka. 1967. Elements of energy budget of *Tribolium castaneum* (Hbst) in its developmental cycle. pp. 859-879 in: Secondary Productivity in Terrestrial Ecosystems. K. Petrusewicz (ed.). Panstwowe Wydawnictwo Nawkowe, Warsaw-Krakow, Poland.

Klekowski, R. Z. and J. Uchmanski. 1980. Simulation model of energy flow through a *Rhyzoglyphus echinopus* (R. et R.) individual during its development. Polish Ecol. Studies 6:61-80.

Knulle, W. 1987. Genetic variability and ecological adaptability of hypopus formation in a stored product mite. Exp. Appl. Acarol. 3:21-32.

Knulle, W. 1991. Genetic and environmental determinants of hypopus duration in the stored-product mite *Lepidoglyphus destructor*. Exp. Appl. Acarol. 10:231-258.

Krantz, G. Q. 1971. A manual of acarology. Oregon State University, Corvallis, OR. 335 pp.

Lacey, J. 1988. Grain storage:the management of ecological change. Biodeterioration 7:614-633.

Lamb, R. J. and S. R. Loschiavo. 1981. Diet, temperature, and the logistic model of developmental rate for *Tribolium confusum* (Coleoptera: Tenebrionidae). Can. Entomol. 113:813-818.

Levinson, H. Z. and A. R. Levinson. 1977. Integrated manipulation of storage insects by pheromones and food attractants—a proposal. Z. Ang. Entomol. 84:337-343.

Levinson, H. Z., A. R. Levinson, and K. Muller. 1991. Functional adaptation of two nitrogenous waste products in evoking attraction and aggregation of flour mites (*Acarus siro* L.). Anz. Schad. Pflanz. Umwelt. 64:55-60.

Linsley, E. G. 1944. Natural sources, habitats and reservoirs of insects associated with stored food products. Hilgardia 16:18-233.

Lippert, G. E. and D. W. Hagstrum. 1987. Detection or estimation of insect populations in bulk-stored wheat with probe traps. J. Econ. Entomol. 80:601-604.

Longstaff, B. C. 1988. A modeling study of the effects of temperature manipulation upon the control of *Sitophilus oryzae* (Coleoptera: Curculionidae) by insecticide. J. Appl. Ecol. 25:163-175.

Longstaff, B. C. 1991. An experimental study of the fitness of susceptible and resistant strains of *Sitophilus oryzae* (L.) (Coleoptera: Curculionidae) exposed to insecticide. J. Stored Prod. Res. 27:75-82.

Loschiavo, S. R., J. Wong, N. D. G. White, H. D. Pierce Jr., J. H. Borden, and A. C. Oehlschlager. 1986. Field evaluation of a pheromone to detect adult rusty grain beetles, *Cryptolestes ferrugineus* (Coleoptera: Cucujidae), in stored grain. Can. Entomol. 118:1-8.

Madrid, F. J., N. D. G. White, and S. R. Loschiavo. 1990. Insects in stored cereals and their association with farming practices in southern Manitoba. Can. Entomol. 122:515-523.

Maga, J. A. 1978. Cereal volatiles:a review. J. Agric. Food Chem. 26:175-178.

Majumder, S. K. 1975. Preharvest and postharvest losses. pp. 181-200 in: Bread. A. Spicer (ed.). Applied Sciences Publ., London.

Matsumoto, K., Y. Wada, and M. Okamoto. 1979. The alarm pheromone of grain mites and its antifungal effect. pp. 243-249 in: Recent Advances in Acarology. J. G. Rodriguez (ed.). Academic Press, New York.

May, R. M. 1976. Simple mathematical models with very complicated dynamics. Nature 261:459-467.

May, R. M. and G. F. Oster. 1976. Bifurcations and dynamic complexity in simple ecological models. Am. Natur. 110:573-599.

McGaughey, W. H. 1985. Insect resistance to the biological insecticide *Bacillus thuringiensis*. Science 229:193-195.

McGaughey, W. H. 1990. Insect resistance to *Bacillus thuringiensis* 2-endotoxin. pp. 583-598 in: New Directions in Biological Control:Alternatives for Suppressing Agricultural Pests and Diseases. Alan R. Liss, New York.

Mensah, G. W. K., F. L. Watters, and G. R. B. Webster. 1979. Insecticide residues in milled fractions of dry or tough wheat treated with malathion, bromophos, iodofenphos, and pirimiphos-methyl. J. Econ. Entomol. 72:728-731.

Mills, K. A. 1983. Resistance to the fumigant hydrogen phosphide in some stored product species associated with repeated inadequate treatments. Mitt. Dtsch. Geo. Allg. Angew. Entomol. 4:98-101.

Mills, K. A., A. L. Clifton, B. Chakrabarti, and N. Sarovidon. 1990. The impact of phosphine resistance on the control of insects in stored grain by phosphine fumigation. pp. 1181-1187 in: Brighton Crop Prot. Conf., Pests and Diseases. Brighton, UK.

Navarro, S., Y. Carmi, Y. Kashanchi, and E. Shaaya. 1986. Malathion resistance of stored-product insects in Israel. Phytoparasitica 14:273-380.

Neumann, R. and H. H. Peter. 1987. Insecticidal organophosphates:Nature made them first. Experimentia 43:1235-1237.

O'Connor, B. M. 1979. Evolutionary origins of Astigmatid mites inhabiting stored products. pp. 273-278 in: Recent Advances in Acarology, Vol. I. J. G. Rodriguez (ed.). Academic Press, New York.

Odum, E. P. 1989. Ecology and our endangered life support systems. Sinauer Assoc., Sunderland, MA.

Oppennorth, F. J. 1985. Biochemistry and genetics of insecticide resistance. pp. 731-774 in: Comprehensive Insect Physiology, Biochemistry, and Pharmacology. Vol. 12. G. A. Kerkut and L. I. Gilbert (eds.). Pergamon Press, Toronto.

Park, T., D. B. Mertz, W. Grodzinski, and T. Prus. 1965. Cannibalistic predation in populations of flour beetles. Physiol. Zool. 37:289-321.

Parkinson, C. L. 1990. Population increase and damage by three species of mites on wheat at 20°C and two humidities. Exp. Appl. Acarol. 8:179-194.

Parsons, P. A. 1982. Adaptive strategies of colonizing animal species. Biol. Rev. 57:117-148.

Patten, B. C. and E. P. Odum. 1981. The cybernetic nature of ecosystems. Am. Natur. 118:886-895

Pierce, A. M., H. D. Pierce Jr., A. C. Oelschlager, and J. H. Borden. 1991a. 1-octen-3-ol, attractive semiochemical for foreign grain beetle, *Ahasverus advena* (Waltl) (Coleoptera: Cucujidae). J. Chem. Ecol. 17:567-580.

Pierce, A. M., H. D. Pierce Jr., A. C. Oelschlager, and J. H. Borden. 1991b. Fungal volatiles:semiochemicals for stored-product beetles (Coleoptera: Cucujidae). J. Chem. Ecol. 17:581-597.

Pimentel, D. 1984. Energy flow in agroecosystems. pp. 121-132 in: Agricultural Ecosystems—Unifying Concepts. P. Lawrance, B. R. Stinner and G. H. House (eds.). John Wiley and Sons Ltd., New York.

Pinniger, D. B. 1974. A laboratory simulation of residual populations of stored product pests and an assessment of their susceptibility to a contact insecticide. J. Stored Prod. Res. 10:217-223.

Pomeranz, U. 1982. Biochemical, functional and nutritive changes during storage. pp. 145-217 in: Storage of Cereal Grains and their Products. C. M. Christensen (ed.). Am. Assoc. Cereal Chem., St. Paul, MN.

Prevett, P. F. 1990. Stored-product entomology—the challenge of the next decade. Bull. Entomol. Res. 80:3-6.

Price, L. A. and K. A. Mills. 1988. The toxicity of phosphine to the immature stages of resistant and susceptible strains of some common stored product beetles, and implications for their control. J. Stored Prod. Res. 24:51-59.

Price, N. R. 1985. The mode of action of fumigants. J. Stored Prod. Res. 21:157-164.

Price, N. R. 1991. Enzymic factors in the resistance of stored-product pests to insecticides. Biochem. Soc. Trans. 19:759-762.

Prickett, A. J., J. Muggleton, and J. E. Llewellin. 1990. Insecticide resistance in populations of *Oryzaephilus surinamensis* and *Cryptolestes ferrugineus* from grain stores in England and Wales. pp. 1189-1194 in: Brighton Crop Prot. Conf. Pests and Diseases. Brighton, UK.

Reed, C. R., V. F. Wright, T. W. Mize, J. R. Pedersen, and J. Brockschmidt Evans. 1991. Pitfall traps and grain samples as indicators of insects in farm-stored wheat. J. Econ. Entomol. 84:1381-1387.

Revsbeck, P. and G. Andersen. 1987. Storage mite allergy among grain elevator workers. Allergy 42:423-429.

Roush, R. T. and J. A. McKenzie. 1987. Ecological genetics of insecticide and acaracide resistance. Annu. Rev. Entomol. 32:361-380.

Rowlands, D. G. 1975. The metabolism of contact insecticides in stored grains. III. 1970-1974. Residue Rev. 58:113-155.

Ryan, M. F. and O. Byrne. 1988. Plant-insect coevolution and inhibition of acetylcholinesterase. J. Chem. Ecol. 14:1965-1975.

Sigaut, F. 1988. A method for identifying grain storage techniques and its application for European agricultural history. Tools and Tillage 6:3-32.

Sinclair, E. R. 1982. Population estimates of insect pests of stored products on farms on the Darling Downs, Queensland. Aust. J. Exp. Agric. Anim. Husb. 22:127-132.

Sinclair, E. R. and J. Adler. 1984. Migration of stored-grain insect pests from a small wheat bulk. Aust. J. Exp. Agric. Anim. Husb. 24:260-266.

Sinclair, E. R. and R. L. Haddrell. 1985. Flight of stored products beetles over a grain farming area in southern Queensland. J. Aust. Entomol. Soc. 24:9-15.

Singh, N. B., A. Campbell, and R. N. Sinha. 1976. An energy budget of *Sitophilus oryzae* (Coleoptera: Curculionidae). Ann. Entomol. Soc. Am. 69:503-512.

Singh, K., S.U. Khan, M.H. Akhtar, S. Kacew, and N.D.G. White. 1993. Nature and bio-availability of nonextractable (bound) residues in stored wheat treated with chlorpyrofos-methyl. J. Agric. Food Chem. 41:2421-2425.

Sinha, A. K. and K. K. Sinha. 1990. Insect pests, *Aspergillus flavus* and aflatoxin contamination in stored wheat:a survey at north Bihar (India). J. Stored Prod. Res. 26:223-226.

Sinha, R. N. 1961. Insects and mites associated with hot spots in farm-stored grain. Can. Entomol. 93:609-621.

Sinha, R. N. 1964. Ecological relationships of stored product mites and seed-borne fungi. Acarologia 6:372-389.

Sinha, R. N. 1965. Development of *Cryptolestes ferrugineus* (Stephens) and *Oryzaephilus mercator* (Fauvel) on seed-borne fungi. Entomol. Exp. Appl. 8:309-313.

Sinha, R. N. 1966. Development and mortality of *Tribolium castaneum* and *T. confusum* (Coleoptera: Tenebrionidae) on seed-borne fungi. Ann. Entomol. Soc. Am. 59:192-201.

Sinha, R. N. 1968. Adaptive significance of mycophagy in stored-product arthropods. Evolution 22:785-798.

Sinha, R. N. 1971. Fungus as a food for some stored-product insects. J. Econ. Entomol. 64:3-6.

Sinha, R. N. 1973a. Ecology of storage. Ann. Tech. Agric. 22:351-369.

Sinha, R. N. 1973b. Interrelations of physical, chemical, and biological variables in the deterioration of stored grains. pp. 15-48 in: Grain Storage:Part of a System. R. N. Sinha and W. E. Muir (eds.). Avi Publ. Co. Inc., Westport, CT.

Sinha, R. N. 1974. Seasonal abundance of insects and mites in small farm granaries. Environ. Entomol. 3:854-862.

Sinha, R. N. 1975. Effect of dockage in the infestation of wheat by some stored-product insects. J. Econ. Entomol. 68:699-703.

Sinha, R. N. 1979. Role of Acarina in the stored grain ecosystem. pp. 263-272 in: Recent Advances in Acarology. Vol. I. J. G. Rodriguez (ed.). Academic Press, New York.

Sinha, R. N. 1982. Food losses through energy transfer from cereal grains to stored-product insects. Food Nutr. Bull., (UN University) 4:13-20.

Sinha, R. N. 1988. Population dynamics of Psocoptera in farm-stored grain and oilseed. Can. J. Zool. 66:2618-2627.

Sinha, R. N. 1990. Stored product mites. Precis d'acarologie, S. A. L. F. France.

Sinha, R. N. 1991. Role of odor volatiles in the stored-grain ecosystem. pp. 3-13 in: Proc. 5th Int. Work. Conf. on Stored Product Prot. F. Fleurat-Lessard and P. Ducom (eds.). Bordeaux, France.

Sinha, R. N. and L. Harasymek. 1974. Survival and reproduction of stored-product mites and beetles on fungal and bacterial diets. Environ. Entomol. 3:243-246.

Sinha, R. N. and H. Kawamoto. 1990. Dynamics and distribution patterns of acarine populations in stored-oat ecosystems. Res. Popul. Ecol. 32:33-46.

Sinha, R. N. and S. Utida. 1967. Climatic areas potentially vulnerable to stored product insects in Japan. Appl. Entomol. Zool. 2:124-132.

Sinha, R. N. and H. A. H. Wallace. 1973. Population dynamics of stored-product mites. Oecologia 12:315-327.

Sinha, R. N. and Watters, F. L. 1985. Insect pests of flour mills, grain elevators, and feed mills and their control. Agric. Canada Publ. 1776E, Canadian Gov't Publ. Centre, Ottawa, Canada. 290 pp.

Sinha, R. N., B. Berck, and H. A. H. Wallace. 1967. Effect of phosphine on mites, insects, and microorganisms. J. Econ. Entomol. 60:125-132.

Sinha, R. N., F. J. Madrid, and N. D. G. White. 1986. Bioenergetics of *Ephestia cautella* (Walker) (Lepidoptera: Phycitidae) feeding on stored wheat. Ann. Entomol. Soc. Am. 79:622-628.

Smith, L. B. and P. S. Barker. 1987. Distribution of insects found in granary residues in the Canadian prairies. Can. Entomol. 119:873-880.

Snelson, J. T. 1987. Grain protectants. ACIAR Monograph No. 3. Aust. Centre Int. Agr. Res. Canberra, Australia. 448 pp.

Solomon, M.E. 1946. Tyroglyphid mites in stored products. Ecological Studies. Ann. Appl. Biol. 33:82-97.

Stables, L. M. 1980. The effectiveness of some recently developed pesticides against stored-product mites. J. Stored Prod. Res. 16:143-146.

Stein, W. 1991. New results about stored-product protection (animal pests). IV. Pfl. Krankh. 98:547-562.

Stepien, Z. A. 1970. Energy budget of *Rhizoglyphus echinopus* (Acarina, Acaridae) during its development. Ph.D. thesis, IEPAN, Dzickanon Lesy, near Warsaw. 103 pp.

Stepien, Z. A., W. Goszcynski, and J. Boczek. 1971. The energy budget of *Tyrophagus putresentiae* (Schr.) (Acaridae). pp. 373-378 in: Proc. 3rd Int. Cong. Acarol., Prague, Czechoslovakia.

Stevens, L. 1989. The genetics and evolution of cannibalism in flour beetles (genus *Tribolium*). Evolution 43:169-179.

Storey, C. L., D. B. Sauer, O. Ecker and D. W. Fulk. 1982. Insect infestations in wheat and corn exported from the United States. J. Econ. Entomol. 75:827-832.

Storey, C. L., D. B. Sauer, and D. Walker. 1983. Insect populations in wheat, corn and oats stored on the farm. J. Econ. Entomol. 76:1323-1330.

Subramanyam, B. J., P. K. Harein, and L. K. Cutkomp. 1989. Organophosphate resistance in adults of red flour beetle (Coleoptera: Tenebrionidae) and sawtoothed grain beetle (Coleoptera: Cucujidae) infesting barley stored on farms in Minnesota. J. Econ. Entomol. 82:989-995.

Taher, M. and L. K. Cutkomp. 1983. Effects of sublethal doses of DDT and three other insecticides on *Tribolium confusum* J. du V. (Coleoptera: Tenebrionidae). J. Stored Prod. Res. 19:43-50.

Tansley, A. G. 1935. The use and abuse of vegetational concepts and terms. Ecology 16:284-307.

Taylor, R. W. D. 1989. Phosphine—a major grain fumigant at risk. Int. Pest Control 31:10-14.

Taylor, T. A. 1971. On the flight activity of *Sitophilus zeamais* Motsch. (Coleoptera, Curculionidae) and some other grain-infesting beetles in the field and a store. J. Stored Prod. Res. 6:295-306.

Tuma, D., R. N. Sinha, W. E. Muir, and D. Abramson. 1989. Odor volatiles associated with microflora in damp ventilated and nonventilated bin-stored bulk wheat. Int. J. Food Microbiol. 8:103-119.

Tuma, D., R. N. Sinha, W. E. Muir, and D. Abramson. 1990. Odor volatiles associated with mite-infested bin-stored bulk wheat. J. Chem. Ecol. 16:713-724.

Van Bronswijk, J. E. M. H. and R. N. Sinha. 1971. Interrelations among physical, biological and chemical variates in stored-grain ecosystems; a descriptive and multivariate study. Ann. Entomol. Soc. Am. 64:789-803.

Vanderwel, D. and A. C. Oelschlager. 1989. Pheromone biosynthesis in Coleoptera. Chem. Scripta 29:401-406.

Van Voris, P., R. V. O'Neill, H. H. Shugart, and W. R. Emanual. 1980. Functional complexity and ecosystem stability. Ecology 61:1352-1360.

Wallace, H. A. H. and R. N. Sinha. 1981. Causal factors operative in distributional patterns and abundance of fungi: a multivariate study. pp. 233-247 in: The Fungal Community:Its Organization and Role in the Ecosystem. D. T. Wicklow and G. C. Carroll (eds.). Marcel Dekker, New York.

Walter, C. M. and N. R. Price. 1989. The uptake and penetration of pirimiphos-methyl into susceptible and resistant strains of the rust red flour beetle, *Tribolium castaneum* Herbst (Coleoptera: Tenebrionidae). Comp. Biochem. Physiol. 94C:419-423.

Waltner-Toews, D. 1991. One ecosystem, one food system:the social and ecological context of food safety strategies. J. Agric. Environ. Ethics 4:49-59.

White, G. G. and T. A. Lambkin. 1988. Damage to wheat grain by larvae of *Tribolium castaneum* (Herbst) (Coleoptera: Tenebrionidae). J. Stored Prod. Res. 24:61-65.

White, N. D. G. 1979. Interrelations in stored-wheat ecosystems infested with multiple species of insects:a descriptive and multivariate analyses. Ph.D. Dissertation, University Manitoba, Winnipeg, MB. 181 pp.

White, N. D. G. 1992. A multidisciplinary approach to stored-grain research. J. Stored Prod. Res. 28:127-137.

White, N. D. G. and R. J. Bell. 1988. Inheritance of malathion resistance in a strain of *Tribolium castaneum* (Coleoptera: Tenebrionidae) and effects of resistance genotypes on oviposition and larval survival in malathion-treated wheat. J. Econ. Entomol. 81:381-386.

White, N. D. G. and R. J. Bell. 1990. Relative fitness of a malathion-resistant strain of *Cryptolestes ferrugineus* (Coleoptera: Cucujidae) in malathion-treated and untreated wheat kernels. J. Stored Prod. Res. 26:23-37.

White, N. D. G. and D. S. Jayas. 1989. Safe storage conditions and infestation potential of canola meal by fungi and insects. J. Stored Prod. Res. 25:105-114.

White, N. D. G. and D. S. Jayas. 1991. Factors affecting the deterioration of stored flaxseed including the potential of insect infestation. Can. J. Plant Sci. 71:327-335.

White, N. D. G. and S. R. Loschiavo. 1985. Testing for malathion resistance in field-collected populations of *Cryptolestes ferrugineus* (Stephens) and factors affecting reliability of the tests. J. Econ. Entomol. 78:511-515.

White, N. D. G. and S. R. Loschiavo. 1986. Effects of insect density, trap depth, and attractants on trapping *Tribolium castaneum* (Coleoptera: Tenebrionidae) and *Cryptolestes ferrugineus* (Coleoptera: Cucujidae) in stored wheat. J. Econ. Entomol. 79:1111-1117.

White, N. D. G. and S. R. Loschiavo. 1988. Oviposition and larval development of the red flour beetle and the rusty grain beetle on ground and ball-milled seeds of various cereal cultivars. Can. J. Plant Sci. 68:617-626.

White, N. D. G. and R. N. Sinha. 1980. Changes in stored wheat ecosystems infested with two combinations of insect species. Can. J. Zool. 58:1524-1534.

White, N. D. G. and R. N. Sinha. 1981. Energy budget for *Oryzaephilus surinamensis* (Coleoptera: Cucujidae) feeding on rolled oats. Environ. Entomol. 10:320-326.

White, N. D. G. and R. N. Sinha. 1987. Bioenergetics of *Cynaeus angustus* (Coleoptera: Tenebrionidae) feeding on stored corn. Ann. Entomol. Soc. Am. 80:184-190.

White, N. D. G. and R. N. Sinha. 1990a. Effects of sublethal exposure to malathion or chlorpyrifos-methyl on offspring production by some stored product beetles. Can. Entomol. 122:1187-1191.

White, N. D. G. and R. N. Sinha. 1990b. Effect of chlorpyrifos-methyl on oat ecosystems in farm granaries. J. Econ. Entomol. 83:1128-1134.

White, N. D. G. and F. L. Watters. 1983. Incidence of malathion resistance in *Tribolium castaneum* and *Cryptolestes ferrugineus* populations collected in Canada. pp. 290-302 in: Proc. 3rd Int. Working Conf. Stored Product Entomol., Manhattan, KS.

White, N. D. G., R. T. Arbogast, P. G. Fields, R. C. Hillman, S. R. Loschiavo, B. Subramanyam, J. E. Throne, and V. F. Wright. 1990. The development and use of pitfall and probe traps for capturing insects in stored grain. J. Kansas Entomol. Soc. 63:506-525.

White, N. D. G., R. N. Sinha, and J. T. Mills. 1986. Long-term effects of an insecticide on a stored-wheat ecosystem. Can. J. Zool. 64:2558-2569.

White, N. D. G., R. N. Sinha, and W. E. Muir. 1982. Intergranular carbon dioxide as an indicator of biological activity associated with the spoilage of stored wheat. Can. Agric. Eng. 24:35-42.

White, N. D. G., V. F. Wright, F. L. Watters, and R. B. Mills. 1983. Future directions and current problems for stored-product entomology. Bull. Entomol. Soc. Am. (Summer) :35-37.

Whitehead, J. R., R. T. Rorish, and B. R. Norment. 1985. Resistance stability and co-adaptation in diazinonresistant house flies (Diptera: Muscidae). J. Econ. Entomol. 78:25-29.

Wiegert, R. G. and C. E. Petersen. 1983. Energy transfer in insects. Annu. Rev. Entomol. 28:445-486.

Wightman, J. A. 1978a. The ecology of *Callosobruchus analis* (Coleoptera: Bruchidae): morphometrics and energetics of the immature stages. J. Anim. Ecol. 47:117-129.

Wightman, J. A. 1978b. The ecology of *Callosobruchus analis* (Coleoptera: Bruchidae):energetics and energy reserves of the adults. J. Anim. Ecol. 47:131-142.

Wilkin, D. R. and J. A. Hope. 1973. Evaluation of pesticides against stored-product mites. J. Stored Prod. Res. 8:323-327.

Wright, V. F., T. W. Mize, and J. R. Pedersen. 1988. Using insect traps to monitor stored-grain pests on the farm. p. 442 in: Proc. XVIII Int. Cong. Entomol. Vancouver, BC.

Yaciuk, G., W. E. Muir, and R. N. Sinha. 1975. A simulation model of temperatures in stored grain. J. Agric. Eng. Res. 20:245-258.

Yoshida, T. 1983. Damage and losses in stored products from insect pests in tropical countries. pp. 9-17 in: Bio-Loss of PostHarvest Food and its Prevention Technology. Proc. National Food Res. Inst. United Nations University Workshop, Japan.

Zdarkova, E. 1967. Stored food mites in Czechoslovakia. J. Stored Prod. Res. 3:155-171.

Zdarkova, E., P. H. Verner, and J. Novosad. 1983. Dispersion and distribution of mites and beetles in stored grain. J. Stored Prod. Res. 19:73-80.

Zettler, J. L. and G. W. Cuperus. 1990. Pesticide resistance in *Tribolium castaneum* (Coleoptera: Tenebrionidae) and *Rhyzopertha dominica* (Coleoptera: Bostrichidae) in wheat. J. Econ. Entomol. 83:1677-1681.

Zettler, J. L. and G. L. LeCato. 1974. Sublethal doses of malathion and dichlorvos:effects on fecundity of the black carpet beetle. J. Econ. Entomol. 67:19-21.

Zettler, J. L., W. R. Halliday, and F. W. Arthur. 1989. Phosphine resistance in insects infesting stored peanuts in the southeastern United States. J. Econ. Enomol. 82:1508-1511.

6

Computer Modeling of the Population Dynamics of Stored-Product Pests

James E. Throne

Agricultural Research Service, United States Department of Agriculture, Savannah, Georgia

SUMMARY

Many of the current quantitative ecological theories were developed using stored-product insects as test organisms. Ecological models have been developed for simulating the population dynamics of *Cadra cautella* (Walker), *Callosobruchus chinensis* (L.), *C. maculatus* (Fabricius), *Cryptolestes ferrugineus* (Stephens), *Tribolium castaneum* (Herbst), and *T. confusum* Jacquelin du Val. An ecological model has also been developed for simulating the effects of the pathogen *Nosema whitei* Weiser on *T. confusum* population growth. Most of this ecological work describes population dynamics in limited environmental conditions and, thus, is not directly applicable to storage ecosystems. Simulation models of microorganism growth in stored grain have not been developed because microorganisms are not discrete organisms with quantifiable population parameters. Population dynamics models have been developed for only one mite pest, *Acarus siro* L., because mites cause far less damage to grain than insects.

Australian scientists pioneered the use of computer models for optimizing management strategies for pests of stored grain. Their work on *Sitophilus oryzae* (L.) inspired the development of models for *C. ferrugineus*, *C. pusillus* (Schönherr), *Oryzaephilus surinamensis* (L.), *Ptinus tectus* Boieldieu, *Rhyzopertha dominica* (Fabricius), and *T. castaneum*. These models are generally single-species models that simulate several basic life processes: duration of immature stages, survivorship, and fecundity. Temperature and moisture content are the major abiotic factors driving the models. Population dynamics models are easily integrated with models that simulate the physical environment.

Models for simulating rates of degradation of residual contact pesticides are readily available and appear to be accurate. Nevertheless, the effects of temperature, moisture content, concentration of pesticide, and duration of exposure to protectant pesticides on mortality of insects have not been well described for any species of stored-product insect. Similar data and models for fumigants (including gases for modified-atmosphere storages) and natural enemies are also lacking.

Computer models that aid in managing stored-grain pests currently are not available commercially. Lack of economic studies on the costs and benefits of control strategies in grain storage will probably impede acceptance of these models. The major area where further research is required is the collection of high-quality, quantitative data for describing the effects of abiotic factors (over the entire range of conditions at which grain is stored) on the life history of stored-grain pests and their natural enemies and the effects of protectant chemicals and fumigants on stored-grain pests. Additionally, the abiotic and biotic factors that affect the storage ecosystem must be characterized and automated monitoring of abiotic factors in grain storages should become a standard practice.

1. INTRODUCTION

1.1. Description and Historical Background of Computer Models

Mathematical models have been used for centuries to describe both observed and theoretical phenomena. A mathematical model is an equation or set of equations that describes observed data or a theory (e.g., $E = mc^2$). Mathematical models are useful because they can be used to simulate and predict the progress of events under conditions different from those for which they were developed. For example, Pearl and Reed (1920) used the logistic equation to describe human population growth in the United States. The equation may be used to predict human population growth in any part of the world or the parameters of the equation can be changed and the modified equation used to predict population growth of species other than humans.

As mathematical models became more sophisticated, the labor involved in calculations restricted the factors that could be included in the models. A simple, single-species population-dynamics model might include the effects of temperature and relative humidity on the rate of development and survivorship of several immature stages of an arthropod, and the effects of those abiotic factors on the longevity and fecundity of the adults (e.g., Berreen 1974). Simulating development of an arthropod for 1 yr with a 1-d time step requires numerous calculations. Daily temperature and relative humidity are used as input to the model. Daily rates of development, survivorship, and fecundity are calculated based on temperature and moisture content. Individuals are advanced through life stages, mortality is applied, and new eggs laid. Individuals in each stage are summed and the summations are output to observe daily progress. Inclusion of other species in the model or of a control method may greatly increase the number of calculations required for the simulation. If calculations are done manually, complex mathematical models for simulating population dynamics may require too many calculations for routine use.

The development of computers that could process calculations with less human input resulted in the development of complex models. Computer programs may integrate: 1) equations describing the effects of the environment on biological processes, 2) a mechanism for inputting information on the environmental conditions and initial state of the population, and 3) a mechanism for outputting the results of the calculations. Hence, a computer model is a method for automating the numerous calculations required in the use of mathematical models.

1.2. Approach to Developing Computer Models

In developing computer models, the systems approach is generally used. The systems approach was initiated as a science by engineers and basically seeks to characterize a system by defining the objects in the system and the interactions of those objects with their environment. The systems approach involves: 1) stating the objective of the study as mathematically as possible (e.g., determining how to keep an insect population below an economic threshold level of two insects per kg of grain); 2) determining the major factors that must be understood to achieve the objective (i.e., defining the system) and differentiating those factors from the environment in which they operate; 3) modeling or mathematically describing the system; and 4) using the model to determine how the system can be modified to achieve the stated objective (Ruesink 1976). The ecosystem approach to grain storage uses the systems approach to optimize grain storage for human use.

Following the systems approach, the objective of a stored-grain ecologist is to manage the environment so that pests are reduced to either an undetectable or low level or to a level that is considered acceptable to the owners of the grain. In some countries (e.g., Australia or Canada) the objective is to maintain grain quality by maintaining the pest populations below a detectable level. In other countries (e.g., the United States) the objective is to maintain the pest populations below a predetermined level. In nations where food is used locally, the objective may be to have enough nutrition so that one's family will survive the year. In this case, acceptable levels of pest damage can be determined based on the quantity and quality of grain harvested and the number of persons in the family.

The system is comprised of the biological organisms in the stored grain. The grain in a storage is generally considered to be of constant nutritive value for the pests, unless altered by the organisms growing in it. Hence, the grain is part of the environment, which also includes the storage structure and climatic conditions. Thus, to describe the grain storage system one must model the environment and the biological organisms living in the grain. The resulting model can be used to determine the optimum methods for achieving the objective of keeping the grain in good condition.

1.3. Potential Application of Storage Ecosystem Models

The process of developing computer models provides several benefits to the researcher. Development of a mathematical model describing a major component of the system provides a tool for synthesizing knowledge about that component and emphasizes areas in which knowledge is lacking. Thus, modeling may help identify research priorities.

Models of the storage ecosystem can be used to synthesize research data in a form that is usable by extension personnel or by grain storage managers. The models may be used in several ways. They may be used by researchers to test various management strategies, or combinations of management strategies, that might normally take years to test in the field. For example, models have been used to determine optimum cooling strategies when used in combination with protectant insecticides to minimize insect damage (Longstaff 1981). The control technologies that are most promising in simulations can then be tested in the field.

Models can also be used in "real time" situations. That is, environmental conditions for a grain storage can be input at regular intervals to the computer model and simulations run to determine the condition of the grain. Such simulations might be particularly useful in any type of deep grain mass (e.g., a 30-m high silo), where sampling the middle of the grain mass is impractical. Placing environmental monitoring equipment in the storage and linking the monitoring equipment to a computer containing appropriate model software offers an automated management-decision system.

Models may also be predictive when environmental and biological conditions in a bin are sampled and used as starting conditions for a simulation. Given historical environmental conditions for the geographical area, optimum control strategies can be determined for the bin. Updated environmental and biological information, based on sampling, should be used to initiate simulations intermittently during the storage season to ensure that the model is accurately simulating storage conditions.

2. REVIEW OF COMPUTER MODELS FOR SIMULATING PEST POPULATION DYNAMICS

2.1. Review of Some Modeling Terms

A *deterministic model* is one that does not include chance events. For example, if 100 eggs are laid by an insect on day 1, then all those eggs will hatch and become larvae on the same day (less any that die as eggs). All simulations with the same initial conditions will have the same output.

A *pseudostochastic model* is a type of deterministic model that includes variation among individuals. For example, if 100 eggs are laid by an insect on day 1, they all will not hatch and become larvae on the same day. Rather, they will hatch over time corresponding to a mathematical distribution. Like the deterministic model, all simulations with the same initial conditions will have the same output.

A stochastic model includes chance events, whose probability of occurring depends on a distribution. For example, if 100 eggs are laid by an insect on the same day, the probability of a particular egg hatching on a following day is based on a value chosen from a random number generator that simulates a mathematical distribution. Simulations with the same initial conditions generally will have diverse results, especially with small population size.

A *table look-up function* is a computer program that linearly interpolates between points in a data set. For example, suppose that we have data on the duration of egg to adult development of an insect for temperatures from 5 to 40°C at 5°C intervals. To determine duration of development at 7°C, one can use a table look-up function to linearly interpolate between values of duration of development for 5 and 10°C.

The results of a model must be compared statistically with observed data to ensure that the model actually simulates reality. This process is called *validation.* In this chapter, the term *anecdotal validation* will be used to refer to the validation process in which a strict mathematical comparison of simulated and observed results has not been made; rather, the simulation results appear to be similar to observed results or the observed results are not strictly quantified.

2.2. Models for Simulating the Population Dynamics of Microorganisms

Models have not been developed for simulating the population dynamics of microorganisms in stored grain because of the difficulty in defining microorganism populations. Life stages of microorganisms are often not readily definable nor discernible which further limits studies on their population dynamics in stored grain.

2.3. Models for Simulating the Population Dynamics of Mites

Management models of mite pests of stored grain generally have not been developed partly because mites are usually a less important pest of stored grain than insects and partly because comprehensive data on life tables of storage mites are scarce. Models have been developed for the grain mite, *Acarus siro* L. (Acari: Acaridae). Berreen (1974) developed a model using the Leslie-type matrix format (Leslie 1945) for describing population growth of the grain mite. The model was not truly a computer model, as it was designed to run on a programmable calculator. This exploratory model was developed to summarize available knowledge on the species. The model simulated the effects of the environment on survival, development rate of five age classes (eggs, larvae, protonymphs, tritonymphs, adults), and age-specific fecundity; these might be considered the basics of a model of the life history of a pest species. The data on which the model was based were collected at 15°C and 70% and 90% relative humidity (rh). The model was validated using anecdotal information. That is, Solomon (1969) indicated that at 15°C and 90% rh, an initial population of 100 adult mites produced a population of about 20,000 juvenile and adult mites in about 120 d. The first model developed by Berreen (1974) underestimated the number of mites after 120 d at 15°C and 90% rh. Further studies indicated that the original 1:1 sex ratio in the model was incorrect for the given environmental conditions and that the proportion of females was closer to 60%. A revised model resulted in 21,044 adults and juvenile mites after 120 d. Development of the model pointed out the shortcomings of the available data, mainly that the effects of temperature and moisture content on population growth of the species were not known and that sex ratio varied with humidity.

To avoid the problem of lack of data describing the effects of abiotic factors on individual stages of *A. siro*, Kawamoto et al. (1991) developed a model describing the effects of temperature and relative humidity on the maximum rate of population growth. When environmental conditions in their simulations were outside the limits for mite development, survivorship under those conditions was predicted as a function of temperature and humidity. The model was validated using data collected from granaries of wheat in Canada, Great Britain, and Czechoslovakia. The model accurately simulated observed grain mite population dynamics when the predatory mite, *Cheyletus eruditis* (Schrank) (Acari: Cheyletidae) was not present in the storages. The number of grain mites found when predatory mites were also present in the bins was less than predicted by the model because it did not account for predation. Although the model appears to be useful for predicting population growth, it still indicates the lack of comprehensive quantitative data for describing the life history of this species. The model cannot be used to simulate stage-specific processes, which include the effects of most control measures.

2.4. Models for Simulating the Population Dynamics of Insects

2.4.1. *Ecological Models*

Much of the quantitative ecological theory that exists today was developed using stored-product insects as the test organisms—for example, the work of L. C. Birch (Andrewartha and Birch 1954) and the numerous publications on *Tribolium* spp. (Sokoloff 1974)—because they can be easily reared and observed. Much of this ecological work intended to establish the basic theory of population dynamics rather than be used for practical application in pest control. Thus, the work described population dynamics in unrealistic situations: long-term experiments with many insects in a small quantity of grain maintained at a moderate temperature and optimum humidity. Much of this ecological work is not directly applicable to the stored-grain ecosystem because the environmental conditions are too restricted or because such high densities of insects rarely occur in a marketable bulk of grain. Therefore, models for simulating the population dynamics of insect pests of stored grain will be discussed in two groups: theoretical ecological models and practical pest-management models.

Ecological models are those that simulate population growth under restricted (usually one) temperature and relative humidity conditions. These ecological models generally were developed to mathematically describe and investigate an ecological principle, such as the effects of density on the rate of oviposition. The second group of models are those that attempt to simulate the stored-grain ecosystem. That is, they generally are intended to simulate pest population dynamics over the entire range of environmental conditions that might be encountered in grain storages and they are generally developed to optimize management strategies.

The first computer model for a stored-grain insect was developed by Niven (1967) for the red flour beetle, *Tribolium castaneum* (Herbst) (Coleoptera: Tenebrionidae). Although numerous mathematical models predate this model (e.g., Landahl 1955a,b; Neyman et al. 1956; Lloyd 1968), this was the first model developed of stored-grain pests to run on a computer. This was a single-species, stochastic model that could simulate population development only at 29°C and 70% rh, and included stage-specific rates of development and survival of the immature stages of four genetic strains of the insect. The model included age-specific mortality, density-dependent fecundity of adults, and rates of cannibalism by adults and larvae. Simulation results were similar to those observed in laboratory experiments (i.e., anecdotally validated). The model was used to investigate long-term population growth for the four strains, the effects of increasing fecundity on population growth, and the effects of decreasing and increasing cannibalism on population growth.

Hardman (1976) compared the results of simulations using deterministic, pseudostochastic, and stochastic models to determine which type of model best simulated the population growth of the confused flour beetle, *T. confusum* Jacquelin du Val. He used a degree-day model to simulate population growth under Malthusian conditions: basically unlimited food, maximum fecundity, and minimal mortality. The deterministic model included the effects of temperature on survivorship, fecundity, and development rate. The pseudostochastic model included variation in development rates among individuals. The stochastic model used Monte Carlo simulation techniques to include chance in sex ratios, mortality rates, development rates, and fecundity. The models were validated using data collected from two experiments. The first test was started with populations of 100 eggs in 1.5 L of flour and incubated at 23.5–24.1°C, and the second was started with nine populations of 80 unsexed adults on 1.5 L of flour and incubated at 31°C. Moisture content of the flour was not reported. Results from both the deterministic and pseudostochastic models compared favorably with observed results, although observed and simulation results were not compared statistically. The stochastic model required too much memory to run for more than 18 d, because data about individual insects had to be maintained separately. In the deterministic and pseudostochastic models, data were maintained for cohorts of insects.

Bellows (1982) developed simulation models for the cowpea weevil, *Callosobruchus maculatus* (Fabricius) (Coleoptera: Curculionidae), and *C. chinensis* (L.). The models simulated population development only at 30°C and 70% rh, and included age-specific fecundity and egg mortality as functions of egg and adult density, age-specific adult mortality that included variation among individuals, density-independent and density-dependent mortality of larval stages, and variation in larval development rates. Validation tests showed that the models predicted fecundity well at low densities, but overestimated fecundity at high densities. Although observed and simulated results were not compared statistically, simulated larval survival appeared to mimic observed survival.

Gordon et al. (1988) developed a model for the almond moth, *Cadra cautella* (Walker) (Lepidoptera: Pyralidae). The model included the amount of food available at a given time, the rate of food ingestion by larvae, mass change of larvae, and the physiological age of the insects (which was based on age and mass). The amount of food ingested depended on the amount of food available, larval mass, and larval density. The model did not include the effects of abiotic factors on population growth. The model appeared to simulate larval development well, except when the food supply deteriorated in validation experiments. As the food deteriorated in validation tests, larvae began to feed on frass and each other. Observed and simulated results were not compared statistically. Jones et al. (1990) expanded on the previous model by allowing larval development to continue by feeding on feces when the food supply was diminishing and, thus, improved the accuracy of simulations. Again, observed and simulated results were not compared statistically.

Campbell and Sinha (1990) developed a model for simulating energy transfer in wheat stored at 30°C and 13.5% moisture content (mc) that was infested by the rusty grain beetle, *Cryptolestes ferrugineus* (Stephens) (Coleoptera: Cucujidae). The model included rates of insect development and survival, fecundity, respiration, egestion, and biomass changes. Simulation results closely agreed with observed population growth, respiration, and food consumption during the population growth phase. Observed and simulated results were not compared statistically. In the simulations, the insect population died after the food was exhausted. In the experimental study, the population continued at a low level even after the food was badly deteriorated.

Onstad and Maddox (1990a,b) developed a model for simulating the effects of the pathogen, *Nosema whitei* Weiser (Microsporida: Nosematidae), on population growth of *T. confusum* at 29°C and 70% rh. The model included rate of immature development, fecundity, longevity, and cannibalism by larvae and adults, and used a distributed-delay (Manetsch 1976) to simulate variation in development rate. Fecundity was influenced by age, adult density, and disease. Disease infection was dependent on spore concentration and larval age. Rates of development of infected larvae were decreased when exposure of larvae to spores exceeded certain limits. Longevity of infected adults was also reduced. Simulation results compared favorably with results from a previous study for the first 30 d of simulation. Conditioning (i.e., the buildup of chemicals secreted or excreted by the insects) of the flour in the validation experiment was responsible for differences between observed and simulated population levels after the first 30 d. Simulation results indicate that at the initial spore concentrations tested, disease may reduce the insect population by 90%.

2.4.2. *Pest-Management Models Developed in Australia*

The first use of computer models for optimizing management strategies in stored grain was for the rice weevil, *Sitophilus oryzae* (L.) (Coleoptera: Curculionidae), on wheat in Australia. Longstaff (1990) reviewed the development of the *S. oryzae* model. Hardman (1978) developed a degree-day model for simulating population growth of the weevil and the effects of the weevil on its environment. This was the first attempt to develop a model for a stored-grain pest that could be used to run simulations over the range of environmental conditions encountered in actual grain storages. Assuming a homogeneous environment, the model included effects of temperature (from 16 to 32°C) and grain moisture content (from 10.5 to 14.5%) on: immature development rate; mortality; oviposition; feeding; oxygen consumption; and production of frass, water, and carbon dioxide. The development threshold was calculated as 12.6°C using median rates of development for weevils reared at 13.5% mc at 16–32°C. On average, 423 degree-days above the threshold temperature were required for development from egg to adult. The effects of moisture content of the grain on duration of the immature development period were simulated using multipliers. For example, duration of development at 10.5% mc was determined by multiplying the duration of development at 13.5% mc by 1.19. Durations of individual immature stages were determined using multipliers calculated from the data of Birch (1945). Hence, the egg stage was 12% of total immature development, the larval stage was 53%, and the pupal stage (included prepupa, pupa, and preemergent adult) was 35%. When more than one weevil was present in a kernel, development was reduced by 9% each day. Mortality varied with age, oxygen concentration, number of larvae per kernel, temperature, and moisture content. All climate-induced mortality occurred on the first day of the larval stage in the model. Adult mortality was not influenced by moisture content or temperature, and was based on data collected at 26°C. Fecundity was based on data collected at 26°C and was influenced by moisture content and density. The insects could influence their environment by consuming oxygen and producing carbon dioxide and water. Rates of respiration, metabolism, food ingestion, and frass production were dependent on oxygen concentration, temperature, and moisture content. Gaseous exchange occurred based on concentrations of gases in the grain and in the air outside the grain.

The model was validated by placing 100 weevils in vials (70 by 230 mm) containing approximately 19900 grains of wheat. Temperature was monitored in each vial. Complete adult censuses were conducted every 7–14 d for 112 d. Simulation results compared favorably with observed data, based on statistical comparison of the data, until near the end of the study when the food was deteriorating. Because the model assumed an unlimited food supply, the simulated number of adults continued to increase while the observed number declined. Simulated moisture contents compared favorably with the observed, based on statistical comparison of the data.

The validated model was used to investigate the benefits of cooling and drying grain by running simulations for 80 d at constant temperatures from 16 to 28°C and constant moisture contents from 10.5 to 14.5%. The populations did not increase at temperatures below 20°C. Above 20°C, population growth increased as temperature and moisture content increased. Given that the model simulates gas exchange, the effects on population growth in a sealed storage were simulated. Airtight storage resulted in the death of all insects within 37 d. Even in bins with 10% leakage, population growth was reduced.

Cuff and Hardman (1980) restructured Hardman's (1978) model transforming it into a Leslie-type matrix format. The Leslie matrix was extended to allow for a heterogeneous environment and to include the effect of the insect on its environment. The effects of moisture content on development rate of the weevils was removed from the model because it was considered inconsequential. The heterogeneous environment was modeled by dividing the grain mass into cells (via a matrix), and having movement of weevils between cells. In simulations, the interior of the grain mass constituted the largest cell and the edges of the bulk constituted smaller individual cells. Although not validated, the model was used to simulate the effects of varying degrees of gas-tightness on population growth under varying conditions of temperature and initial insect density in an underground storage. The simulations showed that as gas-tightness increased, insect damage decreased. Under conditions of 99.9% gas-tightness, population growth was less at higher temperatures (over the storage season) because the oxygen supply was quickly depleted.

Due to the lack of knowledge about heat and moisture transfer in bunker storages, Longstaff et al. (1981) linked Cuff and Hardman's (1980) unvalidated model to that of Sutherland et al. (1971) which describes heat and moisture transfer in aerated grain. The effects on rice weevil population growth of airflow rate, initial grain temperature, initial moisture content, and percentage of time grain was aerated were investigated. Aeration was shown to offset the effects of initial high temperature and moisture content of grain stored in Australia. Optimal aeration strategy was determined by considering both pest-population growth and cost of aeration.

Thorpe et al. (1982) modified Cuff and Hardman's (1980) unvalidated *Sitophilus* model and combined it with the heat and moisture transfer model of Sutherland et al. (1971). The effects of the insect on the environment and the effects of low oxygen concentration on the insect were removed from the model of Cuff and Hardman (1980) because the effects of aeration would be much greater than insect-induced changes in the environment. Simulations were run to determine optimum airflow rates and optimum amount of time to run aeration fans in bins of wheat in Australia. In the simulations, wheat was stored for 6 mo with an initial temperature of 30°C, 12% mc, and 0.1 newly-emerged adults per kilogram of wheat in a part of Australia with a warm, humid climate. Aeration fans were run during the coldest 50% of the first month of storage. The effects of additional cooling were optimized when fans were run for the coldest 15% of each day. Population growth was decreased as airflow rate increased above 0.2 (L/s)/t. It was shown that there would be no insect population growth during the storage period if initial grain moisture content could be reduced to 10.5%, *S. oryzae* being unable to complete development in grain of this moisture content. Population growth was insensitive to initial grain temperature. Weaknesses in the model that were pointed out by the authors were: the physical model was applicable only to the central bulk of the grain mass, the periphery of the mass was subject to external environmental influences, and the biological submodel was limited to a single species and a fairly narrow range of temperatures.

Longstaff and Cuff (1984) restructured the biological model for *S. oryzae* developed by Cuff and Hardman (1980). The temperature domain of the model was extended to 15–33.3°C by using a nonlinear physiological time scale and incorporating new data. The influence of moisture content on development rate, which was removed from the model by Cuff and Hardman (1980), was reintroduced into the model. Oxygen exchange was excluded from the model because the applications for which the model was intended would not include oxygen as a limiting factor. Several table look-up functions in Cuff and Hardman's (1980) model were replaced by continuous functions. Age-specific fertility was described as a function of temperature, moisture content, and density of weevils and included immature mortality and numbers of eggs laid. Additional immature mortality was applied as a function of weevil density. Adult survival was described as a function of temperature and moisture content. The only physical elements retained in the model were dry mass of the wheat, dry mass of frass, and mass of free water. The description of the rate of respiration (upon which rate of food consumption, frass production, and water production depended) was modified to include effects of temperature. Variation in development period was added to the model using an extension of the Leslie matrix (Longstaff 1984), and was assumed to follow a log-normal distribution. Random oviposition, including multiple oviposition in some kernels, was reintroduced to the model. To increase the speed at which simulations ran, the time step in the model was changed from 1 d to 1 wk and the number of immature and adult age classes was reduced by about one half. When compared with the validation data of Hardman (1978), the revised model more accurately described observed numbers of weevils and moisture content than the model of Cuff and Hardman (1980). Observed and simulated results were not compared statistically. The revised model was used to determine the effects of fumigation followed by grain cooling on subsequent insect population growth in wheat stored under Australian climatic conditions. Fumigation was assumed to kill an equal proportion of all stages of the insect. A kill rate of 99.9% was required if a second fumigation was to be avoided during a 40-wk storage period.

Longstaff (1988a) further modified the *S. oryzae* model and used the revised model to investigate the effects of temperature manipulation on *S. oryzae* control using protectant insecticides. The previous model had simulated variation in development rate using data for the closely-related maize weevil, *Sitophilus zeamais* Motschulsky. This component of the model was modified using newly-published data for *S. oryzae*. The Leslie matrix was further modified to reduce the excessive computer memory required and increase the speed at which the simulations ran. Survivorship of insects exposed to insecticide was modeled as a function of time of exposure to the pesticide (in physiological time) and of temperature. The pesticide degradation model used was that of Desmarchelier and Bengston (1979). The revised *S. oryzae* model was not validated. Simulations to determine the effects of aeration on population growth were run with either slow, medium, fast, or no cooling of the grain. Simulations showed that the faster the insecticide-free grain was cooled, the smaller the resulting population level after 1 yr. The pest population, however, increased in size no matter how rapidly the grain was cooled. The model also was used to determine the best control strategies using protectant insecticides in combination with aeration. The pyrethroid insecticide deltamethrin, which is more toxic at low temperatures, was predicted to be most effective when the grain was cooled rapidly. The organophosphate pesticide pirimiphos-methyl, which is less toxic at low temperatures, was less effective when used with cooling even though it breaks down more slowly at low temperatures. The effects of cooling after fumigation were also investigated. Fumigation was simulated by assuming that 99% of adults, 90% of pupae, and 100% of eggs and larvae were killed during fumigation. Cooling after fumigation resulted in extremely low population levels for up to 34 wk of storage, and more rapid cooling was most effective in reducing population growth.

Longstaff (1988b) added a genetic submodel of insecticide resistance to this model and used it to investigate the development of resistance to insecticides under regimes of no cooling and rapid cooling. The genetic model was theoretical and was not validated. The frequency of the resistance gene did not increase, or increased marginally, when rapid cooling regimes were used in simulations. Frequencies of resistance genes increased to as much as 0.99 during a 1-yr storage period when no cooling was used in the simulations. The simulations showed that by reducing the increase in resistance gene frequency, cooling could prolong the period of effective insecticide exposure.

Sinclair and Alder (1985) took an alternative approach to developing a model for the rice weevil in wheat. They developed a pest management model that could be used to investigate the effect of sources of insect infestation on a typical Australian grain farm on resultant insect population levels in the grain marketing system. Based on the life history of *S. oryzae*, they developed a model for a hypothetical pest making certain assumptions about the ecology of *S. oryzae* in the absence of field data. They used the model to investigate the level of infestation of grain delivered to central granaries that was infested by insects breeding in the header (combine) at harvest, insects migrating to a loaded truck, and insects in grain that is bought from other sources. Factors such as removal of infestation sources and killing insects with contact insecticides or fumigants were included in the model. Models for mortality caused by insecticides and fumigants were not based on quantitative data, but on assumptions. The insect model included development time of immature insects as linearly related to temperature, emigration from a grain mass based on temperature and density, age-specific survival as a function of temperature and density, and oviposition as a function of temperature. Two crops per year (wheat and sorghum) were harvested and stored. Cleaning the header had the greatest effect on insect populations on unsprayed farms because the major source of infestation was reduced. The longer the grain was held, the worse the insect infestations became. Cleaning spills had little effect on insect population levels at delivery to central storages because it was assumed that the chance of migration from a spill to stored grain was small. Pest infestation levels in sorghum were low because of cool temperatures after harvest. On farms where insecticides were used, insect populations were minimized by treating grain that was retained for feed. Such treatment also greatly enhanced the chances for development of resistance because the susceptible gene pool was removed. The insect life-history submodels were compared with observed data and seemed reasonable, although observed and simulated data were not compared statistically. The control portions of the model were not validated.

2.4.3. Pest-Management Models Developed in North America and Europe

Hagstrum and Heid (1988) developed a general model for predicting numbers of the lesser grain borer, *Rhyzopertha dominica* (Fabricius) (Coleoptera: Bostrichidae) in wheat stores throughout the United States. They determined the proportions of wheat stored on-farm or in elevators throughout the year and the percentage of the wheat crop that is harvested during the summer. Hence, they could predict the average age and type of storage for wheat stocks in the United States at any time of the year. They determined that the average moisture content of wheat when it was placed into storage was 12%, and that moisture content of the wheat changed little during storage. They assumed that grain temperature was 32°C (actual average for wheat-producing states was 30.7 ± 4.7°C) when initially stored. The weekly rate of increase of *R. dominica* was dependent on temperature and moisture content of the grain. They used the model to predict seasonal marketwide changes in average population levels of *R. dominica.* They determined from iterative simulations that the model best mimicked actual population levels in grain when they assumed that the insect population was reduced by 30% per wk by insect controls, when temperatures dropped in the fall by 0.5°C, and when temperatures increased in the spring by 0.5°C. The model appeared to simulate well the average number of insects in the US wheat-marketing system. Observed and simulated results were not compared statistically.

Hagstrum and Throne (1989) applied the general modeling approach used in Australia to develop management models for insect pests of stored wheat in the United States. The model developed by Throne (1983) was modified to simulate the effects of temperature and moisture content on population growth of *T. castaneum*, *C. ferrugineus*, and *R. dominica* in wheat. The relationships between temperature and development rates when the insects were reared at

10–14% moisture contents were described using the method of Wagner et al. (1984). Linear interpolations were used to determine development rates at intermediate moisture contents. Variation in development was simulated using a time-varying distributed delay (Manetsch 1976). Survivorship for *T. castaneum* and *C. ferrugineus* was 100% in simulations. Daily egg production for *R. dominica*, was dependent on temperature and moisture content but not on age of females; egg production for the other two species was age-specific. Egg production of *R. dominica* was determined using adult emergence data, so the effects of temperature and moisture content on survivorship of immature stages were included. Observed and simulated results were compared statistically. Simulation results for all three species compared favorably with observed insect numbers in laboratory tests and in small grain bins.

Flinn and Hagstrum (1990a) expanded the *R. dominica* model (Hagstrum and Throne 1989) by including subroutines for aeration, fumigation, and protectant insecticides. Aeration was simulated by assuming that it took 12 d to lower the grain temperature to 20°C, regardless of the initial grain temperature. The equation of Desmarchelier and Bengston (1979) was used to simulate the effects of temperature and relative humidity on malathion degradation. Rates of mortality due to malathion were dependent upon time of exposure to the insecticide and concentration of the pesticide during exposure. Fumigation was simulated using the method of Longstaff (1988a), where 100% of eggs and larvae, 90% of pupae, and 99% of adults were assumed killed in 5 d. The effects of various control strategies for managing *R. dominica* populations in wheat were investigated using the model. Simulations were run using initial moisture contents of 10 or 14% and initial grain temperatures of 27° or 32°C. Initial starting dates were 1 June, 1 July, or 1 August depending on the region of the United States for which simulations were run. In unaerated grain, temperature and moisture content were not changed until 1 October, at which time grain temperature was decreased 0.5°C weekly. When grain was aerated beginning 1 October in northern regions, no additional control action was indicated. Further control measures were indicated in other regions of the country. Initial moisture content was more important than initial grain temperature in determining population growth over 1 yr. When fall aeration was delayed after storage on 1 June, each month of delay in aeration resulted in an approximately six-fold increase in population level after 1 yr of storage. Delaying fumigation until just before aeration appeared to be the best control strategy, provided insect population levels did not exceed economic threshold levels earlier. Treatment with malathion provided effective control for 1 yr of storage under all initial temperature and moisture conditions tested. Although the insect model was previously validated (Hagstrum and Throne 1989), the control subroutines were not validated.

Hagstrum and Flinn (1990) developed and statistically validated models for the sawtoothed grain beetle, *Oryzaephilus surinamensis* (L.) (Coleoptera: Cucujidae), and *S. oryzae* in wheat using the model structure and methods of Hagstrum and Throne (1989). They then used these models, and those developed by Hagstrum and Throne (1989) and Flinn and Hagstrum (1990a), to determine whether species differences were important in making management decisions for insect pests of stored wheat. Initial grain temperatures, moisture contents, and aeration regimes in the simulations were as in Flinn and Hagstrum (1990a). The effects of initial storage date on population growth varied with the species; the most rapidly growing populations had a greater reduction in population levels when storage was delayed. As reported by Flinn and Hagstrum (1990a), delaying fumigation until August or September resulted in lower population levels at the end of the storage year. Use of malathion was most effective for the external-feeding species. Due to differences in responses to control strategies among species, the authors concluded that grain pests must be considered separately when optimizing control strategies. Occasionally, the same combination of control measures, however, may be required regardless of which species are present. The five simulation models were subsequently used to develop an expert system for management of insect pests of wheat (Flinn and Hagstrum 1990b, see Chapter 2).

Kawamoto et al. (1989) developed a model for simulating the population dynamics of *C. ferrugineus* in stored wheat. Temperature, relative humidity, and insect density were the factors that controlled population growth in a homogeneous bulk of wheat. Duration of egg development was assumed linear between 15.3 and 35°C, and constant above 35°C. Relative humidity did not affect the duration of the egg stage. Rate of egg mortality was determined by temperature and relative humidity. For the combined immature stages (larva to adult emergence), the effects of temperature on development rate were described using the method of Sharpe and DeMichele (1977). Rate of immature development was adjusted for low relative humidities. Rate of mortality of immature insects was dependent on temperature, relative humidity, and insect density. Rate of oviposition was described using a Weibull function and was dependent on temperature, relative humidity, physiological age of female, and insect density. Adult mortality was dependent on temperature. The model was developed assuming no immigration or emigration. The model was not validated.

Nuttall (1989) developed a model for simulating population dynamics of the spider beetle, *Ptinus tectus* Boieldieu (Coleoptera: Ptinidae). Duration of egg and pupal development were dependent on temperature and relative humidity. Duration of larval development was dependent on temperature, relative humidity, nutritive value of the food, food particle diameter, availability of drinking water, and insect density. Adult longevity was dependent on temperature, relative humidity, nutritive quality of the food, food particle diameter, availability of drinking water, the number of dead bodies on the surface of the food, and food surface area. Rate of fecundity between 0 and 35°C was determined by age, temperature, relative humidity, nutritive quality of the food, the availability of drinking water, food particle diameter, water content of the food, and insect density. Mortality was dependent on age, temperature, relative humidity, nutritive quality of food, availability of drinking water, and insect density. The model was not validated.

I developed a model for simulating the population dynamics of the flat grain beetle, *Cryptolestes pusillus* (Schönherr) (Throne 1989). The effects of temperature on duration of development and mortality of each immature stage and adult longevity and fecundity were described using polynomial equations. Constant high humidity was assumed. Variation in development rate was described using a time-varying distributed delay (Manetsch 1976). The model was anecdotally validated and gave results similar to those observed in small outdoor bins. The model was used to determine the effects of noncatastrophic control technologies (such as resistant plant varieties or sublethal effects of pesticides) that alter life history parameters on subsequent growth of insect populations. Even small delays in larval development or reductions in fecundity or survival were shown to have a large effect on eventual population levels when applied over a storage season using actual bin temperatures. The combination of several noncatastrophic control technologies, none of which can alone keep the insect population below the economic injury levels, when used together may provide adequate control of insect pests. The combination of noncatastrophic control technologies with other methods of control may improve the cost-effectiveness of pest management techniques.

In a similar study, Baker et al. (1991) modified the model of Throne (1989) to simulate population dynamics of *S. oryzae*. The model was used to determine the effects that delays in larval development caused by feeding on varieties of wheat with varying levels of α-amylase inhibitor could have on long-term population growth of *S. oryzae*. Even the small, 0.7-d, delay induced by some varieties of the wheat reduced population growth by 20% in the simulations. The model was not validated.

2.4.4. *Evaluation of Insect Population-Dynamics Models*

All of the models developed for simulating the population dynamics of insect pests of stored grain have used a similar approach. Existing data are used to develop a single-species biological model that may be simple or complex, depending on the data available and the intended use of the model. Basic components of the biological models are duration of immature stages, survivorship of immature insects, and longevity and fecundity of adults. Ecological models generally try to simulate population development in a small quantity of grain and must therefore include factors such as the effects of density and cannibalism on the basic biological components of the model. Pest management models differ in that they generally assume Malthusian conditions (which may be reasonable in a relatively large commercial grain bulk that is managed to even a limited extent) and try to determine the effects of major abiotic factors on insect population dynamics.

The approach used for developing pest management models is reasonable in that it is the approach used for developing any model: to use the simplest model that will adequately describe the data. By far, the most important variable affecting insect population development is temperature. This is the only variable that is included in all of the pest management models for stored-grain pests. Relative humidity or moisture content generally will also have a significant effect on population development, although some biological processes are less sensitive than others. Single-species models are developed because they are the simplest models that might explain population dynamics of a species and because they are easier to validate than multispecies models. Such simple models may explain pest population growth in the field (e.g., Hagstrum and Throne 1989). Other species can be added to the models as need dictates to further explain population dynamics, particularly when several species invade a granary.

The main reason that more pest-management models are not available is that basic biological data are lacking for most pest species. Few scientists are willing to invest the time to conduct complete life table studies on a pest species at the full range of temperatures and humidities at which the pest occurs in grain. For example, long-term studies conducted for pests of corn in the southern United States must be done at temperature ranges of 10–40°C and moisture content ranges of 11–15%. Studies conducted over a narrower range of environmental conditions are of little value when trying to simulate insect population dynamics in actual storages.

Once basic biological data are available, their inclusion into pest management models is relatively simple. The model of Throne (1983) has been easily modified for simulating population dynamics of insect pests of stored grain (Hagstrum and Throne 1989, Throne 1989, Baker et al. 1991). The model is compartmentalized so that all components are separated into individual subroutines so that modification for other species is accomplished easily (i.e., an object-oriented model). Logan (1988) developed an expert system that will build a model from life history data. Thus, the limiting factor to building management models is lack of high-quality quantitative data.

3. INTEGRATION OF PEST-POPULATION DYNAMICS MODELS WITH PHYSICAL MODELS

Temperature and moisture content of grain are the two most important abiotic factors affecting growth of stored-grain insect populations. Any pest-management model must include at least these two factors to adequately describe insect population growth in actual storage conditions. Computer files of historical temperature and moisture content data may be used to run simulations. Availability of such historical data is limited, however, because it is tedious and costly to monitor temperature and moisture throughout a grain mass over a long time.

Historical climatic data do exist for many parts of the world. Therefore, if grain temperature and moisture content could be determined from climatic data, then historical data would be available for use in simulations. Mathematical methods for determining grain temperature and moisture content from climatic data are presented in Chapters 14 and 15. Therefore, only the integration of models of temperature and moisture content with models of insect population dynamics will be discussed here.

The process of integration is straightforward. The environmental model must provide temperatures and moisture contents in constant time intervals that coincide with the time step in the population dynamics model. If the environmental model provides data for different regions of a grain mass (e.g., surface, bin wall, etc.) then the population dynamics model will have to simulate population development separately for each region of the grain mass, which is straightforward. Movement of insects from one region to another is based on behavior related to temperature and moisture content preferences. Input of the environmental data to the population dynamics model is a simple computer programming task.

Few attempts have been made to integrate environmental models with population dynamics models. Longstaff et al. (1981) coupled the environmental model of Sutherland et al. (1971) with the insect population dynamics model of Cuff and Hardman (1980). The model was used to simulate temperature and moisture content change in a bin of aerated wheat in a warm, damp region of Australia. The same models were used together by Thorpe et al. (1982) and Longstaff and Cuff (1984).

4. INTEGRATION OF PEST-POPULATION DYNAMICS MODELS WITH CONTROL MODELS

4.1. Models for Protectant Insecticides

Although they did not develop a computer model, Desmarchelier and Bengston (1979) developed a mathematical model that has been used for determining the effects of temperature and moisture content on rates of degradation of protectant insecticides on stored grain. They provided parameters for the model for simulating degradation of bioresmethrin, bioresmethrin plus piperonyl butoxide, carbaryl, chlorpyrifos-methyl, *d*-phenothrin, *d*-phenothrin plus piperonyl butoxide, fenitrothion, malathion, methacrifos, pirimiphos-methyl, pyrethrum, and pyrethrum plus piperonyl butoxide. They determined that grain type did not affect degradation rate. In general, the equations were developed using data collected in field storages at about 30°C and 50% rh. The equations were anecdotally validated in about 2000 separate tests, some of which were published, and were accurate. However, publication of validation tests over a wide range of environmental conditions would be desirable. The method has been used in models developed by Longstaff (1988a), Flinn and Hagstrum (1990a), and Hagstrum and Flinn (1990).

The second part of developing a model for a protectant insecticide involves simulating the effects of the insecticide on the insect. This has not been done systematically. Presumably temperature, moisture content, insecticide concentration, length of exposure to the insecticide, and stage of the insect will all determine insect mortality. Complete data have not been collected for any stored-grain pest. Therefore, various measures have been used to bypass this lack of data.

Longstaff (1988a) assumed that protectant insecticides did not affect immature stages of *S. oryzae*, because these stages occur within the grain kernels. Adult mortality due to exposure to insecticides was modeled as a function of temperature, length of exposure to the pesticide on a physiological time-scale, and concentration of insecticide. Flinn and Hagstrum (1990a) and Hagstrum and Flinn (1990) predicted adult mortality as a function of length of exposure to the pesticide and concentration of pesticide ignoring environmental effects. The model of Longstaff (1988a) accurately simulated adult mortality due to insecticides under limited conditions. Many questions remain about the quantitative description of the use of insecticides in grain storages. When a protectant insecticide is used on grain, how are the different stages of the insect affected? How do temperature, moisture content, and length of exposure to the pesticide affect the relationship?

One might also question whether collecting such large volumes of data are worthwhile. An examination of current testing methods for protectant chemicals will prove the value of collecting quantitative data. When a new pesticide is being evaluated for controlling stored grain pests, the developer usually tests the chemical at various locations and on various grains for a number of years. If the chemical proves effective, it may eventually be tested in several locations and for several years in each region of each country in which it is intended to be used. What these multiple studies in multiple locations do is to determine how well the pesticide works for average conditions and for the common pests in the region. It has already been pointed out that mortality due to insecticides can be described by two processes: rate of pesticide degradation and effects of the pesticide on the insect. Therefore, if the effects of temperature and moisture content on pesticide degradation are determined for one grain, then those results are applicable to any grain in any country. Thus, based on historical temperature and moisture conditions, one can predict how much pesticide will remain in a grain storage at any point in time in a given region. If the effects of temperature, moisture content, concentration of pesticide, and length of exposure to the pesticide on mortality are known then one can predict from the amount of pesticide in the grain storage whether there is enough pesticide to kill the insects. Thus, wide climatic regions may be eliminated from field testing of the pesticide based on simulations. In regions where use of the pesticide appears promising based on simulations, limited field tests may be conducted to validate model predictions.

4.2. Models for Biological Control

The only model for simulating the interaction of natural enemies of stored-product insects is that for *N. whitei* causing disease in *T. confusum* (Onstad and Maddox 1990a,b). This model was developed using data for only one temperature, so application to actual storages is not practical at this time. Presumably, models for simulating population dynamics of natural enemies of stored-grain pests have not been developed because biological control methods generally are not used in grain storages. However, with the increased interest in making biological control a feasible, commercial control strategy in the United States (as evidenced by government funding of a major pilot test on biological control at several government laboratories), an increase in research on biological control technologies is expected.

Because the bionomics of natural enemies of stored-grain pests have not been well studied thus far, the opportunity exists to collect high-quality quantitative data for these natural enemies. The basic data initially required are the same as those for insect pests of stored grain: the effects of temperature and humidity over the full environmental range encountered in grain storages on longevity and rates of development, mortality, and fecundity. Sound quantitative data are also needed on the effects of abiotic factors on the interaction of natural enemies and pests. Studies conducted at one environmental condition alone will do little to enhance our knowledge of biological control of stored-grain pests. Once developed, models that simulate the population dynamics of natural enemies will be useful for optimizing release strategies for natural enemies and determining which natural enemies will be most effective in reducing population growth of a particular pest.

4.3. Models for Fumigation and Modified-Atmosphere Storages

Fumigation and modified-atmosphere storage are amenable to quantification. Sound quantitative data describing the effects of abiotic factors and concentration and exposure time of fumigants on the mortality of various stages of pests (and natural enemies), however, are lacking. In absence of data, Longstaff (1988a) assumed that a fumigation would kill a certain percentage of each life stage of *S. oryzae* regardless of abiotic conditions. The same method, with minor variation, has been used by Flinn and Hagstrum (1990a) and Hagstrum and Flinn (1990). Researchers working with fumigants to control stored-grain pests should not assume that these "educated guesses" as to how fumigants affect pests relieve them of their responsibility to collect high-quality, quantitative data on the effects of fumigants on pests.

5. USEFULNESS OF PEST POPULATION DYNAMICS MODELS

Thus far, models have not been developed that describe the interrelations of pest species with each other and their grain storage environment. However, the single-species models that have been developed have proven useful for optimizing control strategies in stored grain. The *S. oryzae* models developed in Australia have indicated optimum aeration strategies for control of this pest (which happen to coincide with common aeration practices), optimum use of insecticides in conjunction with aeration for control of *S. oryzae*, and the best use of insecticides to minimize development of resistance to insecticides in *S. oryzae*. The models developed for insect pests of wheat in the United States have again indicated optimum control strategies and have been incorporated into an expert system that has potential for commercial use (see Chapter 2). Perhaps the most important use of the models has been to give structure to research on stored-grain pests and provide a tool for integrating data from diverse disciplines in grain storage. Computer models are not currently being used directly by extension personnel or grain managers to help manage pests. An expert system that is based on rules developed through simulations is being tested by extension personnel and grain managers in the United States (P. W. Flinn, personal communication).

One of the anticipated problems in the commercial use of computer models is acceptance by grain managers. We already have several good methods for controlling insects in stored grain. Unfortunately, growers generally ignore these recommended practices or do a poor job of following them. These methods are sanitation in and around storage structures, and proper drying and use of aeration. If grain is stored at low moisture contents or low temperatures, which is feasible at reasonable cost in many parts of the world, insect population growth could be stopped or retarded. Growers in many countries have little incentive for providing high-quality grain because of low grain prices, hence, they have little incentive for using recommended control practices. Until we can demonstrate to growers the economic benefits of proper storage using relatively inexpensive grain-storage technology, they will continue to ignore storage recommendations.

We also have few reliable data on the costs of insect damage to stored grain. Hence, it is difficult to demonstrate to growers the benefits of good storage practices, thereby reducing the possibility of optimizing control strategies in stored grain. Research is required to quantify damage done by pests to grain (both quantitative and qualitative damage) including the effects of abiotic factors on damage, and the resulting information needs to be incorporated into computer models for pest population dynamics. From these data, economic injury levels can be determined and used to optimize control strategies (e.g., Onstad 1987). The costs of various control strategies and the resulting benefits of management tactics have been determined in a few studies (e.g., Reed et al. 1990), but these are not the type of data that can be used to develop management models and develop dynamic economic injury levels because we cannot relate the resulting costs to known insect infestations.

Computer models will be useful even in countries that have zero tolerance for insects in grain. Some type of control methods will be necessary to avoid insect population growth in grain and computer models will provide the quantitative basis for optimizing these control strategies.

6. FUTURE RESEARCH

One of the benefits in reviewing the literature on a topic is that the deficiencies in our knowledge of that topic are emphasized. Most of the deficiencies in our knowledge of computer modeling of the population dynamics of stored-grain pests have already been pointed out in earlier sections of this chapter but will be reemphasized here as high-priority areas for future research. In addition, new areas for research to further the development of computer models for simulating the population dynamics of stored-grain pests will be pointed out.

1. The biological database for stored-grain pests is incomplete. Further studies are required to determine the effects of abiotic factors on the basic biological processes of the major pests of stored grain. These data must be collected over the complete range of environmental conditions at which grain is stored, and at which the pest can survive and multiply. These data can then be incorporated into computer models for simulating pest population dynamics and optimizing control strategies.

2. We know little about the biology of stored-grain pests outside of storages. We need to identify and quantify where and when these pests breed outside of storages, how and when they move into storages, and the spatial and temporal variation in these processes.

3. The effects of storage pests on their environment must be quantified over the entire range of environmental conditions at which grain is stored and at which the pest can survive and multiply. This includes the effects of the pests on grain quantity, quality, moisture content, temperature, and interstitial gas composition.

4. The database for the effects of abiotic factors on the control of stored-grain pests is scarce. It is no longer justifiable to conduct control studies under some average (often unstated) conditions for a geographic region. Studies must be conducted under controlled environmental conditions and the effects on pest mortality of abiotic factors and concentration and time of exposure to the control factor must be quantified. These data can then be incorporated into computer models and used to make knowledgeable decisions about control strategies.

5. As models are developed, every effort should be made to validate them using field data. If modification of the model is indicated, appropriate data must be collected and incorporated into the model. The model must then be revalidated.

6. The economic costs and benefits of grain storage and pest damage must be determined so that both governmental and storage decisions can be made knowledgeably.

7. The grain marketing system should be characterized so that we know what abiotic and biotic factors influence grain storage, starting on the farm and continuing through the processing plant.

8. Storage structures should be developed in such a way that environmental monitoring equipment is built in. Data from such equipment should be directly transferable to computers to drive computer models to aid in making pest management decisions.

9. In general, single-species models will not be appropriate for defining storage conditions and optimal control strategies. Instead results from simulations for all of the major pests in a grain storage system will need to be combined to optimize storage conditions. Expert systems, such as that developed by Flinn and Hagstrum (1990b), are a tool that can be used to combine the results of simulation studies with heuristic knowledge and make the information available to commercial users. Longstaff (1990) discussed the features that are desirable in an expert system for grain storage.

Although developing the simplest model is the way to start a modeling program, one must eventually include all the abiotic and biotic factors that have a major influence on the system, and the interrelationships among these factors, to be able to successfully model and optimize management of the system. This holistic approach will require cooperation among researchers in all fields of grain storage: cereal chemists, economists, engineers, entomologists, microbiologists, mycologists, and systems analysts.

REFERENCES

Andrewartha, H. G. and L. C. Birch. 1954. The distribution and abundance of animals. The University of Chicago Press, Chicago. 782 pp.

Baker, J. E., S. M. Woo, J. E. Throne, and P. L. Finney. 1991. Correlation of α-amylase inhibitor content in Eastern soft wheats with development parameters of the rice weevil (Coleoptera: Curculionidae). Environ. Entomol. 20:53-60.

Bellows, T. S., Jr. 1982. Simulation models for laboratory populations of *Callosobruchus chinensis* and *C. maculatus*. J. Anim. Ecol. 51:597-623.

Berreen, J. M. 1974. The development and validation of a simple model for population growth in the grain mite, *Acarus siro* L. J. Stored Prod. Res. 10:147-154.

Birch, L. C. 1945. The influence of temperature on the development of the different stages of *Calandra oryzae* L. and *Rhizopertha dominica* Fab. (Coleoptera). Aust. J. Exp. Biol. Med. Sci. 23:29-35.

Campbell, A. and R. N. Sinha. 1990. Analysis and simulation modeling of population dynamics and bioenergetics of *Cryptolestes ferrugineus* (Coleoptera: Cucujidae) in stored wheat. Res. Popul. Ecol. 32:235-254.

Cuff, W. R. and J. M. Hardman. 1980. A development of the Leslie matrix formulation for restructuring and extending an ecosystem model: the infestation of stored wheat by *Sitophilus oryzae*. Ecol. Modelling 9:281-305.

Desmarchelier, J. M. and M. Bengston. 1979. The residual behaviour of chemicals on stored grain. pp. 138-151 in: Proc. Second Int. Working Conf. Stored-Product Entomol. Ibadan, Nigeria.

Flinn, P. W. and D. W. Hagstrum. 1990a. Simulations comparing the effectiveness of various stored-grain management practices used to control *Rhyzopertha dominica* (Coleoptera: Bostrichidae). Environ. Entomol. 19:725-729.

Flinn, P. W. and D. W. Hagstrum. 1990b. Stored grain advisor: A knowledge-based system for management of insect pests of stored grain. AI Appl. Nat. Res. Management 4:44-52.

Gordon, D. M., W. S. C. Gurney, R. M. Nisbet, and R. K. Stewart. 1988. A model of *Cadra cautella* larval growth and development. J. Anim. Ecol. 57:645-658.

Hagstrum, D. W. and P. W. Flinn. 1990. Simulations comparing insect species differences in response to wheat storage conditions and management practices. J. Econ. Entomol. 83:2469-2475.

Hagstrum, D. W. and W. G. Heid, Jr. 1988. U. S. wheat-marketing system: an insect ecosystem. Bull. Entomol. Soc. Am. 34: 33-36.

Hagstrum, D. W. and J. E. Throne. 1989. Predictability of stored-wheat insect population trends from life history traits. Environ. Entomol. 18:660-664.

Hardman, J. M. 1976. Deterministic and stochastic models simulating the growth of insect populations over a range of temperatures under Malthusian conditions. Can. Entomol. 108:907-924.

Hardman, J. M. 1978. A logistic model simulating environmental changes associated with the growth of populations of rice weevils, *Sitophilus oryzae*, reared in small cells of wheat. J. Appl. Ecol. 15:65-87.

Jones, A. E., W. S. C. Gurney, R. M. Nisbet, and D. M. Gordon. 1990. Food degradation as a mechanism of intraspecific competition among the larvae of secondary stored-product pests. Funct. Ecol. 4:629-638.

Kawamoto, H., R. N. Sinha, W. E. Muir, and S. M. Woods. 1991. Simulation model of *Acarus siro* (Acari: Acaridae) in stored wheat. Environ. Entomol. 20:1381-1386.

Kawamoto, H., S. M. Woods, R. N. Sinha, and W. E. Muir. 1989. A simulation model of population dynamics of the rusty grain beetle, *Cryptolestes ferrugineus* in stored wheat. Ecol. Modelling 48:137-157.

Landahl, H. D. 1955a. A mathematical model for the temporal pattern of a population structure, with particular reference to the flour beetle. Bull. Math. Biophysics 17: 63-77.

Landahl, H. D. 1955b. A mathematical model for the temporal pattern of a population structure, with particular reference to the flour beetle. II. Competition between species. Bull. Math. Biophysics 17: 131-140.

Leslie, P. H. 1945. On the use of matrices in certain population mathematics. Biometrika 33:183-212.

Lloyd, M. 1968. Self regulation of adult numbers by cannibalism in two laboratory strains of flour beetles (*Tribolium castaneum*). Ecology 49: 245-259.

Logan, J. A. 1988. Toward an expert system for development of pest simulation models. Environ. Entomol. 17:359-376.

Longstaff, B. C. 1981. The manipulation of the population growth of a pest species: an analytical approach. J. Appl. Ecol. 18:727-736.

Longstaff, B. C. 1984. An extension of the Leslie matrix model to include a variable immature period. Aust. J. Ecol. 9:289-293.

Longstaff, B. C. 1988a. A modelling study of the effects of temperature manipulation upon the control of *Sitophilus oryzae* (Coleoptera: Curculionidae) by insecticide. J. Appl. Ecol. 25:163-175.

Longstaff, B. C. 1988b. Temperature manipulation and the management of insecticide resistance in stored grain pests: a simulation study for the rice weevil, *Sitophilus oryzae*. Ecol. Modelling 43:303-313.

Longstaff, B. C. 1990. The role of modelling in the management of stored-product pests. pp. 1995-2007 in Proc. Fifth Int. Working Conf. Stored-Product Prot. Vol. 3, F. Fleurat-Lessard and P. Ducom (eds.). Bordeaux, France.

Longstaff, B. C. and W. R. Cuff. 1984. An ecosystem model of the infestation of stored wheat by *Sitophilus oryzae*: a reappraisal. Ecol. Modelling 25:97-119.

Longstaff, B. C., W. R. Cuff, and G. R. Thorpe. 1981. The control of *Sitophilus oryzae* (L.) infestations: A computer-based management tool. pp. 6.5-6.12 in P. Williams and T. G. Amos (eds.). Proc. First Aust. Stored Grain Pest Control Conf., Melbourne, Australia.

Manetsch, T. J. 1976. Time-varying distributed delays and their use in aggregative models of large systems. IEEE Trans. Systems Man Cybern. SMC-6:547-553.

Neyman, J., T. Park, and E. L. Scott. 1956. Struggle for existence. The *Tribolium* model: biological and statistical aspects. pp. 41-79 in Proc. Third Berkeley Sympo. Math. Stat. Prob., Berkeley, CA.

Niven, B. S. 1967. The stochastic simulation of *Tribolium* populations. Physiol. Zool. 40:67-82.

Nuttall, R. M. 1989. Simulated population dynamics of a stored-products' pest (*Ptinus tectus*, Coleoptera). Ecol. Modelling 48:291-313.

Onstad, D. W. 1987. Calculation of economic-injury levels and economic thresholds for pest management. J. Econ. Entomol. 80: 297-303.

Onstad, D. W. and J. V. Maddox. 1990a. Epizootiological model of *Nosema whitei* and *Tribolium confusum*. pp. 369-372 in: Proc. Fifth Int. Coll. Invertebrate Pathol. Microbial Control, Adelaide, Australia.

Onstad, D. W. and J. V. Maddox. 1990b. Simulation model of *Tribolium confusum* and its pathogen, *Nosema whitei*. Ecol. Modelling 51:143-160.

Pearl, R. and L. J. Reed. 1920. On the rate of growth of the population of the United States since 1790 and its mathematical representation. Proc. Nat. Acad. Sci. 6: 275-288.

Reed, C., K. Anderson, J. Brockschmidt, V. Wright, and J. Pedersen. 1990. Cost and effectiveness of chemical insect control measures in farm-stored Kansas wheat. J. Kansas Entomol. Soc. 63: 351-360.

Ruesink, W. G. 1976. Status of the systems approach to pest management. Annu. Rev. Entomol. 21:27-44.

Sharpe, P. J. H. and D. W. DeMichele. 1977. Reaction kinetics of poikilotherm development. J. Theor. Biol. 64:649-670.

Sinclair, E. R. and J. Alder. 1985. Development of a computer simulation model of stored product insect populations on grain farms. Agric. Syst. 18:95-113.

Sokoloff, A. 1974. The biology of *Tribolium* with special emphasis on genetic aspects, Volume 2. Clarendon Press, Oxford. 610 pp.

Solomon, M. E. 1969. Establishment, growth and decline of populations of the grain mite *Acarus siro* L. on a handful of wheat. pp. 255-260 in Proc. Second Int. Cong. Acarol. G. O. Evans (ed.)., Sutton Bonington, UK.

Sutherland, J. W., P. J. Banks, and H. J. Griffiths. 1971. Equilibrium heat and moisture transfer in air flow through grain. J. Agric. Eng. Res. 16:368-386.

Thorpe, G. R., W. R. Cuff, and B. C. Longstaff. 1982. Control of *Sitophilus oryzae* infestation of stored wheat: an ecosystem model of the use of aeration. Ecol. Modelling 15:331-351.

Throne, J. E. 1983. A computer model for simulation of seedcorn maggot development. Unpublished Ph.D. Thesis, Cornell University, Ithaca, NY. 126 pp.

Throne, J. E. 1989. Effects of noncatastrophic control technologies that alter life history parameters on insect population growth: a simulation study. Environ. Entomol. 18:1050-1055.

Wagner, T. L., H. I. Wu, P. J. H. Sharpe, R. M. Schoolfield, and R. N. Coulson. 1984. Modeling insect development rates: a literature review and application of a biophysical model. Ann. Entomol. Soc. Am. 77:208-225.

7

The Mycology of Stored Grain: An Ecological Perspective

Donald T. Wicklow

Agricultural Research Service, United States Department of Agriculture, Peoria, Illinois

SUMMARY

As components of the seed microflora, fungi may protect seeds from predators and pathogens, compete with the developing seed for photosynthate, interfere with seed abcission and dispersal, promote or inhibit seed germination, reduce seedling vigor, and cause seed death and decay. The data base for grain-storage mycology becomes enriched when fungal assemblages from seeds are examined for ecological interactions that impact seed survival in nature and which may also have relevance to seed deterioration in storage. We have little understanding of the life history and ecology of storage fungi in crop production fields or their ecological role in native plant communities. Investigations of the chemical ecology of fungus-insect interactions in cereals have led to the discovery of new fungal secondary metabolites that affect insects, antibiotics that cure beetles infected with *Ricketsia*-like bacteria, beetle metabolism of fungal toxins, and yeast-symbiont mediated detoxification. Evidence is offered that a desert rodent granivore (banner-tailed kangaroo rat) manages its underground seed caches to prevent seed losses to fungi and insects. This also invites opportunities for technology transfer of ecological information and potentially useful gene products. Selective elimination of chemical or structural defenses has left the edible seeds of domesticated crop plants more vulnerable to seed deterioration. A summary of the ecologically specialized groups of fungal pathogens reported to rot corn kernels (maize grain) is provided as background for a discussion of the ecology and pathology of *Aspergillus flavus* in corn production and storage. The latter is presented as a model for examining the dynamics of other fungal colonists of stored grain. The seed coat (testa) is identified as the most important barrier preventing fungi from infecting the grain proper (e.g., embryo and endosperm). A hypothesis is offered that major aflatoxin outbreaks occur when corn-belt hybrids are exposed to high evening temperature stress during the interval of grain filling resulting in damage to the testa. There remains a need for critical histopathological studies of fungal infection of grain as it matures in the field or ages physiologically in storage. In the past century the science of grain storage mycology has been shaped by four conceptual advances, each of which is examined with a look to the future.

1. FUNGI AND SEED ECOLOGY

1.1. Seed Dispersal and Survival

Seeds provide plants with an important means of reproduction, dispersal and avoidance of environmental stress (Mayer and Poljakoff-Mayber 1963, Fenner 1985). Seeds are of much higher nutritive value than most other plant material and thus represent a rich food supply for any seed-eating animal or fungal colonist. In nature, the establishment of seedlings may be reduced by vertebrates, insects, and fungal pathogens through seed or seedling mortality before or after seed dispersal. Seeds are the food which animals cache most commonly for the longest times, presumably because seed adaptations to dormancy make them ideal for surviving storage and their concentrated energy reserves make them efficient to carry (Smith and Reichman 1984). Animals have had an active role in the dispersal of the grass seeds they consume. In many small-seeded grasses, the plant's foliage may perform the ecological role of the disperser attractant (Janzen 1984), the seeds being eaten and then voided with the feces of animals at some distance from the parent plant. Dung beetles may then bury seeds in brood or food balls, the feces acting as a fertilizer in initial stages of seedling establishment (Wicklow et al. 1984). Seed dispersal is also accomplished when rodent granivores (e.g., kangaroo rats) that scatter hoard seeds fail to recover a significant portion of these caches (Janzen 1971). One reason that desert seeds are successfully stored by rodents is that so many of them are innately dormant, requiring specific germination cues even in conditions promoting seedling growth (Smith and Reichman 1984).

In nature, a dormant seed bank acts as a long-term insurance against local extinction. The ability to remain dormant for a long period is associated with seeds of species that can survive in arid regions where rainfall is variable or of species that colonize unpredictably disturbed patches of ground. There is no special physiological mechanism preventing seed germination in cereal crops. Dormancy occurs when the seed is deprived of its requirements for germination (i.e., adequate moisture, oxygen, and a suitable temperature). Toole and Brown (1946) found that nearly all crop plants were shown to lack long-term viability during soil burial, presumably because they have been selected for fast synchronous germination. Halloin (1983) has proposed that resistance to seed deterioration may be improved dramatically if greater focus were placed on physiological mechanisms such as the role of flavanols and abscisic acid in preventing premature germination or sprouting damage of grain.

1.2. Seedborne Microflora

Agarwal and Sinclair (1987) recognize "seedborne microflora" as the association of bacteria, fungi, and other microflora with seeds which may or may not have the potential of causing diseases of the seed or plant. As components of this microflora, fungi may protect seeds from predators and pathogens, compete with the developing seed for photosynthate, interfere with seed abcission and dispersal, impact seed germination, reduce seedling vigor, or cause seed death and decay (Baker 1972, Neergaard 1977). Surprisingly little is known about how fungal pathogens may influence seedling success in nature. Only a few such studies have been conducted and these report damping-off mortality to seeds and seedlings attacked by soil-inhabiting pathogenic fungi (e.g., Augspurger 1984). While there has been a great deal of research on seed or seedling losses from seedborne pathogens of crops (Neergaard 1977, Agarwal and Sinclair 1987), I am unaware of similar studies involving seedborne pathogens of native plant species in their natural habitat.

1.3. Seed Germination

Fungi may function to overcome seed dormancy. There is both direct and circumstantial evidence to suggest that seed germination in many grass species may be improved by fungal degradation of the covering palea and lemma (Probert 1981). A fungus identified as a member of the *Penicillium funiculosum* Series (Raper and Thom 1949) was a prevalent colonist of *Oryzopsis miliacea* (L.) Asch. et Schw. seeds that germinated in surprisingly high numbers in laboratory tests. Probert (1981) showed that seed-surface sterilization to remove fungal or microbial colonists significantly reduced germination while germination was significantly increased by inoculation of sterilized seed with the *Penicillium* strain. Fungi might also promote germination by degradation of water-soluble germination inhibitors associated with the outer structures of grass seed (Evenari 1949). Seed germination is stimulated by the production of gibberellin and cytokinin (Kramer and Kozlowski 1979). Sorghum heads infected with *Fusarium moniliforme* Sheldon had a higher incidence of sprouting damage than uninfected heads (Castor and Frederiksen 1980). Does *F. moniliforme* provide a source of gibberellin impacting seed germination in cereals?

1.4. Resistance of Wild versus Cultivated Seeds

Seeds may contain chemicals that are toxic to fungal pathogens and seed-eating insects (Rosenthal and Janzen 1979). In the development of crops as food plants, the level of secondary compounds has been reduced where these are toxic or distasteful to man (van Emden 1990), likely resulting in higher losses of seed to predators and pathogens (Harris and Burns 1973). Halloin (1983) observed that seeds of weeds, wild relatives of cultivated species, and even crops cultivated primarily for portions other than their seeds, tend to have superior resistances to seed deterioration compared with those exhibited by the cultivated, edible seeds. For example, Christensen (1973) has noted that many species of hard-coated weed seeds screened from commercial samples of grain were free from storage fungi but not the seeds of weed grasses which proved as susceptible to fungal infection as cereal grains. Halloin (1983) argues that edible seeds have become more vulnerable to seed deteriorating fungi because "undesirable" seed characteristics (e,g., an impermeable or thick seed coat, the presence of tannins or flavanols, protease inhibitors, lectins, cyclopropene fatty acids, phytic acid, and mechanisms imparting long-term dormancy), have been targets of breeding programs to remove these undesirable properties. He suggests that in regions of the world where food is scarce and storage losses of edible seeds are a serious problem one might reconsider these trade-offs and reincorporate certain of these so-called "undesirable" properties into seeds. Some of the options and problems facing crop breeders who wish to exploit plant secondary metabolites in chemical defense against herbivores and plant pathogens are examined by Wink (1988).

1.5. Natural Biological Control Agents

Do some seed-borne fungi function as natural biological control agents? As examples, *Gonatobotrys simplex* Cda. is included among the more frequently encountered fungal colonists of barley at harvest (Wallace and Sinha 1981). This fungus is a contact mycoparasite of seed-borne fungi such as *Alternaria* (Whaley and Barnett 1963) and Cladosporium (Wallace and Sinha 1981). Phylloplane yeasts reduced corn anthracnose infection by *Colletotrichum graminicola*, a fungus that also infects corn seeds (Nankam and Foley 1988) and presumably competes for nutrients (exploitation competition). Microorganisms antagonistic to *Pyrenophora avenae* Ito and Kuribayashi in oats can inhibit seed transmission of this fungal pathogen (Old 1968). The resistance of an oat cultivar from Brazil to *Drechslera victoriae* (Meehan and Murphy) Subram. and Jain was linked to antagonistic *Chaetomium* spp. on the seed surface (Tveit and Moore 1954). The authors suggest that the lack of any important Victoria blight in Brazil might be attributed, in part, to the natural presence of antagonistic strains of *Chaetomium* on the oat seed or in the soil. Control of disease was obtained when *Chaetomium* inoculum was applied as a seed dressing (dusted) (Tveit and Wood

1955). *Epicoccum nigrum* Link ex Wallr. was antagonistic to other fungal colonists of grain in pairings on malt and wheat extract agar (Magan and Lacey 1984) and was shown to actively antagonize soil-borne plant pathogens (Brown et al. 1987).

Wright (1956a) established that antibiotic production will take place in seeds buried in otherwise unaltered soil. The quantities of toxins were large enough for extraction and characterization. In one example, a fungus producing such toxins was capable of preventing soilborne fungal pathogens from attacking treated seeds (Wright 1956b). *Penicillium breviocompactum* Dierckx produces the antifungal antibiotic mycophenolic acid, the first recorded antibiotic (Dekker 1969), and is the most commonly reported fungicolous *Penicillium* (Hawksworth 1981). *Penicillium brevicompactum* grew slowly yet showed the highest dominance ranking of all other "field fungi" and "storage fungi" tested even though these competing fungi had faster rates of growth (Magan and Lacey 1984). A similar phenomenon was recognized by Wicklow and Hirschfield (1979) who described a competitive hierarchy among fungal species occurring on dung in which the slow-growing *Poronia punctata* (L. ex. Fr.) Fr., produces an antifungal metabolite, later identified as (+)-isoepoxydon (Gloer and Truckenbrod 1988), that inhibits faster-growing colonists of dung.

Information about the interactions of fungal colonists of seeds can be exploited in guiding the formulation of seed dressings requiring a mixed-inoculum. For example, *C. globosum* Kunze applied to corn seeds controlled seedling blight due to *F. moniliforme* and *F. graminearum* Schwabe (Chang and Kommedahl 1968). Baker and Cook (1974) advocated that antagonists used in seed coatings be complimentary, not competitive; that is, they should grow best at different temperatures, water potentials, or pH, or on different nutrients so that some will be operative under most environmental situations. Seed pathologists should be encouraged to carefully investigate the microbial and fungal assemblages on the seeds of native grasses in major grain producing regions.

1.6. Protective Mutualism

Do some weakly pathogenic fungal colonists of seeds function as mutualists by defending plants against more virulent pathogens (sensu McIntyre and Miller 1978, Fokkema 1981, Clay 1990a, Carroll 1992)? The phenomenon of induced plant resistance, where infection by one organism confers resistance to subsequent inoculations of other pathogens, is well known (Dean and Kuc 1987) yet I am unaware of equivalent examples based on seed resistance. Many grass species appear to be ubiquitously infected by asymptomatic *Acremonium* endophytes which are transmitted maternally through the seed of infected plants (Clay 1990a). There is a general lack of cross-compatibility of seedborne endophytes with alien host species. Endophyte-infected grasses are more resistant to fungal pathogens (Stovall and Clay 1991) and to nematodes (West et al. 1988, Pederson et al. 1988). Two studies have found a negative correlation between endophyte and pathogen infection in field situations. Panicum agrostoides Spreng. infected by *Balansia henningsiana* (Moell.) Diehl. had significantly fewer lesions of the leaf spot fungus *Alternaria triticina* Pras. and Prab. than did neighboring endophyte-free plants in a natural population (Clay et al. 1989). Endophyte-infected plots of tall fescue showed reduced levels of crown rust, *Puccinia coronata* Cda. (Ford and Kirkpatrick 1989). The mechanism of resistance may be related to the production in vivo of fungitoxic compounds by the endophytes (White and Cole 1985, Siegel and Latch 1991). Christensen and Latch (1991) and Christensen et al. (1991) have demonstrated in vitro antibiosis against *C. graminicola* (Cesati) Wilson and *D. erythrospila* (Drechsler) Shoemaker for 90% of the *Acremonium* isolates they tested.

Infection by one pathogen may increase susceptibility to a second. For example, in natural stands of *Allium ursinum* L. infected with *Puccinia sessilis* Schneid. substantial mortality followed infection of rust pustules by the necrotroph *Botrytis cinerea* Pers. ex Pers. (Hallett and Paul 1989). Similarly, a combination of fungal antibiosis and mycoparasitism enables certain wood decay fungi to replace the mycelial network of an earlier colonist (Wicklow 1981, Rayner and Boddy 1988). Can hyphae of osmotolerant aspergilli trace the path of endophyte hyphae and enter the seed proper? Nicot (1963) frequently isolated A. glaucus sensu lato from decaying Agaricales but it is not known if this fungus is mycoparasitic or shows any mycoparasitic behavior in grain storage environments.

Patterns of fungal association in individual seeds can point to relevant species interactions (Wicklow 1988b). It would be interesting to compare fungal assemblages on seeds produced by endophyte-infected versus noninfected grasses. Studies of this kind could provide clues as to which fungal pathogens of cereal grains are most negatively impacted by fungal endophytes. Are endophyte-infected seed embryos protected from further attack by seed-infecting fungi?

Endophyte-infected grasses also exhibit enhanced resistance to insect pests (Clay 1990b). Insect feeding trials and food plant preference studies conducted in the laboratory with a variety of grass species have shown that endophyte infection acts as a feeding and oviposition deterrent and reduces the survival, growth, and developmental rates of feeding insects (Clay 1990b). The basis for the observed antiherbivore effects of endophyte-infected grasses appears to be alkaloids produced by fungi (Bacon et al. 1986, Siegel et al. 1987). As a family, grasses are relatively poor in endogenous alkaloids (Culvenor 1973) and Clay (1990b) believes that fungal alkaloids fill an empty defensive niche in the grass family. *Panicum agrostoides* Spreng. is infected by endophytic *Fusarium* species (Leuchtmann and Clay 1988) but neither the fungal isolates nor the grass tissues were investigated chemically for fungal toxins. Laboratory feeding studies with *P. agrostoides* indicated that it was a poor food source for a grass-feeding caterpillar (Lepidoptera). The grass suffered virtually no consumption by insects over an entire growing season (Clay et al. 1989). In tall fescue, the greatest concentration of loline alkaloids is in the seeds, suggesting a role in preventing seed predation (Cheplick and Clay 1988). To determine if the presence of endophytes in grass seeds discouraged predation, seeds of perennial ryegrass and tall fescue infected by *Acremonium* endophytes were pulverized and supplied to flour beetles *Tribolium castaneum* (Herbst). Beetles in vials with endophyte-infected seed had significantly lower initial survival and population growth rate than those supplied with endophyte-free seed (Cheplick and Clay 1988). Clay (1990b) suggests that our entire system of agriculture, based primarily on grasses, may have originated through the accidental elimination of endophytes from grasses that, undefended, became especially nutritious or palatable.

In this section I have attempted to illustrate why the fungi recorded from grain at harvest should not be treated simply as a species list, but as fungal assemblages with a long history of interactions. When grain is harvested and stored, fungal interactions that impact seed survival in nature may have little relevance to safe grain storage or grain deterioration. At the same time, mycologists need to recognize the potential significance of such interactions when determining both fungal ability to infect seeds and fungal impact on grain quality.

2. ORIGINS OF THE STORED GRAIN BIOTA

2.1. Storage Fungi and Arthropods

Interactions among seeds, arthropods, and fungi are not unique to agricultural grain storage environments, nor is the history of those interactions limited to the advent of human grain caching around 7000 yr ago. Stored-grain insects and mites are believed to have evolved as inhabitants of bark; decomposing plants; food caches of other insects especially bees, wasps, and ants; and in the nests and burrows of rodents (Linsley 1944). Many of these arthropod pests are fungus feeders that selectively consume fungal hyphae and spores (Sinha and Watters 1985). Dunkel (1992) suggests that most "storage" insects also occur in natural reservoirs in native ecosystems. She has traced the recent migration (< 80 yr) of a "new" storage insect pest, the larger black flour beetle, *Cynaeus angustus* (LeConte), which has moved from a desert ecosystem to midwestern grain stores (Dunkel et al. 1982).

The so-called "storage fungi" of Christensen and Kaufmann (1965) consist of a few dozen fungal species having conidial structures classified in genera such as *Aspergillus, Penicillium, Paecilomyces*, and *Scopulariopsis* (Samson 1974, Kozakiewicz 1989, Frisvad and Filtenborg 1989, King et al. 1986, Singh et al. 1991, Christensen and Sauer 1982). These fungi are also routinely isolated from a variety of natural substrates and soil microhabitats not associated with animal homes (Domsch et al. 1980, Raper and Thom 1949, Pitt 1979, Raper and Fennell 1965). We have little understanding of the life history and ecology of "storage fungi" in agricultural fields and almost no information about their ecological role in native plant communities. Several of the aspergilli that attack seeds are also frequently recorded from diseased or dead insects (Raper and Fennell 1965, Domsch et al. 1980, Jong and Gantt 1987, Kurtzman et al. 1987), and this may have some bearing on the behaviors of these fungi as seed colonists.

2.2. Pattern of Resource Exploitation

The "*Penicillium* growth pattern" is one in which a fungus densely colonizes small individual substrates with no extension of the mycelium into the surrounding soil and then produces conidia heavily over the surface (Sewell 1959). This pattern is characteristic of the aspergilli and penicillia that invade seeds or other discrete, nutrient-rich substrates, enabling these fungi to exploit the entire resource. Pitt (1979) suggested that the terverticillate penicillia represent an advanced group adapted to growth in cereals, food products, and other substrates of reduced water activity, and to producing maximal numbers of conidia per conidiophore. The ability of *Aspergillus* and *Penicillium* species to be successful seed colonists may also depend on their production of specific kinds of toxic secondary metabolites. Janzen (1977) has argued that fungi are often under strong selection to render seeds as objectionable or unusable to larger seed eating animals in the shortest possible time, as a means of protecting themselves and the substrate they colonize from predation.

2.3. Adaptation to Storage Environments

Storage fungi have the ability to grow on plant materials whose moisture contents are in equilibrium with relative humidities of 70–90% where no free water is present. Some external inoculum of storage fungi is almost always present on freshly harvested seeds of wheat and barley (Christensen 1951, Tuite and Christensen 1955) as "light superficial infections" (Christensen 1957). Hyphae of storage fungi are confined to the pericarp layers in corn that shows no evidence of internal fungal invasion (Russell et al. 1982). Extensive invasion of the grain may occur following storage. During the past 50 years there have been hundreds of studies published that examine the role of different abiotic environmental parameters (e.g., available moisture, temperature, carbon dioxide (CO_2) and oxygen (O_2) concentrations) in influencing fungal colonization and deterioration losses of stored grain (e.g., Koehler 1938, Snow et al. 1944, Semeniuk 1954, Christensen and Kaufmann 1974, Lacey 1989, Sinha 1992). Measures of fungal physiological responses include spore germination, hyphal growth rate, colonizing ability in the presence of competitors, sporulation, biomass accumulation (e.g., ergosterol), and the production of secondary metabolites (e.g., mycotoxins, volatiles). These physiological attributes can be used to define the postharvest niche of individual storage fungi (Wallace and Sinha 1981, Lacey 1986) and therefore are useful taxonomic characters (Wicklow 1985, Frisvad 1986).

2.4. Fungal Deterioration of Grain

Under conditions of commercial grain storage, fungi are the primary cause of loss of germinability and germ damage (Christensen and Kaufmann 1969). Measures of fungal deterioration of seeds have included retention or loss of seed viability, dry matter loss, increased fat acidity, grain heating, and sprouting. Corn free of storage fungi showed high rates of germination (> 97%) after several weeks of storage at 17–19.9% moisture content (mc) and 25–32°C, whereas only 6% of the seeds inoculated with storage fungi germinated (Moreno et al. 1965, Lopez and Christensen 1967). Storage molds such as *Eurotium* spp. (syn. *A. glaucus*), *A. restrictus* G. Smith, *A. candidus* Link, *A. flavus* Link, *A. ochraceus* Wilhelm (syn. *A. alutaceus* Berk. and Curt.), *A. versicolor* (Vuill.) Tiraboschi and *P. viridicatum* Westling (syn. *P. verrucosum* Dierckx) all invade the germ or embryo (Christensen and Kaufmann 1974, Tsuruta et al. 1981). Corn kernels artificially inoculated with *A. candidus* and stored at 18% mc and 25°C develop a dark brown discoloration at the germ ends after 4.5 mo, whereas noninoculated seeds appear normal without loss in germination (Qasem and Christensen 1958). Corn kernels invaded by *A. flavus* or other storage molds prior to storage deteriorated more rapidly under conditions favorable to the fungi than kernels free of storage fungi (Qasem and Christensen 1960). Storage molds can slowly or rapidly kill the embryos of the seeds they invade and therefore should be recognized as facultative parasites (Christensen 1973). The observation that different isolates of *A. candidus* killed wheat seeds at different rates suggested to Christensen (1957) that pathogenic races may occur among these common storage fungi. An interesting study by Griffin (1966) showed that storage molds (aspergilli and penicillia) caused germination losses of wheat seeds buried in nonsterilized soil brought to approximate equilibrium with atmospheres whose relative humidity (rh) is similar to those permitting fungal invasion of the embryos of stored cereal grains. There have been no critical histopathological studies of fungal infection of stored grain as it ages physiologically. When does the testa or other seed tissues cease to become a barrier to hyphal invasion?

The susceptibility of corn to invasion by storage fungi is also related to the amount of damage done to the grain as a result of combining and handling (Flannigan 1978, Tuite et al. 1985, Wieman et al. 1986, Siriacha et al. 1989). Fungi appear sooner on live grains with broken or cracked pericarps (Semeniuk 1954). Loss of germinability in damaged corn is greater than in nondamaged seeds (Qasem and Christensen 1958, Qasem and Christensen 1960). Mechanical damage to kernels allows entry points for fungi not able to penetrate the seed coat.

An increase in fatty acid was correlated with the presence of storage fungi in corn (Nagel and Semeniuk 1947, McGee and Christensen 1970). Fungal lipases are believed to be responsible for fatty acid hydrolysis (FAH) in corn (Goodman and Christensen 1952) and other cereal grains (Dirks et al. 1955). The FAH of sound corn kernels is low but increases with increasing levels of damaged kernels (Baker et al. 1957) and invasion by fungi (Bottomley et al. 1952, Christensen 1967, Sauer and Christensen 1969). Because fat hydrolysis takes place more rapidly than protein or carbohydrate hydrolysis in stored grain, the free fatty acid content of grain has been proposed as a sensitive index of incipient grain deterioration (Pomeranz 1974).

2.5. Critical Moisture Level

The method most commonly used world wide for preventing deterioration losses by fungi and for maintaining "quality" of stored grains is to store at moisture contents too low for fungi to grow (Tuite and Foster 1979). Therefore, the primary research objective of mycologists investigating such problems has been to determine the so-called safe storage conditions for different cereal and legume crops and to understand the conditions which determine the storage life of a grain bulk (Christensen and Kaufmann 1974, Sinha and Muir 1973, Lacey 1989). The importance of water as a limiting factor in fungal spore germination and growth has been the central unifying theme in mycological investigations of grain storage (Ayerst 1986). Such research has focused on the ability of storage fungi to produce mycotoxins at different temperatures and grain moisture contents (Northholt et al. 1977, Cuero et al. 1987a). Growth of individual species of storage fungi from large numbers of surface-disinfected kernels is recognized as evidence that during storage, the moisture content of the kernels must have been at or above the minimum required for the growth of that species. For example, on shelled corn stored at 21°C, *A. glaucus* grew at a minimum moisture of 14.4%, *A. wentii* Wehmer at 15.1%, *Penicillium* sp. at 16.3% and *A. flavus* with *A. niger* van Tieghem at 18.3% (Koehler 1938). This corresponded to the moisture requirements for growth in pure culture on surface-sterilized corn. Koehler noted that parasites or weak parasites such as *Cephalosporium acremonium* Cda., *Nigrospora oryzae* (Berk. and Br.) Petch, *Gibberella zeae* (Schwein.) Petch, and *Diplodia zeae* [syn. *D. maydis* (Berk.) Sacc.] did not grow on shelled corn of lower than 21% mc, although some strains of *F. moniliforme* may grow on corn with 18.4% mc.

Because a relative humidity of 75% is about minimum for the germination of mold spores (Ayerst 1969), there is general agreement that the so-called "critical moisture level" for storing a specific seed type is the percentage moisture at which the seed is in equilibrium with an atmospheric relative humidity of about 70%. Above 70% rh, small increases in the relative humidity of the atmosphere correspond with a large rise in the grain moisture content (Oxley 1948, Pixton 1967). Snow et al. (1944) recommended 13.7% mc (wb) for 2–3 yr safe storage of corn and 14.8% mc for 3 mo safe storage. Rapid moisture equilibration (24 h) between wet and dry corn indicates that seed can be blended with minimal risk of mold damage or aflatoxin contamination if the average moisture content of the blend is low enough to prevent mold growth (Sauer and Burroughs 1980).

Succession of fungi in grain storages (Sinha 1992) typically begins with grain at a moisture content too low to support fungal spore germination or hyphal growth. If the grain moisture content rises, this permits a few xerotolerant storage molds (e.g., *Eurotium* spp.) to grow and invade the germ. The metabolic activities of this initial fungal assemblage produces water which elevates the grain moisture to a level supporting a second wave of fungal colonists. This pattern continues until the now dampened and molded grain begins to heat and sprout. During germination the swelling of hydrophilic colloids in seed tissues produces an "imbibition pressure." In corn seed the imbibition pressure leads to the breaking of the seed coat. Should unharvested grain or stored grain become wet and begin to germinate, breaks in the seed coat would allow fungal access to the germ. Gilman and Barron (1930) first demonstrated that fungi were the cause of heating of stored corn, a fact later substantiated by Christensen and Gordon (1948). By mapping the aflatoxin levels around a naturally-occurring hot spot in an Illinois corn bin, Shotwell et al. (1975) demonstrated that local pockets of deteriorating grain may become contaminated with substantial quantities of this mycotoxin.

The moisture content of grain entering storage determines the initial pattern of fungal colonization. If the grain is well above the minimum moisture requirement for safe storage of the grain, the so-called "later successional species" may appear first and exclude the most xerotolerant fungi. It is not known if secondary successional species actually invade individual grains colonized by earlier successional species. Alternatively, later successional colonists may invade grains not previously colonized by storage molds. Does "succession" result in two or more fungi occupying the same grain with eventual species replacement?

In winter, the upper layers of corn in bins in temperate latitudes may accumulate a moisture content of 20–25%. Semeniuk et al. (1947) observed a stratification of different fungi near the upper surface of bulk corn stored in metal bins. When moisture becomes unevenly distributed or stratified in grain bulks this produces a stratification of the most competitive fungal colonists at each moisture level. *Penicillium rugulosum* Thom, *P. palitans* Westling (syn. *P. verrucosum*) and *P. chrysogenum* Thom were recorded from "Blue Eye" corn kernels collected near the surface of the grain bulk. *Penicillium palitans* was frequently the only fungus matting the corn in the upper 300 mm in these bins. *Aspergillus flavus* was found in a yellow-green band of varying width immediately below the layer of *P. palitans*. *Aspergillus candidus* was abundant in drier corn at the outer edges of the *A. flavus* band and *Eurotium amstelodami* Mangin was the prevalent mold colonist in the corn below. With continued storage through the following summer, these upper stratified zones became difficult to distinguish as they became heavily molded with *Absidia*, *Mucor*, and *Rhizopus* species.

With the advent of the picker-sheller and artifical drying facilities, high-moisture corn (20–35% mc) can be harvested to reduce field losses. Corn at these moistures is subject to mold spoilage and mycotoxin formation unless it is rapidly dried (Friday et al. 1989). High-temperature drying or rapid drying can produce stress cracks in the endosperm and damage the grain in other ways (see Chapter 16). Other methods used to prevent mold growth in high-moisture grain include low-temperature storage, reduced O_2 content of storage atmosphere, and the application of organic acids (Lacey 1989). Ammonia (NH_3) was found to have excellent fungicidal properties at concentrations of 0.5% of the dry mass of corn (Bothast et al. 1973, Nofsinger et al. 1977). Chilled aeration technologies show great promise for controling grain pests and preventing deterioration losses (Maier et al. 1992). At cool temperatures (< 15°C) storage fungi grow slowly and granary insects and mites are dormant.

One of the functions of aeration of stored grain is to maintain a uniform temperature throughout a given bulk and keep moisture migration to a minimum. Lack of uniformity in the distribution of moisture in a grain bulk is an important factor in the initial attack by decay-causing organisms (Christensen 1957). Aeration is also the most commonly used method to uniformly reduce grain temperature throughout the storage bin. High quality has been maintained in corn stored in large bins in tropical Puerto Rico, partly through aeration during periods when the temperature is below 25°C and the relative humidity is below 75% (Christensen and Kaufmann 1975).

2.6. Interactions with Stored-Product Insects

Storage fungi usually accompany or follow insect infestation of grain because insects can increase the water activity and temperature to levels suitable for fungal growth (Dix and All 1987). In addition, insects may serve as internal and external carriers of fungal spores (Sinha and Watters 1985, Dunkel 1988). Dunkel (1992) notes that succession in the insect communities of stored grain ecosystems is not as strictly regulated by the moisture content of the grain as is the pattern of fungal colonization but depends on whether the insect is a parasitoid, predator, primary grain feeder, or scavenger (see Chapter 5). Fungi provide an important nutritional benefit to detritivorous insects (Martin 1987, Ingham 1992), including certain insect pests of stored grain and food products (Sinha 1966, 1971; Hill 1978; Wright et al. 1980; Sinha and Watters 1985). One possible advantage for larvae that develop in molded and toxin-contaminated grain is that they may escape potential competitors and predators, including cannibalistic adults (Dowd 1992a).

Reproduction in the confused flour beetle, *T. confusum*, is mediated by a maternally inherited *Ricketsia*-like bacterium: crosses between infected males and uninfected females are sterile (Wade and Stevens 1985, Stevens and Wade 1988). However, infected individuals were cured of infection and reproduced when raised on flour to which tetracycline antibiotics were added (Wade and Stevens 1985) or flour produced from *Streptomyces*- or *Penicillium*-"molded" wheat grain (Stevens and Wicklow 1992). Likewise, the mushroom toxin amanitin frees a mushroom-breeding species of *Drosophila* fruitfly of internal parasites and thereby increases its fitness (Jaenike 1985).

Insects that feed on moldy grain may be able to metabolize toxic fungal metabolites. For example, the fungus-feeding Nitidulid, *Carpophilus hemipterus* (L.) is unaffected by 25 ppm aflatoxin B1 (Dowd 1992a) and is ten times more efficient in metabolizing the *Fusarium*-produced trichothecene 4-monoacetoxy-scirpenol than nonfungus-feeding caterpillars, the fall armyworm, *Spodoptera frugiperda* (Smith) and the corn earworm, *Heliothis zea* (Boddie) (Dowd and Van Middlesworth 1989). Stored-product beetles (e.g., *T. confusum*) are generally resistant to mycotoxins yet are also selective in their feeding behaviors (Wright et al. 1982). On-farm populations of cereal insect pests might have greater tolerance to toxic fungal metabolites produced by species or strains of cereal fungi common to that farm field or storage bin. A related example is found among *Drosophila* species with different ethanol tolerances occupying separate "niches" around wineries (Holmes et al. 1980, McKenzie and McKechnie 1979). Further research on the interactions of fungi and insects will undoubtedly reveal additional benefits of selective mycophagy and produce novel bioassays for identifying biologically active fungal metabolites (Wicklow 1988a, Stevens and Wicklow 1992).

3. DESERT RODENT GRANIVORES

3.1. The Banner-tailed Kangaroo Rat

The banner-tailed kangaroo rat *Dipodomys spectabilis* Merriam (Heteromyidae) is primarily granivorous and caches seeds below ground within an extensive burrow system (Vorhies and Taylor 1922, Holdenreid 1957, Reichman et al. 1985). Mycological studies have revealed numerous species of storage fungi associated with these rodent seed caches (Reichman et al. 1985, Frisvad et al. 1987) and as contaminants of external fur-lined cheek pouches that the rodent uses to transport seeds (Wicklow and Rebar 1988). However, selection may have guided rodent behavior in manipulating seed caches to promote the growth of less toxigenic fungal strains that are harmless to the rodents when ingested with the seeds (Frisvad et al. 1987, Wicklow and Rebar 1988). *Dipodomys spectabilis* forages at night and harvests mature seeds, typically by removing entire ripened flower heads, and storing these underground in species specific piles. All but an hour or two each day is spent inside of their burrows, but very little is known about the underground activities and caching behaviors of the rodents. Burrow air is close to moisture saturation most of the time (Kay and Whitford 1978) and therefore contributes to a cache environment that promotes the germination, growth, and sporulation of common storage fungi (Reichman et al. 1985). Rodent seed caches may deteriorate if they are not actively managed by the rodents (Rebar and Reichman 1983, Reichman and Rebar 1985, Reichman et al. 1986). While details of management procedures are not known, the rodents appear to prefer slightly molded seeds to nonmolded or heavily molded seeds (Reichman and Rebar 1985). Furthermore, laboratory experiments have shown that *D. spectabilis* will move seeds to locations with specific humidities within an artificial burrow system to selectively promote or prevent mold growth (Reichman et al. 1986). Microarthropods may also be beneficial to this process by dispersing fungal spores throughout individual caches, thus providing a uniform inoculation of potentially beneficial molds (Seastedt et al. 1986). The continued maintenance of these "domesticated" fungal populations would be assisted through rodent and arthropod transfer of fungal spores to freshly harvested seeds and by locating caches in burrow environments favoring the growth of specific molds (Wicklow and Rebar 1988).

3.2. Granivorous Beetles and Rodent Cached Seeds

Curiously, no granivorous beetles or other stored-product insects inhabited *D. spectabilis* seed caches that we recovered in 1982–1983 from several excavated mounds (burrow system) near Portal, Arizona (D.T. Wicklow, personal observation). It was necessary to determine if rodent-cached seeds of native desert plants could support populations of granivorous beetles. Wicklow et al. (1994) contrasted the survival and reproduction of the red flour beetle, *T. castaneum*, and the sawtoothed grain beetle, *Oryzaephilus surinamensis* (L.), on diets consisting of either whole seeds, seed flour, or seed flour containing 5% yeast extract. Few or no adult beetles were alive after 30 d on seed diets of *Astragalus nuttallianus* DC, *Erodium cicutarium* (L.) L'Her., *Monolepis nuttalliana*

(Schult.) Greene, *Prosopis juliflora* Woot., or *Setaria macrostachya* H.B.K. Beetle survival was high with numerous larvae and pupae recorded from seed diets of *Amaranthus palmeri* Wats., *P. sonorum* Beal., *Sporobolus contractus* Hitchc., and *S. cryptandrus* (Torr.) Gray. Interestingly, seeds of *P. sonorum* were also gathered as food by native Americans living in Arizona (Kearney et al. 1960). To prevent potential seed losses, the rodents might first consume those seeds most likely to deteriorate in storage (Smith and Reichman 1984) or with the greatest potential to support populations of insect granivores (Wicklow and McAlpin 1991). I find it significant that *T. confusum* has been used in assays to distinguish the nutritive value of barley varieties (Loschiavo et al. 1969). Cereal biotechnologists and chemical ecologists might target seeds cached and eaten by rodent granivores for potentially useful gene products that would serve to deter stored-product insects while having no or low toxicity to vertebrates.

4. FUNGAL DISEASES OF CORN GRAIN

4.1. Kernel rots and Mycotoxins

Fungal colonists of grain belong to different taxonomic groups and are also classified by plant pathologists according to the type of disease symptoms they produce (i.e., grain replacement diseases, kernel or ear rot, streaked or blotched kernels). Each fungus that rots grain also contaminates the grain with fungal-derived chemicals, including mycotoxins (Wyllie and Morehouse 1977, Betina 1989). The majority of toxigenic fungal species that rot grain is limited to a few orders of ascomycetes (i.e., Eurotiales, Hypocreales, and Pleosporales) (Samuels 1984). A summary of the ecologically specialized groups of fungal pathogens reported to rot corn is provided as background for my discussion of the ecology and pathology of *A. flavus* in corn production and storage. *Aspergillus flavus* coexists or competes with these other fungal colonists of cereals and the outcome of such species interactions is important in determining the extent of *A. flavus* colonization and aflatoxin contamination.

4.2. Seed Replacement Diseases

Seed replacement pathogens invade unfertilized embryos of grasses and develop in place of the grain (Luttrell 1981). *Ustilago maydis* (DC.) Cda. and *Sphacelotheca reiliana* (Kuhn) Clint. produce smut galls that rupture to release brown to black colored teliospores that enable the fungus to overseason in the soil. It is unlikely that smut galls could survive combine harvesting in an intact form. I know of no reports linking corn smut galls to any mycotoxicosis or grain storage problem. *Claviceps gigantea* Fuentes et al. is known only from high cool mountain valleys of central Mexico (Fuentes et al. 1964). The ergot sclerotium is much larger than the grain and resembles a horse's tooth. The sclerotia contain ergot alkaloids (Mantle 1975) but because of their large size they can readily be distinguished and removed from the grain by sieving or hand-sorting.

4.3. Mildly Parasitic Colonists of the Phylloplane

Early phylloplane colonists include fungi that rapidly colonize newly emerging leaves, persist through stages of leaf senescence and litter fall, and remain as saprotrophic colonists of plant residues (Pugh 1972, Dickinson 1981, Blakeman 1985). Among these fungi, *Cladosporium cladosporioides* (Fres.) DeVries and *C. herbarum* Lk.:Fr. produce a kernel rot that is recognized by dark, greenish-black, blotched or streaked kernels (Shurtleff 1980). Late phylloplane colonists include fungi that rapidly colonize living leaf surfaces several weeks after leaf emergence and, along with the early colonists, persist through stages of leaf senescence and litter fall, and remain as saprotrophic colonists of crop residues in the field. On exposed kernels (i.e., not covered by the husk), these mildly parasitic fungi may appear as a greenish-black "soot." The late phylloplane colonists, *A. alternata* (Fr.) Keiss. and *Curvularia lunata* (Wakker) Boed. are weak pathogens that can infect kernels on the ear. *Drechslera maydis* (Nisikado) Shoem., the cause of "southern corn leaf blight," can also produce an ear rot that attacks the husks and cobs (Tuite and Caldwell 1970). However, the mold-rotted ears proved harmless when fed to cattle and swine (Nelson and Christensen 1977). Other leaf spot organisms such as *D. rostrata* (Dreschs.) Richardson and Fraser have been reported to cause symptoms of ear rot (Whitehead and Calvert 1959). *Drechslera zeicola* (Stout) Subram. and Jain causes a minor leaf spot of corn and may be present on corn seed, while *Epicoccum purpurascens* Ehrenb.:Schlecht. produces a kernel rot recognized as "red kernel" (Smith et al. 1988). In corn, the husk shields the grain from weathering and from air-borne spore inoculum of phylloplane colonists and other pathogens that produce spots or blemishes on exposed aerial parts of a cereal crop. Some of these fungi appear if the ear tip is exposed, as in loose-husked varieties developed for rapid dry-down in northern latitudes.

Mycologists have recognized relationships between deteriorative changes in seeds and a reduction in the percentage of grain yielding late phylloplane colonists such as *Alternaria* (Christensen and Kaufmann 1974). Wallace et al. (1976) found that *Alternaria* and associated fungi, *Gonatobotrys, Cochliobolus*, and *Nigrospora*, all of which are colonists of wheat seeds in the field, gradually disappear during grain storage and are replaced by *A. versicolor* (Vuill.) Tiraboschi, *A. glaucus* group, *A. candidus, Penicillium* spp., *Absidia, Chaetomium, Rhizopus*, and *Streptomyces*. Direct parasitism of the "dormant" hyphae and spores of phylloplane colonists by these storage fungi, and selective mycophagy by granary arthropods (Wallace and Sinha 1981) may each contribute to this extinction phenomenon. The disappearance of late phylloplane colonists may also result from cell aging processes which are accelerated by higher storage temperatures.

4.4. Stem-Rot and Ear-Rot Pathogen Complex

Fusarium crookwellense Burgess et al., *F. culmorum* (W.G.Sm.) Sacc. and *Gibberella zeae* cause head blight of wheat and corn (Miller et al. 1991). *Gibberella zeae* is able to rot the ears of plants grown under optimal conditions provided: 1) that the corn variety is susceptible to this pathogen, and 2) there is either natural or experimental wounding of the ears (Sutton 1982, Bennett et al. 1988, Wicklow and Caldwell 1990). Mycotoxicoses associated with *Fusarium*- or *Gibberella*-rotted cereals have been recognized for decades (Joffe 1986). Plant pathologists are giving greater emphasis to investigations of the role of these compounds in disease development and the ability of resistant plant hosts to detoxify these fungal metabolites (Miller et al. 1983, Miller and Young 1985, Chelkowski 1989). Some of the fungi assigned to this group are not considered highly pathogenic but appear to infect corn ears when the plants become debilitated from the effects of disease or drought stress. Included are species classified as Coelomycetes (Sutton 1980) because pycnidia represent the anamorph state [e.g., *Stenocarpella maydis* (Berk.) B. Sutton (syn. *D. maydis*), *Lasiodiplodia theobromae* (Pat.) Griff. and Maubl.]. *Stenocarpella maydis* is the only one of the pycnidial forming Coelomycetes that has been linked to a naturally occurring mycotoxin outbreak (Marasas 1977, Rabie et al. 1987). A suspected mycotoxin, diplodiatoxin, accounted for only a small percentage of the toxicity to animals in feeding experiments (Rabie et al. 1977, Kellerman et al. 1985). *Nigrospora oryzae* has not been linked to any mycotoxicosis. As a pathogen of cotton, *N. oryzae* is transmitted by a mite, *Siteroptes reniformis* (Krantz), which in turn is nutritionally dependent on the fungus (Watkins 1981). Could this suggest a similar vector relationship for *N. oryzae* in corn?

4.5. Endophyte Mutualists

Seed pathologists have long recognized that a fungus identified either as *C. acremonium* Cda. or *A. strictum* W. Gams (e.g., Neergaard 1977, Wicklow 1988b) is a frequent colonist of corn and other cereal grains affecting seed quality (Semeniuk 1954, Koehler 1959). Cephalosporium acremonium is reported to have survived for 7 yr as an internal colonist of corn seeds (Dungan and Koehler, 1944). Gams (1971) characterized Cephalosporium as a nomen confusum and recognized two species of *Acremonium* from corn: *A. kiliense* Grutz, the cause of "black bundle disease," and *A. zeae* W. Gams and Sumn., which is reported to cause a stalk rot and is also isolated from corn grain. *Cephalosporium maydis* Samra, Sabet, and Hingorani is reputed to cause a late wilt of corn in Egypt and India (Shurtleff 1980). This fungus is reported to produce sclerotia, a character that distinguishes it from *A. kiliense* and *A. zeae*. *Acremonium* species from corn may be similar to the *Acremonium* endophytes of wild grasses (Clay 1990a). Is the *Acremonium* species complex infecting domesticated cereals, a source of biologically active chemicals, that might impact corn insects or protect corn grain from certain fungal pathogens? *Acremonium* species have not been linked to any mycotoxicosis in cultivated cereals.

Fusarium moniliforme is the most frequently recorded fungal colonist of freshly harvested corn. This fungus was found in every ear of corn examined from several midwestern states and infected every kernel of 60 representative ears examined that same year (Valleau 1920). In a subseqent survey, 100% of the corn plants sampled from the central and eastern United States were infected with *F. moniliforme* (Leslie et al. 1990). *Fusarium moniliforme* produces symptomless infections of the embryo and may be a systemic pathogen (Kucharek and Kommedahl 1961, Neergaard 1977, Lawrence et al. 1981). Dungan and Koehler (1944) showed that *F. moniliforme* could survive for 8 yr as an internal colonist of corn grain. The fungus commonly reported from corn grain as *F. moniliforme* probably represents a species complex (Gerlach 1982) including *F. sacchari* (Butler) Gams var. *subglutinans* (Wollenw. and Reinking) Nirenberg, *F. verticillioides* (Sacc.) Nirenberg, and *F. proliferatum* (Matsushima) Nirenberg. Plant pathologists are actively investigating the genetics, pathology, and biologically active metabolites (e.g., phytotoxins, mycotoxins, etc.) of pathogen populations comprising this species complex (Leslie et al. 1990, Thiel et al. 1986, 1991).

Wicklow (1988b) and Rheeder et al. (1990) reported that individual corn grains infected with *F. moniliforme* were less likely to be infected with *A. flavus* or other fungal pathogens of corn. Evidence was also offered that *F. moniliforme* and *A. strictum* protected grain from becoming infected by *A. flavus* (Wicklow et al. 1988b). Does *F. moniliforme* function as a mutualist in protecting corn and other cereals from predators and pathogens? In nature, host resistance to infection by seed-borne endophytes would be maladaptive if the relationship is mutualistic. If so, would corn varieties selected for their resistance to *F. moniliforme* in field inoculation trials (King 1981, Ochor et al. 1987) be more susceptible to *A. flavus* infection?

4.6. Saprotrophs in Damaged or Aged Grain

Many of the fungi recorded from corn grain at harvest or from shelled corn in storage are not recognized as true pathogens of corn but opportunistic saprotrophs that colonize and rot grain that has been damaged or has aged significantly in storage (Semeniuk 1954, Koehler 1957, Mislivec and Tuite 1970, Christensen and Kaufman 1974). These saprotrophs include one or more yeasts, numerous species with *Aspergillus* and *Penicillium* anamorph states, and miscellaneous fungal taxa associated with biodeterioration of stored grain, hay, or silage [e.g., *C. globosum* Kunze, *Paecilomyces variotii* Bain. *Scopulariopsis brevicaulis* (Sacc.) Bainier, and *Trichothecium roseum* (Pers.) Link: Gray].

Qasem and Christensen (1958) reported that very few surface-disinfected grains at harvest yielded "storage fungi," even when the plants from which the grain was collected had stood for some time in the field after the normal harvest time. Caldwell et al. (1981) determined the pathogenicity to corn ears of 15 species (81 strains) of *Penicillium* isolated from preharvest and stored corn. *Penicillium oxalicum* Currie and Thom and *P. funiculosum* were the only species that could invade and colonize ears that were sprayed at silk with conidial inoculum or injected in the tip and butt end of the ear. *Penicillium oxalicum* also causes a seedling blight of corn (Johann et al. 1931), and produces the mycotoxin secalonic acid D (Palmgren and Fleischhacker 1987).

Penicillium variabile Sopp was considered a "minor" field species since it infected just 28% of the kernels. *Penicillium tardum* Thom (subgenus *Biverticillium*) colonized *P. oxalicum* suggesting the behavior of a mycoparasite (Caldwell et al. 1981). Malloch (1985) proposed that the teleomorph genera *Eupenicillium* and *Talaromyces*, each having *Penicillium* anamorphs, represent separate lines of evolution involving cleistothecial Ascomycetes. He argues that species in the subgenus *Biverticillium* are more closely related to *Talaromyces* since they can degrade cellulose and were probably derived from species colonizing decayed wood such as *Trichoma* in the subfamily Trichocomoidae. Mycoparasitic ability is an important adaptive behavior for fungi that replace primary wood decay fungi (Rayner and Boddy 1988). Observations that *Penicillium* species classified in subgenus *Biverticillium* can parasitize other seed-infecting fungi such as *A. niger* (Raper and Fennell 1965, pp. 65–66) and reports that *T. flavus* (Klocker) Stolk and Samson parasitizes *Sclerotinia sclerotiorum* (Lib.) deBary (McClaren et al. 1986), and *T. vermiculatus* Dangeard parasitizes the sclerotial phase of *Rhizoctonia solani* Kuhn (Boosalis 1956), might therefore be anticipated given such an evolutionary past.

Malloch (1985) classified *Penicillium* anamorphs with a marked affinity for starchy or oily substrates in the family Dichlaenoideae. The latter would encompass those species in *Eupenicillium* and subgenus *Penicillium* that are commonly isolated from grain and other agricultural products (Pitt 1979). Kernel infection rates following corn ear injection were considerably lower for other species of penicillia tested by Caldwell et al. (1981), including several classified in subgenus *Penicillium:* 10–13% infected kernels—*P. martensii* Biourge (syn. *P. aurantiogriseum* Dierckx), *P. cyclopium* Westling (syn. *P. aurantiogriseum*), and *P. citrinum* Thom; 4–7% infected kernels—*P. expansum* (Link) Thom, *P. viridicatum*, *P. purpurogenum* Stoll, and *P. frequentans* Westling; < 1–2% infected kernels—*P. brevi-compactum* Dierckx, *P. urticae* Bain. (syn. *P. griseofulvum* Dierckx, *P. chrysogenum* Thom, and *P. palitans* Westling (syn. *P. verrucosum*). An equivalent study involving inoculation of corn ears with *Asper-*

gillus species recorded from corn grain could provide us with better contrasts of the abilities of *A.* species to infect and rot corn grain in field versus storage environments. These species would include: *A. flavus*, *A. versicolor*, *A. candidus*, *Eurotium* spp., *A. niger* van Tieghem; *A. terreus* (Bain. and Sart.) Thom and Church; *A. melleus* Yukawa; *A. tamarii* Kita; *A. clavatus* Dezm.; *A. fumigatus* Fres.; *A. wentii* Wehmer, and *Emericella nidulans* (Eidam) Vuill. It would also be useful to learn if ear or kernel rot disease resistance ratings determined for corn varieties in field trials also apply when the mature grain is left standing in the field or exposed to storage environments that permit further growth of the pathogen (e.g., Wicklow et al. 1990).

There have been numerous reports showing that naturally occurring fungal colonists of grain, when grown on autoclaved grain, will prove toxic when fed to laboratory or farm animals (e.g., *Aspergillus*: Wilson 1971a, Semeniuk et al. 1971, Smith and Moss 1985, Moss 1989; *Penicillium*: Wilson 1971b, Carlton et al. 1968). Likewise, toxicity tests with extracts of fungal cultures can also provide evidence of the toxigenic potential of kernel rotting fungi (e.g., Richard et al. 1969). Even so, surprisingly few of these species, or their toxins, have actually been proven to be responsible for naturally occurring mycotoxin outbreaks (e.g., *A. flavus* and aflatoxin: Guthrie 1979, Hall et al. 1989).

Three osmophilic species of *Eurotium* (anamorph = *A. glaucus* sensu lato), *E. amstelodami* Mangin, *E. chevalieri* Mangin, and *E. repens* deBary, are the most commonly recorded fungi from stored grain. However, there are no reports of mycotoxin outbreaks on stored grain that can be attributed to any of these molds. The mycotoxin echinulin, produced by *Eurotium* spp., has been implicated in the refusal of pelletized feed by swine (Vesonder et al. 1988). Vesonder and Horn (1985) showed that *A. versicolor* was the source of sterigmatocystin in moldy corn feed associated with symptoms of bloody diarrhea and death in dairy cattle. *Aspergillus versicolor* and *A. sydowii* (Bain. and Sart.) Thom and Church are not common on corn (Lichtwardt et al. 1958). Ochratoxin A is implicated in natural outbreaks of nephrotoxicosis in the poultry industry (Huff et al. 1975, Hamilton et al. 1982) probably as a result of feeding corn molded with *P. viridicatum* (Ciegler et al. 1973). *Penicillium viridicatum* molded corn has also been implicated in poisoning of swine and horses (Cunningham and Freeman 1953). Shotwell et al. (1969) first reported ochratoxin A as a natural contaminant in a survey of corn. *Aspergillus ochraceus*, another source of ochratoxin A, was a prevalent fungal colonist of partially spoiled corn in Mexico (Lopez and Christensen 1967). Corn contaminated with *P. citrinum* and citrinin was demonstrated to be toxic to broiler chicks (Roberts and Mora 1978, Nelson et al. 1985). *Aspergillus candidus*, which very rapidly kills and discolors germs of grain in storage, is a potential source of citrinin in Canadian grain (Scott et al. 1972, Scott 1978). Penicillic acid has been found in commercial corn samples in the north central United States (Thorpe and Johnson 1974). While there are no reports of mycotoxic disease due to penicillic acid contamination of feeds, it synergizes the toxicity of both citrinin and ochratoxin A (Sansing et al. 1976, Dowd 1989). *Eupenicillium ochrosalmoneum* Scott and Stolk infects preharvest corn and contaminates the grain with the neurotoxic mycotoxin citreoviridin (Wicklow and Cole 1984, Wicklow et al. 1988c). The distribution of citreoviridin was associated with kernel damage. Corn ears infested with *E. ochrosalmoneum* and contaminated with citreoviridin may explain reports of staggers in cattle that ate lodged corn (Cole 1977). Tremorgenic mycotoxins isolated from corn molded by *A. fumigatus* and *P. crustosum* Thom were implicated in a natural intoxication of cattle (Dorner et al. 1984). *Penicillium purpurogenum* (syn. *Penicillium rubrum*) produces rubratoxins implicated in a mycotoxicosis in cattle and swine fed moldy corn (Burnside et al. 1957, Wilson and Wilson 1962). However, as occurs in most of these field cases, the grain was contaminated with other fungi such as *A. flavus* and it is not clear that rubratoxins were entirely responsible for this mycotoxicosis.

Candida guillermondii (Castellani) Langeronet Guerra was the prevalent yeast isolated from insect-damaged and molded corn grain sampled at harvest from fields throughout the state of North Carolina (Wicklow et al. 1980, Horn 1985). The yeast was confined mainly to kernels visibly molded by filamentous fungi such as *F. moniliforme* and *A. flavus*. In a study of the competitive effects of fungal isolates from corn grain to interfere with *A. flavus* growth and aflatoxin production, Wicklow et al. (1980) found that *C. guilliermondii* was not antagonistic to *A. flavus* and did not inhibit aflatoxin production by *A. flavus* when co-inoculated onto autoclaved corn kernels. Both *Hyphopichia burtonii* (Boidin et al.) von Arx and *Bacillus amyloliquefaciens* (Fukumoto) markedly stimulated aflatoxin production at 0.95 and 0.98 water activity (Cuero et al. 1987b).

Yeasts and bacteria may offer an important food source for detritivorous corn insects and may also detoxify both corn allelochemics (Jones 1984, Starmer and Fogleman 1986, Leufven 1991) or toxins produced by a fungus that colonizes the same damaged grain (Toyoda et al. 1988, Dowd 1992b). Other saprotrophs are known to prevent aflatoxin formation or detoxify aflatoxins. For example, *A. niger*, *R. arrhizus* Fischer, and *R. nigricans* Ehrenburg have been found to metabolize aflatoxin B1 to a relatively nontoxic compound, aflatoxin B2a (Ciegler et al. 1966, Cole et al. 1972) or to limit aflatoxin formation on autoclaved or irradiated grain and other culture media (Weckbach and Marth 1977, Wicklow et al. 1980, Horn and Wicklow 1983, Magan and Lacey 1988).

5. SELECTION FOR MYCOTOXIN-PRODUCING ABILITY

In this section, I will attempt to show that the production of specific mycotoxins is of adaptive value to fungi that rot corn grain before harvest. Downed corn ears offer a rich reward to those seed-eating rodents, birds, deer, cattle, or swine found foraging in corn fields following harvest. During the winter, low temperatures limit insect activity and, therefore, the molded kernels that detritivorous beetles might consume in warmer periods are at less risk. By summer of the following year, the sound kernels from most of the exposed ears have either been consumed by vertebrates or have themselves become rotted by molds, with only the most heavily molded kernels remaining on the cob. These mold-rotted kernels are also contaminated with mycotoxins and other fungal metabolites. This is where the ability of the fungal colonist to produce mycotoxins (e.g., tremorgens, refusal factors) would protect it from being consumed with the grain.

These molded ears then become a source of infective inoculum that nitidulid beetles may transmit to the next year's corn crop. Mycotoxin-producing ability is therefore under strong selection in such corn cultivation systems (Wicklow 1989). I do not believe it is an accident that the molds we recorded from nitidulids and downed corn ears in Georgia (listed here) produce neurotoxic metabolites or cause feed refusal (Frisvad 1986). *Aspergillus flavus* produces aflatrem, cyclopiazonic acid, and B-nitropropionic acid, all of which have neurotoxic effects and thus would represent an immediate deterrent to vertebrates (Wilson and Wilson 1964, Bush et al. 1951, Dorner et al. 1983). *Aspergillus flavus* is better known for its production of aflatoxins. However, it is not obvious that aflatoxins are refused by vertebrate seed eaters. *Aspergillus terreus* produces the tremorgens territrem A and B (Ling 1976, Ling et al. 1979) in addition to citrinin, gliotoxin, citreoviridin, and other mycotoxins. *Aspergillus niger* produces malformin C, which is highly toxic to rats (Anderegg et al. 1976). *Aspergillus flavipes* produces sterigmatocystin but has not been implicated in any mycotoxicoses. Because the epidemiology of *A. flavipes* in Georgia corn fields mimics that of the other toxigenic aspergilli, it may also synthesize neurotoxic compounds that function as refusal factors. *Eupenicillium ochrosalmoneum* produces the potent neurotoxin citreoviridin and we have shown that the levels of this toxin in a single corn kernel can exceed the LD_{50} for laboratory mice (Wicklow et al. 1988c). *Fusarium moniliforme* is common on corn wherever it is grown and some of the metabolites it produces have been linked to different mycotoxicoses (Marasas et al. 1984, Gelderblom et al. 1988, Marasas et al. 1989). There is evidence that some strains of *F. moniliforme* produce a feed refusal factor equivalent to 40 mg/kg vomitoxin (Burmeister 1983); this is based on observations of feed refusal of *F. moniliforme* rotted corn by cattle (Beasley et al. 1982) and the results of a mouse drinking water bioassay.

During hand-harvesting of corn in southern Africa, all damaged and conspicuously moldy ears are discarded, either by leaving them on the stalk or dropping on the ground (Marasas 1977). This leaves large numbers of *Diplodia*-molded ears present when livestock are put to graze harvested corn lands during the latter part of winter when outbreaks of diplodiosis occur. The numerous examples of mycotoxicoses associated with *Fusarium* ear rot of overwintered cereals in northern latitudes, such as the dramatic epidemics of alimentary toxic aleukia (ATA) in the Soviet Union occurring between 1941 and 1947 (Joffe 1986), offers relevant perspective. Grain was left in the field unharvested followed by a mild winter and late spring, which left a thick snow cover over the ground. Grain that was infected with *Fusarium* spp. became severely contaminated with trichothecene mycotoxins. As the grain was subsequently harvested and consumed by the human population, there occurred an outbreak of ATA, hemorrhage, vomiting, rejection of food, and dermatitis (Joffe 1986). This same toxin-contaminated grain would also have negatively impacted seed-eating bird and rodent populations. The importance of trichothecenes in feed refusal is demonstrated by mice refusing water contaminated with low levels of these toxins (Burmeister et al. 1980). Dowd (1988) has shown that fusaric acid synergizes the toxicity of trichothecenes to corn earworm and fall armyworm larvae and there may be relevant guides here for the design of experiments involving mycotoxin synergism involving vertebrates. Another *Fusarium* mycotoxin, zearalenone, could be expected to reduce the population size of rodent granivores by causing grossly enlarged uteri, an accepted measure of estrogenic activity, vulvovaginitis, abortion, and infertility (Christensen et al. 1965, Mirocha et al. 1968). In an earlier review (Wicklow 1989), I suggested that *Fusarium* mycotoxicoses may have their ecological origins in natural grassland ecosystems.

6. *ASPERGILLUS FLAVUS*: ECOLOGY AND PATHOLOGY

6.1. Aflatoxin Contamination of Corn

In this section, I shall attempt to characterize the ecology and pathology of *A. flavus* in corn production and storage ecosystems as a model for examining the dynamics of other fungi commonly isolated from grain. The discovery of *A. flavus* and aflatoxin in preharvest corn (Anderson et al. 1975) has invited considerable interdisciplinary research on the ecology and pathology of a fungus that seed pathologists had classified as a "storage mold" (Christensen and Kaufmann 1965). Prior to this discovery, individuals responsible for maintenance of grain quality in storage were being advised that the growth of *A. flavus* from surface-sterilized corn grain was evidence of poor grain storage practices (Christensen and Kaufmann 1974).

Aspergillus flavus colonizes corn as follows (Diener et al. 1987): 1) airborne or insect-transmitted inoculum contaminates the silks and grows into the developing ear; 2) portions of the ear having kernels damaged by insects or birds become infested with *A. flavus* and contaminated with aflatoxin; 3) nitidulid beetles may also vector *A. flavus* inoculum directly from molded crop residues to damaged corn kernels (Lussenhop and Wicklow 1990); and 4) climatic and cultural factors that stress the plant such as high evening temperatures (> 24°C) (Thompson et al. 1980, Smart et al. 1990b) and drought (Payne et al. 1986), may increase susceptibility of corn ears to fungal infection. *Aspergillus flavus* may also be present in grain dust, as spores, and therefore contaminates grain entering storage. Thus, a population or clone representing a single *A. flavus* vegetative compatibility group (VCG) or h-c group according to Papa (1986), may interact with two entirely different assemblages of fungi and insects, first in corn production fields and second in grain storage facilities. The outcome of these interactions determines the degree to which *A. flavus* colonizes grain and contaminates it with aflatoxin.

6.2. Sources of Inoculum

Aspergillus flavus produces yellow-green conidia that function in dispersal and as infective inoculum, in addition to sclerotia (long-lived survival structures). Both types of propagules are associated with damaged corn kernels and are dispersed onto the ground during combine harvesting (Wicklow et al. 1984). Wicklow and Wilson (1986) observed that sclerotium germination (sporogenic) occurred in corn fields just prior to silking. Shade provided by the corn canopy may help to retain moisture at the soil surface and thus promote germination since sclerotia exposed for the same interval on fallow, bare ground adjacent to each field did not germinate. *Aspergillus flavus* is like many other ascomycetes that made the transition from buffered tropical forest habitats to more seasonal, less predictable situations. These fungi appear to have sacrificed their teleomorph, and thus the ability for genetic recombination, in exchange for the massive production of dry conidia (Evans 1988). Evans suggests that for fungal pathogens of insects, conidial production may be correlated with seasonal population explosions of host insects.

In cultivated fields, *A. flavus* may actively colonize the residues of corn and other crops (Zummo and Scott 1990, Shearer et al. 1992). Griffin and Garren (1976) showed that low densities of *A. flavus* in field soil increased several fold around buried rye grass. Any effort to determine the source of storage fungi contaminating grain in corn production fields might begin by examining the fungal populations colonizing crop residues just prior to corn flowering and during the interval of ear maturation. The fate of these residues is important, particularly since they may also be colonized by nitidulids.

6.3. Corn Insects

Insects were immediately implicated as potential vectors of *A. flavus* to corn because the fungus was found sporulating in insect-damaged portions of the ear

(Taubenhaus 1920, Anderson et al. 1975). Initial attempts to identify insects that might vector *A. flavus* to corn proved inconclusive (Diener et al. 1987). Although many kinds of insects collected from corn ears carry *A. flavus* (Fennell et al. 1977), collections have typically been made late in the season when secondary inoculum of *A. flavus* is so abundant that insects are more likely to become contaminated. Moths of the corn earworm, *H. zea* (Boddie), and European corn borer, *Ostrinia nubilalis* (Hubner), have been implicated as vectors of *A. flavus* inoculum to corn ears because earworm larvae and the kernels damaged by larvae may become contaminated with *A. flavus* (Lillehoj et al. 1978) and because adult moths collected during an 8-yr period at a light trap (Coastal Plains Experiment Station, Tifton, GA) were contaminated with *A. flavus* (McMillian 1987, McMillian et al. 1990). Presumably, the moths contaminate the silks with *A. flavus* when depositing their eggs. In this scenario, both the fungus and the developing caterpillar larvae would enter the ear through the silk channel. Where these moths become contaminated with *A. flavus* is not known. Neither insect has actually been found to transport *A. flavus* to corn ears in the field. The corn earworm has also been implicated as providing an entrance into the ear for many insects, including rice weevil, pink scavenger caterpillar, sap beetles and mites (Zuber 1977).

Lussenhop and Wicklow (1990) demonstrated the role of nitidulid beetles in transmitting *A. flavus* infective inoculum to preharvest corn ears in commercial corn fields in the US Georgia Coastal Plain. The nitidulids gain entry to the ears through wounds caused by other insects and birds and are capable of entering some ears on their own, particularly loose-husked varieties. Downed and overwintered corn ears that become infested with *A. flavus* represent an important source of infective inoculum. This is because populations of nitidulid beetles also colonize these ears. Lussenhop and Wicklow (1990) found that 68% of the nitidulids (e.g., *Carpophilus lugubris* Murray, *C. freemani* Dobson) associated with overwintered, *A. flavus*-molded, corn ears were contaminated with *A. flavus* at the time corn was silking. Nitidulids were shown to consume the conidia and mycelium of *A. flavus* (Wicklow et al. 1988a), and subsequent observations have revealed that *A. flavus* conidia can germinate after passage through the nitidulid gut. Nitidulids may preferentially feed on molded plant tissues, using the fungi as food (Miller and Mrak 1954). Dorsey and Leach (1956) suggest that the association between nitidulid and fungus is mutualistic. The vectoring of plant pathogenic fungi by nitidulids provides the beetle with a fungus-enriched plant substrate as food.

There is no information on the feeding choices nitidulid beetles make upon encountering kernel rotting molds or their toxins in nature. Nitidulids are generally more tolerant of individual mycotoxins than strictly herbivorous corn insects and the nutritional benefits of fungivory may outweigh the metabolic cost of detoxification (Martin 1987, Wicklow et al. 1988a, Dowd 1992a). Windels et al. (1976) showed that the nitidulid beetle *Glischrochilus quadrisignatus* Say was associated with *Fusarium* spp. in mold-infested ears that were buried during tillage. *Glischrochilus quadrisignatus* has been shown to transmit conidia and ascospores of *G. zeae* to corn ears experimentally wounded to simulate bird or insect damage (Attwater and Busch 1983). Should nitidulids avoid certain molds, it is less likely that these molds will be vectored to preharvest corn. Both plant volatiles and fungal volatiles from infected or decaying host tissues were shown to synergize the attractancy of the nitidulid (*C. hemipterus*) aggregation pheromone (Dowd and Bartelt 1991).

Aspergillus flavus and aflatoxin are often correlated with insect damage because nitidulid beetles are attracted to wounds made by other insects such as the corn earworm. A significant association between earworm damage, presence of nitidulid beetles, and incidence of aflatoxin contamination in corn has been reported in the southern United States (Lillehoj et al. 1975). However, the authors did not discuss the possibility that nitidulids were potential vectors of *A. flavus*.

Although *A. flavus* was associated with European corn borer damage in 1975, the tracks of vast majority of feeding larvae were not infected with *A. flavus* (Lillehoj and Hesseltine 1977). Nitidulid aggregation patterns in corn fields could explain the irregular distribution of *A. flavus* and aflatoxin among damaged ears from the same field (Wilson et al. 1981).

It is not surprising then that the role of nitidulids in vectoring *A. flavus* or other mycotoxin-producing fungi to corn ears had been underestimated. For example, Widstrom et al. (1976) concluded that an unobserved insect vectored *A. flavus* to ears damaged by corn earworm larvae because insecticides reduced aflatoxin contamination, but contamination was not directly caused by corn earworm larvae. The role of nitidulids in vectoring *A. flavus* also explains why Lillehoj et al. (1984) found that a fungicide sprayed on silks did not affect aflatoxin levels; nitidulids could have carried inoculum directly into the corn earworm wounds.

Lussenhop and Wicklow (1990) described a model where nitidulid beetles overwinter in crop residues. Because some crop residues (e.g., corn ears and cobs) are substrates for *A. flavus*, the nitidulids become contaminated. When *A. flavus*-contaminated nitidulids fly to damaged corn ears, they contaminate the wounds with *A. flavus*. The fungus can then spread to infect the adjacent undamaged kernels (Wicklow et al. 1988b). It is in these damaged kernels (and among kernels 1–2 rows removed from the wound-site) where the bulk of the aflatoxin is produced (Lee et al. 1980, Wicklow et al. 1988b). Some of the kernels will fluoresce a bright greenish-yellow (BGYF) under ultraviolet light (365 nm); the fluorescence is produced by the oxidative action of peroxidase enzymes in living seed on kojic acid, which is formed by *A. flavus*. The cycle is completed by release of conidia and sclerotia from contaminated ears during harvesting; these, plus unharvested ears, fall to the ground and begin the next generation.

6.4. Biocompetitive Strains

Initial evidence from greenhouse studies showed that the nonaflatoxigenic strains of *A. flavus* were effective at reducing aflatoxin contamination of developing cotton seed when inoculated into bolls prior to or simultaneously with virulent aflatoxigenic strains (Cotty 1989, 1990). This conceptual approach is well grounded from an ecological perspective. *Aspergillus flavus* occupies a niche, although only partially defined, within the cotton field. The production of an infective inoculum and the dispersal of this inoculum to the maturing cotton bolls is under some form of environmental control. Therefore, the introduction of a substantial inoculum (conidia, sclerotia) consisting of nonaflatoxigenic strain(s) onto the soil of the cotton field could overwhelm and replace the indigenous aflatoxigenic strains. The nonaflatoxigenic strains would then infect the cotton bolls and seeds instead of indigenous aflatoxigenic strains. I do not advocate this approach with corn cultivation for the following reasons:

1. It has not been established that *A. flavus* populations infecting corn at harvest originate from those same field soils. Wind-dispersed and insect-vectored infective inoculum can originate from cultivated and noncultivated field sites either adjacent to or far removed from the field treated with biocompetitive *A. flavus* inoculum. Because one cannot predict which wound sites in a field of corn will eventually become contaminated, by wind- or nitidulid-dispersed spores, the only way to assure the successful application of a biocompetitive inoculum would be to contaminate the great majority of the wound sites. How does one accomplish this without a massive aerial application of *A. flavus* infective inoculum?
2. Additions of *A. flavus* inoculum (conidia, sclerotia) to field soils in an effort to displace aflatoxigenic strains can only increase the "inoculum potential" of *A. flavus*. This provides many more opportunities for *A. flavus* to contaminate the ears, particulary sites of kernel wounding. Our first line of defense in identifying potential lots of aflatoxin contaminated corn is the detection of visible *A. flavus* sporulation (yellow-green conidial heads) or BGYF in comparatively few of the individual kernels colonized internally by *A. flavus*. Therefore, with the addition of substantial quantities of a biocompetitive inoculum to a corn production field, the frequency of BGYF kernels will surely increase. This would result in greater numbers of false positives in grain bulks that will invite frequent chemical testing for aflatoxin as the grain moves through commercial channels.
3. The issue of toxin-free grain does not disappear with the introduction of nonaflatoxigenic strains. *Aspergillus flavus* makes several secondary meta-

bolites (e.g., aflatrem, cyclopiazonic acid, 3-nitropropionic acid) with known toxicity to animals (Wilson 1966, Cole and Cox 1981, Wicklow and Dowd 1989, Gloer et al. 1988, Liu et al. 1989). No one has quantified these toxins in corn grain at harvest or contrasted such data with toxin profiles from the same lot of grain following storage at temperatures and moisture levels promoting *A. flavus* growth within the grain bulk.

6.5. The Seed Coat in Grain Resistance

Insect feeding on corn ears is not essential to *A. flavus* invasion of the germ or endosperm or to the contamination of those tissues with aflatoxin (Jones et al. 1980). How do "undamaged" corn kernels become infected by *A. flavus* and what accounts for the random pattern of aflatoxin occurrence among undamaged kernels on the same ear (Jones et al. 1980, Lee et al. 1980, Smart et al. 1990a)? The seeds of grasses and cereals are called achenes; these are single-seeded, dry, indehiscent fruit in which the pericarp is fused with the testa and surrounds the embryo (germ) and endosperm. The seed coat (testa) normally develops from the integuments of the ovule. Fungal penetration of hyphae into the seed proper is resisted by the testa which becomes thicker as the grains mature, except over the embryo (Semeniuk 1954). Marsh and Payne (1984a) demonstrated that following silk inoculation, *A. flavus* colonized silks in the honey-brown stage of senescence, the hyphae eventually reaching the kernel. While Marsh and Payne (1984a) found hyphae in the pedicel or tip cap they could find no evidence that the fungus had invaded the seeds internally. Zuber (1977) and others have observed that *A. flavus* is a weak pathogen of corn, infection of the grain usually requires a break in the pericarp. Marsh and Payne (1984b) suggest that the presence of *A. flavus* on kernel surfaces would allow the fungus to rapidly invade kernels that are injured by insect feeding. However, Smart et al. (1990b) showed that *A. flavus* gains entry to the seed proper through random microscopic breaks in the testa when the surrounding pericarp was intact. The fungus first spread from the wound-inoculation site on the ear and colonized tissues of the rachis (cob) entering the rachilla or tip cap region, growing unobstructed through the aerenchyma into the pericarp. Smart et al. (1990b) also examined infected kernels for evidence of hyphal penetration at other suggested sites of internal grain invasion, including the exterior of the pericarp, tip region and black layer as well as the point of silk attachment, but found no evidence of hyphal penetration.

When individual grains are surface disinfected and incubated on agar or moist filter paper it is difficult to determine if the *A. flavus* colonies originate from hyphae within the pericarp arenchyma or from the seed proper (e.g., embryo and endosperm). Aflatoxin is only produced after the fungus has invaded the embryo or endosperm. We have recorded *A. flavus* "infection" of numerous intact kernels from silk-inoculated ears where aflatoxin was not detected (Wicklow et al. 1987, 1988b). It should be obvious then that estimates of percent kernel infection should not be used as the sole guide in identifying corn varieties with aflatoxin resistance (Scott and Zummo 1988). Smart et al. (1990b) suggest that disruption of the vascular system during grain filling could lead to uneven drying-down of the embryo versus endosperm, resulting in disruption of the testa. This study further revealed that vascular bundles (xylem and phloem elements) and aerenchyma cells collapse well before fungal contact, suggesting a phytotoxic effect (Smart et al. 1990b). The role of other *A. flavus* secondary metabolites (Wilson 1966) in producing this phytotoxic effect needs to be investigated. For example, *A. flavus* is known to produce toxins that are lethal to wheat seeds (Thomas 1937). Aflatoxin is probably not the cause of this collapse of vascular bundles because aflatoxin can be taken-up and translocated by plant roots (Mertz et al. 1980, Llewellyn et al. 1982). With the use of kernel culturing techniques (Singletary and Below 1989) one should be able to accurately describe how individual *A. flavus* metabolites impact seed development.

Premature germination of the corn grain while the ears are still on the plants would also produce breaks in the testa. This might be linked to the presence of vivipary genes (Eyster 1931, Neuffer et al. 1968) or as a response to ear infection by *D. zeae* (Calvert et al. 1969) and *F. moniliforme* (Carvajal 1987). The testa is an important barrier to *A. flavus* in other seeds that become contaminated with aflatoxin (e.g., peanuts: LaPrade et al. 1973, Bampton 1963, Mixon and Rogers 1975, Graham 1982, Mehan and McDonald 1984; cotton seeds: Lee et al. 1987). Halloin (1983) presents additional examples showing that resistance to seed deterioration is associated with the seed coat, a mechanical and chemical barrier to potential fungal pathogens. Popcorns and small grains such as wheat, rye, oats, barley, grain sorghum, rice, and millet do not appear to be a significant source of aflatoxin exposure unless damaged in storage (Stoloff 1977). Hence, the seed coats of smaller grains seem to be less susceptible to tearing or stress cracks than large seeded corn hybrids?

Do cuticularized integuments of insect pupae present the same barrier to *A. flavus* as cuticularized seed coats (testa)? *Aspergillus flavus* produces extracellular hydrolytic enzymes that enable it to penetrate the cuticle of some insects, digesting the lipids, protein, and chitin components of the integument (Charnley 1984, Huber 1958). *Aspergillus flavus* can penetrate the integument at points where the eipcuticle becomes naturally abraded (Koidsumi 1957, Sussman 1951). The insect integument represents a formidable barrier to *A. flavus* infection, but the fungus is able to emerge through the cuticle of a diseased insect, presumably because the established mycelium can produce more enzymes (Benz 1963). Although *A. flavus* infected and killed a majority of the larvae and all of the adults of the giant silkworm *Platysamia cecropia* L., pupae were resistant to infection; hyphal attack of the pupae followed the removal of the waxy epicuticle

(Sussman 1951). Koidsumi (1957) found that lipids in the exuviae of *Bombyx mori* (L.) exhibited antifungal activity to *A. flavus* and he suggested that free, medium-chain length, unsaturated fatty acids in the cuticle, presumably caprylic and capric acid, were responsible for this antifungal activity. Furthermore, lipids extracted from *B. mori* strains showing resistance to *A. flavus* had greater antifungal activity than those from susceptible strains (Koidsumi and Wada 1955). Capric and caprylic acid are components of cuticular and seed lipids in many plant species. Do these fatty acids also function in protecting seeds from pathogenic fungi? Seed pathologists should find it interesting that a "phytoalexin-like" response was recorded in which capric acid accumulated as the most fungitoxic of several fatty and aromatic acids in iris bulbs infected with *Penicillium*, but capric acid was not detected in healthy bulbs (Kimura 1980).

Yanagita (1980) found no relationship between the activity of lipase in strains of *A. flavus* and their pathogenicity to Japanese silkworm (B. mori) larvae but chitnolytic activity did parallel pathogenicity. Is insect frass also a source of microbial or insect-derived cutinases and cellulases? If so, it could degrade the cuticle and cell walls of the seed coat, thus removing natural barriers to *A. flavus* infection of the germ. A relevant example is found in the addition of a purified cutinase from *C. gloeosporioides*, a fungal pathogen of papaya (Dickman et al. 1982), to spores of a *Mycosphaerella* sp. which allows this organism to penetrate and cause disease on papaya fruits with intact cuticles. Normally *Mycosphaerella* requires a wound to cause disease on papaya.

The aflatoxins function as entomotoxins and are implicated as having a role in the entomopathogenicity of *A. flavus* (Wright et al. 1982, Charnley 1984). Dowd (1988) has shown that another *A. flavus* metabolite, kojic acid, which is biosynthetically simple to produce, synergized the toxicity of aflatoxin B1 to corn earworm and fall armyworm larvae. *Aspergillus flavus* is a necrotrophic pathogen of corn grain, deriving nutrients from cells that are killed in advance of invading hyphae (Smart et al. 1990b), particularly the lipid-rich nucellar tissues (Fennell et al. 1973). Perhaps *A. flavus* "recognize" oil-rich seeds of some crops as insects and attack them with several entomotoxic metabolites (e.g., aflatoxins, kojic acid, cyclopiazonic acid)?

6.6. Temperature Stress Damage

Physiological stress to the corn plant, especially during grain-filling stages, affects the amount of aflatoxin contaminating the grain. Payne et al. (1986) demonstrated that irrigation reduced kernel infection when silks of a corn hybrid grown on irrigated versus nonirrigated plots were inoculated with *A. flavus* and covered with a bag to restrict insect access. Zuber et al. (1983) observed that high temperatures proved more important than lack of moisture in enhancing aflatoxin levels. High temperatures during grain development have been associated with high levels of aflatoxin in the field (Anderson et al. 1975; Lillehoj et al. 1978, 1980; Widstrom et al. 1990) and greater numbers of infected kernels from silk-inoculated corn during the period of grain filling (Jones et al. 1980, Payne et al. 1988). Aflatoxin levels were significantly greater in kernels from plants grown at the hottest daily temperature in plant growth chambers (Thompson et al. 1980). Jones et al. (1980) suggest that "at high temperatures this fungus (*A. flavus*) has increased parasitic ability." Alternatively, stress injury to maturing grains grown at high temperatures leads to cracks in the testa and allows internal colonization of the grains (Smart et al. 1990b). Stress-related damage to kernels may explain the substantial aflatoxin contamination recorded for silk-inoculated corn grown at elevated (30°C) night temperatures (Jones et al. 1981, Smart et al. 1990b). Unfortunately, Smart et al. (1990b) were unable to distinguish the extent to which the severely wrinkled and damaged kernels may have resulted solely from the effects of high temperature stress. Even so, significant increases in the frequency of kernel abortion and damage to the kernel were found among kernels cultured at high temperatures "in vitro" without any microorganisms present (Hanft and Jones 1986a, 1986b). Temperature stress results in poor kernel filling, vivipary, and kernel abortion, all of which provide entry points for opportunistic kernel rotting fungi such as *A. flavus*.

There is a logical trail of evidence to support the hypothesis that major aflatoxin outbreaks in preharvest corn are the result of physiological stress and damage to the kernel following exposure to "unseasonably" high evening temperatures during kernel filling. Approximately 25 years ago, corn growers in the southern United States stopped planting late-maturing "old southern" lines and began planting early-maturing corn hybrids that were initially developed for the corn belt (Zuber 1977). Zuber argued that corn-belt hybrids are not adapted for southern latitudes; he noted that corn-belt hybrids have inadequate husk coverage which makes them vulnerable to corn insects that damage the kernels and vector *A. flavus*. I suggest that the most important deficiency of the corn-belt hybrids is their inability to tolerate high evening temperatures during the period of grain filling and maturation. When early maturing corn belt hybrids are planted in southern latitudes this often exposes them to "unseasonably" high temperatures during intervals of kernel filling. Consider that in the corn belt, evening temperatures during kernel filling are about 15–20°C while evening temperatures exceeding 24°C are common in regions such as the Georgia coastal plain and the Rio Grande valley of Texas and Mexico, where aflatoxin contamination of preharvest corn is commonly detected. At some locations the problem may be further exacerbated by early planting of these corn belt hybrids. LaPrade and Manwiller (1977) reported that short season hybrids had a much greater build-up of aflatoxin B1 than the long-season hybrids. One study has shown that with progressively later planting dates in southern Georgia the corn had progressively less aflatoxin at harvest (Widstrom et al. 1990). This came as a surprise since Widstrom and his colleagues had predicted increased aflatoxin levels with later plantings because populations of *A. flavus* and numbers of kernel damaging insects both increase during the summer. Widstrom et al. (1990) showed an association between later planting dates and exposure to cooler temperatures during kernel filling. The later-maturing old southern genotypes would normally have ripened when temperatures are less extreme.

Corn has been grown in the American southwest by native Americans for centuries. In the lower Sonoran desert these open pollinated varieties have undergone extensive selection inadvertently for high temperature performance. We need to investigate physiological mechanisms of temperature stress tolerance in corn and to learn if resistance to high temperature stress may also convey some resistance to *A. flavus* infection and aflatoxin contamination.

7. FUTURE RESEARCH

For several decades, two observations have shaped the field of grain storage mycology. First, that a rather predictable fungal assemblage can be shown to dominate the grain when stored at a given moisture-temperature regime (Koehler 1938, Snow et al. 1944, Christensen and Sauer 1982, Lacey 1989). Conditions permitting safe storage of grain could be based upon a knowledge of the minimum moisture requirements of the individual fungi; the latter could be determined experimentally. Stored-product mycologists have begun to design more rigorous ecological experiments. This is best illustrated by Magan and Lacey (1984) who investigated the interaction of temperature and moisture in determining the outcome of competitive encounters between species of cereal fungi. As one example, they were able to show that *E. repens* de Bary, a dominant colonist of stored grain does not "win" a competitive encounter at a water activity of less than 0.85 but avoids competition by exploiting an environment unfavorable to other species.

A second important discovery was that grain heating in storage was a result of fungal and not seed respiration (Darsie et al. 1914, Gilman and Barron 1930). This required that grain-storage science emphasize strategies for control of mold growth apart from strategies that would impact seed respiration. Furthermore, it provided the impetus for research on methods that could accurately measure fungal biomass in grain (e.g., Seitz et al. 1979, 1982; Martin et al. 1989) or identify and quantify fungal volatiles as a measure of fungal deterioration of stored grain (Kaminski et al. 1972, 1974; Abramson et al. 1983). Even so, there have been very few studies of fungal biomass dynamics in preharvest or stored grain (Newell 1992). There is a clear opportunity here for research that would enable one to make useful predictions regarding environmental control of grain decay rates and the potential for population growth of fungus-feeding arthropods, based, of course, on a knowledge of the fungal species and their interactions with other microbes and animals.

The University of Manitoba/Agriculture Canada stored-grain research group recognized that a stored grain bulk could be examined within an ecosystem context, with energy being transfered through the biotic components of a heterotrophic food web (Sinha and Muir 1973). This approach has provided grain storage biologists with a relevant organizational framework for investigating organismal interactions and for evaluating the impact of such interactions on the total process of grain deterioration. Abiotic environmental parameters such as grain moisture content and temperature of the interstitial air have an impact on grain respiration and aging and on the activities of fungi and arthropods. Research organized within an ecosystems context can guide investigators to the more important interactions, suggest novel interdisciplinary studies, and assist in establishing research priorities (Sinha 1992).

Janzen (1977) recognized that fungal colonists of seeds or "seed-eating fungi" also compete with seed-eating animals for the same resource. Hence, these fungi are under strong selection to render seeds objectionable to seed-eating animals in the shortest possible time. Janzen proposes that fungi achieve this through the production of mycotoxins and metabolic conversions of seeds; the fungus both feeds itself and, at the same time, protects its food. This hypothesis identifies an ecological role for mycotoxins and shows how these fungal metabolites impact organisms in different trophic levels. Janzen (1977) also stimulated thinking about the potential role of fungal toxins in defending fungal tissues from arthropod fungivores (Wicklow and Cole 1982, Wicklow 1988a), and offered ecological relevance to studies of the secondary metabolite profiles and chemotaxonomy of these species (Frisvad 1986).

Investigations of fungi in grain storage ecosystems, like all of mycological ecology, are benefiting from numerous technological advances that make analyses of fungal biomass, fungal secondary metabolites and seed composition a more exact science. Methods in molecular ecology can provide us with an accurate assessment of the genetic relatedness among fungal populations and species. Sophisticated laboratory instruments have been developed that enable chemists to quickly and accurately measure various abiotic environmental parameters and to identify and quantify large numbers of known fungal metabolites from a small volume of sample extract. The latter is of practical relevance because it enables accurate identification of known mycotoxins and antibiotics in grain samples or fungal culture media and because it offers the opportunity to investigate the chemical ecology of three-way-interactions involving seeds, arthropods, and fungi. Much interesting research awaits us.

REFERENCES

Abramson, D., R. N. Sinha, and J. T. Mills. 1983. Mycotoxin and odor formation in barley stored at 16 and 20% moisture in Manitoba. Cereal Chem. 60:350-355.

Agarwal, V. K. and J. B. Sinclair. 1987. Principles of Seed Pathology. Vol. I. CRC Press, Inc., Boca Raton, FL. 176 pp.

Anderegg, R. J., K. Biemann, G. Buchi, and M. Cushman. 1976. Malformin C, a new metabolite of *Aspergillus niger*. J. Am. Chem. Soc. 98:3365-3370.

Anderson, H. W., E. W. Nehring, and W. R. Wichser. 1975. Aflatoxin contamination of corn in the field. J. Agric. Food Chem. 23:775-782.

Attwater, W. A. and L. V. Busch. 1983. Role of the sap beetle *Glischrochilus quadrisignathus* in the epidemiology of *Gibberella* corn ear rot. Can. J. Plant Pathol. 5:158-163.

Augspurger, C. K. 1984. Seedling survival of tropical tree species: interactions of dispersal distance, light-gaps, and pathogens. Ecology 65:1705-1712.

Ayerst, G. 1969. The effects of moisture and temperature on growth and spore germination in some fungi. J. Stored Prod. Res. 1:127-141.

Ayerst, G. 1986. Water and fungal ecology in stored products. pp. 359-373 in: Water, Fungi and Plants, P. G. Ayres and L. Boddy (eds.). Cambridge University Press, London.

Bacon, C. W., P. C. Lyons, J. K. Porter, and J. D. Robbins. 1986. Ergot toxicity from endophyte-infected grasses: a review. Agronomy J. 78:108-116.

Baker, K. F. 1972. Seed pathology. pp. 317-416 in: Seed Biology, Vol. II. T. T. Kozlowski (ed). Academic Press, New York.

Baker, K. F. and R. J. Cook. 1974. Biological Control of Plant Pathogens. Freeman, San Francisco. 433 pp.

Baker, D., M. H. Neustadt, and L. Zeleny 1957. Application of the fat acidity test as an index of grain deterioration. Cereal Chem. 34:226-233.

Bampton, S. S. 1963. Growth of *Aspergillus flavus* and production of aflatoxin in groundnuts. Part I. Trop. Sci. 5:74-81.

Beasley, V. R., W. B. Buck, R. F. Vesonder, and J. J. Ellis. 1982. Feed refusal in cattle associated with *Fusarium moniliforme* in corn. Vet. Record 111:393-394.

Bennett, G. A., D. T. Wicklow, R. W. Caldwell, and E. B. Smalley. 1988. Distribution of trichothecenes and zearalenone in *Fusarium graminearum* rotted corn ears grown in a controled environment. J. Agric. Food Chem. 36:639-642.

Benz, G. 1963. Physiopathology and histochemistry. pp. 299-338 in: Insect Pathology an Advanced Treatise, E. A. Steinhaus (ed.). Academic Press, New York.

Betina, V. (ed.) 1989. Mycotoxins—Chemical, Biological and Environmental Aspects. Elsevier Science Publ. Co., New York. 438 pp.

Blakeman, J. P. 1985. Ecological succession of leaf surface microorganisms in relation to biological control. pp. 6-30 in: Biological Control in the Phylloplane, C. E. Windels and S. E. Lindow (eds.). Am. Phytopathological Soc., St. Paul, MN.

Boosalis, M. G. 1956. Effect of soil temperature and green-manure amendment of unsterilized soil on parasitism of Rhizoctonia solani by *Penicillium vermiculatum* and *Trichoderma* sp. Phytopathology 46:473-478.

Bothast, R. J., E. B. Lancaster, and C. W. Hesseltine. 1973. Ammonia kills spoilage molds in corn. J. Dairy Sci. 56:241-245.

Bottomley, R. A., C. M. Christensen, and W. F. Geddes. 1952. Grain storage studies. X. The influence of aeration, time and moisture content on fat acidity, nonreducing sugars and mold flora of stored yellow corn. Cereal Chem. 29:53-64.

Brown, A. E., R. Finlay, and J. S. Ward. 1987. Antifungal compounds produced by *Epicoccum purpurascens* against soil-borne plant pathogenic fungi. Soil Biol. Biochem. 19:657-664.

Burmeister, H. R. 1983. *Fusarium moniliforme* as a producer of a feed refusal factor. Meetings Abstract, Am. Soc. Microbiol. Annual Meeting, New Orleans, LA.

Burmeister, H. R., R. F. Vesonder, and W. F. Kwolek. 1980. Mouse bioassay for *Fusarium* rejection or acceptance when dissolved in drinking water. Appl. Environ. Microbiol. 39:957-961.

Burnside, J. E., W. L. Sippel, J. Forgacs, W. T. Carll, M. B. Atwood, and E. R. Doll. 1957. A disease of swine caused by eating moldy corn. II. Experimental production with pure cultures of molds. Am. J. Vet. Res. 18:817-824.

Bush, D. A., O. Touster, and J. E. Brockman. 1951. The production of B-nitropropionic acid by a strain of *Aspergillus flavus*. J. Biol. Chem. 188:685-693.

Caldwell, R. W., J. F. Tuite, and W. W. Carlton. 1981. Pathogenicity of Penicillia to corn ears. Phytopathology 71:175-180.

Calvert, O. H., M. S. Zuber, and M. G. Neuffer. 1969. Vivipary in Zea mays induced by *Diplodia maydis*. Phytopathology 59:239-240.

Carlton, W. W., J. F. Tuite, and P. Mislivec. 1968. Investigations of the toxic effects in mice of certain species of *Penicillium*. Toxicol. Appl. Pharmacol. 13:372-387.

Carroll, G. C. 1992. Fungal mutualism. pp. 327-354 in: The Fungal Community, Second Edition, G. C. Carroll and D. T. Wicklow (eds.). Marcel Dekker, New York.

Carvajal, M. 1987. Premature germination of maize and the related mycotoxin. pp. 71-73 in: Aflatoxin in Maize: A Proceedings of the Workshop, M. S. Zuber, E. B. Lillehoj and B. L. Renfro (eds.). The International Maize and Wheat Improvement Center (CIMMYT), Mexico D.F.

Castor, L. L. and R. A. Frederiksen. 1980. Fusarium head blight caused by *Fusarium moniliforme* occurrence and effects on sorghum yield and grain characteristics in Texas. Plant Dis. 64:1017-1019.

Chang, I-pin. and T. Kommedahl. 1968. Biological control of seedling blight of corn by coating kernels with antagonistic microorganisms. Phytopathology 58:1395-1401.

Charnley, A. K. 1984. Physiological aspects of destructive pathogenesis in insects by fungi: a speculative review. pp. 229-270 in: Invertebrate-Microbial Interactions, J. M. Anderson, A. D. M. Rayner, and D. W. Walton (eds.). Cambridge University Press, London.

Chelkowski, J. (ed.). 1989. Fusarium: Mycotoxins, Taxonomy and Pathogenicity. Elsevier Science Publishers, Amsterdam. 492 pp.

Cheplick, G. P. and K. Clay. 1988. Acquired chemical defenses of grasses: the role of fungal endophytes. Oikos 52:309-318.

Christensen, C. M. 1951. Fungi on and in wheat seed. Cereal Chem. 28:408-415.

Christensen, C. M. 1957. Deterioration of stored grains by fungi. Bot. Rev. 23:108-134.

Christensen, C. M. 1967. Some changes in No. 2 corn stored two years at moisture contents of 14.5 and 15.2% and temperatures of 12, 20 and 25°C. Cereal Chem. 44:95-99.

Christensen, C. M. 1973. Loss of viability in storage: microflora. Seed Sci. Technol. 1:547-562.

Christensen, C. M. and D. R. Gordon. 1948. The mold flora of stored wheat and corn and its relation to heating of moist grain. Cereal Chem. 25: 40-51.

Christensen, C. M. and H. H. Kaufmann. 1965. Deterioration of stored grains by fungi. Annu. Rev. Phytopathol. 3:69-84.

Christensen, C. M. and H. H. Kaufmann. 1969. Grain storage: The Role of Fungi in Quality Loss. University Minnesota Press, Minneapolis, MN. 153 pp.

Christensen, C. M. and H. H. Kaufmann. 1974. Microflora. pp. 158-192 in: Storage of Cereal grains and their products, C. M. Christensen (ed.). Am. Assoc. Cereal Chem., St. Paul, MN.

Christensen, C. M. and H. H. Kaufmann. 1975. Control of postharvest losses caused by fungi in food, feed grains. Feedstuffs 47(10):47,58.

Christensen, M. J. and G. C. M. Latch. 1991. Variation within isolates of *Acremonium* endophytes (*A. coenophialum* and possibly *A. typhinum*) from tall fescue (*Festuca arundinacea*). Mycol. Res. 95:1123-1126.

Christensen, C. M. and D. H. Sauer. 1982. Microflora. pp. 219-240 in: Storage of Cereal Grains and Their Products, C. M. Christensen (ed.). Am. Assoc. Cereal Chem., St. Paul, MN.

Christensen, M. J., G. C. M. Latch, and B. A. Tapper. 1991. Variation within isolates of *Acremonium* endophytes from perennial rye-grasses. Mycol. Res. 95:918-923.

Christensen, C. M., G. H. Nelson, and C. J. Mirocha. 1965. Effect on the white rat uterus of a toxic substance isolated from *Fusarium*. Appl. Microbiol. 13:653-659.

Ciegler, A., E. B. Lillehoj, R. E. Peterson, and H. H. Hall. 1966. Microbial detoxification of aflatoxin. Appl. Microbiol. 14:934-939.

Ciegler, A., D. I. Fennell, G. A. Sansing, R. W. Detroy, and G. A. Bennett. 1973. Mycotoxin-producing strains of *Penicillium viridicatum:* Classification into subgroups. Appl. Microbiol. 26:271-278.

Clay, K. 1990a. Fungal endophytes of grasses. Annu. Rev. Ecol. Syst. 21:275-297.

Clay, K. 1990b. Insects, endophytic fungi and plants. pp. 111-130 in: Pests, Pathogens and Plant Communities, J. J. Burdon and S. R. Leather (eds.). Blackwell Scientific Pub., Oxford.

Clay, K., G. P. Cheplick, and S. M. Wray. 1989. Impact of the fungus *Balansia henningsiana* in the grass *Panicum agrostoides:* frequency of infection, plant growth and reproduction, and resistance to pests. Oecologia 80:374-380.

Cole, R. J. 1977. Tremorgenic mycotoxins. pp. 583-595 in: Mycotoxins in Human and Animal Health, J. V. Rodricks, C. W. Hesseltine, and M. A. Mehlman (eds.). Pathotox Publishers, Inc, Park Forest South, IL.

Cole, R. J. and R. H. Cox. 1981. Handbook of Toxic Fungal Metabolites. Academic Press, New York. 937 pp.

Cole, R. J., J. W. Kirksey, and B. R. Blankenship. 1972. Conversion of aflatoxin B1 to isomeric hydroxy compounds by *Rhizopus* spp. J. Agr. Food Chem. 20:1100-1102.

Cotty, P. J. 1989. Virulence and cultural characteristics of two *Aspergillus flavus* strains pathogenic on cotton. Phytopathology 79:808-814.

Cotty, P. J. 1990. Effect of aflatoxigenic strains of *Aspergillus flavus* on aflatoxin contamination of developing cottonseed. Plant Dis. 74:233-235.

Cuero, R. G., J. E. Smith, and J. Lacey. 1987a. Interaction of water activity, temperature and substrate on mycotoxin production by *Aspergillus flavus*, *Penicillium viridicatum*, and *Fusarium graminearum* in irradiated grains. Trans. Br. Mycol. Soc. 89:221-226.

Cuero, R. G., J. E. Smith, and J. Lacey. 1987b. Stimulation by *Hyphopichia burtonii* and *Bacillus amyloliquefaciens* of aflatoxin production by *Aspergillus flavus* in irradiated maize and rice grains. Appl. Environ. Microbiol. 53:1142-1146.

Culvenor, C. C. 1973. Alkaloids. pp. 375-446 in: Chemistry and Biochemistry of Herbage, Vol. I., G. W. Butler and R. W. Bailey (eds.). Academic Press, New York.

Cunningham K. G. and G. G. Freeman. 1953. The isolation and some chemical properties of viridicatin, a metabolic product of *Penicillium viridicatum* Westling. Biochem. J. 53:328-332.

Darsie, M. L., C. Elliott, and G. J. Peirce. 1914. Heat evolved in the germination of seeds. Bot. Gaz. 58:101-136.

Dean, R. A. and J. A. Kuc. 1987. Immunization against disease: the plant fights back. pp. 383-410 in: Fungal Infection of Plants, G. F. Pegg and P. G. Ayres (eds.). Cambridge University Press, London.

Dekker, J. 1969. Antibiotics. pp. 579-635 in: Fungicides, Vol. II., D. C. Torgeson (ed.). Academic Press, New York.

Dickinson, C. H. 1981. Biology of *Alternaria alternata*, *Cladosporium cladosporioides*, and *C. herbarum* in respect to their activity on green plants. pp. 169-184 in: Microbial Ecology of the Phylloplane, J. P. Blakeman (ed.). Academic Press, London.

Dickman, M. B., S. S. Patil, and P. E. Kolattukudy. 1982. Purification, characterization and role in infection of an extracellular cutinolytic enzyme from *Colletotrichum gloeosporioides* Penz. on *Carica papaya* L. Physiol. Plant Pathol. 20:333-347.

Diener, U. L., R. J. Cole, T. H. Sanders, G. A. Payne, L. S. Lee, and M. A. Klich. 1987. Epidemiology of aflatoxin formation by *Aspergillus flavus*. Annu. Rev. Phytopathol. 25:249-270.

Dirks, B. M., P. D. Boyer, and W. F. Geddes. 1955. Some properties of fungal lipases and their significance in stored grains. Cereal Chem. 32:356-373.

Dix, D. E. and J. N. All. 1987. Interactions between maize weevil (Coleoptera: Curculionidae) infestations and infection by *Aspergillus flavus* and other fungi in stored corn. J. Entomol. Sci. (The Entomological Soc., Tifton, GA). 22:108-118.

Domsch, K. H., W. Gams, and T. H. Anderson. 1980. A Compendium of Selected Soil Fungi, Vols. I and II. Academic Press, New York. 859 pp.

Dorner, J. W., R. J. Cole, and R. A. Hill. 1984. Tremorgenic mycotoxins produced by *Aspergillus fumigatus* and *Penicillium crustosum* isolated from molded corn implicated in a natural intoxication of cattle. J. Agric. Food Chem. 32:411-413.

Dorner, J. W., R. J. Cole, L. G. Lomax, H. S. Gosser, and U. L. Diener. 1983. Cyclopiazonic acid production by *Aspergillus flavus* and its effects on broiler chickens. Appl. Environ. Microbiol. 46:698-703.

Dorsey, C. K. and J. G. Leach. 1956. The bionomics of certain insects associated with oak wilt, with particular reference to the Nitidulidae. J. Econ. Entomol. 49:219-230.

Dowd, P. F. 1988. Synergism of aflatoxin B1 toxicity with the co-occurring fungal metabolite kojic acid to two caterpillars. Entomol. Exp. Appl. 47:69-71.

Dowd, P. F. 1989. Toxicity of naturally occurring levels of the *Penicillium* mycotoxins citrinin, ochratoxin A, and penicillic acid to the corn earworm, *Heliothis zea*, and the fall armyworm, *Spodoptera frugiperda* (Lepidoptera: Noctuidae). Environ. Entomol. 18:24-29.

Dowd, P. F. 1992a. Insect interactions with mycotoxin-producing fungi and their hosts. pp. 137-155 in: Handbook of Applied Mycology. Vol. 5: Mycotoxins in Ecological Systems, D. Bhatnagar, E. B. Lillehoj and D. K. Arora (eds.). Marcel Dekker, New York.

Dowd, P. F. 1992b. Insect fungal symbionts: a promising source of detoxifying enzymes. J. Indust. Microbiol. 9:149-161.

Dowd, P. F. and R. J. Bartelt. 1991. Host-derived volatiles as attractants and pheromone synergists for dried fruit beetle, *Carpophilus hemipterus*. J. Chem. Ecol. 17:285-308.

Dowd, P. F. and F. L. van Middlesworth. 1989. In vitro metabolism of the trichothecene 4-monoacetoxyscirpenol by fungus- and nonfungus feeding insects. Experimentia 45:393-395.

Dungan, G. H. and B. Koehler. 1944. Age of seed corn in relation to seed infection and yielding capacity. J. Am. Soc. Agron. 36:436-443.

Dunkel, F. V. 1988. The relationship of insects to the deterioration of stored grain by fungi. Int. J. Food Microbiol. 7:227-244.

Dunkel, F. V. 1992. The stored grain ecosystem: A global perspective. J. Stored Prod. Res. 28:73-87.

Dunkel, F. V., A. V. Barak, and P. K. Harein. 1982. Geographical distribution of *Cynaeus angustus* (Le Conte) (Coleoptera: Tenebrionidae) and its association with stored products. J. Biogeogr. 9:345-352.

Eyster, W. H. 1931. Vivipary in maize. Genetics 16:574-590.

Evenari, M. 1949. Germination inhibitors. Bot. Rev. 15:153-194.

Evans, H. C. 1988. Coevolution of entomogenous fungi and their insect hosts. pp. 149-171 in: Coevolution of Fungi with Plants and Animals, K. A. Pirozynski and D. L. Hawksworth (eds.). Academic Press, London.

Fennell, D. I., R. J. Bothast, E. B. Lillehoj, and R. E. Peterson. 1973. Bright greenish-yellow fluorescence and associated fungi in white corn naturally contaminated with aflatoxin. Cereal Chem. 50:404-414.

Fennell, D. I., W. F. Kwolek, E. B. Lillehoj, G. A. Adams, R. J. Bothast, M. S. Zuber, O. H. Calvert, W. D. Guthrie, A. J. Bockholt, A. Manwiller, and M. D. Jellum. 1977. *Aspergillus flavus* presence in silks and insects developing and mature corn ears. Cereal Chem. 54:770-778.

Fenner, M. 1985. Seed Ecology. Chapman and Hall Ltd., London. 151 pp.

Flannigan, B. 1978. Primary contamination of barley and wheat grain by storage fungi. Trans. Br. Mycol. Soc. 71:37-42.

Fokkema, N. J. 1981. Fungal leaf saprophytes, beneficial or detrimental? pp. 433-454 in: Microbial Ecology of the Phylloplane, S. J. P. Blakeman (ed.). Academic Press, London.

Ford, V. L. and T. L. Kirkpatrick. 1989. Effects of *Acremonium coenophialum* in tall fescue on host disease and insect resistance and allelopathy to *Pinus taeda* seedlings. Proc. Ark. Fescue Toxicosis Conf. Special Rep. 140:29-34.

Friday, D., J. F. Tuite, and R. Stroshine. 1989. Effect of hybrid and physical damage on mold development and carbon dioxide production during storage of high-moisture shelled corn. Cereal Chem. 66:422-426.

Frisvad, J. C. 1986. Taxonomic approaches to mycotoxin identification. pp. 415-457 in: Modern Methods in the Analysis and Structure Elucidation of Mycotoxins, R. J. Cole (ed.). Academic Press, New York.

Frisvad, J. C. and O. Filtenborg. 1989. *Terverticillate penicillia:* chemotaxonomy and mycotoxin production. Mycologia 81:837-861.

Frisvad, J. C., Filtenborg, O., and Wicklow, D. T. 1987. *Terverticillate penicillia* isolated from underground seed caches and cheek pouches of banner-tailed kangaroo rats (*Dipodomys spectabilis*). Can. J. Bot. 65:765-773.

Fuentes, S. F., M. de Lourdes de la Isla, A. J. Ullstrup, and A. E. Rodriguez. 1964. *Claviceps gigantea*, a new pathogen of maize in Mexico. Phytopathology 54:379-381.

Gams, W. 1971. *Cephalosporium*-artige Schimmelpilze (Hyphomycetes). Gustav Fischer Verlag, Stuttgart, Germany. 262 pp.

Gelderblom, W. C. A., K. Jaskiewicz, W. F. O. Marasas, P. G. Thiel, R. M. Horak, R. Vleggaar, and N. P. J. Kriek. 1988. Fumonisins-novel mycotoxins with cancer-promoting activity produced by *Fusarium moniliforme*. Appl. Environ. Microbiol. 54:1806-1811.

Gerlach, W. 1982. The Genus *Fusarium*—A Pictorial Atlas, Kommissionsverlag Parey, Berlin, Germany. 406 pp.

Gilman, J. C. and D. H. Barron. 1930. Effect of molds on temperature of stored grain. Plant Physiol. 5:565-573.

Gloer, J. B. and S. M. Truckenbrod. 1988. Interference competition among coprophilous fungi: Production of (+)-isoepoxydon by *Poronia punctata*. Appl. Environ. Microbiol. 54:861-864.

Gloer, J. B., M. R. TePaske, J. R. Sima, D. T. Wicklow, and P. F. Dowd. 1988. Antiinsectan aflavinine derivatives from the sclerotia of *Aspergillus flavus*. J. Org. Chem. 53: 5457-5460.

Goodman, J. J. and C. M. Christensen. 1952. Grain storage studies. XI. Lipolytic activity of fungi isolated from stored corn. Cereal Chem. 29:299-308.

Graham, J. 1982. Aflatoxin in peanuts: occurrence and control. Queensland Agric. J. 108:119-122.

Griffin, D. M. 1966. Fungi attacking seeds in dry seed-beds. Proc. Linnean Soc. N. S. W. 91:84-89.

Griffin, G. J. and K. H. Garren. 1976. Colonization of rye green manure and peanut fruit debris by *Aspergillus flavus* and *Aspergillus niger* group in field soils Appl. Environ. Microbiol. 32:28-32.

Guthrie, L. D. 1979. Effects of aflatoxin in corn on production and reproduction in dairy cattle. J. Dairy Sci. (Meetings Abstract) 62:134.

Hall, R. F., L. R. Harrison, and B. M. Colvin. 1989. Aflatoxicosis in cattle pastured in a field of sweet corn. J. Am. Vet. Med. Assoc. 194:938.

Hallett, S. G. and N. D. Paul. 1989. Rust pustules support invasion of ramsons by *Botrytis cinerea*. The Mycologist 3:174-175.

Halloin, J. M. 1983. Deterioration resistance mechanisms in seeds. Phytopathology 73: 335-339.

Hamilton, P. B., W. E. Huff, J. R. Harris, and R. D. Wyatt. 1982. Natural occurrences of ochratoxicosis in poultry. Poultry Sci. 61:1832-1841.

Hanft, J. M. and R. J. Jones. 1986a. Kernel abortion in maize. I. Carbohydrate concentration patterns and acid invertase activity of maize kernels induced to abort "in vitro". Plant Physiol. 81:503-510.

Hanft, J. M. and R. J. Jones. 1986b. Kernel abortion in maize. II. Distribution of 14C among kernel carbohydrates. Plant Physiol. 81:511-515.

Harris, H. B. and R. E. Burns. 1973. Relationship between tannin content of sorghum grain and preharvest seed molding. Agron. J. 65:957-959.

Hawksworth, D. L. 1981. A survey of the fungicolous conidial fungi. pp. 171-244 in: Biology of Conidial Fungi, Vol. I., G. T. Cole and B. Kendrick (eds.). Academic Press, New York.

Hill, S. T. 1978. Development of *Ahasverus advena* (Coleoptera: Silvanidae) on seven species of *Aspergillus* and on food molded by two of these. J. Stored Prod. Res. 14:227-231.

Holdenreid, R. 1957. Natural history of the banner-tail kangaroo rat in New Mexico. J. Mammal. 38:330-350.

Holmes, R. S., L. N. Moxton, and P. A. Parsons. 1980. Genetic variability of alcohol dehydrogenase among Australian *Drosophila* species: correlation of ADH biochemical phenotype with ethanol resource utilization. J. Exp. Zool. 214:199-204.

Horn, B. W. 1985. Association of *Candida guilliermondii* with amylolytic filamentous fungi on preharvest corn. Can. J. Microbiol. 31:19-23.

Horn, B. W. and D. T. Wicklow. 1983. Factors influencing the inhibition of aflatoxin production in corn by *Aspergillus niger*. Can. J. Microbiol. 29:1087-1091.

Huber, J. 1958. Untersuchungen zur physiologie insektentoetender pilze. Arch. Mikrobiol. 29:257-276.

Huff, W. E., R. D. Wyatt, and P. B. Hamilton. 1975. Nephrotoxicity of dietary ochratoxin A in broiler chickens. Appl. Microbiol. 30:48-51.

Ingham, R. E. 1992. Interactions between invertebrates and fungi: Effects on nutrient availability. pp. 669-690 in: The Fungal Community, Second Edition, G. C. Carroll and D. T. Wicklow (eds.). Marcel Dekker, New York.

Jaenike, J. 1985. Parasite pressure and the evolution of amanitin tolerance in *Drosophila*. Evolution 39:1295-1301.

Janzen, D. H. 1971. Seed predation by animals. Annu. Rev. Ecol. Syst. 2:465-492.

Janzen, D. H. 1977. Why fruits rot, seeds mold and meat spoils. Am. Nat. 111:691-713.

Janzen, D. H. 1984. Dispersal of small seeds by big herbivores: foliage is the fruit. Am. Nat. 123:338-353.

Joffe, A. Z. 1986. *Fusarium* Species: Their Biology and Toxicology. John Wiley and Sons, New York. 588 pp.

Johann, H., J. R. Holbert, and J. G. Dickson. 1931. Further studies on *Penicillium* injury to corn. J. Agric. Res. 56:291-307.

Jones, C. G. 1984. Microorganisms as mediators of plant resource exploitation by insect herbivores. pp. 53-99 in: A New Ecology: Novel Approaches to Interactive Systems, P. W. Price, C. N. Slobodchikoff and W. S. Gaud (eds.). John Wiley and Sons, New York.

Jones, R. K., H. E. Duncan, and P. B. Hamilton. 1981. Planting date, harvest date, and irrigation effects on infection and aflatoxin production by *Apergillus flavus* in field corn. Phytopathology 71:810-816.

Jones, R. K., H. E. Duncan, G. A. Payne, and K. A. Leonard. 1980. Factors influencing infection by Aspergillus flavus in silk-inoculated corn. Plant Dis. 64:859-863.

Jong, S. C. and M. J. Gantt (eds.) 1987. ATCC Catalog of Fungi/Yeasts. 17th edition, American Type Culture Collection, Rockville, MD. 532 pp.

Kaminski, E., L. M. Libbey, S. Stawicki, and E. Wasowicz. 1972. Identification of the predominant volatile compounds produced by *Aspergillus flavus*. Appl. Microbiol. 24:721-726.

Kaminski, E., S. Stawicki, and E. Wasowicz. 1974. Volatile flavor compounds produced by molds of *Aspergillus, Penicillium*, and Fungi Imperfecti. Appl. Microbiol. 27: 1001-1004.

Kay, F. R. and W. G. Whitford. 1978. The burrow environment of the banner-tailed kangaroo rat, *Dipodomys spectabilis*, in south-central New Mexico. Am. Midl. Nat. 99:270-279.

Kearney, T. H., R. H. Peebles, J. T. Howell, and E. McClintock. 1960. Arizona Flora. University California Press, Berkeley, CA. 1085 pp.

Kellerman, T. S., C. J. Rabie, G. C. A. Van-der-Westhuizen, N. P. J. Kriek, and L. Prozesky. 1985. Induction of diplodiosis, a neuromycotoxicosis, in domestic ruminants with cultures of indigenous and exotic isolates of *Diplodia maydis*. Onderstepoort J. Vet. Res. 52:35-42.

Kimura, Y. 1980. Some antifungal components of diseased Blue Ribbon iris bulbs. J. Agric. Food Chem. 28:1022.

King, S. B. 1981. Time of infection of maize kernels by *Fusarium moniliforme* and *Cephalosporium acremonium*. Phytopathology 71:796-799.

King, A. D., J. I. Pitt, L. R. Beuchat, and J. E. L. Corry. (eds.). 1986. Methods for the mycological examination of foods. Plenum Press, New York. 315 pp.

Koehler, B. 1938. Fungus growth in shelled corn as affected by moisture. J. Agric. Res. 56:291-307.

Koehler, B. 1957. Pericarp injuries in seed corn. Ill. Agric. Exp. Stn. Bull. 617, Urbana, IL. 72 pp.

Koehler, B. 1959. Corn ear rots in Illinois. Ill. Agric. Exp. Stn. Bull. 639, Urbana, IL. 87 pp.

Koidsumi, K. 1957. Antifungal action of cuticular lipids in insects. J. Insect Physiol. 1:40-51.

Koidsumi, K. and Y. Wada. 1955. Studies on the antimicrobial function of insect lipids. IV. Racial differences in the antifungal activity in the silkworm integument. Japan J. Appl. Zool. 20:184-190.

Kozakiewicz, Z. 1989. *Aspergillus* species on stored products. Mycol. Papers 161:1-188.

Kramer, P. J. and T. T. Kozlowski. 1979. Physiology of Woody Plants. Academic Press, New York. 811 pp.

Kucharek, T. A. and T. Kommedahl. 1961. Kernel infection and corn stalk rot caused by *Fusarium moniliforme*. Phytopathology 56:983-984.

Kurtzman, C. P., B. W. Horn, and C. W. Hesseltine. 1987. *Aspergillus nomius*, a new aflatoxin-producing species related to *Aspergillus flavus* and *Aspergillus tamarii*. Antonie van Leewenhoek 53:147-158.

Lacey, J. 1986. Water availability and fungal reproduction: patterns of spore production, liberation and dispersal. pp. 65-86 in: Water, Fungi and Plants, P. G. Ayres and L. Boddy (eds.). Cambridge University Press, London.

Lacey, J. 1989. Pre- and postharvest ecology of fungi causing spoilage of foods and other stored products. pp. 11S-22S in: Filamentous Fungi in Foods and Feeds, M. O. Moss, B. Jarvis and F. A. Skinner (eds.). Supplement to J. Appl. Bacteriol. Vol. 67, Blackwell Scientific Pub., Oxford. 149 pp.

LaPrade, J. F., J. A. Bartz, A. J. Norden, and T. J. Demuynk. 1973. Correlation of peanut seed coat surface wax accumulations with tolerance to colonization by *Aspergillus flavus*. J. Am. Peanut Res. Educ. Assoc. 5:89-94.

LaPrade, J. C. and A. Manwiller. 1977. Relation of insect damage, vector, and hybrid reaction to aflatoxin B1 recovery from field corn. Phytopathology 67:544-547.

Lawrence, E. B., P. E. Nelson, and J. E. Ayers. 1981. Histopathology of sweet corn seed and plants infected with *Fusarium moniliforme* and *F. oxysporum*. Phytopathology 71:379-386.

Lee, L. S., P. E. Lacey, and W. R. Goynes. 1987. Aflatoxin in Arizona cottonseed: A model study of insect-vectored entry of cotton bolls by *Aspergillus flavus*. Plant Dis. 71:997-1001.

Lee, L. S., E. B. Lillehoj, and W. F. Kwolek. 1980. Aflatoxin distribution in individual corn kernels from intact ears. Cereal Chem. 57:340-343.

Leslie, J. F., C. A. S. Pearson, P. E. Nelson, and T. A. Tousson. 1990. *Fusarium* spp. from corn, sorghum, and soybean fields in the central and eastern United States. Phytopathology 80:343-350.

Leuchtmann, A. and K. Clay. 1988. *Atkinsonella hypoxylon* and *Balansia cyperi*, epiphytic members of the Balansiae. Mycologia 80:192-199.

Leufven, A. 1991. Role of microorganisms in spruce bark beetle-conifer interactions. pp. 467-483 in: Microbial Mediation of Plant-Herbivore Interactions. P. Barbosa, V. A. Krischik, and C. G. Jones (eds.). John Wiley and Sons, Inc. New York.

Lichtwardt, R. W., G. L. Barron, and L. H. Tiffany. 1958. Mold flora associated with shelled corn in Iowa. Iowa State Coll. J. Sci. 33:1-11.

Lillehoj, E. B. and C. W. Hesseltine. 1977. Aflatoxin control during plant growth and harvest of corn. pp. 107-119 in: Mycotoxins in Human and Animal Health, J. V. Rodricks, C. W. Hesseltine and M. A. Mehlman (eds.). Pathotox Publ., Park Forest South, IL.

Lillehoj, E. B., D. I. Fennell, W. F. Kwolek, G. L. Adams, M. S. Zuber, E. S. Horner, N. Widstrom, H. Warren, W. D. Guthrie, D. B. Sauer, W. R. Findley, A. Manwiller, L. M. Josephson, and A. J. Bockholt. 1978. Aflatoxin contamination of corn before harvest: *Aspergillus flavus* association with insects from developing ears. Crop Sci. 18: 921-924.

Lillehoj, E. B., W. W. McMillian, W. D. Guthrie, and D. Barry. 1980. Aflatoxin-producing fungi in pre-harvest corn: inoculum source in insects and soils. J. Environ. Qual. 9:691-699.

Lillehoj, E. B., W. F. Kwolek, E. E. Vandegraft, M. S. Zuber, O. H. Calvert, N. W. Widstrom, M. C. Futrell, and A. J. Bockholt. 1975. Aflatoxin production in *Aspergillus flavus* inoculated ears grown at diverse locations. Crop Sci. 15:267-270.

Lillehoj, E. B., W. W. McMillian, N. W. Widstrom, W. D. Guthrie, J. L. Jarvis, D. Barry, and W. F. Kwolek. 1984. Aflatoxin contamination of maize kernels before harvest. Mycopathologia 86:77-81.

Ling, K. H. 1976. Study on mycotoxins contaminated in food in Taiwan. II. Tremor-inducing compounds from *Aspergillus terreus*. Proc. Nat. Sci. Council, R. O. C. 9:121-129.

Ling, K. H., C-K Yang, and F. T. Peng. 1979. Territrems, tremorgenic mycotoxins of *Aspergillus terreus*. Appl. Environ. Microbiol. 37:355-357.

Linsley, E. g., 1944. Natural sources, habitats and reservoirs of insects associated with stored food products. Hilgardia 16:187-224.

Liu, X. J., X. Y. Luo, and W. J. Hu. 1989. *Arthrinium* spp. and the etiology of deteriorated sugarcane poisoning. pp. 109-118 in: Mycotoxins and Phycotoxins '88, S. Natori, K. Hashimoto and Y. Ueno (eds.). Elsevier Science Pub., Amsterdam.

Llewellyn, G. C., J. D. Reynolds, L. Hurst, R. A. Vance, and W. V. Dashek. 1982. Uptake of Zn++ and aflatoxin from perlite and liquid culture by Zea mays seedlings. Mycopathologia 77:111-116.

Lopez, L. C. and C. M. Christensen. 1967. Effect of moisture content and temperature on invasion of stored corn by *Aspergillus flavus*. Phytopathology 57:588-590.

Loschiavo, S. R., A. J. McGinnis, and D. R. Metcalfe. 1969. Nutritive value of barley varieties assessed with the confused flour beetle. Nature 224:288.

Lussenhop, J. and D. T. Wicklow. 1990. Nitidulid beetles (Nitidulidae: Coleoptera) as vectors of *Aspergillus flavus* in pre-harvest maize. Trans. Mycol. Soc. Japan 31:63-74.

Luttrell, E. S. 1981. Tissue replacement diseases caused by fungi. Annu. Rev. Phytopathol. 19:373-389.

Lynch, G. P., C. G. Todd, W. T. Shalkop, and L. A. Moore. 1970. Response of dairy calves to aflatoxin contaminated feed. J. Dairy Sci. 53:63-71.

Magan, N. and J. Lacey. 1984. Effect of water activity, temperature and substrate on interactions between field and storage fungi. Trans. Br. Mycol. Soc. 82:83-93.

Magan, N. and J. Lacey. 1988. Mycotoxin formation by *Aspergillus flavus* and *Fusarium graminearum* in irradiated maize grains in the presence of other fungi. J. Food Prot. 51:452-456.

Maier, D. E., R. G. Moreira, and F. W. Bakker-Arkema. 1992. Comparison of conventional and chilled aeration of grains under Texas conditions. Appl. Engin. Agric. 8:661-667

Malloch, D. M. 1985. The Trichocomaceae: Relationships with other Ascomycetes. pp. 365-382 in: Advances in *Penicillium* and *Aspergillus* Systematics, R. A. Samson and J. I. Pitt (eds.). Plenum Press, New York.

Mantle, P. G. 1975. Industrial exploitation of ergot fungi. pp. 281-300 in: The filamentous fungi, Vol. I. Industrial Mycology, J. E. Smith and D. R. Berry (eds.). John Wiley and Sons, New York.

Marasas, W. F. O. 1977. Diplodiosis in Cattle. pp. 163-165. Diplodiosis in Sheep. pp. 218-219 in: Mycotoxic Fungi, Mycotoxins, Mycotoxicoses: An Encyclopedic Handbook, T. D. Wyllie and L. G. Morehouse (eds.). Marcel Dekker, Inc., New York.

Marasas, W. F. O., T. S. Kellerman, W. C. A. Gelderblom, J. A. W. Coetzer, P. G. Thiel, and J. J. Van Der Lugt. 1989. Leukoencephalomalacia in a horse induced by fumonisin B1 isolated from *Fusarium moniliforme*. Onderstepoort J. Vet. Res. 55: 197-203.

Marasas, W. F. O., P. E. Nelson, and T. A. Tousson. 1984. Toxigenic *Fusarium* species: Identity and Mycotoxicology. The Pennsylvania State University Press, University Park, PA. 328 pp.

Marsh, S. F. and G. A. Payne. 1984a. Scanning EM studies on the colonization of dent corn by *Aspergillus flavus*. Phytopathology 74:557-561.

Marsh, S. F. and G. A. Payne. 1984b. Preharvest infection of corn silks and kernels by *Aspergillus flavus*. Phytopathology 74:1284-1289.

Martin, M. M. 1987. Invertebrate-Microbial Interactions. Comstock Pub. Associates, Cornell University Press, Ithaca, NY. 176 pp.

Martin, S. L., J. F. Tuite, and M. A. Diekman. 1989. Inhibition radioimmunoassay for *Aspergillus repens* compared with other indices of fungal growth in stored corn. Cereal Chem. 66:139-144.

Mayer, A. M. and A. Poljakoff-Mayber. 1963. The germination of seeds. The MacMillan Co., New York. 236 pp.

McGee, D. C. and C. M. Christensen. 1970. Storage fungi and fatty acids in seeds held thirty days at moisture contents of 14 and 16%. Phytopathology 60:1775-1777.

McIntyre, J. L. and P. M. Miller. 1978. Protection of tobacco against *Phytopthora parasitica* var. *nicotianae* by cultivar-nonpathogenic races, cell-free sonicates, and *Pratylenchus penetrans*. Phytopathology 68:235-239.

McKenzie, J. A. and S. W. McKechnie. 1979. A comparative study of resource utilization in natural populations of *Drosophila melanogaster* and *D. simulans*. Oecologia (Berlin) 40:299-309.

McLaren, D. L., H. C. Huang, and S. R. Rimmer. 1986. Hyperparasitism of *Sclerotinia sclerotiorum* by Talaromyces flavus. Can. J. Plant Pathol. 8:43-48.

McMillian, W. W. 1987. Relation of insects to aflatoxin contamination in maize grown in the Southeastern the United States. pp. 194-200 in: Aflatoxin in Maize: A Proceedings of the Workshop, M. S. Zuber, E. B. Lillehoj, and B. L. Renfro (eds.). The Int. Maize and Wheat Improvement Center (CIMMYT), Mexico, D. F.

McMillian, W. W., N. W. Widstrom, D. M. Wilson, and B. D. Evans. 1990. Annual contamination of *Heliothis zea* (Lepidoptera: Noctuidae) moths with *Aspergillus flavus* and incidence of aflatoxin contamination in preharvest corn in the Georgia Coastal Plain. J. Entomol. Sci. 25:123-124.

Mehan, V. K. and D. McDonald. 1984. Research on the aflatoxin problem in groundnut at ICRISAT. Plant Soil 79:255-260.

Mertz, D., D. Lee, M. Zuber, and E. B. Lillehoj. 1980. Uptake and metabolism of aflatoxin by Zea mays. J. Agric. Food Chem. 28:963-966.

Miller, M. W. and E. M. Mrak. 1954. Yeasts associated with dried fruit beetles in figs. Appl. Environ. Microbiol. 1:174-178.

Miller, J. D. and J. C. Young. 1985. Deoxynivalenol in an experimental *Fusarium graminearum* infection of wheat. Can. J. Plant Pathol. 7:132-134.

Miller, J. D., R. Greenhalgh, Y. Z. Wang, and M. Lu. 1991. Trichothecene chemotypes of three *Fusarium* species. Mycologia 83:121-130.

Miller, J. D., J. C. Young, and H. L. Trenholm. 1983. *Fusarium* toxins in field corn. I. Time course of fungal growth and production of deoxynivalenol and other mycotoxins. Can. J. Bot. 61:3080-3087.

Mirocha, C. J., C. M. Christensen, and G. H. Nelson. 1968. Estrogenic metabolite produced by *Fusarium graminearum* in stored corn. Appl. Microbiol. 15:497-503.

Mislivec, P. B. and J. F. Tuite. 1970. Species of *Penicillium* occurring in freshly harvested and in stored dent corn kernels. Mycologia 62:67-74.

Mixon, A. C. and K. M. Rogers. 1975. Factors affecting *Aspergillus flavus* Lk. ex. Fr. colonization of resistant and susceptible genotypes of *Arachis hypogaea* L. Peanut Sci. 2:18-22.

Moreno, E., L. C. Lopez, and C. M. Christensen. 1965. Loss of germination of stored corn from invasion by storage fungi. Phytopathology (Meetings Abstract) 55:125.

Moss, M. O. 1989. Mycotoxins of *Aspergillus* and other filamentous fungi. Sym. Ser. Soc. Appl. Bacteriol. (Oxford: Blackwell Scientific Pub.) 18:695-815.

Nagel, C. M. and G. Semeniuk. 1947. Some mold-induced changes in shelled corn. Plant Physiol. 22:20-33.

Nankam, C. and D. C. Foley. 1988. Anthracnose kernel rot of maize caused by *Colletotrichum graminicola* (Ces.) Wils. : mode of entrance into and disease progression in ears. J. Iowa Acad. Sci. 95:79-81.

Neergaard, P. 1977. Seed Pathology, Vol. 1 and 2. MacMillan, London. 1187 pp.

Nelson, G. H. and C. M. Christensen. 1977. Diagnosis of mycotoxicoses in animals and poultry: An overview. pp. 3-8 in: Mycotoxic fungi, Mycotoxins, Mycotoxicoses: An Encyclopedic Handbook, T. D. Wyllie and L. G. Morehouse (eds.). Marcel Dekker, New York.

Nelson, T. S., L. K. Kirby, J. N. Beasley, and Z. B. Johnson. 1985. The effect of drying method and storage time on citrinin activity in corn. Poult. Sci. 64:464-468.

Neuffer, M. G., L. Jones, and M. S. Zuber. 1968. The mutants of maize a pictorial survey in color of the usable mutant genes in maize with gene symbols and linkage map positions, arranged by chromosome location. Crop Sci. Soc. Am., Madison, WI. 74 pp.

Newell, S. Y. 1992. Estimated fungal biomass and productivity in decomposing litter. pp. 521-561 in: The Fungal Community: Its Organization and Role in the Ecosystem, 2nd Ed., G. C. Carroll and D. T. Wicklow (eds.). Marcel Dekker, New York.

Nicot, J. 1963. Les moisissures de champignons superieurs: Liste preliminaire des especes recoltees en 1959, 1960, 1961. Bull. Trimest. Soc. Mycol. Fr. 78:221-238.

Nofsinger, G. W., R. J. Bothast, E. B. Lancaster, and E. B. Bagley. 1977. Ammonia-supplemented ambient temperature drying of high moisture corn. Trans. ASAE (Am. Soc. Agric. Eng.) 20:1151-1159.

Northholt, M. D., H. P. Van Egmond, and W. E. Paulsch. 1977. Difference between *Aspergillus flavus* strains in growth and aflatoxin production in relation to water activity and temperature. J. Food Prot. 40:778-781.

Ochor, T. E., L. E. Trevathan, and S. B. King. 1987. Relationship of harvest date and host genotype to infection of maize kernels by *Fusarium moniliforme*. Plant Dis. 71:311-313.

Old, K. M. 1968. Mercury tolerant *Pyrenophora avenae* in seed oats. Trans. Br. Mycol. Soc. 51:525-534.

Oxley, T. A. 1948. The Scientific Principles of Grain Storage. Northern Pub. Co., Liverpool, UK. 103 pp.

Palmgren, M. S. and D. S. Fleischhacker. 1987. *Penicillium oxalicum* and secalonic acid D in fresh corn. pp. 193-196 in: Biodeterioration Research, G. C. Llewellyn and C. E. O'Rear (eds.). Plenum Press, New York.

Papa, K. E. 1986. Heterokaryon incompatibility in *Aspergillus flavus*. Mycologia 78:98-101.

Payne, G. A., D. K. Cassel, and C. R. Adkins. 1986. Reduction of aflatoxin contamination in corn by irrigation and tillage. Phytopathology 76:697-684.

Payne, G. A., D. L. Thompson, E. B. Lillehoj, M. S. Zuber, and C. R. Adkins. 1988. Effect of temperature on the preharvest infection of maize kernels by *Aspergillus flavus*. Phytopathology 78:1376-1380.

Pederson, J. F., R. Rodriguez-Kabana, and R. A. Shelby. 1988. Ryegrass cultivars and endophyte in tall fescue affect nematodes in grass and succeeding soybean. Agron. J. 80:811-814.

Pitt, J. I. 1979. The Genus *Penicillium* and its Teleomorphic States *Eupenicillium* and *Talaromyces*. Academic Press, London. 634 pp.

Pixton, S. W. 1967. Moisture content—its significance and measurement in stored products. J. Stored Prod. Res. 3:35-47.

Pomeranz, Y. 1974. Biochemical, functional and nutritive changes during storage. pp. 56-114 in: Storage of Cereal Grains and their Products, C. M. Christensen (ed.). Am. Assoc. Cereal Chem., St. Paul, MN.

Probert, R. J. 1981. The promotive effects of a mould, *Penicillium funiculosum* Thom on the germination of *Oryzopsis miliacea* (L.) Asch et Schw. Ann. Bot. 48:85-88.

Pugh, G. J. F. 1972. Saprophytic fungi and seeds. pp. 337-345 in: Seed Ecology, W. Heydecker (ed.). Butterworth, London.

Qasem, S. A. and C. M. Christensen. 1958. Influence of moisture content, temperature, and time on the deterioration of stored corn by fungi. Phytopathology 48:544-549.

Qasem, S. A. and C. M. Christensen. 1960. Influence of various factors on the deterioration of stored corn by fungi. Phytopathology 50:703-709.

Rabie, C. J., J. J. Du-Preez, and J. P. Hayes. 1987. Toxicity of *Diplodia maydis* to broilers, ducklings, and laying chicken hens. Poult. Sci. 66:1123-1128.

Rabie, C. J., S. J. van Rensburg, N. P. J. Kreik, and A. Lubben. 1977. Toxicity of *Diplodia maydis* to laboratory animals. Appl. Environ. Microbiol. 34:111-114.

Raper, K. B. and D. I. Fennell. 1965. The Genus *Aspergillus*. Williams and Wilkins, Baltimore, MD. 686 pp.

Raper, K. B. and C. Thom. 1949. A Manual of the Penicillia. Williams and Wilkins, Baltimore, MD. 875 pp.

Rayner, A. D. M. and L. Boddy. 1988. Fungal Decomposition of Wood: Its Biology and Ecology. John Wiley and Sons, New York. 587 pp.

Rebar, C. and O. J. Reichman. 1983. Ingestion of moldy seeds by heteromyid rodents. J. Mammal. 64:713-715.

Reichman, O. J. and C. Rebar. 1985. Seed preferences by desert rodents based on levels of mouldiness. Animal Behav. 33:726-729.

Reichman, O. J., A. Fattaey, and K. Fattaey. 1986. Management of sterile and mouldy seeds by a desert rodent. Animal Behav. 34:221-225.

Reichman, O. J., D. T. Wicklow, and C. Rebar. 1985. Ecological and mycological characteristics of caches in the mounds of *Dipodomys spectabilis*. J. Mammal. 66:643-651.

Rheeder, J. P., W. F. O. Marasas, and P. S. Van-Wyk. 1990. Fungal associations in corn kernels and effects on germination. Phytopathology 80:131-134.

Richard, J. L., L. H. Tiffany, and A. C. Pier. 1969. Toxigenic fungi associated with stored corn. Mycopathol. Mycol. Appl. 38:313-326.

Roberts, W. T. and E. C. Mora. 1978. Toxicity of *Penicillium citrinum* AUA-532 contaminated corn and citrinin in broiler chicks. Poult. Sci. 57:1221-1226.

Rosenthal, G. A. and D. H. Janzen (eds.). 1979. Herbivores: their Interactions with Secondary Plant Metabolites. Academic Press, New York. 718 pp.

Russell, G. H., M. E. Murray, and P. Berjak. 1982. Storage microflora: on the nature of the host/pathogen relationship in fungal infected maize seeds. Seed Sci. Technol. 10:605-618.

Samson, R. A. 1974. *Paecilomyces* and some allied hyphomycetes. Studies in Mycology No, 6. Centraalbureau voor Schimmelcultures, Baarn. 119 pp.

Samuels, G. J. 1984. Toxigenic fungi as Ascomycetes. pp. 119-128 in: Toxigenic Fungi—Their Toxins and Health Hazzard, H. Kurata and Y. Ueno (eds.). Elsevier Science Pub., Amsterdam.

Sansing, G. A., E. B. Lillehoj, R. W. Detroy, and M. A. Miller. 1976. Synergistic toxic effects of citrinin, ochratoxin A and penicillic acid in mice. Toxicon 14: 213-220.

Sauer, D. B. and R. Burroughs. 1980. Fungal growth, aflatoxin production, and moisture equilibration in mixtures of wet and dry corn. Phytopathology 70:516-521.

Sauer, D. B. and C. M. Christensen. 1969. Some factors affecting increase in fat acidity values in corn. Phytopathology 59:108-110.

Scott, G. E. and N. Zummo. 1988. Sources of resistance in maize to kernel infection by *Aspergillus flavus* in the field. Crop Sci. 28:504-507.

Scott, P. M. 1978. Mycotoxins in feeds and ingredients and their origin. J. Food Prot. 41:385-398.

Scott, P. M., W. van Walbeek, B. Kennedy, and D. Anyeti. 1972. Mycotoxins (ochratoxin A, citrinin, and sterigmatocystin) and toxigenic fungi in grains and other agricultural products. J. Agric. Food Chem. 20:1103-1109.

Seastedt, T. R., O. J. Reichman, and T. C. Todd. 1986. Microarthropods and nematodes in kangaroo rat burrows. Southwest. Nat. 31:114-116.

Seitz, L. M., D. B. Sauer, R. Burroughs, H. E. Mohr, and J. D. Hubbard. 1979. Ergosterol as a measure of fungal growth. Phytopathology 69:1202-1203.

Seitz, L. M., D. B. Sauer, H. E. Mohr, and D. F. Aldis. 1982. Fungal growth and dry matter loss during bin storage of high-moisture corn. Cereal Chem. 59:9-14.

Semeniuk, G. 1954. Microflora. pp. 77-151 in: Storage of Cereal Grains and their Products, J. A. Anderson and A. W. Alcock (eds.). Am. Assoc. Cereal Chem., St. Paul, MN.

Semeniuk, G., G. S. Harshfield, C. W. Carlson, C. W. Hesseltine, and W. F. Kwolek. 1971. Mycotoxins in *Aspergillus*. Mycopathol. Mycol. Appl. 43:137-152.

Semeniuk, G., C. M. Nagel, and J. C. Gilman. 1947. Observations on mold development and on deterioration in stored yellow dent shelled corn. Iowa Agric. Exp. Sta. Res. Bull. 349, Ames, IA. 284 pp.

Sewell, G. W. F. 1959. Studies of fungi in a Calluna-heathland soil. II. perfect states of some Penicillia. Antonie van Leeuwenhoek 33:297-314.

Shearer, J. F., L. E. Sweets, N. K. Baker, and L. H. Tiffany. 1992. A study of *Aspergillus flavus/parasiticus* in Iowa crop fields: 1988-1990. Plant Dis. 76:19-22.

Shotwell, O. L., M. L. Goulden, R. J. Bothast, and C. W. Hesseltine. 1975. Mycotoxins in hot spots in grains. I. Aflatoxin and zearalenone occurrence in stored corn. Cereal Chem. 52:687-697.

Shotwell, O. L., C. W. Hesseltine, and M. L. Goulden. 1969. Note on the natural occurrence of Ochratoxin A. J. Assoc. Off. Anal. Chem. 52:81-83.

Shurtleff, M. (ed.). 1980. Compendium of Corn Diseases, 2nd Ed. Am. Phytopath. Soc., St. Paul, MN. 105 pp.

Siegel, M. R. and G. C. M. Latch. 1991. Expression of antifungal activity in agar culture by isolates of grass endophytes. Mycologia 83:529-537.

Siegel, M. R., G. C. M. Latch, and M. C. Johnson. 1987. Fungal endophytes of grasses. Annu. Rev. Phytopathol. 25:293-315.

Singletary, G. W. and F. E. Below. 1989. Growth and composition of maize kernels cultured "in vitro" with varying supplies of carbon and nitrogen. Plant Physiol. 89:341-346.

Singh, K., J. C. Frisvad, U. Thrane, and S. B. Mathur. 1991. An Illustrated Manual on Identification of some Seed-borne *Aspergilli, Fusaria, Penicillia* and their Mycotoxins. Danish Gov. Inst. of Seed Pathology for Developing Countries, Hellerup, Denmark. 133 pp.

Sinha, R. N. 1966. Association of granary mites and seed-borne fungi in stored grain in outdoor and indoor habitats. Ann. Entomol. Soc. Am. 59:1170-1181.

Sinha, R. N. 1971. Fungus as food for some stored-product insects. J. Econ. Entomol. 64:3-6.

Sinha, R. N. 1992. The fungal community in the stored grain ecosystem. pp. 797-815 in: The Fungal Community, 2nd Ed., G. C. Carroll and D. T. Wicklow (eds.). Marcel Dekker, Inc., New York. 976 pp.

Sinha, R. N. and W. E. Muir (eds.). 1973. Grain Storage: Part of a System. AVI Pub. Co., Westport, CT. 481 pp.

Sinha, R. N. and F. L. Watters. 1985. Insect Pests of Flour Mills, Grain Elevators, and Feed Mills and their Control. Pub. 1776, Res. Branch Agric. Canada, OT, Canada. 290 pp.

Siriacha, P., K. Kawashima, S. Kawasugi, M. Saito, and P. Tonboon-ek. 1989. Postharvest contamination of Thai corn with *Aspergillus flavus*. Cereal Chem. 66:445-448.

Smart, M. G., O. L. Shotwell, and R. W. Caldwell. 1990a. Pathogenesis in *Aspergillus* ear rot of maize: Aflatoxin B1 levels in grains around wound-inoculation sites. Phytopathology 80:1283-1286.

Smart, M. G., D. T. Wicklow, and R. W. Caldwell. 1990b. Pathogenesis in *Aspergillus* ear rot of maize: light microscopy of fungal spread from wounds. Phytopathology 80:1287-1294.

Smith, C. C. and O. J. Reichman. 1984. The evolution of food caching by birds and mammals. Annu. Rev. Ecol. Syst. 15:329-351.

Smith, I. M., J. Dunez, D. H. Phillips, R. A. Lelliott, and S. A. Archer (eds.). 1988. European Handbook of Plant Diseases. Blackwell Scientific Pub. Oxford,. 583 pp.

Smith, J. E. and M. O. Moss. 1985. Mycotoxins: Formation, Analyses and Significance. John Wiley and Sons, Chichester, the United Kingdom. 148 pp.

Snow, D., M. H. G. Crichton, and N. C. Wright. 1944. Mould deterioration of feeding-stuffs in relation to humidity of storage. Part I. The growth of moulds at low humidities. Ann. of Appl. Biol. 31:102-110.

Starmer, W. T. and J. C. Fogleman. 1986. Coadaptation of *Drosophila* and yeasts in their natural habitat. J. Chem. Ecol. 17:1037-1055.

Stevens, L. and M. J. Wade. 1988. Effect of antibiotics on the productivity of genetic strains of *Tribolium confusum* and *Tribolium castaneum* (Coleoptera: Tenebrionidae). Environ. Entomol. 17:115-119.

Stevens, L. and D. T. Wicklow. 1992. Antibiotics in molded grain promote a polymorphism of reproductive incompatibility in *Tribolium* flour beetles. Am. Nat. 140:642-653.

Stoloff, L. 1977. Aflatoxins—An overview. pp. 7-28 in: Mycotoxins in Human and Animal Health. J. Rodricks, C. W. Hesseltine and M. Mehlman (eds.). Pathotox Pub., Park Forest South, IL.

Stovall, M. E. and K. Clay. 1991. Fungitoxic effects of *Balansia* cyperi. Mycologia 83:288-295.

Sussman, A. S. 1951. Studies on an insect mycosis. I. Etiology of the disease. Mycologia 43:338-350.

Sutton, B. C. 1980. The Coelomycetes. Commonwealth Mycological Inst., Kew, Surrey, UK. 696 pp.

Sutton, J. C. 1982. Epidemiology of wheat head blight and maize ear rot caused by *Fusarium graminearum*. Can. J. Plant Pathol. 4:195-209.

Taubenhaus, J. J. 1920. A study of the black and the yellow molds of ear corn. Texas Agric. Exp. Sta. Bull., No. 270, College Station, TX. 38 pp.

Thiel, P. G., W. C. A. Gelderblom, W. F. O. Marasas, P. E. Nelson, and T. M. Wilson. 1986. Natural occurrence of moniliformin and fusarin C in corn screenings known to be hepatocarcinogenic in rats. J. Agric. Food Chem. 34:773-775.

Thiel, P. G., W. F. O. Marasas, E. W. Sydenham, G. S. Shephard, W. C. A. Gelderblom, and J. J. Nieuwenhuis. 1991. Survey of fumonisin production by Fusarium species. Appl. Environ. Microbiol. 57:1089-1093.

Thomas, R. C. 1937. The role of certain fungi in the "sick wheat" problem. Ohio Agric. Exp. Sta. Biomonthly Bull. 22:43-45.

Thompson, D. L., E. B. Lillehoj, K. J. Leonard, W. F. Kwolek, and M. S. Zuber. 1980. Aflatoxin concentration in corn as influenced by kernel development stage and postinoculation temperature in controled environments. Crop Sci. 20:609-612.

Thorpe, C. W. and R. L. Johnson. 1974. Analysis of penicillic acid by gas-liquid chromatography. J. Assoc. Off. Anal. Chem. 57:861-865.

Toole, E. H. and E. Brown. 1946. Final results of the Duvel buried seed experiment. J. Agric. Res. 72:201-210.

Toyoda, H., H. Hashimoto, R. Utsumi, H. Kobayashi, and S. Ouchi. 1988. Detoxification of fusaric acid by a fusaric acid-resistant mutant of *Pseudomonas solanacearum* and its application to biological control of *Fusarium* wilt of tomato. Phytopathology 78:1307-1311.

Tsuruta, O., S. Gohara, and M. Saito. 1981. Scanning electron microscopic observations of a fungal invasion of corn kernels. Trans. Mycol. Soc. Japan 22:121-126.

Tuite, J. F. and R. W. Caldwell. 1970. Infection of corn seed with *Helminthosporium maydis* and other fungi in 1970. Plant Dis. Rep. 55:387-389.

Tuite, J. F. and C. M. Christensen. 1955. Grain storage studies. XVI. Influence of storage conditions upon the fungus flora of barley seed. Cereal Chem. 32:1-11.

Tuite, J. F. and G. H. Foster. 1979. Control of storage disease of grain. Annu. Rev. Phytopathol. 17:343-366.

Tuite, J. F., C. Koh-Knox, R. Stroshine, F. A. Cantone, and L. F. Bauman. 1985. Effect of physical damage to corn kernels on the development of *Penicillium* species and *Aspergillus glaucus* in storage. Phytopathology 75:1137-1140.

Tveit, M. and M. B. Moore. 1954. Isolates of *Chaetomium* that protect oats from *Helminthosporium victoriae*. Phytopathology 44:686—689.

Tveit, M. and R. K. S. Wood. 1955. The control of *Fusarium* blight in oat seedlings with antagonistic species of *Chaetomium*. Ann. Appl. Biol. 43:538-552.

Valleau, W. D. 1920. Seed corn infection with *Fusarium moniliforme* and its relation to the root and stalk rots. Kentucky Agric. Exp. Sta. Bull. 226:27-51.

van Emden, H. F. 1990. Limitations on insect herbivore abundance in natural vegetation. pp. 15-30 in: Pests, Pathogens and Plant Communities, J. J. Burdon and S. R. Leather (eds.). Blackwell Scientific Pub., Oxford.

Vesonder, R. F. and B. W. Horn. 1985. Sterigmatocystin in dairy cattle feed contaminated with *Aspergillus versicolor*. Appl. Environ. Microbiol. 49:234-235.

Vesonder, R. F., R. Lambert, D. T. Wicklow, and M. L. Biehl. 1988. *Eurotium* spp. and echinulin in feed refused by swine. Appl. Environ. Microbiol. 54:830-831.

Vorhies, C. T. and W. P. Taylor. 1922. Life history of the kangaroo rat, *Dipodomys spectabilis spectabilis* Merriam. No. 1091. Bull. US Dept. Agric. Washington, DC. 40 pp.

Wade, M. J. and L. Stevens. 1985. Microorganism mediated reproductive isolation in flour beetles, (Genus *Tribolium*). Science 227:527-528.

Wallace, H. A. H. and R. N. Sinha. 1981. Causal factors operative in distributional patterns and abundance of fungi: A multivariate study. pp. 233-247 in: The Fungal Community—Its Organization and Role in the Ecosystem, D. T. Wicklow and G. C. Carroll (eds.). Marcel Dekker, Inc., New York.

Wallace, H. A. H., R. N. Sinha, and J. T. Mills. 1976. Fungi associated with small wheat bulks during prolonged storage in Manitoba. Can. J. Bot. 54:1332-1343.

Watkins, G. M. 1981. Compendium of Cotton Diseases. Am. Phytopathol. Soc., St. Paul, MN. 87 pp.

Weckbach, L. S. and E. H. Marth. 1977. Aflatoxin production by *Aspergillus parasiticus* in a competitive environment. Mycopathologia 62:39-45.

West, C. P., E. Izekor, D. M. Oosterhuis, and R. T. Robbins. 1988. The effect of *Acremonium coenophialum* on the growth and nematode infestation of tall fescue. Plant Soil 112:3-6.

Whaley, J. W. and H. L. Barnett. 1963. Parasitism and nutrition of *Gonatobotrys simplex*. Mycologia 55:199-210.

White, J. R. and G. T. Cole. 1985. Endophyte-host associations in forage grasses. III. In vitro inhibition of fungi by *Acremonium coenophialum*. Mycologia 77:487-489.

Whitehead, M. D. and O. H. Calvert. 1959. *Helminthosporium rostratum* inciting ear rot of corn and leaf-spot of thirteen grass hosts. Phytopathology 49:817-820.

Wicklow, D. T. 1981. Interference Competition and the Organization of Fungal Communities. pp. 351-375 in: The Fungal Community—Its Organization and Role in the Ecosystem, D. T. Wicklow and G. C. Carroll (eds.). Marcel Dekker, New York.

Wicklow, D. T. 1985. Ecological adaptation and classification in *Aspergillus* and *Penicillium*. pp. 255-265 in: Advances in *Penicillium* and *Aspergillus* Systematics, R. A. Samson and J. I. Pitt (eds.). Plenum Press, New York.

Wicklow, D. T. 1988a. Metabolites in the coevolution of fungal chemical defence systems. pp. 173-201 in: Coevolution of Fungi with Animals and Plants, K. A. Pirozynski and D. Hawksworth (eds.). Academic Press, London.

Wicklow, D. T. 1988b. Patterns of fungal association within maize kernels harvested in North Carolina. Plant Dis. 72:113-115.

Wicklow, D. T. 1989. Maize cultivation selects for mycotoxin-producing ability among kernel-rotting fungi. pp. 263-274 in: Phytochemical Ecology: Allelochemicals, Mycotoxins and Insect Pheromones and Allomones, C. H. Chou and G. R. Waller (eds.). Inst. Botany, Academia Sinica Monograph Series No. 9, Taipei, Taiwan.

Wicklow, D. T., G. Bennett, R. W. Caldwell, and E. B. Smalley. 1990. Changes in the distribution of trichothecenes and zearalenone in *Gibberella zeae*-rotted maize following exposure to cool temperatures. Plant Dis. 74:304-305.

Wicklow, D. T. and R. W. Caldwell. 1990. Noninfection of maize ears inoculated with *Gibberella zeae* ascospores or macroconidia in a controled environment. Trans. Mycol. Soc. Japan. 31:29-34.

Wicklow, D. T. and R. J. Cole. 1982. Tremorgenic indole metabolites and aflatoxins in sclerotia of *Aspergillus flavus* Link: an evolutionary perspective. Can. J. Bot. 60:525-528.

Wicklow, D. T. and R. J. Cole. 1984. Citreoviridin in standing corn infested by *Eupenicillium ochrosalmoneum*. Mycologia 76: 959-961.

Wicklow, D. T. and P. F. Dowd. 1989. Entomotoxigenic potential of wild and domesticated yellow-green aspergilli: toxicity to corn earworm and fall armyworm larvae. Mycologia 81:561-566.

Wicklow, D. T., P. F. Dowd, M. R. TePaske, and J. B. Gloer. 1988a. Sclerotial metabolites of *Aspergillus flavus* toxic to a detritivorous maize insect (*Carpophilus hemipterus*, Nitidulidae). Trans. Br. Mycol. Soc. 91:433-438.

Wicklow, D. T., C. W. Hesseltine, O. L. Shotwell, and G. L. Adams. 1980. Interference competition and aflatoxin levels in corn. Phytopathology 70:761-764.

Wicklow, D. T. and B. J. Hirschfield. 1979. Evidence of a competitive hierarchy among coprophilous fungal populations. Can. J. Microbiol. 25:85-858.

Wicklow, D. T., B. W. Horn, W. R. Burg, and R. J. Cole. 1984. Sclerotium dispersal of *Aspergillus flavus* and *Eupenicillium ochrosalmoneum* from maize during harvest. Trans. Br. Mycol. Soc. 83:299-303.

Wicklow, D. T., B. W. Horn, and O. L. Shotwell. 1987. Aflatoxin formation in preharvest maize ears co-inoculated with *Aspergillus flavus* and *Aspergillus niger*. Mycologia 79:697-682.

Wicklow, D. T., B. W. Horn, O. L. Shotwell, C. W. Hesseltine, and R. W. Caldwell. 1988b. Fungal interference with *Aspergillus flavus* infection and aflatoxin contamination of maize grown in a controled environment. Phytopathology 78:68-74.

Wicklow, D. T., R. Kumar, and J. E. Lloyd. 1984. Germination of blue grama seeds buried by dung beetles (Coleoptera: Scarabaeidae). Environ. Entomol. 13:878-881.

Wicklow, D. T., C. E. McAlpin, and T. C. Nelsen. 1994. Survival and reproduction of stored-product beetles on seeds cached by a desert rodent and by Native Americans. Environ. Entomol. 23(2):(in press).

Wicklow, D. T. and C. Rebar. 1988. Mold inoculum from cheek pouches of a granivorous desert rodent, *Dipodomys spectabilis*. Mycologia 80:750-753.

Wicklow, D. T., R. D. Stubblefield, B. W. Horn, and O. L. Shotwell. 1988c. Citreoviridin levels in *Eupenicillium ochrosalmoneum*-infested maize kernels at harvest. Appl. Environ. Microbiol. 54:1096-1098.

Wicklow, D. T. and D. M. Wilson. 1986. Germination of *Aspergillus flavus* sclerotia in a Georgia maize field. Trans. Brit. Mycol. Soc. 87:651-653.

Widstrom, N. W., E. B. Lillehoj, A. N. Sparks, and W. F. Kwolek. 1976. Corn earworm damage and aflatoxin B1 on corn ears protected with insecticides. J. Econ. Entomol. 69:677-679.

Widstrom, N. W., W. W. McMillian, R. W. Beaver, and D. M. Wilson. 1990. Weather-associated changes in aflatoxin contamination of preharvest maize. J. Prod. Agric. 3:196-199.

Wieman, D. M., G. M. White, J. L. Taraba, I. J. Ross, C. L. Hicks, and B. E. Langlois. 1986. Production of aflatoxin on damaged corn under controlled environmental conditions. Trans. ASAE (Am. Soc. Agric. Eng.) 29:1150-1155.

Wilson, B. J. 1966. Toxins other than aflatoxins produced by *Aspergillus flavus*. Bact. Rev. 30:478-484.

Wilson, B. J. 1971a. Miscellaneous *Aspergillus toxins*. pp. 207-295 in: Microbial toxins. VI. Fungal Toxins, A. Ciegler, S. Kadis, and S. J. Ajl (eds.). Academic Press, New York.

Wilson, B. J. 1971b. Miscellaneous *Penicillium* toxins. pp. 459-521 in: Microbial Toxins, Vol. VI. A. Ciegler, S. Kadis and S. J. Ajl (eds.). Academic Press, New York.

Wilson, B. J. and C. H. Wilson. 1962. Extraction and preliminary characterization of a hepatotoxic substance from cultures of *Penicillium rubrum*. J. Bact. 84:283-290.

Wilson, B. J. and C. H. Wilson. 1964. Toxin from *Aspergillus flavus*: production on food materials of a substance causing tremors in mice. Science 144:177-178.

Wilson, D. M., N. W. Widstrom, L. R. Marti, and B. D. Evans. 1981. *Aspergillus flavus* group, aflatoxin, and bright greenish-yellow fluorescence in insect-damaged corn in Georgia. Cereal Chem. 58:40-42.

Windels, C. E., M. B. Windels, and T. Kommedahl. 1976. Association of *Fusarium* species with picnic beetles on corn ears. Phytopathology 66:328-331.

Wink, M. 1988. Plant breeding: importance of plant secondary metabolites for protection against pathogens and herbivores. Theo. Appl. Genet. 75:225-233.

Wright, J. M. 1956a. The production of antibiotics in soil. IV. Production of antibiotics in coats of seeds sown in soil. Ann. Appl. Biol. 44:561-566.

Wright, J. M. 1956b. Biological control of a soil-borne Pythium infection by seed inoculation. Plant Soil 8:1-9.

Wright, V. F., E. de las Casas, and P. K. Harein. 1980. The nutritional value and toxicity of *Penicillium* isolates from *Tribolium confusum*. Environ. Entomol. 9:204-212.

Wright, V. F., R. F. Vesonder, and A. Ciegler. 1982. Mycotoxins and other fungal metabolites as insecticides. pp. 559-583 in: Microbial and Viral Pesticides, E. Kurstak (ed.). Marcel Dekker, New York.

Wyllie, T. D. and L. G. Morehouse (eds.). 1977. Mycotoxic Fungi, Mycotoxins, Mycotoxicoses—An Encyclopedic Handbook. Marcel Dekker, Inc., New York. 1310 pp.

Yanagita, T. 1980. The formaldehyde resistance of *Aspergillus* fungi attacking silkworm larvae. 4. Relationship between pathogenicity to silkworm larvae and chitinase activity of *Aspergillus flavus* oryzae. J. Sericult. Sci. Japan 49:440-445.

Zuber, M. S. 1977. Influence of plant genetics on toxin production in corn, pp. 173-179 in: Mycotoxins in Human and Animal Health, J. V. Rodricks, C. W. Hesseltine and M. A. Mehlman (eds.). Pathotox Publishers, Park Forest South, IL.

Zuber, M. S., L. L. Darrah, E. B. Lillehoj, L. M. Josephson, A. Manwiller, G. E. Scott, R. T. Gudauskas, E. S. Horner, N. W. Widstrom, D. L. Thompson, A. J. Bockholt, and J. L. Brewbaker. 1983. Comparison of open-pollinated maize varieties and hybrids for preharvest aflatoxin contamination in the southern United States. Plant Dis. 67:185-187.

Zummo, N. and G. E. Scott. 1990. Relative aggressiveness of *Aspergillus flavus* and *A. parasiticus* on maize in Mississippi. Plant Dis. 74:978-981.

8

Mycotoxins and Mycotoxigenic Fungi in Storage

Jens C. Frisvad

Technical University of Denmark, Lyngby, Denmark

SUMMARY

Storage fungi have a strong influence on stored-grain ecosystems, especially through the production of secondary metabolites: mycotoxins, insecticides, antibiotics, plant inhibitors, and volatile insect attractors. Each fungus has a specific profile of secondary metabolites and each commodity has an associated mycoflora depending on the climatic zone and chemical, physical, intrinsic, and extrinsic factors. In hot climates, *Aspergillus flavus*, *A. parasiticus* and possibly *A. nomius* are important mycotoxin producers (aflatoxins, cyclopiazonic acid, and possibly 3-nitropropionic acid and maltoryzin) together with *Fusarium moniliforme* sensu lato (fumonisins). *Penicillium oxalicum* (secalonic acid D), *P. aurantiogriseum* and *P. polonicum* (corn, penicillic acid, verrucosidin, nephrotoxic glycopeptides), *P. viridicatum* (xanthomegnin, viomellein, viridic acid), and *P. citrinum* (citrinin) appear to be important toxigenic storage penicillia in warm climates. In temperate zones, *Fusarium culmorum* and *F. graminearum* (trichothecenes and zearalenone), *P. verrucosum* (ochratoxin A, citrinin and oxalic acid), and *P. freii* and *P. cyclopium* (xanthomegnin and viomellein) are the important toxigenic fungi.

1. INTRODUCTION

Fungal growth and metabolic activity is a major problem in stored foods and feeds throughout the world and may lead to poor appearance of the products because of deterioration of color, texture and taste (Mills 1989, Brooker et al. 1992), loss of seed germination ability (Pelhate 1988a, Sauer 1988), off smell (Abramson 1991, Kaminski and Wasowicz 1991), energy and nutritional loss (Sauer 1988), allergies in humans and domestic animals (Miller 1990, Flannigan et al. 1991, Lacey 1991), infection, and production of secondary metabolites (Lee et al. 1991, Sharma and Salunkhe 1991). Secondary metabolites from fungi include antibiotics, pharmacologically active substances (Masurekar 1991) and mycotoxins (Miller 1991). The presence of any such secondary metabolites from fungi in food and feed is undesirable. It is evident that this multitude of biological effects of fungal growth and metabolism in a man-made environment should be considered in an integrated ecological approach, as has been proposed by Calderon (1981), Multon (1988), and Lillehøj (1991a,b). However, fungal deterioration and mycotoxin production are among the most obvious and significant signs of poor grain quality and several reviews on the ecology of filamentous fungi in cereals have been written (Lacey 1986; Dunkel 1988; Magan and Lacey 1988; Paster and Bullerman 1988; Pelhate 1988a,b,c; Poisson and Cahagnier 1988; Abramson 1991; Chelkowski 1991; Christensen 1991; Frisvad and Samson 1991 a, b; Hocking 1991a; Lacey and Magan 1991; Lacey et al. 1991).

Many authors have considered the aflatoxins, ochratoxin A, the ergot alkaloids, and the trichothecenes as the most important toxins for human health (Richard and Thurston 1986, Krogh 1987, Betina 1989, World Health Organization 1990), but several other toxins may also cause human and animal disease. Of particular importance is the possible synergism between mycotoxins (Miller 1991, Blaney 1991) and the concern that several mycotoxins may impair the animal or human immune system, thereby leading to bacterial infection or neoplastic events (Pestka and Bondy 1990).

Storage fungi have often been considered generalists, being able to grow on all kinds of foods and feeds. This impression, however, has been based on extremely broad species concepts exemplified by outdated names such as *Fusarium roseum* Link and *Penicillium glaucum* Link. Inaccurate mycological methods, avoiding surface decontamination of seeds and grains, would give an impression of a series of species being able to grow on cereals, while they were in some cases only present as surface contaminants (Andrews 1986, Frisvad et al. 1986, Christensen 1991). Because the taxonomy of *Aspergillus, Fusarium* and *Penicillium* is very complex, several isolates have either not been identified, partially identified, or misidentified (Marasas et al. 1984, Frisvad 1989a). Wallace (1973), Frisvad and Samson (1991a,b), and Samson and Frisvad (1991) compiled many of the filamentous fungi reported from cereals. Accurate species identification has shown that each type of food and feed commodity in a certain range of environmental factors (an ecological gradient) contain an actively growing, specific, saprophytic, associated mycoflora even at nonextreme conditions (Frisvad and Filtenborg 1988). Furthermore, each fungal species has a specific profile of secondary metabolites (Frisvad 1986, 1989a, 1992). These "specific connections" indicate that each cereal commodity also has a specific profile of potential mycotoxins and that intrinsic and extrinsic ecological factors are determining this profile (Frisvad 1992). This review will concentrate on the ecological factors determining the profile of species of filamentous fungi, mycotoxins in stored cereals under different environmental conditions, and the role of mycotoxins in man-made ecosystems.

2. MYCOTOXINS IN STORED-CEREAL PRODUCTS

Many mycotoxins have been detected in cereals and some of these may influence both animal and human health (Cole and Cox 1981; Steyn and Vleggaar 1986; Krogh 1987; Scott 1987, 1988, 1989, 1990, 1991, 1992; Marasas and Nelson 1987; Betina 1989; Jelinek et al. 1989; Natori et al. 1989; Chelkowski 1989, 1991; Smith and Henderson 1991; Champ et al. 1991; Abramson 1991; Wilson and Abramson 1992). The mycotoxins which have been found occurring naturally in cereals will be mentioned briefly below. Analytical procedures and regulations for these mycotoxins will not be treated in this review but are discussed elsewhere (Cole 1986; Betina 1989, 1992; van Egmond 1989; Jelenik et al. 1989). Detoxification of foods and feedstuffs containing mycotoxins has been discussed by Trenholm et al. (1991), Wilkins et al. (1992), and others.

Some of the mycotoxins which have been found occurring naturally have been produced by the fungi under field conditions, but may still be present in stored cereals. Several significant mycotoxins can be produced under storage conditions and some mycotoxins may even be produced under both conditions (e.g., aflatoxin and zearalenone).

2.1. Aflatoxins

The aflatoxins are by far the most important mycotoxins. They are only produced by three closely related species: *Aspergillus flavus* Link, *A. parasiticus* Speare, and *A. nomius* Kurtzman, Horn and Hesseltine (Frisvad 1989a). Aflatoxins are produced in all tropical and subtropical zones of the world and are acutely toxic, carcinogenic, teratogenic, and immunosuppressive (Bhat 1991, Coulombe 1991). Aflatoxin B_1, however, may also be produced in temperate regions—especially in cereals treated with formic acid or propionic acid and kept at high humidities (Petterson et al. 1978, Kiessling et al. 1982, Holmberg et al. 1983)—and the aflatoxins also appear in some lots of cereal-based feeds imported from warmer to colder climates. Transportation over the equator often governs the production of aflatoxins (Bhat 1988, Mossel 1988, Milton and Pawsey 1988). Aflatoxins may also be formed in the field, thus complicating preventive measures.

2.2. Sterigmatocystin

Sterigmatocystin is produced by a series of species (Frisvad 1986, 1989a), but the most important potential producer in cereals is *Aspergillus versicolor* (Vuill.) Tiraboschi. The only other storage fungi producing sterigmatocystin are species of *Emericella*, notably *E. nidulans* (Eidam) Vuill. It may be of importance in metabolically-heated cereals. Sterigmatocystin production by *Eurotium* spp. has not been independently confirmed (Frisvad 1986, 1989a).

Possibly, because of rather insensitive analytical methods for sterigmatocystin (Wilson and Abramson 1992), the mycotoxin has not been reported to be a cause of any mycotoxicosis. Sterigmatocystin has been found in mouldy rice (Schmidt et al. 1981), barley (Abramson et al. 1983), and wheat (Scott et al. 1972), and it should be looked for in cereals because it is carcinogenic.

2.3. Cyclopiazonic Acid and 3-Nitropropionic Acid

Efficient producers of cyclopiazonic acid include *P. griseofulvum* Dierckx, *P. commune* Thom, *A. flavus* and *A. tamarii* Kita (Frisvad 1989a). *P. commune* is found on meat and cheese, but practically never on cereals. Similarly, domesticated forms used for production of enzymes and for food fermentations, such as *A. oryzae* (Ahlburg) Cohn (wildtype *A. flavus*) and *P. camemberti* Thom (wildtype *P. commune*) are also efficient producers of cyclopiazonic acid, but are never found on cereals (Frisvad and Filtenborg 1988). *P. griseofulvum*, *A. flavus*, and *A. tamarii*, however, have been found in cereals (Singh et al. 1991). *Penicillium atrovenetum* G. Smith, *A. oryzae*, *A. flavus* and *Arthrinium* species produce 3-nitropropionic acid, but only *A. flavus* is found in stored cereals.

Cyclopiazonic acid has not yet been found occurring naturally in cereals (Urano et al. 1992), but it may very well be a cocontaminant with aflatoxin B_1 because both are produced by *A. flavus*. Most analytical procedures for aflatoxins are not suitable for the analysis of either cyclopiazonic acid (Wilson and Abramson 1992) or other acidic metabolites from *A. flavus* such as 3-nitropropionic acid or kojic acid. Considering the toxicity of cyclopiazonic acid (Cole 1984) and 3-nitropropionic acid (Liu et al. 1989) and the metal-chelating properties of both cyclopiazonic acid and kojic acid, mixtures of these mycotoxins may be much more toxic than aflatoxin B_1 alone (Dowd 1988). The strong influence of cyclopiazonic acid and 3-nitropropionic acid on the central nervous system (Cole 1984, Liu et al. 1989) indicate that both toxins should be considered important candidates for new sensitive analytical/chemical methods in cereals often contaminated by *A. flavus* The domesticated form of *A. flavus*, *A. oryzae*, produce 3-nitropropionic acid on white potato, sweet potato, and banana (Penel and Kosikowski 1990).

2.4. Ochratoxin A

This important mycotoxin is only produced by *P. verrucosum* Dierckx and members of *Aspergillus* subgenus *Circumdati* section *Circumdati* (formerly the *Aspergillus ochraceus* group) (Frisvad 1989a). Until now, only *A. ochraceus* Wilhelm (= *A. alutaceus* Berk. and Curt.) and *A. ostianus* Wehmer (from *Circumdati*) have been reported as having any significance in cereals (Singh et al. 1991, Raper and Fennell 1965). However, ochratoxin A is not produced by *A. ostianus* sensu stricto (Frisvad, unpublished data). Naturally-occurring ochratoxin A in cereals in warm climates has not been linked to *Aspergillus* species yet but it has been found in coffee beans (Ueno et al. 1991).

Being nephrotoxic, teratogenic, and carcinogenic (Steyn 1984, Bendele et al. 1985, Kuiper-Goodman and Scott 1989, Scott 1989), ochratoxin A is a very important mycotoxin especially in temperate regions of the world and it has been found to occur naturally in cereals (barley, rye, wheat, oats, rice, and corn) in temperate countries, especially Norway, Sweden (Holmberg et al. 1991), Denmark (Krogh et al. 1973, Tholstrup and Rasmussen 1990), Great Britain, Poland, Germany, Czechoslovakia (Fukal 1990), and Canada, but also in countries with warmer climates such as Bulgaria (Petkova-Bocharova et al. 1991), Turkey (corn) (Scott 1992), Yugoslavia (Scott et al. 1991), Italy, the United States, Japan (Scott 1992), Senegal (Kane et al. 1991), and Australia (Wilson and Abramson 1992). However, Tholstrup and Rasmussen (1990) reported that the levels of ochratoxin A in Danish rye were higher than in samples of rye imported to Denmark from warmer climates. Reviews and papers on the occurrence and toxicity of ochratoxin A are presented by Castagnero et al. (1991).

2.5. Citrinin

A series of species are reported to produce citrinin and quite a few species are confirmed producers of this mycotoxin (Frisvad 1989a), including: *P. verrucosum*, *P. expansum* Link, *P. albocoremium* Frisvad, *P. lividum* Westling, *P. westlingii* Zaleski, *P. citrinum* Thom, *A. carneus* Blochwitz, *A. niveus* Link and *A. terreus* Thom (Frisvad 1989a). Of these only *P. verrucosum, P. citrinum* and *A. terreus* seem to be of any importance in cereals (Frisvad and Samson 1991b).

Citrinin is produced by *Penicillium verrucosum*—together with two other nephrotoxins, ochratoxin A and oxalic acid—in temperate regions of the world (Frisvad and Filtenborg 1989, Krogh et al. 1970) but it appears that citrinin is found less often than ochratoxin A in cereals. It has been found in cereals in Denmark (Krogh et al. 1973), Sweden (Hoekby et al. 1979), Germany (Dick et al. 1988), United Kingdom (Osborne 1980), Canada (Scott et al. 1972), and the United States (Chalam and Stahr 1979, Nelson et al. 1985). Nelson et al. (1985)

correctly identified the producing organism as *P. citrinum* even though it was received as *P. lanosum* Westling (Frisvad 1989a). *Penicillium citrinum* may also produce citrinin in rice in warmer climates (yellowed rice toxicosis, Saito et al. 1971). Rosa et al. (1985) reported on porcine nephropathy caused by barley grain contaminated with *P. citrinum* in Brazil. The barley sample contained 1000 ppb citrinin. This was the first report of a penicillia other than *P. verrucosum* to cause porcine nephropathy.

2.6. Xanthomegnin, Viomellein, and Vioxanthin

Xanthomegnin, viomellein, and vioxanthin are produced by *P. freii* Frisvad and Samson, *P. cyclopium* Westling, *P. viridicatum* Westling, *P. melanoconidium* (Frisvad) Frisvad and Samson, *P. tricolor* Frisvad, Seifert, Samson and Mills, *A. ochraceus* and other members of *Circumdati*, and dermatophytes such as *Trichophyton rubrum* (Castell.) Sabour. (Frisvad 1989a, Frisvad and Filtenborg 1989, Lund and Frisvad 1994). Of these, only the penicillia and *A. ochraceus* are found in cereals. *Penicillium freii, P. cyclopium, P. viridicatum* and *P. melanoconidium* are very common in cereals and, therefore, more cereal samples should be screened for toxins produced by these species (see later).

Hald et al. (1983) first reported on the natural occurrence of viomellein in Danish barley. Later xanthomegnin and viomellein were found in cereals in the United Kingdom (Scudamore et al. 1986) and in corn in the United States (Garcia and Tuite 1985). Later examinations have shown that *P. freii* was the dominant producer of these hepato- and nephro-toxins in the Danish and British samples and that *P. viridicatum* was responsible for the production of xanthomegnin and viomellein in corn in the United States (Lund and Frisvad 1994).

2.7. Penicillic acid

Fungal producers of penicillic acid include several soil-borne species but the only species found in cereals are *P. aurantiogriseum, P. polonicum* Zaleski, *P. aurantiovirens* Biourge, *P. cyclopium, P. melanoconidium, A. ochraceus* and *A. ostianus* (Frisvad 1989a, Lund and Frisvad 1994). Penicillic acid has only been found occurring naturally in corn from the United States (Thorpe and Johnson 1974, Stoloff 1976, Garcia and Tuite 1985).

2.8. Secalonic Acid D

Secalonic acid D is produced by *P. oxalicum* Currie and Thom, and *Claviceps* spp. (Reddy and Reddy 1991). The report on secalonic acid A production by *A. ochraceus* (Yamazaki et al. 1971) could not be confirmed (Frisvad, unpublished data). Secalonic acid D, a teratogenic and possibly mutagenic mycotoxin (Reddy et al. 1981, Reddy and Reddy 1991), has been found occurring naturally together with aflatoxin B_1 and B_2 in corn dust (Ehrlich et al. 1982, Bullerman 1983) but data from Palmgren and Fleichhacker (1987) indicate that secalonic acid is not produced in the field.

2.9. Citreoviridin

This mycotoxin is produced by *P. smithii* Quintanilla, *P. miczynskii* Zaleski, *P. manginii* Duché and Heim, *P. citreonigrum* Dierckx, *Eupenicillium ochrosalmoneum* Scott and Stolk, and *A. terreus* (Frisvad 1989a), but only the latter three species are found in cereals. Citreoviridin, a neurotoxin (Moreau 1975), was found occurring naturally in corn that was left unharvested for 4 mo in Georgia (Wicklow et al. 1982, 1988; Stubblefield et al. 1988). It was produced by *E. ochrosalmoneum.* This toxin was first found on moldy rice in which *P. citreonigrum* had grown (Hirata 1947).

2.10. Roquefortine C

Roquefortine C is produced by a series of *Penicillium* species, including *P. crustosum* Thom, *P. hordei* Stolk, *P. chrysogenum* Thom, *P. expansum*, *P. griseofulvum* Dierckx, *P. hirsutum* Dierckx, *P. glandicola* (Oud.) Seifert and Samson, *P. coprophilum* (Berk. and Curt.) Seifert and Samson, *P. sclerotigenum* Yamamoto, *P. roqueforti* Thom, *P. carneum* Frisvad, *P. oxalicum*, and *P. vulpinum* (Cooke and Massee) Seifert and Samson. Of these the first five mentioned and *P. roqueforti* seem to be the most important producers of this toxin in cereals. Roquefortine C has only been detected in feed grain once (Häggblom 1990) but other mycotoxins known to be produced by *Penicillium roqueforti*, such as PR-toxin and mycophenolic acid, were not detected in that feed grain sample. The grain heavily infected with *P. roqueforti* was fed to milk cows and caused a loss of appetite, ketosis, mastitis, paralysis, and abortions (Häggblom 1990).

2.11. Trichothecenes

The trichothecenes—notably T-2 toxin, deoxynivalenol, nivalenol and diacetoxyscirpenol—are probably not formed under storage conditions and their producers have been compiled by Marasas et al. (1984) and Thrane (1989). The most important producers include *Fusarium poae* (Peck) Wollenw., *F. sporotrichioides* Sherb., *F. acuminatum* Ell. and Everh., *F. equiseti* (Corda) Sacc., *F. crookwellense* Burgess, Nelson and Toussoun, *F. culmorum* (W.G. Smith) Sacc., *F. graminearum* Schwabe, and *F. sambucinum* Fuckel (Thrane 1989). The trichothecenes, principally T-2 toxin, HT-2 toxin, diacetoxyscirpenol, deoxynivalenol and nivalenol, have been found in cereals in both temperate regions and warmer climates. Trichothecenes have been found in Denmark, Sweden, Norway, Finland, Poland (Perkowski et al. 1990), Germany, Austria, Russia, United Kingdom, Canada, the United States, France, Italy, Greece, Hungary, Bulgaria, Yugoslavia, Japan, Korea (Park et al. 1991), Australia (Williams et al. 1992), South Africa, Argentina (Chulze et al. 1989, Torres et al. 1990), India, Nepal, China, Taiwan, and Yemen (Wilson and Abramson 1992, Lew et al. 1991).

2.12. Zearalenone

Zearalenone has been found worldwide, mostly in corn but also in wheat, barley, oats, rye, rice, and sorghum. It has been found occurring naturally in the United Kingdom, Germany (Ranfft et al. 1990), Poland (Perkowski et al. 1990), Austria (Lew et al. 1991), and Canada; more zearalenone has been found in warmer climates: Italy, Portugal, Spain, the United States, Argentina (Chulze et al. 1989), Australia (Williams et al. 1992), China, Taiwan, Japan, Korea, India, Nepal, South Africa, Zambia, and Yemen (Lee et al. 1991, Wilson and Abramson 1992). Zearalenone is primarily produced by *F. sporotrichioides, F. pallidoroseum* (Cooke) Sacc., *F. equiseti, F. crookwellense, F. culmorum*, and *F. graminearum* (Thrane 1989).

2.13. Fusarochromanone

Fusarochromanone has been found occurring naturally in cereal feed (a mixture of wheat, corn, and peas) causing tibial dyschondroplasia in chickens (Krogh et al. 1989). The mycotoxin was produced by *F. equiseti.*

2.14. Fumonisins

Fumonisin B_1 and B_2 have been shown to cause equine leukoencephalomalacia, to have cancer promoting activity, and to be associated with porcine pulmonary edema (Nelson et al. 1992, Ross et al. 1992). These polar toxins naturally occur in corn in the United States (Plattner et al. 1990, Ross et al. 1990), Poland (Lew et al. 1991), and South Africa (Sydenham et al. 1990, Thiel et al. 1991). Producers of the fumonisins include the broad species *Gibberella fujikuroi* (Sawada) Ito and the related anamorphs *F. verticillioides* (Sacc.) Nirenberg (= *F. moniliforme* Sheldon), *F. anthophilum* (A. Braun) Wollenw., *F. dlamini* Marasas, Nelson, and Toussoun, *F. napiforme* Marasas, Nelson and Rabie, *F. nygamai* Burgess and Trimboli, and *F. proliferatum* (Mastsushima) Nirenberg, the latter being the most important (Nelson et al. 1992).

2.15. Moniliformin

Thiel et al. (1982, 1986) reported the natural occurrence of moniliformin in corn from South Africa; Scott and Lawrence (1987) only found moniliformin in two corn samples from Canada. Thalmann et al. (1985) found low levels of moniliformin in German grown corn but Sharman et al. (1991) found moniliformin in several samples of corn from Great Britain, the Netherlands, the United States, France, Italy, Thailand, South Africa, Zimbabwe, Nigeria, Gambia, Tanzania, Kenya, and Malawi. They also found moniliformin, produced by *F. avenaceum* (Corda:Fr.) Sacc., in hand-selected samples of wheat, rye, and oats; and moniliformin produced by *F. subglutinans* (Wollenw. and Reinking) Nelson, Toussoun and Marasas in corn from Poland. Lew et al. (1991) found that the latter species produced large amounts of moniliformin in Austrian corn. Other producers of moniliformin include *F. verticillioides, F. chlamydosporum* Wollenw. and Reinking, *F. pallidoroseum, F. acuminatum, F. anthophilum*, and *F. proliferatum* (Thrane 1989).

2.16. Fusarin C

Fusarin C, a mutagenic mycotoxin, has been found occurring naturally in corn from South Africa (Gelderblom et al. 1984), the United States (Thiel et al. 1986), and China (Cheng et al. 1985). It is primarily produced by *F. poae, F. tricinctum* (Corda) Sacc., *F. sporotrichioides, F. avenaceum, F. crookwellense, F. culmorum, F. graminearum, F. sambucinum*, and *F. verticillioides* (Thrane 1989).

2.17. Alternariols and Tenuazonic Acid

Alternariol monomethylether, alternariol, altenuene, and altertoxin I have been found occurring naturally in poor-quality sorghum from the United States and Australia (Seitz et al. 1975, Sauer et al. 1978, Andrews 1986), and in Polish wheat and rye (Grabarkiewicz-Szczesna et al. 1989). The most widespread producer of these toxins and tenuazonic acid is *Alternaria alternata*; however, the taxonomy of this species and related taxa is still under revision (Simmons 1992).

2.18. Other Mycotoxins

A series of other mycotoxins have been isolated from pure laboratory cultures of filamentous fungi that have been isolated from cases of suspected mycotoxicoses. The unavailability of standards or insensitive analytical methods may have caused underestimation of the significance of such mycotoxins or the mycotoxins may still be unknown. Furthermore, known and unknown mycotoxins may have synergistic toxic effects (Miller 1991). Therefore, it is very important to evaluate the actively growing mycoflora and to identify the species present accurately. Furthermore, the fungal isolates should be examined for production of mycotoxins in pure culture and in cereals; methods for prevention of growth and mycotoxin production should be investigated.

3. THE ASSOCIATED MYCOFLORA OF STORED CEREALS

3.1. Methods for Mycological Examination of Cereals

Two international workshops have been devoted to development and standardization of methods in food and feed mycology (King et al. 1986, Samson et al. 1992) and several methods for the examination of cereals were compared and certain methods recommended. Cereal grains should be surface disinfected to remove the superficial fungal spores which may not be able to germinate and grow in the cereals at the given conditions. The recommended treatment with sodium hypochlorite will not remove all superficial contaminants (Christensen 1991) but at least the picture of the actively growing mycoflora will be less disturbed by too many surface contaminants. The surface-disinfected grains should be directly plated on a suitable medium (Christensen 1991, Frisvad et al. 1986, Andrews 1986, Trojanowska 1991). A good indication of actual growth of certain fungi is specific mycotoxin production. Andrews (1986) examined two sorghum samples suspected of causing a mycotoxicosis in broiler chickens. Direct plating without surface "disinfection" showed 100% infection with *A. flavus*. Aflatoxins, however, could not be detected in the sample. Dilution plating gave a similar result; *A. flavus* dominated the sample. Nevertheless, when the grains were surface disinfected, *Alternaria alternata* and *Drechslera* species were detected on 55% and 38% of the kernels, respectively, and indeed alternariol and alternariol monomethylether were present in both samples. Frisvad et al. (1986) examined two samples of barley that had been placed uncovered in two buckets in the same barn for some weeks. One of the samples had been stored for 1.5 yr in an airtight silo. Direct plating showed that the mycoflora on both samples were dominated by species typical of airtight storage: *P. roqueforti*, *A. candidus*, and *Hyphopichia burtonii* (Boedin et al.) von Arx and van der Walt. After surface disinfection, only the sample which had been actually stored in the airtight silo contained the species listed above. The other sample contained species typical of fresh cereals: *Alternaria, Drechslera*, and *Fusarium* spp. Thus, surface disinfection is highly recommended.

Dilution plating is not particularly suitable for analysis of grains because species which cannot actually grow in the cereals may be over represented. Products such as cereal flour, however, are not easily plated directly. For such products, dilution plating is recommended in addition to direct plating of several plates with 30 mg flour each. In both dilution and direct plating, the identity of the fungi is much more important than the actual number of colony forming units or percentage of infected grains.

For storage fungi, the medium DG18 (Hocking and Pitt 1980) compares favorably with traditional media such as malt-salt agar (King et al. 1986, Samson et al. 1992). Advantages of DG18 (compared with malt salt agar) include better control of fast-growing fungi, higher recovery of fungi, less general inhibition of important fungi, better electivity, less intralaboratory differences (differences between malt extracts may be considerable), better resistance to autoclaving, and better transparency. Several other media traditionally used for examination of cereals (Christensen 1991)—such as acidified potato dextrose agar or malt extract agars—compare very poorly with DG18 (King et al. 1986, Samson et al. 1992), both because antibiotics are less inhibitory than acid and because xerophilic species recover much better on DG18. Even though DG18 was developed for xerophilic fungi, it also compares favorably with other media for less xerophilic species (Samson et al. 1992). However, sporulation of field fungi such as *Drechslera*, *Fusarium*, *Botrytis*, and *Alternaria* is poor on DG18, making direct identification to genus level difficult.

Other media, such as dichloran chloramphenicol peptone agar (DCPA) (Andrews and Pitt 1986) and dichloran rose bengal chloramphenicol (DRBC) agar (King et al. 1979), are quite effective for recovering field fungi and give favorable counts compared with traditional media (King et al. 1986, Samson et al. 1992). Neither DCPA nor DRBC are ideal for recovering, isolating, and identifying to the genus level; fungi on DCPA will develop ammonia after prolonged incubation and late-developing fungi on DCPA may show signs of deterioration if isolated from this medium (Andrews and Pitt 1986). Rose bengal (present in DRBC) will develop singlet oxygen when irradiated and may severely inhibit several filamentous fungi (Gams and Laar 1982; Madelin 1987; Frisvad et al. 1992a,b). Black light alternating with darkness is necessary for good sporulation of several field fungi. Several media for field fungi are compiled by Tuite (1969). A series of media was developed for *Fusarium* species (e.g., Nishikawa and Kohgo 1975, Gyllang et al. 1981, Abildgren et al. 1987, Thrane et al. 1992) including pentachloronitrobenzene, dichloran (= botran), or iprodione as selective principles.

Only a few selective and indicative media have been developed for particularly important toxigenic fungi. Bothast and Fennell (1974) developed a basal medium for the detection of *A. flavus* and *A. parasiticus*. This medium was later modified by Hamsa and Ayers (1977) to include dichloran as recommended by Bell and Crawford (1967); Pitt et al. (1983) further modified the medium for practical use. It is now commercially available as AFPA (*Aspergillus flavus* and *parasiticus* agar) and can be used to directly identify the three species *A. flavus*, *A. parasiticus* and *A. nomius* (orange reverse on AFPA), while *A. tamarii* has a dark brown reverse (Samson and Frisvad 1991). Other media have also been suggested for the screening of aflatoxin-producing species, including: a coconut basal medium (Lin and Dianese 1976, Davis et al. 1987, Lemke et al. 1989), silica gel (Torrey and Marth 1976) or yeast extract sucrose (YES) broth or agar, or similar media (Davis et al. 1966; Hara et al. 1974; Yabe et al. 1987, 1988). The simplest method seems to be that of Filtenborg et al. (1983, 1990) using agar plugs for direct thin layer chromatography (TLC) analysis of the aflatoxins.

Dichloran rose bengal yeast extract sucrose (DRYES) agar was developed for the selective and indicative detection primarily of *P. verrucosum* (violet to red brown reverse) but also *P. aurantiogriseum* (strong yellow reverse) (Frisvad 1983). Other species of *Penicillium* have yellow (orange) reverses on DRYES agar, among them *P. solitum* Westling, *P. crustosum* and *P. expansum*, but the latter species are uncommon in cereals. DRYES agar was later modified to include magnesium sulphate to enhance the production of the violet brown (nondiffusible) colony reverse color of *P. verrucosum* (Filtenborg et al. 1990, 1992). In certain cases, species of *Penicillium* and *Aspergillus* can be identified directly on DRYES by macromorphological features and production of specific secondary metabolites (Filtenborg and Frisvad 1990, Filtenborg et al. 1992, Frisvad et al. 1992b) using the simple agar plug TLC method developed by Filtenborg and Frisvad (1980) and Filtenborg et al. (1983). Ochratoxin A, citrinin, and several other mycotoxins can be screened directly in mixed cultures using this agar plug TLC method and the DRYES medium.

3.2. Association Between Mycoflora and Environmental Factors

3.2.1. Associated Mycoflora of Cereals under Extreme Conditions

A large number of storage fungi have been detected in cereals and these species have been compiled by Wallace (1973), Lacey et al. (1991), Samson and Frisvad (1991) and Frisvad and Samson (1991a,b). Many of these species, primarily belonging to *Penicillium* and *Aspergillus*, are superficial contaminants or their identity may be less certain because of changing taxonomies. This was realized by Christensen (1991) and others who reported only on the really severe storage fungi that were able to spoil cereals and sometimes produce mycotoxins. The most important of these species include *Eurotium* spp. (*Aspergillus glaucus* "group" = *Aspergillus* subgenus *Aspergillus* section *Aspergillus*), *A. restrictus* G. Smith, and *A. penicillioides* Speg. (subgenus *Aspergillus* section *Restricti*), *A. candidus*, *A. flavus*, and *Penicillium* spp. Other less dominating storage fungi include *A. versicolor, A. ochraceus, A. fumigatus, A. niger* van Tieghem, *A. terreus, Emericella nidulans, Chrysosporium fastidium* Pitt, and *Wallemia sebi* (Fr.) von Arx (Christensen 1991; Lacey et al. 1991, Frisvad and Samson 1991a,b). Studies of Danish, Norwegian, Swedish, British, and Canadian cereals (Frisvad, unpublished) have confirmed that *Eurotium*, *Aspergillus* section *Restricti*, *A. flavus*, *A. candidus*, and *Penicillium* spp. are the genuinely dominating storage fungi in cereals in temperate regions.

Aspergillus section *Restricti* will dominate cereals stored for a long period at low water activity (a_w < 0.75) but may lead the way for *Eurotium* spp. (*E.*

repens de Bary, *E. rubrum* Koenig, Spieckermann, and Bremer, *E. herbariorum* (Wiggers:Fr.) Link, *E. amstelodami* Mangin, *E. chevalieri* Mangin) which require water activities between 0.75 and 0.85. At water activities between 0.85 and 0.90, species of *Penicillium* subgenus *Penicillium* will dominate. At temperatures higher than 25°C, *Penicillium* subgenus *Biverticillium* and a few from the subgenus *Furcatum* and *Aspergilloides* will dominate. The latter penicillia will also dominate in warm climates. At very high temperatures or in the course of heating of cereals *A. candidus, A. flavus, A. fumigatus* Fres. and *E. nidulans*, certain *Talaromyces*, and *Penicillium* subgenus *Biverticillium* species will dominate (Lacey et al. 1991).

Under special circumstances, cereal samples will be dominated by species of *Penicillium* and *Aspergillus* that are adapted to extreme conditions, including: *A. fumigatus, P. roqueforti, P. carneum, P. glandicola, Monascus ruber* van Tieghem, *Byssochlamys fulva* Olliver and G. Smith, *B. nivea* Westling, and *Paecilomyces variotii* Bain. These species can tolerate high levels of acetic acid, propionic acid, and other acids and may, therefore, dominate silage or cereals treated with propionic acid (Escoula 1973, Pelhate 1974/75, Cole et al. 1977, Gedek et al. 1981, Woolford 1990). *Aspergillus flavus* may grow and produce aflatoxin in cereals insufficiently treated with propionic or formic acid, provided the water activity is sufficiently high (Pettersson et al. 1978).

In modified atmospheres, especially at very low oxygen tensions, *A. terreus, A. candidus, Pae. variotii, Pen. roqueforti, P. griseofulvum*, and *Scopulariopsis brevicaulis* (Sacc.) Bain. (Ceynowa 1986, Clarke and Hill 1981, Hocking 1991b) appear to be the most important fungi. It is interesting to note that nearly all the species listed above are good producers of patulin: *B. fulva, B. nivea, Pae. variotii, A. terreus, Pen. glandicola* and *P. carneum* (Frisvad and Filtenborg 1989; Frisvad and Samson 1991a,b).

Apart from *A. fumigatus* (Cole et al. 1977), *A. flavus, A. parasiticus*, and *A. nomius* (aflatoxins and probably several other toxins), little is known on mycotoxins actually produced by the fungi mentioned above at more extreme conditions. Thus, the mesophilic mycoflora growing at water activities between 0.85 and 0.98 appears to be the most important toxigenic fungi in stored cereals.

3.2.2. Important Storage Fungi in Warm Climates

In warm climates and especially in cereals containing a high percentage of lipid such as corn, *A. flavus, A. niger, A. tamarii, P. aethiopicum* Frisvad, *P. viridicatum, P. oxalicum, P. citrinum, P. pinophilum* Hedgcock, *P. funiculosum* Thom, and *P. crateriforme* Gilman and Abbott are the dominating filamentous storage fungi (Frisvad and Filtenborg 1989, Singh et al. 1991, Wheeler et al. 1991). Natural occurrence of aflatoxin and secalonic acid D shows that *A. flavus* and *A. parasiticus* and *P. oxalicum* are capable of producing some of their toxins in corn. *P. aethiopicum* (formerly called *P. viridicatum* or *P. expansum*, Frisvad 1989a) was isolated from corn and was able to produce the mycotoxin viridicatumtoxin on wet sterile corn meal (Hutchinson et al. 1973). It is not known, however, whether this mycotoxin is produced under natural conditions. All of 23 strains examined of *P. aethiopicum* produced viridicatumtoxin (Frisvad and Filtenborg 1989).

Rubratoxin production has been established in cultures of *P. crateriforme* (formerly *P. rubrum* Stoll), but again, rubratoxin B has not been found as a natural contaminant. *P. crateriforme* has been isolated from toxic moldy corn (Moss 1971). More examinations of the dominating mycoflora and naturally occurring mycotoxins of stored tropical and subtropical cereals are needed to evaluate the role of *Penicillium* and *Aspergillus* toxins, apart from aflatoxin B_1, which is of paramount importance (Dawson 1991).

The mycoflora of rice in the far east is quite similar to the flora of corn, but *P. citrinum* (citrinin), *P. islandicum* Sopp (luteoskyrin, erythroskyrin, rugulosin, cyclochlorotine), and *P. citreonigrum* (citreoviridin) seem to be associated with rice and may have been involved in yellowed rice toxicosis (Saito et al. 1971). Several cases of mycotoxicoses seemed to be caused by growth and toxin production of the penicillia mentioned above and possibly also by other species of *Biverticillium* such as *P. rugulosum* Thom (rugulosin) and *P. tardum* Thom (rugulosin). Other mycotoxicosis in Japan appear to have been caused by growth of *P. griseofulvum* in malt from Kobe (patulin, roquefortine C, cyclopiazonic acid are known mycotoxins from this fungus) and growth of *A. flavus* (called *A. oryzae* var. *microsporus* Yamamoto) in "commercial food" from Tokyo and Chiba Prefecture (maltoryzin) (Saito et al. 1971).

3.2.3. Important Storage Fungi in Temperate Climates

3.2.3.1. Classification and Identification Problems

Even though a large number of filamentous fungi have been reported from cereals from temperate regions of the world, most of these were superficial surface contaminants. The actively growing *Penicillium* mycoflora of most cereal samples is restricted to surprisingly few species. Failure to surface-disinfect cereal grains being examined, different and difficult taxonomic systems, and misidentifications have obscured the association between cereals and storage fungi in temperate regions. According to the taxonomic nomenclature of Frisvad and Filtenborg (1989) and Lund and Frisvad (1994), the only consistently found penicillia in cereals grown in northern Europe are *P. verrucosum*, *P. freii*, *P. cyclopium*, *P. aurantiovirens*, *P. aurantiogriseum*, *P. polonicum*, and *P. hordei* (Frisvad and Filtenborg 1988, 1989). In addition to those species, *P. chrysogenum* and *P. tricolor* have been found in Canadian grown cereals (Mills and Abramson 1982, Mills et al. 1989).

3.2.3.2. *Penicillium verrucosum*

Penicillium verrucosum sensu stricto is the only known producer of ochratoxin A in *Penicillium* (Frisvad 1989a, Frisvad and Filtenborg 1989). All other claimed producers have been re-identified to be *P. verrucosum* (Table 1 and Frisvad 1989a) or the detected chemical was a secondary metabolite with "similar" chemical characteristics to ochratoxin A. Ciegler et al. (1973) realized that ochratoxin A and citrinin producing strains called *P. viridicatum* were different from the neotype culture and called them *P. viridicatum* "group" II (strains from cereals) and "group" III (strains from meat); only group II species produced citrinin. Samson et al. (1976) placed most strains now called *P. verrucosum, P. viridicatum, P. aurantiogriseum, P. solitum, P. commune, P. hirsutum*, and *P. crustosum* in a broad micromorphogically-based species *P. verrucosum.* Their concept of *P. verrucosum* var. *verrucosum* included strains now identified as either *P. verrucosum, P. viridicatum* or *P. commune* (i.e., the three subgroups of Ciegler et al. (1973) were again placed together and certain strains of *P. commune* were included). Pitt (1979) placed *P. viridicatum* "group" II together with strains typical of the culture ex type of *P. viridicatum* and some strains of *P. commune* and *P. solitum.* Pitt's original concept of *P. verrucosum* included *P. viridicatum* "group" III, *P. commune*, and a strain of *P. echinulatum.* Thus, both the concept of Samson et al. (1976) and Pitt (1979) were at variance with that of Ciegler et al. (1973) and their taxa did not produce consistent mycotoxin profiles. Frisvad (1981) concurred with Ciegler et al. (1973) and even proposed that a new taxon should be described for "*P. viridicatum* o-c". This was based on several physiological characters which separated the ochratoxin producers from *P. aurantiogriseum* and *P. commune. Penicillium verrucosum* sensu stricto (= "*P. viridicatum* o-c") grows on nitrite (as the sole nitrogen source) and grows very poorly (and does not produce acid) on creatine-sucrose agar (Frisvad 1981, 1985a). Frisvad and Filtenborg (1983) proposed that "*P. viridicatum* o-c" (= *P. viridicatum* II + III) should be called *P. verrucosum.* The problem was finally solved when Frisvad (1985a,b) proposed the name *Penicillium verrucosum* in the strict sense only for slow-growing, nitrite-positive, creatine-negative, ochratoxin, anacine, and verrucolone producing penicillia, agreeing with the culture ex neotype. Later, Pitt (1987) and Samson and coworkers (Stolk et al. 1990) also followed this concept. As mentioned previously, the medium DRYES was developed for the indicative selection of cereal-borne *P. verrucosum* (Frisvad 1983); production of ochratoxin A, citrinin, and verrucolone can be directly monitored on this medium by using the simple agar plug TLC method (Filtenborg et al. 1990, 1992). Using this medium, Frisvad and Viuf (1986) found that 67 of 70 barley samples contained *P. verrucosum.* These samples were collected from all parts of Denmark from barley that had been fed to pigs which were discarded because of too high levels of ochratoxin A in their kidneys. Of these 70 samples, 66 contained ochratoxin A. A large number of the 514 isolates of

Table 1 Examples of Isolates of *Penicillium verrucosum* from Cereals Producing Ochratoxin A but Originally Identified as Other Species

Original identification	Substrate
P. palitans CBS 222.71[1]	White beans, Canada
P. palitans CBS 223.71[1]	White beans, Canada
P. palitans CBS 224.71[1]	White beans, Canada
P. viridicatum CBS 226.71F[1]	Poulty feed, Canada
P. viridicatum M93[2]	Cereal, Sweden
P. viridicatum NRRL 3710[3]	Wheat, Canada
P. viridicatum NRRL 3712[3]	Wheat, Canada
P. viridicatum NRRL 5571[3]	Wheat, United States
P. viridicatum NRRL 5572[3]	Wheat, Canada
P. viridicatum NRRL 5583[3]	Beans, Canada
P. cyclopium NRRL 6065[4]	United States
P. verrucosum IMI 285522[5]	Barley, Denmark
P. verrucosum IMI 293203[5]	Barley, Denmark

Sources: [1]Scott et al. (1972), [2]Häggblom (1982), Häggblom and Ghosh (1985), [3]Ciegler et al. (1973), [4]Northolt et al. (1979), [5]Frisvad and Filtenborg 1989.

P. verrucosum reported by Frisvad and Filtenborg (1989) were isolated from Danish barley and wheat, but some were also from Swedish, Norwegian, and British cereals and beans. Of the 514 isolates, all produced the small lactone verrucolone whereas 470 produced ochratoxin A and 493 produced citrinin. Holmberg et al. (1991) used the DRYES medium for screening of Swedish isolates in cereal samples containing ochratoxin A; they used the agar plug TLC method of Filtenborg and Frisvad (1980) to screen for ochratoxin A producers. They also found *P. verrucosum* to be the only producer of ochratoxin A.

Growth medium and environmental factors (water activity and temperature) influence different strains of *P. verrucosum* rather differently (Harwig and Chen 1974, Northolt et al. 1979, Chelkowski et al. 1981, Häggblom 1982, Häggblom and Ghosh 1985, Damoglou et al. 1984, Boley and Müller 1986, Madhyastha et al. 1990, Müller and Boley 1991). Thus, all strains of *P. verrucosum* may be able to produce ochratoxin A and citrinin at "optimal" conditions. The substrate has a marked influence on ochratoxin A and citrinin production (Filtenborg et al. 1990, Madhyastha et al. 1990). Filtenborg et al. (1990) found that one brand of yeast extract from Sigma Chemical Co. (in YES agar) supported very strong ochratoxin A and citrinin production, while another brand from the same company completely inhibited production of these two nephrotoxins. As YES agar has been used very often for screening of ochratoxin A production (Scott et al. 1972, Frisvad and Filtenborg 1983), more strains of *P. verrucosum* than reported may be able to produce ochratoxin A and citrinin. The strains of *Penicillium* used by Northolt et al. (1979) in their examination of the influence of substrate, temperature and water activity on ochratoxin A production were called *P. cyclopium* and *P. viridicatum*, but they were all *P. verrucosum* (Frisvad 1989a). Therefore, it can be concluded that environmental factors influence different strains of the same species differently and that prediction of ochratoxin A and citrinin production is still very difficult. Many more strains of *P. verrucosum*, from different habitats and geographic regions, should be examined to find general trends in the production of mycotoxins under realistic conditions, preferably using statistical design. A multivariate statistical high performance liquid chromatographic (HPLC) study of secondary metabolites of five species of *Penicillium* showed that the qualitative presence of the secondary metabolites were species specific but there were quantitative differences between strains (Frisvad 1992).

3.2.3.3. *Penicillium aurantiogriseum, P. viridicatum,* and Related Taxa

Penicillium aurantiogriseum and allied species are the other important species found in cereals. Frisvad and Filtenborg (1989) subdivided this species in several varieties and chemotypes. *Penicillium aurantiogriseum*, in this sense, is the same as the "*P. cyclopium* p" concept of Frisvad (1981) (i.e., including *P. viridicatum*). Strains of this species were called *P. cyclopium, P. aurantiovirens, P. martensii, P. puberulum* pro parte, *P. olivinoviride*, and *P. aurantiocandidum* by Raper and Thom (1949). Samson et al. (1976) included such strains in *P. verrucosum* var. *cyclopium* together with strains of *P. aetiopicum, P. solitum, P. crustosum*, and certain strains of *P. commune* and *P. concentricum*. Pitt (1979) included *P. aethiopicum* and *P. solitum* in his concept of *P. aurantiogriseum*. However, there has been agreement of the concept of *P. aurantiogriseum* (Stolk et al. 1990, Pitt and Cruickshank 1990, Frisvad and Filtenborg 1989, Lund and Frisvad 1994). In practise, creatine sucrose agar (Frisvad 1985a) has been valuable in separating species occurring on meat, nuts, and cheese (e.g., *P. solitum, P. crustosum* and *P. commune* from *P. aurantiogriseum*) and this medium is recommended to confirm identifications. Strains of *P. aurantiogriseum* grow poorly on creatine sucrose agar but produce large amounts of acid (which is not followed by production of base) and produce yellow colors in colony reverse and obverse on YES agar. The total profile of known mycotoxin produced by *P. aurantiogriseum* and related species includes penicillic acid, terrestric acid, xanthomegnin, viomellein, verrucosidin, viridic acid, penitrem A, and some nephrotoxic glycopeptides (Frisvad and Filtenborg 1989). An overview of the mycotoxins and other secondary metabolites produced by the most important

Table 2 Important Species of *Penicillium* and Their Production of Known Secondary Metabolites

Species	Mycotoxins[1]	Other secondary metabolites
P. aurantiogriseum	Penicillic acid	Anacine
	Terrestric acid	Aurantiamine
	Verrucosidin	Auranthine
	Nephrotoxic glycopeptides	
P. aurantiovirens	Penicillic acid	Puberulic acid[2]
		3-methoxyviridicatin[3]
		Verrucofortine[4]
P. cyclopium	Xanthomegnin[5]	Verrucofortine[4]
	Penicillic acid	3-methoxyviridicatin[3]
P. freii	Xanthomegnin[5]	Aurantiamine
	Penicillic acid	3-methoxyviridicatin[3]
P. hordei	Roquefortine C	
	Terrestric acid	
P. melanoconidium	Penitrem A	Meleagrin[6]
	Penicillic acid	
	Verrucosidin	
	Xanthomegnin[5]	
P. polonicum	Penicillic acid	3-methoxyviridicatin[3]
	Verrucosidin	Anacine
	Nephrotoxic glycopeptides	Verrucofortine[4]
P. tricolor	Asteltoxin	Verrucofortine
	Xanthomegnin[5]	
	Terrestric acid	
P. verrucosum	Ochratoxin A	Verrucolone
	Citrinin	Anacine
P. viridicatum	Viridic acid	Brevianamide A
	Xanthomegnin[5]	Viridamine
	Penicillic acid	

[1] Only vertebrate toxins included.
[2] The puberulic acid family also includes puberulonic acid.
[3] The 3-methoxyviridicatin biosynthetic family also includes cyclopeptin, dehydrocyclopeptin, cyclopenol, cyclopenin, viridicatol, and viridicatin.
[4] Verrucofortine is often coproduced with puberuline and the precursors leucyltryptophanyldiketopiperazine and rugulosuvine.
[5] The hepato- and nephro-toxins xanthomegnin, viomellein and vioxanthin are always coproduced. Other derivatives are also produced, at least by *P. viridicatum:* rubrosulphin, viopurpurin, viocristin, and xanthoviridicatin D and G.5.
[6] Meleagrin and oxaline are coproduced.

cereal-borne species is listed in Table 2 (Lund and Frisvad 1994). Of these toxins, only xanthomegnin and viomellein have been found occurring naturally in wheat or barley; however, other mycotoxins may prove to be important. The most important mycotoxins, apart from xanthomegnin and viomellein, may prove to be the nephrotoxic glycopeptides isolated from a strain (strains) of *P. aurantiogriseum* (Yeulet et al. 1988, Macgeorge and Mantle 1990, Mantle et al. 1991, Hard and Greig 1991). The original fungus strain examined (IMI 180922) for nephrotoxicity (Barnes et al. 1977) was first correctly identified as *P. aurantiogriseum* (Yeulet et al. 1988), but was later identified as *P. commune* (Macgeorge and Mantle 1990). However, all taxonomic characters currently used (conidial size, colony diameters on different substrates, growth on creatine-sucrose agar, profiles of secondary metabolites, substrate), dictate that IMI 180922 should be allocated to *P. aurantiogriseum.*

Penicillium freii, *P. cyclopium*, *P. aurantiovirens*, *P. hordei*, and *P. verrucosum* are the most widespread species of *Penicillium* in Danish cereals, while *P. aurantiogriseum*, *P. polonicum* and *P. viridicatum* are less common (Frisvad and Filtenborg, unpublished data). *P. freii* and *P. cyclopium* are very common and both produce xanthomegnin, viomellein, and vioxanthin. It is possible that these naphthoquinones are important cereal-borne mycotoxins in Denmark. Most of the samples examined by Frisvad and Viuf (1986) for *P. verrucosum* also contained *P. freii* and *P. cyclopium.* Both *P. aurantiogriseum* and *P. freii* produce the fluorescing alkaloid aurantiamine (Larsen et al. 1992) but it is not known whether this secondary metabolite is toxic.

A strain involved in a neurological disease of cattle (Wilson et al. 1981) was later shown to produce the tremorgenic mycotoxin verrucosidin (Burka et al. 1983). This strain (PREM 47750 = IBT 3447) also produces normethylverrucosidin, cyclopenin, cyclopenol, 3-methoxyviridicatin, viridicatol, and verrucofortine (Hodges et al. 1988, Frisvad and Filtenborg 1989, Lund and Frisvad, 1994). The results of a repeated analysis of *P. polonicum* PREM 47750 by HPLC using the method of Frisvad and Thrane (1987) is shown in Figure 1. PREM 47750 was grown on laboratory substrates, rye, oats, barley, and wheat (Christensen and Frisvad, unpublished data). A very large amount of penicillic acid is produced on the laboratory media but penicillic acid was not produced at all on any of the four types of cereals. Different members of the cyclopenin biosynthetic family (i.e., cyclopenin, cyclopenol, cyclopeptin, dehydrocyclopeptin, viridicatol, and 3-methoxyviridicatin are produced on laboratory substrates and the cereals, whereas viridicatin is only produced on wheat, oats, and barley. Verrucosidin, anacine, and verrucofortine are produced on all substrates. This indicates that verrucosidin should be among the mycotoxins analyzed for in cereals. It is not known whether anacine or verrucofortine are toxic to vertebrates.

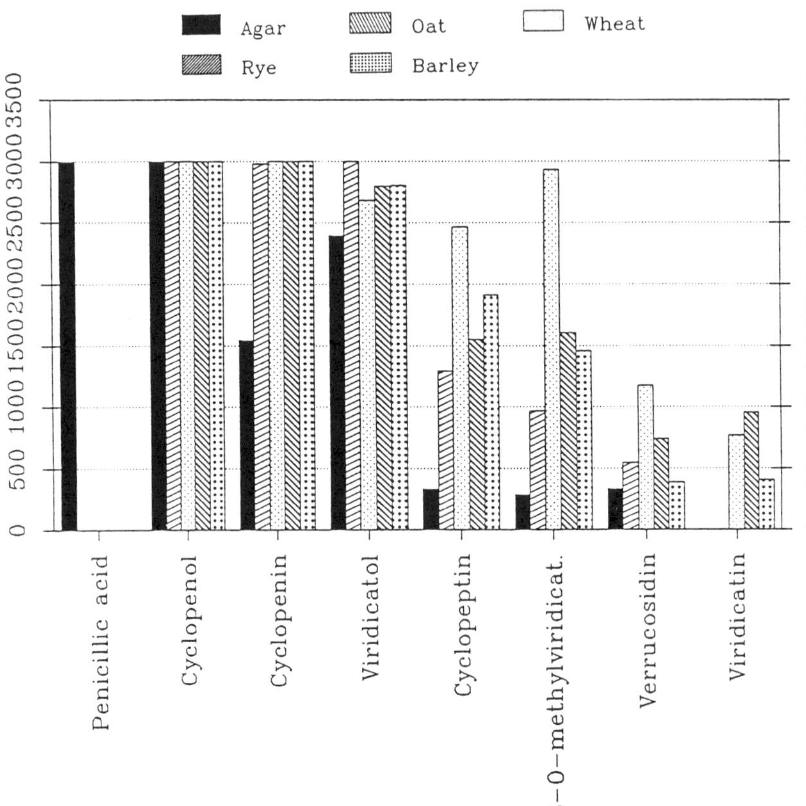

Figure 1 Production of secondary metabolites of *Penicillium polonicum* (IBT 3447) on laboratory substrates (Czapek yeast extract agar and YES agar) compared with production on rye, oat, barley, and wheat as determined by HPLC with diode array detection (Frisvad and Thrane 1987) after 14 days of growth at 25°C (from L. Christensen and Frisvad, unpublished data). The y-axis is the amount of the different secondary metabolites in milli-absorbtion units. Verrucosidin is the only known mycotoxin produced on cereals. The nephrotoxic glycopeptides partially characterized by Yeulet et al. (1988) are apparently produced by several strains of *P. polonicum*, however (P.G. Mantle, personal communication).

Penicillium hordei and *P. chrysogenum* both produce the mycotoxin roquefortine C (Frisvad and Filtenborg 1989) but it is not known if this or any other mycotoxins are produced by these species on cereals.

4. THE ROLE OF SECONDARY METABOLITES ON A STORED-GRAIN ECOSYSTEM

4.1. Arthropods

Dowd (1991) reviewed the interactions between mycotoxins and insects and showed that important mycotoxins produced in cereals are toxic to moth larvae. The mycotoxins that are produced simultaneously by *P. verrucosum*—ochratoxin A and citrinin—even have synergistic toxic effects on *Spodoptera frugiperda* J.E. Smith at realistic levels (Dowd 1989a). However, Dowd (1991) also noted that most stored-products beetles appeared to be relatively resistant to known mycotoxins even though other secondary metabolites may have specific effects on certain insects. The relatively restricted *Penicillium* mycoflora of cereals in temperate regions indicate that secondary metabolites may play a significant ecological role in the insect-fungus-cereal system. Several of the secondary metabolites produced by *P. verrucosum* (ochratoxin A and citrinin) and *P. aurantiogriseum* and related species (brevianamide A, cyclopenol, viomellein, and penicillic acid) were toxic to other insects that do not occur in stored cereals (i.e., *Spodoptera littoralis* and *Drosophila melanogaster*; Paterson et al. 1987). This shows that both mycotoxins and other secondary metabolites can be toxic to several types of insects.

Several species of *Aspergillus, Fusarium,* and *Penicillium* occurring in cereals produce metal-chelating metabolites (Dowd 1988, 1989b; Frisvad 1989b) and Dowd (1988, 1989b) has shown that such chelators (kojic acid from *A. flavus* and fusaric acid from *Fusarium* species) synergize the toxicity of other secondary metabolites from the fungi or plant host allelochemicals. These examples and the clear profiles of secondary metabolites in cereal strains of *Penicillium* and other genera (Frisvad 1989b, Frisvad and Filtenborg 1989) encourage further experiments to show the influence of more than one secondary metabolite at a time on the interactions between fungi, insects, and cereals.

Volatiles produced by cereal fungi are important in plant-fungus-insect interactions (Lussenhop and Wicklow 1990, Dunkel 1988). For instance, 1-octen-3-ol is produced by several fungi in cereals (Abramson 1991), while other volatiles produced by these fungi may be more specific. 1-octen-3-ol, 3-methyl-1-butanol, 3-octanone, and 3-octanol, all produced in cereals (Abramson 1991, Kaminski and Wasowicz 1991), are attractive semiochemicals for the foreign grain beetle (*Ahasverus advena* Waltl) and other stored-product beetles

(Pierce et al. 1991a, b). Again this should be advantageous to consider both nonvolatile and volatile secondary metabolites in ecological studies. This requires more data on the full profile of secondary metabolites of several strains of each species from different origins. Such data on the full profile allow differentiation even between closely related fungi (Frisvad 1992).

4.2. Other Microflora

Cyclopenol is not considered to be a mycotoxin but it has been shown to be both insecticidal and inhibitory to the fungus *Phytophthora infestans* (Mont.) de Bary (Cutler et al. 1988, Adams 1988). Furthermore, the related cyclopenin shows weak antibiotic activity against *Staphylococcus aureus* Rosenbach and *Escherichia coli* (Migula) Castellani and Chalmers (Cutler et al. 1988), while viridicatin at 5 ppm inhibited several gram-positive bacteria including *Bacillus subtilis* (Ehrenberg) Cohn and *Staphylococcus aureus* (Taniguchi and Satomura 1970). As shown in Figure 1, cyclopenin and other related metabolites can be produced by *P. aurantiogriseum* in artificially inoculated cereals and thus may play an important role in cereal ecosystems.

4.3. Seeds

Cyclopenin inhibits the growth of wheat, beans and corn, while cyclopenol inhibits the tobacco plant (Cutler et al. 1988). Viridicatin inhibits the roots and buds of rice, except at low concentrations (3 ppm) where growth is actually stimulated (Taniguchi and Satomura 1970, 1972). Another example is that citrinin inhibits *Allium cepa* root cells (Sinha et al. 1992) and a citrinin-producing fungus like *P. hirsutum* var. *albocoremium* Frisvad is common on onions (Frisvad and Filtenborg 1989). Common secondary metabolites from fungi may also interact with the plants they grow on and this is also well known for *Fusarium* species and their secondary metabolites (Desjardins 1991). It has been shown that kojic acid strongly inhibits polyphenol oxidases of several plants (Chen et al. 1991a,b) and this metabolite may, thus, also play an important ecological role in the interaction between cereals and *A. flavus*, *A. parasiticus*, or *A. nomius*. Thus, secondary metabolites play a role in the interaction between the living plant and the fungi. Data are meager, however, on the interaction of stored seeds and fungal secondary metabolites.

5. CONCLUSIONS

Stored cereals may be invaded by storage fungi when they are not properly dried or growth may start from locally-moistened areas. The most important toxigenic storage fungi in warm climates are *A. flavus* (aflatoxin B_1, cyclopiazonic acid, 3-nitropropionic acid), *A. parasiticus* (aflatoxin B_1 and G_1), and *P. oxalicum* (secalonic acid D). In temperate regions, the most important stored product fungi are *P. verrucosum, P. aurantiogriseum*, and *P. hordei. Penicillium verrucosum* produces oxalic acid, verrucolone, citrinin, ochratoxin A, and several unknown alkaloids together with volatile compounds. *Penicillium aurantiogriseum* and related species produce different combinations of the following mycotoxins: penicillic acid, terrestric acid, xanthomegnin, viomellein verrucosidin, penitrem A, the nephrotoxic glycopeptides, and viridic acid. Furthermore, other secondary metabolites are produced such as cyclopenin, cyclopenol, brevianamide A, verrucofortine, auranthine, aurantiamine, oxaline, and several unknown metabolites plus several volatiles.

The most important mycotoxins in stored cereals in warmer climates are aflatoxins, secalonic acid D, and possibly the yellowed rice toxins. In cooler climates, ochratoxin A, citrinin, xanthomegnin, viomellein, and possibly the nephrotoxic glycopeptides described by Mantle et al. (1991) and verrucosidin play the dominant role. Apart from these toxins *Fusarium* toxins (trichothecenes, fumonisins, fusarochromanone, fusarin C, moniliformin, and zearalenone) may be formed in the field and remain in the stored cereals.

6. FUTURE RESEARCH NEEDS

Many secondary metabolites may have a strong influence on the ecosystem of stored grains, insects, and filamentous fungi. Given the known profile of secondary metabolites in storage fungi, more research emphasis should be given to the interactions between fungi, their secondary metabolites (including volatiles), insects, and grains under natural conditions. Possibilities to monitor and prevent growth and mycotoxin production may be the result from such research. This future field of predictive mycology requires improved multisecondary metabolite screenings methods in cereals and accurate and reliable identification methods for the fungi involved.

REFERENCES

Abildgren, M. P., F. Lund, U. Thrane, and S. Elmholt. 1987. Czapek-Dox agar containing iprodione and dicloran as a selective medium for the isolation of *Fusarium* species. Lett. Appl. Microbiol. 5:83-86.

Abramson, D. 1991. Development of molds, mycotoxins and odors in moist cereals during storage. pp. 119-147 in: Cereal grain. Mycotoxins, fungi and quality in drying and storage. J. Chelkowski (ed.). Elsevier, Amsterdam.

Abramson, D., R. N. Sinha, and J. T. Mills. 1983. Mycotoxin and odor formation in moist cereal grain during granary storage. Cereal Chem. 60:350-355.

Adams, S. 1988. Fungi are key future sources of safe chemicals. Agric. Res. 36:13.

Andrews, S. 1986. Dilution plating versus direct plating of various cereal samples. pp. 40-45 in: Methods for the mycological examination of foods. A. D. King, J. I. Pitt, L. R. Beuchat and J. E. L. Corry (eds.). Plenum Press, New York.

Andrews, S. and J. I. Pitt. 1986. Selective medium for isolation of *Fusarium* species and dematiaceous Hyphomycetes from cereals. Appl. Environ. Microbiol. 51:1235-1238.

Anonymous. 1990. International program on chemical safety. Environmental heath criteria 105. Selected mycotoxins: ochratoxins, trichothecenes, ergot. World Health Organization, Geneva. 263 pp.

Barnes, J. M., R. L. Carter, G. C. Peristianis, P. K. C. Austwick, F. V. Flynn, and W. N. Aldridge. 1977. Balkan (endemic) nephropathy and a toxin-producing strain of *P. verrucosum* var. *cyclopium*: an experimental model in rats. Lancet i:671-675.

Bell, D. K. and J. L. Crawford. 1967. A botran-amended medium for isolating *Aspergillus flavus* from peanuts and soil. Phytopathology 57:939-941.

Bendele, A. M., W. W. Carlton, P. Krogh, and E. B. Lillehøj. 1985. Ochratoxin A carcinogenesis in the (C57BL/6JxC3H)F1 mouse. J. Natl. Cancer Inst. 75:733-742.

Betina, V. 1989. Mycotoxins - chemical, biological and environmental aspects. Elsevier, Amsterdam. 438 pp.

Betina, V. (ed.). 1992. Chromatography of mycotoxins: techniques and applications. Elsevier, Amsterdam. 436 pp.

Bhat, R. V. 1988, Mould deterioration of agricultural commodities during transit: problems faced by developing countries. Int. J. Food Microbiol. 7:219-225.

Bhat, R. V. 1991. Aflatoxins: Successes and failures of three decades of research. pp. 80-85 in: Fungi and mycotoxins in stored products. B. R. Champ, E. Highley, A. D. Hocking, and J. I. Pitt (eds.). ACIAR Proceedings No. 36. Aust. Centre Int. Agric. Research, Canberra, Australia.

Blaney, B. J. 1991. *Fusarium* and Alternaria toxins pp. 86-98 in: Fungi and mycotoxins in stored products. B. R. Champ, E. Highley, A. D. Hocking, and J. I. Pitt (eds.). ACIAR Proceedings No. 36. Aust. Centre Int. Agric. Research, Canberra, Australia.

Boley, A. and H. -M. Müller. 1986. Production of ochratoxin A, citrinin, and ergosterol by *Penicillium viridicatum* in autoclaved and nonautoclaved wheat at low temperature. Mycotoxin Res. 2:99-104.

Bothast, R. J. and D. I. Fennell. 1974. A medium for rapid identification and enumeration of *Aspergillus flavus* and related organisms. Mycologia 66:365-369.

Brooker, D. B., F. W. Bakker-Arkema, and C. W. Hall. 1992. Drying and storage of grains and oilseeds. AVI Van Nostrand Reinhold, New York. 450 pp.

Bullerman, L. B. 1983. Health hazards of dust and pesticides reported at annual meetings. Grain Qual. Newslett. 5:3-6.

Burka, L. T., M. Ganguli, and B. J. Wilson. 1983. Verrucosidin, a tremorgen from *Penicillium verrucosum* var. *cyclopium*. J. Chem. Soc. Chem. Commun. 1983:544-545.

Calderon, M. 1981. The ecosystem approach for apprehending the extent of postharvest grain losses. Phytoparasitica 9:157-167.

Castagnero, M., R. Plestina, G. Dirheimer, I. N. Chernozemsky, and H. Bartsch (eds.). 1991. Mycotoxins, endemic nephropathy and urinary tract tumors. IARC Scientific Publications No. 115. Int. Agency Res. Cancer, Lyon, France. 340 pp.

Ceynowa, J. 1986. Mykologische Untersuchungen an luftdicht gelagertem Getreide. Dissertation, Agrarwissenschaftliche Fakultät der Christian-Albrechts-Universität, Kiel, Germany. 211 pp.

Chalam, R. V. and H. M. Stahr. 1979. Thin layer chromatographic determination of citrinin. J. Assoc. Off. Anal. Chem. 62:570-572.

Champ, B. R., E. Highley, A. D. Hocking, and J. I. Pitt (Eds.). 1991. Fungi and mycotoxins in stored products, ACIAR Proc. No. 36, Aust. Centre Agric. Res., Canberra, Australia. 270 pp.

Chelkowski, J. (ed.). 1989. *Fusarium*—Mycotoxins, taxonomy and pathogenicity. Elsevier, Amsterdam. 492 pp.

Chelkowski, J. (ed.). 1991. Cereal grain. Mycotoxins, fungi and quality in drying and storage. Elsevier, Amsterdam. 607 pp.

Chelkowski, J., G. Dopierala, B. Godlewska, W. Radomska, and K. Szebiotko. 1981. Mycotoxins in ceral grains. Part III. Production of ochratoxin A in different varieties of wheat, rye and barley. Nahrung 25:525-629.

Chen, J. S., C. Wei and M. R. Marshall. 1991a. Inhibition mechanism of kojic acid in polyphenol oxidase. J. Agric. Food Chem. 39:1897-1901.

Chen, J. S., C. Wei, R. S. Rolle, W. S. Otwell, M. O. Balaban, and M. R. Marshall. 1991b. Inhibitory effect of kojic acid on some plant and crustacean polyphenol oxidases. J. Agric. Food Chem. 39:1396-1401.

Cheng, S. J., Y. Z. Jiang, M. H. Li, and H. Z. Lo. 1985. A mutagenic metabolite produced by *Fusarium moniliforme* isolated from Linxian county, China. Carcinogenesis 6: 903-905.

Christensen, C. M. 1991. Fungi and seed quality. pp. 99-120 in: Handbook of applied mycology. Vol. 3. Foods and feeds. D. K. Arora, K. G. Mukerji, and E. H. Marth (eds.). Marcel Dekker Inc., New York.

Chulze, S., C. Bertinetti, A. Dalcero, M. Etcheverry, C. Farnochi, A. Torres, I. Rizzo, and E. Varsavsky. 1989. Incidence of aflatoxin, zearalenone, and deoxynivalenol on corn in Argentine. Mycotoxin Res. 5:9-12.

Ciegler, A., D. I. Fennell, G. A. Sansing, R. W. Detroy, and G. A. Bennett. 1973. Mycotoxin-producing strains of *Penicillium viridicatum*: classification into subgroups. Appl. Microbiol. 26:271-278.

Clarke, J. H. and S. T. Hill. 1981. Mycofloras of moist barley during sealed storage in farm and laboratory silos. Trans. Brit. Mycol. Soc. 77:557-565.

Cole, R. J. and R. H. Cox. 1981. Handbook of toxic fungal metabolites. Academic Press, New York. 935 pp.

Cole, R. J. 1984. Cyclopiazonic acid and related toxins. pp. 405-414 in: Mycotoxins—production, isolation, separation, and purification. V. Betina (ed.). Elsevier, Amsterdam.

Cole, R. J. (Ed.). 1986. Modern methods in the analysis and structural elucidation of mycotoxins. Academic Press, New York. 471 pp.

Cole, R. J., J. W. Kirksey, J. W. Dorner, D. M. Wilson, J. C. Johnson, A. Neill Johnson, D. M. Bedell, J. P. Springer, K. K. Chexal, J. C. Clardy, and R. H. Cox. 1977.

Mycotoxins produced by *Aspergillus fumigatus* species isolated from molded silage. J. Agric. Food Chem. 25:826-830.

Coulombe, R. A. 1991. Aflatoxins. pp. 103-143 in: Mycotoxins and phytoalexins. R. P. Sharma and D. K. Salunkhe (eds.). CRC Press, Boca Raton, FL.

Cutler, H. G., E. Ammermann, and J. P. Springer. 1988. Diverse but specific biological activities of four natural products from three fungi. pp. 79-90 in: Biologically active natural products. Potential use in agriculture. H. G. Cutler (ed.). Am. Chem. Soc., Washington.

Damoglou, A. P., G. A. Downey, and W. Shannon. 1984. The production of ochratoxin A and citrinin in barley. J. Sci. Food Agric. 35:395-400.

Davis, N. D., U. L. Diener, and D. W. Eldridge. 1966. Production of aflatoxins B_1 and G_1 by *Aspergillus flavus* in a semisynthetic medium. Appl. Microbiol. 14:378-380.

Davis, N. D., S. K. Iyer, and U. L. Diener. 1987. Improved method of screening for aflatoxin with a coconut agar medium. Appl. Environ. Microbiol. 53:1593-1595.

Dawson, R. J. 1991. A global view of the mycotoxin problem. pp. 22-28 in: Fungi and mycotoxins in stored products. ACIAR Proc. No. 36. B. R. Champ, E. Highley, A. D. Hocking, and J. I. Pitt (eds.). Aust. Centre Int. Agric. Res., Canberra, Australia.

Desjardins, A. E. 1991. Genetic approaches to the chemical ecology of phytopathogenic *Fusarium* species. pp. 333-357 in: Handbook of applied mycology Vol. 5. Mycotoxins in ecological systems. D. Bhatnagar, E. B. Lillehøj, and D. K. Arora (eds.). Marcel Dekker, New York.

Dick, R., U. Baumann, and B. Zimmerli. 1988. Zum vorkommen von citrinin in cerealien. Mitt. Gebiete Lebensm. Hyg. 79:159-164.

Dowd, P. 1988. Synergism of aflatoxin B_1 toxicity with the co-occurring fungal metabolite kojic acid to two caterpillars. Entomol. Exp. Appl. 47:69-71.

Dowd, P. 1989a. Toxicity of naturally occurring levels of the *Penicillium* mycotoxins citrinin, ochratoxin A, and penicillic acid to the corn earworm, *Helothis zea*, and the fall armyworm, *Spodoptera frugiperda* (*Lepidoptera: Noctuidae*). Environ. Entomol. 18:24-29.

Dowd, P. 1989b. Fusaric acid. A secondary fungal metabolite that synergizes toxicity of co-occurring host allelochemicals to the corn earworm, *Helothis zea* (*Lepidoptera*). J. Chem. Ecol. 15:249-254.

Dowd, P. 1991. Insect interactions with mycotoxin-producing fungi and their hosts. pp. 137-155 in: Handbook of applied mycology Vol. 5. Mycotoxins in ecological systems. D. Bhatnagar, E. B. Lillehøj, and D. K. Arora (eds). Marcel Dekker, New York.

Dunkel, F. V. 1988. The relationship of insects to the deterioration of stored grain by fungi. Int. J. Food Microbiol. 7:227-244.

Ehrlich, K. C., L. S. Lee, A. Ciegler, and M. S. Palmgren. 1982. Secalonic acid D: natural contaminant of corn dust. Appl. Environ. Microbiol. 44:1007-1008.

Escoula, L. 1973. Moissisures toxinogène des fourrages ensilè. Ann. Rech. Vétér. 6:303-310, 315-324.

Filtenborg, O. and J. C. Frisvad. 1980. A simple screening method for toxigenic fungi in pure cultures. Lebensm. Wiss. Technol. 13:128-130.

Filtenborg, O. and J. C. Frisvad. 1990. Identification of *Penicillium* and *Aspergillus* species in mixed cultures in petri dishes using secondary metabolite profile. pp.

27-37 in: Modern concepts in *Penicillium* and *Aspergillus* classification. R. A. Samson and J. I. Pitt (eds.). Plenum Press, New York.

Filtenborg, O., J. C. Frisvad, and J. A. Svendsen. 1983. Simple screening method for molds producing intracellular mycotoxins in pure cultures. Appl. Environ. Microbiol. 45:581-585.

Filtenborg, O., J. C. Frisvad and U. Thrane. 1990. The significance of yeast extract composition on metabolite production in *Penicillium*. pp. 433-441 in: Modern concepts in *Penicillium* and *Aspergillus* classification. R. A. Samson and J. I. Pitt (eds.). Plenum Press, New York.

Filtenborg, O., J. C. Frisvad, F. Lund, and U. Thrane. 1992. Simple identification procedure for spoilage and toxigenic mycoflora of foods. pp. 248-258 in: Modern methods in food mycology. R. A. Samson, A. D. Hocking, J. I. Pitt, and A. D. King (eds.). Elsevier, Amsterdam.

Flannigan, B., E. M. McCabe, and F. McGarry. 1991. Allergenic and toxigenic microorganisms in houses. J. Appl. Bact. 70:61S-73S.

Frisvad, J. C. 1981. Physiological criteria and mycotoxin production as aids in identification of commmon asymmetric penicillia. Appl. Environ. Microbiol. 41:568-579.

Frisvad, J. C. 1983. A selective and indicative medium for groups of *Penicillium viridicatum* producing different mycotoxins in cereals. J. Appl. Bacteriol. 54:409-416.

Frisvad, J. C. 1985a. Creatine-sucrose agar, a differential medium for mycotoxin producing terverticillate *Penicillium* species. Lett. Appl. Microbiol. 1:109-113.

Frisvad, J. C. 1985b. Classification of asymmetric penicillia using expressions of differentiation. pp. 327-333 in: Advances in *Penicillium* and *Aspergillus* systematics. R. A. Samson and J. I. Pitt (eds.), Plenum Press, New York.

Frisvad, J. C. 1986. Taxonomic approaches to mycotoxin identification. pp. 415-457 in: Modern methods in the analysis and structure elucidation of mycotoxins. R. J. Cole (ed.). Academic Press, New York.

Frisvad, J. C. 1989a. The connection between the penicillia and aspergilli and mycotoxins with emphasis on misidentified isolates. Arch. Environ. Contam. Toxicol. 18:452-467.

Frisvad, J. C. 1989b. The use of high-performance liquid chromatography and diode array detection in fungal chemotaxonomy based on profiles of secondary metabolites. Bot. J. Linn. Soc. 99:81-95.

Frisvad, J. C. 1992. Chemometrics and chemotaxonomy: a comparison of multivariate statistical methods for the evaluation of binary fungal secondary metabolite data. Chemom. Intell. Lab. Syst. 12:253-269.

Frisvad, J. C. and O. Filtenborg. 1983. Classification of terverticillate penicillia based on profiles of mycotoxins and other secondary metabolites. Appl. Environ. Microbiol. 46:1301-1310.

Frisvad, J. C. and O. Filtenborg. 1988. Specific mycotoxin producing *Penicillium* and *Aspergillus* mycoflora of different foods. Proc. Jpn. Assoc. Mycotoxicol. Suppl. 1:163-166.

Frisvad, J. C. and O. Filtenborg. 1989. Terverticillate penicillia: chemotaxonomy and mycotoxin production. Mycologia 81:837-861.

Frisvad, J. C. and R. A. Samson. 1991a. Filamentous fungi in foods and feeds: Ecology, spoilage and mycotoxin production. pp. 31-68 in: Handbook of applied mycology.

Vol. 3. Foods and feeds. D. K. Arora, K. G. Mukerji, and E. H. Marth (eds.). Marcel Dekker, New York.

Frisvad, J. C. and R. A. Samson. 1991b. Mycotoxins produced by species of *Penicillium* and *Aspergillus* occurring in cereals. pp. 441-476 in: Cereal grain. Mycotoxins, fungi and quality in drying and storage. J. Chelkowski (ed.). Elsevier, Amsterdam.

Frisvad, J. C. and U. Thrane. 1987. Standardized high-performance liquid chromatography of 182 mycotoxins and other fungal metabolites based on alkylphenone retention indices and UV-VIS spectra (diode array detection). J. Chromatogr. 404:195-214.

Frisvad, J. C. and B. T. Viuf. 1986. Comparison of direct and dilution plating for detecting *Penicillium viridicatum* in barley containing ochratoxin. pp. 45-47 in: Methods for the mycological examination of foods. A. D. King, J. I. Pitt, L. R. Beuchat, and J. E. L. Corry (eds.). Plenum Press, New York.

Frisvad, J. C., A. B. Kristensen, and O. Filtenborg. 1986. Comparison of methods used for surface desinfection of food and feed commodities before mycological analysis. pp. 32-40 in: Methods for the mycological examination of foods A. D. King, J. I. Pitt, L. R. Beuchat, and J. E. L. Corry (eds.). Plenum Press, New York

Frisvad, J. C., O. Filtenborg, U. Thrane, and P. V. Nielsen. 1992a. Collaborative study on media for detecting and enumerating toxigenic *Penicillium* and *Aspergillus* species. pp. 241-247 in: Modern methods in food mycology. R. A. Samson, A. D. Hocking, J. I. Pitt, and A. D. King (eds.). Elsevier, Amsterdam.

Frisvad, J. C., O. Filtenborg, F. Lund, and U. Thrane. 1992b. New selective media for the detection of toxigenic fungi in cereal products, meat and cheese. pp. 259-268 in: Modern methods in food mycology. R. A. Samson, A. D. Hocking, J. I. Pitt, and A. D. King (eds.). Elsevier, Amsterdam.

Fukal, L. 1990. A survey of cereals, cereal products, feedstuffs and porcine kidneys for ochratoxin A by radioimmuneassay. Food Addit. Contam. 7:253-258.

Gams, W. and W. van Laar. 1982. The use of solacol (validamycin) as a growth retardant in the isolation of soil fungi. Neth. J. Plant Pathol. 88:39-45.

Garcia, R. P. and J. Tuite. 1985. Screening of *Penicillium* isolates from shelled corn for production of mycotoxins. Phillip. Agric. 68:453-459.

Gedek, B., J. Bauer, and H. Schreibner. 1981. Zur Mykotoxinbildung Silage-verderbender Schimmelpilze. Wien. tierärzl. Mschr. 68:299-301.

Gelderblom, W. C. A., P. G. Thiel, W. F. O. Marasas, and K. J. van der Merwe. 1984. Natural occurrence of fusarin C, a mutagen produced by *Fusarium moniliforme*, in corn. J. Agric. Food Chem. 32:1064-1067.

Grabarkiewicz-Szczesna, J., J. Chelkowski, and P. Zajkowski. 1989. Natural occurrence of *Alternaria* mycotoxins in the grain and chaff of cereals. Mycotoxin Res. 5:77-80.

Gyllang, H., K. Kjellén, A. Haikara, and P. Sigsgaard. 1981. Evaluation of fungal contamination on barley and malt. J. Inst. Brew. 87:248-251.

Häggblom, P. 1982. Production of ochratoxin A in barley by *Aspergillus ochraceus* and *Penicillium viridicatum*: effect of fungal growth, time, temperature, and inoculum size. Appl. Environ. Microbiol. 43:1205-1207.

Häggblom, P. 1990. Isolation of roquefortine C from feed grain. Appl. Environ. Microbiol. 56:2924-2926.

Häggblom, P. and J. Ghosh. 1985. Postharvest production of ochratoxin A by *Aspergillus ochraceus* and *Penicillium viridicatum* in barley with different protein levels. Appl. Environ. Microbiol. 49:787-790.

Hald, B., D. H. Christensen, and P. Krogh. 1983. Natural occurrence of the mycotoxin viomellein in barley and associated quinone-producing penicillia. Appl. Environ. Microbiol. 46:1311-1317.

Hamsa, T. A. P. and J. C. Ayers. 1977. A differential medium for the isolation of *Aspergillus flavus* from cotton seed. J. Food Sci. 42:449-453.

Hara, S., D. I. Fennell, and C. W. Hesseltine. 1974. Aflatoxin-producing strains of *Aspergillus flavus* detected by fluorescence of agar medium under ultraviolet light. Appl. Microbiol. 27:1118-1123.

Hard, G. C. and J. B. Greig. 1991. Comparative acute nephrotoxicity of *Penicillium aurantiogriseum* in rats and hamsters. pp. 113-117 in: Mycotoxins, endemic nephropathy and urinary tract tumours. M. Castagnero, R. Plestina, G. Dirheimer, I. N. Chernozemsky, and H. Bartsch (eds.). Int. Agency Res. Cancer, Lyon, France.

Harwig, J. and Y. -K. Chen. 1974. Some conditions favoring production of ochratoxin A and citrinin by *Penicillium viridicatum* on barley. Can. J. Plant Sci. 54:17-24.

Hirata, Y. 1947. On the product of mold. I. Poisonous substance from moldy rice (Part I). Extraction. J. Japan. Chem. Soc. 68:63,74.

Hocking, A. D. 1991a. Xerophilic fungi in intermediate and low moisture foods. pp. 69-97 in: Handbook of applied mycology. Vol. 3. Foods and feeds. D. K. Arora, K. G. Mukerji, and E. H. Marth (eds.). Marcel Dekker Inc., New York.

Hocking, A. D. 1991b. Effects of fumigation and modified atmosphere storage on growth of fungi and production of mycotoxins in stored grains. pp. 145-156 in: Fungi and mycotoxins in stored products. ACIAR Proc. No. 36. B. R. Champ, E. Highley, A. D. Hocking, and J. I. Pitt (eds.). Aust. Centre Int. Agric. Res., Canberra, Australia.

Hocking, A. D. and J. I. Pitt. 1980. Dichloran-glycerol medium for enumeration of xerophilic fungi from low-moisture foods. Appl. Environ. Microbiol. 39:488-492.

Hodges, R. P., C. M. Harris, and T. M. Harris. 1988. Verrucofortine, a major metabolite of *Penicillium verrucosum* var. *cyclopium*, the fungus that produces the mycotoxin verrucosidin. J. Nat. Prod. 51:66-73.

Hoekby, E., K. Hult, S. Gatenbeck, and L. Rutquist. 1979. Ochratoxin and citrinin in 1976 crop of barley stored on farms in Sweden. Acta Agric. Scand. 29:174-178.

Holmberg, T., H. Pettersson, N. G. Nilsson, B. Göransson, and R. Grossman. 1983. A case of aflatoxicosis in fattening calves caused by aflatoxin formation in inadequate formic acid treated grain. Zbl. Vet. Med. A 30:656-663.

Holmberg, T., A. Breitholtz-Emanuelson, P. Haggblom, O. Schwan, and K. Hult. 1991. *Penicillium verrucosum* in feed of ochratoxin A positive swine herds. Mycopathologia 116:169-176.

Hutchinson, R. D., P. S. Steyn, and S. J. van Reensburg. 1973. Viridicatumtoxin, a new mycotoxin from *Penicillium viridicatum* Westling. Toxicol. Appl. Pharmacol. 24: 507-509.

Jelinek, C. F., A. E. Pohland, and G. E. Wood. 1989. Worldwide occurrence of mycotoxins in foods and feeds. J. Assoc. Off. Anal. Chem. 72:223-230.

Kaminski, E. and E. Wasowicz. 1991. The usage of volatile compounds produced by moulds as indicators of grain deterioration. pp. 229-258 in: Cereal grain. Mycotoxins, fungi and quality in drying and storage. J. Chelkowski (ed.). Elsevier, Amsterdam.

Kane, A., N. Diop, and T. S. Diack. 1991. Natural occurrence of ochratoxin A in food and feed in Senegal. pp. 93-96 in: Mycotoxins, endemic nephropathy and urinary tract tumours. M. Castagnero, R. Plestina, G. Dirheimer, I. N. Chernozemsky, and H. Bartsch (eds.). Int. Agency Res. Cancer, Lyon, France.

Kiessling, K. -H., H. Pettersson, K. Tideman, and I. -L. Andersson. 1982. A survey of aflatoxin and *Aspergillus flavus/parasiticus* in acid treated Swedish grain. Swedish J. Agric. Res. 12:63-67.

King, A. D., A. D. Hocking, and J. I. Pitt. 1979. Dichloran-rose bengal medium for enumeration and isolation of molds from foods. Appl. Environ. Microbiol. 37:959-964.

King, A. D., J. I. Pitt, L. R. Beuchat, and J. E. L. Corry (eds.). 1986. Methods for the mycological examination of foods. Plenum Press, New York. 315 pp.

Krogh, P. (ed.) 1987. Mycotoxins in food. Academic Press, New York. 263 pp.

Krogh, P., B. Hald, and E. J. Pedersen. 1973. Occurrence of ochratoxin A and citrinin in cereals associated with mycotoxin porcine nephropathy. Acta Pathol. Microbiol. Scand. Sect. B 81:689-695.

Krogh, P., D. H. Christensen, B. Hald, B. Harlou, C. Larsen, E. J. Pedersen, and U. Thrane. 1989. Natural occurrence of the mycotoxin fusarochromanone, a metabolite of *Fusarium equiseti*, in cereal feed associated with tibial dyschrondroplasia. Appl. Environ. Microbiol. 55:3184-3188.

Krogh, P., E. Hasselager, and P. Friis. 1970. Studies on fungal nephrotoxicity. II. Isolation of two nephrotoxic compounds from *Penicillium viridicatum* Westling. Acta Pathol. Microbiol. Scand. Sect. B 78:401-413.

Kuiper-Goodman, T. and P. M. Scott. 1989. Risk assessment of the mycotoxin ochratoxin A. Biomed. Environ. Sci. 2:179-248.

Lacey, J. 1986. Factors affecting fungal colonization of grain. pp. 29-34 in: Spoilage and mycotoxins of cereals and other stored products B. Flannigan (ed.). Int. Biodeterior. 22 (Supplement). CAB International, Farnham Royal, UK.

Lacey, J. 1991. Aerobiology and health: the role of airborne fungal spores in respiratory disease. pp. 157-185 in: Frontiers in mycology. D. L. Hawksworth (ed.). CAB International, Farnham Royal, UK.

Lacey, J. and N. Magan. 1991. Fungi in cereal grains: their occurrence and water and temperature relationships. pp. 77-118 in: Cereal grain. Mycotoxins, fungi and quality in drying and storage. J. Chelkowski (ed.). Elsevier, Amsterdam.

Lacey, J., N. Ramakrishna, A. Hamer, N. Magan, and I. C. Marfleet. 1991. Grain fungi. pp. 121-177 in: Handbook of applied mycology. Vol. 3. Foods and feeds. D. K. Arora, K. G. Mukerji, and E. H. Marth (eds.). Marcel Dekker, New York.

Larsen, T. O., J. C. Frisvad, and S. R. Jensen. 1992. Aurantiamine, a diketopiperazine from two varieties of *Penicillium aurantiogriseum*. Phytochemistry 31:1613-1615.

Lee, L. S., P. Bayman, and J. W. Bennett. 1991. Mycotoxins. pp. 463-503 in: Biotechnology of filamentous fungi. Technology and products D. B. Finkelstein and C. Ball (eds.). Butterworth-Heineman, Boston.

Lemke, P. A., N. D. Davis, and G. W. Creech. 1989. Direct visual detection of aflatoxin synthesis by minicolonies of *Aspergillus* species. Appl. Environ. Microbiol. 55: 1808-1810.

Lew, H., A. Adler, and W. Edinger. 1991. Moniliformin and the european corn borer (*Ostrinia nubilalis*). Mycotoxin Res. 7A:71-76.

Lillehøj, E. B. 1991a. Aflatoxin, an ecologically-elicited, genetic activation signal. pp. 1-17 in: Mycotoxins and animal foods J. E. Smith and R. S. Henderson (eds.). CRC Press, Boca Raton, FL.

Lillehøj, E. B. 1991b. Aflatoxin: genetic mobilization agent. pp. 1-22 in: Handbook of applied mycology. Vol. 5. Mycotoxins in ecological systems. D. Bhatnagar, E. B. Lillehøj, and D. K. Arora (eds.). Marcel Dekker, New York.

Lin, M. T. and J. C. Dianese. 1976. A coconut-agar medium for rapid detection of aflatoxin production by *Aspergillus* spp. Phytopathology 66:1466-1469.

Liu, X. J., X. Y. Luo, and W. J. Hu. 1989. *Arthrinium* spp. and ethiology of deteriorated sugarcane poisoning. pp. 109-118 in: Mycotoxins and phycotoxins '88. S. Natori, K. Hashimoto, and Y. Ueno (eds.). Elsevier, Amsterdam.

Lund, F. and J. C. Frisvad. 1994. Chemotaxonomy of *Penicillium aurantiogriseum* and related species. Mycol. Res. 98(5):(in press).

Lussenhop, J. and D. T. Wicklow. 1990. Nitidulid beetles (Nitudulidae: Coleoptera) as vectors of *Aspergillus flavus* in pre-harvest maize. Trans Mycol. Soc. Japan. 31:63-74.

Macgeorge, K. M. and P. G. Mantle. 1990. Nephrotoxicity of *Penicillium aurantiogriseum* and *P. commune* from an endemic nephropathy area of Yugoslavia. Mycopathologia 112:139-145.

Madelin, T. M. 1987. The effect of surfactant in media for enumeration, growth and identification of airborne fungi. J. Appl. Bacteriol. 63:47-52.

Madhyastha, S. M., R. R. Marquardt, A. A. Frohlich, G. Platford, and D. Abramson. 1990. Effects of different cereal and oilseed substrates on the growth and production of toxins by *Aspergillus alutaceus* and *Penicillium verrucosum*. J. Agric. Food Chem. 38:1506-1510.

Magan, N. and J. Lacey. 1988. Ecological determinants of mould growth in stored grain. Int. J. Food Microbiol. 7:245-256.

Mantle, P. G., K. M. McHugh, R. Adatia, J. M. Heaton, T. Gray, and D. R. Rurner. 1991. *Penicillium aurantiogriseum*-induced persistent renal histopathological changes in rats; an experimental model for balkan endemic nephropathy competitive with ochratoxin A. pp. 119-127 in: Mycotoxins, endemic nephropathy and urinary tract tumours. M. Castagnero, R. Plestina, G. Dirheimer, I. N. Chernozemsky, and H. Bartsch (eds.). Int. Agency Res. Cancer, Lyon, France.

Marasas, W. F. O. and P. E. Nelson. 1987. Mycotoxicology. Pennsylvania State University Press, University Park, PA. 102 pp.

Marasas, W. F. O., P. E. Nelson, and T. A. Toussoun. 1984. Toxigenic *Fusarium* species. Identity and mycotoxicology. Pennsylvania State University Press, University Park, PA. 328 pp.

Masurekar, P. S. 1991. Therapeutic metabolites. pp. 241-301 in: Biotechnology of filamentous fungi. Technology and products D. B. Finkelstein and C. Ball (eds.). Butterworth-Heineman, Boston.

Miller, J. D. 1990. Fungi as contaminants in indoor air. pp. 51-63 in: Indoor air '90. Proc. 5th Int. Conf. Indoor Air Quality Clim. Vol. 5., D. S. Walkinshaw (ed.), Canada Mortgage and Housing Co., Ottawa.

Miller, J. D. 1991. Significance of grain mycotoxins for health and nutrition. pp. 126-135 in: Fungi and mycotoxins in stored products B. R. Champ, E. Highley, A. D. Hocking, and J. I. Pitt (eds.). ACIAR Proc. No. 36., Aust. Centre Int. Agric. Res., Canberra, Australia.

Mills, J. T. 1989. Spoilage and heating of stored agricultural products. Prevention, detection and control. Agr. Can. Publ. #1823. Canadian Gov. Pub. Center, Ottawa. 101 pp.

Mills, J. T. and D. Abramson. 1982. Ochratoxigenic potential of *Penicillium* spp. isolated from stored rapeseed and cereals in western Canada. Can. J. Plant Pathol. 4:37-41.

Mills, J. T., D. Abramson, A. A. Frohlich, and R. R. Marquardt. 1989. Citrinin and ochratoxin A production by *Penicillium* spp. from stored durum wheat. Can. J. Plant Pathol. 11:357-360.

Milton, R. F. and R. K. Pawsey. 1988. Spoilage relating to the storage and transport of cereals and oil seeds. Int. J. Food Microbiol. 7:211-217.

Moreau, C. 1975. Le *Penicillium ochrosalmoneum* Udagawa, moisissure de chataignes et la neurotoxicose de moutons en Ardeche. Bull. Soc. Mycol. Fr. 91:413-421.

Moss, M. 1971. The rubratoxins, toxic metabolites of *Penicillium rubrum* Stoll. pp. 381-407 in: Microbial toxins. Vol. VI. Fungal toxins. A. Ciegler, S. Kadis, and S. J. Ajl. (eds.). Academic Press, New York.

Mossel, D. A. A. 1988. Mould spoilage of cereals during transportation by sea from Latin America to Europe, mechanisms, impact and management. Int. J. Food Microbiol. 7:205-209.

Müller, H. -M. and A. Boley. 1991. Einfluss von Autoklavierung und Oberflächendesinfektion von Weizen (Triticum aestivum) auf die Produktion von Ergostrin, Ochratoxin A und Citrinin durch *Penicillium verrucosum*. Zentralb. Mikrobiol. 146: 445-451.

Multon, J. L. 1988. Spoilage mechanisms of grains and seeds in the postharvest ecosystem. pp. 3-63 in: Preservation of grains, seeds and their by-products J. L. Multon (ed.). Lavoisier Publishing, New York.

Natori, S., K. Hashimoto, and Y. Ueno (eds.). 1989. Mycotoxins and phycotoxins 1988. Elsevier, Amsterdam. 443 pp.

Nelson, T. S., L. K. Kirby, J. N. Beasley, Z. B. Johnson, and A. Ciegler. 1985. The effect of drying method and storage time on citrinin activity in corn. Poultry Sci. 64:464-468.

Nelson, P. E., R. D. Plattner, D. D. Shackelford, and A. E. Desjardins. 1992. Fumonisin B1 production by *Fusarium* species other than *F. moniliforme* in section Liseola and by some related species. Appl. Environ. Microbiol. 58:984-989.

Nishikawa, N. and M. Kohgo. 1975. A selective medium for genera *Fusarium* and *Gibberella* contaminating on barley and malt grains. Bull. Brew. Sci. 21:9-17.

Northolt, M. D., H. P. van Egmond, and W. E. Paulsch. 1979. Ochratoxin A production by some fungal species in relation to water activity and temperature. J. Food Prot. 42:485-490.

Osborne, B. G. 1980. The occurrence of ochratoxin A in mouldy bread and retail flour. Food Cosmet. Toxicol. 18:615-617.

Palmgren, M. S. and D. S. Fleischhacker. 1987. *Penicillium oxalicum* and secalonic acid D in fresh corn. Biodeterior. Res. 1:193.

Park, J., M. S. Zong, and I. -M. Chang. 1991. Survey of the presence of the *Fusarium* mycotoxins nivalenol, deoxynivalenol and T-2 toxin in Korean cereals of the 1989 harvest. Food Addit. Contam. 8:447-451.

Paster, N. and L. B. Bullerman. 1988. Mould spoilage and mycotoxin formation in grains as controlled by physical means. Int. J. Food Microbiol. 7:257-265.

Paterson, R. R. M., M. S. J. Simmmonds, and W. M. Blaney. 1987. Mycopesticidal effects of characterized extracts of *Penicillium* isolates and purified secondary metabolites (including mycotoxins) on *Drosophila melanogaster* and *Spodoptera littorals*. J. Invert. Pathol. 50:124-133.

Pelhate, J. 1974/1975. Mycoflore des mais-fourrages ensils. Rev. Mycol. 24:65-95.

Pelhate, J. 1988a. Microbiology of seeds in relation to conditioning: its influence on germinative capacity. pp. 310-327 in: Preservation of grains, seeds and their by-products. J. L. Multon (ed.). Lavoisier Publishing, New York.

Pelhate, J. 1988b. Ecology of the microflora of grains and seeds. pp. 244-262 in: Preservation of grains, seeds and their by-products. J. L. Multon (ed.). Lavoisier Publishing, New York.

Pelhate, J. 1988c. Microbiology of moist grains. pp. 328-346 in: Preservation of grains, seeds and their by-products J. L. Multon (ed.). Lavoisier Publishing, New York.

Penel, A. J. and F. V. Kosikowski. 1990. Beta-nitropropionic acid production by *Aspergillus oryzae* in selected high protein and carbohydrate-rich foods. J. Food Prot. 53:321-323.

Perkowski, J., R. D. Plattner, P. Golinski, R. F. Vesonder, and J. Chelkowski. 1990. Natural occurrence of deoxynivalenol, 3-acetyl-deoxynivalenol, 15-acetyl-deoxynivalenol, nivalenol, 4,7-dideoxynivalenol, and zearalenone on Polish wheat. Mycotoxin Res. 6:7-12.

Pestka, J. J. and G. S. Bondy. 1990. Alteration of immune function following dietary mycotoxin exposure. Can. J. Physiol. Pharmacol. 68:1009-1016.

Petkova-Bocharova, T., M. Castagnero, J. Michelon, and V. Maru. 1991. Ochratoxin A and other mycotoxins in cereals from an area of Balkan endemic nephropathy and renal tumours in Bulgaria. pp. 83-87 in: Mycotoxins, endemic nephropathy and urinary tract tumours. M. Castagnero, R. Plestina, G. Dirheimer, I. N. Chernozemsky, and H. Bartsch (eds.). Int. Agency Res. Cancer, Lyon, France.

Petterson, H., B. Göransson, K. -H. Kiessling, K. Tideman, and T. Johansson. 1978. Aflatoxin in a Swedish grain sample. Nord. Vet. Med. 30:482-485.

Pierce, A. M., H. D. Pierce, A. C. Oehlschlager, and J. H. Borden. 1991a. 1-octen-3-ol, atractive semiochemical for foreign grain beetle, *Ahasverus advena* (Waltl) (Coleoptera: Cucujidae). J. Chem. Ecol. 17:567-580.

Pierce, A. M., H. D. Pierce, J. H. Borden, and A. C. Oehlschlager. 1991b. Fungal volatiles: semiochemical for stored-product beetles (Coleoptera: Cucujicae). J. Chem. Ecol. 17:581-597.

Pitt, J. I. 1979. The genus *Penicillium* and its teleomorphic states *Eupenicillium* and *Talaromyces*. Academic Press, London. 634 pp.

Pitt, J. I. 1987. *Penicillium viridicatum, P. verrucosum* and the production of ochratoxin A. Appl. Environ. Microbiol. 53:266-269.

Pitt, J. I. and R. H. Cruickshank. 1990. Speciation and synonomy in *Penicillium* subgenus *Penicillium*—towards a definitive taxonomy. pp. 103-119 in: Modern concepts in *Penicillium* and *Aspergillus* classification. R. A. Samson and J. I. Pitt (eds.). Plenum Press, New York.

Pitt, J. I., A. D. Hocking, and D. R. Glenn. 1983. An improved medium for the detection of *Aspergillus flavus* and *A. parasiticus*. J. Appl. Bacteriol. 54:109-114.

Plattner, R. D., W. P. Norred, C. W. Bacon, K. A. Voss, R. Peterson, D. D. Shackelford, and D. Weisleder. 1990. A method of detection of fumonisins in corn samples associated with field cases of equine leukoencephalomalacia. Mycologia 82:698-702.

Poisson, J. and B. Cahagnier. 1988. Effects of grain "stabilization" processes. pp. 263-289 in: Preservation of grains, seeds and their by-products. J. L. Multon (ed.). Lavoisier Publishing, New York.

Ranfft, K., R. Gerstl, and G. Mayer. 1990. Bestimmung und Vorkommen von Zearalenon in Getreide und Mischfuttermitteln. Z. Lebensm. Unters. Forsch. 191:449-453.

Raper, K. B. and D. I. Fennell. 1965. The genus *Aspergillus*. Williams and Wilkins Co., Baltimore, MD. 686 pp.

Raper, K. B. and C. Thom. 1949. A manual of the penicillia. Williams and Wilkins Co., Baltimore, MD. 874 pp.

Reddy, C. S. and R. V. Reddy. 1991. Secalonic acids. pp. 167-190 in: Mycotoxins and phytoalexins. R. P. Sharma and D. K. Salunkhe (eds.). CRC Press, Boca Raton, FL.

Reddy, C. S., R. V. Reddy, and A. W. Hayes. 1981. High-performance liquid chromatographic analysis of the mycotoxin secalonic acid D and its application to biological fluids. J. Chromatogr. 208:17-26.

Richard, J. L. and J. R. Thurston (eds.). 1986. Diagnosis of mycotoxicosis. Martinus Nijhoff Publishers, Dordrecht, South Africa. 411 pp.

Rodricks, J. V. (ed.). 1976. Mycotoxins and other fungal related food problems. American Chemical Society, Washington. 409 pp.

Rosa, C. A. da R., L. C. H. da Cruz, W. A. Chagas, and C. E. M. de O. Veiga. 1985. Ocorrencia natural de nefropatia micotoxica suina causada pela ingestao de ceveda contaminada com citrina. Rev. Bras. Med. Vet. 7:87-90.

Ross, P. F., L. G. Rice, R. D. Plattner, G. D. Osweiler, T. M. Wilson, D. L. Owens, P. A. Nelson, and J. L. Richard. 1990. Concentrations of fumonisin B_1 in feeds associated with animal health problems. Mycopathologia 114:129-135.

Ross, P. F., L. G. Rice, G. D. Osweiler, P. E. Nelson, J. L. Richard, and T. M. Wilson. 1992. A review and update of animal toxicoses associated with fumonisin-contaminated foods and production of fumonisins by *Fusarium* isolates. Mycopathologia 117:109-114.

Saito, M., M. Enomoto, and T. Tatsuno. 1971. Yellowed rice toxins. pp. 299-380 in: Microbial toxins. Vol. VI. Fungal toxins. A. Ciegler, S. Kadis and S. J. Ajl. (eds.). Academic Press, New York.

Samson, R. A. and J. C. Frisvad. 1991. Current taxonomic concepts in *Penicillium* and *Aspergillus*. pp. 405-439 in:Cereal grain. Mycotoxins, fungi and quality in drying and storage. J. Chelkowski (ed.). Elsevier, Amsterdam.

Samson, R. A., A. C. Stolk, and R. Hadlok. 1976. Revision of subsection Fasciculata of *Penicillium* and some allied species. Stud. Mycol. (Baarn) 11:1-47.

Samson, R. A., A. D. Hocking, J. I. Pitt, and A. D. King (eds.). 1992. Modern methods in food mycology. Elsevier, Amsterdam. 388 pp.

Sauer, D. B. 1988. Effects of fungal deterioration on grain:nutritional value, toxicity, germination. Int. J. Food Microbiol. 7:267-275.

Sauer, D. B., L. M. Seitz, R. Borroughs, H. E. Mohr, J. L. West, L. J. Milleret, and H. D. Anthony. 1978. Toxicity of *Alternaria* metabolites found in weathered sorghum grain at harvest. J. Agric. Food Chem. 26:1380-1383.

Schmidt, R., J. Modani, E. Ziegenhagen, and K. Dose. 1981. High performance liquid chromatography of the mycotoxin sterigmatocystin and its application to the analysis of mouldy rice for sterigmatocystin. J. Chromatogr. 207:435-438.

Scott, P. M. 1987. Mycotoxins. J. Assoc. Off. Anal. Chem. 70:276-280.

Scott, P. M. 1988. Mycotoxins. J. Assoc. Off. Anal. Chem. 71:70-76.

Scott, P. M. 1989. Mycotoxins. J. Assoc. Off. Anal. Chem. 72:75-80.

Scott, P. M. 1990. Mycotoxins. J. Assoc. Off. Anal. Chem. 73:98-105.

Scott, P. M. 1991. Mycotoxins. J. Assoc. Off. Anal. Chem. 74:120-128.

Scott, P. M. 1992. Mycotoxins. J. Assoc. Off. Anal. Chem. 75:95-102.

Scott, P. M. and G. A. Lawrence. 1987. Liquid chromatographic determination and stability of the *Fusarium* mycotoxin moniliformin in cereal grain. J. Assoc. Off. Anal. Chem. 70:850-853.

Scott, P. M., S. R. Kanhere, R. Canela, G. L. Lombaert, and S. Bacler. 1991. Determination of ochratoxin A in meat by liquid chromatography. Prehrambeno-technol. Biotechnol. Rev. 29:61-64.

Scott, P. M., W. van Walbeek, B. Kennedy, and D. Ayeti. 1972. Mycotoxins (ochratoxin A, citrinin, and sterigmatocystin) and toxigenic fungi in grains and other agricultural products. J. Agric. Food Chem. 20:1103-1109.

Scudamore, K. A., P. M. Atkin, and A. E. Buckle. 1986. Natural occurrence of the napthoquinone mycotoxins xanthomegnin and viomellein, and vioxanthin in cereals and animal feedstuffs. J. Stored Prod. Res. 22:81-84.

Seitz, L. M., D. B. Sauer, H. E. Mohr, and R. Borroughs. 1975. Weathered grain sorghum: natural occurrence of alternariols and storability of the grain. Phytopathology 65: 1259-1263.

Sharma, R. P. and D. K. Salunkhe. 1991. Occurrence of mycotoxins in foods and feeds. pp. 13-32 in: Mycotoxins and phytoalexins. R. P. Sharma and D. K. Salunkhe (eds.). CRC Press, Boca Raton, FL.

Sharman, M, J. Gilbert, and J. Chelkowski. 1991. A survey of the occurrence of the mycotoxin moniliformin in cereal samples from sources worldwide. Food Addit. Contam. 8:459-466.

Sinha, R. K., A. Dubey, and K. K. Sinha. 1992. Cytotoxic effects of citrinin on relative division rate of Allium cepa root cells. Lett. Appl. Microbiol. 14:118-120.

Simmons, E. G., 1992. *Alternaria* taxonomy: current status, viewpoint, challenge. pp. 1-35 in: *Alternaria*. Biology, plant diseases and metabolites. J. Chelkowski and A. Visconti (eds.). Elsevier, Amsterdam.

Singh, K., J. C. Frisvad, U. Thrane, and S. B. Mathur. 1991. An illustrated manual on identification of some seed-borne aspergilli, fusaria, penicillia and their mycotoxins. Jordbrugsforlaget, Copenhagen. 133 pp.

Smith, J. E. and R. S. Henderson (eds.). 1991. Mycotoxins and animal foods.CRC Press, Boca Raton, FL. 875 pp.

Steyn, P. S. 1984. Ochratoxin A and related dihydroisocoumarins. pp. 183-216 in: Mycotoxins—Production, isolation and purification. V. Betina (ed.). Elsevier, Amsterdam.

Steyn, P. S. and R. Vleggaar (eds.). 1986. Mycotoxin and phycotoxins '86. Elsevier, Amsterdam. 546 pp.

Stoloff, L. 1976. Occurrence of mycotoxins in foods and feeds. pp. 23-50, in: Mycotoxins and other fungal related food problems. J. V. Rodricks (ed.), Am. Chem. Soc., Washington.

Stolk, A. C., R. A. Samson, J. C. Frisvad, and O. Filtenborg. 1990. The systematics of the terverticillate penicillia. pp. 121-136 in: Modern concepts in *Penicillium* and *Aspergillus* classification. R. A. Samson and J. I. Pitt (eds.). Plenum Press, New York.

Stubblefield, R. D., J. I. Greer, and O. L. Shotwell. 1988. Liquid chromatographic method for determination of citreoviridin in corn and rice. J. Assoc. Off. Anal. Chem. 71:721-724.

Sydenham, E. W., P. G. Thiel, W. F. O. Marasas, G. S. Shephard, D. J. Van Schalkwyk, and K. R. Koch. 1990. Natural occurrence of some *Fusarium* mycotoxins in corn from low and high esophagal cancer prevalence areas of the Transkei, Southern Africa. J. Agric. Food Chem. 38:1900-1903.

Taniguchi, M. and Y. Satomura. 1970. Isolation of viridicatin from *Penicillium crustosum*, and the physiological activity of viridicatin and its 3-carboxymethylene derivative on microorganisms and plants. Agric. Biol. Chem. 34:506-510.

Taniguchi, M. and Y. Satomura. 1972. Structure and physiological activity of carbostyril compounds. Agric. Biol. Chem. 36:2169-2175.

Thalmann, A, S. Metzenauer, and S. Gruber-Schley. 1985. Survey of occurrence of fusariotoxins in cereals Berich. Landwirtsch. 63:257-272.

Thiel, P. G., C. J. Meyer, and W. F. O. Marasas. 1982. Natural occurrence of moniliformin together with deoxynivalenol and zearalenone in Transkeian corn. J. Agric. Food Chem. 30:308-312.

Thiel, P. G., W. C. A. Gelderblom, W. F. O. Marasas, P. E. Nelson, and T. M. Wilson. 1986. Natural occurrence of moniliformin and fusarin C in corn screenings known to be hepatocarcinogenic in rats. J. Agric. Food Chem. 34:773-775.

Thiel, P. G., C. J. Meyer, and W. F. O. Marasas. 1982. Natural occurrence of moniliformin together with deoxynivalenol and zearalenone in Transkeian corn. J. Agric. Food Chem. 30:308-312.

Thiel, P. G., G. S. Shephard, E. W. Sydenham, W. F. O. Marasas, P. E. Nelson, and T. M. Wilson. 1991. Levels of fumonisin B_1 and B_2 in feeds associated with confirmed cases of equine leukoencephalomalacia. J. Agric. Food Chem. 39:109-111.

Tholstrup, B. and G. Rasmussen. 1990. Ochratoxin A in cereals 1986-1989. Levnedsmidelstyrelsen, Copenhagen, Denmark. 50 pp.

Thorpe, C. W. and R. L. Johnson. 1974. Analysis of penicillic acid by gas-liquid chromatography. J. Assoc. Off. Anal. Chem. 57:861-865.

Thrane, U. 1989. *Fusarium* species and their specific production of secondary metabolites. pp. 199-225 in: *Fusarium.* Mycotoxins, taxonomy and pathogenicity. J. Chelkowski (ed.). Elsevier, Amsterdam.

Thrane, U., O. Filtenborg, J. C. Frisvad, and F. Lund. 1992. Improved methods for detection and identification of toxigenic *Fusarium* species. pp. 231-240 in: Modern methods in food mycology. R. A. Samson, A. D. Hocking, J. I. Pitt, and A. D. King (eds.). Elsevier, Amsterdam.

Torres, A., S. Chulze, A. Dalcero, M. Etcheverry, and C. Farnochi. 1990. Deoxynivalenol and nivalenol in wheat and by-products in Argentine. Mycotoxin Res. 6:89-92.

Torrey, G. S. and E. H. Marth. 1976. Silica gel medium to detect molds that produce aflatoxin. Appl. Environ. Microbiol. 32:376-380.

Trenholm, H. L., L. L. Charmley, D. B. Prelusky, and W. L. Bryden. 1991. Safety of mycotoxins in animal feeds and approaches to detoxification. pp. 136-144 in: Fungi and mycotoxins in stored products. ACIAR Proc. No. 36. B. R. Champ, E. Highley, A. D. Hocking, and J. I. Pitt (eds.). Aust. Centre Int. Agric. Res., Canberra, Australia

Trojanowska, K. 1991. Evaluation of cereal grain quality using mycological methods. pp. 185-215 in: Cereal grain. Mycotoxins, fungi and quality in storage. J. Chelkowski (ed.). Elsevier, Amsterdam.

Tuite, J. 1969. Plant pathological methods. Burgess, Minneapolis, MN. 239 pp.

Ueno, Y., O. Kawamura, Y. Sugiura, K. Horiguchi, M. Nakajima, K. Yamamoto, and S. Sato. 1991. Use of monoclonal antibodies, enzyme-linked immunosorbent assay and immunoaffinity column chromatography to determine ochratoxin A in porcine sera, coffee products and toxin-producing fungi. pp. 71-75 in: Mycotoxins, endemic nephropathy and urinary tract tumours. M. Castagnero, R. Plestina, G. Dirheimer, I. N. Chernozemsky, and H. Bartsch (eds.). Int. Agency Res. Cancer, Lyon, France.

Urano, T., M. W. Trucksess, J. Matusik, and J. W. Dorner. 1992. Liquid chromatographic determination of cyclopiazonic acid in corn and peanuts. J. Assoc. Off. Anal. Chem. Int. 75:319-322.

Van Egmond, H. P. 1989. Current situation on regulations for mycotoxins. Overview of tolerances and status of standard methods of sampling and analysis. Food Addit. Contam. 6:139-188.

Wallace, H. A. H. 1973. Fungi and other organisms associated with stored grain. pp. 71-98 in: Grain storage: part of a system. R. N. Sinha and W. E. Muir (eds.). AVI Publishing Co., Westport, CT.

Wheeler, K. A., B. F. Miscamble, M. Ng, A. D. Hocking and K. Bhudhasamai. 1991. Survey of fungi and mycotoxins associated with maize and other commodities in Thailand. pp. 214-216 in: Fungi and mycotoxins in stored products. ACIAR Proc. No. 36. B. R. Champ, E. Highley, A. D. Hocking, and J. I. Pitt (eds.). Aust. Centre Int. Agric. Res., Canberra, Australia.

Wicklow, D. T., B. W. Horn, and R. J. Cole. 1982. Cleistothecia of *Eupenicillium ochrosalmoneum* form naturally within corn kernels. Can. J. Bot. 60:1050-1053.

Wicklow, D. T., R. D. Stubblefield, B. W. Horn, and O. L. Shotwell. 1988. Citreoviridin levels in *Eupenicillium ochrosalmoneum*-infected maize kernels at harvest. Appl. Environ. Microbiol. 54:1096-1098.

Wilkins, C. K., C. Tørngren, O. Filtenborg, and K. P. Poulsen. 1992. Destruction of mold spores, mycotoxins and insects in grain by HTST (high temperature short time) treatment. pp. 263-272 in: Advances on Food Engineering. R. P. Singh and M. A. Wirakartakusumah (eds.). CRC Press, Boca Raton, FL.

Williams, K. C., B. J. Blaney, R. L. Dodman, and C. L. Palmer. 1992. Assessment for animal food of maize kernels naturally intected predominantly with *Fusarium moniliforme* and *Diplodia maydis*. 1. Fungal isolations and changes in chemical composition. Aust. J. Agric. Res. 43:773-782.

Wilson, D. M. and D. Abramson. 1992. Mycotoxins. pp. 341-391 in: Storage of cereal grain and its products. D. B. Sauer (ed.). Am. Assoc. Cereal Chem., St. Paul, MN.

Wilson, B. J., C. S. Byerly, and L. T. Burka. 1981. Neurological disease of fungal origin in three herds of cattle. J. Amer. Vet. Med. Assoc. 179:480-481.

Woolford, M. K. 1990. A review. The detrimental effects of air on silage. J. Appl. Bacteriol. 68:101-116.

Yabe, K., Y. Ando, M. Ito, and N. Terakado. 1987. Simple method for screening aflatoxin-producing molds by UV photography. Appl. Environ. Microbiol. 53:230-234.

Yabe, K., H. Nakamura, Y. Ando, N. Terakado, H. Nakajima, and T. Hamasaki. 1988. Isolation and characterization of *Aspergillus parasiticus* mutants with impaired aflatoxin production by a novel tip culture method. Appl. Environ. Microbiol. 54:2096-2100.

Yamazaki, M., Y. Maebayashi, and K. Miyaki. 1971. The isolation of secalonic acid A from *Aspergillus ochraceus* cultured on rice. Chem. Pharm. Bull. 19:199-202.

Yeulet, S. E., P. G. Mantle, M. S. Rudge, and J. B. Greig. 1988. Nephrotoxicity of *Penicillium aurantiogriseum*, a possible factor in the ethiology of Balkan Endemic Nephropathy. Mycopathologia 102:21-30.

9

Rodents and Birds as Invaders of Stored-Grain Ecosystems

Robert H. Smith

University of Leicester, Leicester, United Kingdom

SUMMARY

Birds and rodents are associated with larger magnitudes of scale in space and time than the nonvertebrate pests of stored grain and the vertebrates tend to be invaders rather than residents of stored-grain ecosystems. Environmental aspects of vertebrate pest control involve spatial and temporal changes in populations, evolutionary change, and ecotoxicological effects. One impact of vertebrate pests is to carry insects, mites, and fungi between stores, linking local populations into an assemblage of connected populations or metapopulation. Direct losses due to consumption of grain by vertebrates are small (of the order of 1%) but indirect losses are greater, including damage to structures, contamination, and spread of pathogens and pests. The major bird pests are sparrows and pigeons, but many bird species are opportunistic pests of stored grain. Apart from the three main commensal rodent pests (Norway rat, roof rat, and house mouse), many field pests invade grain stores from time to time. Integrated Pest Management (IPM) requires an appreciation of density-dependent compensation and of the spatial scale of the pest population relative to the grain store. The most effective long-term measure in IPM is to prevent access of both birds and rodents but complete exclusion is only feasible in purpose-built stores and any effective exclusion is difficult in on-farm storage in the tropics. Repellents (birds) and chemosterilants (rodents) may play a role in IPM. Lethal methods have limited value against birds but have been widely developed for rodent control. Because of the development of bait shyness, the most effective rodenticides have a delayed action. Delayed-effect, multiple-feed rodenticides include first-generation anticoagulants and calciferols. The most effective rodenticides are second-generation anticoagulants, which are delayed-effect, single-feed poisons. Even with small losses (i.e., of the order of 1%), control can be very cost-effective. Future needs include better data on bird losses, improved store construction, resistance management, and reduction of environmental hazards.

1. INTRODUCTION

1.1. Rodents and Birds as Invaders

The stored-grain ecosystem is a community of interacting species (arthropods, fungi, and microorganisms) that inhabit stored grain and interact with the physical environment. To describe rodents and birds as "invaders" implies that they are not really a part of that ecosystem and to some extent, this is true because stored grain does not provide all of their needs. Birds and most rodents need to drink water daily and therefore would have to go outside the stored-grain ecosystem for this basic requirement (though the house mouse *Mus musculus* is able to survive on metabolic water alone and can live all of its life within an appropriate unit of stored grain). In general, therefore, we may think of rodents and birds as opportunistic marauders rather than colonizing settlers, perhaps more akin to the Vikings in Europe in the Middle Ages rather than to the Pilgrim Fathers or other settlers. To this extent, the effect of the invading vertebrates on the more permanent residents (e.g., fungi and arthropods) is unpredictable in time and patchy in space but may be important where it occurs, for example where rat urine forms a focus for *Aspergillus* growth.

1.2. Vertebrate Pests and Local Ecosystem Dynamics

The main impact of vertebrate pests on the stored-grain ecosystem, then, seems to be in the way that the physical environment is changed. The physical changes have an impact on suitability for population growth of the permanent residents (arthropods, fungi, and microorganisms) and on the condition of the energy source (the grain, which may germinate in store or be prevented from later germination because of the vertebrate invaders). As consumers of the grain, however, the direct effect of the vertebrates in terms of energy flow (by means of food consumed) is rather small in most instances. Thus, in terms of the conventional descriptions of ecosystem structure considering energetics and species interactions, rodents and birds seem to be relatively unimportant.

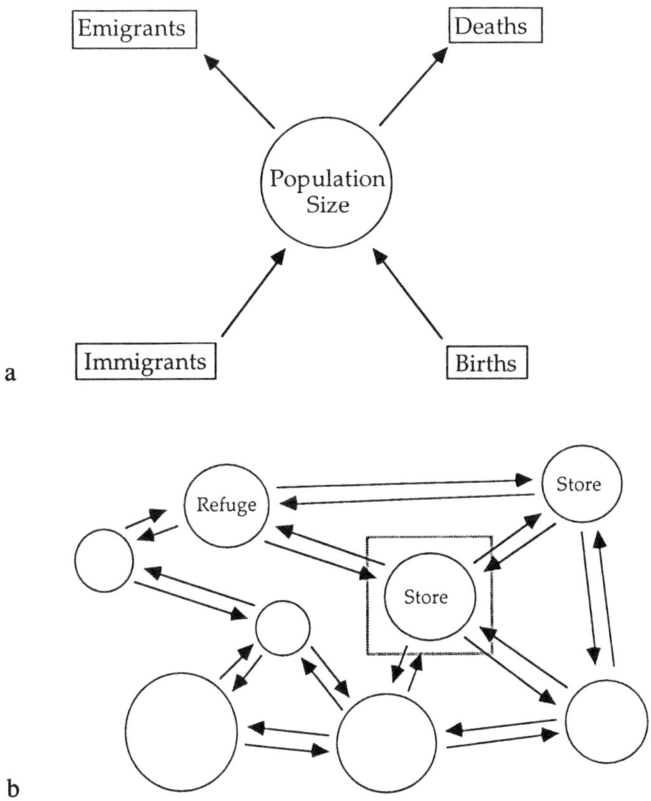

Figure 1 Schematic representation of local and metapopulation dynamics: (a) the four processes that contribute to changes in population size, (b) a metapopulation structure resulting from migration between local populations inhabiting discrete patches of environments.

1.3. Vertebrate Pests and Metapopulations

There is a newer view of population and ecosystem dynamics in which rodents and birds will be seen to assume greater importance. Conventionally, population dynamics have been described in terms of the dynamics of local populations of an organism, where a population is those organisms present in one place at one time (Fig. 1a). Births, deaths, immigration and emigration are the processes that lead to changes in population size, and these four processes will depend on the physical environment, intraspecific competition (density dependence), and interspecific interactions (e.g., competition, predation). Ecologists used to focus on births and deaths, but lately the role of migration in determining population stability has been shown to be important. Ecosystems are now seen as assemblages of *metapopulations of species* (Fig. 1b), a metapopulation being a group of local populations linked by migration. An essential feature of metapopulation dynamics is that aggregations of organisms in local populations (often occupying discrete patches of suitable environment such as grain stores) can lead to refuges in space that have a stabilizing effect on population dynamics; aggregation in space means that intraspecific competition assumes relatively greater importance than interspecific interactions and thereby confers population stability (see Shorrocks and Swingland 1990). Now the stabilizing effect depends on there being a certain amount of migration between patches, but not so much that population density becomes homogeneous throughout the metapopulation. Rodents, and more particularly birds, move between grain stores as well as between "patches" (e.g., bags, stacks, bins) within a store and they carry with them the smaller, more permanent residents. Thus, the main effect of rodents and birds on stored-grain ecosystem dynamics may be in linking local populations into a metapopulation structure in the other species through the passive migration that they induce. The magnitude of the effect has not been measured, as far as I know, but it would not have to be large to have a significant effect on the dynamics of the species transported. The practical importance of rodents and birds in leaving behind arthropods and microbial spores (as well as hair and feces) is already well recognized, although again not quantified, but the qualitative effect of vertebrates on the dynamics of the nonvertebrate pests may turn out to be more than just starting a local infestation of the nonvertebrates.

2. VERTEBRATE PEST SPECIES

2.1. Major Bird Pests: Sparrows and Pigeons

Many species of bird and rodent are capable of attacking stored grain and grain products. There are rather few major pests (i.e., cosmopolitan species that regularly attack stored grain) and they will be described below, along with notes on some minor pests.

Bird pests really can be described as invaders because they reside for a very short period of time in stores. Sometimes birds may nest in the upper parts of a structure and they will frequently roost in roof structures, defecating on the stored commodities below. The major bird pests of stored grain belong to the Orders Passeriformes (perching birds) and Columbiformes (pigeons and doves). Of these, the house sparrow *Passer domesticus* (L.) and the rock pigeon *Columba livia* Gmelin are generally recognized to be the most common and widespread pests (e.g., Shuyler 1970, Meyer 1989, Semple, undated).

There are more than 300 species of *Passer* named in the literature and the taxonomy of sparrows is such that it is not always clear which species or subspecies is present. Some species of sparrow nest singly while a few nest in small colonies, and nests may either be in buildings or in trees, though usually associated with human habitation. A pair of *P. domesticus* may rear three to five young on two or three occasions a year and the young birds form feeding flocks after they leave the nest. The activities of feeding flocks, which forage over distances up to 2–3 km, are responsible for most of the problems.

Pigeons are larger than sparrows and they also live and feed in flocks for some of the year, mostly feeding on seeds and foraging over distances up to around 8 km. Again, there are many species of *Columba*, several of which occur all over the world as agricultural pests. The rock pigeon *C. livia* has been domesticated (e.g., for food in China and for racing in Britain). Pigeons from domestic stock may become feral and have little fear of man, often causing problems by producing large quantities of feces as well as consuming substantial amounts of food. Pigeons rear less offspring than sparrows, but even in a temperate country like Britain it is normal for pigeons to produce three broods of two young each year. Pigeons and sparrows, then, share the following features:

- the need to eat and drink daily,
- formation of feeding flocks during at least part of the year,
- defecation from a roost or perching point,
- consumption of seed and grain products, and
- flight which allows daily movement up to several kilometers.

2.2. Minor Bird Pests

Many species are opportunistic grain feeders and can be local pests. The Passeriformes contain species that perch on twigs or branches in and around fields and many will attack ripe grain immediately before harvest. Apart from sparrows and weavers (Family Ploceidae), some of the finches (Family Fringillidae) also cause damage to ripening crops. Within the Columbiformes, various doves (*Streptopelia* spp.) also damage stored grain in the tropics and, like *Columba* species, are often associated with habitation and built-up areas of large towns and cities. The biology of the major bird pests and some of the minor ones is summarized very succinctly by Brooks and Ahmad (1990).

2.3. Major Rodent Pests: Rats and Mice

Rats and mice belong to the family Muridae which constitutes the main group of rodents that damage stored grain and grain products. The three cosmopolitan species that cause most problems are:

- house mouse *M. musculus* (L.)
- Norway rat *Rattus norvegicus* (Berkenhout)
- roof rat *R. rattus* (L.)

While *R. norvegicus* is a quite distinct species, the taxonomy of *M. musculus* and *R. rattus* has been a source of debate that still continues. For example, in Europe there are two species or subspecies of house mouse, *M. musculus* and *M. domesticus* Schwarz, though earlier literature does not distinguish between them (Lund 1994). *Mus musculus* was taken to Latin America and California by early Portuguese and Spanish explorers, while Dutch, French and British ships probably took *M. domesticus* to the northern United States. The *Mus* species complex originated in the steppes of central Asia where a wild form *M. wagnerii* can still be found. The commensal habit of *Mus* spp. probably began in the area of the border between Iran and the former USSR where man first cultivated cereals and stored grain. *Mus* was well known to the ancient Egyptians and presumably spread alongside human trade and settlement.

The two main *Rattus* pest species probably also radiated from Asia, though *R. rattus* is thought to have evolved in south-east Asia and *R. norvegicus* farther north in eastern Asia, maybe in China, from where it has moved westwards across Europe during the last two centuries (Brooks 1973).

The main contrast between the house mouse and the two cosmopolitan rats is that house mice may settle and breed for several generations entirely within a store as they do not need to drink free water provided the moisture content of their food is about 15% or more. In contrast, rats must drink free water at least once a day. Although a source of water is not uncommon in or around a store (e.g., a rain-water gully, a dripping tap or a cistern), rats frequently live outside stores, which they visit to feed, and they may forage away from stores even when their nests are inside or under the store structure.

All three commensal species have a preference for grain and grain products, though they will balance their diets with various growing plants, invertebrates, and fish and meat if they are able. Because of their different drinking requirements, house mice may have a home range as small as 4–5 m^2 (e.g., a stack of animal-feed bags on a pallet) while rats may forage over a distance of 1 km or more in a night. However, the behavior of rats and mice is opportunistic and can change substantially with local conditions. For example, mice migrate en masse during mouse plagues in Australia though such behavior may never be seen in most parts of the world.

Important aspects of the biology of the three species are reviewed in Meehan (1984), Macdonald and Fenn (1994), and Lund (1994). Some behavior relevant to control will be discussed in later sections.

2.4. Minor Rodent Pests

Many species normally thought of as field (as opposed to commensal) rodents may cause damage to stored grain, though usually on a sporadic, localized basis. In sub-Saharan Africa, two of the most important species complexes are the multimammate rat *Mastomys/Praeomys natalensis* (Smith) and the Nile rat *Arvicanthis niloticus* (Desmarest) (reviewed by Mbise 1983 and Farrag 1983, respectively). Brooks and Rowe (1979) highlight two Asian species as developing into commensal pests in addition to the African multimammate rat; these other "developing commensals" are the lesser bandicoot rat *Bandicota bengalensis* (Gray) and the Polynesian rat *R. exulans* (Peale). Although primarily field species, *B. bengalensis* and *R. exulans* have been associated with human building and store structures in both urban and rural areas and, according to Brooks and Rowe (1979), both species now tend to invade and live within human structures.

Many other *Rattus* spp. (often formerly referred to as subspecies of *R. rattus*) attack on-farm food stores, and sometimes security stocks in government warehouses, as do the rice rats (*Oryzomys* spp.) of South and Central America and *Nesokia* spp. in India. Apart from the Muridae, some voles (Family Cricetidae) damage stored grain in rural areas; these include jirds (*Meriones* spp.) in northern Africa and the Middle East, cane rats (*Holochilus* spp.) in South America, gerbils (*Meriones* spp.) in north Africa and field voles (*Microtus* spp.) in Europe and North America. Squirrels (Family Sciuridae) may also damage stored grain (e.g., *Xerus*, *Funisciurus*, and *Helosciurus* spp. in West Africa) (Grolleau and Gautun 1991).

3. IMPACT OF PESTS IN STORED-GRAIN ECOSYSTEMS

3.1. Societal Perception of Losses

The impact of rodents and birds (and our response to them) depends on the perception in a particular society of how important the problem is. For example, four stages of development of storage can be recognized in Britain since 1945 (Barnes 1987).

1. *Subsistence*: the only real concern is direct losses. During the 1940s, Britain had a food shortage and food was rationed because of the problems of importing food caused by the Second World War. Because food in any form was in short supply, a high degree of contamination was accepted and there was very little rejection. Interestingly, this period of shortage is now recognized as a high spot in the nutritional health of the nation!

2. *Shortages*: allowances are made for some contamination. By the 1950s, food rationing was over but there were still shortages of some commodities and food was rejected only if perceived to be dangerous.
3. *Sufficiency*: with increasing affluence, food of all types can be readily purchased and consumers become concerned about hygiene and quality. During the 1960s and 1970s in Britain, contamination was no longer accepted as normal and the value of stored grain and products was reduced if contaminants were detected, to the point where batches might be rejected for sale.
4. *Sophistication*: although there is a zero tolerance of visible contamination, consumers assume that pests are not a problem and become increasingly concerned about environmental issues. Pesticide residues became a big issue in Britain in the 1980s and 1990s (in part, because lower levels of pesticides could be detected) and we arrive at the contradiction in many of the affluent, industrialized nations where people expect clean food but do not like pesticides to be used. Thus, the economic impact of rodent and bird pests has to be judged in relation to the expectations of the consumer. The problems that may be caused by any pest are:

   a. loss of usable stored grain or product,
   b. contamination of final product, and
   c. aversion of customers (and staff) to signs of pests.

In addition, two problems that apply particularly to birds and rodents are:

   d. spread of pathogens and pests, and
   e. structural damage.

3.2. Direct Losses

3.2.1. Birds

There are two types of direct loss: spillage and consumption. Birds as well as rodents may cause spillage by damaging bags (especially pigeons). There is no doubt that losses caused by vertebrates are greater in bag storage than in bulk storage, as is also true of invertebrate pests. There is no easy way, however, of changing over to bulk storage in developing countries (see Chapter 3).

Much consumption of grain by birds occurs immediately preharvest, and during threshing and drying in developing countries. Inadequately bird-proofed grain stores are also attacked. In Sri Lanka, Meyer (1989) found pigeon populations of up to several hundred birds associated with individual stores of the Sri Lankan Food Department. An adult pigeon eats around 35g/d of seed, thus 80 pigeons consume more than 1 t/yr of dry food and 20 pigeons eat about as much as a single human. A store that supports 500 pigeons may therefore lose enough grain by direct consumption to feed 25 people.

3.2.2. *Rodents*

An adult Norway rat eats about 10% of its body mass each day (i.e., 28 g/d on average). Thus 100 rats would consume more than 1 t/yr of dry food (Meyer 1994) and 25 rats eat about as much as a single human. A badly infested store supporting 500 Norway rats could then be losing enough grain by direct consumption to feed 20 people.

3.2.3. *Loss Estimates*

There is a shortage of good data for storage losses caused by birds, though Khan and Ahmad (1990) give methods for estimating losses to ripe grain immediately preharvest. Shuyler (1970) gives a figure of 4–5% loss caused by birds to wheat postharvest in the United Arab Republic, though he does not say how this figure was determined.

Rodent damage to farm and village storage in the tropics and subtropics was estimated by Hopf et al. (1976) to be generally in the range 1–10%, though direct losses are probably nearer 1% on average (J. H. Greaves, personal communication). Occasionally, up to 50% loss has been reported. In an experimental study, Barnett (1951) found that small populations of Norway rats (10–26 rats) caused a loss in mass of stored grain of 4–5% over a period of 4–6 mo, but there was a total financial loss of more than 18% of the value of the damaged wheat and sacks.

It is difficult to generalize, but rats and mice frequently cause significant losses in the order of 1% due to direct consumption of stored grain. Thus the size of direct losses is not usually spectacular. However, the rejection level (because of contamination) may be much higher. Greaves (1978) discusses methods for the determination of postharvest losses caused by rodents.

3.3. Spread of Pathogens and Pests

3.3.1. *Diseases*

Diseases known to be spread to other animals and to man by rodents and birds (zoonoses) are numerous and the lists are without doubt incomplete. The main route of transmission of the pathogens responsible for zoonoses is contamination by urine and feces which is then spread by ingestion (e.g., salmonella, toxoplasmosis) or direct entry through a mucous membrane (leptospirosis, tuberculosis). Direct transfer of blood parasites is unlikely, though it can occur infrequently if animals are bitten. However, arthropod vectors may transmit some dangerous diseases (e.g., plague, tick-borne fever, leishmaniasis) carried in blood. Brooks and Rowe (1979) reviewed the more important rodent-borne diseases and Gratz has presented up-to-date comprehensive reviews in a number of publications (e.g., Gratz 1988, 1994). Apart from the cost in terms of human misery when people are debilitated or die through zoonoses, there is a substantial economic cost because of work lost in the case of humans and loss of production in the case of domesticated animals.

Feed mills (see Chapter 18) are particularly attractive to both rodents and birds because food is easily available and of high nutritional value; contamination at the feed preparation stage leading to disease transmission can be a major problem in animal production units. Opitz et al. (1991) identified rodents as the main cause of spread of Salmonella in poultry houses.

3.3.2. Grain Pests

The extent to which insects, mites, bacteria and fungi are carried on the feet, fur, and feathers of vertebrates between and within stores is not known. Clearly, grain that is uncontaminated at entry needs only a small number of pests carried onto it to be infested.

3.4. Damage to Materials, Equipment, and Storage Structures

3.4.1. Gnawing, Pecking, and Defecation

Containers such as jute bags and cardboard boxes are easily damaged by both rodents and the larger birds (e.g., pigeons). Sparrows often do a tremendous amount of damage by pecking directly into soft insulation in walls. Bird feces deposited on a cardboard box of food will render that food unsaleable, even though the box could often be cleaned without risk of contamination of the contents. Feces, and more particularly dead bodies, of birds and rodents may reduce the efficiency of or severely damage expensive machinery in food handling and processing. Nests and feces may clog water-drainage systems and cause significant expense through water damage to structures and equipment.

The main damage to structures is caused by the gnawing of rodents. All rodents have single pairs of upper and lower incisor teeth that grow continuously and are separated by a large gap (diastema) from the molar teeth in the cheeks. The back of the incisors (dentine) is softer than the front (enamel). The enamel is harder than steel and continuous wear through gnawing or grinding of the teeth gives the incisors sharp tips like chisels. Cogelia et al. (1976) suggest that Norway rats spend about 2% of their time gnawing and can exert a pressure of 49 MPa with an average bite rate of 1.5 bites/s and a maximum of 6 bites/s (Meehan 1984).

3.4.2. *Damage to Electrical and Water Supplies*

Serious damage occurs when electricity cables or water pipes are gnawed. The consequences may be fire, flood, or failure of cooling/aeration systems with consequent problems of grain spoilage. It is rarely possible to pin down the cause of an unexplained fire to rodents, though it is thought in Great Britain and the United States that around one quarter of the fires caused by electrical faults are rodent related. Sophisticated electrical control systems are now a part of grain storage in the industrialized world (see Chapter 16) and rodent damage to these systems could be very costly. Meyer (1994) refers to an egg production facility in the United Kingdom where an electrician is available every hour of every day to repair electrical faults caused by House mice.

3.4.3. *Damage to Building Structures*

Doors, roofs, walls, and floors are all vulnerable to rodent damage unless they are made of appropriate materials or rodents are absent. Rodent entry holes reduce both efficiency and security of storage, they may let in water (hence causing mold growth) and they can prevent successful fumigation. The increased use of fumigation sheets to control insects and fungi selectively in small stacks of stored grain is a practice that can be rendered useless if rodents chew through the plastic sheeting material.

3.5. Contamination

Contamination of grain or grain products can occur with any of the following:

- urine, blood, and saliva
- hair and feathers
- feces
- whole bodies and body fragments (e.g., digits, tail pieces, teeth).

The contaminants are listed in order of size and ease of detection by the consumer, though the ranking in terms of the real health hazard might be different. Contamination may infringe the law in some countries but the legal position is rarely clear. In Great Britain, for example, strictly speaking there is probably a nil tolerance by law (i.e., no level of contamination of food is acceptable). In practice, a local Environmental Health Department might only take action if there is an extreme breach of an accepted norm or previous warnings have been ignored; this is because some contamination is recognized as inevitable. For example, the US Food and Drug Administration requires samples of imported food to be tested for the presence of pest activity ("filth test"). Food is rejected for import if contamination is above a certain level (e.g., 1–3 rodent hairs/100 g chocolate or 1 rodent hair/50 g wheat flour, depending on frequency and size of samples). Thus, it is possible that a loaf of bread could contain perhaps half a dozen rat hairs! Most consumers, of course, would be horrified by the thought of any rat hairs in their toasted bread at breakfast but in practice there is a minimal hazard to health.

The real cost to food producers, however, is not the force of the law (which might result in a fine of a few hundred dollars for accidental contamination above an accepted norm) but the adverse publicity and the effect on sales. Contamination of any sort is bad economic news and, in countries where storage is beyond the subsistence and shortage stages, contamination is much more costly than direct losses. During the 1950s in the United States, more than 90% of corn samples were reckoned to be contaminated by rodents (Dykstra 1954). By the late 1970s in the United States, the level of contamination of wheat was estimated to be 7% (Gecan et al. 1980).

4. CONTROL OF VERTEBRATE PESTS

4.1. Population Equilibrium and Compensation

Pest management is often thought of as killing pests but lethal control may be neither the most effective nor the most economic approach in every case. An equilibrium population size is achieved when births, deaths, immigration, and emigration balance out. If one or more of these population processes is density dependent because of some limiting factor in the environment, the equilibrium population may be stable such that an immediate reduction in pest numbers will only be short-lived because of density-dependent compensation, unless the population reduction continues. In general, lethal methods achieve the most rapid results but nonlethal methods usually provide a more cost-effective and acceptable alternative, especially in grain stores, because they reduce the carrying capacity or equilibrium population size.

4.2. Bird Pests

A sociological difficulty that hampers bird-pest control in industrialized countries is that many people act as if wild animals live forever and will not countenance lethal control, especially of attractive animals such as rabbits and birds. However, the main problem with bird-pest control is the mobility of birds (see Section 5.1). For example, the major bird pest in the world is *Quelea quelea* (L.) and there is still no effective means of controlling *Quelea* damage to field cereals in Africa, largely because of the distances that *Quelea* fly in order to feed (Bruggers and Elliott 1989). Thus, lethal control of field pests may mean finding and destroying bird pests over massive areas. In grain storage, the problem is less acute because the control measures are more easily targeted on those birds causing damage. However, nonlethal exclusion methods as part of IPM are feasible in modern storage structures and probably represent the most cost-effective approach.

4.2.1. *Nonlethal Methods*

The best way to prevent damage to stored grain is to prevent access, in other words exclude the birds from the store. In on-farm stores in the tropics and subtropics, exclusion is hardly ever feasible. But, in a purpose-built warehouse constructed to proper specifications exclusion is attainable.

Exclusion methods include:

- clear plastic overlapping flaps in doorways,
- no structural apertures allowing access at any height,
- netting below beams and cross-girders to prevent access to perches if birds do enter, and
- screening of open ventilators and windows.

These bird-proofing measures are routinely implemented in food storage and processing premises in Europe, North America and Australasia by professional pest management or pest control operators (PCOs) (Cornwell 1979). Exclusion of pigeons requires a net mesh no greater than 4.4 cm, 1.25 cm for sparrows.

Scaring is one of the oldest methods of preventing damage, usually achieved by making a noise or providing a visual stimulus to alarm birds (reviewed by Hussain 1990). Apart from passive scarecrows, which are rarely effective for long, scaring methods tend to be time-consuming and only work for a short time; birds soon learn that there is no real danger. Electronic devices that reproduce recorded alarm calls are available but they work best when moved around and are therefore also labor-intensive. There is a chemical that is used for scaring (4-aminopyridine, marketed as Avitrol) though it does kill some birds. The principle is that a few birds in a flock eat treated bait and these birds then show frightening symptoms (distress calls, disorientation, tremors, convulsions) that cause the rest of the flock to leave. Avitrol, however, is toxic to both birds and mammals (acute LD_{50} around 5–10 mg/kg) though hazards to nontarget animals are probably minimal. As far as any possible risk to humans is concerned, Avitrol (and other pesticides) pose little threat to professional PCOs if used according to directions. But it is hard to recommend the use of any toxicant in close proximity to stored grain where scattering and regurgitation of poisoned grain bait might occur, potentially leading to chemical contamination of grain. The question as to whether Avitrol (or any other chemical used in pest control) is humane is an issue of concern that may never be resolved satisfactorily. Use of any chemical that causes distress and pain is likely to be questioned by the public (and now by registration authorities) in industrialized countries, yet nonchemical alternatives (including predators) may be no more humane.

Taste repellents have been developed to deter birds from consuming particular foods. The carbamate Methiocarb causes physiological distress which leads to bait avoidance in some species. Other chemicals (mostly fungicides) have also been used (Brooks and Hussain 1990). A basic problem is that birds have a poor sense of taste compared with, for example, rodents and therefore they may not respond to taste repellents.

Sterilization by chemosterilants (e.g., Ornitrol) has been used against pigeons. Population increase may be reduced or prevented but the problem birds are still present.

Removal following trapping or sedation is widely used in Great Britain and other industrialized countries, largely for cosmetic reasons, but also to prevent mortality of nontarget birds. The sedative alphachlorase is often used. Although trapping and sedation are not in themselves lethal, the birds that are removed are usually killed.

Sanitation is an important element of IPM around stores. Cleaning up spilled grain will help to reduce problems caused by rodents and nonvertebrate pests as well as birds. If spilled grain cannot easily be removed, it might be made unpalatable as an alternative.

4.2.2. Lethal Methods

Lethal methods are used to control bird pests but they are rarely part of a long-term IPM approach and their effectiveness is very dependent on local circumstances. Avicide may not be acceptable to the general public in industrialized countries. The compound Starlicide has been used in grain baits to reduce consumption and contamination of cattle feed by starlings in the United States. Death occurs slowly (3–50 h), apparently painlessly, and there is no secondary poisoning hazard to scavengers (Brooks and Hussain 1990). Two contact poisons have also been used but they are less acceptable on humaneness grounds; these are the insecticides Endrin and Fenthion which are applied on wicks attached to perches. Both are used at a very high concentration (around 10% in solution) and neither can be recommended for use near stored grain. Indeed, there are legal restrictions on their use in most industrialized countries and Endrin is no longer registered in the United States, but both may be found in many tropical and subtropical countries.

Euthanasia following sedation or trapping has been mentioned in Section 4.2.1. Large-scale incendiary slaughter of roosting or nesting birds with napalm has been used against *Quelea*, with little apparent success. It may sometimes be possible to decimate a local problem population provided there is a time when the birds flock together. Shooting is feasible if there is a small, local problem (e.g., a few birds in a warehouse) but is ineffective as a large-scale measure.

4.3. Rodent Pests

4.3.1. Nonlethal Methods

The technology of controlling rodent pests is more developed and successful than that of controlling birds, though most developments concern lethal control. There is an extensive literature, comprehensively reviewed by Meehan (1984). Prakash (1988a) includes chapters of variable length and coverage on many aspects of rodent pest management. The most recent and comprehensive coverage is given by Buckle and Smith (1994). The Food and Agriculture Organization (FAO) World Food Program Food Storage Manual includes a useful chapter on rodent behavior and control relevant to storage (FAO 1983). Because all these texts are available, the treatment here will highlight important principles and developments rather than give a comprehensive review or provide a manual of methodology.

Smith (1994) reviews the feasibility of a range of nonlethal methods in an ecological context. Nonlethal methods may affect three of the four population processes in Figure 1a: immigration, emigration, and births. Table 1 summarizes the options available and whether they might be used in stores. Some methods are listed as "Field only" but they may be used to protect on-farm stores from

Table 1 Summary of Nonlethal Methods Available for Control of Rodent Pests

Population process	Method	Applications
Immigration	Rodent proofing	Buildings in good condition
	Barriers	Limited to high-value units
	Electric fences	Limited to high-value units
	Ultrasound	Limited, and doubtful efficacy
	Chemical repellents	Theoretical only, at present
Emigration	Clear cultivation	Field only, after harvest
	Hygiene/sanitation	Stores and food processing
	Ultrasound	As above
Births	Remove nest opportunities	Buildings at design stage
	Disrupt behavior	Largely theoretical
	Sterilization	Field only

Source: Smith 1994

Table 2 The Common Points of Rodent Entry and the Action Needed to Prevent Entry to Grain Stores

Entry point	Action to prevent entry
Earth floor	Install concrete floor
Wooden walls	Line walls with sheet metal
Sliding doors	Movable 1 m metal barrier inside doors
Windows and ventilators	Expanded metal screens
Eaves	Seal with concrete
Pipes and cables	Metal baffles projecting at least 250 mm
Rough walls	Smooth paint band on mortar rendering (at least 150 mm deep, 1 m high)
Overhanging trees	Cut back branches

Source: FAO 1983.

depredation by pests living in the fields nearby. While Table 1 classifies methods according to their effect on population processes, the four sections that follow are based on the technology used.

4.3.2. Mechanical Methods

Exclusion from storage structures is feasible in large, modern stores (Jenson 1979) but more difficult in on-farm storage. In purpose-built stores, the common points of entry and the action needed to prevent entry are listed in Table 2.

4.3.3. Hygiene and Cultural Methods

Stacks of commodities provide shelter as well as food for rodents, especially bags and boxes. Stacks should be elevated on pallets, kept about 1 m away from walls and roof structures (beams and girders) and moved regularly. Small stacks with a passage around them to allow inspection are best. Bulk storage is generally less vulnerable to rodent infestation and storage in metal containers with lids closed will prevent access though other considerations (e.g., condensation) may make this unacceptable.

Spillage must be cleaned up and all areas around stacks should be swept frequently. Hygiene is important in nonfood areas such as offices as well as around buildings. Most species of rodent are unwilling to cross-open areas such as roads (Wilkins 1982) and an area clear of weeds and harborage extending for at least 20 m around a building will provide a behavioral barrier. For on-farm storage, clean farming methods reduce local rodent populations, namely: eliminate nearby noncrop vegetation and crop trash, harvest and cultivate as soon as possible, use large fields with minimal boundaries, reduce the bund to paddy ratio in rice, and synchronize planting and harvesting in different plots (Fitzwater 1988).

4.3.4. *Repellents*

The uses of various types of repellent have been reviewed by the following: chemical (Meehan 1984, Stoddart 1988), electrical (McKillop and Sibly 1988, Schumake et al. 1979), ultrasonic (Meehan 1984, Lund 1988a) or electromagnetic (Fitzwater 1978). For a variety of reasons (Smith 1994), none of these methods has more than theoretical interest in the context of grain stores at present, though potentially some of them could be developed. Electromagnetic repellents have not shown any efficacy and, from a practical standpoint, it seems unlikely that ultrasonic repellents will be effective.

4.3.5. *Chemosterilants*

Because of density-dependent compensation (Sinclair 1989), a reduction in birth rate needs to be substantial to have any significant effect on population size. Spontaneous abortion and suppression of female reproduction because of the presence of a strange male is known in laboratory mice (Bruce 1960) and other mice and voles (Meehan 1984) but is unlikely to have any practical application (Stoddart 1988). Poor nutrition will reduce reproductive output and it is well established that nutritional status as affected by rainfall and vegetation succession has an important role in the population dynamics of the multimammate rat in East Africa (Leirs 1992). Thus, hygiene methods that prevent access to high quality food (e.g., animal feed concentrate) can help to reduce birth rate. The presence of an insect infestation can enhance nutrition and reproduction of rodents, although the effect has not, as far as I know, been measured experimentally. For example, house mice cannot live on processed white flour and water alone in laboratory cages, but in a store containing only white flour infested with insects, mice can thrive and reproduce (J. H. Greaves, personal communication). Thus, effective control of insects may lead to poor health and reduced reproduction in rodents.

Chemical inhibitors of reproduction are invariably orally presented and enough animals must consume sufficient bait for there to be a significant effect at the population level. The development of the contraceptive pill in humans lead to oestrogenic steroids being tested against rodents (Brooks and Bowerman 1971, Kendle et al. 1973, Marsh 1988). The synthetic oestrogen BDH 10131 looked promising because a single exposure to bait made female rats infertile for almost one year (Kendle et al. 1973) but palatability and bait attractiveness present problems. A high proportion of females must stop breeding if an effect at the population level is to be achieved.

Sterilization of males requires an even higher level of success than sterilization of females in polygamous species such as rats. There is a male antifertility compound (alphachlorohydrin) marketed as Epibloc that has given variable results in different species (Marsh 1988) and is toxic at higher doses, causing mortality up to 50%. Ericsson (1982) felt able to describe alphachlorohydrin as a toxicant-sterilant, though the compound has not made much impact in rodent control.

Singleton and Redhead (1991) discuss the use of an hepatic nematode as a means of reducing fertility in house mice. The chronic effect of the parasite may regulate mouse populations below plague levels (McCallum and Singleton 1989). A new approach based on developments in molecular biology is to introduce immunosterilants delivered by a virus vector, though this method is still at an early stage of development (Singleton and Redhead 1991).

Overall, it seems that reproductive management may have potential in rodent IPM, for example in resistance management (Smith and Greaves 1987) but potential users still need to be convinced of its value. In food processing premises, where there is normally a nil tolerance of rodents, reproductive management alone is unlikely to be of much use (Smith 1994).

4.3.6. Lethal Methods

A variety of lethal methods is available of which the following could be used in grain stores: trapping, predators, disease, and chemical rodenticides. The drawback with trapping is that it is labor-intensive, basically because there is a limit on the number of rodents that can be trapped in conventional traps (single-catch, break-back traps). Sticky traps or glue boards may be used in buildings where other methods fail, though this method is generally considered to be inhumane. Although trapping is unsuitable as a means of rapidly controlling large infestations—Greaves (1982) recommends a ratio of traps to rodents of 2–3:1!—traps are valuable when only a small number of rodents is present in a store, in part because there is zero risk of accidental poisoning. Multiple-catch mouse traps are said to be quite effective and cost-effective for controlling mice in areas where poison bait cannot be used and single-catch traps are too labor intensive. However, traps do need regular attention otherwise the bodies will smell, attract blowflies, and cause secondary health hazards.

Predators may regulate rodent populations under certain circumstances (Newsome 1990), in particular if the spatial dynamics allow the predators to respond effectively to increases in prey numbers (Duckett 1991, Smith 1994). In stored-grain ecosystems, predators may assist by deterring some immigration of rodents and cats have traditionally played this preventive role (Elton 1953) rather than that of controlling infestations (Tolar 1991). But in general, predators are not a practical means of controlling rodents around grain stores. Agents of disease are only feasible if there is no risk of spreading the disease to humans or livestock. McCallum and Singleton (1989) have considered how a parasite may regulate house mouse populations, though this approach is useful for reducing the probability of immigration into rather than for controlling an infestation inside a grain store.

In general, rapid control of an infestation requires the use of a chemical rodenticide (Buckle 1994) used either as a fumigant, contact powder, or poison bait.

4.3.7. *Chemical Control: Fast-Acting Poisons*

Fumigation of warehouses with phosphine or methyl bromide in order to control insects will, incidentally, kill rodents because rats and mice are more susceptible to most fumigant gases than are insects. However, fumigation to kill rodents alone is unlikely to be cost-effective, except by small-scale application to burrows in damp soil around grain stores, where cyanides or phosphides may be used with extreme care.

Fast-acting or "acute" poisons were the only chemical rodenticides available until the middle of the twentieth century. Their efficacy depends on palatability and speed of action as well as intrinsic toxicity. Over much of the world, acute poisons such as zinc phosphide are the most widely used poisons because they are cheap and readily available. There are three main drawbacks: safety to humans and livestock (no antidote), nontarget hazards (primary poisoning), and bait shyness. Most rodenticides are toxic to all mammals and there is no antidote to fast-acting poisons other than treatment of symptoms. Nontarget animals that eat bait may also be poisoned, though the risks to larger animals are minimized by the use of tamper-resistant, enclosed bait stations. Generally there is little risk to scavengers or predators of rodents because of rapid metabolism of zinc phosphide to nontoxic compounds.

The major problem with fast-acting poisons is induction of bait shyness or conditioned aversion, at least in the commensal *Rattus* species. The Norway rat and other species develop bait shyness as follows. When rats approach new food (e.g., poison bait), they tend to sample a small amount and wait to discover the effect; this behavior is almost certainly related to the anatomical inability of rodents to vomit because they cannot quickly get rid of any noxious substance once it has been consumed. During poison baiting, therefore, some rodents in a population only consume a sublethal dose of poison. Acute poisons that act within minutes or hours of ingestion may cause stressful symptoms. Survivors remember the stressful experience and associate it with the taste or smell of the poison, or even the carrier bait, which they will then avoid, maybe for many months (Prakash 1988b). The problem of bait shyness is made worse because of the cautious reaction of many rodents to any new object or food that they

Table 3 Properties of the Most Commonly Used Acute Rodenticides.

Compound	% in bait (w/w)	LD_{50} (Norway rat) (mg/kg)	Notes
Zinc phosphide	1–5	46	Releases phosphine in stomach. Causes pulmonary oedema and kidney damage.
Sodium fluoroacetate (1080)	0.2–1	5	Causes convulsions and disturbance of central nervous system and heart (blocks Krebs cycle).
Fluoroacetamide (1081)	0.2–1	15	Converts to 1080 in vivo.
Scilliroside	0.1–0.5	0.4–0.7	Cardiac glycoside. Nontargets eliminate by vomiting. Causes convulsions and heart failure.

encounter (neophobia) which may in extreme cases become a widespread bait avoidance (Quy et al. 1992). Prakash (1988b) lists many species in which bait shyness has been demonstrated but *Rattus* spp., especially *R. norvegicus*, seem to show bait shyness to the greatest extent.

Development of bait shyness may be prevented by persuading the pests that the bait is harmless, presenting unpoisoned bait for at least two (Prakash 1988b) or preferably 4–8 d (FAO 1983) before placing poisoned baits for no more than 2 d. This "prebaiting" is essential if satisfactory control is to be achieved, though it does increase the amount of bait and labor used. Even with prebaiting, an experienced operator is likely to achieve only up to 90% control (Shell Research 1985) and the survivors will later be very difficult to control with poison bait. Nevertheless, their relatively low cost and easy availability make acute rodenticides (especially zinc phosphide) the most widely used poisons in rodent control worldwide (Buckle 1994) and their single-feed action and the immediate results (dead bodies) make them attractive to users.

Because of problems of efficacy and safety, many older compounds are no longer available. Table 3 summarises the properties of those acute poisons that are the most widely used and acceptable. However, anticoagulant poisons are invariably more effective than even the best acute poisons. Acute poisons ought really only to be used under the following conditions:

- where there is absolutely no risk of accidental poisoning (e.g., in sewers, such as compounds 1080 and 1081),
- where there is resistance to anticoagulants,
- where rapid and economical knockdown of a large infestation is required, before treatment with a more expensive anticoagulant, and
- where there is no alternative available.

4.3.8. *Delayed-Effect, Multiple-Feed Poisons: First-Generation Anticoagulants*

The major revolution in rodent control was the introduction of the anticoagulants around 1950. Because the history of the development of anticoagulants is based on a problem with stored animal feed and elucidates their mode of action, it will be summarized below.

It was known in the second quarter of the twentieth century that cattle fed "spoiled" (moldy) sweet clover hay sometimes developed hemorrhages, which were attributed to a deficiency of prothrombin without which blood will not clot (Roderick 1929). The anticoagulant was isolated in 1939 (Link 1944) and found to be a dicoumarin compound. It had also been found that the hemorrhagic disease was prevented if dried alfalfa was added to the spoiled sweet clover hay (Quick 1937) and that adding coumarin to dried clover induced toxicity (Smith and Brink 1938). Since alfalfa contains high levels of vitamin K, the stage was set for the development of anticoagulants and antidotal treatment of anticoagulant poisoning 10 yr before their potential as rodenticides was first noted in the literature (O'Connor 1948).

Sweet clover disease was a problem in the cattle rearing areas of the northern United States, and much of the research was carried out by the Wisconsin Alumni Research Foundation (WARF). A dicoumarin compound (Dicumarol) was patented by WARF in 1941 and patented as a rodenticide in 1952. A more toxic compound was patented in 1947 and registered by the US Department of Agriculture in 1950 for sale as a rodenticide under the name Warfarin. Prior to that, O'Connor (1948) described the potential use of multiple-dose rodenticides as follows: "The most satisfactory technique so far evolved has been to ensure that the poisoned bait is constantly and freely available to rodents."

Anticoagulants acted more slowly than previously developed rodenticides because of their distinctive mode of action. In the vertebrate liver, some of the factors involved in blood clotting are formed by the operation of the vitamin K cycle (Fig. 2). Put simply, anticoagulants are thought to bind vitamin K epoxide reductase, causing the cycle to stop at vitamin K epoxide which builds up while the levels of vitamin K (specifically vitamin K_1) and its reduced form

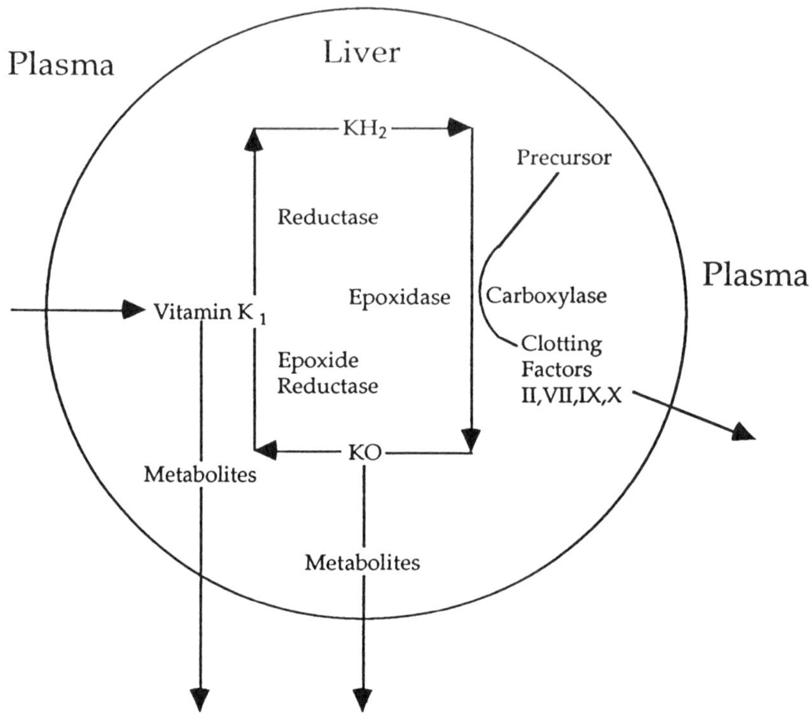

Figure 2 The vitamin K cycle in the mammalian liver. Vitamin K_1 is first converted to a reduced form (KH_2) and then oxidized to Vitamin K epoxide (KO).

decline. The antidotal effect of dietary vitamin K occurs because part of the cycle is pushed around and clotting factors are formed before the block. Without the antidote, clotting factors are used up without being replaced and the clotting mechanism ceases to function two or more days after inhibition of the epoxide reductase. Death follows after 4 or 5 d, or longer in some animals. Hemorrhage and clotting are normal processes in the organs and tissues of active animals, and time to death in social groups of enclosed wild rats has been found to be significantly shorter than in individually caged wild rats (Cox and Smith 1992). Because animals do not associate ingestion of the poison bait with the delayed symptoms, there is no development of bait shyness.

Other anticoagulants based on 4-hydroxycoumarin were developed and marketed in the 1950s and the equivalent anticoagulant effect of another class of compounds (derivatives of 1,3-indandione) was also discovered. All of these compounds are known as "first-generation" anticoagulants and all require multiple-feed application because toxicity at palatable bait concentrations is not sufficiently high for a single-feed effect (though the range of tolerances in a population is so wide that a few animals may succumb after only one or two feeds). The best known coumarins that were marketed are Warfarin, Coumachlor and Coumatetralyl, while Diphacinone and Chlorophacinone are the best known indandiones. Concentrations in bait are 50–300 mg/kg.

In summary, the advantages of the first-generation anticoagulants are: delayed action, therefore no bait shyness; antidote is vitamin K (reverses anticoagulation, not hemorrhage); and low hazard to nontarget animals, especially birds. The disadvantages are: expensive in bait and labor (because of multiple-feed requirement); extended treatment time (3–5 wk); lower efficacy than second-generation anticoagulants (see Section 4.3.9); and development of resistance since late 1950s in some locations.

4.3.9. Delayed-Effect, Single-Feed Poisons: Second-Generation Anticoagulants

A series of halogenated hydroxycoumarin derivatives was developed as a response to resistance to first-generation anticoagulants; these were first marketed in the late 1970s. Two of those that were marketed (difenacoum and bromadiolone) are toxic enough to kill a rat after a single feed but they are normally used as multiple-feed poisons at a concentration of 50 mg/kg to ensure effectiveness. Difenacoum is marketed under various names, including Neosorexa and Ratak; bromadiolone as Rodine, Slaymor, Deadline, Maki, and Contrac. Both were effective against rats resistant to first-generation anticoagulants, though cross-resistance has now developed in southern Great Britain (Greaves 1994, Smith and Greaves 1987).

There are three true, single-feed anticoagulants: brodifacoum (e.g., Klerat, Talon, Matikus, Ratak Super, Volak), flocoumafen (Storm), and difenthialone. Because of their high toxicity, even at concentrations of 20–50 mg/kg, they do not require multiple feeds to be effective. In order to save on bait and labor and to reduce hazard, recommended use is by pulsed baiting—many, small bait points renewed at intervals of 7–10 d—rather than by surplus baiting with larger bait points renewed every 3–4 d (Shell Research 1985, Lund 1988b). Although some tolerance to low doses of these most toxic compounds is known, resistance to field-strength baits has not been documented (Gill and MacNicoll 1991, Greaves 1994).

4.3.10. Delayed-effect, Multiple-Feed Poisons: Calciferols

Apart from bromethalin (marketed as Despatch, Vengeance), which is a relatively new delayed-effect poison not much used outside the United States, the main alternative slow-acting poisons to anticoagulants are calciferols.

Calciferols are forms of vitamin D, which at normal levels is necessary for normal calcium metabolism in order to prevent rickets and osteomalacia. The two forms are ergocalciferol (vitamin D_2; marketed as Sorexa C.R. and Rodine C) and cholecalciferol (vitamin D_3; marketed as Quintox). Concentrations in bait are 75–100 mg/kg and at this level calcium is resorbed from the skeleton in 10–12 h, hypercalcemia is induced, and there is calcification of the blood vessels and organs leading to death in 3–14 d. There is no specific antidote though the thyroid hormone calcitonin reduces hypercalcemia. In small mammals, calcification of the blood vessels in the heart leads to heart failure while in large mammals, calcification of the kidney vessels may increase blood pressure and lead to a brain hemorrhage (stroke).

Calciferol treatments average 7–14 d and bait must be supplied in excess (surplus baiting) to prevent development of bait shyness, which may occur because of the fairly rapid onset of physiological disturbances (Shell Research 1985, Prescott et al. 1992). Palatability and stability of ergocalciferol have been found to be variable (Meehan 1984). Some nontarget animals such as dogs may be at risk of accidental primary poisoning if baits are used carelessly, but there is little risk of secondary poisoning to predators and scavengers.

Ergocalciferol is marketed only as a mouse poison in the United Kingdom. The problem of instability in damp conditions does not arise in grain stores and calciferols are fully effective against animals resistant to anticoagulants. Indeed, ergocalciferol was first introduced in the 1970s to circumvent the resistance to first-generation anticoagulants that had arisen during the 1960s. The main application of calciferols is in food stores and food-processing premises, though the introduction of "second-generation" anticoagulants has lead to a decline in the use of calciferols.

4.3.11. Formulations and Baiting

In order to persuade a rodent pest to take in a lethal dose of poison (or a nonlethal chemical such as a chemosterilant), an appropriate formulation must be chosen and applied correctly. The choice of formulations is: contact powder (tracking powder in the United States) or gel, liquid bait, or edible bait.

Contact preparations rely on a rodent grooming poison from its fur. For safety reasons, contact powders should not be used in food stores though gels may have a place. Contact preparations use high concentrations of toxins (e.g., 25% lindane) and may be repellent to rodents as well as dangerous if misplaced.

Liquid baits are often overlooked but can be very useful in grain stores and food processing premises because rodents will drink water baits if available in dry stores and edible baits may be less attractive than alternative foods. Liquid bait containers need frequent attention to top-up liquid that has evaporated or else they may become unpalatable as concentration increases.

Edible baits are the most widely used and may be offered as cereal bait, pellets, block baits, or packet style (known as "throw packs" in Europe). The most appropriate formulation will depend on the situation, and attractiveness of bait as an alternative to other food in a store will depend on placement as well as intrinsic palatability. Lund (1988b) discusses the selection and distribution of baits. Baiting strategies have been discussed in previous sections and are described in more detail by Meehan (1984) and Shell Research (1985).

5. ENVIRONMENTAL PERSPECTIVE

5.1. Population Dynamics Scales in Space and Time

Having considered at some length the technology of rodent and bird pest control, I will now return to the environmental perspective introduced in Section 1 and summarize the biological features that make birds and rodents different from insects, mites and fungi. The relevant spatial and temporal scales of the pests need to be examined in relation to the scales associated with the stored grain, namely: the size (and movement) of individuals relative to the size of the store and the generation time of the pest relative to the store turnover time. Clearly, both the size of individuals and the scale of movement increase through fungi, mites and insects, to rodents and birds.

5.1.1. Rodents

The population dynamics of rodents may be stable or irruptive (Macdonald and Fenn 1994). Commensal species tend to breed continuously and population numbers increase towards an equilibrium, while field species often breed seasonally and are irruptive. While many rodent species hold individual territories, the main pests defend group ("clan") rather than individual territories (Macdonald and Fenn 1994). Individuals respond to food availability and some rats may travel kilometers or tens of kilometers in search of food. Even the commensal Norway rat moves from farm buildings to the field for the summer in Great Britain, moving back to farm buildings in the autumn. In general, population size is limited by availability of food and harborage and only rarely by natural enemies or disease. Hence it is possible to reduce damage by local modification of the environment (clean farming, sanitation, and hygiene). In grain stores, food is abundant and harborage is often the main factor limiting population growth. Although rodents can live for many months, generation time in the wild is of the order of weeks rather than months for most pest species. Thus, genetic changes occur quite rapidly on a human time scale, for instance the evolution and spread of resistance in some parts of the world (see Section 5.2).

5.1.2. Birds

The main pest species are only weakly territorial or nonterritorial; like rodents, individual territoriality would limit population density and reduce the pest problem. Bird pests often feed in flocks and may travel tens or hundreds of kilometers for food, responding opportunistically and rapidly to variations in food supply. Natural enemies are unlikely to be population-regulating factors. Because of the scale of movement of birds, sanitation will reduce the chances of invasion (your neighbors' messy stores may be more attractive) but bird proofing whenever practical is the best line of defence.

5.2. Evolution of Resistance

Physiological resistance to warfarin, and to all other first-generation anticoagulants, evolved remarkably quickly, and was first documented in the United Kingdom in Scotland in 1958 (Boyle 1960) and on the England/Wales border in 1960 (Drummond 1966), within 10 yr of the introduction of anticoagulants. In both cases, resistance was associated with a single, dominant gene (Greaves and Ayres 1967, 1976). Resistance in some strains seems to be due to a modified vitamin K epoxide reductase; the modified enzyme is not bound by first-generation anticoagulants but neither is it very efficient at reducing vitamin K epoxide. Consequently, resistant homozygotes in a Welsh strain suffer from vitamin K deficiency unless their food is supplemented, and there is evidence of a balanced polymorphism in the presence of anticoagulant, where the heterozygote is fitter than either homozygote (Greaves et al. 1977, Smith and Greaves 1987). Additional work has indicated that homozygous resistant rats suffer a growth-rate reduction in addition to (or perhaps as a consequence of) vitamin K deficiency (Smith et al. 1991) but that there may be heterozygous advantage even when anticoagulants are not used (Smith et al. 1993).

The second-generation anticoagulants overcame the problem of resistance until cross-resistance to difenacoum was recorded in southern Great Britain (Greaves et al. 1982, Greaves and Cullen-Ayres 1988). Neither difenacoum nor bromadiolone are effective against certain rat populations in Berkshire, England (Smith, unpublished observations) and a real practical problem exists there because the more powerful compounds brodifacoum and flocoumafen are not registered for outdoor use (Cox and Smith 1990, Smith 1993). It should be stressed, however, that the problem of resistance to difenacoum and bromadiolone is not widespread.

Behavioral resistance (avoidance of all rodenticide baits) is also a theoretical possibility (Greaves 1994, Macdonald and Fenn 1994). However, the evidence for behavioral resistance as opposed to problems with bait-take associated with habitat differences is not compelling (Quy et al. 1992).

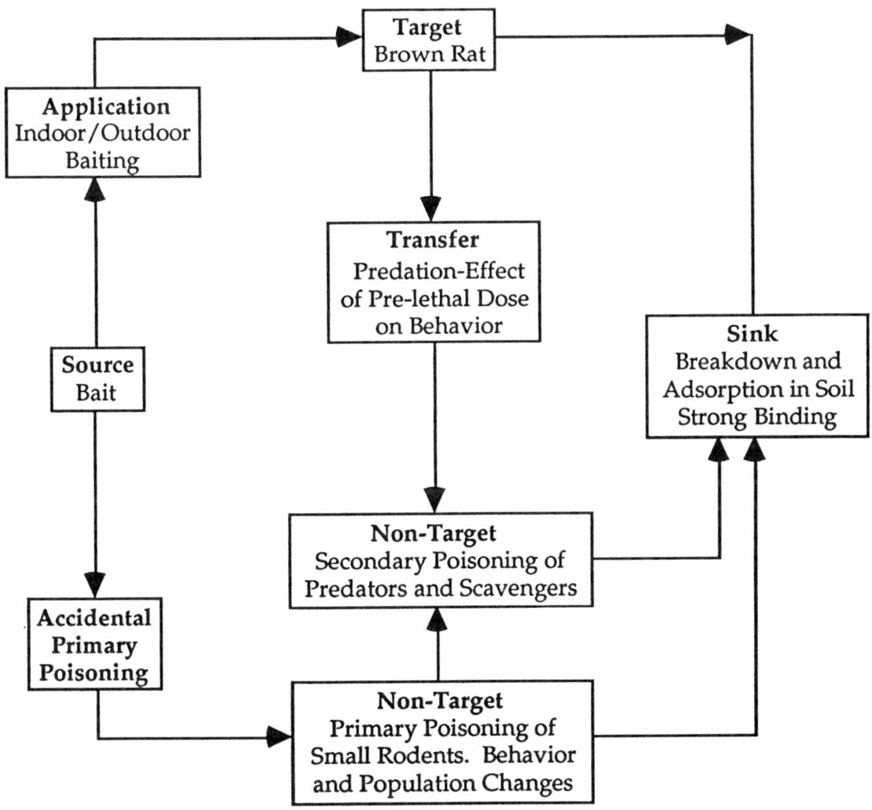

Figure 3 A compartment model of rodenticide ecotoxicology showing the routes by which a persistent rodenticide can move from source (bait point) to sink (soil).

5.3. Rodenticide Ecotoxicology

Rodenticides generally have high toxicity to most mammals, and are often also toxic to birds. In some countries, registration authorities have restricted the use of some of the more toxic and persistent second-generation anticoagulants because of possible secondary-poisoning hazards to predators and scavengers, especially predatory birds (Smith 1993). Experimental studies relating to assessment of the real hazard as opposed to theoretical risks (Brown 1994) are reported by Cox and Smith (1990 1992), who also carried out a computer simulation study (Smith et al. 1990). Figure 3 shows a model of how persistent rodenticides move between application and the final "sink" in the soil. Too little information is available yet to be able to predict quantitatively how rodenticides are partitioned between compartments in the model. However, it is clear that the hazard to nontarget, wild animals when controlling store pests is less than when controlling field pests and that careful positioning and covering of bait points followed up by removal of uneaten bait and dead bodies can help to minimize any adverse environmental effects.

6. COST-BENEFIT ANALYSIS

In principle, the optimal solution to a pest problem is obtained by working through the following sequence:

1. Define the problem. What is (are) the:	nature of damage?
	boundaries of problem?
	distribution of damage?
	pest species?
2. Define the constraints. Which options are:	available?
	realistic?
	acceptable?
3. Define the solution. Which option(s)	maximize gain?

The optimal solution will maximize gain (benefits minus costs), subject to the defined constraints, but in practice it may be more difficult to estimate the benefit in stores than in field crops where small plots can be compared, treated with untreated (Richards 1988). Therefore, a more practical optimization criterion might be "choose the procedure that will achieve effective control with minimal cost," again subject to constraints.

The only published estimate of the cost-effectiveness of reducing damage in food stores is based on a study of warehouses in Cuba (Hernandez and Drummond 1984). The essential features of this example are:

- Initial losses were low (<1%)
- Warfarin treatment gave a high ratio of benefits to costs (29:1)
- Some of the assumptions are questionable (e.g., ignoring cost of a second control phase because it did not last for long).

Nevertheless, it is clear that a total rodent control program in stores as part of routine store management can be, and usually will be, cost effective. A total IPM program will consider all approaches, including sanitation, trapping, and rodenticides as part of store management (Meyer 1994).

7. FUTURE NEEDS

It is evident from the uneven coverage of birds and rodents that more is known and published about rodent control than bird control in the stored grain ecosystem. In part, the difference reflects the relative importance of the two but the greater migratory potential of birds renders them more difficult both to assess and to control as pests. Ideally, then, more data should be obtained on losses caused by bird pests and on methods of managing bird-pest problems in grain stores.

The only effective way to prevent losses caused by birds is to prevent them from invading, by effective proofing of stores. There are ways of controlling rodent infestations but improved store construction and maintenance could greatly reduce dependence on chemical rodenticides which sensibly ought to be kept away from stored food if at all possible. Ideally, there should be effective repellents that can be applied to structures to prevent gnawing damage. But improved design and construction of stores according to well-known principles (Jenson 1979) could prevent many invasions and infestations by both birds and rodents. If lethal chemicals are used, they clearly need to be formulated effectively to operate in the presence of abundant food and this may mean more use of liquid baits.

Inherited physiological resistance to rodenticides is not a widespread problem but it can be serious where it occurs. The rodenticide market has probably reached a plateau and there is little commercial incentive to develop new products for the relatively small rodenticide market (Smith and Buckle 1994), and no guarantee that cross-resistance would not rapidly evolve. There are several options for managing resistance, including judicious use of different rodenticides, sanitation, mopping up or chemosterilizing survivors of anticoagulant control, and using natural selection against resistant genotypes (Smith and Greaves 1987). Resistance management should be a part of rodent IPM, even when resistance is anticipated rather than proven.

Vertebrate poisons could potentially cause tremendous environmental damage by decimating populations of sensitive species such as predatory birds. In practice, the risk of using rodenticides has rarely manifested itself as a real hazard because rodenticides are normally used selectively and effectively (Smith et al. 1990). Widespread application of bromadiolone to control water voles in Switzerland, however, did result in accidental poisoning of a number of predatory birds (Shawyer 1985). Improvements in formulation technology may help to minimize environmental hazards; for example, a wax-block formulation of flocoumafen presented no observable primary hazard to birds in field trials reported by Johnson and Scott (1986). Thus, in addition to monitoring, experimental studies are needed to determine patterns of use that minimize nontarget hazards while maintaining efficacy.

Finally, at least for rodents, it should be emphasized that there is no real technological problem of control. New technological tools may help in monitoring entry of rodents into sophisticated grain facilities but more important is the implementation of operational storage practices that reduce the likelihood of rodent immigration and damage (Meyer 1994). What is required most of all is improved organization to use the technology effectively. There are two levels at which organization needs to be improved. The individual store manager should have hygiene and maintenance of proofing as part of his normal work routine. In addition, regular monitoring for signs of invasion and maintenance of any control measures (including preventive baiting at permanent bait points) should be part of his delegated responsibilities (Meyer 1994). But coordinated control is also important; if your neighbors are rat-infested, there will always be a queue of invaders waiting to gain entry into your store! It is difficult to achieve effective control campaigns without governmental intervention because neither birds nor rodents recognize the boundaries of individual stores, communities, or even states. Unfortunately, the tide of opinion in many countries has turned against intervention, cooperation, and coordination to such an extent that progressive IPM of vertebrates may become more and more difficult to achieve. While the simple operation of market forces may encourage effective local control of the smaller store pests like insects, the larger, more mobile pests require coordination of efforts over larger areas. Rat-free campaigns have proved successful in cities and even a whole Canadian province (Alberta); once established, a rat-free zone is relatively cheap to maintain. Unfortunately, large-scale programs are not effective if they are badly designed. Apart from resources wasted on badly-designed programs, J. H. Greaves (personal communication) has identified the following as primary problems in many parts of Africa, Asia, and South America: absence of effective municipal programs, unavailability of technical knowledge at local level, and unavailability of rodenticides at local level for farmers. Thus, the extension of existing technology to the level of individuals responsible for managing grain stores is the most pressing need for effective IPM of vertebrate pests.

REFERENCES

Barnes, G. 1987. Food industry standards—their trends and significance in stored product pest control. Pages 3-8 in: Stored Products Pest Control T. J. Lawson, (ed.). BCPC Monograph No. 37. British Crop Protect. Council, Thornton Heath, UK.

Barnett, S. A. 1951. Damage to wheat by enclosed populations of *Rattus norvegicus*. J. Hyg., Cambridge 49:22-25.

Boyle, C. M. 1960. Case of apparent resistance of *Rattus norvegicus* Berkenhaut to anticoagulant poisons. Nature 188:517.

Brooks, J. E. 1973. A review of commensal rodents and their control. CRC Crit. Rev. Environ. Control. 3:405-453.

Brooks, J. E. and E. Ahmad. 1990. Pest birds of Pakistan: identification and distribution. pp. 173-180 in: Vertebrate Pest Management J. E. Brooks, E. Ahmad, I., Hussain, S. Munir, and A. A. Khan, (eds.). GOP/USAID/DWRC, Islamabad, Pakistan.

Brooks, J. E. and A. M. Bowermann. 1971. Estrogenic steroid used to inhibit reproduction in wild Norway rats. J. Wildlife Management. 35:444-449.

Brooks, J. E. and I. Hussain. 1990. Chemicals for bird control. pp. 193-195 in: Vertebrate Pest Management, J. E. Brooks, E. Ahmad, I. Hussain, S. Munir, and A. A. Khan, (eds.). GOP/the United States ID/DWRC, Islamabad, Pakistan.

Brooks, J. E. and F. P. Rowe. 1979. Commensal rodent control. WHO Report WHO/VBC/79. 726. World Health Org., Geneva, Switzerland. 106 pp.

Brown, R. A. 1994. Rodenticides and the environment. pp. 363-380 in: Rodent Pests and Their Control, A. P. Buckle and R. H. Smith, (eds.). CAB Int., Wallingford, UK.

Bruce, H. M. 1960. A block to pregnancy in the mouse caused by the proximity of strange males. J. Rep. Fert. 1:96-103.

Bruggers, R. L. and C. C. H. Elliott (eds.) 1989. *Quelea quelea*: Africa's bird pest. Oxford Univ. Press, Oxford.

Buckle, A. P. 1994. Rodent control methods: chemical. pp. 127-160 in: Rodent Pests and Their Control, A. P. Buckle and R. H. Smith, (eds.). CAB Int., Wallingford, UK.

Buckle, A. P. and R. H. Smith (eds.) 1994. Rodent Pests and their Control. CAB Int., Wallingford, UK. 405 pp.

Cogelia, N. J., G. K. Lavoie, and J. F. Glahn. 1976. Rodent biting pressure and chewing action and their effects on wire and cable sheath. pp. 16-18 in: Proc. 25th Int. Wire and Cable Sympo., Cherry Hill, NJ.

Cornwell, P. D. 1979. Pest Control in Buildings (2nd ed.). Rentokil Ltd., East Grinstead, UK.

Cox, P. R. and R. H. Smith. 1990. Rodenticide ecotoxicology: assessing nontarget population effects. Functional Ecol. 4:315-320.

Cox, P. R. and R. H. Smith. 1992. Rodenticide ecotoxicology: prelethal effects of anticoagulants on rat behaviour. pp. 165-170 in: Proc. 15th Vertebrate Pest Conf., Los Angeles, CA. University of California, Davis, CA.

Drummond, D. C. 1966. Rats resistant to warfarin. New Scientist 30:771-772.

Duckett, J. E. 1991. The evaluation of management techniques in the biological control of rats in oil palm plantations. MPhil thesis, Cranfield Univ. Technology, Silsoe, UK. 147 pp.

Dykstra, W. W. 1954. Rodent filth in food. Pest Control. 22:9-14.

Elton, C. S. 1953. The use of cats in farm rat control. Brit. J. Anim. Behaviour. 1:151-155.

Ericsson, R. J. 1982. Alpha-chlorohydrin (Epibloc^R): a toxicant-sterilant as an alternative rodent control. pp. 6-9 in: Pro. 10th Vertebrate Pest Conf., Davis, CA. University of California, Davis, CA.

FAO. 1983. Food Storage Manual. Trop. Dev. Res. Inst., London. 263 pp.

Farrag, H. A. 1983. Reproductive biology and social behaviour in the Nile rat (*Arvicanthis niloticus*). PhD thesis, Univ. Reading, Reading, UK. 175 pp.

Fitzwater, W. D. 1978. Electromagnetic repellers—fact or fiction? pp. 88-92 in: Proc. 8th Vertebrate Pest Conf., University of California, Davis, CA.

Fitzwater, W. D. 1988. Nonlethal methods in rodent control. pp. 421-426 in: Rodent Pest Management I. Prakash (ed.). CRC Press Inc., Boca Raton, FL.

Gecan, J. S., J. L. Thrasher, W. Eisenberg, and P. M. Brickey. 1980. Rodent excreta contamination and insect damage of wheat. J. Food Prot. 43:203-204.

Gill, J. E. and A. D. MacNicoll. 1991. Determination of the susceptibility of wild populations of the Norway rat (*Rattus norvegicus*) to the anticoagulant rodenticide brodifacoum. Zeitschrift für angewandte Zoologie. 78:101-117.

Gratz, N. G. 1988. Rodents and human disease: a global appreciation. pp. 101-169 in: Rodent Pest Management I. Prakash (ed.). CRC Press, Boca Raton, FL.

Gratz, N. G. 1994. Rodents as carriers of disease. pp. 85-108 in: Rodent Pests and their Control, A. P. Buckle and R. H. Smith (eds.). CAB Int., Wallingford, UK.

Greaves, J. H. 1978. A manual of methods for the evaluation of postharvest losses. pp. 109-115 in: Postharvest Grain Loss Assessment Methods, K. L. Harris, and C. J. Linblad (eds.). Am Assoc. Cereal Chem., Minneapolis, MN.

Greaves, J. H. 1982. Rodent Control in Agriculture. FAO Plant Production and Protection Paper No 40. Food and Agric. Org. United Nations, Rome, Italy. 88 pp.

Greaves, J. H. 1994. Resistance to anticoagulant rodenticides. pp. 197-217 in: Rodent Pests and their Control A. P. Buckle, and R. H. Smith (eds.). CAB Int., Wallingford, UK.

Greaves, J. H. and P. B. Ayres. 1967. Heritable resistance to warfarin in rats. Nature. 215:877-878.

Greaves, J. H. and P. B. Ayres. 1976. Inheritance of Scottish-type resistance to warfarin in the Norway rat. Genet. Res. 28:231-239.

Greaves, J. H. and P. B. Cullen-Ayres. 1988. Genetics of difenacoum resistance in the rat. pp. 389-397 in: Current Advances in Vitamin K Research J. W. Suttie (ed.). Elsevier, Amsterdam.

Greaves, J. H., R. Redfern, P. B. Ayres, and J. E. Gill. 1977. Warfarin resistance: a balanced polymorphism in the Norway rat. Genet. Res. 30:257-263.

Greaves, J. H., D. S. Shepherd, and J. E. Gill. 1982. An investigation of difenacoum resistance in Norway rat populations in Hampshire. Ann. Appl. Biol. 100:581-587.

Grolleau, G. and J. Gautun. 1991. Protection des denrées stockées contre les degats de vertebres ravageurs. pp. 1469-1478 in: Proc. 5th Int. Working Conf. Stored-Prod. Protect. III., Fleurat-Lessard and P. Ducom (eds.). INRA, Bordeaux, France.

Hernandez, A. and D. C. Drummond. 1984. A study of rodent damage to food in some Cuban warehouses and the cost of preventing it. J. Stored Prod. Res. 20:83-86.

Hopf H. S., G. E. J. Morley, and J. R. O. Humphries (eds.). 1976. Rodent damage to growing crops and to farm and village storage in tropical and subtropical regions. Centre for Overseas Pest Res. and Trop. Prod. Inst., London. 115 pp.

Hussain, I. 1990. Trapping, netting and scaring techniques for bird control. pp. 187-191 in: Vertebrate Pest Management J. E. Brooks, E. Ahmad, I., Hussain, S. Munir, and A. A. Khan, (eds.). GOP/USAID/DWRC, Islamabad, Pakistan.

Jenson, A. G. 1979. Proofing of buildings against rats, mice and other pests. ADAS advisory leaflet. HMSO, London. 33 pp.

Johnson, R. A. and R. M. Scott. 1986. Flocoumafen—a new second generation anti-coagulant. 7th British Pest Control Conf., Session 5, Paper 3, Guernsey, Channel Islands, UK. 20 pp.

Kendle, K. E., A. Lazarus, F. P. Rowe, J. M. Telford, and D. K. Vallance. 1973. Sterilization of rodent and other pests using a synthetic oestrogen. Nature 244:105-108.

Khan, A. A. and E. Ahmad. 1990. Damage patterns of pest birds and assessment methods. pp. 181-186 in: Vertebrate Pest Management J. E. Brooks, E. Ahmad, I., Hussain, S. Munir, and A. A. Khan, (eds.). GOP/USAID/DWRC, Islamabad, Pakistan.

Leirs, H. 1992. Population ecology of *Mastomys natalensis* (Smith, 1834) multimammate rats: possible implications for rodent control in Africa. PhD thesis, Univ. Antwerp., Belgium. 263 pp.

Link, K. P. 1944. The anticoagulant from spoiled sweet clover hay. The Harvey Lectures. Series 39:162-216.

Lund, M. 1988a. Ultrasound devices. pp. 407-409 in: Rodent Pest Management I. Prakash (ed.). CRC Press, Boca Raton, FL.

Lund, M. 1988b. Selection of baits and their distribution. pp. 261-268 in: Rodent Pest Management I. Prakash (ed.). CRC Press, Boca Raton, FL.

Lund, M. 1994. Rodents as commensal pests. pp. 23-43 in: Rodent Pests and their Control A. P. Buckle and R. H. Smith (eds.). CAB Int., Wallingford, UK.

Macdonald, D. W. and M. G. P. Fenn. 1994. The natural history of rodents: pre-adaptations to pestilence. pp. 1-21 in: Rodent Pests and their Control A. P. Buckle, and R. H. Smith (eds.). CAB Int., Wallingford, UK.

Marsh, R. E. 1988. Chemosterilants for rodent control. pp. 101-169 in: Rodent Pest Management I. Prakash (ed.). CRC Press, Boca Raton, FL.

Mbise, T. 1983. The biology of *Mastomys natalensis* (Smith 1834) and its control in Tanzania. MSc. thesis, Univ. Reading, Reading, UK. 70 pp.

McCallum, H. I. and G. R. Singleton. 1989. Models to assess the potential of *Capillaria hepatica* to control population outbreaks of house mice. Parasitology 98:425-437.

McKillop, I. G. and R. M. Sibly. 1988. Animal behaviour at electric fences and the implications for management. Mammal Rev. 18:91-103.

Meehan, A. P. 1984. Rats and Mice: their biology and control. Rentokil Ltd., East Grinstead, UK. 383 pp.

Meyer, A. N. 1989. Vertebrate pest control report on a visit to the food department, Sri Lanka. ODNRI Report R1581 (R). Overseas Dev. Nat. Resources Inst., Chatham, UK. 37 pp.

Meyer, A. N. 1994. Rodent control in practice: food stores. pp. 273-290 in: Rodent Pests and their Control A. P. Buckle and R. H. Smith (eds.). CAB Int., Wallingford, UK.

Newsome, A. 1990. The control of vertebrate pests by vertebrate predators. Trends in Ecology and Evolution. 5:187-191.

O'Connor, J. A. 1948. Use of blood anticoagulants for rodent control. Research, London, 1:334-336.

Opitz, H. M., D. J. Henzler, J. A. Singer, and D. J. Bean. 1991. *Salmonella entereditis*: a three-year epidemiological study of laying flocks. Paper presented at the 63rd Annual Meeting Northeastern Conf. Avian Diseases, Cornell Univ., Ithaca, NY.

Prakash, I. (ed.). 1988a. Rodent Pest Management. CRC Press Inc., Boca Raton, FL. 480 pp.

Prakash, I. 1988b. Bait shyness and poison aversion. pp. 101-169 in: Rodent Pest Management I. Prakash (ed.). CRC Press, Boca Raton, FL.

Prescott, C. V., M. El-Amin, and R. H. Smith. 1992. Calciferols and bait shyness in the laboratory rat. pp. 218-223 in: Proc. 15th Vertebrate Pest Conf., Los Angeles. University of California, Davis, CA.

Quick, A. J. 1937. The coagulation defect in sweet clover disease and in the haemorrhagic chick disease of dietary origin. Am. J. Physiol. 188:260-271.

Quy, R. J., D. S. Shepherd, and I. R. Inglis. 1992. Bait avoidance and effectiveness of anticoagulant rodenticides against warfarin- and difenacoum-resistant populations of Norway rats (*Rattus norvegicus*). Crop Prot. 11:14-20.

Richards, C. G. J. 1988. Large-scale evaluation of rodent control techniques. pp. 269-284 in: Rodent Pest Management I. Prakash (ed.). CRC Press, Boca Raton, FL.

Roderick, L. M. 1929. The pathology of sweet clover disease in cattle. J. Am. Vet. Med. Assoc. 74:314-326.

Schumake, S. A., A. L. Kolz, R. F. Reidinger, and M. W. Fall. 1979. Evaluation of nonlethal electric barriers for crop protection against rodent damage. pp. 29-38 in: Vertebrate Pest Control and Management Materials J. R. Beck (ed.). ASTM STP 680. Am. Soc. Testing Mat. Philadelphia.

Semple, R. L. (undated). Problems related to pest control and use of pesticides in grain storage: the current situation in ASEAN and future requirements. Report to ASEAN Crops Post-Harvest program. National Postharvest Inst. Res and Ext., Manila, The Philippines. 72 pp.

Shawyer, C. R. 1985. Rodenticides - a review and assessment of their potential hazard to nontarget wildlife with special reference to the barn owl (*Tyto alba*). The Hawk Trust, London.

Shell Research. 1985. Rodenticides: Profile of a New Product. Shell Research, Sittingbourne, UK. 37 pp.

Shorrocks, B. and I. Swingland. 1990. Living in a Patchy Environment. Oxford Univ. Press, Oxford.

Shuyler, H. R. 1970. Control of rodents and grain-eating birds: report to the government of the United Arab Republic. FAO Report No. RP7. Food and Agric. Org. United Nations, Rome, Italy. 246 pp.

Sinclair, A. R. E. 1989. Population regulation in animals. pp. 197-241 in: Ecological Concepts J. M. Cherrett (ed.). Blackwell, Oxford.

Singleton, G. and T. Redhead. 1991. Future prospects for biological control of rodents using micro- and macro-parasites. pp. 75-82 in: Rodents and Rice G. R. Quick (ed.). Int. Rice Res. Inst., Los Baños, Philippines.

Smith, P., M. Berdoy, R. H. Smith, and D. W. Macdonald. 1993. A new aspect of warfarin resistance in wild rats: benefits in the absence of poison. Functional Ecol. 7:190-194.

Smith, P., M. Townsend, and R. H. Smith. 1991. A cost of resistance in the brown rat: reduced growth rate in warfarin-resistant lines. Functional Ecol. 5:441-447.

Smith, R. H. 1993. Chapter 19: Mammals. pp. 339-352 in: Handbook of Ecotoxicology. Vol. 1 Techniques P. Calow (ed.). Blackwell, Oxford.

Smith, R. H. 1994. Rodent control methods: nonchemical and nonlethal chemical. pp. 109-125 in: Rodent Pests and Their Control, A. P. Buckle and R. H. Smith, (eds.). CAB Int., Wallingford, UK.

Smith, R. H. and A. P. Buckle. 1994. Rodent control: back to the future. pp. 381-390 in: Rodent Pests and Their Control, A. P. Buckle and R. H. Smith (eds.). CAB Int., Wallingford, UK.

Smith, R. H. and J. H. Greaves. 1987. Resistance to anticoagulant rodenticides: the problem and its management. pp. 302-315 in: Proc. 4th Int. Working Conf. Stored-Prod. Protect., E. Donahaye and S. Navarro (eds.). Agricultural Research Organisation, Bet Dugan, Israel.

Smith, R. H., P. R. Cox, and M. Rampaud. 1990. Rodenticide ecotoxicology: systems analysis and simulation. pp. 47-54 in: Proc. 14th Vertebrate Pest Conf., University of California, Davis, CA.

Smith, W. K. and R. A. Brink. 1938. Relation of bitterness to the toxic principle in sweet clover. J. Agric. Res. 57:145-154.

Stoddart, D. M. 1988. The potential for pheromonal involvement in rodent control programs. pp. 101-169 in: Rodent Pest Management I. Prakash (ed.). CRC Press, Boca Raton, FL.

Tolar, V. 1991. The use of domestic cats in the control of the black rat, *Rattus rattus*, in a swine-farm: a case history. pp. 1535-1540 in: Proc. 5th Int. Working Conf. Stored-Prod. Protect. III., F. Fleurat-Lessard and P. Ducom (eds.). INRA, Bordeaux, France..

Wilkins, K. T. 1982. Highways as barriers to rodent dispersal. Southwestern Naturalist. 27:459-460.

10

Quality and Nutritional Changes in Stored Grain

Keith H. Tipples

Grain Research Laboratory, Canadian Grain Commission, Winnipeg, Manitoba, Canada

SUMMARY

Quality and nutritive changes that take place in stored grain are the net result of interactions within a complex ecological system; four major categories of factors that influence potential storage changes are grain condition, environmental conditions, invasive forces, and grain treatment. Under favorable storage conditions, two technologically important types of change involve germination and germination capacity, particulary for malting barley, and subtle "aging" effects on freshly harvested wheat. High moisture content is by far the most important factor influencing the deterioration of stored grain but other factors include condition of the grain, in particular its temperature when put into storage and its physical condition. Cracked or broken kernels will allow microorganisms easy access to internal tissues and also will promote autolytic enzymatic activity. Presence of weed seeds and immature kernels encourage increased respiration within the grain mass, while impurities (dockage) harbor fungal and bacterial spores which can develop under adverse conditions. Insects and molds can interact to produce different degrees of deterioration depending on grain condition. Mold development, insect infestation, or both can lead to grain heating with concomitant darkening of color, loss of germination, reduction in quality, and in extreme cases self-ignition and total loss. The most serious consequence of mold development is production of mycotoxins by toxigenic species. Canada goes to great lengths to maintain the integrity of commercial shipments of grain including ensuring freedom from unaccepptable levels of mycotoxins. Other mold-related changes include development of musty odors causing feed refusal by animals, and grade loss, which is also associated with germ discoloration and visible mold. Oilseeds are particulary sensitive to deterioration caused by lipolytic or oxidative changes in the oil. Apart from promoting heating and mold development, the most obvious adverse effects of insect infestation relate to customer rejection. Several types of grain treatment can be applied to decrease the probability of storage deterioration. They include drying, cleaning, fumigating, aerating, and controlled atmosphere storage. Storage changes must be considered from the point of view of a total integrated system to the extent that discussions based on simple cause and effect may be misleading or invalid.

1. INTRODUCTION

Annual world production of cereal grains and oilseeds has averaged around 2 Gt during the past decade. Because most of this grain is intended for human or animal consumption and because the time between harvest and end processing may be considerable, the relation of storage to changes in quality and nutritional characteristics is of paramount importance.

Postharvest losses of grain are a significant factor in the world food supply and may represent as much as 5–10% of world production of cereal grains and oilseeds (Richard-Molard 1988). While most losses result from infestation by insects, microorganisms, rodents, and birds, a smaller but quite significant proportion of total losses results from respiration and gradual deterioration of viability, nutritive quality, and end-use properties during storage under commercial conditions (Pomeranz 1992).

Storage of cereal grains and their products has been considered sufficiently important by the American Association of Cereal Chemists to have published four editions of a major monograph on the subject. The latest edition appeared in 1992 (Sauer 1992) and followed previous editions published in 1982, 1974, and 1954. These books contain major reviews of biochemical, functional, and nutritive changes during storage (Pomeranz 1992, 1982) which was a valuable reference source for this paper. A second recent major review on grain storage was edited by Multon (1988). These and other authoritative publications on the subject (e.g., Mills 1989) make it unnecessary to present an exhaustive review. Instead, my aim is to summarize the available information on quality and nutritional changes and to add a Canadian perspective.

Although the most serious deleterious changes in grains and oilseeds occur under adverse storage conditions and are associated with invasive forces, even under the most favorable conditions, biochemical and physiological changes occur in grain because the seeds are living, albeit quiescent, organisms that respire and age. Therefore, I will discuss not only development of such major catastrophes as microbial spoilage and mycotoxin development but also the occurrence of some of the more subtle changes that can affect end processing acceptability and hence value or "quality." Since the Canada Grain Act specifically prohibits delivery into the licensed elevator system of grain that is "out of condition," it is these more subtle changes that can be of major interest in markets for high quality grain.

2. THE TOTAL STORAGE EFFECT: A COMPLEX ECOLOGY

2.1. Grain Condition

When considering storage changes, it is important to consider the total ecological system. Figure 1 shows a simplistic and arbitrary subdivision of the basic elements influencing potential storage changes into four categories: grain condition, environmental conditions, invasive elements, and grain treatments.

Factors governing grain condition include the type of grain and its composition (especially oil content), moisture content, temperature when put into storage, and its physical condition. Cracked or broken kernels will allow microorganisms easy access to internal tissues and also promote autolytic enzymatic activity. Presence of weed seeds and immature kernels encourages increased respiration within the grain mass and impurities (e.g., dust, dirt, and trash) harbor fungal and bacterial spores which can develop under poor storage conditions.

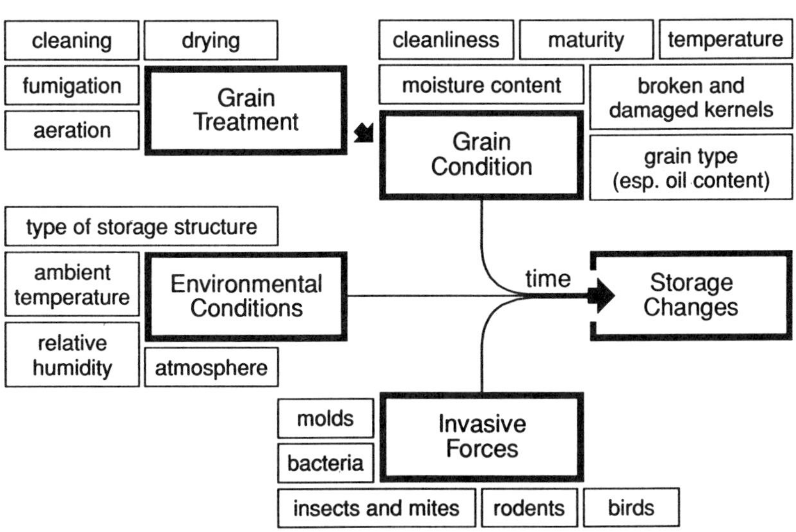

Figure 1 The total storage effect: a complex ecology.

2.2. Moisture and Temperature

Of all the various factors influencing the deterioration of stored grain, moisture is by far the most important. If moisture content is maintained at a sufficiently low level, grain can be stored for many years with little adverse effect even under conditions that may otherwise be unfavorable. Grain moisture content, water activity, and relative humidity of the interstitial atmosphere all affect the rates at which grain will respire and molds and insects will flourish. Microflora and insects require a certain minimum relative humidity to survive and flourish, while enzymes are inhibited, if not destroyed, by lack of it. It is the moisture associated with the nonfatty component which is critical with respect to the development of microflora.

At the same time, temperature is inextricably linked to moisture in determining the rate at which deteriorative changes will occur. In general terms, insects will not develop in cereal grains below 15°C or 10% moisture content (mc) on a wet mass basis (wb). Fungi will not grow below 13.5% mc (wb) or 0°C, and mites are inhibited below 13% mc (wb) or 5°C. Bacteria normally only grow under anaerobic conditions at moisture contents in excess of 20% (wb).

2.3. Grain Treatment

Several types of treatment can be applied to decrease the probability of storage deterioration. They include drying, cleaning, fumigation, aeration, and controlled atmosphere storage. These will be discussed briefly later. All the above factors are intimately interrelated and often interdependent. They must be considered as a total integrated system when discussing the implications and effects of storage on grain quality.

Before dealing with the causes of major spoilage and deterioration that occur under unfavorable storage conditions, we will first consider some of the "natural" changes that occur under favorable conditions during short- and long-term storage.

3. CHANGES IN GRAIN QUALITY OCCURRING UNDER FAVORABLE STORAGE CONDITIONS

3.1. Germination-Related Changes

3.1.1. Seed Dormancy

In an ideal storage situation, the grain or oilseed in question will have been grown and harvested under near perfect conditions. The grain is low in moisture content (e.g., 12% mc (wb) or less for cereal grains and 8% (wb) or less for small seeded oilseeds) and will store almost indefinitely under good storage conditions. It is free of damaged kernels, broken grain, and weed seeds, and is clean and cool when put into storage.

Two categories of "natural changes" occur with storage. One is related to subsequent ability of the seed to germinate and the other to subtle chemical and biochemical changes that may affect processing quality and acceptability.

Nature designed seeds to be largely impervious to outside influences until such time as they germinate and produce a new plant. All the food stores necessary for future growth are carefully packaged and protected from the precursors and stimulators of chemical reactions by cell walls and an outer seed coat. As well, the germ is located safely away from the polysaccharides, proteins, and other storage molecules present in the endosperm cells. Provided a seed remains dry and intact it can continue to be inactive yet viable for long periods of time under cool, favorable storage conditions.

Freshly harvested seeds will not necessarily germinate if subjected to moisture and temperature conditions that normally promote seed growth. Dormancy is the mechanism built in by nature to ensure that seeds will not germinate until such time as external conditions are suitable to allow plant growth to proceed to full maturity. Dormancy of most harvested cereals is slowly lost in dry storage. In nature, however, temperature and light are most important in breaking dormancy. Alternating temperatures (e.g., several hours at 10°C followed by several hours at 20°C) is often effective, although the mechanism of the cyclic effect is not well understood (Pomeranz 1992). Often chilling imbibed seeds at 0–5°C for several days help them to overcome dormancy and to germinate during later exposure to temperatures favorable to growth. This chilling requirement apparently prevents germination in late fall and regulates it in spring.

The causes of variation in dormancy are extremely diverse (Come 1988). They are primarily of genetic origin and the type and intensity of dormancy vary according to species and variety. Among cereals, red-seeded varieties of wheat are more dormant than white-seeded varieties, and winter barley is often more dormant than spring barley. The dormancy state of seeds of the same variety changes from year to year depending on location. For cereals, a cool, rainy period during planting reinforces dormancy.

Soundness is considered an important quality factor in both milling wheat and malting barley. Buyers of high quality wheat demand sound wheat with low α-amylase activity. Since one severely sprouted kernel may contain several thousand times as much α-amylase activity as a sound kernel (Kruger and Tipples 1983), even a small percentage of germinated kernels in a sample of milling wheat may be enough to make it unacceptable for certain end uses. Wet harvest conditions occur approximately once every 3 yr in western Canada, so resistance to sprouting under conditions that normally favor germination is a desirable feature. Breeders of red wheat have been more successful in maintaining and improving sprouting resistance. The red spring wheat variety Columbus, registered in 1980, was the first of a new generation of Canadian wheat varieties with genetically-introduced superior resistance to weathering and sprouting. Breeders aiming for increased levels of sprout resistance have to ensure that good germination capacity and vigor is still shown by seed planted the following season.

Similarly with malting barley, sprouting in the ear during harvest is highly undesirable since kernels that do begin to germinate may not germinate or modify properly when malted. The requirements of growers and maltsters may therefore be in conflict. Especially in countries that frequently experience wet harvest conditions, growers prefer grain with some dormancy to resist field sprouting. Maltsters, by contrast, prefer barley with a very limited dormancy that will germinate rapidly and uniformly and produce optimum modification during malting.

3.1.2. Seed Viability

For barley that is to be used for malting, or for seed of any grain that is intended for planting, it is of paramount importance that a high percentage of seeds will germinate when subjected to conditions conducive to germination. Furthermore, it is important that germination be vigorous and uniform for all seeds.

A wide range of survival times has been noted for stored barley including up to 96% germination after 32 yr storage (Briggs 1978). Of the barley grains found sealed in the foundation stone of the Nuremburg Theatre after 123 yr, 3 of the 25 tested germinated. This record of longevity, however, does not come close to rivalling the finding by museum botanists in Canada who reported germination of lupine seeds buried in a peat bog for about 10,000 yr (Pomeranz 1992).

Viability of cereal seeds is prolonged by storing at low moisture contents and temperatures. Several authors have presented model equations based on moisture content and temperature to determine theoretical longevity (Renard 1988).

Apart from storage conditions, other factors influencing seed longevity include variety and harvest conditions. Survival after a wet harvest is poorer than after a dry harvest, likely due to an initial heavy microbial contamination (Briggs 1978).

For preserving special, relatively small stocks of cereal seed in "gene banks" with the intention of preserving viability unchanged for 10 yr or more while also minimizing genetic deterioration, several steps must be taken. These include harvesting the grain at full maturity under dry harvest conditions, carefully threshing it to minimize physical damage, drying carefully, and storing under carbon dioxide (CO_2) or nitrogen (N_2) in sealed containers at low temperature in the dark (Briggs 1978). For really prolonged storage, the moisture content should possibly be reduced to about 8% and the sample, sealed under an inert gas, should be stored at about –20°C (Briggs 1978).

3.2. Changes in Milling and Baking Quality During Storage of Freshly Harvested Wheat

3.2.1. Short-Term Storage

For many years commercial millers and bakers have believed that freshly harvested wheat does not perform as well as wheat that has been stored (aged) for a period of several weeks or months (Zeleny 1954). While much of this belief may be based on their own experience or on anecdotal evidence, there are a number of reports in the literature which substantiate an improvement in milling and baking quality during short term storage. The literature reviewed by Pomeranz (1992) generally confirms an improvement of breadmaking quality of flour milled from freshly harvested wheat that has undergone short-term storage. Subsequently, a point is reached where further aging no longer improves baking potential and longer storage may even be accompanied by a gradual decline in breadmaking quality.

Postharvest maturation is reportedly manifested by increases in milling extraction or loaf volume and overall baking quality. These improvements in technological processing quality appear to reach a maximum around 2–4 mo following harvest depending on wheat type and variety. They are accompanied by increases in gluten strength and elasticity that result in an improved balance of gluten properties and higher gas-retention capacity in baking. At the molecular level, the postharvest aging is reported to be associated with an interchange between thiol (-SH) and disulphide (S-S) groups, a slow oxidation of thiol groups to disulphide, an aggregation of low molecular-mass gliadins (Rao et al. 1978) and an increase in protein molecular mass (Ariyama and Khan 1990).

It is not entirely clear whether postharvest aging involves an improvement in baking potential or whether it is largely an oxidation-related effect that can be rectified in the bakery by judicious use of the correct chemical oxidants and optimization of their level to suit the particular breadmaking process and product. In some ways, the distinction may be immaterial since, by law, many bakers are unable to use certain chemical oxidizing agents required to fully overcome the lack of maturity of freshly milled flours or flours milled from freshly harvested wheat. Furthermore, bakers do not like to have to modify any of their processing conditions or formula ingredients once they find a suitable combination of flour supply and processing conditions that will consistently produce bread of the required specification.

Millers and bakers worldwide become very nervous when the time of year approaches that they have to use new harvest wheat. This nervousness has been exacerbated because potassium bromate is in the process of being banned worldwide (Ranum 1992). Potassium bromate is a slow-acting oxidant that has for many years been regarded by many bakers as an essential ingredient having the unique ability to maximize oven spring and loaf volume when used at an optimum level (e.g., 10–50 ppm, based on flour mass, depending on the baking process).

Interpretation of published milling and baking data is complicated because baking tests may involve a wide range of formulations and processing conditions (Tipples 1979) including long or short fermentation, high-speed or slow-speed mixing, variations in oxidant type and level, and inclusion of "rich" formula ingredients such as fat and milk solids. Furthermore, it is notoriously difficult to obtain repeatable results when the same flour sample is rebaked on different days, particularly after several weeks or months.

Laboratory milling is also difficult to standardize so as to obtain consistent results in terms of milling yield and flour properties, particularly ash, color, starch damage, and water absorption. As with test baking, it is hard to establish the statistical significance of test result differences obtained weeks or months apart. As well, any change in flour extraction caused by variation in milling conditions can affect baking quality, particularly chemical oxidation requirements in baking.

Potential varieties of bread wheat in Canada must pass strict statutory requirements of quality before they can be supported for registration. They are selected on their ability to do well over a wide range of baking conditions and tend not to be overly strong or to have unduly long requirements for dough mixing (Tipples 1986). Because of its good, all-round quality, Canadian wheat does not have a high oxidation requirement (i.e., it will perform reasonably well even in situations where no chemical oxidant is used). The Canadian Grain Commission's Grain Research Laboratory has for many years monitored the milling and baking quality of composite samples representing Canadian wheat produced in each new harvest and subsequently exported in cargo shipments. Results are published in technical bulletins. In most years, little difference is observed in milling or baking performance between samples tested 1–2 mo after harvest and cargo shipments tested up to a year later.

Further indirect evidence for Canadian wheat not showing a large aging effect of postharvest short-term storage was obtained in an unpublished study in my laboratory. The varieties Neepawa and Manitou, which have been representative of Canadian wheat quality for over 20 yr, were harvested at different stages of field maturity, then milled immediately after harvest or after varying periods of subsequent storage. Flours were baked after varying periods of short-term storage by a number of different baking procedures with variable chemical oxidation. I found no evidence of an aging effect on baking performance for this set of samples.

3.2.2. *Long-Term Storage*

Long-term studies on wheat stored under favorable conditions have generally indicated that potential breadmaking quality can remain relatively unimpaired despite severely reduced germination capacity and large increases in free fatty acids. In one study, Pixton (1980) stored samples of Canadian and English wheat at about 12% mc (wb) and under variable conditions of temperature and oxygen availability for 18 yr.

Flour yields obtained on a Buhler laboratory mill with constant setting varied irregularly with the highest yield in the last 2 yr of storage (Pixton 1980). Water absorption remained fairly constant during the storage period. Doughs prepared from both types of wheat flour increased in resistance to extension and decreased in extensibility for up to 16 yr of storage but tended to become softer and weaker during the last 2 yr, possibly because of enzyme activity.

Throughout the storage period, the flour milled from Canadian wheat responded to the addition of potassium bromate in a baking test carried out with variable levels of bromate addition of 0–30 ppm (Pixton 1980). The volume of "pup" loaves made from Canadian flour with 20 ppm bromate was initially 715 cm^3 when the wheat was first stored. After low-temperature storage (4.5°C) for 18 yr, the loaf volume was 816 cm^3 and after storage at "ambient" temperatures (2–20°C), was 760 cm^3. Over the last 2 yr of storage there was a slight decrease in loaf volume, consistent with the observed weakening of dough properties. Fungal α-amylase had to be added to both types of flour to ensure adequate gas production since natural α-amylase activity decreased after the first 8 yr of storage.

4. DETERIORATIVE STORAGE CHANGES

4.1. Invasive Forces

4.1.1. *Storage Fungi*

Microorganisms represent one of the major causes of deterioration in stored grain. The major types of losses caused by fungi growing in stored grains are: decrease in germination capacity; discoloration of part (usually the germ) or all of the kernel; heating and mustiness; various biochemical changes, notably to the lipid components of grain; production of mycotoxins; and dry matter loss. These deteriorative changes affect the grade and price of grain and contribute to buyer dissatisfaction when grain is marketed. In extreme cases, grain will be rendered completely unfit for human or animal consumption.

Spoilage molds exist as spores in soil, on decaying debris, and in harvesting equipment. The spores may be gathered by the combine harvester and distributed among the grains; they may also be present in the storage structure. Growth of spoilage molds on the surface of seeds often results in a dull rather than a bright appearance as in normal seeds. Dull appearance is sometimes considered a degrading factor (Mills 1989). Under the right moisture conditions, spoilage fungi can invade the germs without being visibly evident, weaken the seeds, and eventually cause seed death. Some species such as *Aspergillus restrictus* group species (gr. spp.), *A. candidus* Link, and *A. flavus* Link can cause severe damage and, the latter two species at least, quickly kill the germs. As fungal invasion continues, the germs become brown, then black and friable, making them difficult to separate from the endosperm during flour milling. This increases speckiness and imparts a rancid taste to the wheat flour. This type of discoloration results in lower grades in both Canada and the United States.

One type of damage in wheat associated with storage deterioration is the condition known to the grain trade as "sick wheat." It is manifested by kernels with a dull appearance in which the germs are dead and exhibit various degrees of darkening. Mold growth is usually present in commercial samples of such grain (Pomeranz 1992).

A variety of chemical changes accompany fungal invasion of grain. These include increases in reducing sugars and free fatty acids; protein changes including changes in enzyme activity and the production of fungal constituents or metabolites such as chitin, ergosterol, antibiotics, and mycotoxins. Mycotoxins will be discussed in more detail later.

The lipids of grains and seeds, in particular the triglycerides, are very sensitive to degradation by microorganisms (Richard-Molard 1988). Breakdown begins with the hydrolysis of glycerides by microbial lipases, leading to an increase in free fatty acid content. Numerous molds, including *Penicillium* and *Aspergillus* species, are capable of high lipolytic activity, often affecting the germ which is rich in lipids. Another consequence of lipid breakdown by microorganisms is the formation of various volatile substances which causes the stored grain to have disagreeable, rancid, earthy, and moldy odors (Richard-Molard 1988). Free unsaturated fatty acids are sensitive to oxidation, autoxidation, or enzymatic oxidation, giving rise to unstable hydroperoxides which slowly decompose into short-chained acids, aldehydes, etc., which is characteristic of rancidity.

In reviewing the effects of fungal deterioration on grain, Sauer (1988) noted that the storage fungi each have specific moisture and temperature requirements. The minimum moisture content for growth in starchy cereals is about 13.5%. At this level, xerophilic group species such as *A. restrictus* and then *A. glaucus* gr. spp. can begin to grow and because of their metabolic processes, an increase in moisture and perhaps temperature may occur. These changing abiotic variables stimulate the growth of other species such as members of the *A. candidus* and *A. flavus* gr. spp. and some *Penicillium* spp. *Aspergillus candidus* and *A.*

flavus gr. spp. are especially important in the death and discoloration of the germs and the heating of the grain. Increasing moisture and temperature results in more rapid growth of storage fungi and, as the temperature rises past the optimum of the mesophyllic species, thermophilic group species such as *A. fumigatus* Fresenius increase. The importance of this increased temperature can be illustrated by the fact that deterioration is about 10 times faster at 25°C than 3°C. Fungi can eventually raise the temperature of the grain bulk to 57°C, at which time death of the fungi would result. Further heating may occur but would be due to growth of other thermophilic organisms and finally to purely chemical processes.

Mycotoxin production by storage fungi may be associated with the latter stages of grain deterioration. The most important, potentially-toxigenic storage-fungi in Canada are the *Penicillium* spp. and, to a much lesser extent, the *A. versicolor* gr. spp. The ability of some *Penicillium* spp. to grow, although very slowly, at low temperatures is one of, if not the, primary reason for their supremacy over the toxigenic *Aspergillus* spp. *Aspergillus flavus* gr. spp. are the most notorious of the storage fungi due to the ability of some members to produce aflatoxin. Aflatoxin, however, is more commonly produced in the field than in storage and is primarily a problem in climates warmer than found in Canada.

In any lot of stored grain, certain events can be observed. Mold spore count increases slowly and temperature rises slowly. At about 32°C odors, some germ discoloration, and small increases in fatty acids become evident. At about 43°C strong odors (usually sour), browning of kernels, and sharp increases in fatty acids and moisture increases are measurable. At about 57°C, there is a pause in the temperature rise but kernels are likely to be black, highly rancid, and collapsing.

Fungi are usually strongly dependent on a supply of O_2 (Briggs 1978) for growth, which is slowed by 10% CO_2 in the atmosphere and strongly checked by 20% CO_2. Below 1% O_2, even for very moist grains, no mold activity is observed (Richard-Molard 1988) even though spores may survive for a considerable length of time.

Malting barley quality is particularly sensitive to storage mold activity. Microorganisms may kill grains, reduce the vigor of seedlings, or may be responsible for water sensitivity (a type of water-induced dormancy). When heavily infected barley is malted the progress of growth, rootlet production, and modification may be altered (Briggs 1978). Beer made from such malt may have quality defects such as "gushing" (overfoaming) and poor flavor.

Since 1985, the Canadian Grain Commission's Grain Research Laboratory has routinely monitored export cargoes of Canadian grain for the presence of various fungi, including those normally associated with spoilage. This program is the initial screening step in determining which mycotoxins, if any, will be assayed in which classes and grades of Canadian grain. It is aimed especially

Table 1 Average Percentage of Seed Infection by Storage Fungi in Canadian Export Grain 1986–90

Crop	Number of samples	*Aspergillus* *glaucus*	*Aspergillus* flavus	All *Aspergillus*	*Penicillium*
CWRW[1]	72	3.1	0.1	3.4	1.6
CWRS[2]	774	5.1	< 0.1	5.3	2.0
CWAD[3]	339	7.3	0.1	7.6	2.7
CFW[4]	211	10.7	0.1	11.2	3.9
CEWW[5]	339	10.5	0.4	11.2	3.3
Barley	643	9.9	0.4	10.8	9.6
Corn	94	28.5	1.9	30.6	7.0

Note: CWRS wheats were of grade #3 in 4 crop years, all grades in 1 crop year. CWAD wheats were of grades #3, 4, and 5 in 4 crop years, all grades in 1 crop year.
[1] Canada Western Red Winter wheat
[2] Canada Western Red Spring wheat
[3] Canada Western Amber Durum wheat
[4] Canada Feed wheat
[5] Canada Eastern White Winter wheat

at lower grades of export grain which are more likely to show evidence of degrading factors related to field or storage fungi.

A total of 12 *Aspergillus* gr. spp. have been recovered from surface disinfected grain cultured on both potato dextrose agar and wet filter paper with a 7.5% NaCl solution (method of Mills et al. 1978). *Penicillium* spp. have also been reported, although the exact species has not been determined. *Aspergillis glaucus* gr. spp. and *Penicillium* spp. are the storage fungi most frequently isolated from Canadian grain (R. Clear, personal communication). *Aspergillus glaucus* gr. spp. accounted for 91–97% of the Aspergilli, depending on the type of grain, followed by *A. flavus* gr. spp. at <1–6%. The other *Aspergillus* gr. spp. detected were *A. candidus*, *A. clavatus* Desmazieres, *A. fumigatus*, *A. nidulans* (Eidam) Wint., *A. niger* van Tieghem, *A. ochraceus* Wilhelm, *A. restrictus* Smith, *A. terreus* Thom, *A. versicolor* (Vuill.) Tiraboschi, and *A. wentii* Wehmer.

As indicated in Table 1, the numbers of seeds infected by the storage fungi (*Aspergillus* spp.) were the lowest for western red winter wheat, followed by western red spring wheat, western amber durum wheat, western feed barley, western feed wheat, eastern white winter wheat, and finally, eastern corn. Several other crops have been tested but numbers of samples and years involved do not allow direct comparison with the aforementioned grains. Nevertheless, the levels of infection by storage fungi observed to date in lentils, canola (rapeseed), and soybeans appear to be below those shown by the cereals.

Corn had the greatest number of infected seeds. On average, 28.5% of the seeds were infected by *A. glaucus* gr. spp., 1.9% by *A. flavus* gr. spp., and 7% by *Penicillium* spp. All these spp. could occur on the same seed. Although the number of infected corn seeds appear substantial, it should be noted that the main species involved were those that develop at low osmotic potentials. These early invaders (*A. glaucus* gr. spp., some *Penicillium* spp.) can cause the degrading factor in corn known as blue-eye but usually their presence is not readily detectable by the unaided eye. Sauer et al. (1992) states that samples with 40% or more seed infection by *A. glaucus* gr. spp. have questionable storability. The small numbers of seeds infected by the other *Aspergillus* spp. indicates that at time of export any spoilage that might have been initiated during storage of individual lots making up the corn shipments, had not progressed very far. The low numbers of *A. flavus* also help explain the absence of aflatoxin in Canadian export grain. It is interesting to note, based on the greater abundance of *Penicillium* spp. on corn seeds plated onto PDA than salt filter paper, that many of the *Penicillium* spp. were less xerophilic than those from other seeds. This is the opposite of all the other crops in Table 1 and may be related to the fact that corn is being stored at a higher moisture content than other cereals.

Penicillium spp. were more common on barley than any other crop tested. It is not clear the extent to which this may be partially due to the lemma and palea of the barley seed protecting spores from surface disinfection, as opposed to truly reflecting greater colonization of barley seeds during storage. Perhaps this is part of the reason why ochratoxin A, produced by some *Penicillium* spp., is more often associated with moldy barley feed than other feeds.

4.1.2. Insects and Mites

More than 60 species of insects can occur in stored grain and grain products in Canada alone (Mills 1989) (see Chapter 5). Fortunately, Canada's cold winters are very effective in minimizing insect infestations. Worldwide, literally hundreds of species of insects and mites occur in stored grain. Whereas many are adventitious, some 50 species are considered to be major pests (Briggs 1978). About 5% of the world's grain is destroyed in store by insects and much more than this is partially spoiled.

Insects and mites are a hazard to stored grain in several ways. Some destroy whole kernels, others consume broken kernels and dust; all cause rises in grain temperature and moisture. All contaminate the grain and insect fragments may end up in grain products where, though harmless, they are aesthetically objectionable (Bailey 1982). Live and dead insects, their body fragments, and fecal matter constitute "filth" and may cause adverse psychological or allergic reactions. Not surprisingly, many purchasers will not accept infested grain. Finally, mites and some insects can induce allergies and cause digestive upsets in farm animals (Briggs 1978). Insects also spread microorganisms and create conditions suitable for their multiplication.

Some stored-grain insects consume all parts of the kernel whereas others feed primarily on the germ. Such damage reduces germination and seed vigor. Several species of grain-infesting insects develop inside kernels and comprise "hidden infestation" in a grain mass (Cotton and Wilbur 1982). Weevils deposit their eggs inside the kernels while lesser grain borers and angoumois grain moths deposit eggs outside but their newly hatched larvae promptly tunnel into the kernels; there is little telltale evidence of their presence until they emerge as beetles or moths.

Grain-infesting insects are very sensitive to temperature. They multiply slowly or not at all below 15°C and they cannot survive in temperatures of 42°C or above (Bailey 1982). They appear to thrive best at about 30°C and at that temperature their life cycles may be as short as 30 d. A grain temperature of 21°C is considered favorable for the insects and constitutes the danger line (Cotton and Wilbur 1982). Above this temperature, severe damage to stored grain from insects may be expected. With few exceptions, temperatures above 35°C are unfavorable for the reproduction of most grain-infesting insects; oviposition ceases and the adults are short-lived. Insect metabolic activity within grain bulks at 15% mc or less can result in heating up to 42°C (Mills 1989). Insect-induced hot spots are common in places where grain is often binned at a high ambient temperature (e.g., 30°C). Localized insect metabolic activity can cause an increase in moisture content above 15% in the vicinity of the hot spot, thereby permitting spoilage molds to grow and sometimes producing temperatures up to 57°C.

Mites are difficult to see and more easily detected by the strong, minty odor some species impart to grain which, when grain is heavily infested, becomes unpalatable as animal feed. About eight kinds of mites are common in stored grain in Canada (Mills 1989), all of which can withstand low temperatures. Mites feed on broken grain, weed seeds, and molds present in the grain and they thrive in moist grain. They spread mold spores on and in their bodies and, like insects, encourage development of spoilage molds through their metabolic activity.

The Canadian Grain Commission goes to great lengths to detect and control any insect infestation in the commercial elevator system. Grain delivered in rail cars from primary elevators and producers located across western Canada is tested at terminal elevators for overt or hidden infestation before binning. Primary elevators found to deliver insect-infested grain are instructed to clean up their elevator and could be shut down if the infestation is not eliminated. Terminal elevators are inspected for phytosanitary certification and all binned commodities are routinely sampled for infestation. Grain leaving terminals is thoroughly sampled to ensure that no infested grain is exported from Canada.

Most infested grain originates at farm and primary elevator positions on the prairies and the rusty grain beetle is the most common insect pest encountered. Infestation rates vary according to overall conditions such as length of storage and weather. The average infestation rate for inspected railcars carrying grain from the prairies increased from 3.1% in 1989–90 to 3.5% in 1990–91 (Canadian Grain Commission 1992). The entomology section together with research scientists from the Agriculture Canada Winnipeg Research Station are currently engaged in a nation-wide survey for lesser grain borers and rice weevils to monitor their potential migration into Canada from the United States.

4.1.3. Bacteria

Under normal moisture conditions, bacteria do not develop on grain even though spores are always present and can survive during storage. By contrast, bacterial flora, especially lactic bacteria, play a significant role during the storage of moist grain under anaerobic conditions. This is most notably seen during preparation of silage (Richard-Molard 1988).

The genera and species found on grain include two species whose spores can persist into flour and survive the baking process to cause problems. They are *Bacillus mesentericus* Trevisan and *Serratia marcescens*, which are responsible for the conditions "ropey bread" and "bleeding bread," respectively. The species of bacteria found on grains and seeds vary in frequency and abundance depending on climate conditions, but are essentially present everywhere and in particular, in the soil.

4.1.4. Rodents and Birds

Of all invasive agents, rodents cause the most destruction of grain worldwide (Briggs 1978) (see Chapter 9). It has been estimated that rodents are responsible for the spoilage of about 2% of total world production. They spoil grain by eating or partially eating it (particularly mice, who nibble at four or five times as much grain as they actually eat), by urinating and defecating on grain, by contaminating the grain with hair and other extraneous debris, and by leaving unpleasant odors.

Rats eat about 10% of their mass in food each day and contaminate a great deal more with their droppings and urine thereby rendering it unfit for human consumption (Harris and Bauer 1982). Presence of rodent droppings in wheat is very troublesome to flour millers. Even though great pains are taken to screen, wash, scour, aspirate, and otherwise clean wheat prior to milling, it is very difficult to remove rodent filth completely. Mouse pellets, being approximately the same size as wheat kernels are particularly hard to deal with. Rodent hair can find their way into milled products where they may be detected by sensitive "filth" tests.

Birds are not generally regarded as a serious threat to stored grain, except in some developing countries where grain is stored relatively unprotected (Williams 1991). Birds (e.g., starlings and sparrows) invade granaries through holes in windows and under eaves. Apart from the physical losses incurred when birds feed on grain, the main effect on quality is through contamination with droppings. Unlike rodents, birds consume whole grains rather than nibbling at significantly more than they eat.

4.2. Grain Respiration and Heating

Respiration occurs in all living cells. Aerobic respiration involves oxidative breakdown of plant storage materials into carbon dioxide, water, and energy. The energy liberated during aerobic respiration is used by the cells to fuel metabolic processes and is then released as heat (Mills 1989). Dry, mature seeds generally are quiescent and have a very low respiration in storage. Increased moisture content causes seeds to respire at a much faster rate and also promotes active respiration of molds that are present on the surface and within the seed coats. The heat produced by both seed and mold respiration causes an increase in grain temperature.

Mills (1989) in discussing self-heating of grain, distinguished between biological heating and chemical heating. The first phase of grain heating—biological heating—normally occurs up to 55°C and up to 75°C in exceptional cases involving bacteria. The second phase—chemical heating—occurs from above 75°C to at least 150°C and may, in extreme situations, lead to self-ignition of the grain mass. Biological heating is caused by the activity and respiration of plant cells, molds, bacteria, insects, and mites. Chemical heating is caused by oxidation.

Heating causes severe damage to grain in proportion to the temperature reached. Elevated temperatures result in denaturation of proteins, internal browning or blackening of seeds, and destruction of seed quality and germination capacity. The severity of end-use quality damage caused by heated seeds is recognized in the grade tolerances imposed in Canada. The presence of only 2% heated, internally brown seeds in a sample of canola is enough to lower the grade from No. 1 Canada to No. 3 Canada. Higher levels cause further degrading to "sample account heated." Presence of heated seeds causes problems for processors since oil from heat-damaged oilseeds requires costly extra decolorizing procedures. Kernels of both cereals and oilseeds that have been chemically heat-damaged to an advanced stage are externally blackened and are classified as either bin-burnt or fire-burnt, depending on the severity of the heating. Grade tolerances for such kernels are very low.

It should be noted that not all blackened seeds have been exposed to high temperatures. Indeed, Christensen and Meronuck (1986) indicate that in most cases blackened kernels presumed to be "heat damaged" have probably not been subjected to particularly high temperatures.

4.3. Storage Mycotoxins

4.3.1. Aflatoxins

Mycotoxins are metabolites produced by fungi growing in feed or feedstuffs. They are toxic when ingested by man or other animals. Mycotoxins have been identified only relatively recently even though incidences of mycotoxicoses in man and animals have likely occurred from time immemorial. Certain strains of fungi can produce mycotoxins in the field if the conditions are favorable to their growth and to synthesis of secondary natural products. Although the development of fungi on stored grain may not result in mycotoxin production, the presence of significant levels of mycotoxins is almost always associated with a high degree of fungal invasion (see Chapter 8).

Molds are generally very specific in terms of the mycotoxins that they may produce in natural infections. Many fungi can be made to produce mycotoxins and other metabolites under the right conditions in the laboratory. In the field, however, they are not so diverse in their metabolites. Around the world, a large number of storage mycotoxins have been detected on grains. Those of major importance from the point of view of quality assurance of international grain shipments are aflatoxins and ochratoxins.

The aflatoxin family are compounds primarily derived from some members of the *Aspergillus flavus* gr. spp. and are more of a problem in warmer climates. Those important to quality assurance of grains include the B1, B2, G1, and G2 metabolites. Aflatoxins are regarded as being highly carcinogenic as well as mutagenic, teratogenic, and acutely toxic with B1 being the most potent.

Corn is the most susceptible of the cereal grains to aflatoxin contamination. Although *A. flavus* gr. spp. are generally regarded as storage fungi, most of the aflatoxin present in contaminated corn is field incurred.

The strictest international tolerance limits for aflatoxins in grain are generally in the ranges of 10–30 parts per billion (ppb) and 5–20 ppb for total aflatoxin and aflatoxin B1, respectively. Ongoing monitoring by the Canadian Grain Commission of cargo shipments of Canadian grain, particularly Ontario corn, have never indicated detectable levels (> 1 ppb) of aflatoxins.

4.3.2. Ochratoxins

Although they were first characterized as metabolites of *A. ochraceus*, hence their name, ochratoxins are synthesized by at least seven strains of *Aspergillus* spp. and six strains of *Penicillium* spp. (Frayssinet 1988). Ochratoxins may affect grains in countries with temperate or continental climates. Of the six closely related metabolites, the most important is ochratoxin A. The presence of ochratoxin A in grains is associated with storage at grain moisture levels above 20% and the toxin is apparently produced at low temperatures. International tolerance limits for ochratoxin A in cereal grains are generally in the range of 20–300 ppb.

4.3.3. *Other Mycotoxins*

Zearalenone is a metabolite synthesized by molds of *Fusarium* spp., notably *F. graminearum*, contaminating cereals during storage. It is found principally on corn. It is not a toxin in the strict sense of the term but its estrogenic action can provoke serious physiopathological disorders (Frayssinet 1988). Citrinin is produced by at least 14 species of *Penicillium* and *Aspergillus.* The principal producer in cereals appears to be *P. viridicatum* Westling, the same organism that produces ochratoxins. Citrinin is not carcinogenic but promotes kidney cancer, particularly that caused by nitrosamines, and it has a teratogenic effect in the chicken embryo. Citrinin can contaminate cereals: corn, wheat, oats, and, above all, barley where high concentrations have been noted.

4.4. Toxic Substances and Food Safety

It is now appropriate to consider some of the food safety issues that have emerged in relation to toxic substances. The importance of food safety to grain marketing has been increasing steadily since the mid 1970s, evolving into a major factor in international grain sales. Interest has been fueled by a constantly rising level of consumer awareness and concern over the possible presence of toxic substances (notably pesticide residues and mycotoxins) in food. Governments have responded by introducing tough legislation pertaining to toxicants in foods. As a consequence, grain buyers have been forced to become more discriminating in their purchases of foreign grain and have increased their demands to suppliers for quality assurance of commercial shipments.

There are two implications of these increased demands that relate to grain storage. One is the importance of ensuring that any fumigation carried out to deal with insect infestations does not result in unacceptable levels of chemical residues in commercial shipments. The other is the need to maintain a close awareness of what mycotoxins and toxic chemical contaminants might possibly be present in stored grain.

In Canada, maintenance of the integrity of commercial shipments of grain is greatly assisted by the provisions of the Canada Grain Act which ensures that infested, contaminated, or out-of-condition grain is not allowed into commercial channels. Specifically the Canada Grain Act (1985) prohibits operators of licensed elevators from receiving any grain that they have "reason to believe is infested or contaminated." The Act further states that "no operator of a licensed elevator is required to receive into the elevator any grain that has gone or is likely to go out of condition." The elevator operator must also "exercise reasonable care and diligence to prevent any grain in the elevator from suffering damage or from deteriorating or going out of condition". Aluminum phosphide which produces phosphine gas is the only fumigant allowed in Canada for direct application on grain. Malathion is used only under some circumstances for bin or equipment treatment and rarely as a grain protectant.

4.5. Grain Treatment

4.5.1. Cleaning

Several methods can be used to either reduce the risk of adverse quality changes during grain storage or to slow down or arrest deterioration that has already begun. These include cleaning, drying, fumigation, aeration, and storage under controlled atmosphere. In Canada, almost all grain received at terminal elevators is cleaned to export specifications to remove separable impurities or "dockage." Dockage components include small or large seeds of weeds or other grains, soil pellets, chaff, straw, dust, and other trash. Green seeds and fines in particular, are areas where spoilage initiates because aeration is less effective.

The majority of grain is stored on farms in Canada. Grain stored on farms should be cleaned because cleaning grain before storage could remove much of the source of microbiological contamination and retard insect or mite development. It also makes the grain much more porous and less liable to contain regions of high moisture content which can lead to development of hot spots.

4.5.2. Drying

It has been stressed many times that grain moisture content is the single most important factor determining whether or not deterioration will occur during storage. In Canada, grain is usually artificially dried if moisture content of harvested grain is above so-called "straight-grade" limits. Great care must be taken to ensure that grain quality is not impaired through use of too high a drying temperature (see Chapters 15 and 16). This is particularly important in the case of malting barley or seed grain as germination capacity is easily reduced by improper drying conditions. Too high a temperature in milling wheat causes denaturation of gluten proteins which seriously damages baking quality. Because of the effect of temperature on deterioration in storage, it is important that artificially dried grain be properly cooled before it is stored.

4.5.3. Fumigation

Phosphine gas is used to control insect pests in storages in Canada. Worldwide consumer concerns with food safety have led to the banning or voluntary withdrawal of several other fumigants that were previously effective in treating grain or empty storage facilities.

4.5.4. Aeration

Aeration is the practice of forcing unheated air, by means of a fan, through grain. If ambient temperature is "suitably" low, this is effective in lowering the temperature of the bulk of the grain and maintaining a uniform temperature throughout the grain mass. This, in turn, reduces or eliminates moisture migration, removes hot spots, reduces mold and insect growth, and removes storage odors (Mills 1989).

4.5.5. Controlled Atmosphere Storage

The effects of O_2 on seed longevity have been the subject of many studies (as compared the number of studies on CO_2 and especially N_2) (Renard 1988). Under anaerobic storage conditions for over a year, no deterioration of wheat quality occurs and mold counts are significantly lower than under normal atmospheric storage at higher grain moisture contents. In a well-sealed airtight storage structure, respiration of the organisms present uses up the available O_2 and most pests are killed (Patterson 1989). Some anaerobic activity may continue and is more obvious with grain at higher moisture contents. Such activity can lead to sour off-flavors and odors. When largely inert gas mixtures are used to fill the storage atmosphere, the effectiveness of killing one pest or another is largely related to the extent to which O_2 content is reduced. Molds appear more sensitive to lack of O_2 than to the presence of CO_2. In general, oilseed storage in N_2 is better than storage in air. For mold inhibition, O_2 must be absent for even a fraction of 1% of O_2 allows mold growth.

Most studies have supported the use of either $< 1\%$ O_2 in N_2 or $> 60\%$ CO_2 in air for the control of insects and inhibition of mold growth in cereal grain (Jayas et al. 1991). This is especially true if it is necessary to maintain a high level of germination (as with malting barley) or where high temperature or high moisture content makes grain stored in air prone to mold growth.

4.6. Changes in Grain Components During Storage

4.6.1. Lipids

Deteriorative changes in grain lipids may be either oxidative, resulting in typical rancid flavors and odors, or hydrolytic, resulting in the production of free fatty acids. The development of oxidative rancidity is rarely a problem in grain storage because grains contain fairly active antioxidants and the fat or oil in unbroken kernels of grain is well protected against the effects of oxygen in the air (Pomeranz 1982).

Fats in grain are readily broken down by lipases into free fatty acids and glycerol, particularly when the temperature and moisture content are high. Free fatty acid production is also greatly accelerated by the growth of molds which have high lipolytic activity. Because fat hydrolysis takes place much more rapidly than protein or carbohydrate hydrolysis, the free fatty acid content of grain is a sensitive indicator of in-storage deterioration. Changes in the composition of triglycerides have been noted in stored wheat but losses of triglycerides are proportionally much less than losses of polar lipids (Morrison 1988). Grain deterioration associated with mold activity in high-moisture wheat is accompanied by rapid disappearance of glycolipids and phospholipids (Pomeranz 1992).

4.6.2. *Carbohydrates*

Deterioration of wheat and other grains during storage is accompanied by a decrease in the amount of nonreducing sugars and total sugars present (Lineback and Rasper 1988). Conditions affecting these changes include moisture content, temperature, and growth of molds. Under normal storage conditions changes in carbohydrates appear to be largely due to mold growth. Under good storage conditions the concentrations of disaccharides (sucrose and maltose) and trisaccharides in wheat remain relatively unchanged except for a slight decrease in sucrose content. In wheat stored at high moisture contents and high temperatures sucrose, glucodifructose, and raffinose decrease. Long term storage of sound wheat for 6 yr showed a reduction in total sugars, especially nonreducing sugars (Pixton and Hill 1967).

4.6.3. *Proteins*

Crude protein remains unchanged during long-term storage under favorable conditions but protein content may increase in mold-damaged samples due to a respiratory loss of carbohydrates (Pomeranz 1992). Studies on the fractional composition of the proteins of high quality wheat have indicated little change over 3 to 4 yr. Longer storage decreased hydrophillic characteristics and aggregation of the protein molecules resulting in a decrease in water-solubles. Proteolytic activity and production of free amino acids occur only when storage damage is advanced.

4.6.4. *Nutrition and Feed Quality*

Cereal grains are important sources of certain vitamins in food and feed, notably thiamine, niacin, pyridoxine, inositol, biotin, and vitamin E (Pomeranz 1992). Cereals also contain significant quantities of pantothenic acid. Vitamin A activity occurs in yellow corn but is practically absent in all other cereal grains. Thiamine appears to be quite stable during storage at normal safe moisture levels but losses of up to 30% can occur during storage at high moisture contents. Although there is little supporting evidence, it is generally believed that the other B vitamins found in grain, with the possible exception of pantothenic acid, are rather stable and not readily destroyed in unbroken grain under normal storage conditions. However, considerable losses of vitamin A have been shown to occur in yellow corn during storage even at low temperatures. Vitamin A activity of yellow corn is considered to be of importance in animal feeding and may also be of significance in human nutrition where cornmeal is an important item of the diet.

Tocopherol losses are accelerated in grain stored under adverse conditions with losses correlated with decreases in unsaturated fatty acids of cereal lipids. No major reduction in tocopherol or other antioxidants occur even after considerable aging of wheat.

Although inorganic mineral matter is seldom gained or lost in storage, the availability of phosphorus, nutritionally important to animals and man, appears to increase on storage (Pomeranz 1992). Most of the phosphorus in grain is present in the form of phytin, a calcium-magnesium salt of inositol phosphoric acid. Phosphorus in this form is not well utilized in the animal body and over half of it is excreted unchanged by man. During grain storage phytin is gradually hydrolysed by the enzyme phytase to yield water-soluble, assimilable phosphorus compounds.

Grain having good feeding quality may be characterized as having protein with high digestibility and biological value, being free from toxic substances, having lipids that have not been excessively hydrolyzed or oxidized, and unchanged in water-soluble and fat-soluble vitamins (Pomeranz 1982). Feeding trials with large animals have generally shown that storage under good dry conditions does not change the value of cereal grains as a source of energy or as a source of proteins in a well-balanced diet (Pomeranz 1982).

The effects of feeding deteriorated-stored grain are often difficult to demonstrate and have not been clearly documented. Consumption of cereals infested by insects or mites may cause digestive or similar ailments. Certain nutritional factors are destroyed in moldy cereals and oxidation of unsaturated fatty acids in damaged grain may cause muscular dystrophy in pigs. Mycotoxicoses of animals is associated with the formation of mycotoxins in moldy grain.

When mold-damaged grain is being considered for animal feed the effect on nutritional value is the major concern. Several possible scenarios were discussed by Sauer (1988).

1. Overall improvement in feed value. Palatability is not a problem, vitamins and antibiotics may be produced and constituents in grain may be broken down to a digestible form.
2. Reduced palatability associated with off-flavors and odor but no other problems. Feed consumption and rate of gain may be reduced but feed efficiency may be acceptable.

3. Reduced palatability and reduced digestibility. Both rate of gain and efficiency are reduced but it may be possible to compensate with dietary supplements.
4. Palatability satisfactory, but digestibility is reduced. Feed consumption is at or above normal but feed to weight gain ratio is poor.
5. Some degree of toxicity. Various kinds of toxic effects can be overlaid on any of the above situations.

4.7. Relative Sensitivity of Different Grains and Oilseeds to Storage Deterioration

4.7.1. Cereal Grains

The effects of storage on wheat milling and baking quality and on barley malting quality have been covered in other sections. In the wet milling of corn it is reportedly more difficult to obtain a good separation of starch from other constituents when the corn has been stored under unfavorable conditions, even though no apparent damage can be observed (Pomeranz 1992). Raw rice stored for 11–15 yr at low temperatures was of acceptable quality, even though germinability was basically lost (Pomeranz 1992). However, rice stored at room temperature during the summer was less palatable after cooking than cooked rice from grain stored at 4°C. The amounts of *n*-hexanal in cooked rice increased with increasing storage time at room temperature and the amounts of acetone or propionaldehyde increased with storage at 4°C.

4.7.2. Oilseeds

Because of their high oil content, oilseeds are more susceptible to storage deterioration than are cereal grains. This is particularly the case with oilseeds containing higher proportions of unsaturated fatty acids which are prone to oxidative rancidity. Confectionery sunflower kernels have a relatively short shelf life because of their high oil content and high proportion of polyunsaturated fatty acids.

Moisture levels for safe storage of oilseeds are lower than for cereal grains in direct proportion to the lower content of oil-free matter. Hence, in Canada, maximum moisture content levels for straight-grade seeds are 14.5% for wheat, but only 10.0% for flax and canola.

Deterioration of soybeans at elevated temperatures and relative humidities causes a darkening of the beans and an increase in acidity of the beans and of the extracted oil (Saio et al. 1980). Extractability of protein and other solids from soybean milk decreased rapidly after storage at higher temperatures and relative humidities. Color of soybean milk darkened and hardness of a tofu-like gel from stored beans decreased. There was an increase in total solids, N-compounds, sugars, and mineral components in aqueous extracts.

Canola (rapeseed) goes through a period of active respiration after binning. If the heat and moisture of respiration are not quickly removed, mold growth and continued high respiration of the seeds can occur (Mills 1989). Molding and heating can occur very quickly in moist canola and where this happens the seeds are likely to stick together. In such cases, the value of the stored product for processing is greatly reduced because of a marked increase in the level of free fatty acids.

Flaxseed harvested damp can produce deadly hydrogen cyanide gas, a fast-acting poison which can be absorbed through the skin. Harvested seed must be stored under dry conditions because flaxseed is coated with a mucilaginous substance that becomes sticky when wet. If flaxseed is stored in a tough or damp condition or is exposed to rain or snow, severe caking of the seed can occur, rendering it unfit for sale. Flaxseed respires much more vigorously than do cereals in the 11–17% mc range and when binned in a moist condition it will heat within a few days. Flaxseed is degraded when it contains fire-burnt or heated kernels or has a fire-burnt or heated odor.

5. FUTURE RESEARCH DIRECTIONS

Figure 2 attempts to summarize some of the interrelationships among the major deteriorative changes that can take place in grain during storage and their implications for quality and nutritional value. This figure is of necessity overly simplistic but it does serve to reinforce the earlier statement that storage must be considered from the point of view of a total integrated system. Discussion of effects of storage on grain quality on the basis of simple cause-and-effect may be misleading or invalid.

Future research should endeavour to maintain an holistic approach to the deteriorative changes occurring during storage. There is clearly a need to carry out further research to determine safe storage levels, particulary with respect to oilseeds and Canada's changing export grain market patterns. What may be safe upper ("straight-grade") moisture content limits for temperate climates may be unacceptable where grain is stored for any appreciable period of time after unloading in tropical countries.

Given the trend of consumers around the world to demand food that is free from chemical residues, there is an increasing urgency to develop alternatives to fumigation and other chemical treatments for grain protection. Therefore, it is important that approaches such as controlled atmosphere storage be investigated fully with respect to potential application in the Canadian grain industry.

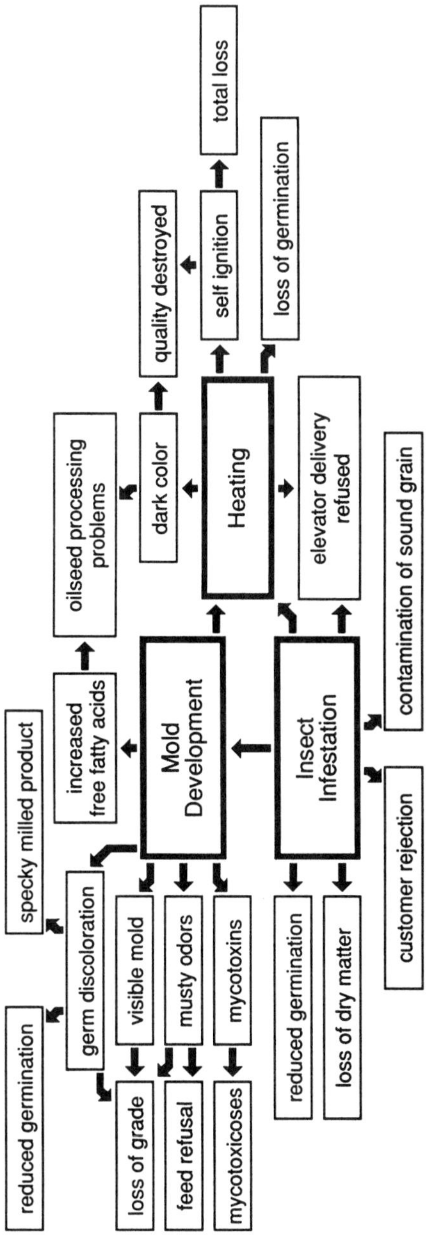

Figure 2 Simplified representation of the major effects of spoilage on grain quality.

ACKNOWLEDGEMENTS

Contribution No. M188 of the Canadian Grain Commission, Grain Research Laboratory, 1404-303 Main Street, Winnipeg, MB, R3C 3G8, Canada.

REFERENCES

Ariyama, T. and K. Khan. 1990. Effect of laboratory sprouting and storage on physicochemical and breadmaking properties of hard red Spring wheat. Cereal Chem. 67:53-58.

Bailey, J. E. 1982. Whole grain storage. pp. 53-78 in: Storage of cereal grains and their products, 3rd Edition C. M. Christensen (ed.). Am. Assoc. Cereal Chem., St. Paul, MN.

Briggs, D. E. 1978. Barley. Chapman and Hall, London.

Canada Grain Act. 1985. Revised Statutes of Canada, Chapter G-10. The Queen's Printer, Ottawa, ON.

Canadian Grain Commission. 1992. Annual Report-1991. Canadian Grain Commission, Winnipeg, MB. 57 pp.

Christensen, C. M. and R. A. Meronuck. 1986. Quality maintenance in stored grains and seeds. University of Minnesota Press, Minneapolis, MN. 138 pp.

Clear, R. 1992. Unpublished data on presence of storage fungi in Canadian export grain. Canadian Grain Commission, Grain Research Laboratory, Winnipeg, MB.

Come, D. 1988. Seeds, organs of survival. pp. 205-225 in: Preservation and storage of grains, seeds and their byproducts. J. L. Multon (ed.). Lavoisier Publishing, New York.

Cotton, R. J. and D. A. Wilbur. 1982. Insects. pp. 281-318 in: Storage of cereal grains and their products, 3rd Edition C. M. Christensen. Am. Assoc. Cereal Chem., St. Paul, MN.

Frayssinet, C. 1988. Classification and chemical characteristics of fungal toxins in grains and seeds. pp. 290-309 in: Preservation and storage of grains, seeds and their byproducts. J. L. Multon (ed.). Lavoisier Publishing Inc., New York.

Harris, K. L. and F. J. Bauer. 1982. Rodents. pp. 363-405 in: Storage of cereal grains and their products, 3rd Edition C. M. Christensen (ed.) Am. Assoc. Cereal Chem., St. Paul, MN.

Jayas, D. S., B. Khangura, and N. D. G. White. 1991. Controlled atmosphere storage of grains. Postharvest News and Information 2:423-427.

Kruger, J. E. and K. H. Tipples. 1983. Some experiences with monitoring alpha-amylase levels in Canadian wheat. pp. 125-131 in: Third International Symposium on Preharvest Sprouting in Cereals J. E. Kruger and D. E. LaBerge (eds.). Westview Press, Boulder, CO.

Lineback, D. R. and V. F. Rasper. 1988. Wheat carbohydrates. pp. 277-392 in: Wheat Chemistry and Technology Y. Pomeranz (ed.). Am. Assoc. Cereal Chem., St. Paul, MN.

Mills, J. T. 1989. Spoilage and heating of stored agricultural products. Publication 1823E. Research Branch, Agriculture Canada, Ottawa, ON. 101 pp.

Mills, J. T., R. N. Sinha, and A. H. Wallace. 1978. Multivariate evaluation of isolation techniques for fungi associated with stored rapeseed. Phytopathology 68:1520-1525.

Morrison, W. R. 1988. Technological aspects of lipids. pp. 417-439 in: Wheat: Chemistry and technology. Y. Pomeranz (ed.) Am. Assoc. Cereal Chem., St. Paul, MN.

Multon, J. L. (ed.). 1988. Preservation and storage of grains, seeds and their byproducts. Lavoisier Publishing, New York.

Patterson, H. B. W. 1989. Handling and storage of oilseeds, oils, fats and meal. Elsevier Science Publishers, Ltd. New York.

Pixton, S. W. 1980. Changes in quality of wheat during 18 years storage. pp. 301-310 in: Controlled atmosphere storage of grains, Shejbal (ed.). Elsevier Scientific Pub. Co., Amsterdam.

Pixton, S. W. and S. T. Hill. 1967. Long term storage of wheat II. J. Sci. Fd. Agric. 18:94-98.

Pomeranz, Y. 1982. Biochemical, functional and nutritive changes during storage. pp. 145-217 in: Storage of cereal grains and their products, 3rd Edition C. M. Christensen (ed.). Am. Assoc. Cereal Chem., St. Paul, MN

Pomeranz, Y. 1992. Biochemical, functional, and nutritive changes during storage. pp. 55-118 in: Storage of cereal grains and their products, 4th edition D. B. Sauer (ed.). Am. Assoc. Cereal Chem., St. Paul, MN.

Ranum, P. M. 1992. Potassium bromate in bread baking. Cereal Foods World 37:253-258.

Rao, V. S., U. K. Vakil, and A. Sreenivasan. 1978. Comparative studies on physico-chemical and baking properties of newly harvested and stored Indian varieties of wheat. J. Sci. Fd. Agric. 29:155-164.

Renard, H. A. 1988. The germinative value of seeds. pp. 1011-1048 in: Preservation and storage of grains, seeds and their byproducts J. L. Multon (ed.). Lavoisier Publishing, New York.

Richard-Molard, D. 1988. General characteristics of the microflora of grains and seeds and the principal resulting spoilages. pp. 226-243 in: Preservation and storage of grains, seeds and their byproducts J. L. Multon (ed.). Lavoisier Publishing, New York.

Saio, K., I. Nikkuni, Y. Ando, M. Otsuru, Y. Terauchi, and M. Kito. 1980. Soybean quality changes during model storage studies. Cereal Chem. 57:77-82.

Sauer, D. B. 1988. Effects of fungal deterioration on grain, nutritional value, toxicity, germination. Int. J. Food Microbiology 7:267-275.

Sauer, D. B. (ed.) 1992. Storage of cereal grains and their products, 4th edition. Am. Assoc. Cereal Chem., St. Paul, MN. 630 pp.

Sauer, D. B., R. A. Meronuck, and C. M. Christensen. 1992. Microflora. pp. 313-340 in: Storage of Cereal Grains and their Products, 4th Edition D. B. Sauer (ed.). Am. Assoc. Cereal Chem., St. Paul, MN.

Tipples, K. H. 1972. Unpublished data on effect of short-term storage on milling and baking quality of freshly harvested Canadian wheat varieties. Canadian Grain Commission, Grain Research Laboratory, Winnipeg, MB.

Tipples, K. H. 1979. The baking test: Fact or fiction. Cereal Foods World 24:15-18.

Tipples, K. H. 1986. Testing candidate wheat cultivars for quality. pp. 500-511 in: Wheat production in Canada—A review A. E. Slinkard and D. B. Fowler (eds.). Saskatoon, SK.

Williams, P. C. 1991. Storage of grains and seeds. pp. 721-746 in: Mycotoxins and animal foods J. E. Smith and R. S. Henderson (eds.) CRC Press, Inc., Boca Raton, FL.

Zeleny, L. 1954. Chemical, physical and nutritive changes during storage. pp. 46-76 in: Storage of Cereal Grains and their Products J. A. Anderson and A. W. Alcock (eds.). Am. Assoc. Cereal Chem., St. Paul, MN.

11

Physical Methods for Insect Control in Stored-Grain Ecosystems

Jonathan Banks

CSIRO Division of Entomology, Canberra, Australia

Paul Fields

Agriculture and Agri-Food Canada, Winnipeg, Manitoba, Canada

SUMMARY

Physical control of insect pests involves the manipulation of physical factors to eliminate pests or reduce their populations to a tolerable level. Temperature, relative humidity, atmospheric composition, impact, desiccation, physical exclusion, removal, and ionizing radiation all may be employed separately or in combination.

Reducing temperatures from 20 to 14°C will slow population growth of most insect pests of stored products. Temperatures below 14°C will result in death, particularly of immature stages. Durations required to kill insects decrease with lower temperatures, with death occurring rapidly at the freezing point (−10 to −20°C) of the insects. High temperatures (40–65°C) are also lethal to stored-product insect pests. Species, stage of development, thermal history (acclimation), relative humidity, and sex can affect the sensitivity of insects to temperature control methods. Models describing insect mortality at extreme temperatures are discussed. Low temperatures are usually obtained by aeration with cool ambient air but refrigeration systems can also cool bulk grain. Methods to obtain high grain temperatures are more diverse, including: microwaves, infrared, hot air, and dielectric heating.

Controlled atmosphere techniques involve changing the CO_2, O_2, and N_2 content of storage atmospheres to render them lethal to insects. Response data for 16 stored-product insects for high CO_2 and low O_2 are given and the effects of temperature, relative humidity, and exposure time are outlined.

Relative humidities below 55% reduce population growth of many stored-product pests but some can survive and reproduce even below 20% relative humidity. Inert dusts damage the insect's cuticle, causing death through desiccation. Diatomaceous earth, the most widely used inert dust, is used for grain protection in several countries either as an admixture or as a structural spray. It has low mammalian toxicity but can affect grain handling characteristics and grade of the grain.

Movement of grain can reduce insect populations significantly. Laboratory studies and the few field studies in the literature are reviewed. Use of impact devices (e.g., Entoleters) is discussed. Excluding pests from a product through packaging and sealing is often an effective method of protection with a range of application from small retail packs to very large grain stores. The effects of ionizing radiation, either using electron beam or γ-irradiation, on 24 stored-product insects and mites are reviewed.

Combinations of methods frequently give better results than a single process. Examples of beneficial combinations of physical methods are summarized. These include use of ionizing irradiation with percussion or controlled atmospheres and controlled atmospheres with heat.

The resurgence in interest in physical control is a result of the increasing restrictions on the use of chemically-based control programs. Before the use of chemical insecticides, physical control methods were the main form of protection of stored-products. These physical control methods are now being re-evaluated and adapted to meet the demands of modern grain storage systems.

---

1. INTRODUCTION

Physical methods of pest control in grain storage are becoming increasingly important. In the near future, it is likely that they will be the predominant, if not exclusive, processes in advanced systems of grain handling and storage. The main driving force behind the change from the use of chemically-based control techniques (i.e., the use of fumigants and grain protectants, including "biorationals" such as insect hormone mimics) is the progressively severe restrictions on the use of these chemicals. At present, two grain fumigants—phosphine and methyl bromide—remain in common use yet Bond (1984) listed 13 such materials only a few years ago. Even methyl bromide use may soon be severely restricted or discontinued because of its contribution to stratospheric ozone depletion (Watson et al. 1992). Protectants, typically organophosphates and pyrethroids, are under pressure from several sources, including increasing aversion to chemical residues in grain in domestic and international trade and development of resistance by the target pests.

Physically-based technologies for control of stored grain pests, particularly insect pests, are based on a diverse set of forces and activities that can be used to manipulate the storage environment or ecosystem to provide conditions inimical to pests (Table 1). Physical control techniques were the main methods for grain protection before the extensive use of synthetic chemicals. Pest control in stored grain in many systems of traditional and subsistence-level agriculture has always been based largely on physical control techniques. Techniques include sun drying (use of heat and aridity; Murdock and Shade 1991), pit storage (use of controlled atmospheres, CA, and exclusion; Hyde et al. 1973, Levinson and Levinson 1989), admixture of ground rock and ash (inert dusts; Ebeling 1971), and winnowing and sieving (physical removal; Banks 1987). Winterbottom (1922) describes grain storage in Australia in the 1914–1918 period when large grain stocks were held for long periods due to interruption of export by war. All the categories of physical control given in Table 1, except irradiation, were either in use or under research at that time. Use of physical control techniques will therefore be a process of rediscovery and adaptation, more than development of new processes.

The return to physical control methods will require their modification to suit modern systems of grain handling and storage, notably to suit large bulks of grain, high rates of handling, with possibly an increased tolerance to low levels of insect infestations. Unlike the fumigants, many of the methods are "soft." Thus, unlike the use of chemically-based pest control where usually high levels of mortality (over 99.9%) and rapid action (a few days or less) is expected, physical control methods tend to be slow and some may not give high levels of mortality even when well managed. Heat disinfestation is one of the few rapid, physically-based processes. Many provide no post-treatment protection of the disinfested product against reinvasion. Adaptation by the grain industry of physical methods will therefore require a much higher level of management and understanding than chemical control methods if they are to be as successful. Often several control measures in combination will be required to achieve an adequate level of control. This may also involve tactical use of chemically-based methods. Overall successful use of physical control measures will involve adoption of Integrated Commodity Management (ICM) (Evans and van S. Graver 1989), with a detailed understanding of the stored-grain ecosystem and how this is influenced by physical changes, such as alteration in temperature or atmospheric composition.

This review covers the main areas of progress in understanding and adoption of physical control measures for stored-grain disinfestation and protection since 1970. This period is remarkable for the general lack of progress in research and achievement in physical control measures, largely because of the outstanding success of chemical control of insects in grain. Where possible, generalizations are developed for the various measures. Such rules will become more important

Table 1 Comparison of General Features of Physical Methods for Disinfestation of Grain

Force or action	Typical embodiment	Does it give 100% kill	Speed of action	In use for bulk grain?
Heat	Heat to 63°C	Yes	A few seconds	Experimental
Cold	Cool to 12°C	Yes, eventually	Weeks or months	Widespread
CA (Low O_2)	Reduce to < 1% O_2	Yes	Weeks or months	Few commercial installations
CA (high CO_2)	Raise to > 60% CO_2	Yes, for most species	Weeks	Some large scale use
Irradiation	Expose to 0.3 kGy	Yes	Weeks or months	Commercial plant prototype
Exclusion	Sealing	No	n/a	Widespread
Inert dusts	Add dust	Yes, for most species	A few days	Limited commercial use
Aridity	Dry to < 11% mc	Yes, for some species	A few days or weeks	Widespread
Impact	Entoleter, pneumatics	Not usually	A few days	Not specifically used
Removal	Sieving	Not usually	A few seconds	Little specific use

as management of stored-grain pests become more complex and computer-based decision systems ("expert systems," see Chapter 2) are created to assist this management. It has been an unfortunate feature of research in physical control that results are generally presented in raw and even anecdotal form without analysis to reveal underlying trends and principles, and systematic or directly comparable studies are scarce. An objective of this review is to indicate where new studies are required.

There have been several general reviews of physical control of stored-grain pests (Banks 1976, Evans 1987a, Fleurat-Lessard 1987, Watters 1991, Bell and Armitage 1992). Specific processes have also been reviewed: effects of temperature (Howe 1965, Fields 1992), controlled atmospheres (Ripp et al. 1984, Annis 1987, Jayas et al. 1993), impact (Banks 1987), inert dusts (Ebeling 1971, Quarles 1992a,b). Most of these reviews have concentrated on presenting the effects of particular methods in isolation without consideration of their interaction and how a particular tactic or action may affect the effectiveness and economics of others. While this may be because of lack of relevant data, the combined effects of various methods are nevertheless an important aspect of the study of physical control of pests.

2. TEMPERATURE MANIPULATION

2.1. Low Temperature

Stored-grain insect pests can survive and multiply only over a narrow range of temperature (Howe 1965, Fields 1992; see Chapter 5). Outside this range insect pest populations either remain static or die out at a faster or slower rate (Table 2). Unlike mammals and birds, insects are ectotherms, their body temperature is in large part determined by the temperature of the environment.

Lowering the temperature of commodities to prevent spoilage is an ancient technique. Part of the success of underground stores used in ancient Egypt (Levinson and Levinson 1989) compared with above-ground stores was due to cooler temperatures of the underground stores. Bulk storage in temperate climates extensively use low temperature control through ambient air aeration. In tropical climates or in the summer in temperate areas, low temperature aeration supplied by chillers or refrigerated aeration units makes low temperature control possible (Navarro et al. 1973, Maier 1993). There are two basic effects of low temperature: 1) reducing the development rate, feeding, and fecundity, and 2) decreasing survival.

Howe (1965) reviewed temperatures that slow the development of insects. It should be noted that temperatures listed in his review are not developmental thresholds, as often quoted, but temperatures that allow only two generations

Table 2 Response of Stored-Product Insect Pests to Temperature.

Zone	Temperature range (°C)	Effects
Lethal	> 62	Death in less than 1 min
	50–62	Death in less than 1 h
	45–50	Death in less than 1 d
	35–42	Populations die out, mobile insects seek cooler environment
Suboptimum	35	Maximum temperature for reproduction
	32–35	Slow population increase
Optimum	25–32	Maximum rate of population increase
Suboptimum	13–25	Slow population increase
Lethal	5–13	Slowly lethal
	3–5	Movement ceases
	−10 – −5	Death in weeks, or months if acclimated
	−25 – −15	Death in less than 1 h

Source: adapted from Fields (1992)

per year. For most major stored-product insects, temperatures of about 14°C reduce the rate of multiplication to below replacement rate, causing the eventual elimination of the population.

Temperatures at which insect and mite populations have a zero finite rate of increase, r_m, are strongly dependent on the humidity of their environment. For example, mites in damp grain only stop developing below 2°C (Howe 1965). Desmarchelier (1988) found that the combined influence of humidity and temperature can be conveniently described using wet-bulb temperature as a parameter (Fig. 1). The temperature at which population increase is zero is important to pest control as it defines the target temperature at which a grain mass must be maintained to prevent an infestation from increasing. With a grain bulk infested below detectable levels, if it is held at or below the target temperature (and there is no immigration) then it is effectively protected indefinitely. Specifying a target wet-bulb temperature is also convenient since optimum aeration strategy and control is dependent on wet-bulb temperature of the incoming air. Although no pest development occurs at these temperatures, the insects and mites survive for long periods—more than a year in some cases (Wohlgemuth 1989)—and there will be a resurgence of infestation if the temperature or moisture content of the commodity rises.

The relationship between duration of exposure and temperature for a given mortality level is usually an approximately hyperbolic curve (Evans 1987b; Figure 2). Several models have been developed to describe cold related mortality

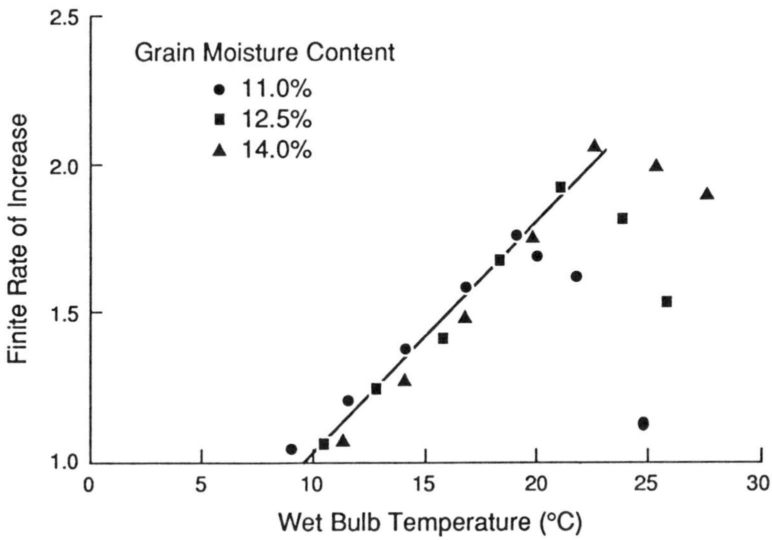

Figure 1 Rate of increase of *S. oryzae* as a function of wet bulb temperature with trend line fitted to data for below optimum development temperatures. *Source*: Desmarchelier (1988).

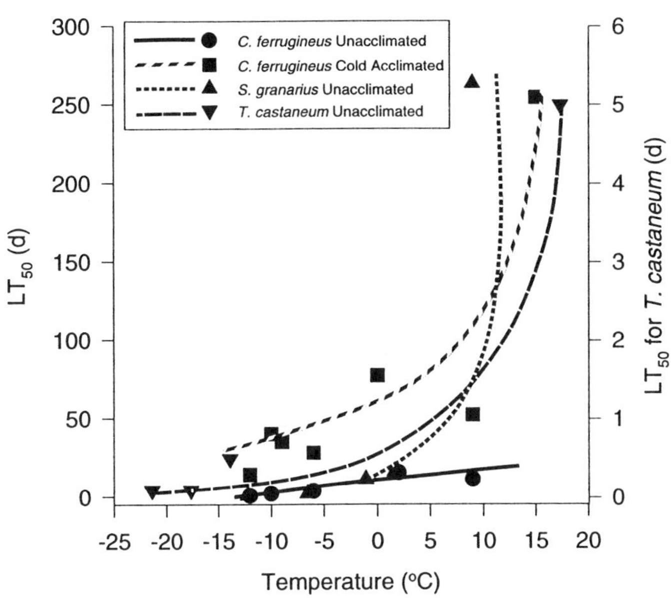

Figure 2 Lethal time for 50% of the population (LT_{50}) for some stored-product insects at temperatures below their developmental thresholds. *Source*: Fields (1992).

(Kawamoto et al. 1989, Turnock et al. 1983, Brokerhof et al. 1992), though temperature responses of not all species are described by these models (Stratil and Reichmuth 1984).

Stored-grain pests have a wide range of tolerances to cold. In general, *Tribolium castaneum* (Herbst), *T. confusum* (J. du Val), and *Oryzaephilus mercator* (Fauvel), are the most cold susceptible, whereas *Trogoderma granarium* (Everts), *Sitophilus granarius* (L.), *Ephestia elutella* (Hübner), *E. kuehniella*, and *Plodia interpunctella* (Hübner) are the most cold tolerant species. Durations of exposure required to control species at a given temperature have been summarized by Fields (1992).

The stage of development (egg, larva, pupa, or adult), will affect ability to withstand low temperatures. For many species the egg is most susceptible (Fields 1992) and, for a few species, there have been detailed work showing that age of the egg can affect susceptibility (Watters 1966, Cline 1970, Daumal et al. 1974, Johnson and Wofford 1991). Larvae are the most cold-tolerant stages for *Rhyzopertha dominica* (F.) and *S. oryzae* (L.), (David et al. 1977). Diapausing larvae are the most cold-hardy stage of *E. kuehniella* (Cox 1987) and this may be true for other pyralid moth pests; *P. interpunctella*, *E. cautella* (Walker), and *E. elutella* (Fields 1992). The adult is the most cold-hardy stage for *Cryptolestes ferrugineus* (Stephens) (Smith 1970) and *T. confusum* (Nagel and Sheppard 1934). The larvae of some species are as cold hardy as the adults (Howe and Hole 1968, David et al. 1977).

In most field applications of cold temperature to control stored-product pests, insects are exposed to between 10 and 20°C for some days before being exposed to lethal cold temperatures. Exceptions to this rule are commodities that are rapidly cooled in freezers to control insect populations. When insects are exposed to cool temperatures (20–10°C) they become acclimated and their cold-hardiness increases by two- to ten-fold (Smith 1970; Evans 1983; Fields 1990, 1992). There are only two studies, both with *S. granarius*, that demonstrate that some species do not increase in cold-hardiness if exposed to cool temperatures before being exposed to cold temperatures (Robinson 1926, Evans 1983). The implications for low-temperature control are that laboratory studies often underestimate cold-hardiness of insects because of inadequate acclimation before cold treatment. Rapid cooling of commodities from 25°C also should prevent cold-acclimation and may be a useful tool to more rapidly control insects with low temperatures.

2.2. High Temperature

Exposure to temperatures only about 5°C above the optimum for population increase will stop development of stored-product insects (Tables 2, 3). The temperature tolerance of stored-product insects, have been grouped into two

Table 3 Exposures Producing Over 90% Mortality to Stored-Product Pests at 42–50°C

Species	Stage†	Temperature (°C)	Exposure time	Mortality (%)
Ephestia cautella	P	50	2 h	100
Ephestia elutella	L	43	24 h	100
		45	16 h	100
Oryzaephilus surinamensis	A	44	4.4 s	90
	A	49	20 s	97
Plodia interpunctella	P	45	2 h	100
Rhyzopertha dominica	A	43	4 d	99.7
Sitophilus oryzae	A	47	4 s	90
Tribolium confusum	L, P, A	44	29 h	> 95
		46	2.7 h	> 95
Tribolium castaneum	E, L, P, A	50	155 s	100
Tribolium confusum	E, L, P, A	48	155 s	100
Sitophilus granarius	A	48	155 s	100
Oryzaephilus surinamensis	E, L, P, A	50	155 s	100

†L = larvae, P = pupae, A = adult, E = egg.
Source: from Fields (1992) and Lapp et al. (1986)

categories: those needing high temperatures for optimal rate of increase, typically about 30°C, and those thriving at more moderate temperatures, typically around 26°C. The first category includes most common beetle pests and the moths *P. interpunctella* and *E. cautella*, while the latter include common moth and mite pests (Howe 1965).

The lethal effect of temperatures from 42–50°C has been subject to little study (Gonen 1977a,b). There is little survival (over 95% mortality) of most species when exposed for some hours to this range of temperatures (Table 3). On balance, the available information suggests that exposure of 2 h at 50°C should eliminate all pests but studies on heat-tolerant species and stages (e.g., *R. dominica* pupae) are lacking. The data of Lapp et al. (1986) appears to give anomalously brief exposure times to achieve complete kill.

The temperature of an insect-produced "hot spot" in dry grain is about 42°C. At this temperature the adult insect pests are driven from the hot region to the cooler periphery and immobile stages within the region are killed. The process of hot spot development is well known and has not been subject to recent study. Wheat is intentionally held at high temperatures (over 40°C) for some weeks immediately after harvest in parts of China to eliminate infestation and to mature the grain (Ren Yong Lin, personal communication, Liu et al. 1983).

Table 4 Temperatures Requiring Less Than 5 min to Disinfest (Over LD_{99}) Grain

Species	Stage†	Peak temperature (°C)	Exposure (sec)	Heating system	Reference
Lasioderma serricorne	A	66	40	Infrared	Kirkpatrick and Tilton (1972)
Oryzaephilus surinamensis	A	66	40	Infrared	Kirkpatrick and Tilton (1972)
	E, L, P	80	120	Microwave	Locatelli and Traversa (1989)
Rhyzopertha dominica	A	66	40	Infrared	Kirkpatrick and Tilton (1972)
	E, L, P, A	67	290	Fluidized bed	Vardell and Tilton (1981)
	E, L, P	64	16	Fluidized bed	Fleurat-Lessard (1985)
	E, L, P	68	25	Fluidized bed	Evans (1987a)
	E, L, P	82	120	Microwave	Locatelli and Traversa (1989)
Sitophilus oryzae	A	72	12	Microwave	Baker et al. (1956)
	E, L, P	83	120	Microwave	Locatelli and Traversa (1989)
	A	66	40	Infrared	Kirkpatrick and Tilton (1972)
	E, L, P, A	64	16	Fluidized bed	Fleurat-Lessard (1985)
Tribolium castaneum	A	66	40	Infrared	Kirkpatrick and Tilton (1972)
Trogoderma granarium	L	70	16	Fluidized bed	Fleurat-Lessard (1985)

†E = eggs; L = larvae; P = pupae; A = adults.

The studies of Nelson and Kantack (1966) and Kirkpatrick and Tilton (1972) appear anomalous. They give over 90% kill for only a few seconds exposure at about 45°C, using high frequency and infrared heating, respectively, as heat sources. The same studies show that under the same conditions, complete disinfestation requires much higher temperatures (over 62°C). Furthermore, other studies (Gonen 1977a,b; Table 3) indicate that an hour or more are required to give over 90% control. This suggests that true exposure temperatures, measured by Nelson and Kantack (1966) and Kirkpatrick and Tilton (1972), may have been higher than those recorded.

Present research on effects of high temperatures has concentrated on obtaining rapid disinfestation without damage to the treated grain (Table 4). Evans (1987a) concluded that a grain temperature of 65°C must be attained briefly to control *R. dominica*, the most common heat-tolerant pest. Studies with a variety of means of heating grain—hot air fluidized bed, infrared, high frequency (dielectric), and microwave heating—all appear to require similar grain temperatures to achieve the same lethal effect. It is predicted that insect tissue may be heated selectively and more rapidly than grain with dielectric heating but that microwave heating will not exhibit this differential (Nelson 1986). Differences in rate of heating and differential heating have been observed in practice with dielectric heating (Nelson and Stetson 1974, Kato and Yamashita 1987) but this has yet to be shown to be economically useful. Bedi and Singh (1992) claim some difference in mortality when using microwave frequencies of 12, 15, and 18 GHz. Comparisons of the heat susceptibility of stored-product pests at temperatures above 50°C are confounded by the difficulties in accurate measurement of the temperature to which the target insect was actually exposed and the lack of a system that allows different temperature-exposure time combinations to be evaluated (Fields 1992).

The quality of treated material can be reduced if temperatures are too high or if durations are too long (Fleurat-Lessard 1985, Evans 1987a, More et al. 1992). When any heat disinfestation procedure is examined, the effects on end use quality must also be tested. In general, the conditions needed to control insects do not affect quality but the margin of safety before damage occurs is narrow, necessitating close control of the heating process and subsequent cooling.

2.3. Commercialization

Several different approaches have been used to adapt heat disinfestation for use with modern systems of grain storage (Tables 4, 5); they all heat the grain by some process during conveyance. The target commodity-temperature to achieve disinfestation appears to be similar for all approaches, except possibly dielectric heating. Thus, they can all be viewed as systems for heating grain. The choice of which process is appropriate will depend on local costs of energy, construc-

Table 5 Heat Disinfestation of Grain: Processes and Stage of Development

Heat transfer process	Type of system	Stage of development	References
Conduction	Conduction with mixing	Small scale pilot demonstration	Lapp et al. (1986)
Hot air	Fluidized bed	Laboratory scale investigation	Fleurat-Lessard (1985) Evans and Dermott (1981)
		Full scale commercial prototype	Evans et al. (1983) Thorpe et al. (1984)
	Pneumatic conveying	Pilot scale demonstration	Fleurat-Lessard (1985) Sutherland et al. (1989)
	Spouted bed	Laboratory scale	Claflin et al. (1986)
Radiation	Infrared	Laboratory scale investigation	Kirkpatrick and Cagle (1978)
	Dielectric (RF) heating	Laboratory scale investigation	Zhang Cong et al. (1981)
		Plans for pilot scale unit	Hafner 1975
	Microwave	Laboratory scale investigation	Zhang Cong et al. (1981)
		Full scale demonstration	Fleurat-Lessard (1985)

tion, and operating costs. Fluidized bed heating is the only heat disinfestation process developed to the full scale prototype stage for treatment of over 150 t/h of grain (Fig. 3). The others have all been demonstrated successfully at the laboratory or pilot plant stage of development (Sutherland et al. 1986).

Hot air systems, such as the fluid bed approach, have the advantage that they can easily be adapted for cooling the treated grain after exposure. Cooling is necessary as grain held for more than brief periods at temperatures that are rapidly insecticidal (over 60°C) loses important qualities (e.g., germination, baking quality). Approaches using radiative heating will require grain cooling after treatment, adding to the cost and complexity of the system. A fluidized bed cooling system is well suited to the rate of cooling and throughput required for large scale use, and it has been suggested for use after dielectric heating (Fleurat-Lessard 1987).

All heat disinfestation technologies of the types so far proposed are both capital and energy intensive. Direct comparative cost and power usage figures for full-scale plants are not available. However, running costs for fluidized-bed

Figure 3 Prototype of a fluidized bed for heat disinfestation at a rate of 150 t/h, Dunolly, Australia. *Source*: Thorpe (1983).

heat-disinfestation of $0.75/t were achieved under Australian conditions (Sutherland et al. 1986) but this did not include the cost of the equipment. This is similar to the cost of using protectants (fenitrothion/bioresmethrin combination) under Australian conditions.

Of the physical control methods given in Table 1, heat-based systems are the only ones that can provide complete in-line disinfestation at the high grain transfer rates (several hundred tonnes/hour) found typically in large modern storage facilities.

2.4. Modeling

Two approaches to the modeling of lethal high temperature effects on stored grain insects have been proposed. In one, Dermott and Evans (1978, Equation 1) describes the combination of exposure time and temperature producing a particular level of mortality in a way similar to the concentration-time product concept used for assessing effects of fumigations. The other (Brokerhof et al. 1992, Equation 2), is based on the Arrhenius equation, a model describing the variation in rate of some simple chemical reactions with temperature. Thus,

$$t_{x\%}\,(T - T_o) = k \tag{1}$$

and

$$\frac{1}{t_{x\%}} = Ae^{-b/T} \tag{2}$$

where $t_{x\%}$ is the time to give $x\%$ mortality, T_o is the threshold lethal temperature and A, b and k are constants.

At present there is insufficient data available to distinguish which of the two models is better; however, Equation (1) is likely to be more useful in practice. It provides a convenient measure (k) of the combined effect of temperature (T) and exposure time, above a threshold temperature (T_o), and may be easily used to integrate effects under changing temperatures as occur in the heating and cooling phases of heat treatments of grain for either disinfestation or drying.

Table 6 gives data for heat treatment of three species of stored-product pests in a fluidized bed system, fitted to Equation 1. The fitted threshold temperature is close to the maximum temperature at which insects do not die directly from acute effects of heat. Information from studies on insect reproduction at high temperatures (Fields 1992, Table 3) and the temperature of insect-induced hot spots in bulk grain indicate this temperature to be about 42°C. Brokerhof et al. (1992) gives data on heat treatments fitted to Equation 2.

Most of these studies do not measure the temperature to which pests are actually exposed but measure the temperature of the air used to heat the grain. Direct correlation of these temperatures with true exposure temperature is not usually possible. Evans (1987a) and Fleurat-Lessard (1985) use:

Table 6 Values of Threshold Lethal Temperature, T_O, and Temperature-Time Product k (Eq. 1) for 50 and 99% Mortality

	LT_{50}		LT_{99}	
Species	T_o (°C)	k (min.°C)	T_o (°C)	k (min.°C)
Sitophilus oryzae	45.3	75.4	44.8	101.8
Rhyzopertha dominica	49.2	89.4	47.6	113.5
Sitotroga cerealella	48.5	74.7	46.3	108.8

Source: recalculated from Dermott and Evans (1978).

$$\frac{1}{t_{x\%}} = T + C \tag{3}$$

and

$$T_{max} - T = ab^{t_{x\%}} \tag{4}$$

respectively to relate temperature of input air in fluidized bed systems to the mortality attained, where a, b, and C are constants, T is the grain temperature (°C), and T_{max} is the air inlet temperature. Equation 4 apparently supersedes the model given earlier by Fleurat-Lessard (1980).

Over the range of values used for inlet air, both approaches give good models. There is an apparent discontinuity in the slope of the line at about 130°C inlet temperature in Evans' (1987a) data when fitted to Equation 1; this is attributed to a change in mechanism of action. The change is not apparent in the data of Fleurat-Lessard (1987).

3. CONTROLLED ATMOSPHERES

Controlled atmosphere (CA) disinfestation technology involves the alteration of the natural storage gases—CO_2, O_2, and N_2— to render the atmosphere in the stores inimical to pests. Several combinations can be used (Table 7). As normally understood, CA does not include either alteration of the storage atmosphere by addition of toxic gases such as phosphine or methyl bromide, or regulation or alteration of the atmospheric water content. The controlled atmosphere may be achieved in several ways (Table 8): adding gaseous or solid CO_2, adding a gas of low O_2 content (e.g., pure N_2 or output from a hydrocarbon burner), or allowing metabolic processes within a storage to remove O_2, usually with associated release of CO_2. Such atmospheres are referred to as "high-CO_2," "low-O_2," and "hermetic storage" atmospheres, respectively. They are sometimes known collectively as "modified atmospheres" or, inaccurately, as "inert atmospheres" and these terms may also be applied to specific atmospheres or relate to how they are created or controlled. The relative advantages and disadvantages of using CO_2-based or N_2-based atmospheres have been discussed (Banks and Annis 1990). In general, the optimum process will depend on many factors, including the storage structures to be treated, logistic conditions, and the relative costs of gases (e.g., Spillman 1989).

Controlled atmosphere technology for control of grain pests has been extensively studied in the laboratory and the field.. Reviews and books on CA technology for grain are listed in Table 7. Despite the large quantity of research on CA disinfestation for grains, there are important areas of uncertainty remain-

Table 7 Reviews of Aspects of Controlled Atmosphere (CA) Technology for Stored Grain Protection

Subject	References
Overview of current state of CA technology	Jayas et al. (1993)
Differences between CO_2- and N_2-based CA technologies	Banks and Annis (1990)
Effects of CA on insect pests of grain, particularly at > 20°C	Annis (1987)
Effects of CA on insect pests of grain, particularly from 10–20°C	Bell and Armitage (1992)
Australian field trials using CA	Banks et al. (1980)
General CA technology for grain, particularly in temperate climates	Shejbal (1980)
General CA technology for grain, particularly for dry, warm grain	Ripp et al. (1984)
General CA technology for grain, with emphasis on use in humid tropics	Champ et al. (1990)
Effects of CA on grain mycoflora	Richard-Molard et al. (1987), Hocking (1990)
Effects on CA on insect pests of grain, including sublethal effects	Bailey and Banks (1980)
Hermetic storage systems	Hyde et al. (1973)
Effects of CA on end-use properties of grain	Gras and Bason (1990), Banks (1981)
General CA technology, including grain	Calderon and Barkai-Golan (1990)

Table 8 Composition of Typical Controlled Atmospheres (CA) Easily Created in Sealed Grain Storages

		Composition (%)			
Type of CA	Atmosphere source	O_2	CO_2	N_2	Ar
Low O_2	Nitrogen from liquid or other source (< 0.1% O_2)	0.5		99.4	
Low O_2	Propane burner	0.5	13.4	85.1	1.0
Low O_2	Producer gas burner	0.5	20.8	78.2	0.5
Low O_2	CO_2 from liquid or other source (< 0.1% O_2)	0.5	97.5	2.0	
Hermetic storage	Metabolism within the storage	2.0	18.0	81.0	1.0
High-CO_2	CO_2 from liquid or other source (> 98% pure)	4.2	80.0	15.6	0.2
High-CO_2	CO_2 from liquid or other source (> 98% pure)	8.4	60.0	31.2	0.4
High-CO_2	CO_2 from liquid or other source (> 98% pure)	12.6	40.0	46.9	0.5

Source: modified from Banks and Annis (1990).

ing in the effects on pests, notably the most appropriate gas composition and exposure period required for given situations and the best means of creating and maintaining the desired atmospheres. Compositions for typical atmospheres effective against insect pests of grain are given in Table 8. These atmospheres are also effective against mites and may restrict growth of some molds (Richard-Molard et al. 1987, Hocking 1990).

3.1. Acute Mortality

Both the species and the developmental stage have a large influence on the response of insects to CAs (Table 9). The general impression given by the data in Table 9 is that (except for *R. dominica*, *T. granarium*, and *Sitophilus* spp.) these pests are controlled (i.e., over 95% mortality) in less than 14 d by typical CAs of either high-CO_2 or low-O_2 at between 14 and 29°C. In the few instances where there is comparable information for both temperatures, more rapid mortality is observed at the higher temperatures.

Usually a 14 d exposure to either atmosphere gives a technically satisfactory level of control. A 95% level of mortality, however, provides only a brief

Table 9 Exposure Times (d) for > 95% Mortality of Insect Pests to High-CO_2 or Low-O_2 CA

	50–70% CO_2		<1% O_2	
	14–17°C	20–29°C	14–17°C	20–29°C
Cryptolestes ferrugineus	7 A	—	—	< 2 A
Ephestia cautella	7 E,L	< 5 E,L	6 E,L	1.5 E,L
Ephestia kuehniella	7	—	6 L	< 1.5 E
Cryptolestes pusillus	—	—	—	2.5 A
Oryzaephilus surinamensis	5 E,L,A	< 3 A	10 A	< 1.0 A
Oryzaephilus mercator	—	—	—	3
Plodia interpunctella	7 L	< 7	14 L	< 7
Rhyzopertha dominica	28	1.5 A	28	> 4 E,A
Sitophilus granarius	42–56	6.5 A	12 A	16 A
Sitophilus oryzae	21	6	28	> 14
Tribolium castaneum	5–6 L,A	< 7	> 14	7
Tribolium confusum	—	3.5 A	7	> 7
Trogoderma glabrum	7 L	< 3 L	14 L	< 3.5 L
Trogoderma granarium	—	—	12 L	—

E = eggs, L = larvae, A = adults. Exposure times for > 95% mortality of all stages except where stages are indicated.
Source: condensed and updated from Annis (1987) and Bell and Armitage (1992).

time—less than one generation time for many species—before infestation returns to the level before treatment. It can be argued that such partially effective treatments are undesirable because of the probability of selection for resistance. Development of resistance has already been demonstrated under particular laboratory conditions and selection (Bond and Buckland 1979, Donahaye 1990).

Annis (1987) comprehensively reviewed the available data on lethal response of stored-product insect and mite pests to CAs at above 20°C. He attempted to simplify the large body of published data for the temperature range 20–29°C, by producing equimortality plots for the best worked examples and by showing the variation in insect response to exposure times and concentrations of either O_2 or CO_2 (e.g., Figure 4). Other species and stages generally produced qualitatively similar plots. The review of Annis (1987) concluded that the following CA dosage regimes would be effective against all stages of common stored-product pests:

- O_2 levels maintained at less than 1.0% for longer than 20 d; or

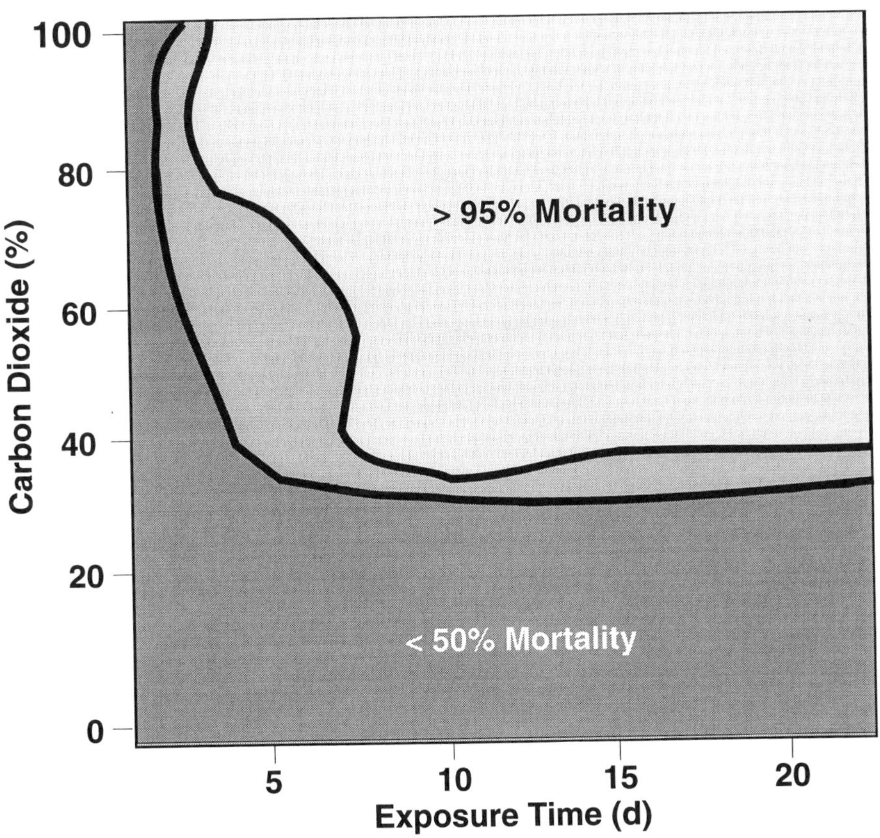

Figure 4 Response of *S. granarius* to modified atmospheres containing various properties of CO_2 in air as a function of exposure time. *Source*: Annis (1987).

- CO_2 levels held at 80% for 8.5 d, 60% for 11 d, or 40% for 17 d; or
- CO_2 levels, initially above 70% declining over at least 15 d to not less than 35%.

Only the low-O_2 dosage effectively controls larvae of *Trogoderma* spp. in diapause. Specific pests may require shorter exposure periods. Annis' (1987) recommendations were targeted principally at the most tolerant species, *S. oryzae* and *R. dominica*, in their tolerant pupal and egg stages.

The O_2 or CO_2 levels that must be maintained indefinitely to prevent population increase of insect pests, as opposed to eliminating the pests, have not been well defined nor subject to much research. Bailey (1955, 1965) concluded that the "lethal level" to which O_2 in hermetic storage systems must fall is about 2%. He also noted that young immature stages of *S. oryzae* died at around 4% O_2, suggesting that populations would die out if this level was maintained. Reichmuth (1987) found that there was a pronounced increase in tolerance to the species and stages he tested above 3% O_2, suggesting that this level is close to the limit for effective control. Bailey and Banks (1980), however, found significant survival (over 5%) in *S. granarius* cultures exposed for 12 wk at 18°C to 1.3% O_2. Complete mortality occurred at 24°C under the same conditions in only 3 wk.

Spratt (1979) showed a decrease in reproduction rate of *Sitophilus zeamais* Motsch. under atmospheres containing 10% CO_2, 10% O_2, and 80% N_2 compared with that in air. The intrinsic rate of increase fell from 0.32 to 0.12 per 6 d. It can be expected that the rate of increase of species less susceptible to CA than *S. zeamais* with regard to acute mortality (e.g., *S. oryzae*, *R. dominica*) may also be less affected.

There is still debate as to whether CO_2- or N_2-based CAs are better. The decision is likely to be made on technical and economic grounds rather than on whether one or the other is faster acting in particular circumstances. In general, atmospheres containing about 60% CO_2 with about 8% O_2 act more rapidly on species that develop within the grain (e.g., *R. dominica* and *Sitophilus* spp.) than on external feeders which are better controlled with low-O_2 atmospheres (Banks and Annis 1990). *Trogoderma* spp. are particularly tolerant of high-CO_2 atmospheres but are susceptible to low-O_2 ones (Jay 1984a, Spratt et al. 1985).

The presence of CO_2 increases the effectiveness of low-O_2 atmospheres for some species and developmental stages but appears to have little influence on others. Low-O_2 atmospheres containing 10–20% CO_2, typically produced by burners, are more effective than mixtures of N_2 and O_2 against adults of several species (Calderon and Navarro 1979, Krishnamurthy et al. 1986, Reichmuth 1987) at the same O_2 content but there is little difference with *S. granarius* (Krishnamurthy et al. 1986, Figure 4). Bailey and Banks (1980) concluded that there is some practical advantage in using a proportion (e.g., 10%) of CO_2 in low-O_2 atmospheres but the reduction in exposure time allowable will not be substantial if *Sitophilus* spp. are present. There is a need to investigate the action of CO_2 in low-O_2 atmospheres in species and stages (e.g., *S. oryzae* pupae) that are particularly tolerant of this type of CA and to determine if the results observed in easily studied adults are also exhibited by the stages that would often be targeted in practice.

Reduced pressure appears to speed the action of high-CO_2 atmospheres. For example, 100% kill of the normally tolerant pupae and eggs of *S. oryzae* was obtained within 36 h at 20°C in an atmosphere of 98% CO_2, containing less than 0.5% O_2, at 35 kPa absolute. All stages of *R. dominica* were killed under the same conditions after 48 h (Locatelli and Daolio 1993). Low pressure appears to have a similar effect to low-O_2 systems with its effect directly related to the residual concentration of O_2 and no added specific effect of reduced pressure. Raised O_2 concentrations are also lethal to stored-product insects (Calderon et al. 1991) but the fire hazard associated with high O_2 makes this control method impractical for bulk storages.

Controlled atmosphere treatment times can also be substantially reduced where it is possible to apply high pressures. High pressure CO_2 (over 1.5 MPa) reduces the exposure time for *Sitophilus* spp. to a few hours from 5–18 d for pure CO_2 at normal atmospheric pressure (Le Torc'h and Fleurat-Lessard 1991). Similar results were found by Prozell and Reichmuth (1990). High pressure CO_2 is also effective against mites (Mitsura et al. 1973). Due to the difficulties of sealing and maintaining large storage structures at pressures different than atmospheric pressure, these control methods are not useful for large-scale stored-grain protection at present.

3.2. Temperature Effects

At either high-CO_2 or low-O_2 concentrations, increasing temperature augments the rate of mortality of stored-product insects (Table 9). Banks and Annis (1990), using data from several sources, showed that for *S. oryzae* adults exposed to CAs, the logarithm of time to 95% mortality was approximately proportional to the exposure temperature. There is inadequate data to provide such an analysis for other species or for the tolerant developmental stages, such as eggs and pupae, whose control is critical to the effectiveness of a CA treatment.

It has been suggested that the high-CO_2 CA is less sensitive to temperature than low-O_2 CA. This may be based on the anomalous observations of Harein and Press (1968) who found little change in *T. castaneum* larvae mortality at 16, 27, or 39°C to 48% CO_2 in air. Zakladnoi (1976) also found little temperature dependence in the response of adults of several species to 100% CO_2 at 20, 25, and 35°C, but a marked positive temperature dependence in tests with 100% N_2. Zakladnoi's (1976) observations contrast with those of AliNiazee (1972) who found a strong temperature effect with 100% CO_2 against *Tribolium* spp. Further studies and clarification of these results are required.

Under temperatures low enough to be lethal or to prevent rapid development, CAs may exert some protective effect as shown in Table 10 where reduced mortalities were observed under both types. The effect was apparent with both species tested and at both temperatures with low-O_2 CA but only with *S. oryzae* at low temperatures with high-CO_2 CA. This effect deserves further study as it may limit the utility of CA treatments to moderate or high storage temperatures (i.e., above 15°C) or where specific pests are to be treated at low temperatures.

Table 10 Percentage Reduction in Emergence of Adults of *Sitophilus oryzae* and *Rhyzopertha dominica* After 1 wk of Exposure of Immature Stages to Air or a CA Showing the Protective Effects of Particular Atmospheres at Low Temperatures

	Sitophilus oryzae		*Rhyzopertha dominica*	
Atmosphere	2°C	16°C	2°C	16°C
Air	99	67	73	25
60% CO_2 in air	96	81	89	91
1% O_2 in nitrogen	95	42	63	60

Source: Jay (1980b, 1984b).

3.3. Relative Humidity

Controlled atmospheres achieve better control at lower relative humidity (rh) (Pearman and Jay 1970, Jay et al. 1971, Spratt 1975, Zakladnoi 1976, Navarro 1978). Navarro and Calderon (1973, 1974) showed that there was little difference in response of *E. cautella* pupae to high-CO_2 atmospheres at either 21 or 55% rh (no grain present) but that greater mortality was observed at 95% rh. Navarro (1978) found that by reducing the relative humidity as the O_2 concentration was increased in a low-O_2 CA system, the mortality of *T. castaneum* adults and *E. cautella* pupae could be maintained. The influence of relative humidity has not been incorporated into recommendations for dosage rates for CAs (e.g., Annis 1987) beyond specifying that they refer to "dry grain." Where CAs are used for insect control in grain with equilibrium relative humidities over 70%, an allowance may be required for the slower action.

3.4. Comparative Effects of CO_2 and N_2

Banks and Annis (1990) have reviewed the comparative effects of N_2- and CO_2-based atmospheres. Atmospheres of about 60% CO_2 in air are comparatively more rapidly effective against internal feeders (e.g., *R. dominica*), while low-O_2 atmospheres are more effective against external feeders.

Atmospheres containing 1% or less O_2, with the balance either N_2 or CO_2, are usually equally effective (Lindgren and Vincent 1970, Verma and Wahdi 1978, Calderon and Navarro 1979) but some exceptions are known. Thus, Bell et al. (1980) found 100% CO_2 to be more effective than 100% N_2 against *Ephestia* eggs and Zakladnoi (1976) found different orders of susceptibility of pests to 100% N_2 or CO_2. The differences in rate of response to optimum high-CO_2 and low-O_2 atmospheres are small so that in practice the appropriate atmosphere will usually be selected on grounds such as cost, availability, and suitability of the structure for the gas, not because of different speeds of action. Again there are special cases. Low-O_2 atmospheres are more effective than high-CO_2 ones against *Trogoderma* spp. as larvae, particularly in diapause (Jay 1986, Spratt et al. 1985) and high-CO_2 atmospheres are substantially more effective against *Sitophilus* spp., notably as pupae.

3.5. Commercialization

Controlled atmosphere technology for pest control in grains has now been developed to a level where it can be applied in commercial practice. Table 7 gives examples of applications that have been published in the scientific literature.

Controlled atmosphere technology has been subject to numerous pilot- and full-scale trials (Shejbal 1980, Ripp et al. 1984). These trials have covered a wide range of storage structures or enclosures and atmosphere sources. In general, it is now known how to generate and maintain CAs of the compositions given in Table 8 in almost all grain storage enclosures. Large horizontal stores, bunkers, and bolted metal bins are notable exceptions. Uncertainties remain as to how long the particular atmosphere should be maintained to give the required level of disinfestation, the degree of modification (e.g., sealing) required to render an enclosure suitable for CA, the required gas supply rate, and the economics of providing the CA gases under specific conditions.

The technology for generating a controlled atmosphere with liquified gases supplied by a tanker-truck has been described for the low-O_2 type by Banks and Annis (1977) and Tranchino et al. (1980), and for high-CO_2 atmospheres by Jay (1971, 1980a). Essentially, the liquified gas is vaporized by passage through a heat exchanger and piped into the base or top of the storage enclosure. The applied gas displaces or dilutes the existing interstitial atmosphere and air in the free space within the enclosure, such as in the headspace in bulk stores or spaces between bags in bag stacks. After target atmospheric composition is achieved, further gas can be added to maintain the composition as air ingress and leakage occur. Addition of maintenance gas is always required with large storages, except where they are adequately sealed to retain CO_2 at insecticidal levels for more than 15 d. Small containers and flexible enclosures (e.g., drums, packs of commodity in barrier film) may not need added gas if they are well sealed. It has been assumed that the CA atmosphere created in retail packs of about 1 kg of grain by gas flushing with CO_2 or O_2 absorbing compounds, such as Ageless (Ohguchi et al. 1983), is insecticidal but no thorough tests involving both adult and immature insects have been published. The N_2-based CA system for treating stacks of bags of milled rice in a polyethylene envelope, formerly used in Shanghai, China (Champ et al. 1981), was also sufficiently well sealed not to require addition of maintenance gas, though respiration associated with the high-moisture milled rice may have assisted in maintaining the low O_2 level.

Most commercial installations for CA preservation of grain are based on atmospheres created by gas derived from industrial sources. In many countries, such sources may be distant from the grain storage and it may be more economic to generate the CA directly on site; this approach does not appear to be suited to high-CO_2 atmospheres but there are several alternative systems that can produce from air N_2 or mixtures of N_2 and CO_2 that are potentially insecticidal. Potential systems for on-site generation have been reviewed by Banks (1984) and Zanon (1980). Hydrocarbon burners, membrane separation systems and pressure-swing absorption (PSA) systems are all of interest because of their level of development and commercialization (Banks et al. 1990). They are already in use for generation of CAs for preservation of fresh fruit but may require some adaptation for disinfestation of grain because of the lower O_2 requirement. The optimum method of generation of N_2-CA will be site-specific.

Insect control with CA generated from burner systems have been successfully demonstrated in grain on pilot scale and in large storages (e.g., Storey 1973, Fleurat-Lessard and Le Torc'h 1987). There have been no studies published using membrane- and PSA-generated atmospheres. Because the output rate and cost of these machines (Spillman 1989) is dependent on the O_2 content required, a tendency can be expected to run CA storages at higher contents of residual O_2 than when liquid N_2 is used as an atmosphere source. Available data are insufficient to define precisely the usable upper limit but increased risk of failure and increased exposure times can both be expected as the O_2 content in the store is allowed to rise. Small-capacity machines may also create the target atmosphere in the store more slowly than when liquid N_2 is used, decreasing purging efficiency and subjecting the infestation to a slower change in O_2 content. There are no data to show whether the rate of change is important to the lethal effect of CAs.

3.6. Site Preparation

Unless the storage enclosures are purpose-built, enclosures will normally require some sealing and modification to be suitable for CA use. Banks and Annis (1984) discussed the influence of environmental forces on gas loss from grain storage enclosures and provided a theoretical justification for the 5-min pressure half-life for CA storages. This is used in Australia as a minimum standard for considering an enclosure to be suitably sealed for CA or fumigation. There is evidence that half-life greater than 10 min may be more appropriate for tall concrete cells and bag stacks enclosed in sheets intended for treatment with high-CO_2 atmospheres where the "one-shot" system is used (Wilson et al. 1984, Annis and van S. Graver 1990).

The 5-min pressure test standard corresponds to a level of sealing where gas losses from wind and the chimney effect are of a similar magnitude to those from temperature changes within the enclosure (Banks and Annis 1984). If gas to maintain the CA can be supplied cheaply, it may be feasible to relax this standard, possibly avoiding the additional costs of sealing while accepting an increased cost of gas. Trials with low-O_2 atmospheres have been carried out in leaky storages (farm bins, McGaughey and Akins 1989; and silo bins, Storey 1973) using a high rate of flow of maintenance gas to compensate for a high leak rate. Jay (1980a) recommends the use of a continuous addition of CO_2 to maintain insecticidal compositions in leaky silo bins but at the cost of using several times as much gas as would have been required in a well-sealed structure. White and Jayas (1991) showed a similar approach was feasible in poorly sealed bins.

The technology of sealing storages is now well developed (Ripp et al. 1984). Sealing of permanent structures involves modification of accesses and grain handling chutes and application of sealants such as thick emulsions of modified acrylic, silicone rubbers with reinforcement, polyurethane foams to fill gaps and solid materials as appropriate. Many designs and types of existing store can now be sealed to a high level at acceptable cost and well-sealed temporary enclosures can be made routinely. Bag stacks are routinely held within well-sealed polyvinyl chloride (PVC) enclosures in Indonesia (Annis et al. 1984, Annis and van S. Graver 1990, Nataredja and Hodges 1990) and China (Liang 1990). Various types of permanent storage structures, horizontal storages, concrete silo bins, and farm bins have all been sealed as part of normal commercial practice in Australia (Ripp et al. 1984). Capacities range from typical farm stores of about 50 t to large horizontal stores of more than 250 000 t.

4. ARIDITY

Many stored-product pests cannot survive or reproduce in grain of low moisture content. Longevity of the adults may also be reduced. The effect of aridity on insect pests is seldom studied in isolation but usually as a component of the combined influence of temperature and moisture. This combined effect can be described as the influence of prevailing wet bulb temperature on pest biology (Desmarchelier 1988). Aridity is also useful as an insecticidal principle on its own.

The controlling influence of aridity on molds (Chelkowski 1991) and mites (Cunnington 1965) is well known. Both are unable to proliferate at less than about 70% rh, even at otherwise optimum temperatures, but molds and some mites can enter long-lived resting stages, spores or hypopi, which can withstand long exposure to low relative humidity.

Table 11 Effect of Low Relative Humidity (rh) on Adult Survival (Weeks to 99% Mortality) for Adults of Six Common Species of Grain Pest

	Weeks to 99% mortality			
	Held at 30 or 32°C		Cooled to 13.5°C	
Species	45% rh	70% rh	45% rh	70% rh
Cryptolestes ferrugineus	42	60	42	>52
Oryzaephilus surinamensis	29	31	40	>50
Rhyzopertha dominica	30	31	24	48
Sitophilus granarius	16	23	47	>55
Sitophilus oryzae	19	30	53	>52
Tribolium castaneum	41	47	20	28

Source: condensed from Evans (1983).

Howe (1965) listed the many species of insect grain pests that are dependent on high relative humidity for reproduction. Most insects that are important stored-grain pests can multiply at less than 70% rh, equivalent to about 14% moisture content (mc) for wheat at 25°C (Brooker et al. 1974). Other species, however, notably *Sitophilus* spp., are severely affected by drier environments. Evans (1983) compared survival of several grain pests at 45 and 70% rh (Table 11). He noted a slight adverse effect of the lower rh on various species at optimum temperatures but for some species the influence was more pronounced at 13.5°C, which is close to the limit of reproduction. Evans (1982) found that a mean adult life span of unacclimated *S. oryzae* was decreased from more than 24 wk at 15°C in 12.5% mc wheat (about 50% rh) to only about 11.5 wk at 10.3% mc (about 35% rh). Khan (1983) studied adult survival of 10 species of stored-product insects without food under various relative humidities at 25°C. All species tested showed increasing longevity as the rh increased to about 80%. For most species there was a reduction in survival at still higher levels but *S. granarius* and *S. zeamais* showed increased survival even at "100%" rh. None of the species studied survived more than 6 wk, much less than that observed by Evans (1983), showing that Khan (1983) was observing the combined effect of aridity and starvation. Le Patourel (1986) compared survival of *S. granarius*, *Cryptolestes pusillus* (Schön), and *O. surinamensis* with and without food and found that the presence of food substantially increased their survival (e.g., Figure 5). Food is very important with stored-product insects as it provides water through the metabolic pathways (O'Donnell and Machin 1991).

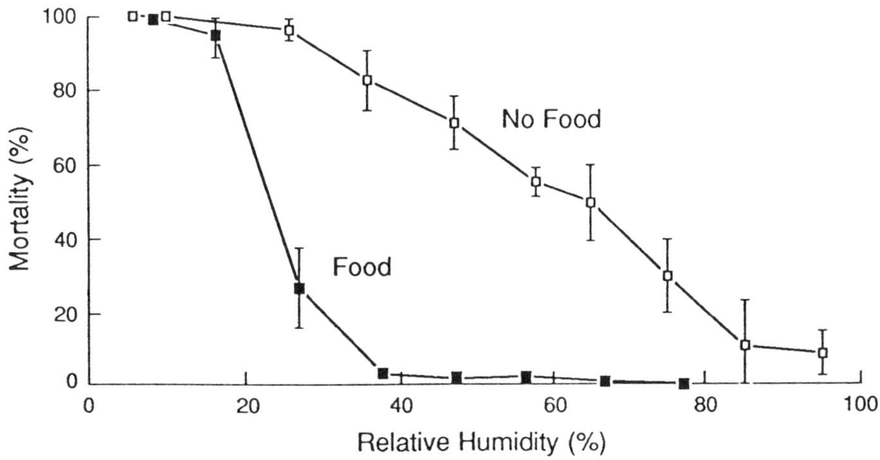

Figure 5 Relationship between mortality and relative humidity over a 7-d period for *Sitophilus granarius*. *Source*: Le Patourel (1986).

5. INERT DUSTS

Inert dusts—such as ash, lime, various ground minerals and clays—have a long history of use for grain protection (Ebling 1971; Golob and Webley 1980; Ross 1981; Quarles 1992a,b). There are several advantages to using inert dusts to control stored-product insect pests. In the United States, amorphous silicon dioxide (diatomaceous earth) is considered "Generally Recognized as Safe" (GRAS) by the US Food and Drug Administration and is a registered food additive (21 CFR 182.90, 182.1711). Silica aerogels have been shown to have low mammalian toxicity (3160 mg/kg LD_{50}, rat oral; Ebeling 1971). Also, inert dusts do not affect baking quality (Desmarchelier and Dines 1987, Aldryhim 1990) and provide long lasting protection (La Hue 1965, 1978; White et al. 1975). The main disadvantages are decreased flowability of grain, visible residues that affect the grading of grain and decreased bulk density.

Interest has revived in the use of inert dusts after neglect resulting from the outstanding success of chemical protectants. There are four basic types of inert dusts:

1. *Clays, sands, and earth* have been traditional insecticides that are used as a layer on top of stored seed. These materials were often mixed at a mass ratio to grain of 10 kg/t or more, a proportion regarded as excessive by modern standards despite its effectiveness (Golob and Webley 1980).

2. *Diatomaceous earth* is the fossilized remains of diatoms—microscopic, unicellular, aquatic-plants—that have a fine shell made of amorphous hydrated silica. Deposits of diatomaceous earth can be hundreds of meters thick (Ross 1981). The main constituent of these deposits is silica though there are small amounts of other minerals (aluminum, iron oxide, lime, magnesium, and sodium). Many proprietary insecticidal formulations are available, some of which contain additional materials (e.g., ammonium fluorosilicate or attractants) that are said to enhance their potency. These dusts are effective against a wide range of pests when admixed to grain at rates of 1 kg/t or less.
3. *Silica aerogels* are produced by drying an aqueous solution of sodium silicate (Quarles 1992b). These are very light, nonhygroscopic powders which are effective at lower rates than diatomaceous earth formulations.
4. *Nonsilica dusts* such as pulverised phosphate rock have been used in traditional stores in Egypt (Fam et al. 1974). Ground rock also has some activity (Golob and Webley 1980). Some farmers in Australia still use lime (calcium hydroxide) for successfully protecting grain to be used as animal feed.

La Hue (1978) showed a diatomaceous earth, Kenite, added at 0.5 kg/t at 33°C, 40% rh reduced multiplication of *S. oryzae* and *S. zeamais* to one-quarter of that on untreated grain over a 6 mo storage period. White et al. (1975) compared the performance of some silica aerogels and diatomaceous earths as protectants for wheat under field conditions in the United States. They concluded that the brands of diatomaceous earth tested were superior to the silica aerogels under their conditions when applied at equal volumes. Due to differences in density, however, this was equivalent to mass rates of 3 and 0.5 kg/t, respectively. Residual infestation was 1.8 insects/kg after 3 yr storage with Perma-Guard, a diatomaceous earth and the best material tested, compared with over 100 insects per kg in untreated grain. Treated grain was still protected against *S. oryzae* after the 3 yr storage. Desmarchelier and Dines (1987) found a diatomaceous earth and silica aerogel mixture, Dryacide, gave complete mortality of adults and immature stages of *S. granarius*, the most tolerant of the five pest species tested, at 2–3 kg/t and suppression of increases in population at 1–2 kg/t at 25°C and 65% rh. Maceljski and Korunic (1971, 1972) observed rapid mortality (less than 12 d) of adults of *S. granarius*, *S. oryzae*, *O. surinamensis*, and *Cryptolestes* spp. exposed to wheat dusted with diatomaceous earth at 2 kg/t or silica aerogel at 0.5 kg/t. They noted that the same dosage applied in water suspension to wheat, initially at 10.7% mc, was equally effective against *O. surinamensis* and *Cryptolestes* spp. but the activity of the diatomaceous earth, not the silica aerogel, was reduced against *Sitophilus* spp. Diatomaceous earth at 3.5–7 kg/t controlled *S. oryzae* and *T. castaneum* in milled and brown rice at 27°C and 60% rh but more than 56 d exposure was required in brown rice to affect complete mortality of *T. castaneum* (McGaughey 1972). The slow action was attributed to deactivation of the dust by rice oils.

The available data on responses of immature stages of grain pests is very limited although it can be inferred from the success of inert dusts in suppressing population growth that a strong effect on free living immature stages is probable. Desmarchelier and Dines (1987) noted that the effective dosage required to suppress the development of progeny in *R. dominica* and *Sitophilus* spp. was the same or greater than that for 100% mortality of adults, whereas only a low dosage (0.25–0.50 kg/t at 25°C and 65% rh) was needed to suppress development of *T. castaneum* and *E. cautella*. Permual and Le Patourel (1990) showed acid-activated kaolin at 5 kg/t suppressed progeny production of a range of stored-grain pests including *Sitophilus* spp. and *R. dominica*. *Sitotroga cerealella* (Olivier) was particularly sensitive and high larval mortality was observed in 0.75 kg/t.

The activity of inert dusts is strongly dependent on the prevailing relative humidity, as to be expected from their presumed mode of action. This is believed to be through removal of the cuticular wax layer by absorption on to the dust, leaving the insect prone to desiccation. Insects die when they lose about 60% of their water or about 30% of their total body mass (Ebeling 1971). Silica aerogels can absorb as much as 3 times their own mass in oils. It is believed that as insects move through the grains the silica aerogels absorb waxes off the insect's cuticle (Le Patourel et al. 1989). Besides the absorption of cuticular waxes, diatomaceous earth abrades the cuticle. Maceljski and Korunic (1971) found that *Sitophilus* spp. and *T. castaneum* adults treated with silica aerogel and held at 90–100% rh can regenerate the wax layer of the cuticle and survive, whereas adults held at 40% rh rapidly die. Le Patourel (1986), investigating the action of silica dusts in wheat at various moisture contents, showed a 37-fold increase in the dosage required to give 50% mortality of *S. granarius* adults in 16% mc wheat compared with that at 11% mc. Humidity had a lesser influence on the response of the other three species studied. Aldryhim (1990) and Desmarchelier and Dines (1987) also found a strong effect of relative humidity on response of *Sitophilus* spp. to diatomaceous earth (Table 12). Permual and Le Patourel (1990) noted a similar effect on the response of adults of several common rice pests to acid-activated kaolin.

Systematic studies on the effect of temperature on the activity of modern inert dusts are still required. There is a notable absence of laboratory data at both high (above 30°C) and low (less than 20°C) storage temperatures. Almost all contemporary publications give data for one temperature only. Aldryhim (1990, 1993), however, studied the responses of adult *R. dominica*, *S. granarius*, and *T. confusum* at 20 and 30°C. *Rhyzopertha dominica* and *S. granarius* are more susceptible to a diatomaceous earth (Dryacide) at the higher temperatures whereas *T. confusum* is less susceptible at higher temperatures. Diatomaceous earth suppressed progeny completely at 20°C after 7 d exposure but there was some progeny production at 30°C. The effect of temperature on response to inert dusts is likely to be complex as it will depend on several factors, including:

Table 12 Effect of Relative Humidity on the Toxicity of a Diatomaceous Earth (Dryacide) at 30°C

Concentration (kg/t)	Relative humidity (%)	Mortality (%) *Sitophilus oryzae* 7 d	*Sitophilus oryzae* 28 d	*Sitophilus granius* 7 d	*Sitophilus granius* 28 d
0.5	45	93	100	59	98
1		100	100	100	100
2		100	100	100	100
0.5	55	63	96	38	94
1		100	100	85	100
2		100	100	100	100
0.5	65	49	96	2	12
1		84	100	22	87
2		100	100	71	100

Source: Desmarchelier and Dines (1987).

the net rate of pickup of the dust (Gowers and Le Patourel 1984, Le Patourel et al. 1989), the rate of repair of the wax layer (Maceljski and Korunic 1971, Le Patourel et al. 1989), the influence of temperature on desiccation resulting from the dust, and how much water loss is tolerable at a particular temperature.

Maceljski and Korunic (1971) found that it was not necessary for insects to be completely covered with the inert dust for it to be active. They applied 15 mg of a silica aerogel, Dri-Die, directly onto *S. granarius* adults. Complete mortality occurred within 5 d. White and Loschiavo (1989) showed that access to food greatly increased the chances of survival of adult *O. mercator* and *T. confusum* after exposure to a surface treated with silica aerogel. Apparently, food itself rather than mechanical removal of the aerogel favored survival because exposure to cellulose after the aerogel did not increase survival over insects exposed and then kept in empty containers. Larvae of both species also survived better if allowed access to food after treatment.

Inert dusts can be quite rapid in their lethal action under favorable conditions with complete mortality of adults achievable within 7 d at low dosage rates. Response times, however, are generally poorly defined, with most studies observing mortality at one fixed assessment time, typically less than 7 d. Long exposures are usually more effective than short ones when assessed by the concentration-exposure product (effectiveness ratio, Table 13).

Table 13 Lethal Concentrations for 50% of the Population (LC_{50}) for Diatomaceous Earth at 60% rh for Two Exposure Periods

Species	Temperature (°C)	LC_{50} (g/kg) 2 d	7 d	Effectiveness ratio†
Sitophilus granarius	20	1.7	0.26	1.81
	30	0.97	0.37	0.75
Tribolium confusum	20	2.6	0.42	1.74
	30	2.6	0.31	2.46

† Ratio of LC_{50}s for 2- and 7-d exposures multiplied by ratio of exposure times.
Source: calculated from Aldryhim 1990.

For grain pests that develop within the grain kernel (e.g., *Sitophilus* spp. and *R. dominica*), it is presumably the lethal response time of adults that largely determines the level of dust that prevents progeny formation because the adults must be killed before they can mate and lay eggs. Concentrations of modern dusts required to eliminate free-living pests with long exposures is yet to be adequately defined, with current knowledge restricted to dosages found to be effective under field conditions with dry grain. It is important to define the dosage required to eliminate pests at high relative humidity and, in effect, to define the end-point mortality for combinations of dosage and conditions so that the upper limit of utility of inert dusts can be determined.

The main problems with the use of inert dusts are that they change the flowability of grain, decrease the bulk density of grain, are dusty to apply, and are ineffective in some situations. Because inert dusts adhere to the surface of the kernels and increase friction between grains, grain does not flow as easily, thereby increasing angles of repose and decreasing bulk density. Diatomaceous earth at 2 kg/t caused a 4.4 kg/hL in corn and 6.2 kg/hL in wheat decrease in bulk density (La Hue 1965). As desiccation is the mode of action, diatomaceous earth does not control insects in moist grain as well as in dry grain (La Hue 1965). Application of inert dusts can be undesirable because of the dust generated. To alleviate this, aqueous applications for surface treatments are used in Australia, although this somewhat reduces the effectiveness of the inert dusts (Maceljski and Korunic 1971). Silica aerogel glued onto cards can kill insects that walk across the cards (Loschiavo 1988a); however, mortality was reduced when insects were given access to food (Loschiavo 1988b). Finally, diatomaceous earth is an abrasive mineral and there is concern over increased wear on grain handling machinery.

6. IMPACT AND REMOVAL

6.1. Laboratory Studies

The adverse effect of shock and disturbance, though well known, has been little studied despite its potential as a simple, residue-free means of controlling grain pests. Banks (1987) provided a detailed review on the subject.

Loschiavo (1978) extended the studies of Bailey (1969) on the effects of dropping stored product insects on to surfaces. He found that dropping of adult insects in wheat filled sacks or in free flowing grain caused substantial mortality. Multiple impacts were more effective than single impacts (Figs. 6, 7). The experiments were designed to simulate effects of loading grain into a silo bin.

Gentle tumbling of beans infested with *Acanthoscelides obtectus* Say every 8 h over a 2 wk period reduced populations by 97% relative to undisturbed cultures (Quentin et al. 1991). It was hypothesised that the effect was caused

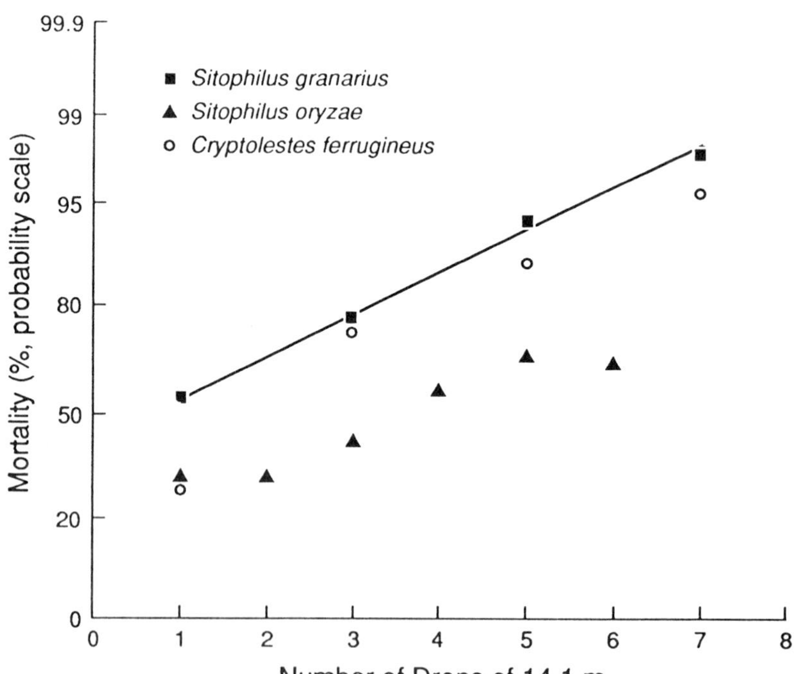

Figure 6 Mortality of adult insects in wheat-filled sacks (400 g) dropped repeatedly onto a concrete floor. *Source*: data of Loschiavo (1978).

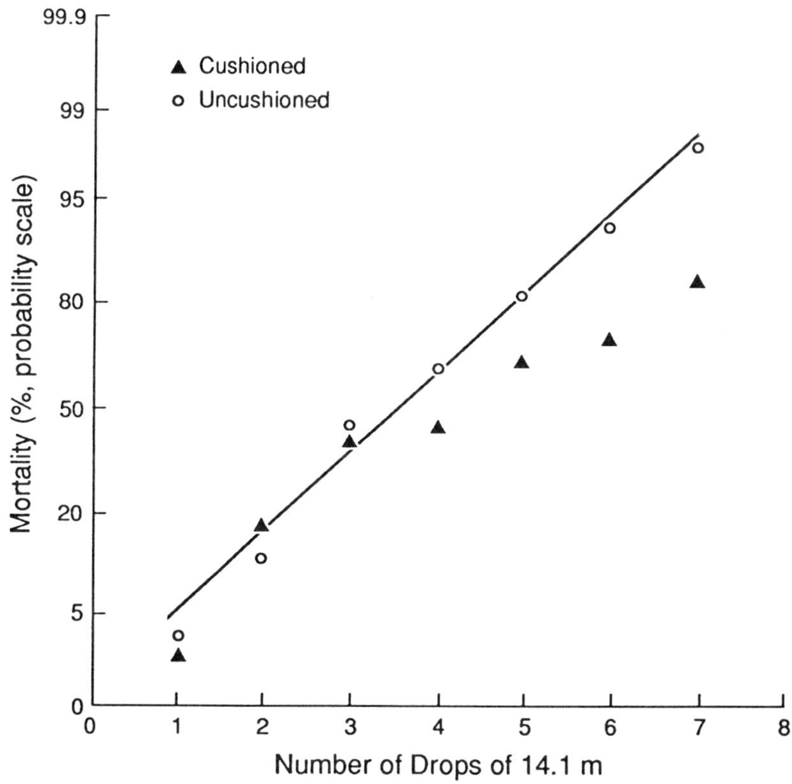

Figure 7 Mortality of *Cryptolestes ferrugineus* adults in grain poured repeatedly on to a metal surface ("uncushioned") or a metal surface covered with several centimeters of wheat ("cushioned"). *Source*: data of Loschiavo (1978).

by preventing the first instar from boring an entrance hole in a bean. This level of control was close to that predicted (96%) from calculations of the chances of a partially completed hole being in a favorable position for further boring after each tumbling. Mechanical disturbance also reduces the productivity of *Sitophilus* spp. cultures (Ungsunantwiwat and Mills 1979).

Stratil et al. (1987) investigated the insecticidal effect of Entoleters. These machines subject grain or, more normally, flour to severe impact. Seeds or flour are dropped on to a rotating disk with studs. Insects are killed or injured when they collide with the studs or the walls of the discharge hopper. They found, as Bailey (1962) had, that slow speeds of about 20 m/s produced significant mortality (30–60%) but high speeds (64 m/s) were required to disinfest grain completely of adult *S. granarius*, *O. surinamensis*, and *T. castaneum*. At speeds producing even moderate mortality there was unacceptable levels of breakage (3–8%) with dry, storable wheat, showing that impact machines are unlikely to be useful in their present form for disinfestation of whole kernels, except where it is to be processed soon thereafter.

Shock and disturbance of grain can make it more susceptible to insect attack. Many insects (e.g., *C. ferrugineus*, *Oryzaephilus* spp., and *T. castaneum*) are classified as secondary pests because they require a break in the seed coat to infest cereal grains. These species have higher populations when the seed coat is damaged (Tuff and Telford 1964, Li and Arbogast 1991). Mechanized harvesting can cause 50% of wheat kernels to be cracked or broken (Tuff and Telford 1964, Kline 1973, Bourgeois 1993). Movement of grain and effectiveness of cleaners will also affect the amount of dockage present and the integrity of the seed coat. Corn is especially sensitive to breaking during movement (Baker et al. 1986).

6.2. Field Studies

Modern field studies investigating the effect of shock on insect pests directly are restricted to the work of Bahr (summarized in Bahr 1990). He showed that some species were particularly susceptible to damage during pneumatic conveying of grain. Thus, conveying of severely infested rye at 38 t/h produced 99.8% and 96% mortality of all stages of *O. surinamensis* and *C. ferrugineus*, respectively. Conveying of wheat through an industrial vacuum system gave high mortality (over 90%) of several species of stored-grain pests (Bahr 1990). It was particularly effective against *T. granarium* and *O. surinamensis* where 99.6 and 100% kill was obtained after four passes through the system (Table 14).

Several studies intended to show the effects of other factors have demonstrated the highly adverse effects grain conveying has on infestation. Cogburn et al. (1972), investigating radiation disinfestation of grain, observed substantial mortality of several pest species by grain transfer including pneumatic conveying. Over 99% of adults of *R. dominica*, *S. oryzae*, and *Cryptolestes* spp., and over 80% of their immature stages were killed. Kirkpatrick and Cagle (1978), studying disinfestation of wheat by infrared heating, observed a similar effect but with less mortality to the immature stages (*S. oryzae* 94 and 63%; *R. dominica* 100 and 66%; *C. ferrugineus* 100 and 77% kill of adults and immatures, respectively). Green and Tyler (1966) had to modify their experiment, investigating the effect of an insecticide, because of a 95% mortality of test insects resulting from mechanical handling of the grain (a pneumatic conveyor, elevators, and conveyors). Muir et al. (1977) noted a mortality of 80 and 60% (adults and larvae, respectively) of *C. ferrugineus* on two transfers of the grain through screw conveyors. The observed effects were presumably a result of the combined

Table 14 Mortality of Stored-Grain Pests During Pneumatic Conveying

	Mortality (%)		
	Ship discharge†		Cleaning system*
Species	All stages	Immatures	All stages
Sitotroga cerealella	65	65	72
Sitophilus granarius	—	—	73
Sitophilus oryzae	48	46	75
Cryptolestes ferrugineus	62	24	91
Tribolium castaneum	88	59	—
Rhyzopertha dominica	89	70	94
Trogoderma granarium	—	—	100
Oryzaephilus surinamensis	95	81	100

†Lowest observed mortality resulting from pneumatic discharge of grain from ship.
*Observed mortality after 4 passages through an industrial vacuum cleaning system (6 m, 7 cm diameter).
Source: Bahr (1990).

effects of shock, disturbance, and changes in temperature distribution resulting from the grain turning.

Sutherland et al. (1989) obtained 34% mortality of immature *R. dominica* by pneumatic conveying during their studies on heat disinfestation. Fleurat-Lessard (1980, 1985) noted 80% reduction in populations of *S. oryzae* (all stages) in a smaller pneumatic conveyor, with over 97% reduction in all stages except eggs. Bahr (1973, 1975, 1990) rated insects in order of most to least susceptibility to mortality during pneumatic conveying: *Typhaea stercorea* (L.), *O. surinamensis*, *T. granarium*, *T. castaneum*, *R. dominica*, *C. ferrugineus*, *S. oryzae*, *S. granarius*, and *S. cerealella*. Even pneumatic sampling systems for grain may cause appreciable kill (Bryan and Elvidge 1977).

Physical removal of pests from grain and grain storages can be accomplished by several techniques; these include sieving and aspiration for bulk grain, and various forms of cleaning for grain stores. Banks (1987) reviewed some of the very limited data on physical removal as a pest control process.

It is normally held to be self-evident that various forms of store hygiene, such as vacuum cleaning and sweeping up of residues, are useful in restricting infestation of grain, though quantitative studies on the usefulness of such measures are generally lacking. Apart from store hygiene, physical removal as a control process is seldom carried out deliberately in modern grain storage.

7. PACKAGING AND EXCLUSION

Insect pests may be prevented from entering large or small quantities of stored grain by sealing or packaging. The level of sealing required for effective exclusion of grain insects from large bulks of grain has not been well researched. It will probably depend on several factors including pest pressure, size of holes in the barrier, and tolerance of infestation. Nevertheless, physical exclusion plays an important role in the overall effectiveness of sealed storages. Banks and Ripp (1984) discuss the effectiveness of a large-scale sealed storage in western Australia in preventing insect reinvasion of disinfested stocks of wheat and conclude a substantial period of protection is afforded by a store sealed to meet the local pressure test specification (pressure decay of 500 to 250 Pa in not less than 5 min in a full store). Nataredja and Hodges (1990) report successful insect-free storage of bagged rice over 8 mo in sealed PVC enclosures under the severe pest pressure of the humid tropics.

For small bulks of up to a few kilograms, packaging and sealing are widely used to prevent invasion of insect pests into cereal grains and products. There has been extensive work on the capacity of packaging to prevent insects from infesting food products and there are several reviews on this topic (Wohlgemuth 1979; Newton 1988; Highland 1978, 1981, 1991). Packaging provides a physical barrier that prevents or impedes infestation by insects. This is important as manufacturers are held responsible for insects in their products, even if infestation occurs in stores that are outside their control. There are three factors that determine if a commodity will be infested: insect species, packaging material, and commodity packaged.

Insects vary in their capacity to penetrate packaging. *Lasioderma serricorne* (F.), *Stegobium paniceum* (L.), *P. interpunctella*, *E. cautella*, *Corcyra cephalonica* (Stainton), and *Trogoderma variable* (Ballion) are all capable of penetrating intact packages. *Rhyzopertha dominica* is also capable of penetrating but is rarely found in packaged foods. Other stored-product insects (e.g., *T. castaneum*, *T. confusum*, *C. ferrugineus*, *C. pusillus*, *O. mercator*, and *O. surinamensis*) must have an opening to enter most packaged goods (Highland 1991). However *T. castaneum*, *O. mercator*, and *C. pusillus* can penetrate the less resistant packing materials (Cline 1978). Holes larger than 2 mm^2 will allow most stored-product adult insects to enter packages, whereas holes smaller than 0.3 mm^2 will prevent entry (Cline and Highland 1981). Neonate larvae can enter small holes and imperfections as their head capsules width is often less than 0.3 mm (Khan 1983). Wohlgemuth (1979) found young *P. interpunctella* can locate and pass through a 0.25 mm diameter (0.39 mm^2) hole. A convenient measure of package integrity is the pressure packages can sustain and is related to the susceptibility of the packaging to insect infestation (Yerington 1978).

Packaging materials differ in their capacity to prevent penetration. Listed in order of decreasing ease of penetration they are: cellophane, polyethylene, paper, polyvinyl chloride, aluminum foil, polyester, polypropylene, and polycarbonate (Cline 1978). Most packages, however, have small holes that allow entry. Packaging materials also vary in slipperiness that affects the capacity of insects to move from package to package. In general, packaging materials in order of increasing difficulty for adult beetles to climb are paper, ethylenetetrafluoroethylene (ETFE), polyester (PE), aluminum foil, cellophane, fluorinated ethylenepropylene (FEP), glass, PVC, and polypropylene (PP). Packaging materials rank differently for larvae than for adults with the easiest to most difficult to climb being paper, glass, PET, ETFE, PVC, cellophane, FEP, foil, PE, and PP (Cline 1978).

Insecticides can be inserted into the packaging materials to render them more insect resistant. Currently, in the United States, the only insecticide registered for this use is synergized pyrethrum in 23 kg bags of grain used for export (Newton 1988). Multiwall packages are necessary to prevent the migration of the pesticide into the commodity.

Packaging is also a factor when considering heat sterilization of a packaged product. Using longwave irradiation, commodities can be heated at high enough temperatures to kill any insects within a package. Packaging must transmit the longwave radiation and be adequately resistant to the heat and pressure created during heating (Fleurat-Lessard 1989). Airtight packaging with modified atmospheres are used to maintain freshness of high-value products and these atmospheres will also prevent development of insects (Cline and Highland 1987).

Combining the parasitoid, *Bracon hebetor* Say, with insect resistant packaging reduced infestations by *E. cautella*. The proposed mechanism is that the larvae were exposed to the parasitoid for longer periods of times in treatments when insect resistant packaging was used (Cline and Press 1990).

8. IONIZING IRRADIATION

8.1. Types

Two types of ionizing irradiation have been considered for insect control in grain; γ-radiation produced from ^{60}Co or ^{137}Cs sources and accelerated electrons. The radiosensitivity of stored-product pests has been extensively studied (Urbain 1986, Tilton and Brower 1987, Watters 1991). Tolerance increases progressively from egg to larval to pupal to adult stage of insects (Shipp 1983) so dosages that control or kill adults can be taken to be the maximum dosage required for grain treatment. It is often assumed that irradiation by accelerated electrons and γ-rays are equally effective, though both Bull and Cornwell (1966) and Adem et al. (1978) found γ-irradiation to be more effective for the same dosage. In these studies, the rate at which the dosage was applied was different for the two forms of treatment and strictly-controlled comparative studies are still needed. Also, the higher energy γ-irradiation penetrates at least ten times as far into solid foods as does electron-beam irradiation (Anderson 1989), thus grain must be in shallow layers during electron-beam irradiation.

Irradiation at 10 kGy is regarded as not producing toxins in treated foods and the World Health Organization recommends that foods treated at this or lower doses be exempt from toxicological testing (Urbain 1986). At lower dosages of 3–5 kGy that cause death of stored-product pests within 24 h, there can be significant damage to processing quality of wheat and other cereals. Irradiation can reduce vitamins A, C, E, B_1 (thiamin), and K. The amount of reduction is dependent on the food irradiated, dose, and other factors (Urbain 1986). Bread quality is affected only after wheat was irradiated at doses above 2.5 kGy (Lee 1959) but some starch damage may be apparent at 0.5 kGy (Ng et al. 1989).

The maximum dosage may be limited to avoid these problems. In Canada, for instance, 0.75 kGy is the maximum dosage that wheat and flour can be given. Consequently, studies have concentrated on dosages required for death after several weeks and for sterilization. These are typically between 0.2 and 0.5 kGy for coleopterous and mite pests, but more than 1.0 kGy for moths. The times-to-complete-mortality after a given dosage vary widely among species and there are sometimes also notable discrepancies between results of different experimenters. Table 15 summarizes data available for action of γ-irradiation against adult stored-product pests and time-to-death for 0.3 and 0.5 kGy. These dosages are those required to ensure sterilization of the commonest pests of stored wheat (*R. dominica*, *T. castaneum*, *S. oryzae*, *S. granarius*, *O. surinamensis*, and *C. ferrugineus*) and were advocated by Tilton and Burditt (1983) for control of grain pests in the United States. At 0.5 kGy, beetles and the immature forms of mites and moth pests are sterilized but some adult moths may remain fertile. The few progeny they produce, however, will be sterile due to inherited genetic damage (Tilton and Brower 1987). The doses of irradiation needed to kill insects can also affect germination making this type of control unsuitable for malting barley (Watters and MacQueen 1967) or seed stocks.

The use of a low intensity radiation source placed permanently in a grain store has been discussed but rejected because of the possibility of continuous reinfestation. Erdman (1973) showed that constant irradiation at 1.5 Gy/d caused slow decline of *Tribolium* populations.

8.2. Commercialization

Disinfestation of bulk grain by radiation from either γ-ray or accelerated electron sources has been technically feasible for several decades but its full-scale routine use is limited to the installation at Krasnodar, Ukraine, commissioned in 1980 (Zakladnoi et al. 1982). This plant has two accelerated-electron units each operating at 1.4 MeV (20 kW) with a capacity of 200 t/h. The grain after treatment is aspirated to remove dust and ozone produced during the irradiation.

A pilot-scale grain irradiator operating with a ^{60}Co source was constructed in the late 1960s in the United States for research and demonstration purposes (Tilton et al. 1971, Cogburn et al. 1972) but is now dismantled. This had a treatment capacity of about 1 t/h at 0.4 kGy. A research scale corn irradiator was constructed in Mexico; this operated at 0.2 t/h using 1.5 MeV electrons (Watters 1979).

The main limitations on the use of ionizing radiation for insect control are capital cost of the installations, public and market acceptance of the process, and the inability to apply an acutely lethal dosage to pests without commodity damage. The slow speed of action has prompted several studies to seek a more rapid kill by combining irradiation with other physical or chemical processes (Cogburn et al. 1972, Tilton and Brower 1985, Buscarlet et al. 1987).

9. LIGHT AND SOUND

Rees (1985) reviewed the use of light for control of stored-grain insects. Light may find some use with other processes when used to lure flying insects into traps and thereby remove them from the commodities but is unlikely to find general use on its own.

Attempts have been made to control stored-grain pests with sound waves. Mullen (1973) showed low-frequency sound adversely affected development of *P. interpunctella*. Pradzynska (1982) reviewed previous attempts to use ultrasound as a control measure. He showed a 5-min exposure to 1 MHz sound at 14.5 W/cm^2 killed all stages of *S. granarius* at 26°C in wheat. There seems little promise, however, that sound can be used in commercial situations (Andrieu et al. 1977, 1978).

10. COMBINED TREATMENTS

Some physical control measures on their own are not effective against the complete range of pests in grain, though they do have a high level of activity against particular species or developmental stages. They may be limited by a narrow margin of safety between killing the most tolerant pest or stage and appreciable damage to grain quality. Examples of the former include cooling, with long survival of adult pests; percussion, with limited effectiveness against eggs; sieving and aspiration, largely ineffective against internal stages; and high-CO_2 CA with excessive exposure times required against diapausing *T.*

Table 15 Response of Adult Stored-Grain Pests to γ-Radiation

Species	Sterilization dose:† (kGy)	Time to attain 100% mortality (d)		References
		0.3 kGy	0.5 kGy	
Acarina				
Acarus siro L.	4.5	175	203	Burkholder et al. (1966)
Coleoptera				
Attagenus piceus Oliv.	1.75	35	35	Tilton et al. (1966a)
Cryptolestes ferrugineus	0.625	21	14	Watters and MacQueen (1967)
Oryzaephilus mercator	2.0	14	14	Brower and Tilton (1972a)
Oryzaephilus surinamensis	2.0	M>70 F21	M>70 F14	Brower and Tilton (1972a)
Rhyzopertha dominica	1.32	42	28	Tilton et al. (1966b)
	1.25	70	28	Watters and MacQueen (1967)
	5.6	0	>28	Cornwell et al. (1957)
Sitophilus granarius	10	28	>49	Brown et al. (1972)
	1.25	14	14	Watters and MacQueen (1967)
Sitophilus oryzae	1.75	21	21	Tilton et al. (1966b)
				Cornwell et al. (1957)

Sitophilus zeamais	1.0	28	28	Brown et al. (1972)
Tenebrio molitor (L.)	M1.5 F.5	21	21	Brower (1973)
Tenebrio obscurus	1.0	7	7	Brower (1973)
Tribolium castaneum	2.0	28	28	Brower and Tilton (1973)
	1.25	21	21	Watters and MacQueen (1967)
Tribolium confusum	1.75	21	21	Tilton et al. (1966b)
	0.625	14	14	Watters and MacQueen (1967)
Trogoderma glabrum	M2.5 F1.3	—	—	Tilton et al. (1966a)
Trogoderma granarium	M>1.5 F0.5	—	—	Carney (1959)
Trogoderma inclusum	M3.0 F2.0	35	42	Brower and Tilton (1972b)
Tribolium madens	M3.0 F1.0	M28 F35	28	Brower and Tilton (1973)
Tribolium variabile	M3.0 F1.0	28	M28 F35	Brower and Tilton (1972b)
Lepidoptera				
Ephestia cautella	M10 F3.0	—	—	Cogburn et al. (1973)
	—	M4.5 F4.7	M3.2 F4.7	Burkholder et al. (1966)
Ephestia elutella	M45 F30	—	—	Brower (1979)
Plodia interpunctella	>10	—	—	Cogburn et al. (1966)
Sitotroga cerealella	M>10	—	—	Cogburn et al. (1966)

†M is for males and F for females.

Source: compiled from Tilton and Brower (1987) and D.E. Evans (unpublished).

granarium larvae. In the latter category, late larvae or pupae of *R. dominica* limit effectiveness of heat disinfestation and Lepidoptera require high dosages of irradiation treatment. Banks (1987) discussed the advantages of combining the strengths of the physical control measures—percussion and physical removal by sieving and aspiration—with several physical and chemical control measures to enhance their effectiveness. Table 16 summarizes the processes suggested or tested that in combination are likely to be much more effective than in isolation.

Banks (1987) concluded there was considerable scope for development of processes based on removal, particularly in combination with other physically-based methods. He noted, in particular, that the hirsute larvae of *Trogoderma* spp., tolerant of many control processes, may be particularly susceptible to removal from grain by aspiration. Mathlein (1971) showed that sieving to remove adult insects could be beneficially combined with aeration to give good insect control.

Physical exclusion, through sealing or packaging, is likely to be complementary to any disinfestation process that does not leave residual protection (e.g., heat treatment, ionizing radiation, and fumigation).

The actions of some physical control methods can be directly antagonistic to others. Irradiation of grain pests in inert gas decreases the effectiveness of irradiation, although pre- or post-treatment with N_2 atmospheres enhances speed of death (Buscarlet et al. 1987). Cooling of grain reduces the effectiveness of fumigants (Bond 1984) and controlled atmospheres (Tables 9, 10). The requirement to have a sealed system to use fumigants or controlled atmospheres is difficult to integrate with the requirement for ventilation to cool with aeration. Generally cooling needs to be applied after a disinfestation technique in order to control any surviving insects and to protect the grain from residual pests, subsequent reinvasion, and slow loss of quality during storage.

11. FUTURE RESEARCH

Historically, physical control methods such as sun-drying, pit storage, winnowing, and inert dusts were the mainstay of stored-product insect control. These methods were replaced by chemicals as the control method of choice. However, with the increased concern of chemical residues on grain and the development of pest resistance to chemicals, physical control methods have gained renewed interest. Much of the work in development of physical methods has been a process of extension or even re-invention of old methods; there are, however, a few new approaches. For instance, ice nucleating bacteria make low-temperature control possible at higher, but still low, temperatures, (Fields 1990, Lee et al. 1992, Fields 1993). Knowledge of the underlying mechanisms that cause insects to die during physical control (desiccation, cold, heat, modified atmo-

Table 16 Beneficial Combinations of Control Measures Against Grain Pests

Main method	Limiting species of stage	Complementary method	Status†	References
Cold disinfestation	Adult beetle pests	Sieving and aspiration	F	Mathlein (1971)
Heat disinfestation	*Rhyzopertha dominica* late larvae and pupae	Percussion	F	Sutherland et al. (1989)
CA treatments	*Sitophilus* spp.	Heat	L	Jay (1986)
Low-O_2, CA	*Sitophilus* pupae	Percussion	S	Banks (1987)
High-CO_2 CA	*Trogoderma* larvae	Sieving and aspiration	S	Banks (1987)
Irradiation	Lepidoptera	Percussion	F	Tilton and Brower (1985), Cogburn et al. (1972)
Irradiation	Adults	Low-O_2, CA	L	Buscarlet et al (1987)
Fenitrothion protectant	Resistant insects *Oryzaephilus surinamensis*	Percussion	S	Banks (1987)
Methyl bromide	*Trogoderma* larvae	Sieving and aspiration	S	Banks (1987)
Phosphine	*Sitophilus* pupae	Percussion	S	Banks (1987)

†F = field demonstration, L = lab demonstration, S = suggested.

spheres, and irradiation) may lead to other synergists that would enhance these methods of control.

Processes using high-pressure CO_2 appear to promise a rapid method for disinfestation of commodities. There is obvious potential for development of nonchemical insect control methods using disturbance and impact.

In the future, there must be a recognition by persons managing grain storages that they are dealing with an ecosystem. Furthermore, storage ecosystems can be manipulated to the detriment of the pests. With this in mind, the technology of grain storage should develop the rational use of methods to alter the physical environment of the pests. A combination of different approaches will be used to give an optimum outcome in many situations. Overall, the demands of the marketplace to provide residue-free and insect-free grain will be met. Physical methods of pest control are the most likely processes to be able to deliver grain to this specification in the near future.

REFERENCES

Adem, E., F. L. Watters, R. Uribe-Rendn, and A. de la Piedad. 1978. Comparison of ^{60}Co gamma radiation and accelerated electrons for suppressing emergence of *Sitophilus* spp. in stored maize. J. Stored Prod. Res. 14:135-142.

Aldryhim, Y. N. 1990. Efficacy of the amorphous silica dust, Dryacide, against *Tribolium confusum* Duv. and *Sitophilus granarius* (L.) (Coleoptera: Tenebrionidae and Curculionidae). J. Stored Prod. Res. 26:207-210.

Aldryhim, Y. N. 1993. Combination of classes of wheat and environmental factors affecting the efficacy of amorphous silica dust, Dryacide, against *Rhyzopertha dominica*. J. Stored Prod. Res. 29:271-275.

AliNiazee, M. T. 1972. Susceptibility of confused and red flour beetles to anoxia produced by helium and nitrogen at various temperatures. J. Econ. Entomol. 65:60-64.

Anderson, J. 1989. Food Irradiation: An alternative food processing technology. Food Development Division, Agriculture Canada, Ottawa, ON. 17 pp.

Andrieu, A. J., F. Fleurat-Lessard, and R. G. Busnel. 1977. Effect of high intensity sound waves on stored flour insect pests. Int. Cong. Acoustics, Madrid, 1977, Paper J-11.

Andrieu, A. J., F. Fleurat-Lessard, and R. G. Busnel. 1978. Disinsectisation of stored grain by high powered sound waves. pp. 249-265 in: Effects of noise on wildlife, J. L. Flectcher and R. G. Busnel (eds.). Academic Press, New York.

Annis, P. C. 1987. Towards rational controlled atmosphere dosage schedules: a review of current knowledge. pp. 128-148 in: Proc. Fourth Int. Working Conf. Stored-Product Prot., E. Donahaye and S. Navarro (eds.). Tel Aviv, Israel.

Annis, P. C. and J. van S. Graver. 1990. Suggested recommendations for the fumigation of grain in the ASEAN region. Part 2. Carbon dioxide fumigation of bag-stacks sealed in plastic enclosures: an operation manual. ASEAN Food Handling Bureau, Kuala Lumpur. 58 pp.

Annis, P. C., H. J. Banks, and Sukardi. 1984. Insect control in stacks of bagged rice using CO_2 treatment and an experimental PVC-membrane enclosure. CSIRO (Aust.) Div. Entomol. Tech. Pap. No. 22, 38 pp.

Bahr, I. 1973. Untersuchungen uber die Verminderung des Schadlingsbesatzes im Getreide durch pneumatische Forderung. Nachr. Pflanz. DDR 27:232-237.

Bahr, I. 1975. Uber das Schadauftreten des Getreidekapuziners (*Rhyzopertha dominica* F.) und die Wirkung eines Saug- und Druck geblases auf den Befall im Getreide. Nachr. Pflanz. DDR 29:228-231.

Bahr, I. 1990. Reduction of stored product insects during pneumatic unloading of ship cargoes. pp. 1135-1144 in: Proc. Fifth Int. Working Conf. Stored-Product Prot., F. Fleurat-Lessard and P. Ducom (eds.). Bordeaux, France.

Bailey, S. W. 1955. Air-tight storage of grain; its effect on insect pests. I. *Calandra granaria* L. (Coleoptera: Curculiondae). Aust. J. Agric. Res. 6:33-51.

Bailey, S. W. 1962. The effects of percussion on insect pests of grain. J. Econ. Entomol. 55:301-304.

Bailey, S. W. 1965. Air-tight storage of grain; its effect on insect pests. IV. *Rhyzopertha dominica* (F.) and some other Coleoptera that infest stored grain. J. Stored Prod. Res. 1:25-33.

Bailey, S. W. 1969. The effects of physical stress in the grain weevil *Sitophilus granarius*. J. Stored Prod. Res. 5:311-324.

Bailey, S. W. and H. J. Banks. 1980. A review of recent studies of the effects of controlled atmospheres on stored product pests. pp. 101-118 in: Controlled Atmosphere Storage of Grains: an International Symposium, J. Shejbal (ed.). Elsevier, Amsterdam.

Baker, K. D., R. L. Stroshine, K. J. Magee, G. H. Foster, and R. B. Jacko. 1986. Grain damage and dust generation in a pressure pneumatic conveying system. Trans. Am. Soc. Agric. Eng. 29:840-847.

Baker, V. H., D. E. Wiant, and G. Taboada. 1956. Some effects of microwaves on certain insects which infest wheat and flour. J. Econ. Entomol. 49:33-37.

Banks, H. J. 1976. Physical control of insects—recent developments. J. Aust. Entomol. Soc. 15:89-100.

Banks, H. J. 1981. Effects of controlled atmosphere storage on grain quality: a review. Food Technol. Aust. 33:335-340.

Banks, H. J. 1984. Current methods and potential systems for production of controlled atmospheres for grain storage. pp. 523-542 in: Controlled Atmosphere and Fumigation in Grain Storages, B. E. Ripp, H. J. Banks, E. J. Bond, D. J. Calverley, E. G. Jay, and S. Navarro (eds.). Elsevier, Amsterdam, Netherlands.

Banks, H. J. 1987. Impact, physical removal and exclusion for insect control in stored products. pp. 165-184 in: Proc. Fourth Int. Working Conf. Stored-Product Prot., E. Donahaye and S. Navarro (eds.). Tel Aviv, Israel.

Banks, H. J. and P. C. Annis. 1977. Suggested procedures for controlled atmosphere storage of dry grain. CSIRO (Aust.) Div. Entomol. Tech. Pap. No. 13. 23 pp.

Banks, H. J. and P. C. Annis. 1984. On criteria for success of phosphine fumigations based on observation of gas distribution patterns. pp. 327-341 in: Controlled Atmosphere and Fumigation in Grain Storages, B. E. Ripp, H. J. Banks, E. J. Bond, D. J. Calverley, E. G. Jay, and S. Navarro (eds.). Elsevier, Amsterdam.

Banks, H. J. and P. C. Annis. 1990. Comparative advantages of high CO_2 and low O_2 types of controlled atmospheres for grain storage. pp. 93-122 in: Food Preservation by Modified Atmospheres, M. Calderon and R. Barkai Golan (eds.). CRC Press, Roca Baton, FL.

Banks, H. J. and B. E. Ripp. 1984. Sealing of grain storages for use with fumigants and controlled atmospheres. pp. 375-390 in: Proc. Third Int. Working Conf. Stored-Product Entomol. Manhattan, KS.

Banks, H. J., P. C. Annis, R. C. Henning, and A. D. Wilson. 1980. Experimental and commercial modified atmosphere treatments of stored grain in Australia. pp. 207-224 in: Controlled Atmosphere Storage of Grains, J. Shejbal (ed.). Elsevier, Amsterdam.

Banks, H. J., P. C. Annis, and G. R. Rigby. 1990. Controlled atmosphere storage of grain: the known and the future. pp. 695-706 in: Proc. Fifth Int. Working Conf. Stored-Product Prot., F. Fleurat-Lessard and P. Ducom (eds.). Bordeaux, France.

Bedi, S. S. and Singh, M. 1992. Microwaves for control of stored grain insects. Nat. Acad. Sci. Letters 15: 195-197.

Bell, C. H. and D. M. Armitage. 1992. Alternative storage practices. pp. 249-311 in: Storage of Cereal Grains and their Products, D. B. Sauer (ed.). Am. Assoc. Cereal Chem., St. Paul, MN.

Bell, C. H., E. C. Spratt, and D. J. Mitchell. 1980. The effect of nitrogen and carbon dioxide on eggs of *Ephestia cautella* (Walker) and *E. kuehniella* Zeller (Lepidoptera: Pyramidae). Bull. Entomol. Res. 70:293-298.

Bond, J. 1984. Manual of fumigation for insect control. FAO Plant Production and Protection Paper No. 54. Rome, Italy. 432 pp.

Bond, J. and C. T. Buckland. 1979. Development of resistance to carbon dioxide in the granary weevil. J. Econ. Entomol. 72:770-771.

Bourgeois, L. 1993. Vigour loss of wheat caused by threshing. M.Sc. Thesis, University of Manitoba, Winnipeg, MB. 115 pp.

Brokerhof, A. W., H. J. Banks, and R. Morton. 1992. A model for time-temperature-mortality relationships for eggs of the webbing clothes moth, *Tineola bisselliella* (Lepidoptera: Tineidae), exposed to cold. J. Stored Prod. Res. 28:269-277.

Brooker, D. B., F. W. Bakker-Arkema, and C. A. Hall. 1974. Drying cereal grains. AVI Pub. Co., Westport, CT. 266 pp.

Brower, J. H. 1973. Sensitivity of *Tenebrio molitor* and *T. obscurus* to gamma irradiation. J. Econ. Entomol. 66:1175-1179.

Brower, J. H. 1979. Effects of irradiation on reproductive biology of adult tobacco moth, *Ephestia elutella*. J. Georgia Entomol. Soc. 14:201-209.

Brower, J. H. and E. W. Tilton. 1972a. Comparative gamma-radiation sensitivity of the sawtoothed grain beetle and the merchant grain beetle. Environ. Entomol. 1:735-738.

Brower, J. H. and E. W. Tilton. 1972b. Gamma-radiation effects on *Trogoderma inclusum* and *T. variable*. J. Econ. Entomol. 65:250-254.

Brower, J. H. and E. W. Tilton. 1973. Comparative gamma-radiation sensitivity of *Tribolium madens* (Charpentier) and *T. castaneum* (Herbst). J. Stored Prod. Res. 9:93-100.

Brown, G. A., J. H. Brower, and E. W. Tilton. 1972. Gamma radiation effects on *Sitophilus zeamais* and *S. granarius*. J. Econ. Entomol. 65:203-205.

Bryan, J. M. and J. Elvidge. 1977. Mortality of adult grain beetles in sample delivery systems used in terminal grain elevators. Can. Entomol. 109:209-213.

Buscarlet, L. A., B. Aminian, and C. Bali. 1987. Effect of irradiation and exposure to nitrogen on mortality of adults of *Tribolium confusum* J. du Val. pp. 186-193 in: Proc. Fourth Int. Working Conf. Stored-Product Prot., E. Donahaye and S. Navarro (eds.). Tel Aviv, Israel.

Bull, J. O. and P. B. Cornwell. 1966. A comparison of the susceptibility of the grain weevil *Sitophilus granarius* (L.) to accelerated electrons and ^{60}Co gamma radiation. pp. 157-175 in: The Entomology of Radiation Disinfestation of Grain, P. B. Cornwall (ed.). Pergamon Press, Oxford.

Burkholder, W. E., E. W. Tilton, and R. R. Cogburn. 1966. Effects of gamma radiation on the grain mite, *Acarus siro*. J. Econ. Entomol. 59:976-980.

Calderon, M. and R. Barkai-Golan. 1990. Food Preservation in Modified Atmospheres. CRC Press, Roca Baton, FL. 402 pp.

Calderon, M. and S. Navarro. 1979. Increased toxicity of low oxygen atmospheres supplemented with carbon dioxide on *Tribolium castaneum* adults. Entomol. Exp. App. 25:39-44.

Calderon, M., J. G. Leesch, and E. G. Jay. 1991. Toxicity of high oxygen atmospheres to *Tribolium castaneum* (Coleoptera: Tenebrionidae). J. Entomol. Sci. 26: 197-204.

Carney, G. C. 1959. Differential response of male and female adults of *Trogoderma granarium* Everts towards sterilizing doses of gamma-radiation. Nature 183: 338-339.

Champ, B. R., E. Highley, and H. J. Banks (eds.). 1990. Fumigation and Controlled Atmosphere Storage of Grain. ACIAR, Proc. No. 25, Aust. Centre Inst. Agric. Res., Canberra, Australia. 301 pp.

Champ, B. R., W. S. Murray, and M. G. Jefferies. 1981. Grain storage and pest problems in China. Bulk. Wheat 15:21-26.

Chelkowski, S. 1991. Mycological quality of mixed seeds and ingredients. Dev. Food. Sci. 26:217-227.

Claflin, J. K., D. E. Evans, A. G. Fane, and R. J. Hill. 1986. Thermal disinfestation of wheat in a spouted bed. J. Stored Prod. Res. 22:153-161.

Cline, L. D. 1970. Indian-meal moth egg hatch and subsequent larval survival after short exposures to low temperature. J. Econ. Entomol. 63:1081-1083.

Cline, L. D. 1978. Clinging and climbing ability of larvae of eleven species of stored-product insects on nine flexible packaging materials and glass. J. Econ. Entomol. 71:689-691.

Cline, L. D. and H. A. Highland. 1981. Minimum size of holes allowing passage of adults of stored-product Coleoptera. J. Georgia Entomol. Soc. 16:525-531.

Cline, L. D. and H. A. Highland. 1987. Survival of four species of stored-product insects confined with food in vacuumized and unvacuumized film pouches. J. Econ. Entomol. 80:73-76.

Cline, L. D. and J. W. Press. 1990. Reduction in almond moth (Lepidoptera: Pyralidae) infestations using commercial packaging of foods in combination with the parasitic wasp, *Bracon hebetor* (Hymenoptera: Braconidae). J. Econ. Entomol. 83:1110-1113.

Cogburn, R. R., E. W. Tilton, and W. E. Burkholder. 1966. Gross effects of gamma radiation on the Indian-meal moth and the Angoumois grain moth. J. Econ. Entomol. 59:682-685.

Cogburn, R. R., E. W. Tilton, and J. H. Brower. 1972. Bulk-grain gamma irradiation for control of insects infesting wheat. J. Econ. Entomol. 65:818-821.

Cogburn, R. R., E. W. Tilton, and J. H. Brower. 1973. Almond moth: gamma radiation effects on the life stages. J. Econ. Entomol. 66:745-751.

Cornwell, P. B., L. J. Crook, and J. O. Bell. 1957. Lethal and sterilizing effects of gamma radiation on insects infesting cereal commodities. Nature. 179:670-672.

Cox, P. D. 1987. Cold tolerance and factors affecting the duration of diapause in *Ephestia kuehniella* Zeller (Lepidoptera: Pyralidae). J. Stored Prod. Res. 23:163-168.

Cunnington, A. M. 1965. Physical limits for complete development of the grain mite, *Acarus siro* L. (Acarina, Acaridae), in relation to its world distribution. J. Appl. Ecol. 2: 295-306.

Daumal, J., P. Jourdheuil, and R. Tomassone. 1974. Variabilité des effects letaux des basses temperatures en fonction du stade de developpement embryonnaire chez la pyrale de la farine (*Anagasta kuhniella* Zell., Lepid., Pyralidae). Ann. Zool. -Ecol. Anim. 6:229-244.

David, M. H., R. B. Mills, and G. D. White. 1977. Effects of low temperature acclimation on developmental stages of stored-product insects. Environ. Entomol. 6:181-184.

Dermott, T. and D. E. Evans. 1978. An evaluation of fluidized-bed heating as a means of disinfesting wheat. J. Stored Prod. Res. 14:1-12.

Desmarchelier, J. M. 1988. The relationship between wet-bulb temperature and intrinsic rate of increase of eight species of stored-product Coleoptera. J. Stored Prod. Res. 24:107-113.

Desmarchelier, J. M. and J. C. Dines. 1987. Dryacide treatment of stored wheat: its efficacy against insects, and after processing. Aust. J. Exp. Agric. 27:309-312.

Donahaye, E. 1990. The potential for stored-products insects to develop resistance to modified atmospheres. pp. 989-996 in: Proc. Fifth Int. Working Conf. Stored Prod. Prot., F. Fleurat-Lessard and P. Ducom (eds.). Bordeaux, France.

Ebeling, W. 1971. Sorptive dusts for pest control. Ann. Rev. Entomol. 16:123-158.

Erdman, H. E. 1973. Effects of chronic ^{60}Co (gamma) radiation, dose rate, temperature and cohabitation on population performance of flour beetles. Environ. Entomol. 2:41-48.

Evans, D. E. 1982. The influence of temperature and grain moisture content on the intrinsic rate of increase of *Sitophilus oryzae* (L.) (Coleoptera: Curculionidae). J. Stored Prod. Res. 18:55-66.

Evans, D. E. 1983. The influence of relative humidity and thermal acclimation on the survival of adult grain beetles in cooled grain. J. Stored Prod. Res. 19:173-180.

Evans, D. E. 1987a. Some biological and physical constraints to the use of heat and cold for disinfesting and preserving stored products. pp. 149-164 in: Proc. Fourth Int. Working Conf. Stored-Product Prot., E. Donahaye and S. Navarro (eds.). Tel Aviv, Israel.

Evans, D. E. 1987b. The survival of immature grain beetles at low temperatures. J. Stored Prod. Res. 23:79-83.

Evans, D. E. and T. Dermott. 1981. Dosage-mortality relationships for *Rhyzopertha dominica* (F.) (Coleoptera: Bostrychidae) exposed to heat in a fluidized-bed. J. Stored Prod. Res. 17:53-64.

Evans, D. E. and J. van S. Graver. 1989. Integrated pest and commodity management: putting first things first. pp. 206-217 in: Grain Postharvest systems, B. M. de Mesa (ed.). ASEAN Postharvest Program, Bangkok, Thailand.

Evans, D. E., G. R. Thorpe, and T. Dermott. 1983. The disinfestation of wheat in a continuous-flow fluidized bed. J. Stored Prod. Res. 19:125-137.

Fam, E. Z., A. K. M. El-Nahal, and H. Fahmy. 1974. Influence of grain moisture on the efficiency of silica aerogel and katelsous used as grain protectants. Bull. Entomol. Soc. Egypt Econ. Ser. 8:105-114.

Fields, P. G. 1990. The cold-hardiness of *Cryptolestes ferrugineus* and the use of ice nucleation active bacteria as a cold-synergist. pp. 1183-1191 in: Proc. Fifth Int. Working Conf. Stored-Product Prot., F. Fleurat-Lessard and P. Ducom (eds.). Bordeaux, France.

Fields, P. G. 1992. The control of stored-product insects and mites with extreme temperatures. J. Stored Prod. Res. 28:89-118.

Fields, P. G. 1993. Reduction of cold tolerance of stored-product insects by ice-nucleating-active bacteria. Environ. Entomol. 22:470-476.

Fleurat-Lessard, F. 1980. Lutte physique par l'air chaud ou les hautes fréquences contre les insectes des grains et des produits céréaliers. Bull. Tech. d'Inf. Cer. Stockées 349:345-352.

Fleurat-Lessard, F. 1985. Les traitements thermiques de désinfestation des céréales et des produits céréaliers: possibilité d'utilisation pratique et domain d'application. Bull. OEPP 15:109-118.

Fleurat-Lessard, F. 1987. Control of storage insects by physical means and modified environmental conditions. Feasibility and application. pp. 209-218 in: Stored Products Pest Control, T. J. Lawson (ed.). British Crop Protection Council, London.

Fleurat-Lessard, F. 1989. La désinsectisation des stocks de farine. Industries des Céréales 60:17-22.

Fleurat-Lessard, F. and J. M. Le Torc'h. 1987. Practical approach to purging grain with low oxygen atmosphere for disinfestation of large wheat bins against the granary weevil, *Sitophilus granarius*. pp. 208-217 in: Proc. Fourth Int. Working Conf. Stored-Product Prot., E. Donahaye and S. Navarro (eds.). Tel Aviv, Israel.

Golob, P. and D. J. Webley. 1980. The use of plants and minerals as traditional protectants of stored products. Report Trop. Prod. Inst. G138, 32 pp.

Gonen, M. 1977a. Susceptibility of *Sitophilus granarius* and *S. oryzae* (Coleoptera: Curculionidae) to high temperature after exposure to supra-optimal temperature. Entomol. Exp. Appl. 21:243-248.

Gonen, M. 1977b. Survival and reproduction of heat-acclimated *Sitophilus granarius* (Coleoptera, Curculionidae) at a moderately high temperature. Entomol. Exp. Appl. 21:249-253.

Gowers, S. L. and G. N. J. Le Patourel. 1984. Toxicity of deposits of an amorphous silica dust on different surfaces and their pick-up by *Sitophilus granarius* (L.) (Coleoptera: Curculionidae). J. Stored Prod. Res. 20:25-29.

Gras, P. W. and M. L. Bason. 1990. Biochemical effects of storage atmospheres in grain and grain quality. pp. 83-91 in: Fumigation and Controlled Atmosphere Storage of

Grain: B. R. Champ, E. Highley, and H. J. Banks (eds.). ACIAR, Proc. No. 25, Aust Centre Int. Agric. Res., Canberra, Australia.

Green, A. A. and P. S. Tyler. 1966. A field comparison of malathion, dichlorvos and fenitrothion for control of *Oryzaephilus surinamensis* (L.) (Coleoptera: Silvanidae) infesting stored barley. J. Stored Prod. Res. 1:273-285.

Hafner, T. 1975. Traitment sans dommage par chauffage HF des denrées alimentaires à stocker. Revue Brunn Boveri 1/2: 52-54.

Harein, P. K. and A. F. Press. 1968. Mortality of stored-peanut insects exposed to mixtures of atmospheric gases at various temperatures. J. Stored Prod. Res. 4:77-82.

Highland, H. A. 1978. Insect resistance of food packages—a review. J. Food Process. Preserv. 2:123-130.

Highland, H. A. 1981. Resistant barriers for stored-product insects. pp. 41-46 in: CRC Handbook of Transportation and Marketing in Agriculture, E. E. Finney (ed.). Vol. 1. CRC Press, Boca Raton, FL.

Highland, H. A. 1991. Protecting packages against insects. pp. 345-350 in: Ecology and Management of Food-Industry Pests, J. R. Gorham (ed.). Assoc. Official Anal. Chem. Arlington, VA.

Hooking, A. D. 1990. Responses of fungi to modified atmospheres. pp. 70-82 in: Fumigation and Controlled Atmosphere Storage of Grain: B. R. Champ, E. Highley, and H. J. Banks (eds.). ACIAR, Proc. No. 25, Aust. Centre Inst. Agric. Res., Canberra, Australia.

Howe, R. W. 1965. A summary of estimates of optimal and minimal conditions for population increase of some stored products insects. J. Stored Prod. Res. 1:177-184.

Howe, R. W. and B. D. Hole. 1968. The susceptibility of developmental stages of *Sitophilus granarius* (L.) (Coleoptera: Curculionidae) to moderately low temperatures. J. Stored Prod. Res. 4:147-156.

Hyde, M. B., A. A. Baker, A. C. Ross, and C. O. Lopez. 1973. Airtight grain storage. FAO Agric. Serv. Bull. 17. 71 pp.

Jay, E. G. 1971. Suggested conditions and procedures for using carbon dioxide to control insects in grain storage facilities. USDA-ARS 51-46, Sept. 1971. 6 pp.

Jay, E. G. 1980a. Methods of applying carbon dioxide for insect control in stored grain. USDA-SEA, Advances in Agricultural Technology, Southern Series, S-13. 7 pp.

Jay, E. G. 1980b. Low temperatures: Effects on control of *Sitophilus oryzae* (L.) with modified atmospheres. pp. 65-71 in: Controlled Atmosphere Storage of Grains. J. Shejbal (ed.). Elsevier, Amsterdam.

Jay, E. G. 1984a. Imperfections in our current knowledge of insect biology as related to their response to controlled atmospheres. pp. 493-508 in: Controlled Atmosphere and Fumigation in Grain Storages, B. E. Ripp (eds.). Elsevier, Amsterdam.

Jay, E. 1984b. Control of *Rhyzopertha dominica* with modified atmospheres at low temperatures. J. Agric. Entomol. 1:155-160.

Jay, E. G. 1986. Factors affecting the use of carbon dioxide for treating raw and processed agricultural products. pp. 173-189 in: GASGA Seminar on Fumigation Technology in Developing Countries, V. Howe (ed.). Slough, UK.

Jay, E. G., R. T. Arbogast, and G. C. Pearman. 1971. Relative humidity: its importance in the control of stored-product insects with modified atmospheric gas concentrations. J. Stored Prod. Res. 6:325-329.

Jayas, D. S., N. D. G. White, W. E. Muir, and R. N. Sinha. 1993. Controlled atmosphere storage of cereals and oilseeds. J. Appl. Zool. Res. 4:1-12.

Johnson, J. A. and P. L. Wofford. 1991. Effects of age on response of eggs of Indian meal moth and Navel orangeworm (Lepidoptera: Pyralidae) to subfreezing temperatures. J. Econ. Entomol. 84:202-205.

Khan, M. A. 1983. Untersuchungen über die Invasion von Eilaven von vorratsschädlichen Insekten durch verschieden grosse Poren des Verpackungsmaterials. Anzeiger für Schädlingskunde, Pflanzenschutz, Umweltschutz 56:65-67.

Kato, K. and R. Yamashita. 1987. Stored-grain insect control by dielectric heating. J. Japan Soc. Agric. Machinery 49:443-450.

Kawamoto, H., S. M. Woods, R. N. Sinha, and W. E. Muir. 1989. A simulation model of population dynamics of the rusty grain beetle, *Cryptolestes ferrugineus*, in stored wheat. Ecol. Model. 48:137-157.

Kirkpatrick, R. L. and A. Cagle. 1978. Controlling insects in bulk wheat with infrared radiation. J. Kansas Entomol. Soc. 51:386-393.

Kirkpatrick, R. L. and E. W. Tilton. 1972. Infrared radiation to control adult stored-product Coleoptera. J. Georgia Entomol. Soc. 7:73-75.

Kline, G. L. 1973. Mechanical damage levels in shelled corn from farms. Paper 73-331. Am. Soc. Agric. Eng., St. Joseph, MI. 7 pp.

Krishnamurthy, T. S., E. C. Spratt, and C. H. Bell. 1986. The toxicity of carbon dioxide to adult beetles in low oxygen atmospheres. J. Stored Prod. Res. 22:145-151.

La Hue, D. W. 1965. Evaluation of malathion, synergized pyrethrum, and diatomaceous earth as wheat protectants in small bins. Mktg. Res. Rep. USDA No. 726, Washington. 13 pp.

La Hue, D. W. 1978. Insecticidal dusts: grain protectants during high temperature—low humidity storage. J. Econ. Entomol. 71:230-232.

Lapp, H. M., F. J. Madrid, and L. B. Smith. 1986. A continuous thermal treatment to eradicate insects in stored wheat. Paper 86-3008. Am. Soc. Agric. Eng., St. Joseph, MI. 14 pp.

Lee, C. 1959. Baking quality and maltose value of flour irradiated with cobalt-60 γ rays. Cereal Chem. 36:70-77.

Lee, R. E., J. M. Strong-Gunderson, M. R. Lee, and E. C. Davidson. 1992. Ice nucleating active bacteria decrease the coldhardiness of stored-grain insects. J. Econ. Entomol. 85:371-374.

Le Patourel, G. N. J. 1986. The effect of grain moisture content on the toxicity of a sorptive silica dust to four species of grain beetle. J. Stored Prod. Res. 22:63-69.

Le Patourel, G. N. J. and J. Singh. 1984. Toxicity of amorphous silicas and silica pyrethroid mixtures to *Tribolium castaneum* (Herbst) (Coleoptera: Tenebrionidae). J. Stored Prod. Res. 20:183-190.

Le Patourel, G. N. J., M. Shawir, and F. I. Moustafa. 1989. Accumulation of mineral dusts from wheat by *Sitophilus oryzae* (L.) (Coleoptera: Curculionidae). J. Stored Prod. Res. 25:65-72.

Le Torc'h, J. M. and F. Fleurat-Lessard. 1991. Effet des fortes pressions sur l'efficacité insecticide des atmosphères modifées par CO_2 contre *Sitophilus granarius* (L.) et *S.*

oryzae (Coleoptera: Curculionidae). pp. 847-857 in: Proc. Fifth Int. Working Conf. Stored-Product Prot. F. Fleurat-Lessard and P. Dacom (eds.). Bordeaux, France.

Levinson, H. Z. and A. R. Levinson. 1989. Food storage and storage protection in ancient Egypt. Boletin de Sanidad Vegetal 17:475-482.

Li, L. and R. T. Arbogast. 1991. The effect of grain breakage on fecundity, development, survival, and population increase in maize of *Tribolium castaneum* (Herbst) (Coleoptera, Tenebrionidae). J. Stored Prod. Res. 27:87-94.

Liu, R., Y. Xiong, and C. Wang. 1983. Studies on heat treatment of wheat grains in China. pp. 443-452 in: Chemistry and World Food Supplies: The New Frontiers, L. W. Sheritt (ed.). Pergamon Press, Oxford.

Liang, Q. 1990. The current status of fumigation and controlled atmosphere storage technologies in China. pp. 166-173 in: Fumigation and Controlled Atmosphere storage of Grain, B. R. Champ, E. Highley, and H. J. Banks (eds.). ACIAR, Proc. No. 25, Aust. Centre Int. Agric. Res., Canberra, Australia.

Lindgren, D. L. and L. E. Vincent. 1970. Effect of atmospheric gases alone or in combination on the mortality of granary and rice weevils. J. Econ. Entomol. 63: 1926-1929.

Locatelli, D. and E. Daolio. 1993. Effectiveness of carbon dioxide under reduced pressure against some insects infesting packaged rice. J. Stored Prod. Res. 29:81-87.

Locatelli, D. P. and S. Traversa. 1989. Microwaves in the control of rice infestations. Ital. J. Food Sci. 2:53-62.

Loschiavo, S. 1978. Effect of disturbance of wheat on four species of stored-product insects. J. Econ. Entomol. 71:888-893.

Loschiavo, S. R. 1988a. Safe method of using silica aerogel to control stored-product beetles in dwellings. J. Econ. Entomol. 81:1231-1236.

Loschiavo, S. R. 1988b. Availability of food as a factor in effectiveness of a silica aerogel against the merchant grain beetle (Coleoptera: Cucujidae). J. Econ. Entomol. 81: 1237-1240.

Maceljski, M. and Z. Korunic. 1971. The results of investigation of the use of inert dusts in water suspensions against stored-product insects. Zast. Bilja 23:376-387.

Maceljski, M. and Z. Korunic. 1972. A contribution to the knowledge of the mechanism of action of the inert dusts on insects. Zast. Bilja 23:49-64.

Maier, D. E. 1993. Grain chilling: a nonchemical storage technology. pp. 64-65 in: Practical Use of Fumigants and Pheromones. An International Technical Conference and Workshop. December 1-3, Lübeck, Germany.

Mathlein, R. 1971. Mechanical cleaning of infested grain as a control method against some insect pests. Meddn St. VäxtstAnst. 152:209-227.

McGaughey, W. H. 1972. Diatomaceous earth for confused flour beetle and rice weevil control in rough, brown, and milled rice. J. Econ. Entomol. 65:1427-1428.

McGaughey, W. H. and R. G. Akins. 1989. Application of modified atmospheres in farm grain storage bins. J. Stored Prod. Res. 25:201-210.

Mitsura, A., R. Amano, and H. Tanabe. 1973. The acacricidal effects of compressed gas treatments on the grain mite, *Tyrophagus putrescentiae*. Fd. Hyg. Soc. Japan 14:511-516.

More, H. G., B. C. Stenning, and N. Magan. 1992. Effect of high temperature treatment on disinfestation and quality characteristics of sorghum. Ann. Appl. Biol. 120:161-171.

Muir, W. E., G. Yacuik, and R. N. Sinha. 1977. Effects on temperature and insect and mite population of turning and transferring farm-stored wheat. Can. Agric. Eng. 19:25-28.

Mullen, M. A. 1973. Low frequency sound affecting the development of the Indian Meal moth. J. Georgia Entomol. Soc. 8:320-321.

Murdock, L. L. and R. W. Shade. 1991. Eradication of cowpea weevil (Coleoptera: Bruchidae) in cowpeas by solar heating. Am. Entomol. 37:228-231.

Nagel, R. H. and H. H. Sheppard. 1934. The lethal effect of low temperatures on the various stages of the confused flour beetle. J. Agric. Res. 48:1009-1016.

Nataredja, Y. C. and R. J. Hodges. 1990. Commercial experience of sealed storage of bag stacks in Indonesia. pp. 197-202 in: Fumigation and Controlled Atmosphere storage of Grain, B. R. Champ, E. Highley, and H. J. Banks (eds.). ACIAR, Proc. No. 25, Aust. Centre Int. Agric. Res., Canberra, Australia.

Navarro, S. 1978. The effects of low oxygen tensions on three stored-product insect pests. Phytopar. 6:51-58.

Navarro, S. and M. Calderon. 1973. Carbon dioxide and relative humidity: interrelated factors affecting the loss of water and mortality of *Ephestia cautella* (Walker) (Lepidoptera: Phyciticae). Israel J. Entomol. 8:143-152.

Navarro, S. and M. Calderon. 1974. Exposure of *Ephestia cautella* (Wlk.) pupae to carbon dioxide concentrations at different relative humidities: the effect on adult emergence and loss of weight. J. Stored Prod. Res. 10:237-241.

Navarro, S., E. Donahaye, and M. Calderon. 1973. Studies on aeration with refrigerated air—I. Chilling of wheat in a concrete elevator. J. Stored Prod. Res. 9:253-259.

Nelson, S. O. 1986. Potential agricultural application for RF and microwave energy. Paper 86-6539. Am. Soc. Agric. Eng., St. Joseph, MI. 11 pp.

Nelson, S. O. and B. H. Kantack. 1966. Stored-grain insect control studies with radio-frequency energy. J. Econ. Entomol. 59:588-594.

Nelson, S. O. and L. E. Stetson. 1974. Comparative effectiveness of 39- and 2450-MHz electric fields for control of rice weevils in wheat. J. Econ. Entomol. 67:592-595.

Newman, C. J. E. 1990. Specification and design of enclosures for gas treatment. pp. 107-130 in: Fumigation and Controlled Atmosphere storage of Grain, B. R. Champ, E. Highley, and H. J. Banks (eds.). ACIAR, Proc. No. 25, Aust. Centre Int. Agric. Res., Canberra, Australia.

Newton, J. 1988. Insects and packaging—a review. Int. Biodet. 24:175-187.

Ng, P. K. W., W. Bushuk, and J. Borsa. 1989. Effect of gamma ray and high-energy electron irradiations on breadmaking quality of two Canadian wheat cultivars. Can. Inst. Food Sci. Technol. J. 22:173-176.

O'Donnell, M. J. and J. Machin. 1991. Ion activities and electrochemical gradients in the mealworm rectal complex. J. Exp. Biol. 155:375-402.

Ohguchi, Y., H. Suzuki, S. Tatsuki, and J. Fukami. 1983. The lethal effect of oxygen absorber (Ageless®) on several stored grain and clothes pest insects. Japan J. Appl. Entomol. Zool. 27:270-275.

Pearman, G. C. and E. G. Jay. 1970. The effect of relative humidity on the toxicity of carbon dioxide to *Tribolium castaneum* in peanuts. J. Georgia Entomol. Soc. 5:61-64.

Permual, D. and G. Le Patourel. 1990. Laboratory evaluation of acid-activated kaolin to protect stored paddy against infestation by stored product insects. J. Stored Prod. Res. 26:149-153.

Pradzynska, A. 1982. The suitableness of ultrasounds in control of storage pests. II. The effectiveness of treatment with ultrasonic waves on all stages of granary weevil (*Sitophilus granarius* L.). Prace naukowe Rustytutu Ochrony Roslin. 24:77-90.

Prozell, S. and C. Reichmuth. 1990. Response of the granary weevil *Sitophilus granarius* (L.) (Coleoptera: Curculionidae) to controlled atmospheres under high pressure. pp. 911-918 in: Proc. Fifth Int. Working Conf. Stored-Product Prot., F. Fleurat-Lessard and P. Ducom (eds.). Bordeaux, France.

Quarles, W. 1992a. Diatomaceous earth for pest control. IPM Pract. 14:1-11.

Quarles, W. 1992b. Silica gel for pest control. IPM Pract. 15:1-11.

Quentin, M. E., J. L. Spencer, and J. R. Miller. 1991. Bean tumbling as a control measure for the common bean weevil, *Acanthoscelides obtectus*. Ent. Exp. Appl. 60:105-109.

Rees, D. P. 1985. Review of the response of stored product insects to light of various wavelengths, with particular reference to the design and use of light traps for population monitoring. Trop. Sci. 25:197-213.

Reichmuth, C. 1987. Low oxygen content to control stored product insects. pp. 194-207 in: Proc. Fourth Int. Working Conf. Stored-Product Prot., E. Donahaye and S. Navarro (eds.). Tel Aviv, Israel.

Richard-Molard, D., B. Diawara, and B. Cahagnier. 1987. Susceptibility of cereal microflora to oxygen deficiency and carbon dioxide concentration. pp. 85-92 in: Proc. Fourth Int. Working Conf. Stored-Product Prot., E. Donahaye and S. Navarro (eds.). Tel Aviv, Israel.

Ripp, B. E., H. J. Banks, E. J. Bond, D. J. Calverley, E. G. Jay, and S. Navarro (eds.). 1984. Controlled Atmosphere and Fumigation in Grain Storages. Proc. Int. Symp., Elsevier, Amsterdam. 798 pp.

Robinson, W. 1926. Low temperature and moisture as factors in the ecology of Rice weevil, *Sitophilus oryza* L. and the granary weevil, *Sitophilus granarius* L. Univ. Minn. Agric. Exp. Stat. Tech. Bull. 41:40 pp.

Ross, T. E. 1981. Diatomaceous earth as a possible alternative to chemical insecticides. Agric. Environ. 6:43-51.

Shejbal, J. (ed.). 1980. Controlled Atmosphere Storage of Grains. An Int. Symp., Elsevier, Amsterdam. 608 pp.

Shipp, E. 1983. Radiation entomology in food preservation. pp. 27-31 in: Ionizing Energy Treatment of Foods. Nat. Symp. Proc., Sydney, Australia.

Smith, L. B. 1970. Effects of cold-acclimation on supercooling and survival of the rusty grain beetle, *Cryptolestes ferrugineus* (Stephens) (Coleoptera: Cucujidae), at subzero temperatures. Can. J. Zool. 48:853-858.

Spillman, R. W. 1989. Economics of gas separation membranes. Chem. Eng. Prog. 85:42-62.

Spratt, E. C. 1975. Some effects of the carbon dioxide absorbency of humidity controlling solutions on the results of life history experiments with stored products insects. J. Stored. Prod. Res. 11:127-134.

Spratt, E. C. 1979. Some effects of a mixture of oxygen, carbon dioxide and nitrogen in ratio 1:1:8 on the oviposition and development of *Sitophilus zeamais* Mots. (Coleoptera: Curculionidae). J. Stored Prod. Res. 15:73-80.

Spratt, E. C., G. Dignan, and H. J. Banks. 1985. The effects of high concentration of carbon dioxide in air an *Trogoderma granarium* Everts. (Coleoptera: Dermestidae). J. Stored Prod. Res. 4:41-46.

Storey, C. L. 1973. Exothermic inert atmosphere generators for control of insects in stored wheat. J. Econ. Entomol. 66:511-514

Storey, C. L., C. R. Martin, and D. R. Sukkestad. 1981. Concentrations of 80:20 grain fumigant (CCL_4-CS_2) in handling equipment during transfers of fumigated wheat. J. Econ. Entomol. 74:188-190.

Stratil, H. H. and C. H. Reichmuth. 1984. Uberlebensdauer von Eiern der vorratsschadlichen Motten *Ephestia cautella* (Wlk.) und *Ephestia elutella* (Hbn.) (Lepidoptera, Pyraloidea) bei Temperaturen unterhalb ihres Entwicklungsminimums. Z. ang. Entomol. 97:63-70.

Stratil, H., R. Wohlgemuth, H. Bolling, and H. Zwingelberg. 1987. Optimierung des Prallmaschinenverfahrens zum Abtoten und Entfernen von vorratsschadlichen Insekten aus Nahrungsmitteln unter besonderer Berucksichtigung der Qualitat der Mahlerzeugnisse. Getreide, Mehl und Brot 41:294-302.

Sutherland, J. W., D. E. Evans, A. G. Fane, and G. R. Thorpe. 1987. Disinfestation of grain with heated air. pp. 261-274 in: Proc. Fourth Int. Working Conf. Stored-Product Prot., E. Donahaye and S. Navarro (eds.). Tel Aviv, Israel.

Sutherland, J. W., P. W. Fricke, and R. J. Hill. 1989. The entomological and thermodynamic performance of pneumatic conveyor wheat disinfestor using heated air. J. Agric. Eng. Res. 44:113-124.

Sutherland, J. W., G. R. Thorpe, and P. W. Fricke. 1986. Grain disinfestation by heating in a pneumatic conveyor. pp. 419-425 in: Proc. Conf. Agric. Eng., Adelaide. Canberra, Institution of Engineers, Melbourne, Australia.

Thorpe, G. E. 1983. High temperature disinfestation of grain. CSIRO (Australia) Research Review 1983:41-47.

Thorpe, G. E., D. E. Evans, and J. W. Sutherland. 1984. The development of a continuous-flow fluidized-bed high-temperature grain disinfestation process. pp. 617-622 in: Controlled Atmosphere and Fumigation in Grain Storages, B. E. Ripp, H. J. Banks, E. J. Bond, D. J. Calverley, E. G. Jay, and S. Navarro (eds.). Elsevier, Amsterdam.

Tilton, E. W. and J. H. Brower. 1985. Supplemental treatments for increasing the mortality of insects during irradiation of grain. Food Technol. 39:75-79.

Tilton, E. W. and J. H. Brower. 1987. Ionizing radiation for insect control in grain and grain products. Cereal Foods World 32:330-333.

Tilton, E. W. and A. K. Burditt. 1983. Insects disinfestation of grain and fruit. pp. 215-229 in: Preservation of Food by Ionizing Radiation. E. S. Josephson and M. S. Peterson (eds.). Vol. III. CRC Press, Boca Raton, FL.

Tilton, E. W., J. H. Brower, and R. R. Cogburn. 1971. Critical evaluation of an operational bulk grain and packaged product irradiator. Int. J. Radiat. Eng. 1:49-50.

Tilton, E. W., W. E. Burkholder, and R. R. Cogburn. 1966a. Effect of gamma radiation on *Trogoderma glabrum* and *Attagenus piceus*. J. Econ. Entomol. 59:944-948.

Tilton, E. W., W. E. Burkholder, and R. R. Cogburn. 1966b. Effect of gamma radiation on *Rhyzopertha dominica*, *Sitophilus oryzae*, *Tribolium confusum*, and *Lassioderma serricorne*. J. Econ. Entomol. 59:1363-1368.

Tranchino, L., P. Agostinello, A. Constantini, and J. Shejbal. 1980. The first Italian large scale facilities for the storage of cereal grains in nitrogen. pp. 445-459 in: Controlled Atmosphere Storage of grain, J. Shejbal (ed.). Elsevier, Amsterdam.

Tuff, D. W. and H. S. Telford. 1964. Wheat fracturing as affecting infestation by *Cryptolestes ferrugineus*. J. Econ. Entomol. 57:513-516.

Turnock, W. J., R. J. Lamb, and R. P. Bodnaryk. 1983. Effects of cold stress during pupal diapause on the survival and development of *Mamestra configurata* (Lepidoptera: Noctuidae). Oecologia 56:185-192.

Ungsunantwiwat, A. and R. B. Mills. 1979. Influence of medium and physical disturbances during rearing on development and numbers of *Sitophilus* progeny. J. Stored Prod. Res. 15:37-42.

Urbain, W. M. 1986. Food irradiation. Academic Press, New York. 351 pp.

Vardell, H. H. and E. W. Tilton. 1981. Control of the lesser grain borer, *Rhyzopertha dominica* (F.) and the rice weevil , *Sitophilus oryzae* (L.) in rough rice with a heated fluidized bed. J. Georgia. Entomol. Soc. 16:521-524.

Verma, A. K. and S. R. Wadhi. 1978. Susceptibility of walnut pests to carbon dioxide and nitrogen and effect of gas storage on keeping quality of walnut kernels. Ind. J. Entomol. 40:290-298.

Watson, R. T., D. L. Albritton, S. O. Andersen, and L. Lee-Bapty. 1992. Methyl bromide: its atmospheric science, technology and economics. United Nations Environment Program, Nairobi. 42 pp.

Watters, F. L. 1966. The effects of short exposures to subthreshold temperatures on subsequent hatching and development of eggs of *Tribolium confusum* Duval (Coleoptera: Tenebrionidae). J. Stored Prod. Res. 2:81-90.

Watters, F. L. 1979. Potential of accelerated electrons for insect control in stored grains. pp. 278-286 in: Proc. Second Int. Working Conf. Stored Product Entomol. Ibadan, Nigeria.

Watters, F. L. 1991. Physical methods to manage stored-food pests. pp. 399-414 in: Ecology and Management of Food-Industry Pests, J. R. Gorham (ed.). Assoc. Off. Anal. Chem., Arlington, VA.

Watters, F. L. and K. F. MacQueen. 1967. Effectiveness of gamma irradiation for control of five species of stored-product insects. J. Stored Prod. Res. 3:223-234.

White, N. D. G. and D. S. Jayas. 1991. Control of insects and mites with carbon dioxide in wheat stored at cool temperatures in nonairtight bins. J. Econ. Entomol. 84:1933-1942.

White, N. D. G. and S. R. Loschiavo. 1989. Factors affecting survival of the merchant grain beetle (Coleoptera: Cucujidae) and the confused flour beetle (Coleoptera: Tenebrionidae) exposed to silica aerogel. J. Econ. Entomol. 82:960-969.

White, G. D., W. L. Berndt, and J. L. Wilson. 1975. Evaluating diatomaceous earth, silica-aerogel dusts, and malathion to protect stored wheat from insects. Mktg. Res. Rep. USDA No. 1038, Washington. 18 pp.

Wilson, A. D., S. A. Buchanan, and A. G. Sharp. 1984. The use in and recirculation of carbon dioxide in welded steel bins in Victoria. pp. 257-268 in: Controlled Atmosphere and Fumigation in Grain Storages, B. E. Ripp, H. J. Banks, E. J. Bond, D. J. Calverley, E. G. Jay, and S. Navarro (eds.). Elsevier, Amsterdam.

Winterbottom, D. C. 1922. Weevils in wheat and storage of grain in bags: A record of Australian experience during the war period (1915 to 1919). Govt. Printer, Adelaide. 122 pp.

Wohlgemuth, R. 1979. Protection of stored foodstuffs against insect infestation by packaging. Chem. Ind. 5:330-334.

Wohlgemuth, R. 1989. Longevity of stored products insects in grain in cold-storage warehouses. Anz. Schadlingskunde, Pflanzschutz, Umweltschutz, 62:114-119.

Yerington, A. P. 1978. Insects and package seal quality. Modern Pack. 51:41-42.

Zakladnoi, G. A. 1976. Regulation of the gas composition of the atmosphere for eliminating insects in grain. USSR Ministry of Procurement, Moscow. 3 pp.

Zakladnoi, G. A., A. I. Men'shenin, E. S. Pertsovskii, R. A. Salimov, V. G. Cherepkov, and V. S. Krsheminskii. 1982. Industrial application of radiation deinsectification of grain. Soviet Atomic Energy 52:74-78.

Zanon, K. 1980. Systems of supply of nitrogen for the storage of grains in controlled atmosphere. pp. 507-516 in: Controlled Atmosphere Storage of Grains, J. Shejbal, (ed.). Elsevier, Amsterdam.

Zhang Cong, Z., Y. Ye Bing, C. Cui Er, and G. Ding Han. 1981. Studies on the stored-grain pest control with high frequency and microwave dielectric heating. Tech. Bull. Plant Quarantine Res. China 6:1-17.

12

Engineering Properties of Grains and Oilseeds

Stefan Cenkowski and Qiang Zhang

University of Manitoba, Winnipeg, Manitoba, Canada

SUMMARY

This chapter reviews and summarizes the engineering properties of grains and oilseeds. These properties include geometrical, thermal, hygroscopic, and mechanical properties of both bulk grain and single grain kernels. Measurement techniques and laboratory procedures for the determination of these properties are presented and some theoretical and empirical models for characterizing these properties are discussed. The discussion is extended to interactions between grain and containing structures, and the effects of additives (e.g., oil for dust suppression) on engineering properties of grain. The trends of the future research activities in the areas concerning engineering properties of grain and oilseeds are addressed.

1. INTRODUCTION

The design and evaluation of facilities for processing, handling, and storing grain and oilseeds requires knowledge of the physical, thermal, and mechanical behavior of grain and oilseeds. Mathematical modeling and computer simulation are the most effective methods used by researchers to study the behavior of grains and oilseeds. Numerous mathematical models have bee developed which have been used to:

- control and optimize aeration or drying processes to minimize the energy usage cost (O'Callaghan et al. 1971, Bakker-Arkema 1986, Ryniecki et al. 1993; see Chapter 16).
- study the susceptibility of grain and oilseeds to thermal or mechanical stresses to reduce grain spoilage and preserve the quality during storage (Gunasekaran and Paulsen 1985, Tang and Sokhansanj 1990; see Chapter 14).
- predict mechanical behavior of bulk grain and interactions between grain and structures for use when designing grain storage and handling facilities (Zhang et al. 1986,1988; Thompson et al. 1988; Xu et al., 1993).

One of the most important factors which limits the accuracy of the model simulations and predictions is the availability of adequate data on the engineering properties of grain and oilseeds (Cenkowski et al. 1990, Jaros et al. 1992). For example, bulk density is a fundamental property of grain and oilseeds. Different techniques for determining bulk density (e.g., standard test weight and dielectric property test where a small coaxial sample holder is used) may give different results (Nelson 1980). Moreover, in a practical storage facility, the bulk density is difficult to interpret because the degree of packing depends on the method used to fill the container (Jayas et al. 1989).

It is generally agreed that the mechanism of moisture movement during the falling-rate period of grain drying is by diffusion. This can be described by Fick's Law. It has not yet, however, been proven conclusively whether the moisture moves in the liquid or vapor state. Drying may be caused by the diffusion of liquid, vapor, or both. Furthermore, one of these phenomena may be dominant during certain phases of drying. The differences among diffusion coefficients calculated from different formulae are substantial (Zaremba 1977, Pabis 1982, Jaros et al. 1992). One explanation for the discrepancy is that different methods are being used to measure the coefficient of diffusion. Also, assumed kernel geometry when computing the coefficient of diffusion can affect its magnitude (Pabis and Henderson 1961, Chittenden and Hustrulid 1966). Many of the factors that govern the diffusion coefficient are still unknown. There is a need for a unifying theory which could serve as the basis for development of a comprehensive model of the diffusion process.

There are a number of equations published in the literature for determination of convective heat transfer coefficient which is one of the most critical parameters required for analysis of drying and thermal processing. Selecting formulae from different authors causes substantial discrepancies in the final results (Boyce 1965, Pabis 1967, O'Callaghan et al. 1971, Bruce 1983).

The purpose of this chapter is to summarize and discuss the engineering properties of grain and oilseeds, as well as techniques and procedures for determining these properties.

2. PHYSICAL PROPERTIES

2.1. Geometrical Properties of Grain Kernels

The geometrical properties of grain kernels (e.g., shape, size, volume, sphericity, and surface area) are fundamental physical characteristics of grain. To a large extent, they determine the interactions among and between particles, and with the surrounding air and structures. These interactions, in turn, influence almost all engineering properties of grain, including thermal conductivity and diffusivity, surface and volume heat transfer coefficients, bulk density, resistance to airflow, internal friction, contact stresses between kernels, and friction between grain and the surrounding structure. These engineering properties must be considered in the design and evaluation of grain storage and handling systems.

Grain kernels are usually irregular in shape. A complete specification of their form theoretically requires an infinite number of dimensional measurements. Griffith and Smith (1964) attempted to establish a criterion for determining the number of measurements required to define the form of an irregular-shaped object. From a practical point of view, a measurement of three mutually perpendicular axes is usually sufficient for a quantitative description of the size and shape of a grain kernel (Nelson 1980, Fortes and Okos 1980, Sheperd and Bhardwaj 1986, Mohsenin 1986).

The shape of a grain kernel may be described by its sphericity. A widely-accepted definition of sphericity is the ratio of the surface area of a sphere, which has the same volume as that of a kernel, to the actual surface area of the kernel. Surface area and surface to volume ratio affect drying rate (Bakker-Arkema et al. 1983). Several methods for determination of surface area of selected agricultural products have been reviewed by Jindal et al. (1974) and Mohsenin (1986). Because it is difficult to measure the surface area of a kernel, the sphericity is estimated as the ratio of the diameter of a sphere of the same volume as the object (d_e; equivalent diameter) to the diameter of the smallest circumscribing sphere (d_c; the longest diameter of the object) (Mohsenin 1986). Another simple definition of sphericity is the ratio of the geometric mean of three mutually perpendicular dimensions to the longest dimension: $(\ell_1 \ell_2 \ell_3)^{1/3}/\ell_1$ where ℓ_1 = longest intercept; ℓ_2 = longest intercept normal to ℓ_1; and ℓ_3 = the longest intercept normal to ℓ_1 and ℓ_3. The intercepts need not to intersect each other at a common point.

The size of a spherical body can be uniquely defined by its diameter. For a nonspherical grain kernel, the size is often expressed in terms of the equivalent diameter (d_e) which is the diameter of a sphere of the same volume (V) as the grain kernel ($d_e = (6V/\pi)^{1/3}$). Henderson and Pabis (1962) used the geometric mean of the three principal dimensions as the equivalent diameter

$((\ell_1 \ell_2 \ell_3)^{1/3})$. Later, Pabis (1982) defined an average volume of a single kernel, V_{ke}, from the grain mass, number of kernels in the mass and the kernel density:

$$V_{ke} = \frac{S_w}{\rho_k N} \tag{1}$$

where:

S_w = wet mass of grain kernels (kg)
ρ_k = kernel density of wet grain (kg/m^3)
N = number of grain kernels.

Usually, the grain mass for 1000 kernels (seeds) can be found in appropriate tables (e.g., Pabis 1982). The wet grain mass is related to its dry mass (S_d) and moisture content (M; decimal fraction dry basis) so that $S_w = S_d(1 + M)$. Thus, replacing the average volume of a grain kernel (considered as a sphere) in Equation 1 by $V_{ke} = \pi d_e^3/6$, the equivalent diameter can be expressed as a function of the grain moisture content (mc):

$$d_e = \sqrt[3]{\frac{6S_d(1+M)}{\pi \rho_k N}} \tag{2}$$

However, for the same kernel, different methods of measuring the equivalent diameter may not necessarily yield the same results.

Image processing techniques have been introduced to measure shape and size of kernels (Casady and Sinclar 1990, Ding and Morey 1990, Kim et al. 1990). Brogan and Edison (1974) developed a pattern recognition system using several line scanners to record length, width, depth, and area of grain kernels. The extraction of size features such as area and perimeter are relatively straightforward operations. The measurement of perimeter or contour length is a common indication of object size and is easily extracted from information contained in the chain code (Saperstein et al. 1987). The method of moments is another way of determining the area, position, orientation, and other noninterpretive parameters of an object (Saffari 1989). The shape of any object may be described by a number of different descriptors. The most widely used description of shape in image processing is the shape factor or roundness (Pratt 1978). Travis and Draper (1985) used the shape factor to distinguish between 49 crops and wheat cultivars.

2.2. Kernel Density and Bulk Density

Density of a kernel is a ratio of its mass to its volume: $\rho_k = S_w/V_k$. The kernel density is usually determined by weighing a certain amount of grain and dividing the measured grain mass by the grain volume excluding the intergranular spaces. The true grain volume may be determined with an air comparison pycnometer (Beckman Instruments Inc., Mohsenin 1986).

The bulk density is the grain mass divided by the volume occupied by the grain kernels and intergranular spaces: $\rho_b = S_w/V$. Bulk density of grain varies with kernel density and spatial arrangement of the kernels (porosity) (see Section 2.2.3) which is a quantitative measure of kernel packing. Also, kernel density is dependent on the moisture content, and porosity is influenced by factors such as the method of filling and the period of storage.

Many researchers have attempted to describe the relationship between density and grain moisture content. Mühlbauer and Scherer (1977) assumed a linear relationship between the bulk density of shelled corn and its moisture content:

$$\rho_k = \rho_{kd} - c_1 M$$
$$\rho_b = \rho_{bd} - c_2 M \tag{3}$$

where: c_1, c_2 are empirical coefficients depending on corn cultivar. For example, for Velox corn cultivar ρ_{kd} = 1400 kg/m^3, ρ_{bd} = 892 kg/m^3, c_1 = 230 kg/m^3, and c_2 = 373 kg/m^3 (Mühlbauer and Scherer 1977).

Nelson (1980) developed polynomial expressions for both bulk and kernel densities of yellow-dent corn and hard red winter wheat:

$$\rho_{b\text{or}k} = c_1 + c_2 M + c_3 M_w^2 + c_4 M_w^3 + c_5 M_w^4 \tag{4}$$

where:

M_w = moisture content (% wet basis)
c_1–c_2 = empirical coefficients.

Gadaj and Suprunowicz (1971) determined the bulk density as a function of dry basis moisture content (M) and the ratio of wet kernel volume to dry kernel volume:

$$\frac{\rho_b}{\rho_{bd}} = \frac{1 + M}{V_k/V_{kd}} \tag{5}$$

The ratio of V_k/V_{kd} may be expressed as a polynomial function of moisture (Gadaj and Suprunowicz 1971):

$$\frac{V_k}{V_{kd}} = 1 + \sum_{i=1}^{n} a_i M^i \tag{6}$$

A fourth-order polynomial gave them the best fit for legume seeds. Empirical coefficients a_i for selected agricultural materials were reported by Pabis (1982).

Substituting Equation 6 in Equation 5 yields the relationship for calculating the bulk density of kernels as a function of dry mass bulk density and kernel moisture content:

$$\frac{\rho_b}{\rho_{bd}} = \frac{1+M}{1+\sum_{i=1}^{n} a_i M^i} \tag{7}$$

where M in Equation 7 is in decimal fraction of dry basis.

Pabis (1982) used Equation 7 to compare densities of fababeans, lupin, peas, and beans. They pointed out that for seeds tested in the moisture content range from 7 to 30% dry basis (db). The ratio of ρ_b to ρ_{bd} is less than 0.05 except for lupin. It was suggested that for approximate calculations in this range of moisture, the bulk densities of fababeans, peas, and beans can be assumed equal to their dry mass bulk densities with a relative error of 5% or less.

2.3. Porosity

Porosity (ξ) of bulk grain is defined as the ratio of the intergranular space to the total space occupied by the grain (kernel volume plus the volume of intergranular space) and it can be calculated from kernel and bulk density of grain:

$$\xi = \left(1 - \frac{\rho_b}{\rho_k}\right) \times 100\% \tag{8}$$

According to Equation 8, if the kernel density is constant, the plot of ξ versus ρ_b should be a straight line with an intercept of 1 and a slope of $1/\rho_k$. For changing moisture contents, Chung and Converse (1971) used a linear regression to describe porosity of wheat and corn in relation to their bulk densities (ρ_b in kg/m^3): $\xi = 100(14.4 - 0.149\ \rho_b)$, for wheat; and $\xi = 100(16.2 - 0.20\ \rho_b)$, for shelled corn.

3. THERMAL PROPERTIES

3.1. Specific Heat

The properties of grain that are necessary for analysis of thermal processes are specific heat (c) thermal conductivity (k) and thermal diffusivity (α). In addition, because these processes involve convective heat transfer between air and grain kernels in a packed bed, the fluid-particle heat transfer coefficient (h) is required. Some of thermal properties for grain and grain products are listed in ASAE Standards (1991).

Specific heat indicates how much heat (Q) is required to change the temperature (t; °C) of a material. The specific heat (c) of moist agricultural products can be related to their dry mass (S_d) or their wet mass (S_w):

$$c_{db} = \frac{1}{S_d}\frac{dQ}{dt}$$

or

$$c_{wb} = \frac{1}{S_w}\frac{dQ}{dt} \tag{9}$$

Where subscripts *db* and *wb* indicate dry basis and wet basis, respectively.

From the known relationships between S_d and S_w it follows that:

$$c_{db} = c_{wb}(1 + M) \tag{10}$$

Based on many empirical investigations, the specific heat of moist agricultural products, above the freezing point, can be estimated as the sum of the specific heat of its dry mass (c_d) and the specific heat of water held in the product (c_w). For 1 kg dry mass, of a given product, the heat capacity will be:

$$c = c_d + c_w M$$

where c_w = 4186 J/(kg K) for the temperature range of 0–70°C and finally:

$$c = c_d + 4186M \tag{11}$$

where M in Equation 11 is in decimal fraction of dry basis. Pabis (1982) summarized measured values of c_d and mathematical formulas for specific heat as a function of moisture content reported by other researchers.

Various methods of measuring thermal properties of grains are reviewed by Reidy and Rippen (1971) and Mohsenin (1980). The intent of this chapter is to recommend methods and strategies that may be employed for measuring thermal properties of crops. Precision and accuracy of measurement are important factors to consider. More precise values will improve the accuracy of heat transfer models such as those developed using finite element methods. Variations in composition, size, and shape of agricultural products, however, preclude the need for accuracies greater than ± 2–5%. These models often require specification of the effects of temperature, density, and moisture content on thermal properties.

Mohsenin (1980) briefly described methods used by researchers for measuring the specific heat of agricultural products. Among them are:

1. the method of mixtures which is the most common method reported in the literature for specific heat determination of seeds and grains (Muir and Viravanichai 1972, Dutta et al. 1988);
2. the guarded-plate method where heat loss from a specimen is prevented by surrounding it with electrically-heated thermal guards maintained at the same temperature as the specimen;
3. the calculation method in which specific heat is determined from other thermal properties such as thermal conductivity and diffusivity (Gordon and Thorne 1990);
4. the adiabatic method where a measured quantity of heat is added by means of heating cables buried in the bulk of the agricultural material. The sample is placed in a container which is placed in the test chamber. Heat energy raises the temperature of the material, the material container, the chamber wall, the air in the heater, and any auxiliary equipment in the chamber. To maintain adiabatic conditions the temperature of the outside and inside walls of the test chamber must be kept equal;
5. differential scanning calorimetry (DSC) in which very small thermal effects produced in thermal processes are measured. This method is well suited for determining the effect of temperature on specific heat (Tang et al. 1991). Any gain or loss of energy is recorded as the equipment scans the sample by heating it at a given rate of temperature rise over a selected temperature interval. The disadvantages of this method are that: a) it requires a small sample size (5–15 mg), which makes it difficult to obtain a homogeneous sample that is representative of the product tested, and b) it is a comparative device and must be calibrated.

Many of the published data on thermal properties are of limited value because insufficient supporting data are included. Information needed include detailed product description, temperature, and errors in measurement. This description should include the cultivar, the size of the individual kernels, maturity, and pretreatment. Details of the experiments should include sample size, surface conditions, porosity, temperature, relative humidity, atmospheric pressure, and sampling procedure. The equipment description should provide sufficient detail so that the experiment can be precisely duplicated. Comparison of the mathematical formulae given by different authors shows that the extreme values of specific heat of wheat can differ by 15% in the moisture content range of 1–15% wb.

Many researchers have used the method of mixtures for determining specific heat of grains. In this method, a sample is placed in water and the temperature change of the water is recorded. Muir and Viravanichai (1972) used this method to study the specific heat of hard red spring wheat harvested in different years. They showed that there was a linear relationship between specific heat and moisture content and that the specific heat was not affected by the year of harvest. They also found that in the temperature range of –33.5–21.8°C the specific heat of wheat was a linear function of temperature for up to 20% mc wb. Above the range of 23–30% mc wb, the specific heat of wheat was not a linear function of moisture content or temperature. The results indicate that above approximately 23% mc wb, water begins to exist as free water which undergoes a change of phase at temperatures below 0°C.

When low-moisture grain is used in the method of mixtures, the absorption of water by the grain generates heat of hydration. This heat increases the rate at which the water temperature rises. Muir and Viravanichai (1972) reported that the heat of hydration was not significant. Sharma and Thompson (1973) stated that this heat generation in low moisture grain can create erroneous results. Therefore, to minimize the error they incorporated a correction factor.

3.2. Thermal Conductivity

Thermal conductivity is described by the following formula:

$$q = -k\,\nabla t$$

For solids, the heat flux (q) is directly proportional to the temperature gradient (∇t) and thermal conductivity of the body (k). The minus sign refers to the direction of the flux, which is in the direction opposite to the temperature gradient. Therefore, if the temperature gradient is unity (for a unit time interval), thermal conductivity indicates the amount of thermal energy transmitted through a unit area within the unit time interval. The area is perpendicular to the direction of flow.

The mechanism of heat transfer is usually interpreted as propagation of the kinetic energy of oscillating molecules, with kinetic energy increasing with an increase in temperature. This energy is transferred to surrounding molecules.

Thermal conductivity of solids varies with chemical composition, physical structure, the state of the substance, its temperature, and moisture content. Biological materials are not homogenous and vary in cell structure, composition, and air content. Therefore, their thermal conductivities vary even more than thermal conductivity of nonbiological materials. The bulk density of granular products also influences thermal conductivity. For example, the method by which a bin of grain is filled influences thermal conductivity. Thermal conductivity of the air in the intergranular space is much smaller than thermal conductivity of most agricultural products. Therefore, thermal conductivity increases as porosity decreases. Thus, at a constant moisture content, thermal conductivity can be expressed as a linear function of bulk density:

$$k = a + b\,\rho_b$$

where a and b are coefficients. Chang (1986) gives values of a and b for wheat, corn, and grain sorghum at 22°C and various moisture contents.

Thermal conductivities of grains and oilseeds are linearly dependent on dry basis moisture content (Pabis 1982):

$$k(M) = k_d + a_1 M \tag{12}$$

where:

k_d = thermal conductivity of a layer of dry mass seeds ($W{\cdot}m^{-1}{\cdot}K^{-1}$)
a_1 = coefficient which depends on the material

Thermal conductivities of most biomaterials, including grain, vary slightly with temperature. The influence of temperature on thermal conductivity of hard red spring wheat was investigated by Chandra and Muir (1971). They determined that thermal conductivity between –6°C and 20°C was a linear function of only moisture content.

Thermal conductivity measurement presents more of a challenge than specific heat measurement because, as it was indicated before, thermal conductivity depends on the physical structure, chemical composition, and physical arrangement of a sample. Factors influencing the choice of method for determination of thermal conductivity were discussed by Nesvadba (1982). Measurement techniques have been discussed by Mohsenin (1980). Basically, methods are divided into two categories: those based on steady-state flow of heat through the media and those based on transient flow. Many researchers recommend a line heat-source, a thermal-conductivity probe (Mohsenin 1980, Sweat 1986). The line heat-source method is simple and fast and requires relatively small sample sizes. The basic theory of this technique was summarized by Sweat (1986). In brief, a probe is inserted into a sample, which is initially at a uniform temperature. The probe is heated at a constant rate and the temperature adjacent to the line heat-source is recorded. Approximate analytical and numerical methods are used in conjunction with temperatures measured as a function of time to determine thermal conductivity of the medium. While the analytical method has the advantage of being quick, however, it applies only to a perfect line source which can only be approximated by a small diameter tube enclosing a thin wire. This problem has been partially overcome by introducing a time correction procedure (Alagusundaram et al. 1991, Papadakis et al. 1990).

3.3. Thermal Diffusivity

Unsteady-state or transient heat conduction usually occurs during heating or cooling of agricultural products. It involves the accumulation or depletion of heat within a body which results in temperature changes with time, θ. The rate at which heat diffuses out of or into a body or porous layer of agricultural material is called thermal diffusivity.

The effect of water evaporation on thermal diffusivity can be evaluated using the following relationships (Luikov 1968):

$$\text{for a dry body:} \quad \frac{\partial t}{\partial \theta} = \alpha(M=0)\nabla^2 t$$

$$\text{for a wet body:} \quad \frac{\partial t}{\partial \theta} = \alpha(M)\nabla^2 t + h_{fg}^{*}\,\varepsilon^{*}\,\frac{\partial M}{\partial \theta} \tag{13}$$

where:

ε^* = coefficient of internal evaporation (decimal fraction)
h_{fg}^* = latent heat of vaporization of water from a solid body (kJ/kg H_2O)
α = coefficient of thermal diffusivity (m^2/s)

The coefficient of thermal diffusivity can be described as follows:

$$\text{for a dry body:} \quad \alpha(M=0) = \frac{k_d}{c_d\,\rho_d}$$

$$\text{for a wet body:} \quad \alpha(M) = \frac{k(M)}{c(M)\,\rho(M)} = \frac{k}{c\,\rho} \tag{14}$$

According to Equation 14, thermal diffusivity is not a linear function of moisture content. This was confirmed by Kazarian and Hall's (1965) experimental studies. Based on Equation 13, we conclude that the magnitude of thermal diffusion coefficient influences the drying kinetics.

Usually, thermal diffusivity is calculated from Equation 14 using experimentally determined values of thermal conductivity, specific heat, and mass density. This gives an approximate value with a relative error which is difficult to estimate. The true value can only be determined by direct measurement. This was demonstrated by thermal properties for wheat and corn reported by Kazarian and Hall (1965) and for single soybeans reported by Watts and Bilanski (1973).

Kazarian and Hall (1965) found that, over a range of moisture contents, thermal diffusivities calculated from k, c, and ρ were approximately 6–12% higher than the measured values for wheat; for corn, the difference was approximately 10–16%. The reason given for this discrepancy was experimental error. Kazarian and Hall (1965) estimated that if all possible sources of error (including temperature measurement errors (± 1%) and accidental errors) were considered, the maximum error for such a comparison could be greater than 17%. This means that the difference between measured and calculated diffusivity which they observed was acceptable.

Thermal diffusivity can be also measured directly using transient heat conduction. The method uses the Fourier heat conduction equation (the basic differential equation) which governs heat conduction in an isotropic body. The equation has been solved for three basic geometries: an infinite slab (Kazarian and Hall 1965), an infinite cylinder (Poulsen 1982), and a sphere (Watts and Bilanski 1973).

Results of experiments (Mohsenin 1980, Muir and Viravanichai 1972, Kazarian and Hall 1965) have confirmed the relationship between thermal conductivity and kernel moisture content described by Equation 12. In the same way specific heat can be described as a function of moisture content:

$$c(M) = c_d + b\,M \tag{15}$$

where b is experimentally determined and not necessarily equal to 4186 J/(kg·K) as in Equation 11. Sharma and Thompson (1973) and Pabis (1980) have reported relationships similar to Equation 15. As previously discussed, the bulk density of crops in the moisture content range of 0–25% wb is within 5% of the bulk density of the dry product. Assuming Equation 15 is valid, Equation 14 can be rearranged to give the following equation for thermal diffusivity as a function of moisture and bulk density:

$$\alpha(M) = \frac{k(M)}{\rho_b\, c(M)} = \frac{k_d + a_1 M}{\rho_b\,(c_d + b\,M)} \tag{16}$$

Equation 16 can only be used for moisture contents up to 25% wb because of the effect of changes in the bulk density.

Some researchers have improved the fit of their empirical formulae by incorporating the temperature effect into thermal conductivity and specific heat (Chandra and Muir 1971, Dutta et al. 1988, Alagusundaram et al. 1991).

3.4. Convective Heat Transfer Coefficient

The convective heat transfer coefficient is one of the most critical parameters required for analysis of heating, cooling, or drying processes. The heat transfer between air (a fluid) and particles of agricultural products can be due to conduction, natural convection, or forced convection. Natural convection occurs within stored grain as a result of temperature gradients induced by ambient weather conditions. Solar radiation has a major effect on storage temperature (Jiang and Jofriet 1987). Thus, the color of buildings and the material from which they are constructed can have a significant effect on temperature distribution inside a bin (Muir 1973). Most models which simulate the temperature of stored products assume that conduction is the main form of heat transport (Yaciuk et al. 1975). Smith and Sokhansanj (1990) showed theoretically that natural convection has strong effects on temperature distribution in packed beds formed of particles with diameters larger than those of small cereal grains such as wheat.

The main mechanism of air-particle heat transfer during heating, cooling, or drying of agricultural products is forced convection. For such convection the heat transfer coefficient (h) is determined from the Nusselt number (Nu). The Nu number is determined from empirical correlations between Reynolds (Re) and Prandtl (Pr) numbers. Perhaps the simplest type of relation to describe the relationship between Nu, Re, and Pr is an exponential function. Constants of this relationship are determined from the experiments. For the temperature range used for drying most agricultural products (except perhaps in high temperature drum dryers) the Prandtl number is assumed to be constant and its value is usually incorporated into a constant (C). This simplifies the exponential function so that it has only one variable.

Relationships between the Nusselt and Reynolds numbers are in the literature (Pabis 1982, Löf and Hawley 1948). Researchers give different relationships and the differences are particularly evident for Reynolds numbers less than 200. Pabis (1967) summarized the data of many researchers for cylindrical and spherical shapes using the following equations:

$$Nu = 1.40\ \mathrm{Re}^{0.214} \quad \text{for} \quad 10 \le \mathrm{Re} \le 40$$
$$Nu = 0.338\ \mathrm{Re}^{0.6} \quad \text{for} \quad 40 < \mathrm{Re} < 1000 \tag{17}$$

Löf and Hawley (1948) gave an empirical relationship which can be used to determine the volumetric heat transfer coefficient (h_v) for gravel of equivalent diameter (d_e) in the range of 0.0048–0.0381 m (see Equations 1 and 2) and air mass flux per unit area ($\dot{m}_a$) between 0.09 to 0.49 kg/(m^{2}s):

$$h_v = 0.655 \left(\frac{\dot{m}_a}{d_e} \right)^{0.7} \tag{18}$$

where h_v is in kJ/(s·m^3·K).

Bruce (1983) used the formula:

$$h_v = 0.022606\, R_{sv}\, \dot{m}_a^{0.59} \left[\frac{0.0006\,(1.72 + 0.0463\, t_a)}{d_e} \right]^{0.41} \tag{19}$$

where R_{sv} is the ratio of a kernel surface area to its volume (m2/m^3), t_a is drying air temperature (°C), d_e is the equivalent diameter expressed in m, and air mass flux per unit area ($\dot{m}_a$) is in kg/(m^2·s).

Equation 19 gives good results for a stationary bed of wheat (Bruce 1983). Boyce (1965) studied drying of a thin-layer of barley. He determined the convective heat transfer coefficient based on a heat balance for a thin layer where no moisture transfer takes place and proposed the following formula:

$$h_v = \frac{c_g \rho_b}{\theta} \ln\left(\frac{t_a - t_g}{t_a - t_{go}}\right) \tag{20}$$

where:

c_g = specific heat of grain ($kJ \cdot kg^{-1} \cdot K^{-1}$)
ρ_b = bulk density (kg/m3)
θ = time (s)
t_a = drying air temperature (°C)
t_{go} = initial grain temperature (°C)
t_g = grain temperature (°C).

The slope of the logarithm of the temperature ratio versus time determines the heat transfer coefficient for specific values of time (θ) and grain temperature (t_g).

O'Callaghan et al. (1971) used the following formula:

$$h_v = 856.8 \left[\frac{\dot{m}_a (273 + t_a)}{p_a}\right]^{0.6011} \tag{21}$$

where: p_a= atmospheric pressure (Pa).

Direct relationships such as Equation 20 and 21 are easy to use for determination of the heat transfer coefficient, and they were created for a porous bed of grain.

4. WATER DIFFUSION

A wet solid body can be described as a binary (two-component) system in which water molecules can be transported within the boundary of the solid. By analogy to thermal diffusivity which influences the rate of temperature equalization within the boundary of a solid body or a porous layer, mass diffusion governs the rate at which water concentration equalizes (assuming all other significant parameters are held constant).

It is generally agreed that the mechanism of moisture movement during the falling rate period of grain drying is by diffusion. This can be described by Fick's Law. It has not yet, however, been proven conclusively whether the moisture moves in the liquid or vapor state. Drying may be caused by the diffusion of liquid, vapor, or both. Furthermore, one of these phenomena may be dominant during certain phases of drying. In the case of mass diffusion, individual molecules move through a wet solid body by means of their random, individual movements. Molecules travel only in straight lines and change directions when they bounce off other molecules. This movement is known as molecular diffusion or the random-walk process. Depending on the size, shape, and connections of pores, however, diffusion in porous solids may take place via mechanisms other then molecular diffusion. Comparison of the liquid and vapor diffusion models given by Chinnan and Young (1977) indicate that diffusivities through peanut kernels and hulls for both models are exponential functions of the reciprocal of absolute dry-bulb temperature, which is a trend that can be predicted theoretically. Whitaker and Young (1972b), however, showed an opposite trend for vapor diffusivities. Chinnan and Young (1977) concluded that neither vapor nor liquid diffusivities are affected by changes in dew point temperatures.

For modeling thin-layer drying of different grains, many researchers solved the equation based on Fick's Law of diffusion. Pabis and Henderson (1961) proposed a model for internal water diffusion in corn kernels. Becker and Sallans (1955) presented the diffusion model for wheat, Chu and Hustrulid (1968) for corn, and Young and Whitaker (1971) for peanuts. Whitaker and Young (1972a) concluded that in most cases, the diffusion equation predicted the experimental drying rate for peanut kernels more accurately than for peanut pods; probably because peanut kernels are more homogeneous than pods. They indicated that further study was needed to evaluate whether the diffusion equation characterizes moisture movement in a composite body.

The diffusion coefficient during soaking of long-grain white and brown rice was obtained by solving the diffusion equation for an infinite cylinder (Engels et al. 1986). They used the following Arrhenius-type equation to describe the relationship between diffusion coefficient (D, m^2/s) and temperature, t_a (K):

$$D = D_o \exp\left(-\frac{E}{R\,T_a}\right) \tag{22}$$

where:

D_o = diffusion coefficient at a reference temperature (m^2/s),
E = activation energy of the diffusion coefficient (kJ/mol)
R = universal gas constant (kJ·mol^{-1}·K^{-1}).

Engels et al. (1986) used linear regression to determine the constant D_o and the activation energy (E) for white rice in the air temperature range of 30–50°C and a moisture range of 14–35% (db) (with a correlation coefficient r = 0.998). Engels et al. (1986) considered brown-rice kernels as a two-layer system (hull and endosperm). The diffusion coefficient of the endosperm was considered to be equal to the diffusion coefficient of white rice and the diffusion of the hull was again expressed in the form of an Arrhenius-type equation (Eq.

22) with $r = 0.997$. The constants D_o and E for both white long-grain rice and the endosperm of brown rice were 4.01×10^{-7} m^2/s and 22.5 kJ/mol, respectively. The constants D_o and E for the testa were 205.0 m^2/s and 64.6 kJ/mol, respectively.

Steffe and Singh (1980) developed a diffusion-type model for rough medium-grain rice in which three different diffusion coefficients (i.e., for endosperm, bran, and hull) were employed. Diffusivity values were calculated for temperatures ranging from 35 to 55°C, initial moisture contents from 27 to 31% db, and final grain moisture contents above 15.6% db. Diffusivity was related to the temperature using a modified form of Equation 22 where E/R in the exponential function was replaced by a coefficient B:

$$D_m = A \exp\left(-\frac{B}{T_a}\right) \tag{23}$$

The coefficients for the rice-grain components were: starchy endosperm, $A = 7.13 \times 10^{-7}$ m^2/s, $B = 2.88 \times 10^3$ K; bran, $A = 2.2 \times 10^{-4}$ m^2/s, $B = 5.11 \times 10^3$ K; hull, $A = 0.13$ m^2/s, $B = 7.38 \times 10^3$ K.

On the basis of Equation 23 and empirically defined A and B coefficients, Steffe and Singh (1980) concluded that bran and hull have greater resistance to moisture movement than the starchy endosperm; likewise, the hull is more limiting to moisture movement than the bran.

Husain et al. (1973) modeled a kernel of a rough rice as an infinite cylinder and solved the coupled heat and moisture diffusion equation. They reported that the diffusion coefficient (D_m) was a function of temperature (t_a) and moisture content (M):

$$D_m = D_o(t_a) \exp\,[P(t_a)M]$$

where the function $D_o(t_a)$ and the parameter $P(t_a)$ were represented by regression equations. They reported values of diffusion coefficients for a range of temperatures and moisture contents. The function $D_o(t_a)$ appears to follow the Arrhenius equation (Eq. 23).

The Arrhenius type equation has been used by many researchers for expressing diffusivity (Pabis and Henderson 1961, Steffe and Singh 1980) but it failed to give good predictions in some cases. For example, Husain et al. (1972) found that it gave good agreement with the experimental results when used to predict the moisture independent diffusivity for drying of potato slabs during the first falling rate period. The correlation deteriorated, however, as the drying entered the second falling rate period. The diffusion coefficient for this period was a function of both moisture and temperature:

$$D(M,t_a) = a \exp\,(b\,t_a) \exp\,(c\,t_a - d\,M) \tag{24}$$

where: a, b, c, d = coefficients.

An attempt was made to find a single exponential function for diffusivity that could predict drying behavior during the entire period of drying. The correlation between experimental and predicted results was not as good as obtained by two diffusivity models.

During drying of agricultural products shrinkage occurs simultaneously with moisture diffusion. Shrinkage may have a significant effect on the mass diffusivity and thus on the moisture removal rate. The effect of shrinkage on moisture diffusion in soybeans was studied by Misra and Young (1980). The authors described moisture diffusion by introducing an apparent liquid diffusivity (D_F):

$$D_F = D_o \left(\frac{\rho_o}{\rho} \right)^n \exp\left[n_2 \left(C - C_1 \right) \right] \tag{25}$$

where:

D_o = constant mass diffusivity at a given dry bulb temperature (m^2/s)
ρ_0 = initial dry mass density (kg/m^3)
ρ = instantaneous dry mass density (kg/m^3)
C = instantaneous liquid concentration (kg/m^3)
C_1 = initial liquid concentration (kg/m^3)
n, n_1 = constants with appropriate dimensions.

Dry-bulb temperature had a significant effect on the constant diffusivity (D_o) for soybeans whereas dew point temperature did not have a significant effect. Therefore, Misra and Young (1980) modeled D_o at different dry-bulb temperatures using an Arrhenius-type relationship (Eq. 23) with D_m replaced by D_o. The constants for soybeans were estimated to be: $A = 1.3 \times 10^{-7}$ m^2/s and B = 3437 K.

Crank and Park (1968) described five general methods for determining the diffusion coefficient: 1) permeation method, 2) sorption and desorption method, 3) concentration-distance curves method, 4) radioactive-tracer method, and 5) nuclear magnetic resonance and self-diffusion method. The most commonly used method for determining the moisture diffusion coefficient in grain is by fitting the diffusion-based drying equations to the experimental data in the falling rate period.

Tolaba et al. (1990) demonstrated that Fick's diffusion equation satisfactorily describes the drying behavior of corn kernels for finite Biot numbers when the resistance of the pericarp is included as part of the external resistance. They calculated the mass diffusion coefficient from the Fourier number by fitting the experimental drying curves to the diffusion model. The mass diffusion coefficient (D_m) for the interior of the kernel (endosperm and germ) at 50 and 60°C was 7.7×10^{-11} and 9.8×10^{-11} m^2/s, respectively. The D_m value for the pericarp was 1.11×10^{-11} and 1.14×10^{-11} m^2/s for the same two temperatures (50, 60°C), respectively.

A new method for determining diffusion coefficients of corn components (germ, endosperm, and pericarp)—thermogravimetric analysis—was described by Syarief et al. (1987). The diffusion coefficient D_m(M) was calculated as a function of moisture content using an exponential equation:

$$D_m(M) = A_o \exp(0.086\, M)$$

where M = moisture content in % db. The values for A_o were: germ = 6.4×10^{-11}, floury endosperm = 1.8×10^{-11}, horny endosperm = 1.3×10^{-11} m^2/s. The diffusion coefficient for the germ was 3.6 times that of floury endosperm and 4.9 times that of the horny endosperm. The value of A_o for the pericarp varied with location as follows: over the germ = 6.5×10^{-12}, over the floury endosperm = 5.6×10^{-12}, and over the horny endosperm = 3.8×10^{-12} m^2/s. This could partially be explained by the fact that the pericarp thickness varies according to position on the corn kernel (Wolf et al. 1969). Wolf et al. (1952) reported the pericarp covering the abgerminal side of the corn kernel to be thicker than the germinal side. Permeability of the pericarp, however, also varies among hybrids due to the different physical structures of their pericarps (Purdy and Crane 1967).

An empirical formula for calculation of the mass diffusion coefficient in wheat kernels was given by Pabis (1982):

$$D_m(M,t_a) = 10^{-11}\,(-0.88 + 18.7M - 64.6M^2) \exp(0.0330\, t_g) \tag{26}$$

where M in Equation 26 is in kg H_2O/kg db. Equation 26 was developed for moisture contents $6 \leq M \leq 23\%$ db and kernel temperatures $30 \leq t_g \leq 72$°C. The correlation coefficients for D_m versus M and D_m versus t_g were between 0.989 to 0.998, respectively.

Pabis (1982) made two important observations regarding the formulation of Equation 26. First, determination is indirect since it is based on data from drying experiments. Diffusion coefficients determined by this method are of practical importance in applying drying theory but the formulas are empirical and cannot be used for other materials which differ in structure. Secondly, although there is a logical justification for the empirical formula, the model was not based on a unifying theory. Therefore, the formula can only be used in the ranges of moisture and temperature for which experiments were conducted.

The differences among diffusion coefficients calculated from different formulae are substantial. The formula of Pabis (1967) for wheat kernels at 50°C in the moisture content range of 16–25% db, gives an average diffusion coefficient D_m of 8.3×10^{-11} m^2/s while Zaremba (1977) reported D_m to be 2.5×10^{-11} m^2/s, one-third of Pabis's results. One explanation for the discrepancy is that different methods were used to measure the coefficients. Also, kernel geometry can affect the prediction. Assuming a brick shape for the corn kernel, Pabis and Henderson (1961) found that D_m was approximately 2.17×10^{-9} m^2/s at a drying temperature of approximately 38°C and a moisture content of 30% db. For the same drying conditions Chittenden and Hustrulid (1966) assumed a spherical shape and determined the diffusion coefficient to be 6.70×10^{-10} m^2/s.

For the falling rate drying period a mass diffusion coefficient which is related to the geometric shape of the solid can be calculated by assuming that the moisture ratio (MR) at time θ is a function of Fo_m, the Fourier number for mass transfer:

$$MR(\theta) = \frac{M(\theta) - M_e}{M_o - M_e} = f(Fo_m) \tag{27}$$

Pabis (1982) gave $f(Fo_m)$ for three basic geometrical shapes: sphere, infinite cylinder, and infinite plate.

Methods for experimentally determining the mass diffusion coefficient for grains were discussed by Ginzburg et al. (1969), and Jaros et al. (1992). Jaros et al. (1992) determined the overall coefficient of diffusion (liquid and vapor) as a function of temperature and moisture content. They modeled a single wheat kernel as a sphere, used the series solution of the diffusion equation and developed the following empirical relationship for D_m in the moisture content range for M of 5.5–33.0% db:

$$D_m = A(t)\, M^{B(t)} \exp\,[C(t)\, M] \tag{28}$$

where for calculations, Equation 28 requires M in kg H_2O/kg db. The coefficients A, B, and C were related to the air temperature t_a (52–82°C) by the following equations:

$$A(t) = 0.99 \times 10^{-11} \exp\,(0.064\, t_a)$$

$$B(t) = -0.05\, C(t)$$

$$C(t) = -(0.124\, t_a + 0.09)$$

Although Jaros et al. (1992) did not attempt to separate liquid from vapor diffusion, they concluded that in the falling-rate of drying of a single wheat kernel, water moves mainly in the form of water vapor and that the proportion of water that moves as vapor becomes higher at lower moisture contents. The conclusion was based on the effect of shape of the curves for diffusion as a function of moisture content along with the results of Fortes et al. (1981) and the explanation given by Keey (1972).

All the factors that govern diffusion coefficient are still unknown. There is a need for a unifying theory which could serve as the basis for development of a comprehensive model of the diffusion process.

5. EQUILIBRIUM MOISTURE CONTENT

Prolonged drying, ventilation, or storage of grains or oilseeds (even without forced ventilation) always causes the crop to reach a certain equilibrium in moisture. This equilibrium depends on the following:

1. the partial pressure of the water vapor (p_v) of the surrounding air which can be expressed in terms of air relative humidity (rh) and temperature; (this assumes that the air temperature is equal to the kernel temperature);
2. the method of reaching the equilibrium (sorption or desorption, multiple drying and rewetting);
3. cultivar; and
4. biological stage of maturity.

The equilibrium moisture content (M_e) can be expressed by the following relationship:

$$M_e = f(p_v, t)$$

At constant partial vapor pressure (p_v) the equilibrium moisture content can be considered an isobar and at constant temperature it is an isotherm. In drying of agricultural products the equilibrium moisture data are presented as isotherms and the partial vapor pressure of water is replaced by the air relative humidity. Thus:

$$M_e = f(rh, t = \text{constant})$$

For agricultural products the shapes of the equilibrium isotherms are sigmoidal and correspond to isotherms of capillary-porous colloidal materials. The isotherm can be divided into three regions. Monomolecular adsorption takes place in the region below the inflection point of the curve. Usually, for agricultural products at ambient temperatures, this region occurs for relative humidities up to 15%. According to Becker and Sallans (1956), the transition from monolayer desorption occurs at 0.078 and 0.060 db for 25 and 50°C isotherms, respectively. In the intermediate region, multilayer adsorption occurs when additional layers of water are adsorbed over the monolayer. Multilayer adsorption usually takes place at relative humidities between 15 and 90%. Adsorption occurs in both the mono- or multilayer regions of the curve. During molecular adsorption (mono- and multilayer) a certain amount of heat is emitted and the physical properties of the bound water are altered. For example, the density, specific heat, and freezing temperature of water molecules that are bound in a monolayer are 1790 kg/m^3, 2930.8 J/kg, and –78°C, respectively (Pabis 1982).

The amount of osmotically bound water (absorbed) can be several times greater than the amount of adsorbed water. The physical characteristics of the osmotically bound water do not change and there is no heat emitted during the absorption process. Capillary absorption refers to the free water held in the pores of a solid. This absorption takes place in the region between 90 and 100% relative humidity where the water is held in capillaries of a diameter less than 10^{-4} mm. For wheat, the transition from bound to free water occurs at 32% db (Becker and Sallans 1956). At rh = 100% when the moisture content of a solid body exceeds its hygroscopic moisture content, free water appears also on the surface of the body.

Many researchers have attempted to develop mathematical models of the equilibrium isotherms on the basis of theory. So far, none of the proposed models has been adequate over the entire range of relative humidities.

Langmuir (1918) developed one of the first equations to describe equilibrium isotherms. It describes the adsorption of a monolayer of water vapor on the internal surfaces of a solid.

Utilizing the assumption that van der Waals forces account for the adsorption of water molecules on a product's surface, Brunauer et al. (1938) developed an expression for multilayer molecular adsorption which is known as the BET (Brunauer, Emmett, and Teller) equation. The BET equation gives a good fit for a variety of foods over the region of 5–45% rh (Chirife and Iglesias 1978). Therefore, it can be used for dehydrated foods. Becker and Sallans (1956) applied this equation to describe desorption isotherms of wheat. They found good agreement with their isotherms for temperatures of 25 and 50°C and relative humidities of 4–30%. Desorption data can be obtained when equilibrium is reached by lowering the moisture content of a product (for example, during drying). Conversely, adsorption or absorption can be obtained when equilibrium is reached by adding moisture to a product (for example, aeration of grain with 100% rh air).

Henderson (1952) developed a formula based on Gibb's adsorption equation:

$$\ln(1 - rh) = -C\,t\,M_e^n$$

In this equation, C and n are constants. This equation has been questioned by some researchers (Brooker et al. 1974). Thompson et al. (1968) modified the Henderson equation by adding another constant to the temperature. The modified Henderson equation was adopted by the American Society of Agricultural Engineers (ASAE 1991, Standard D245.4) as one of the two standard equations to describe moisture versus relative humidity relationships for agricultural grains:

$$M_e = \left[\frac{\ln (1 - \mathrm{rh})}{-K\,(t + C)} \right]^{1/N} \tag{29}$$

where K, C, and N are constants which are tabulated for several grains in the ASAE Standard D245.4 (ASAE 1991).

Chung and Pfost (1967) developed a two-parameter isotherm model for cereal grains. The model is based on thermodynamic relationships and the equation for free energy change during sorption:

$$\ln (\mathrm{rh}) = [-A/(R\,t)] \exp (-B/M_e)$$

In their equation, A and B are constants and R is the universal gas constant. Pfost et al. (1976) modified the equation by adding a new parameter to the temperature term. It was the second equation included in the ASAE Standard D245.4.

$$M_e = 100\{E - F \ln [-(t + C) \ln (rh)]\} \tag{30}$$

where E, F, and C are constants which are listed for several grains in the ASAE Standard D245.4.

Halsey (1948) developed a two-parameter equation to describe the condensation which occurs in the multilayer region. Iglesias and Chirife (1976) modified the Halsey equation by relating one parameter of the original Halsey equation to temperature by an empirical exponential function. Their equation is:

$$M_e = \left[\frac{-\exp (A + B\,t)}{\ln (rh)} \right]^{1/-C} \tag{31}$$

where: A, B, and C = constants.

Oswin (1946) presented an equation for sigmoid-shaped curves:

$$M_e = A\,[\mathrm{rh}/(1 - \mathrm{rh})]^n$$

In this equation A and n are constants. Chen and Morey (1989) found that the parameter A was a linear function of temperature and modified Oswin's equation:

$$M_e = (B + C) \left(\frac{\mathrm{rh}}{1 - \mathrm{rh}} \right)^n \tag{32}$$

for which values of parameters B, C, and n are given by Chen and Morey (1989). Based on the residual plots Chen and Morey (1989) concluded that the Modified-Oswin equation is a good model for popcorn, corn cobs, whole pods of peanuts, and some varieties of corn and wheat.

The three parameter Guggenheim-Anderson-de Boer (GAB) equation was first proposed as a modification of the BET equation by Anderson (1946):

$$M_e = \frac{M_o\, C\, K\, \text{rh}}{(1 - k\, \text{rh})\,(1 - K\, \text{rh} + C\, K\, \text{rh})} \tag{33}$$

where M_o is the monolayer moisture content (% db). Constants K and C are related to the energies of interaction between sorbed layers of water molecules. When K is unity, the GAB equation becomes the BET equation.

The GAB equation was later derived kinetically by de Boer (1953) and by Guggenheim (1966) using statistical mechanics; it is recognized as the most satisfactory of theoretical isotherm equations (Multon 1988, Kapsalis 1987, Bakker-Arkema 1986). Jayas and Mazza (1991) incorporated the temperature effect into Equation 33 by dividing the coefficient C by the temperature (t). This gives a relatively simple three parameter model:

$$M_e = \frac{M_o\,(C/t)\, K\, \text{rh}}{(1 - \text{rh})\,[1 - k\, \text{rh} + (C/t)\, K\, \text{rh}]} \tag{34}$$

Jayas and Mazza (1991) evaluated the modified GAB equation against four other commonly used three-parameter equations: the modified Henderson, Chung-Pfost, Hasley, and Oswin equations. For the desorption (M_e) of oats, the modified GAB equation was the second best among the five with the best being the modified Chung-Pfost equation.

There are many methods used to obtain sorption isotherms. Gal (1981) indicated that the experimental data on equilibrium moisture content (emc) should be accompanied by a detailed description of the material, experimental procedures, and apparatus. He stressed that while taking M_e data, the temperature and water vapor pressure (rh) in the space around the sample should be precisely maintained constant. He proposed that the temperature should be kept constant within ± 0.2°C for routine work and ± 0.02°C for reference purposes. He also indicated that only moderate accuracy could be attained using methods in which the water vapor is maintained constant by controlling the vapor content in the space around the sample. On the other hand the accuracy of the relative humidity is approximately ± 0.5% when saturated or unsaturated solutions of various salts are used to maintain constant vapor pressures.

Osborn et al. (1989) used the equilibrium relative humidity (erh) method (described in their study as a closed-loop, dew point method), to obtain moisture content data for soybeans. They indicated that relatively little time was required for the small mass of air to reach equilibrium with the grain in the erh method. Much more time was required for the grain to reach equilibrium with the air when the saturated salt method was used.

Cenkowski et al. (1989) used both an erh and an equilibrium moisture method to measure the emc of lentils. The erh method was based on equilibrating a small mass of air with a relatively large mass of grain in an airtight system. The emc method was based on equilibrating a thin layer of grain with air of a constant relative humidity. Gal (1981) described a method in which final air relative humidities were attained by mixing dry and moist quantities of a substance. In this case, the points on the isotherms lay inside the hysteresis loop (the difference between adsorption and desorption data obtained at the same rh). Multon et al. (1981) obtained a hysteresis loop for different relative humidities by mixing predetermined quantities of dry and wet grain (treated with dye) and placing the mixture in sealed containers. After the equilibrium was reached, moist (dyed) and dry (not dyed) samples were hand-separated and the moisture contents were measured. Yang and Cenkowski (1991, 1993) placed moist and dry layers of kernels on separate tray sections. The air was circulated in a closed loop through the alternating dry and wet layers. This accelerated moisture exchange.

6. LATENT HEAT OF VAPORIZATION

When the sorption of water by agricultural products was discussed, heat emission during the adsorption process was mentioned. This is a result of a decrease in the free energy of the surface layer of the adsorbent. During desorption heat input is needed to overcome the attraction forces between the adsorbed water molecules and the surface of a solid (sorbent) and in addition, heat is required for the change in phase from liquid to vapor.

Moisture is held in hygroscopic materials by physical and chemical forces. The exact mechanism is not fully understood. In general, the affinity of a biological material for water is a function of the relative numbers of polar and nonpolar groups, their steric accessibility in the molecules, and their orientations. This means that sorption, solution, hydration, chemical combination, and capillarity are factors influencing the bonding of water in a product.

For each degree of dehydration the forces of attraction between water molecules and the products change. In free water, the hydrogen bonds are the major factors affecting the exchange of water with the atmosphere. Therefore, energy is required to bring about a phase change from liquid to vapor. As moisture content decreases, an additional increase in the amount of energy required to separate the adsorbed water molecules can be expected. Less energy will be needed to separate water adsorbed in a multilayer than water adsorbed in a monolayer while even less energy will be needed to separate water in microcapillaries.

Gallaher (1951) established the following formula for the dependence of the latent heat of vaporization of water from the product (h^*_{fg}) on its moisture content (M):

$$h^*_{fg} = h_{fg} [1 + a \exp(b M)] \tag{35}$$

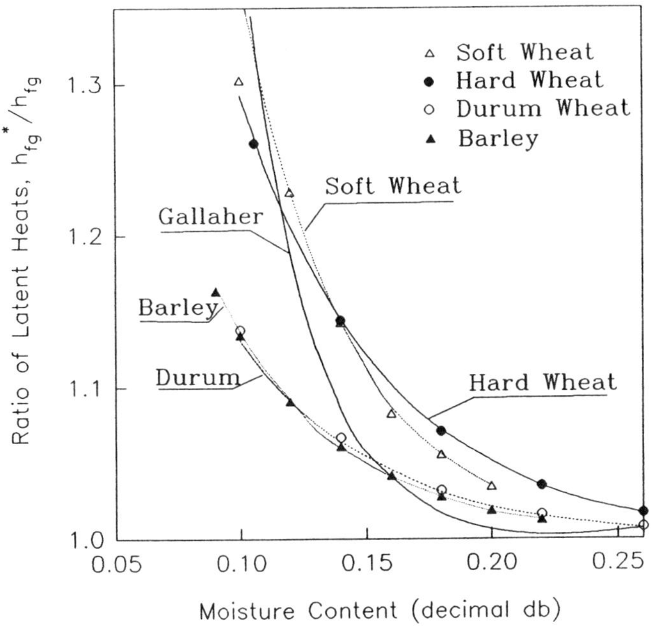

Figure 1 Effect of moisture content on the latent heat of vaporization of different wheat cultivars and barley compared with Gallaher's (1951) model.

where:

h_{fg} = latent heat of vaporization of free water (kJ/kgH_2O),
a, b = coefficients.

Values of a and b for selected agricultural products were reported by Cenkowski et al. (1992a).

The moisture dependency of latent heats of barley and three varieties of wheat are shown in Figure 1. Values of h_{fg}^*/h_{fg} were calculated on the basis of the moisture content predicted from equilibrium moisture content data (ASAE 1991). The curves represent the best fit line for Equation 35. Figure 1 also shows predictions using coefficients for wheat given by Gallaher (1951).

The values of h_{fg}^*/h_{fg} for hard and soft wheat are similar, but there is a significant difference in the ratio for durum wheat compared with soft or hard wheat. At a moisture content of 0.10 db, for example, the ratio h_{fg}^*/h_{fg} for durum wheat is 1.138 and for soft wheat it is 1.302, approximately 15% higher. At low moisture contents (below 0.12 db) Gallaher's curve is closer to the curves for soft or hard wheat. Above 0.14 mc db, it is closer to the curve for durum wheat. This could be attributed to differences in the chemical composition among wheat cultivars (Lorenz and Kulp 1991). Starch and protein play a significant role in water adsorption and the starch to protein ratio varies with cultivar. As wheat contains approximately 59–65% db starch, the differences in the latent heat of vaporization could probably be attributed to starch content. Since the ASAE data (ASAE 1991) do not indicate chemical composition of the selected grains, it is difficult to explain the behavior of Gallaher's curve. Environment has a strong effect on the chemical composition and, for the same cultivar grown in different locations or different years, differences up to 3% db in starch and protein content are possible (Martens and Hlynka 1969). The variation of starch in barley is even greater. For the same barley types, starch ranges from 21 to 64% db (Lorenz and Kulp 1991). In Figure 1, the h_{fg}^*/h_{fg} ratio for barley is similar to the ratio for durum wheat. This was probably due to the fact they have similar starch contents.

There is a substantial body of data on equilibrium moisture content in the literature. Even when the same methods were used for determining water sorption isotherms, there is a large variation among laboratories for the same material. For instance, Brooker et al. (1974) summarized the equilibrium moisture contents of cereal grains, oilseeds, and similar products from 22 sources. They concluded that variations in the values reported for one grain at the same temperature and humidity are common. As explained previously, a variation in the M_e data influences the determination of the latent heat of vaporization.

Another problem associated with determinations of the equilibrium moisture content is that the sorption behavior of a given material is influenced by prior treatments. High or low drying temperatures may alter sorption capacity. For example, Tuite and Foster (1963) reported that artificial drying at temperatures above 60°C lowers the equilibrium moisture content of corn by 0.5–1.0 percentage point. Chung et al. (1972) reported higher M_e values for broken corn kernels than for sound kernels and higher M_e values for sound kernels than for heat-damaged kernels. An important implication of these observations is that prior treatment influences the shape of the latent heat of vaporization curve.

7. GRAIN MOISTURE MEASUREMENT

Accurate determination of grain moisture content is desirable in grain storage and marketing. Various methods have been used to measure the moisture content of bulk grain and single kernels including: conventional oven drying, distillation, chemical titration, electrical moisture meters, nuclear magnetic resonance (NMR), microwave oven drying, near-infrared spectrophotometry, and acoustic devices. Among these methods, the oven drying and the electrical methods are most commonly used in grain storage management and the grain trade.

Various oven drying procedures have been used as standard methods by many government agencies and professional associations (e.g., ASAE 1991), whereas electrical moisture meters provide fast and reasonably accurate moisture measurements for commerce. Electrical moisture meters are usually calibrated against one of the standard oven drying methods. The performance of electrical moisture meters has been investigated extensively (Hlynka et al. 1949, Stevens and Hughes 1966, Paulsen et al. 1983, Hurburgh et al. 1985).

Early electrical moisture meters were based on direct current conductance measurements. They were later superseded by alternating current conductance measurement techniques (Kraszewski 1980). Electrical moisture meters are based on the correlation between moisture content of grain and its dielectric properties, such as the dielectric constant and loss factor (Nelson 1973). Although there exists a high correlation between grain moisture and its dielectric properties, factors other than moisture content also influence the dielectric properties of grain (Nelson 1982). To improve the accuracy of measurements, microwave attenuation (microwaves are electromagnetic waves with frequencies between 10^{10} and 10^{12} Hz) has been used to sense grain moisture (Okabe et al. 1973, Miyai 1978, Völgyi 1989). At high frequencies, variations in dielectric properties due to ionic conductivity and bound water relaxation do not interfere with the detection of the amount of water in the grain (Nelson and Kraszewski 1990). Therefore, measurements carried out at a high frequency are virtually independent of the material salinity or variations in chemical composition (Kraszewski et al. 1991). To minimize the effect of density on the dielectric properties when measuring moisture content of bulk grain, some commercial instruments use microwave attenuation for moisture sensing and gamma-rays for density correction (Nelson and Kraszewski 1990). Kraszewski and Kulinski (1976) proposed a technique which eliminates the density effect through simultaneous measurements of both attenuation and phase-shift. The technique was shown to be adequate for measuring moisture contents of bulk wheat (Kraszewski 1988).

A quick and inexpensive method for determining grain moisture content is microwave-oven drying (Hanna and Sharma 1988, Zhang and Brusewitz 1990). Brusewitz (1984) observed that if the final sample temperature during microwave drying is the same as that during convection oven drying, there was little difference between the moisture measurements by microwave oven and those by convection ovens.

Nuclear magnetic resonance (NMR) provides a nondestructive way of measuring transient moisture profiles of grain kernels during drying (Song and

Litchfield 1990, Ruan and Litchfield 1992, Sun and Litchfield 1991). The method has been shown to be a useful research tool in studying stress cracking of grain kernels during drying.

8. RHEOLOGICAL PROPERTIES OF SINGLE GRAIN KERNELS

A knowledge of rheological properties of grain kernels (e.g., modulus of elasticity, Poisson's ratio, relaxation and retardation time, shear modulus, and coefficient of thermal expansion) is essential when designing and evaluating grain processing facilities. Fundamental tests allow the measurement of properties that are inherent to the material and do not depend on the geometry of the sample tested, the conditions of loading, or the apparatus. These tests may be classified into two different groups: static or quasi-static loading tests, and dynamic tests. Various types of tests include a compressive stress-strain test, a stress relaxation test, and a creep-recovery test.

In the compressive stress-strain test, the load is increased slowly while deformation of the specimen is usually measured until the specimen fractures. Such tests for individual kernels, at different moisture contents, have been conducted by Bilanski (1966) and Liu et al. (1990) for soybeans; Waananen and Okos (1988) for shelled corn kernels; Davison et al. (1975) for rapeseed; and Cenkowski et al. (1991, 1992b) for canola. When necessary, these researchers used Hertz's equation of contact stress (Mohsenin 1986) to interpret their data. Cenkowski et al. (1992b) compared Hertz theory calculations of modulus of elasticity for canola kernels with the modulus of elasticity obtained from the creep-recovery tests. They indicated that the discrepancy in the moduli were between 0.4 to 19.6% and varied with kernel maturity.

In a stress relaxation test, the specimen is deformed instantly to a predetermined strain which is within the elastic limits for the specimen. The decrease or relaxation in stress is recorded as a function of time. Expressing dissipating stress as a function of time gives a relaxation curve. A generalized Maxwell model (Mohsenin 1986) is often used to represent agricultural products subjected to stress relaxation. This model consists of a number of Maxwell elements (a spring and a dashpot in series) in parallel with each other. An additional residual spring is also placed in parallel to the Maxwell elements. A better fit to the relaxation data may be obtained by adding more Maxwell elements. Pappas and Rao (1989) characterized the stress relaxation behavior of cowpeas using Maxwell elements in parallel without a residual spring. The relaxation studies conducted by Liu et al. (1990) demonstrated that the soybean cotyledon can be treated as a simple viscoelastic material and can be represented by a two-term Maxwell model.

In a creep-recovery test, the load is suddenly applied and held constant, and deformation is measured as a function of time. Creep is analogous to in-bin storage where a bottom layer of kernels is exposed to a dead load from the kernels above. The rheological model which represents the creep behavior is known as the four-element Burgers model (Mohsenin 1986) and consists of a Maxwell (a spring and a dashpot in parallel) and Kelvin model (a spring and a dashpot in series) in series.

The creep and recovery tests are carried out using an apparatus consisting of a beam supported at its midpoint (Mohsenin 1986). The sample is placed on one side while a dead load is applied on the other side of the beam. With this method, it is difficult to accurately measure recovery characteristics since the smallest imbalance in the beam can influence the reading. Also, tilting of the beam influences the precision in reading the deformation.

9. STRESS-STRAIN BEHAVIOR OF BULK GRAIN

9.1. Classification of Constitutive Models

Stress-strain behavior of bulk grain has been shown to be highly inelastic and nonlinear (Smith 1981, Manbeck and Nelson 1975, Zhang et al. 1989, Li et al. 1989, Hardin et al. 1990). Bulk grain is composed of individual kernels and pores filled with air. Forces imposed on the grain bulk are transmitted across the contact points of grain kernels, while deformations of grain bulk result from both inter-kernel slip (or rotation) and deformations of individual kernels. Apparently the load response of bulk grain depends on the properties of individual kernels as well as interactions between kernels. No constitutive model for a grain bulk, however, has been developed using the fundamental characteristics of grain kernels and the interactions between kernels. Almost all existing models have been founded on the assumption that a grain bulk is a continuum. These continuum models can be categorized as: 1) nonlinear elastic (Manbeck and Nelson 1975, Smith 1981, Thompson and Ross 1984, Hardin et al. 1990), 2) elastoplastic (Zhang et al. 1986, 1989), and 3) endochronic (Xu et al. 1993). Time dependent models, such as viscoelastic and elastic-viscoplastic models, have also been used (Bock et al. 1989, Li et al. 1990).

9.2. Nonlinear Elastic Model

A commonly used method of nonlinear analysis of the stress-strain behavior of granular materials is the tangent modulus analysis. In this analysis, material is assumed to be nonlinearly elastic and stresses are related to strains incrementally through the tangent modulus of elasticity. The tangent modulus is generally a function of stresses. Smith (1981) applied a hyperbolic model to wheat, barley, corn, and oats. The model approximates the stress-strain curve of grain by a hyperbola (Duncan and Chang 1970):

$$(\sigma_1 - \sigma_3) = \frac{\varepsilon_1}{1/E_i + (\varepsilon_1 R_f)/(\sigma_1 - \sigma_3)_f} \tag{36}$$

where:

σ_1, σ_3 = major and minor principal stresses (kPa)
$(\sigma_1 - \sigma_3)_f$ = stress difference at failure (kPa)
ε_1 = major principal strain (m/m)
E_i = initial modulus of elasticity (kPa)
R_f = failure stress ratio (kPa/kPa)

The initial modulus of elasticity E_i may be expressed as a power function of the minor principal stress:

$$E_i = Mp_a \left(\frac{\sigma_3}{p_a} \right)^a \tag{37}$$

where:

p_a = atmospheric pressure (kPa)
M = elastic modulus number (dimensionless)
n = elastic modulus exponent (dimensionless).

Assuming that grain is a cohesionless material, the Mohr-Coulomb failure criterion can be used to relate the stress difference at failure to the angle of internal friction (φ) as follows:

$$(\sigma_1 - \sigma_3)_f = \frac{2\sigma_3 \sin\varphi}{1 - \sin\varphi} \tag{38}$$

The tangent modulus of elasticity E_t can be determined from Equations 36–38:

$$E_t = \frac{d\sigma_1}{d\varepsilon_1} = Mp_a \left[1 - \frac{R_f(1 - \sin\varphi)(\sigma_1 - \sigma_3)}{2\sigma_3 \sin\varphi} \right]^2 \left(\frac{\sigma_3}{p_a} \right)^n \tag{39}$$

Smith (1981) determined hyperbolic model parameters for corn, wheat, barley, and oats. The results are summarized in Table 1.

9.3. Elastoplastic Model

In an elastoplastic model, strains are related to stresses incrementally. For each stress increment, the corresponding strain increment is composed of elastic and plastic components. The elastic component is determined by using Hooke's law

Table 1 Hyperbolic Model Parameters for Corn, Wheat, Barley, and Oats

Parameter	Corn	Wheat	Barley	Oats
Moisture content, (% db)	15.2	10.3	13.3	12.2
Bulk density, (kg/m^3)	764–821	836–898	691–752	565–597
Failure stress ratio, R_f	0.931	0.966	0.875	0.776
Modulus number, M	316	992	125	33
Modulus exponent, n	0.853	0.947	0.825	0.570
Internal friction, φ	25.5°	25.8°	39.6°	31.9°

Source: Smith (1981).

(Timoshenko and Goodier 1969), whereas the determination of plastic strain components is based on theory of plasticity (e.g., Hill 1950). A key concept in the elastoplastic theory is the yielding surface of the material, which is a geometric surface in the stress space. The material behaves elastically if the stress state is within the yielding surface, whereas both elastic and plastic deformations occur when the stress state is on the yielding surface. A material may have many yielding surfaces of different shapes and sizes. These yielding surfaces may expand or translate as loading proceeds.

Lade (1977) proposed an elastoplastic model with two yielding surfaces, i.e., a collapse yielding surface and an expansive yielding surface given by:

$$\begin{aligned} &\text{Collapse:} \quad f_c - \kappa_c = 0 \\ &\text{Expansive:} \quad f_p - \kappa_p = 0 \end{aligned} \tag{40}$$

Loading functions f_c and f_p and hardening functions κ_c and κ_p are defined as:

$$\begin{aligned} f_c &= I_1^2 + 2I_2 \\ \kappa_c &= p_a^2\left(\frac{1}{c_1}\frac{W_c}{p_a}\right)^{1/c_2} \end{aligned} \tag{41}$$

$$\begin{aligned} f_p &= \left(\frac{I_1^3}{I_3} - 27\right)\left(\frac{I_1}{p_a}\right)^{Y_2} \\ \kappa_p &= ac^{-bW_p}\left(\frac{W_p}{p_a}\right)^{1/q} \end{aligned} \tag{42}$$

where:

I_1, I_2, I_3	= the first, second, and third stress invariants, respectively
W_c	= plastic collapse work per unit volume (product of collapse strain and stress) (kJ/m^3)
W_p	= plastic expansive work per unit volume (product of expansive strain and stress) (kJ/m^3)
c_1	= collapse modulus (dimensionless)
c_2	= collapse exponent (dimensionless)
Y_1	= yield constant (dimensionless)
Y_2	= yield exponent (dimensionless)
a	= $Y_1(ep_a/W_{pp})^{(1/q)}$, intermediate variable
e	= base of natural log
b	= $(qW_{pp})^{-1}$, intermediate variable
W_{pp}	= $w_3 p_a(\sigma_3/p_a)^{w4}$, intermediate variable
q	= $w_1 + w_2(\sigma_3/p_a)$, intermediate variable
w_1, w_2, w_3, w_4	= work-hardening constants (dimensionless)

There are two plastic strain components that correspond to the two yielding surfaces, namely the plastic collapse strain and the plastic expansive strain. The total strain increment is the sum of the elastic and plastic strain components:

$$\{d\varepsilon\} = \{d\varepsilon_e\} + \{d\varepsilon_c\} + \{d\varepsilon_p\} \tag{43}$$

where:

$\{d\varepsilon\}$ = total strain increment vector (m/m)
$\{d\varepsilon_e\}$ = elastic strain increment vector (m/m)
$\{d\varepsilon_c\}$ = plastic collapse increment vector (m/m)
$\{d\varepsilon_p\}$ = plastic expansive increment vector (m/m).

The elastic strain increment is calculated using the generalized Hooke's law:

$$\{d\sigma\} = [D_e]\,\{d\varepsilon^e\}$$

where $[D_e]$ is the elastic stress-strain matrix (Timoshenko and Goodier 1969). Two elastic parameters in $[D_e]$—the modulus of elasticity (MOE) and Poisson's ratio—are needed when calculating the elastic strain increment. Lade (1977) suggested that the MOE may be determined using Equation 37 and Poisson's ratio can be determined experimentally.

The two plastic strain components are calculated from the flow rules (Lade 1977):

$$\{d\varepsilon_c\} = d\lambda_c\left(\frac{\partial g_c}{\partial\{\sigma\}}\right)$$

$$\{d\varepsilon_p\} = d\lambda_p \left(\frac{\partial g_p}{\partial\{\sigma\}} \right) \tag{44}$$

where:

g_c, g_p = plastic collapse and expansive potentials, respectively
$d\lambda_c, d\lambda_p$ = positive differential multipliers (functions of stresses)

The two plastic potentials are given as:

$$g_c = I_1^2 + 2I_2$$

$$g_p = I_1^3 - \left[27 - \zeta \left(\frac{p_a}{I_1} \right) \right] I_3$$

$$\zeta = P_1 \left(\frac{I_1^3}{I_3} - 27 \right) \left(\frac{I_1}{p_a} \right)^{Y_2} + P_2 \left(\frac{\sigma_3}{p_a} \right)^{1/2} + P_3 \tag{45}$$

where:

P_1, P_2, P_3 = plastic expansive potential constants (dimensionless)
ζ = intermediate variable

From Equations 40–45, an overall stress-strain equation is obtained:

$$\{d\sigma\} = [D_{ep}]\,\{d\varepsilon\} \tag{46}$$

$$[D_{ep}] = [D_e] - \frac{[D_e]}{A} \left(\frac{\partial g_c}{\partial\{\sigma\}} \{b_c\}^{\mathrm{t}} + \frac{\partial g_p}{\partial\{\sigma\}} \{b_p\}^{\mathrm{t}} \right) [D_e] \tag{47}$$

$$A = L_{11} L_{22} - L_{12} L_{21} \tag{48}$$

$$\{b_c\} = L_{22} \frac{\partial f_c}{\partial\{\sigma\}} - L_{12} \frac{\partial f_p}{\partial\{\sigma\}} \tag{49}$$

$$\{b_p\} = L_{11} \frac{\partial f_p}{\partial\{\sigma\}} - L_{21} \frac{\partial f_c}{\partial\{\sigma\}} \tag{50}$$

$$L_{11} = \left\{ \frac{\partial f_c}{\partial\{\sigma\}} \right\}^t [D_e] \frac{\partial g_c}{\partial\{\sigma\}} + \frac{\partial \kappa_c}{\partial W_c} \{\sigma\}^{\mathrm{t}} \frac{\partial g_c}{\partial\{\sigma\}} \tag{51}$$

$$L_{22} = \left\{ \frac{\partial f_p}{\partial\{\sigma\}} \right\}^t [D_e] \frac{\partial g_p}{\partial\{\sigma\}} + \frac{\partial \kappa_p}{\partial W_p} \{\sigma\}^{\mathrm{t}} \frac{\partial g_p}{\partial\{\sigma\}} \tag{52}$$

$$L_{12} = \left\{ \frac{\partial f_c}{\partial\{\sigma\}} \right\}^t [D_e] \frac{\partial g_p}{\partial\{\sigma\}} \tag{53}$$

Table 2 Elastoplastic Parameters for Wheat at 8.1% mc (db) and Density of 817 kg/m^3

Parameter	Value
Modulus number and exponent, M, n	204.17, 0.54
Poisson's ratio, μ	0.29
Collapse modulus and exponent, c_1, c_2	0.0037, 0.81
Yield constant and exponent, Y_1, Y_2	9.91, 0.14
Potential constants, P_1, P_2, P_3	0.37, –0.033, 0.37
Work-hardening constants, w_1, w_2, w_3, w_4	3.43, –0.13, 0.13, 1.0

Source: Zhang et al. (1986).

$$L_{21} = \left\{ \frac{\partial f_p}{\partial\{\sigma\}} \right\}^t [D_e] \frac{\partial g_c}{\partial\{\sigma\}} \tag{54}$$

Zhang et al. (1986) determined elastoplastic model parameters for wheat at 8.1% mc (db) and density of 817 kg/m^3 (Table 2). They have shown that under triaxial loading conditions, Lade's model adequately predicts the stress-strain behavior of bulk wheat.

9.4. Endochronic Model

For dissipative materials such as bulk grain, the material behavior is dependent on the entire history of deformation which can be measured by an intrinsic time. The intrinsic time is a monotonically increasing "time-like" parameter of the material. Valanis (1971) originally defined the intrinsic time as the length of the total strain path. Valanis and Peters (1991) modified the definition to be an equivalent length of plastic strains:

$$dZ^2 = \| de_{ij}^p \|^2 + \kappa^2 (d\varepsilon_v^p)^2 \tag{55}$$

where:

dZ = intrinsic time increment (dimensionless)
$d\varepsilon_v^p$ = plastic volumetric strain increment (m^3/m^3)
de_{ij}^p = plastic deviatoric strain increment tensor (m/m)
κ = deviatoric-hydrostatic deformation coupling parameter (dimensionless)
$\| \|$ = norm of tensor.

The plastic volumetric and deviatoric strain increments are given as:

$$d\varepsilon_v^p = d\varepsilon_v - \frac{d\sigma_h}{K}$$

$$de_{ij}^p = de_{ij} - \frac{ds_{ij}}{2G} \tag{56}$$

where:

$d\sigma_h$ = hydrostatic stress increment (kPa)
ds_{ij} = deviatoric stress increment tensor (kPa)
$d\varepsilon_v$ = total volumetric strain increment (m^3/m^3)
de_{ij} = total deviatoric strain increment tensor (m/m)
K = initial bulk modulus of elasticity (kPa)
G = initial elastic shear modulus (kPa)

The formulation of endochronic models is based on thermodynamic principle of internal variables and the dissipation inequality (Valanis 1971, 1980; Wu and Wang 1983). Under isothermal conditions, the free energy density of the material may be expressed in terms of plastic strains and a set of internal variables (Valanis 1971, Wu and Wang 1983, Peters 1988). The stress tensor (σ_{ij}) is related to the strain tensor (ε_{ij}) through the free energy density (ψ) (Valanis 1980):

$$\sigma_{ij} = \partial\psi/\partial\varepsilon_{ij}$$

Various forms of the endochronic model may be obtained if different formulations of free energy and different definitions of intrinsic time are used.

Valanis and Peters (1991) expressed the total free energy as the sum of a deviatoric and a hydrostatic component. Corresponding to the two components of free energy density, two constitutive equations were obtained. One defined the relationship between the deviatoric stress and the deviatoric strains, and the other related the hydrostatic stress to both hydrostatic and deviatoric strains:

$$s_{ij} = \int_o^{Z_d} \Phi_d\,(Z_d = Z')\,\frac{de_{ij}^p}{dZ'}\,dZ'$$

$$\sigma_h = \frac{d\varepsilon_v^p}{dZ_h} + \Gamma_0\, s_{ij}\,\frac{de_{ij}^p}{dZ} \tag{57}$$

where:

s_{ij} = deviatoric stress tensor (kPa)

Table 3 Endochronic Model Parameters for Wheat at 8.1% mc (db) and a Bulk Density of 817 kg/m^3.

Elastic bulk modulus, K	2067 kPa
Elastic shear modulus, G	2750 kPa
Coupling parameters, Γ_0, κ	1.3, 1.13
Deviatoric kernel function parameters, k_1, k_2, k_3	0.54, 979.7, 3637
Hydrostatic hardening parameters, h_1, h_2, h_3	0.7, 0.000088, 73.2

Source: Xu et al. (1991).

σ_h = hydrostatic stress (kPa)
Φ_d = $k_1\delta(Z_d) + k_2e^{-k_3Z_d}$, deviatoric kernel function
k_1, k_2, k_3 = model parameters associated with the deviatoric kernel function (dimensionless)
Γ_0 = coupling constant (dimensionless)
Z' = intrinsic time variable (dimensionless)
Z_d = deviatoric component of intrinsic time (dimensionless)
Z_h = hydrostatic component of intrinsic time (dimensionless)
δ = delta function.

The deviatoric and hydrostatic components of intrinsic times are determined from the total intrinsic time as follows:

$$dZ_d = \frac{dZ}{F_d}$$

$$dZ_h = \frac{dZ}{\kappa F_h}$$

$$F_d = \sigma_h \tag{58}$$

$$F_h = (1 - h_1)\, h_2 e^{h_3[1-(v/v_o)+h_1\sigma_h]}$$

where:

F_h = hydrostatic hardening function
F_d = deviatoric hardening function
h_1, h_2, h_3 = hydrostatic hardening parameters (dimensionless)
v, v_0 = specific volume and initial specific volume, respectively (m^3/kg)

Xu et al. (1993) applied the endochronic model developed by Valanis and Peters (1991) to bulk wheat. Model parameters were determined for wheat at

8.1% mc (db) having a bulk density of 817 kg/m^3 (Table 3). The predicted stress-strain-dilatancy relationship compared favorably with the triaxial test data.

10. DETERMINING MECHANICAL PROPERTIES OF BULK GRAIN

A number of testing methods have been developed for determining mechanical properties of granular materials. These methods include cylindrical triaxial test, cubic (true) triaxial test, direct shear test, simple shear test, and hollow cylinder test. The following sections review testing methods which have been used for bulk grain.

10.1. Cylindrical Triaxial Tests

The cylindrical triaxial test has been considered a standard method for evaluating soil properties (ASTM 1993a). Many researchers have adapted the test for the studies of behavior of bulk grain (Smith 1981, Zhang et al. 1986, Li et al. 1989, Hardin et al. 1990). In a cylindrical triaxial tester, a cylindrical sample is enclosed in a thin rubber membrane which is housed inside a cylindrical plastic chamber. The chamber is filled with fluid which applies a confining pressure to the specimen. An axial stress is applied through a vertical loading ram. In a simple compression test, the vertical axial stress and strain are measured while the confining pressure is kept constant. Isotropic compression tests may be performed by increasing the confining pressure without any axial stresses being applied through the loading ram.

Because the size of grain kernels is relatively large compared with soil particles, large specimens should be used to minimize particle size effects (Bock et al. 1991). Specimens with a height-to-diameter ratio of 2.0 produced more consistent results when testing granular materials having large particles.

In soil tests, water is often used to provide confining pressures. The depth of water in the triaxial cell causes a pressure difference between the top and bottom of the specimen. In grain tests, this pressure difference may significantly affect test results because the specimen is large and the confining pressures are low. Therefore, compressed air is commonly used to provide the confining pressures (Bock et al. 1991).

10.2. Cubic Triaxial Tests

A cubic triaxial tester provides for independent and simultaneous measurements of stresses and strains in three orthogonal directions. Manbeck and Nelson (1972) used a cubical tester to study the behavior of bulk wheat. The apparatus that they used consisted of a cubical membrane (latex rubber, 0.4 mm thick) enclosed in larger aluminum container. A space was provided between the concentric wall of the container and the membrane. This concentric space was divided into six facial regions into which fluid pressure was applied independently to stress the sample, which was contained in the membrane, in three orthogonal directions. Deformations in each direction were measured by monitoring the volume of fluid flowing into or out from the concentric space.

10.3. Resonant Column Tests

The resonant column method is recommended by ASTM (1993b) as a standard method for testing soil properties. Ziolkowski et al. (1985) used the method to determine the shear modulus of bulk wheat. The method is founded on theory of wave propagation in prismatic rods. When an excitation is applied to a confined column of grain, a compression or shear wave is induced and propagates through the column. The frequency of the excitation is adjusted until the system reaches resonance. The modulus of elasticity (if compression excitation is used) or the shear modulus (if shear excitation is used) may be calculated from the resonant frequency. Details of the calculation procedures are illustrated in ASTM Standard D4015-92 (ASTM 1993b).

11. FRICTION BETWEEN GRAIN AND STRUCTURAL SURFACES

The interaction between grain and the surfaces of a structure must be taken into account when designing grain storage and handling facilities. Commonly, the classical friction (Coulomb) theory has been used. This theory states that the ratio of the tangential (frictional) force to the normal force on the contacting surfaces is constant. The ratio is termed the coefficient of friction. Research has shown that the coefficient of friction between grain and a structural surface is dependent on many factors, including grain moisture content, normal pressure, and number of passes made over the surface.

As grain moisture content increases, the coefficient of friction also increases (Lawton 1980, Thompson and Ross 1983). Conflicting results have been reported about the effect of normal pressure on friction. Some researchers have reported that the normal pressure has no effect on the coefficient of friction (Bikert and Bakker-Arkema 1968, Lawton 1980) while others have observed that the coefficient of friction decreased with normal pressure (Thompson and Ross 1983, Moore et al. 1984, Zhang et al. 1988). Friction between grain and a structural surface is very different from the friction between two solid surfaces in contact. Under high normal pressures, grain kernels will deform substantially and they may even break. It is more likely that the coefficient of friction is lower when the normal pressure is high. Thompson and Ross (1983) explained that while the bonds between the asperities of the grain and a structural surface become stronger as normal pressure increases, the frictional force, which acts between the bonds in proportion to the normal force, would be smaller.

As grain repeatedly passes over a surface, the friction between grain and the surface decreases (Thompson et al. 1988, Bucklin et al. 1989). The coefficient of friction becomes constant after sufficient number of passes. Thompson et al. (1988) studied the deposits on the galvanized steel surface after wheat had passed over it. The residue included fatty acids (2–4% of the residue), triphenol acetate (23–27%), and lipid (70%). They concluded that the coefficient of friction decreased after repeated movement of grain over the surface as a result of the deposition of a waxy-paraffin coating on the surface.

12. EFFECTS OF OIL ADDITIVES ON MECHANICAL PROPERTIES

Grain dust in grain handling and storage facilities can cause explosions and can be hazardous to the health of workers. A cost-effective way of suppressing grain dust is to add oil to the grain (Lai et al. 1984). The oil acts as a lubricant between grain kernels and therefore alters the mechanical properties of the grain bulk. Jayas et al. (1992) investigated the effects of oil addition on properties of wheat. They found that both emptying and filling angles of repose increased by 10–30% when either food grade canola oil or mineral oil (5 N-F) was added at rates up to 0.2% by mass (Table 4). The greatest increases occurred at a level of 0.05% oil.

Table 4 Effect of Canola Oil on Mechanical Properties of Wheat

	Amount of oil, % by mass				
Mechanical property	0	0.05	0.10	0.15	0.20
Emptying angle of repose(°)	32.1	34.8	33.3	35.2	35.2
Filling angle of repose (°)	30.1	35.2	34.9	36.5	35.9
Friction coefficients on:					
Parallel grain plywood	0.41	0.39	0.36	0.37	0.34
Perpendicular grain plywood	0.44	0.45	0.42	0.43	0.44
Steel-troweled concrete	0.47	0.49	0.47	0.51	0.50
Wood-floated concrete	0.48	0.47	0.45	0.48	0.48
Smooth galvanized steel	0.34	0.28	0.26	0.28	0.24
Corrugated steel	0.59	0.64	0.63	0.66	0.62

Source: Jayas et al. (992).

Oil also affects the interactions between grain and structural surfaces with which it is in contact. Jayas et al. (1992) studied the effects of oil additives on coefficients of friction between grain and various structural materials, including flat galvanized steel, corrugated galvanized steel, plywood, steel-troweled concrete, and wood-floated concrete (Table 4).

13. FUTURE RESEARCH DIRECTIONS

Precision and accuracy of measurement of engineering properties of grain and oilseeds are important factors to consider. More precise values will improve the accuracy of heat transfer models such as those developed using finite element methods. Variations in composition, size, and shape of agricultural products, however, preclude the need for accuracies greater than ± 2–5%. These models often require specification of the effects of temperature, density, and moisture content on thermal properties.

There is a substantial body of data of physical, thermal, and hygroscopic properties, however, there is also a large variation among laboratories for the same material. Many of the published data on these properties are of limited value because insufficient supporting data are included. In future research, information needed should include detailed product description, sampling procedure, and errors in measurement. This description should include the cultivar, the size of individual kernels, maturity, and pretreatment. Details of the experiment should include sample size, surface conditions, porosity, temperature, relative humidity, atmospheric pressure, and sampling procedure. The equipment description should provide sufficient detail so that the experiment can be precisely duplicated.

Knowledge of physical and mechanical properties of bulk grain is essential in designing and evaluating grain handling and storage facilities. Grain bulks are composed of individual kernels and pores filled with air. Bulk properties, therefore, depend on both individual kernels and interactions among kernels. For example, bulk density reflects the kernel density and spacial arrangement of kernels. Another example is that the stress-strain behavior of bulk grain is governed by forces transmitted across the contact points of grain kernels, interkernel slip, and deformations of individual kernels. However, description of bulk properties for grain and oilseeds has been phenomenological. Research is needed to define bulk properties using the fundamental microscopic parameters, such as the size and shape of grain kernels, surface friction of kernels, orientation of kernels, contact strength between kernels, and number of contacts between kernels.

14. NOMENCLATURE

A	= intermediate variable or empirical constant
a	= intermediate variable or empirical constant
a_1	= empirical constant
B	= empirical constant
b	= intermediate variable or empirical constant
$\{b_c\}, \{b_p\}$	= intermediate variable
C	= instantaneous liquid concentration (kg/m^3)
C	= intermediate variable or empirical constant
C_1	= initial liquid concentration (kg/m^3)
C_e	= coefficient of internal evaporation (dimensionless)
$C_{\ell,}$	= instantaneous liquid concentration (kg/m^3)
$C_{\ell,0}$	= initial liquid concentration (kg/m^3)
c	= specific heat ($kJ \cdot kg^{-1} \cdot K^{-1}$)
c_1	= collapse modulus (dimensionless)
c_2	= collapse exponent (dimensionless)
c_1–c_5	= intermediate variable or empirical constant
c_d	= specific heat of dry grain mass ($kJ \cdot kg^{-1} \cdot K^{-1}$)
c_g	= specific heat of grain ($kJ \cdot kg^{-1} \cdot K^{-1}$)
c_w	= specific heat of water held in the product ($kJ \cdot kg^{-1} \cdot K^{-1}$)
$[D_e]$	= elastic stress-strain matrix
$[D_{ep}]$	= elastoplastic stress-strain matrix
D_F	= apparent liquid diffusivity (m^2/s)
D_m	= mass diffusion coefficient (m^2/s)
D_o	= constant mass diffusivity at a given dry bulb temperature (m^2/s)
d	= intermediate variable or empirical constant
d_c	= diameter of the smallest circumscribing sphere (m)
d_e	= equivalent diameter of grain kernel (m)
ds_{ij}	= deviatoric stress increment tensor (kPa)
dZ	= intrinsic time increment (dimensionless)
dZ_d	= increment of deviatoric component of intrinsic time (dimensionless)
dZ_h	= increment of hydrostatic component of intrinsic time (dimensionless)
$d\varepsilon_v^p$	= plastic volumetric strain increment (m^3/m^3)
$\{d\varepsilon\}$	= strain increment vector (m/m)
$\{d\varepsilon_c\}$	= plastic collapse increment vector (m/m)

$\{d_e\}$ = elastic strain increment vector (m/m)
$\{d\varepsilon_p\}$ = plastic expansive increment vector (m/m)
$d\lambda_c$ = differential multiplier for plastic collapse strain
$d\lambda_p$ = differential multiplier for plastic expansive strain
$d\sigma_h$ = hydrostatic stress increment (kPa)
E = intermediate variable or empirical constant
E = activation energy of the diffusion coefficient (kJ/mol)
E_0 = instantaneous modulus of elasticity or modulus of elasticity at zero time (kPa)
E_i = initial modulus of elasticity (kPa)
E_t = tangent modulus of elasticity (kPa)
e = base of natural log
erh = equilibrium relative humidity (decimal fraction)
de_{ij} = total deviatoric strain increment tensor (m/m)
de_{ij}^p = plastic deviatoric strain increment tensor (m/m)
F = intermediate variable or empirical constant
F_d = deviatoric hardening function
F_h = hydrostatic hardening function
Fo_m = Fourier number for mass transfer
f_c = plastic collapse loading function
f_p = plastic expansive loading function
G = initial elastic shear modulus (kPa)
g_c = plastic collapse potential function
g_p = plastic expensive potential function
h_1, h_2, h_3 = hydrostatic hardening parameters (dimensionless)
h_v = volumetric heat transfer coefficient ($kJ{\cdot}s^{-1}{\cdot}m^{-3}{\cdot}K$)
h_{fg} = latent heat of vaporization of free water (kJ/kgH_2O)
h_{fg}^* = latent heat of vaporization of water from a solid body (kJ/kgH_2O)
I_1, I_2, I_3 = first
K = intermediate variable or empirical constant
K = initial bulk modulus of elasticity (kPa)
k = thermal conductivity ($W{\cdot}m^{-1}{\cdot}K^{-1}$)
k_1, k_2, k_3 = deviatoric kernel parameters (dimensionless)
k_d = thermal conductivity of a layer of dry grain ($W{\cdot}m^{-1}{\cdot}K^{-1}$)
k_p = thermal conductivity of grain kernel ($W{\cdot}m^{-1}{\cdot}K^{-1}$)
$L_{11}, L_{12}, L_{22}, L_{21}$ = intermediate variables
ℓ_1 = the longest dimension (m)

ℓ_2	= the longest dimension perpendicular to ℓ_1 (m)
ℓ_3	= the longest intercept perpendicular to ℓ_1 and ℓ_2 (m)
ℓ_i	= measurements of grain kernel dimension (m)
M	= elastic modulus number (dimensionless)
M	= moisture content (decimal fraction)
M_d	= dry basis moisture content (% db)
M_e	= equilibrium moisture content (% db)
M_o	= initial moisture content (% db)
M_w	= wet basis moisture content (% wb)
MR	= moisture ratio (decimal fraction)
m_a	= air mass flux per unit area ($kg \cdot m^{-2} \cdot s^{-1}$)
N	= number of grain kernels
Nu	= Nusselt number
n	= constant
n	= elastic modulus exponent (dimensionless)
n_2	= constant
P_1, P_2, P_3	= plastic expansive potential constants (dimensionless)
p_a	= atmospheric pressure (kPa)
p_v	= partial pressure of vapor (kPa)
Q	= heat (J)
q	= intermediate variable
q	= heat flux (W/m^2)
R	= universal gas constant (8.314 $kJ \cdot kmol^{-1} \cdot K^{-1}$)
Re	= Reynolds number
R_f	= failure stress ratio (kPa/kPa)
rh	= relative humidity of air (decimal fraction)
R_{sv}	= ratio of a kernel surface area to its volume (m^2/m^3)
S_d	= dry mass of grain kernel (kg)
S_{d1000}	= dry mass of 1000 grain kernels (kg)
s_{ij}	= deviatoric stress tensor (kPa)
S_w	= wet mass of grain kernel (kg)
S_{w1000}	= wet mass of 1000 grain kernels (kg)
t_a	= drying air temperature (K)
t_a	= drying air temperature (°C)
t_g	= grain temperature (°C)
t_{go}	= initial grain temperature (°C)
V	= volume (m^3)
V_k	= volume of a grain kernel (m^3)

V_{kd}	= volume of a dry grain kernel (m^3)
v	= specific volume
v_0	= initial specific volume
W	= water mass (kg)
W_c	= plastic collapse work per unit volume (kJ/m^3)
W_p	= plastic expansive work per unit volume (kJ/m^3)
W_{pp}	= intermediate variable
w_1, w_2, w_3, w_4	= work hardening constants (dimensionless)
Y_1	= yield constant (dimensionless)
Y_2	= yield exponent (dimensionless)
Z'	= intrinsic time variable (dimensionless)
Z_d	= deviatoric component of intrinsic time (dimensionless)
Z_h	= hydrostatic component of intrinsic time (dimensionless)
$\|\| \ \|\|$	= norm of tensor
α	= thermal diffusivity (m^2/s)
Γ_0	= coupling constant (dimensionless)
δ	= delta function
ε	= strain (m/m)
ε^*	= coefficient of internal evaporation
ε_1	= major principal strain (m/m)
ε_{ij}	= strain tensor (m/m)
∇t	= temperature gradient
ζ	= intermediate variable or empirical constant
θ	= time (s)
κ	= deviatoric-hydrostatic deformation coupling parameter (dimensionless)
κ_c	= collapse hardening function
κ_p	= expansive hardening function
μ	= Poisson's ratio
ξ	= porosity (%)
ρ	= instantaneous dry mass density (kg/m^3)
ρ_0	= initial dry mass density (kg/m^3)
ρ_{1000}	= bulk density of 1000 grain kernels (kg/m^3)
ρ_b	= bulk density (kg/m^3)
ρ_{bd}	= bulk density of dry grain (kg/m^3)
ρ_k	= kernel density (kg/m^3)
ρ_{kd}	= kernel density of dry grain (kg/m^3)
σ_0	= constant stress (kPa)

σ_1, σ_3 = major and minor principal stresses (kPa)
$(\sigma_1 - \sigma_3)_f$ = stress difference at failure (kPa)
σ_h = hydrostatic stress (kPa)
σ_{ij} = stress tensor (kPa)
$\{\sigma\}$ = stress vector (kPa)
φ = angle of internal friction (degree)
Φ_d = deviatoric kernel function
ψ = free energy density function

REFERENCES

Alagusundaram, K., D. S. Jayas, W. E. Muir, and N. D. G. White. 1991. Thermal conductivity of bulk barley, lentils, and peas. Trans. ASAE (Am. Soc. Agric. Eng.) 34:1784-1788.

Anderson, R. B. 1946. Modifications of the Brunauer, Emmet and Teller equation. J. Am. Chem. Soc. 68:686-691.

ASAE. 1991. ASAE Standards, 38th edition, Engineering Practices, and Data. Am. Soc. Agric. Eng. St. Joseph, MI. 745 pp.

ASTM. 1993a. Standard test method for unconsolidated—undrained triaxial compressive test on cohesive soils. pp. 1071-1081 in: Annual Book of Standards, Vol 04. 08. Am. Soc. Testing and Materials, Philadelphia.

ASTM. 1993b. Standard test method for modulus and damping of soils by the resonant column method. pp. 581-593 in: Annual Book of Standards, Vol 04. 08. Am. Soc. Testing and Materials, Philadelphia, PA.

Bakker-Arkema, F. W. 1986. Modeling of forced convection in-stored grain drying: the state of the art. pp. 89-95 in: Preserving Grain Quality by Aeration and in-Storage Drying: Proceedings of an International Seminar, B. R. Champ and E. Highley (eds). ACIAR Proc. No. 15. Aust. Centre Int. Agric. Res., Canberra, Australia.

Bakker-Arkema, F. W., C. Fontana, G. L. Pedawa, and I. P. Schisler. 1983. A comparison of drying rates of different grains. Paper 83-3009. Am. Soc. Agric. Eng., St. Joseph, MI. 16 pp.

Becker, H. A. and H. R. Sallans. 1955. A study of internal moisture movement in the drying of wheat kernel. Cereal Chem. 32:212-226.

Becker, H. A. and H. R. Sallans. 1956. A study of the desorption isotherms of wheat at 25°C and 50°C. Cereal Chem. 33:79-91.

Bickert, W. G. and F. W. Bakker-Arkema. 1968. Inconsistencies among friction data for grains. Quarterly Bull. Michigan Agric. Exp. Station 50:591-598.

Bilanski W. K. 1966. Damage resistance of grains. Trans. ASAE (Am. Soc. Agric. Eng.) 9:360-363.

Bock, R. G., V. M. Puri, and H. B. Manbeck. 1989. Stress relaxation response of wheat en masse. Trans. ASAE (Am. Soc. Agric. Eng.) 32:1701-1708.

Bock, R. G., V. M. Puri, and H. B. Manbeck. 1991. Triaxial test sample size effect on stress relaxation of wheat en masse. Trans. ASAE (Am. Soc. Agric. Eng.) 34:996-971.

Boyce, D. S. 1965. Grain moisture and temperature changes with position and time during through drying. J. Agric. Eng. Res. 10:333-341.

Brogan, W. L. and A. R. Edison. 1974. Automatic classification of grains via pattern recognition. Pattern Recognition 6:97-103.

Brooker, D. B., F. W. Bakker-Arkema, and C. W. Hall. 1974. Drying Cereal Grains. AVI Pub. Co. Inc. Westport, CT. 265 pp.

Bruce, D. M. 1983. A simulation of multibed, concurrent-flow and counter flow grain driers. Part I. The model. Div. Note DN 1171, Nat. Inst. Agric. Eng., Silsoe, the United Kingdom. 30 pp.

Brunauer, S., P. H. Emmett, and E. Teller. 1938. Adsorption of gases in multimolecular layers. J. Am. Chem. Soc. 60:309-319.

Brusewitz, G. H. 1984. Microwave drying for moisture determination with accuracy related to temperature. Trans. ASAE (Am. Soc. Agric. Eng.) 27:1217-1221.

Bucklin, R. A., S. A. Thompson, I. J. Ross, and R. H. Biggs. 1989. Apparent coefficient of friction of wheat on bin wall material. Trans. ASAE (Am. Soc. Agric. Eng.) 32:1769-1773.

Casady, W. W. and J. B. Sinclair. 1990. Optical properties of damaged soybean seeds. Paper 90-6578. Am. Soc. Agric. Eng., St. Joseph, MI. 10 pp.

Cenkowski, S., D. S. Jayas, and S. Pabis. 1993. Deep-bed drying—a review of particular theories. Drying Technol. 11:1553-1581.

Cenkowski, S., W. E. Muir, D. S. Jayas. 1990. Simulation of canola and barley drying in a deep bed. J. Food Process Eng. 12:171-190.

Cenkowski, S., S. Sokhansanj and F. W. Sosulski. 1989. Equilibrium moisture content of lentils. Can. Agric. Eng. 31:159-162.

Cenkowski, S., J. Bielewicz, and M. G. Britton. 1991. A single kernel creep and recovery test. Trans. ASAE (Am. Soc. Agric. Eng.) 34:2484-2490.

Cenkowski, S., D. S. Jayas, and D. Hao. 1992a. Latent heat of vaporization for selected foods and crops. Can Agric. Eng. 34:280-285.

Cenkowski, S., Q. Zhang, J. Bielewicz, and M. G. Britton. 1992b. Effect of maturity stage on mechanical properties of canola seeds. Trans. ASAE (Am. Soc. Agric. Eng.) 35:1243-1248.

Chandra, S. and W. E. Muir. 1971. Thermal conductivity of spring wheat at low temperature. Trans. ASAE (Am. Soc. Agric. Eng.) 14:644-646.

Chang, C. S. 1986. Thermal conductivity of wheat, corn, and grain sorghum as affected by bulk densities and moisture content. Trans. ASAE (Am. Soc. Agric. Eng.) 29:1447-1450.

Chen, C. C. and R. V. Morey. 1989. Comparison of four EMC/ERH equations. Trans. ASAE (Am. Soc. Agric. Eng.) 32:983-990.

Chinnan, M. S. and J. H. Young. 1977. A study of diffusion equations describing moisture movement in peanuts pods—I. Comparison of vapor and liquid diffusion equations. Trans. ASAE (Am. Soc. Agric. Eng.) 20:539-546.

Chirife, J. and H. A. Iglesias. 1978. Equation for fitting water sorption isotherm of foods. J. Food Technol. 13:159-174.

Chittenden, D. H. and A. Hustrulid. 1966. Determining drying constants for shelled corn. Trans. ASAE (Am. Soc. Agric. Eng.) 9:52-55.

Chu, S. T. and A. Hustrulid. 1968. General characteristics of variable diffusivity process and the dynamic equilibrium moisture content. Trans. ASAE (Am. Soc. Agric. Eng.) 11:709-710, 715.

Chung, D. S. and H. H. Converse. 1971. Effect of moisture content on some physical properties of grains. Trans. ASAE (Am. Soc. Agric. Eng.) 14:612-614, 620.

Chung, D. S. and H. B. Pfost. 1967. Adsorption and desorption of water vapor by cereal grains and their products. Part II. Trans. ASAE (Am. Soc. Agric. Eng.) 10:552-555.

Chung, D. S., S. W. Park, W. J. Hoover, and C. A. Watson. 1972. Sorption kinetics of water vapor by yellow dent corn. II—Analysis of kinetic data for damaged corn. Cereal Chem. 49:593-604.

Crank, J. and G. S. Park. 1968. Diffusion in Polymers. 2nd Ed. Academic Press, New York. 452 pp.

Davison, E., F. J. Middendorf, and W. K. Bilanski. 1975. Mechanical properties of rapeseed. Can. Agric. Eng. 17:50-54.

de Boer, J. H. 1953. The Dynamical Character of Adsorption. Clarendon Press, Oxford, the United Kingdom. 240 pp.

Ding, K. and R. V. Morey. 1990. Corn quality evaluation with computer vision. Paper 90-3532. Am. Soc. Agric. Eng., St. Joseph, MI. 11 pp.

Duncan, J. M. and C. Y. Chang. 1970. Nonlinear analysis of stress and strain in soils. J. Soil Mech. Found. ASCE 96:1629-1651.

Dutta, S. K., V. K. Nema, and R. K. Bhardwaj. 1988. Thermal properties of gram. J. Agric. Eng. Res. 39:269-275.

Engels, C., M. Hendrickx, S. De Samblanx, I. De Gryze, and P. Tobback. 1986. Modeling of water diffusion during long-grain rice soaking. J. Food Eng. 5:55-73.

Fortes, M. and M. R. Okos. 1980. Change in physical properties of corn during drying. Trans. ASAE (Am. Soc. Agric. Eng.) 23:1004-1008.

Fortes, M., M. R. Okos, and J. B. Barrett. 1981. Heat and mass transfer analysis of intra-kernel wheat drying and rewetting. J. Agric. Eng. Res. 26:109-125.

Gadaj, S. and R. Suprunowicz. 1971. Physico-mathematical description of the relationship for bulk density as a function of moisture content for selected legumes (in Polish). Biuletyn Informacyjny IMER, No. 80. pp. 21-32.

Gal, S. 1981. Recent development for obtaining full range moisture sorption isotherms. pp. 89-110 in: Water Activity: Influences on Food Quality, L. B. Rockland and G. F. Stewart (eds). Academic Press, New York.

Gallaher, G. L. 1951. A method of determining the latent heat of agricultural crops. Agric. Eng. 32:34, 38.

Ginzburg, A. S., W. P. Dubrovski, E. D. Kozakov, G. S. Okunand, and W. A. Rezezikov. 1969. Moisture in Grain (in Russian). Kolos, Moscow. 223 pp.

Gordon, C. and S. Thorne. 1990. A computerized method for determining thermal conductivity and specific heat of foods from temperature measurements during cooling. J. Food Eng. 11:175-185.

Griffith, J. C. and C. M. Smith. 1964. Relationship between volume and axes of some quartzite pebbles from the Olean conglomerate Rock City, New York. Am. J. Sci. 262:497-512.

Guggenheim, E. A. 1966. Applications of Statistical Mechanics. Clarendon Press, Oxford. 211 pp.

Gunasekaran, S. and M. R. Paulsen. 1985. Breake resistance of corn as a function of drying rates. Trans. ASAE (Am. Soc. Agric. Eng.) 28:2071-2076.

Halsey, G. 1948. Physical adsorption on nonuniform surfaces. J. Chem. Phys. 16:931-937.

Hanna, M. A. and N. Sharma. 1988. Oven drying for soybean moisture determination. Trans. ASAE (Am. Soc. Agric. Eng.) 31:1851-1854.

Hardin, B. O., K. O. Hardin, I. J. Ross, and C. V. Schwab. 1990. Triaxial compression, simple shear, and strength of wheat en masse. Trans. ASAE (Am. Soc. Agric. Eng.) 33:933-943.

Henderson, S. M. 1952. A basic concept of equilibrium moisture. Agric. Eng. 33:29-31.

Henderson, S. M. and Pabis, S. 1962. Grain drying theory IV: The effect of airflow rate on the drying index. J. Agric. Eng. Res. 7:85-89.

Hill, R. 1950. The Mathematical Theory of Plasticity. Oxford Univ. Press, Oxford. 356 pp.

Hlynka, W. H., V. Martens, and J. A. Anderson. 1949. A comparative study of ten electrical meters for determining moisture content of wheat. Can. J. Res. 27:382-397.

Hurburgh, C. R., Jr., T. E. Hazen, and C. J. Bern. 1985. Corn moisture measurement accuracy. Trans. ASAE (Am. Soc. Agric. Eng.) 28:634-640.

Husain, A., C. S. Chen, J. T. and Clayton. 1973. Simultaneous heat and mass diffusion in biological materials. J. Agric. Eng. Res. 18:343-354.

Husain, A., C. S. Chen, J. T. Clayton, and L. F. Whitney. 1972. Mathematical simulation of mass and heat transfer in high moisture foods. Trans. ASAE (Am. Soc. Agric. Eng.) 15:732-736.

Iglesias, H. A. and J. Chirife. 1976. Prediction of the effect of temperature on water sorption isotherms of food material. J. Food Technol. 11:109-116.

Jaros, M. and S. Cenkowski, D. S. Jayas, and S. Pabis. 1992. A method of determination of the diffusion coefficient based on kernel moisture content and its temperature. Drying Technol. 10:213-222.

Jayas, D. S. and J. Mazza. 1991. Comparison of modified GAB equation with four other three-parameter equations for the description of sorption data of oats. Paper MNSK91-115. Am Soc. Agric. Eng., St. Joseph, MI. 13 pp.

Jayas, D. S., S. Sokhansanj, and N. D. G. White. 1989. Bulk density and porosity of two canola species. Trans. ASAE (Am. Soc. Agric. Eng.) 32:291-294.

Jayas, D. S., N. D. G. White, M. G. Britton, and J. T. Mills. 1992. Effect of oil used for duct control on engineering properties of stored wheat. Trans. ASAE (Am. Soc. Agric. Eng.) 35:659-664.

Jiang, S. and J. C. Jofriet. 1987. Finite element prediction of silage temperature in tower silos. Trans. ASAE (Am. Soc. Agric. Eng.) 30:1744-1750.

Jindal, V. K., N. N Mohsenin, and J. V. Husted. 1974. Surface area of selected agricultural seeds and grains. Trans. ASAE (Am. Soc. Agric. Eng.) 17:720-725,728.

Kapsalis, J. G. 1987. Influence of hysteresis and temperature on moisture sorption isotherms. pp. 173-213 in: Water activity: Theory and Application to Food. L. B. Rockland and L. R. Beuchat (eds). Marcel Dekker, New York.

Kazarian, E. A. and C. W. Hall. 1965. Thermal properties of grains. Trans. ASAE (Am. Soc. Agric. Eng.) 8:33-37.

Keey, R. B. 1972. Drying: Principle and Practice. Pergamon Press, New York. 358 pp.

Kim, C., J. F. Reid, and M. R. Paulsen. 1990. Evaluation of a machine vision system for corn stress-crack detection. Paper 90-7051. Am. Soc. Agric. Eng., St. Joseph, MI. 12 pp.

Kraszewski, A. W. 1980. Microwave aquametry—A review. J. Mirco. Power 15:209-220.

Kraszewski, A. W. 1988. Microwave monitoring of moisture content in grain—Further considerations. J. Micro. Power 23:236-246.

Kraszewski, A. W. and S. Kulinski. 1976. An improved microwave method of moisture content measurement and control. IEEE Trans. Industr. Electronics and Control Instr., IECI-23:364-370.

Kraszewski, A. W., S. O. Nelson, and T. S. You. 1991. Moisture content determination in single corn kernel by microwave resonator techniques. J. Agric. Eng. Res. 48:77-87.

Lade, P. V. 1977. Elastoplastic stress-strain theory for cohesionless soil with curved yield surfaces. I. J. Solids Struct. 13:1019-1035.

Lai, F. S., C. R. Martin, and B. S. Miller. 1984. Examining the use of additives to control dust in a commercial grain elevator. Paper 84-3575. Am. Soc. Agric. Eng., St. Joseph, MI. 43 pp.

Langmuir, I. 1918. The adsorption of gases on plane surfaces of glass, mica and platinum. J. Am. Chem. Soc. 40:1361-1402.

Lawton, P. J. 1980. Coefficients of friction between cereal grain and various silo wall materials. J. Agric. Eng. Res. 25:75-86.

Li, Y., V. M. Puri, and H. B. Manbeck. 1990. Elastic-viscoplastic cyclic constitutive model parameter determination and evaluation for wheat en masse. Trans. ASAE (Am. Soc. Agric. Eng.) 33:1984-1995.

Li, Y., Q. Zhang, V. M. Puri, and H. B. Manbeck. 1989. Physical properties effect on stress-strain behavior of wheat en masse—Part I. Load response dependence on initial bulk density and moisture content. Trans. ASAE (Am. Soc. Agric. Eng.) 32:203-209.

Liu, M., K. Haghighi, R. L. Stroshine, and E. C. Ting. 1990. Mechanical properties of the soybean cotyledon and failure strength of soybean kernels. Trans. ASAE (Am. Soc. Agric. Eng.) 33:559-566.

Luikov, A. W. 1968. Drying Theory (in Russian). Energia, Moscow. 472 pp.

Löf, G. O. G. and R. W. Hawley. 1948. Unsteady-state heat transfer between air and loose solids. Ind. Eng. Chem. 40:1061-1070.

Lorenz K. J. and K. Kulp (eds). 1991. Handbook of Cereal Science and Technology. Marcel Dekker, New York. 882 pp.

Manbeck, H. B. and G. L. Nelson. 1972. Methods and instrumentation for evaluating the stress strain behavior of wheat en masse. Trans. ASAE (Am. Soc. Agric. Eng.) 15:919-923.

Manbeck, H. B. and G. L. Nelson. 1975. Three dimensional constitutive equations for wheat en masse. Trans. ASAE (Am. Soc. Agric. Eng.) 18:1122-1127.

Martens, V. and I. Hlynka. 1969. Protein Content of Canadian Wheat. Agric. Canada, Winnipeg, MB. 137 pp.

Misra, R. M. and J. H. Young. 1980. Numerical solution of simultaneous moisture diffusion and shrinkage during soybean drying. Trans. ASAE (Am. Soc. Agric. Eng.) 23:1277-1282.

Miyai, Y. 1978. A new microwave moisture meter for grains. J. Microwave Power 13:163-166.

Mohsenin, N. N. 1980. Thermal Properties of Foods and Agricultural Materials. Gordon and Breach Sci. Pub., New York. 407 pp.

Mohsenin, N. N. 1986. Physical Properties of Plant and Animal Materials. Gordon and Breach Sci. Pub., New York. 891 pp.

Moore, D. W., G. M. White, and I. J Ross. 1984. Friction of wheat on corrugated metal surfaces. Trans. ASAE (Am. Soc. Agric. Eng.) 27:1842-1847.

Mühlbauer, W. and R. Scherer. 1977. Die spezifische Wärme von Körnerfrüchten. Grundl. Landtechnik, 27:33-40.

Muir, W. E. 1973. Temperature and moisture in grain stores. pp. 49-70 in: Grain Storage: Part of a System, R. N. Sinha and W. E. Muir (eds). The AVI Pub. Co., Westport, CT.

Muir, W. E. and S. Viravanichai. 1972. Specific heat of wheat. J. Agric. Eng. Res. 17:338-342.

Multon, J. L. 1988. Interaction between water and the constituents of grains, seeds and by-products. pp. 89-159 in: Preservation and Storage of Grains, Seeds and Their By-products. J. L. Multon (ed). Lavoisier Pub., New York.

Multon, J. L., H. Bizot, J. L. Doublier, J. Lefebre and D. C. Abbot. 1981. Rheological behavior of wheat kernel. pp. 179-198 in: Water activity: Influences on Food Quality. L. B. Rockland and G. F. Steward (eds). Academic Press, New York.

Nelson, S. O. 1973. Electrical properties of agricultural products—A critical review. Trans. ASAE (Am. Soc. Agric. Eng.) 16:384-400.

Nelson, S. O. 1980. Moisture dependent kernel and bulk density relationship for wheat and corn. Trans. ASAE (Am. Soc. Agric. Eng.) 203:139-143.

Nelson, S. O. 1982. Factors affecting the dielectric properties of grain. Trans. ASAE (Am. Soc. Agric. Eng.) 25:1045-1049, 1056.

Nelson, S. O. and A. W. Kraszewski. 1990. Grain moisture content determination by microwave measurement. Trans. ASAE (Am. Soc. Agric. Eng.) 33:1303-1307.

Nesvadba, P. 1982. Methods for the measurement of thermal conductivity and diffusivity of foodstuffs. J. Food Eng. 1:93-113.

O'Callaghan, J. R., D. J. Menzies, P. H. Bailey. 1971. Digital simulation of agricultural drier performance. J. Agric. Eng. Res. 16:223-244.

Okabe, T., M. T. Huang, and Okamura. 1973. A new method for the measurement of grain moisture content by the use of microwaves. J. Agric. Eng. Res. 18:59-66.

Osborn, G. S., G. M. White, A. H. Sulaiman, and L. R. Walton. 1989. Predicting equilibrium moisture properties of soybeans. Trans. ASAE (Am. Soc. Agric. Eng.) 32:2106-2113.

Oswin, C. R. 1946. The kinetics of package life. III. Isotherm. J. Soc. Chem. Ind. (London) 65:419-421.

Pabis, S. 1967. Grain drying in thin layers. Paper No. 1/C/4. Presented at the Ag. Eng. Symp. Inst. Agric. Eng., Nat. College Agric. Eng., Silsoe, Bedford, United Kingdom. 12 pp.

Pabis, S. 1982. Theory of the Convective Drying of Agricultural Products (in Polish). PWRiL, Warsaw. 229 pp.

Pabis, S. and S. M. Henderson. 1961. Grain drying theory: II A critical analysis of the drying curve for shelled maize. J. Agric. Eng. Res. 6:272-277.

Papadakis, G., P. Giaglaras, and S. Kyritsis. 1990. A numerical method for determining thermal conductivity of porous media from in-situ measurements using a cylindrical heat source. J. Agric. Eng. Res. 45:281-293.

Pappas, G. and V. N. M. Rao. 1989. Effects of temperature and moisture content on the viscoelastic behavior of cowpeas. J. Texture Studies 20:393-407.

Paulsen, M. R., L. D. Hill, and B. L. Dixon. 1983. Moisture meter-to-oven comparisons for Illinois corn. Trans. ASAE (Am. Soc. Agric. Eng.) 26:576-583.

Peters, J. F. 1988. Internal variable model for frictional materials. pp. 551-569 in: Proc. Int. Workshop Constitutive Equations Granular Noncohesive Soils. Case Western Univ., Cleveland, OH.

Pfost, H. B., S. G. Mourer, D. S. Chung, and G. A. Milliken, 1976. Summarizing and reporting equilibrium moisture data for grains. Paper 76-3520. Am. Soc. Ag. Eng., St. Joseph, MI. 11 pp.

Poulsen, K. P. 1982. Thermal diffusivity of foods measured by simple equipment. J. Food Eng. 1:115-122.

Pratt, W. K. 1978. Digital Image Processing. Wiley-Interscience, New York. 750 pp.

Purdy, J. L. and P. L. Crane. 1967. Influence of pericarp on differential drying rate in "mature" corn (Zea mays L.). Crop Sci. 7:379-381.

Reidy, G. A. and A. L. Rippen. 1971. Methods for determining thermal conductivity in foods. Trans. ASAE (Am. Soc. Agric. Eng.) 14:248-254.

Ruan, R. and J. B. Litchfield. 1992. Determination of water distribution and mobility inside maize kernels during steeping using magnetic resonance imaging. Cereal Chem. 69:13-17.

Ryniecki, A. D. S. Jayas, and W. E. Muir. 1993. A generalized controll-strategy for near-ambient drying of wheat under Canadian-Prairie climate. Trans. ASAE (Am. Soc. Agric. Eng.) 36:1175-1183.

Saffari, B. 1989. Putting DSPs to work. Byte Magazine 14:268-269.

Saperstein, H. D. M. Neuman, E. H. Wright, E. Shedyk, and W. Bushuk. 1987. An instrumental system of cereal grain classification using digital image analysis. J. Cereal Sci. 6:3-14.

Sharma, D. K. and T. L. Thompson. 1973. Specific heat and thermal conductivity of sorghum. Trans. ASAE (Am. Soc. Agric. Eng.) 16:114-117.

Shepherd, H. and R. K. Bhardwaj. 1986. Moisture dependent physical properties of pigeon pea. J. Agric. Eng. Res. 35:227-234.

Smith, D. L. O. 1981. The triaxial load response of grain. Ph. D thesis. Iowa State Univ., Ames, IA. 99 pp.

Smith, E. A. and S. Sokhansanj. 1990. Natural convection and temperature of stored products—A theoretical analysis. Can. Agric. Eng. 32:91-97.

Song, H. and J. B. Litchfield. 1990. Nondestructive measurement of transient moisture profiles in ear corn during drying using NMR imaging. Trans. ASAE (Am. Soc. Agric. Eng.) 33:1286-1290.

Steffe, J. F. and R. P. Singh. 1980. Liquid diffusivity of rough rice components. Trans. ASAE (Am. Soc. Agric. Eng.) 23:767-774, 782.

Stevens, G. N. and M. Hughes. 1966. Moisture meter performance in filed and laboratory. J. Agric. Eng. Res. 11:210-217.

Sun, X. and J. B. Litchfield. 1991. Temperature mapping in a model food with magnetic resonance imaging. Paper 91-6525. Am. Soc. Agric. Eng., St. Joseph, MI. 26 pp.

Sweat, V. E. 1986. Thermal properties of foods. pp. 49-87 in: Engineering Properties of Foods. M. A. Roa and S. S Rizvi (eds.) Marcel Dekker, New York.

Syarief, A. M., R. J. Gustafson, and R. V. Morey. 1987. Moisture diffusion coefficients for yellow-dent corn components. Trans. ASAE (Am. Soc. Agric. Eng.) 30:522-528.

Tang,J. and S. Sokhansanj. 1990. Postharvest technology for lentils. Paper NC90-103. Am. Soc. Agric. Eng. St. Joseph, MI. 16 pp.

Tang, J., S. Sokhansanj, S. Yannacopoulus, and S. O. Kasap. 1991. Specific heat of lentil by differential scanning calorimetry. Trans. ASAE (Am. Soc. Agric. Eng.) 34:517-522.

Thompson, S. A., R. A. Bucklin, C. D. Batich and I. J. Ross. 1988. Variation in apparent coefficient of fiction of wheat on galvanized steel. Trans. ASAE (Am. Soc. Agric. Eng.) 31:1518-1524.

Thompson, T. L., R. M. Peart, and G. H. Foster. 1968. Mathematical simulation of corn drying—A new model. Trans. ASAE (Am. Soc. Agric. Eng.) 11:582-586.

Thompson, S. A. and I. J. Ross. 1983. Compressibility and frictional coefficients of wheat. Trans. ASAE (Am. Soc. Agric. Eng.) 26:1171-1176.

Thompson, S. A. and I. J. Ross. 1984. Using the tangent modulus of grain. Trans. ASAE (Am. Soc. Agric. Eng.) 27:165-168.

Timoshenko, S and J. N Goodier. 1969. Theory of elasticity. 3rd Ed., McGraw-Hill, New York. 567 pp.

Tolaba, M. P., C. Suarez, C., and P. E. Viollaz. 1990. The use of a diffusional model in determining the permeability of corn pericarp. J. Food Eng. 12:53-66.

Travis, A. J. and S. R. Draper. 1985. A computer based system for the recognition of seed shape. Seed Sci. Technol. 13:813-820.

Tuite, J. and G. H. Foster. 1963. Effect of artificial drying on the hygroscopic properties of corn. Cereal Chem. 40:630-637.

Valanis, K. C. 1971. A theory of viscoplasticity without a yield surface, part I: General theory. Arch. Mech. 23:517-533.

Valanis, K. C. 1980. Fundamental consequences of a new intrinsic time measure, plasticity as a limit of the endochronic theory. Arch. Mech. 32:171-191.

Valanis, K. C. and J. F. Peters. 1991. An endochronic plasticity theory with shear-volumetric coupling. I. J. Num. Analy. Meth. Geomech. 15:77-102.

Völgyi, F. 1989. Versatile microwave moisture sensor. Proc. SBMO Int. Micro. Symp., Sao Paulo, Brazil, II:3457-4622.

Waananen, K. M. and M. R. Okos. 1988. Failure properties of yellow-dent corn kernels. Trans. ASAE (Am. Soc. Agric. Eng.) 31:1816-1827.

Watts, K. C. and W. K. Bilanski. 1973. Methods for estimating thermal diffusivity of whole soybeans. Trans. ASAE (Am. Soc. Agric. Eng.) 16:143-1145.

Whitaker, T. B. and J. H. Young. 1972a. Simulation of moisture movement in peanut kernels: Evaluation of the diffusion equation. Trans. ASAE (Am. Soc. Agric. Eng.) 15:163-166.

Whitaker, T. B. and J. H. Young. 1972b. Application of the vapor-diffusion equation for concentric spheres in describing moisture movement in peanut pods. Trans. ASAE (Am. Soc. Agric. Eng.) 15:167-171.

Wolf, M. J., C. L. Buzan, M. M. MacMaster, and C. E. Rist. 1952. Structure of the mature corn kernel. II. Microscopic structure of pericarp, seed coat, and hilar layer of dent corn. Cereal Chem. 29:334-348.

Wolf, M. J., I. M. Cull, J. L. Helm, and M. S. Zuber. 1969. Measuring thickness of excised mature corn pericarp. Agronomy J. 61:777-779.

Wu, H. C. and T. P. Wang. 1983. Endochronic description of sand response to static loading. J. Eng. Mech. 109:970-989.

Xu, S., Q. Zhang, and M. G. Britton. 1993. An endochronic finite element model for predicting loads in grain stroage structures. Trans. ASAE (Am. Soc. Agric. Eng.) 36(4):1191-1199. 18 pp.

Yaciuk, G., W. E. Muir, and R. N. Sinha. 1975. A simulation model of temperatures in stored grain. J. Agric. Eng. Res. 20:245-258.

Yang, W. H. and S. Cenkowski. 1991. Moisture sorption hysteresis of canola upon multiple drying and rewetting cycles. Paper MNSK91-116. Am. Soc. Agric. Eng., St. Joseph, MI. 23 pp.

Yang, W. H. and S. Cenkowski. 1993. Effect of successive adsorption-desorption cycles and drying temperature on hygroscopic equilibrium of canola. Can. Agric. Eng. 35:119-126.

Young, J. H. and T. B. Whitaker. 1971. Evaluation of the diffusion equation for describing thin-layer drying of peanuts in the hull. Trans. ASAE (Am. Soc. Agric. Eng.) 14:309-312.

Zaremba, A. 1977. A method of average temperature and moisture content changes calculation in deep bed drying of wheat (in Polish). Ph. D. thesis. Agric. Univ., Cracow, Poland. 72 pp.

Zhang, Q., Y. Li, V. M. Puri, and H. B. Manbeck. 1989. Physical properties effect on stress-strain behavior of wheat en masse—Part II: constitutive elastoplastic parameter dependence on initial bulk density and moisture content. Trans. ASAE (Am. Soc. Agric. Eng.) 32:203-209.

Zhang, Q., V. M. Puri, and H. B. Manbeck. 1986. Determination of elastoplastic constitutive parameters for wheat en masse. Trans. ASAE (Am. Soc. Agric. Eng.) 29:1739-1746.

Zhang, Q., V. M. Puri, and H. B. Manbeck. 1988. Model for frictional behavior of wheat on structural materials. Trans. ASAE (Am. Soc. Agric. Eng.) 31:898-903.

Zhang, X. and G. H. Brusewitz. 1990. Corn moisture measurement by infrared thermometry. Trans. ASAE (Am. Soc. Agric. Eng.) 33:553-558.

Ziolkowski. D. P., B. O. Hardin, C. V. Schwab, and I. J. Ross. 1985. Shear modulus of wheat at low strain amplitude. Trans. ASAE (Am. Soc. Agric. Eng.) 28:884-888.

13

Structural Loads in Grain Storages

Harvey B. Manbeck and Virendra M. Puri

The Pennsylvania State University, University Park, Pennsylvania

Myron G. Britton

University of Manitoba, Winnipeg, Manitoba, Canada

SUMMARY

This chapter summarizes the primary theories and codes used to predict the magnitude of grain loads acting on grain bins used to store whole grains. The fundamental limit state theories of Rankine and Coulomb are derived and discussed as are two of the commonly used (Janssen and Reimbert) grain bin pressure theories. Janssen's theory is the foundation for most grain load specifications. Indeed, most modern load codes use Janssen's equation, either directly or modified, to calculate static lateral and vertical pressures on bin walls. Dynamic emptying loads are almost universally estimated by applying overpressure factors to the Janssen static pressures. The overpressure factors are based on measurements of wall pressures in model- and full-scale bins during emptying. Thermally-induced pressures likewise are calculated as percentages of Janssen's static pressures in some codes; in others thermal pressures are ignored. Pressure induced by moisture content change, as well as vibration-induced pressures, are not well defined and are generally ignored.

1. INTRODUCTION

Stored grains are generally noncohesive particulate systems in which the cohesive bonds between individual grain kernels are negligible. The loads, both lateral and vertical, that the grains exert on the grain bin walls, floors, and footers are critical to the safe and economical structural design of grain bins.

There are other loads, such as snow, wind, and earthquake, which the grain bin must carry but are not unique to grain bins. Such loads are covered adequately in other engineering practices and codes such as ASCE (1990), ASAE (1992a), and BOCA (1990). Structural design details of bin walls, roofs, and footers are also not the focal point of this chapter. Publications such as those by Gaylord and Gaylord (1984) and Reimbert and Reimbert (1976, 1987), discuss structural analysis and design of bins in detail. This chapter will focus exclusively on the loads exerted on the bin structure by the granular material. Granular materials are often called "semifluids" because their load-deformation (constitutive) behavior is neither like that of solids nor fluids. Thus, their behavior under load cannot be adequately described by either solid mechanics or fluid mechanics principles. This chapter begins with a discussion of two fundamental pressure theories, Rankine and Coulomb, for granular materials followed by discussion of the basic Janssen grain storage pressure theory.

2. BASIC PRESSURE THEORIES: STATIC LOADS

Two commonly used pressure theories for noncohesive particulate systems are the Rankine theory and the Coulomb theory. Brief developments of both theories, along with discussion of their applicability and limitations, are presented in this section.

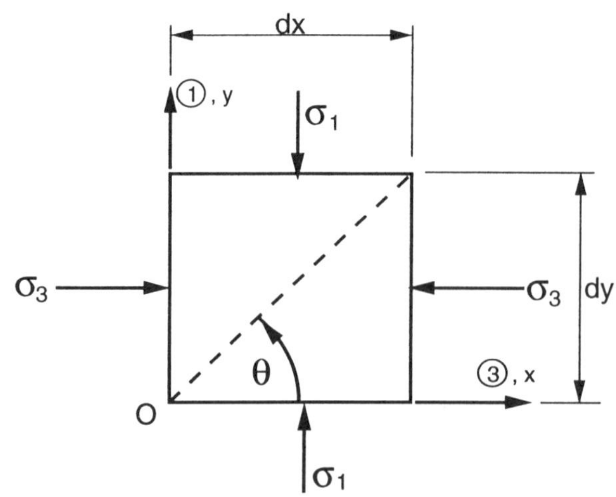

Figure 1 Two-dimensional differential stress element at point O.

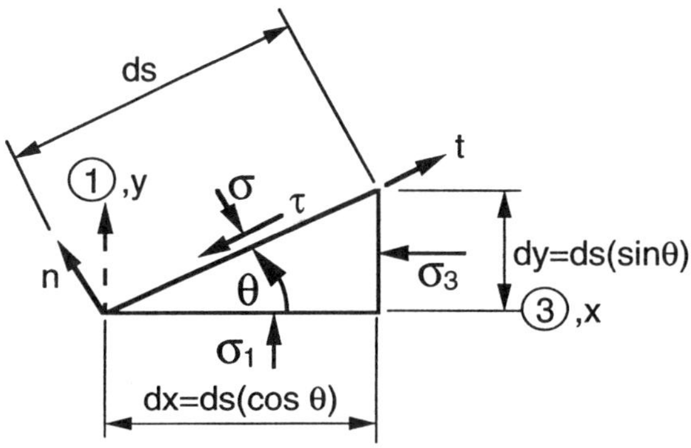

Figure 2 Differential stress element with stresses in the *t-n* directions.

2.1. Rankine Pressure Theory

The development of the Rankine pressure theory (Rankine 1857, Ketchum 1919) begins by considering a differential element of unit height of grain taken at any point within the stored mass subjected to normal compressive stresses (σ_1 and σ_3; Fig. 1). From the free body diagram of the differential element with a cutting plane at an arbitrary angle θ (measured counter-clockwise) from the *x*-axis (Fig. 2), the following equilibrium equations can be written:

$$\sum F_n = 0; \qquad \sigma\, ds - \sigma_1\, ds\, (\cos^2\theta) - \sigma_3\, ds\, (\sin^2\theta) = 0 \tag{1}$$

$$\sum F_t = 0; \qquad \tau ds - \sigma_1\, ds\, (\sin\theta\cos\theta) + \sigma_3\, ds\, (\sin\theta\cos\theta) = 0 \tag{2}$$

Reducing Equations 1 and 2 and solving for the normal and shear stresses, σ and τ, on the *n-t* plane yields:

$$\sigma = \frac{\sigma_1 + \sigma_3}{2} + \frac{\sigma_1 - \sigma_3}{2}\cos 2\theta \tag{3}$$

$$\tau = \frac{\sigma_1 - \sigma_3}{2}\sin 2\theta \tag{4}$$

When plotted in $\sigma - \tau$ space, Equations 3 and 4 yield the familiar Mohr's circle of stresses centered at point O (Fig. 3). Mohr's circle has its center at $(\sigma_1 + \sigma_3)/2$ and a radius equal to $(\sigma_1 - \sigma_3)/2$. The stresses, σ and τ, acting on the *n-t* plane oriented at a counter-clockwise angle θ from the *x*-axis (Fig. 2) are the circle's coordinates at a counter-clockwise angle 2θ from the radial line $\overline{OP}$ (line $\overline{OP}$ corresponds to the σ_1-plane).

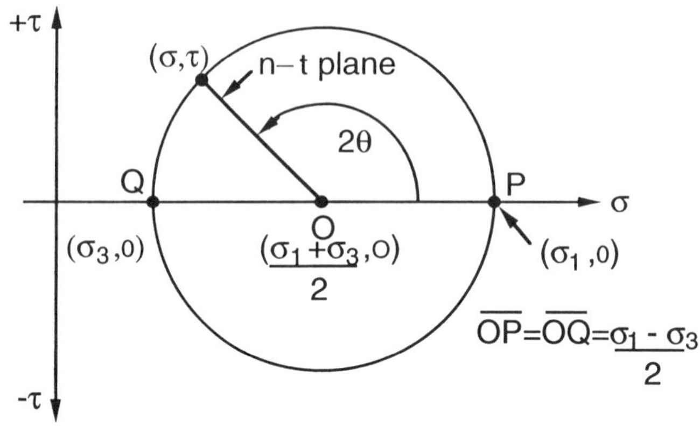

Figure 3 Mohr's circle of stresses at point O in the grain mass.

Coulomb's equation (Eq. 5) defines the relationship between the shear and normal stresses at failure:

$$\tau = \sigma \tan \varphi \tag{5}$$

where φ = angle of internal friction (degrees). The angle of internal friction is unique for each granular material and can be measured by either direct shear tests (Jenike shear tests) or by triaxial shear tests. These methods are presented in Chapter 12. Typical values of φ can be found in Mohsenin (1986) and Gaylord and Gaylord (1984). For many cereal grains, in the absence of shear tests, the angle of internal friction can be approximated by the emptying angle of repose of the grain. This approximation, however, can result in nonconservative estimates for materials such as flaxseed (Moysey and Hiltz 1985).

Plotting Equation 5 in σ–τ space defines the limit states for the granular material (Fig. 4). Any state of stress within the two lines defined by

$$\tau = \pm \sigma \tan \varphi$$

is a stable, and a possible, state of stress. States of stress outside the boundary are not possible. The two Mohr's circles in Figure 4 represent the active (*A*) and passive (*P*) failure states of stress.

The active failure state is obtained by decreasing σ_3 (while holding σ_1 constant) until the Mohr's circle is tangent to the Coulomb limit state boundary. The failure plane is defined by plane $\overline{AC}$ at an angle $2\theta_{fa}$ from the σ_1-plane. The angle $2\theta_{fa}$ equals 90° + φ, so

$$\theta_{fa} = 45° + \varphi/2$$

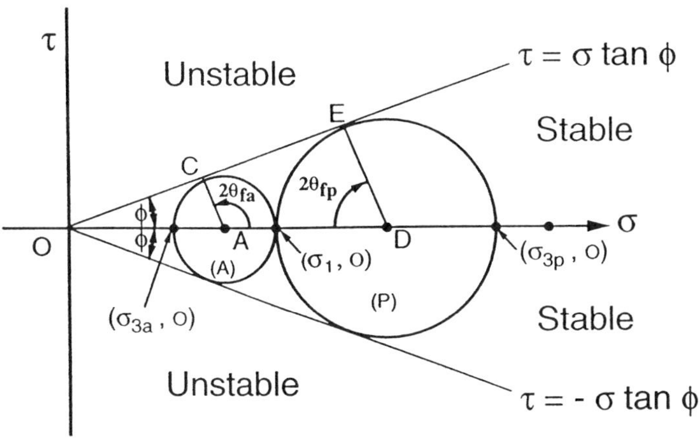

Figure 4 Mohr's circles for the Rankine active (*A*) and passive (*P*) stress states.

This means that the active failure plane is at an angle 45° + φ/2 from the horizontal (Fig. 5).

The active state of horizontal stress (σ_{3a}) is obtained from the following geometric and trigonometric relationships:

$$\sin\theta = \frac{\overline{AC}}{\overline{AO}} = \frac{(\sigma_1 - \sigma_3)/2}{(\sigma_1 + \sigma_3)/2} = \frac{\sigma_1 - \sigma_3}{\sigma_1 + \sigma_3}$$

or

$$\frac{\sigma_3}{\sigma_1} = \frac{1 - \sin\varphi}{1 + \sin\varphi} = \tan^2\left(45^\circ - \frac{\varphi}{2}\right)$$

and

$$\sigma_3 = \sigma_1 \tan^2\left(45^\circ - \frac{\varphi}{2}\right) \tag{6}$$

In shallow grain storages, the vertical pressure (σ_1) in the grain mass equals $\rho g y$; thus the active state lateral pressure, σ_3, in the grain mass is expressed by Equation 7:

$$\begin{aligned}\sigma_3 &= \rho \frac{g'}{g_c} y \tan^2\left(45^\circ - \frac{\varphi}{2}\right) = \rho\, g y \tan^2\left(45^\circ - \frac{\varphi}{2}\right) \\ &= K_a \rho\, g y\end{aligned} \tag{7}$$

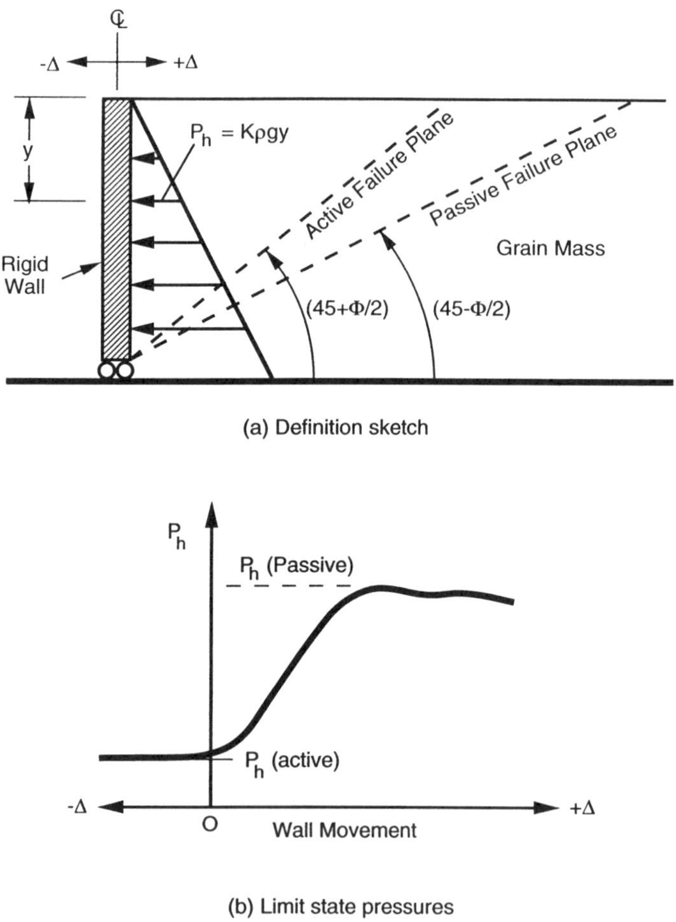

Figure 5 Active and passive states.

where:

σ_3 = active lateral pressure (Pa)
ρ =mass density of grain (kg/m^3)
g = mass to force conversion factor, g'/g_c (9.8N/kg)
g' = acceleration due to gravity (9.8 m/s^2)
g_c = gravitational constant (1 m · kg · N^{-1} · s^{-2})
K_a = Rankine active pressure coefficient, defined as: $\tan^2(45° - \varphi/2)$
y = distance below surface of the grain (m).

The passive failure state is obtained by increasing σ_3, while holding σ_1 constant, until the Mohr's circle is tangent to the Coulomb limit state boundary (Fig. 4). The failure plane is defined by plane $\overline{DE}$ at an angle $2\theta_{fp}$ from the σ_1-plane; angle θ_{fp} equals 45° − φ/2 measured from the horizontal plane. From geometric and trigonometric relationships, the passive state horizontal failure stress (σ_3) is defined by Equation 8:

$$\sigma_3 = \sigma_1 \tan^2\left(45° + \frac{\varphi}{2}\right) \tag{8}$$

or upon substituting $\sigma_1 = \rho\, gy$ into Equation 8:

$$\sigma_3 = \rho\, gy \tan^2\left(45° + \frac{\varphi}{2}\right)$$
$$= K_p\, \rho\, gy \tag{9}$$

where K_p = Rankine passive pressure coefficient, defined as: $\tan^2(45° + \varphi/2)$.

The states of lateral stress defined in Equations 7 and 9 are grain stresses. The loads exerted on a bin wall are:

$$P_h = \sigma_1 \tag{10}$$

and

$$P_w = \mu\, P_h \tag{11}$$

where:

P_h = lateral pressure on a vertical bin wall (Pa)
P_w = vertical pressure on the bin wall (Pa)
μ = coefficient of friction between the grain and bin wall

The Rankine lateral and vertical pressure distributions vary linearly with depth below the grain surface. The pressure distributions are valid for vertical wall surfaces and for the case of a semi-infinite mass of grain behind the wall. Active lateral pressures are normally used for properly designed and managed bins. The active state may be visualized as the minimum limit state pressures acting on the rigid wall in Figure 5 as the wall moves away from the grain mass. Grain failure occurs along a plane oriented 45° + φ/2 to the horizontal. The passive pressure limit state may be visualized as the maximum limit state pressures acting on the rigid wall in Figure 5a as the wall moves towards the grain mass. Grain failure occurs along the plane oriented at 45° – φ/2 to the horizontal. Figure 5b illustrates the relative wall movements required to achieve the limit state pressures. Thus, active pressures are those experienced as the grain medium contracts or is allowed to expand unrestrained. Passive pressures are those experienced as the grain medium expands or is restrained from expansion. Pressures exerted by grain during filling and at-rest are examples of active pressures. Pressures exerted by grain in a bin during contraction of the walls of a steel bin during periods of ambient temperature decline or pressures exerted by grain in a bin when grain expands as moisture content increases are examples of, at least, partial passive pressures.

2.2. Coulomb Pressure Theory

Coulomb's theory of granular pressure (also referred to as the Sliding Wedge Theory) is based on the assumption that a potential failure plane, which originates at the toe of the wall, exists within the granular mass. This plane defines a wedge of material which is kept in place by friction forces along the plane and the reaction of the wall. An illustration of the forces considered for active and passive pressures is given in Figure 6. Essentially, Coulomb's equation is the static analysis of the forces required to maintain equilibrium.

Since the size and shape of the assumed wedge of material are not restricted, this equation permits consideration of both a sloped grain surface and a sloped

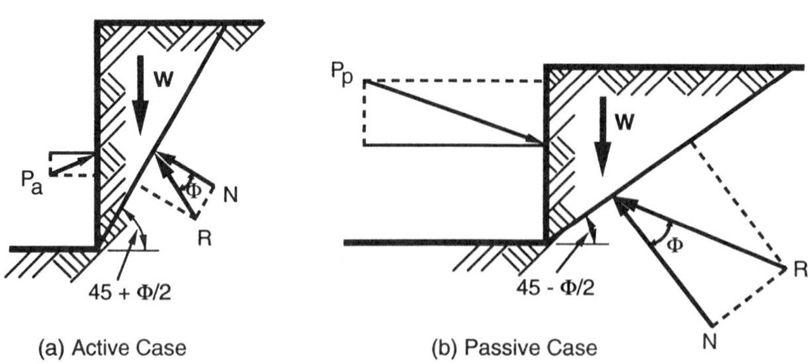

Figure 6 Coloumb active and passive pressure assumptions. (Hough 1957).

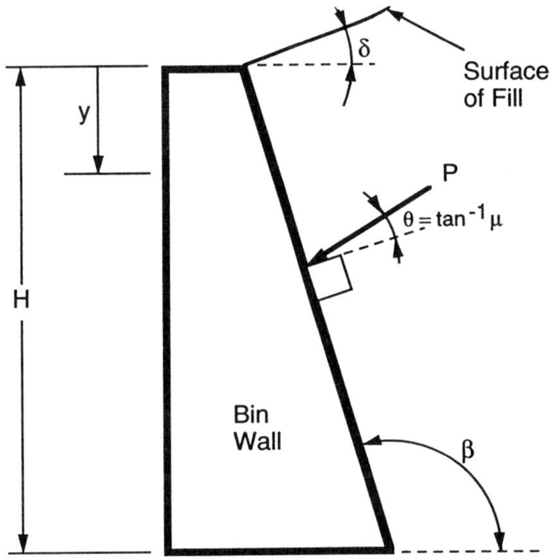

Figure 7 Definition sketch for variables in the Coloumb pressure equation (Eq. 12). *Source*: CFBC (1983).

wall. The resultant force against the wall, the force required to maintain equilibrium, is the resolution of the horizontal and the vertical forces produced from the static analysis. Figure 7 illustrates how the resultant force, the grain surface, and the wall slope are defined. Typically it is assumed that the resultant force (P) acts at an angle defined by the friction between the wall and the grain ($\theta = \tan^{-1} \mu$).

Equation 12 is the most general form, including the effects of both a sloped grain surface and a sloped wall surface, of Coulomb's equation for active pressure:

$$P_h = \rho\, gy \left[\frac{\sin^2(\beta - \varphi)}{\sin^2\beta \sin(\beta + \theta)\left(1 + \sqrt{\dfrac{\sin(\varphi + \theta)\sin(\varphi - \delta)}{\sin(\beta - \delta)\sin(\beta - \theta)}}\right)^2} \right] \tag{12}$$

This complex equation reduces to a simpler form when the bin wall is vertical ($\beta = 90°$) and the grain surface is assumed to be level ($\delta = 0°$). Normally, the active case is used to estimate grain bin loads with the point of action of P being $2H/3$ from the surface.

Like Rankine, Coulomb assumed that the material under consideration was noncohesive and that it existed as a semi-infinite mass. The latter assumption is a serious limitation when considering grain bins because the existence of bin walls implies a finite horizontal dimension. In fact, it is conceivable that a bin can exist in such a manner that the wedges from opposite walls interfere with one another and cause an over-conservative estimate of loads.

All of the calculations upon which Coulomb's equation is based assume a straight wall and a wedge which is bounded by two parallel surfaces, normal to that wall. In the case of a bin with a circular cross-section, the outer limits of any given wedge would, in fact, form a triangle with the apex at the center of the bin (assuming center opening). This would reduce both the volume of the wedge and the resultant force necessary to keep it in a state of equilibrium. Therefore, in circular bins, the Coulomb Pressure Theory overestimates loads.

Unlike other approaches, Coulomb's equation predicts a resultant force. If this force vector is to be reduced to a series of pressure values under various depths of grain, a linear distribution of pressure from zero (stress free) at the surface to a maximum at the base of the wall is normally assumed. As long as the bin is shallow, this linear variation is not a serious problem. If grain depth exceeds the bin diameter, Coulomb's equation will seriously overestimate pressures at the lower levels of the wall. (See Section 3.1 for discussion of shallow versus deep bin criteria).

3. GRAIN BIN THEORIES: STATIC CASE

3.1. Shallow versus Deep Bins

The development of two methods for the prediction of lateral and vertical pressures in grain bins will be described for deep bins. The first method uses Janssen's equation (Janssen 1896, Ketchum 1919) and the second method utilizes Reimbert's pressure theory. The Rankine and Coulomb pressure theories are suited only for the prediction of lateral and vertical pressures in shallow bins.

The Rankine and Coulomb pressure theories assume semi-infinite masses of granular material behind the bin wall. This assumption is reasonable if the limit-state failure-plane intercepts the grain surface before intercepting the opposite bin wall. For no flow cases, the failure plane is oriented at an angle ($45° + \varphi/2$) from the horizontal and originates at the wall-floor intersection (Fig. 8). Thus, for a bin width or diameter of D, the bin is shallow if the bin height (H) is less than $D \cdot \tan(45° + \varphi/2)$. For emptying conditions the failure plane emanates from the edge of the emptying port (Fig. 9). Again, if the failure plane intersects the grain surface before the bin wall, it is classified as shallow; otherwise, it is considered a deep bin.

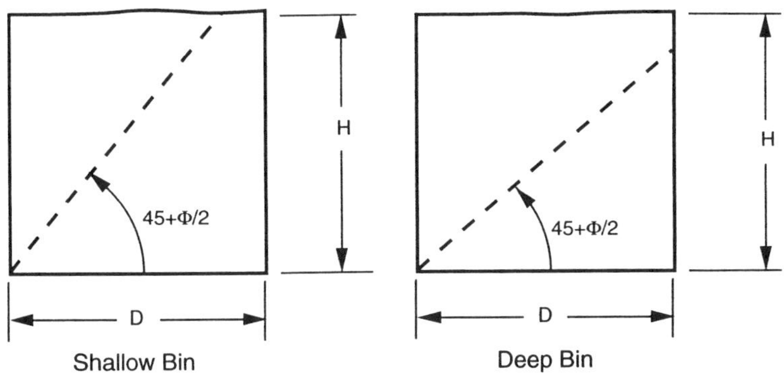

Figure 8 Definition of deep and shallow bins for static pressures.

3.2. Shallow Bin Pressures: Static Case

Static vertical and lateral pressures on bin walls may be estimated using either the Rankine or the Coulomb Theories. Emptying pressures are generally not significantly higher than at-rest pressures in shallow bins although some design specifications still require modest pressure increases during emptying. These provisions will be discussed in more detail in Section 5.

3.3. Deep Bins: Janssen's Equations

The first and most widely used equation for predicting lateral and vertical pressures in grain bins is Janssen's equation (Janssen 1896, Ketchum 1919). Indeed most design codes use Janssen's equation as the fundamental static case.

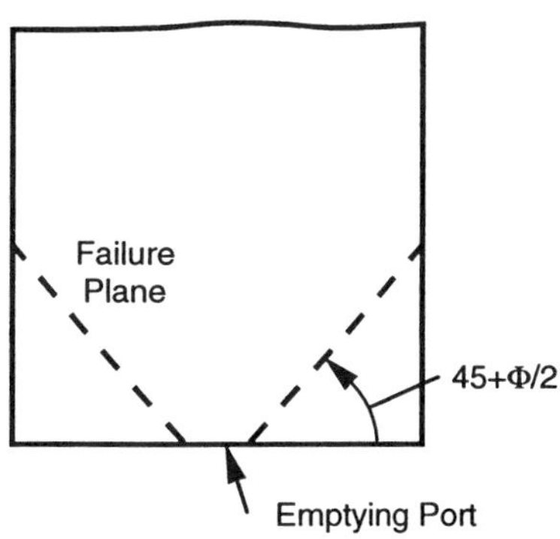

Figure 9 Definition of deep bin for emptying conditions.

For dynamic cases (i.e., during emptying), overpressure factors are applied to Janssen's static pressures. The magnitude of the overpressure factor is determined from observed pressures in laboratory model studies or in field studies during emptying of bins. Since Janssen's equation is fundamental to many design codes and specifications for grain loads, its derivation will be presented herein for a circular cylindrical bin.

Janssen's solution assumes the bulk density of grain (ρ), internal angle of friction (φ), and coefficient of friction between the bin wall and grain (μ) are constant throughout the grain mass. The solution also assumes axisymmetric load distributions; thus, pressure distributions vary with depth but not with location around the circumference. In addition, pressure is assumed not to vary with radius. Figure 10 is a schematic of the grain bin along with a free body diagram of a differential element of grain mass at depth y below the grain surface.

Summing forces acting on the differential element in the vertical direction yields:

$$(P_v + dP_v)\,A + P_w\,C\,dy - P_v\,A - \rho\,g\,A\,dy = 0 \tag{13}$$

where:

A = cross-sectional area of the bin (m^2)
C = bin circumference (m)
P_w = vertical friction pressure on the wall (Pa)
P_v = vertical pressure in the grain mass (Pa)

Defining the ratio of lateral to vertical pressure (as in Rankine's development) within the grain mass (k) as:

$$k = \frac{P_h}{P_v} = \frac{1 - \sin\varphi}{1 + \sin\varphi} = \tan^2\left(45^\circ - \frac{\varphi}{2}\right) \tag{14}$$

and defining the hydraulic radius (R) as:

$$R = \frac{A}{C} \tag{15}$$

and solving Equation 13 yields the expression for the vertical pressure in the grain mass:

$$P_v(y) = \frac{\rho\,g\,R}{k\,\mu}\left(1 - e^{-k\,\mu\,y/R}\right) \tag{16}$$

For grain mass close to the bin wall, the pressure ratio is influenced by grain-wall friction. Gaylord and Gaylord (1984) provides an additional detailed dis-

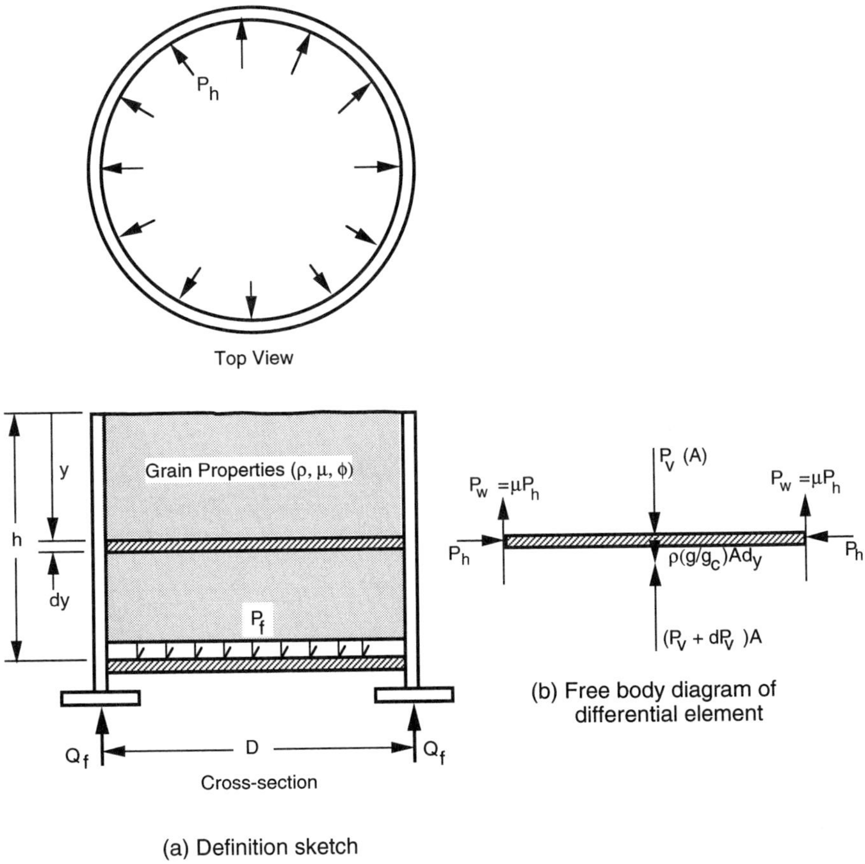

Figure 10 Schematics of circular cylindrical bin for Janssen's equation.

cussion of k. The lateral pressure within the grain mass, which also equals the lateral pressure acting on the bin wall, is defined by Equation 17.

$$P_h = k\,P_v = \frac{\rho\, g\, R}{\mu}\,(1 - e^{-k\mu y/R}) \tag{17}$$

The vertical frictional pressure acting on the bin wall is defined by Equation 18.

$$P_w = \mu\, P_h = \rho\, g\, R\,(1 - e^{-k\mu y/R}) \tag{18}$$

Floor loads (P_f in Figure 10) include the mass of stored grain minus the vertical frictional loads carried by the walls. The frictional load carried by a unit arc

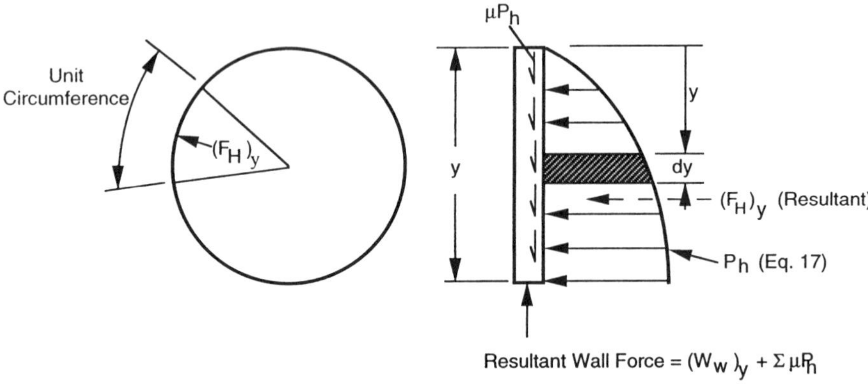

Figure 11 Free body diagram of a unit arc of bin wall to depth y below grain surface.

length of bin wall circumference to a depth y below the surface is illustrated in Figure 11. The resultant horizontal load carried by the unit arc element $(F_H)_y$ is found by integrating P_h from zero to y.

$$(F_H)_y = \int_0^y P_h (1)\, dy' \tag{19}$$

Substituting Equation 17 into Equation 19 and integrating yields:

$$(F_H)_y = \frac{R \rho g}{\mu}\left[y - \frac{R}{k \mu}(1 - e^{-k\mu y/R})\right] \tag{20}$$

The resultant vertical force acting on the unit arc element is defined by Equation 21.

$$(F_v)_y = \mu\, (F_H)_y \tag{21}$$

The vertical footer load per unit circumference (Q_f) is the sum of the force of gravity on the wall plus $(F_v)_y$ evaluated at $y = H$ (Eq. 22).

$$Q_f = W_w + (F_v)_{y=H} \tag{22}$$

where:

Q_f = footer load per unit circumference (N/m)
W_w = total force of gravity on the wall per unit circumference (N/m)

The vertical floor load ($P_f A$) is defined by Equation 23.

$$P_f A = \rho\, g\, AH - C(F_v)_{y=H} \tag{23}$$

from which the floor pressure equals

$$P_f = \rho\, g\, H - \frac{(F_v)_{y=H}}{R} \tag{24}$$

Equations 17, 18, 20, 22, and 24 are the basic equations for estimating lateral and vertical grain loads acting on the bin walls and vertical loads on the footer and floor in a circular bin under static loads, for active states of stress, and for a bin filled with grain of uniform properties. The equations can be extended to bins with rectangular cross-sections by using the hydraulic radius of an equivalent circular bin (i.e., a circular bin with the same area as the rectangular bin). Nonstatic and nonactive pressure states are considered in Section 4. The influences of material properties, which indeed vary within the grain mass, have been investigated and reported by several investigators including Isaacson and Boyd (1965), Loewer et al. (1977), Ross et al. (1979), Thompson and Ross (1981), and Clower et al. (1973). Variable properties, however, are usually not considered in design codes for grain loads. The solutions to the Janssen's equations are quite sensitive to the grain properties used in the solution. This represents a challenge because property values are quite different among grains and some properties are reported as a range of values. Such variations emphasize the importance of estimating the bin loads for the material(s) which produce the maximum lateral or vertical loads over the lifetime of the bin. It is also important to note that the material yielding the highest lateral load may not yield the maximum vertical design loads.

As an example, for a 10 m diameter by 15 m deep, flat-bottom steel-bin filled with granular material having ρ = 800 kg/m^3, φ = 20°, and μ = 0.2, the lateral pressures at 3, 8, and 13 m below the grain surface are 11.1, 26.8, and 39.7 kPa, respectively; the vertical pressures in the grain mass are 22.6, 54.8, and 81.1 kPa, respectively; and the vertical frictional pressure acting on the bin walls at the same depths are 2.2, 5.4, and 7.9 kPa, respectively.

3.4. Deep Bins: Reimbert's Pressure Theory

The deep-bin grain pressure theory known as Reimbert's theory has evolved from work conducted in France by M. L. Reimbert and A. M. Reimbert, beginning in the 1950s (Reimbert and Reimbert 1987). Based on limited experimental evidence which suggested that Janssen's equation underestimated loads in grain bins, they undertook an extensive series of tests using many different shapes and sizes of bins, ranging from models to full-scale structures. Upon completion of these tests, they concluded that:

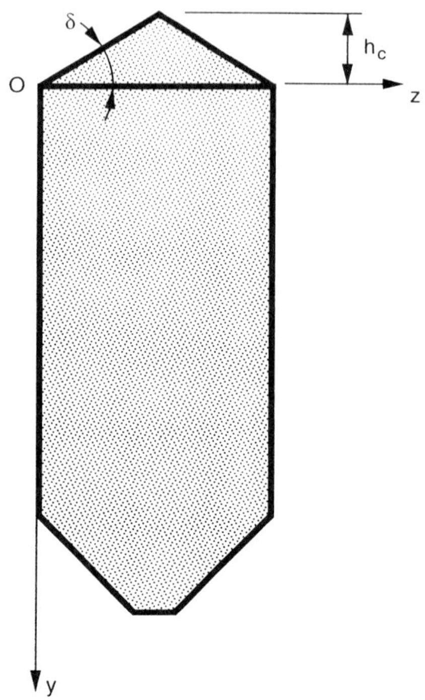

Figure 12 Definition sketch for Riembert pressure theory.

- The ratio of vertical to lateral pressure is not constant but rather varies with height of fill and shape of bin.
- In a rectangular bin, the lateral pressure is not the same on adjacent walls.
- In polygonal and circular cross-section bins, the hydraulic radius is not an adequate definition of the shape of the bin.

This original series of test, supplemented by additional experimental work, forms the base upon which their prediction equations are founded.

The theoretical development was founded on the assumption that by defining the load carried by the walls, both the lateral and vertical pressures in the bin can be calculated. Prediction equations for lateral and vertical pressures were defined with respect to the z (the lateral and/or vertical pressure) and y (depth below the top of the bin wall) axes (Fig. 12).

The height of the cone of grain (h_c) above the z axis was defined as:

$$h_c = R \tan \delta \qquad (25)$$

where:

R = the circular bin cross-section radius (m)
δ = the at-rest angle of grain in the cone (degrees)

The force of gravity on this cone of grain is:

$$Q_0 = \rho\, g\, A \frac{h_c}{3} \tag{26}$$

Assuming that there is no friction between the grain and the bin wall, the maximum vertical load at depth y is calculated using:

$$Q_y = \rho\, g\, Ay + Q_0$$

or

$$Q_y = \rho\, g\, A\left(y + \frac{h_c}{3}\right) \tag{27}$$

Based on experimental observation, the load balanced by friction on the walls at depth y for a horizontal slice having thickness Δy is estimated using:

$$P_h = C\, \mu'\, \Delta y \tag{28}$$

By assuming that the vertical load on the walls due to friction can be defined as $F(y)$, the differential of the load is defined as:

$$\Delta F(y) = P_h\, C\, \mu\, (\Delta y) \tag{29}$$

The limit $(\Delta F(y)/\Delta y)$ as Δy approaches 0 yields:

$$P_h = \frac{F'(y)}{C\, \mu} \tag{30}$$

where

$$F'(y) = \frac{dF}{dy}$$

It follows that the vertical load can be defined as:

$$Q_y = F'(y)$$

and the vertical pressure, therefore, becomes:

$$P_v = \rho\, g\left(y + \frac{h_c}{3}\right) - \frac{F'(y)}{A} \tag{31}$$

Experimental pressure versus depth results were fitted to a hyperbolic curve of the form:

$$F_y = \frac{A_0 y^2 + By + C_0}{Dy + E} \tag{32}$$

By considering limits and various physical constraints on the system, the constants in Equation 32 become:

$$\begin{aligned} A_0 &= \rho g A \\ B &= 0 \\ C_0 &= 0 \\ D &= 1.0 \\ E &= (Q_{max} - Q_0)/(\rho A) \end{aligned}$$

Therefore Equation 31 reduces to the form:

$$F(y) = \frac{\rho g A y^2}{y + [(Q_{max} - Q_0)/(\rho A)]} \tag{33}$$

Normally, this is presented in the general form:

$$F(y) = \frac{\rho g A y^2}{y + A_1} \tag{34}$$

where the A_1 term is defined as:

$$A_1 = \frac{Q_{max} - Q_0}{\rho A} \tag{35}$$

However, the lateral pressure (P_h) was defined earlier as a function of $F'(y)$ in Equation 30, therefore:

$$P_h = \frac{\rho g A\{1 - [(y/A_1) + 1]^{-2}\}}{C \mu} \tag{36}$$

But

$$\rho g A C \mu = P_{max}$$

is the expression for the maximum lateral pressure, so the expression reduces to:

$$P_h = P_{max}\left[1 - \left(\frac{y}{A_1} + 1\right)^{-2}\right] \tag{37}$$

Similarly, substituting the value for $F(y)$ (Eq. 34) into Equation 31 yields:

$$P_v(y) = \rho\, g \left[y \left(\frac{y}{A_1} + 1 \right)^{-1} + \frac{h_c}{3} \right] \tag{38}$$

To utilize the equations it is necessary to define the "characteristic abscissa" (A_1) for specific cases. For cylindrical bins:

$$A_1 = \frac{D}{4\,\mu \tan^2 [45^\circ - (\varphi/2)]} - \frac{h_c}{3} \tag{39}$$

Characteristic abscissas are available for other bin cross-sections as well. Readers are referred to Reimbert and Reimbert (1987) for tables which contain these values.

The sample calculations for the model bin with D = 10 m, H = 15 m, ρ = 800 kg/m^3, φ = 20°, and μ = 0.20 are presented in Sec. 5.2 (also see Table 8).

4. GRAIN LOADS: NONSTATIC CASES

4.1. Types of Loads

Dynamic conditions in grain storages create loads which must be considered in addition to the static loads predicted by the methods described in Section 3. These loads occur when the bins are filled or emptied, or when environmental conditions inside or outside the bin change. The following discussion addresses nonstatic loads in both shallow and deep storage bins.

Numerous nonstatic grain loading conditions may occur during the service life of a storage structure. Chief among these are filling loads, emptying loads, thermally-induced loads, vibrational loads, loads induced by change in grain moisture content, and loads induced by temperature cables. Rational solutions are still in the developmental stages for most of these load cases. Rational solutions for predicting thermally-induced loads are the most developed; whereas, rational solutions for predicting loads induced by changes in grain moisture content and vibrational loads are nonexistent. Thus, most of the nonstatic grain loads are based on empirical equations, many of which are factored levels of one of the static pressure theories such as those of Janssen, Rankine, or Coulomb. The factors used are often based on observed levels of lateral loads in either field or laboratory model studies.

4.2. Filling Pressures

Model- and full-size bin studies have shown that lateral and vertical wall grain loads during filling are the same as the static, or at-rest, grain pressures. Indeed, many grain-load codes recommend filling pressures be the same as at-rest pressures (ACI 1991; DIN 1987; ASAE 1992b).

4.3. Emptying Pressures

For shallow bins with height to diameter ratios less than approximately 1.25, lateral wall loads during emptying are not significantly greater than static grain loads. This is primarily a consequence of the bin emptying in funnel flow mode (Fig. 13a). In funnel flow mode, the grain at the surface forms an inverted cone into which the grains near the surface gradually flow from the sidewalls to the center before flowing down a central "grain funnel" to the exit port. The grain at the top surface is the first to exit the bin in this flow mode.

Deep bins ($H/D > 1.25$) are sufficiently deep that the funnel cannot develop at the beginning of flow. Indeed when the depth of grain is lowered during emptying such that the depth of grain to diameter ratio approaches 1.25, funnel flow will commence. Nguyen et al. (1980) observed the transition between funnel and plug flow at H/D ratios near 1.3 in model bins. However, Usry and Thompson (1986) observed that the transition from funnel to plug flow in large diameter bins occurs at H/D ratios nearer 2.0. The discharge of grain from a deep bin is by plug (mass) flow (Fig. 13b).

There have been some attempts to theoretically define the state of pressure in bins during emptying (e.g., see the review article by Puri and Manbeck 1991); however, the majority of work has been directed toward experimental measurement of bin wall pressures in both model and prototype bins (e.g., Caughey et al.

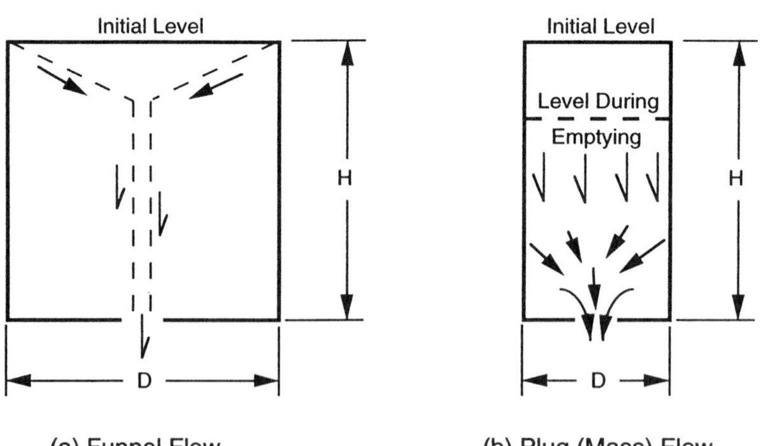

Figure 13 Emptying flow regimes.

1951; Turitzin 1963; Jenike and Johanson 1969; Smid 1975; Reimbert and Reimbert 1976; Manbeck et al. 1977; Thompson et al. 1986; Schwab 1989; Zhang et al. 1991). Many factors have a significant effect on the maximum dynamic pressures including particle size, discharge rate, height to diameter ratio, bin wall corrugation, hopper design, discharge eccentricity, and bin wall stiffness.

There is general agreement that bin wall pressures during emptying exceed the static wall pressures predicted by Janssen's equations (see Section 3.3). There is, however, disagreement about the magnitude and location of the dynamic overpressures during emptying. Dynamic overpressures are those lateral pressures in excess of the at-rest bin wall pressure predicted by Janssen's equations (Section 3.3). The results of numerous research reports vary considerably, as do the recommendations expressed in several design codes.

The measurement of dynamic overpressures in both full-size and model bins has been reported by numerous researchers since about 1939. Takhtamishev (Bernstein 1947, cited by Turitzin 1963) reported emptying overpressures two to three times the static pressures. After conducting tests on model concrete bins, Takhtamishev concluded that the dynamic overpressure factor, defined as the ratio of the dynamic pressure during emptying to the static pressure, was 1.65 for sand and 1.35 for wheat. He also noted that much higher overpressures occurred during mass flow than during funnel flow. Kim (1959), in tests with model and full-size bins, observed that dynamic overpressures only occurred during mass flow. He concluded that dynamic overpressures were greater during emptying than during filling and that during centric emptying the distribution of pressure around the circumference of the bin is not constant at a given elevation. Dynamic overpressure factors between 1.64 and 2.32 were measured by Platonov and Kovtun (1959, cited by Turitzin 1963) in a circular bin with a height to diameter ratio (H/D) of 6.67. The maximum overpressure factor occurred at 0.8 H below the top of the bin wall.

Reimbert and Reimbert (1959, 1963), after having measured the wall pressures in many bins during emptying, concluded that the dynamic overpressure factor varies with the type and the depth of material. They measured wall pressures in bins 4.1 × 4.1 m in cross-section by 10.0 m high filled with wheat. Overpressure factors between 1.76 and 2.39 were observed, but a definite method for computing dynamic pressures could not be advanced because the pressure readings were erratic. They did recommend that funnel flow in bins can be achieved by installing perforated tubes in the bin center to achieve funnel- instead of mass-flow during emptying. Tests showed that no overpressures occur when the perforated tubes are installed in a bin.

Pieper's (1969) observations on dynamic pressures in both square and round model deep-bins filled with granular materials support the DIN (1965) standard specification's procedure, in which the angle of wall friction is assumed to be 0.75 and 0.60 times the angle of internal friction of the material during filling and emptying, respectively. Pieper also found that ten-fold variations in discharge rates from a centric opening had no influence upon silo wall pressures. During eccentric emptying of the bins, Pieper concluded that the maximum lateral pressure occurred on the wall nearest the outlet and that eccentric discharge outlets nearer the center, rather than eccentric outlets near the wall, produced the largest lateral pressures during emptying. Theimer (1969) favors the use of Janssen's equation for estimating lateral pressures during emptying. He recommends the use of an angle of wall friction 0.75 and 0.50 times the at-rest angle of friction for concrete and steel silos, respectively.

Myers (1974) measured the wall pressures in a full-size metal bin, 15.2 m high, 18.2 m in diameter and filled to a depth of 9.7 m. He reported no overpressures during centric emptying of corn at a rate of 2.44×10^{-2} m^3/s.

Deutsch and Schmidt (1969) conducted studies of bin overpressures with a Plexiglas model with an H/D ratio of 4.62 and filled with sand. They recommended overpressure factors of 3.75 for the lower 2/7 of the bin wall height, a linearly varying value between 3.8 and 3.0 for the middle 2/7 of the wall height, and a linearly varying value between 3.0 and 0 for the top 3/7 of the wall height. Yamahara and Takahashi (1967) observed dynamic overpressure factors of 2.5 in a concrete bin with a height-to-diameter ratio of 3.6. They have recommended use of the Soviet Code (1965) for estimating overpressures.

McCabe (1974) studied the flow patterns in granular fill materials during emptying deep, circular, flat-bottomed bins of various sizes filled with three gradings of fine gravel. He observed four separate zones of flow during emptying, namely: 1) plug flow, 2) dynamic equilibrium, 3) dead zone, and 4) free fall. The zone of dynamic equilibrium, which appears to be the zone in which maximum lateral pressures occur, commences at a distance equal to the bin diameter (D) above the discharge outlet and extends to $2.3D$, $2.1D$, and $1.8D$ above the bottom for bins filled with 1–3 mm, 2–5 mm, and 5–8 mm diameter particles, respectively. The upper limit of the zone of dynamic equilibrium is called the crater formation level. It is in the dynamic equilibrium zone that the pressure dome breaks down and plug flow transforms into free-fall or funnel-type flow.

Manbeck et al. (1977) conducted emptying experiments on deep ($H/D = 5$) and shallow ($H/D = 1.25$) flat-bottom model-bins. They concluded that: 1) there are no overpressures in shallow flat-bottom grain bins with $H/D < 1.25$; 2) significant overpressures (overpressure ratios in the range of 0.7–2.4) develop in deep grain bins with $H/D = 5.0$; 3) overpressures were not significantly influenced by a 34:1 variation in discharge rate; and 4) the Soviet Code (1965) best fit the observed overpressures in the middle third of the deep bin whereas the German Code (DIN 1965) specification best fit the observed overpressures in the remainder of the deep bin.

In a model study of deep bins ($H/D = 5.0$), Tyson and Manbeck (1980) measured emptying overpressures over a range of bin wall stiffness. They found that an eight-fold increase in the wall stiffness did not significantly influence the magnitude of overpressures during emptying. They found the ACI Code (ACI 1977) to best fit the observed overpressure data, the Soviet Code (1965) to be conservative, and the German Code (DIN 1965) specification to underestimate observed overpressures by approximately 50%.

Schwab (1989) investigated the grain pressure phenomenon in a cylindrical flat-bottom bin (4.08 m diameter with working height of up to 20 m). He studied the influence of the following parameters: 1) unloading flow rate, 2) flow regime, 3) starting *H*/*D* ratio, and 4) detention time. Some of the key conclusions were: 1) rate of discharge did not significantly influence the total floor load for an *H*/*D* ratio of 1.8 but the influence was significant for an *H*/*D* ratio of 3.0; 2) a complete cone inversion and no movement along bin wall occurred at an average *H*/*D* ratio of 1.6; 3) different starting *H*/*D* ratios of 1.8 to 3.0 did not influence bin wall loads; and 4) the amount of detention time of grain in the bin statistically influenced the static loads.

Zhang et al. (1991) conducted experiments to measure dynamic wall loads during discharge. They reported that bin wall construction had no significant effect on lateral loads but produced significant differences in vertical wall loads.

When plug flow conditions exist, lateral wall pressures have been observed in model and full-scale bins to increase significantly over static pressures. The increases are most significant immediately after the initiation of discharge. Although overpressure factors (ratio of emptying to static loads) have been universally reported, the magnitude of the overpressure factors vary considerably from lows of 1.25–1.50 to highs of 3.5–4.0 (as discussed in the previous paragraphs). Furthermore, the overpressure factors vary with depth below the surface of the grain. Thus, most codes (e.g., ACI 1991; DIN 1987; ASAE 1992b) either define a continuously increasing overpressure factor with depth (except near the bin floor) or define overpressure factors as stepwise increases for different depth zones in the bin.

Several researchers have suggested analytical equations for estimating bin wall loads under mass flow conditions for hopper bins, for example, Walker (1966), Walker and Blanchard (1967), Jenike and Johanson (1969), Jenike et al. (1973a,b). In addition, procedures for calculating the peak pressure during funnel flow conditions have also been outlined by Jenike et al. (1973c).

4.4. Thermally-Induced Loads

4.4.1. Background

Thermally-induced loads in grain bins are significant in thin-walled metal bins. The loads are those induced when a filled bin is subjected to a rapid decrease in ambient air temperature and thus bin wall temperature. Thermal contraction of the bin wall is partially restrained by the grain mass. Thermal mass of the grain is sufficiently large that the grain temperature remains nearly constant. The level of the static plus thermal pressures will be somewhere between the active and passive limit-state pressures (Fig. 5).

When ambient temperature declines rapidly, the thin metal wall of the grain bin contracts. The bin wall contraction, however, is incomplete compared with the free (zero resistance) contraction because of the presence of the compressible stored material. This induces loads in excess of the at-rest loads. Such loads are referred to as thermally-induced loads. Weiland (1964) commented on the significant number of bin failures occurring during periods of rapid decreases in ambient temperatures. It was Anderson (1966) who first presented a semi-empirical mathematical model for circular bins to estimate thermally-induced loads on bin walls. This formulation assumed the mechanical response of the stored material to be elastic. The equation recommended by Anderson for load calculations is given in Equation 40. Because of the wide range of bulk-elastic moduli between and within grains (e.g., Thompson and Ross 1984, Manbeck 1984) the calculated lateral pressures can differ by as much as a factor of five.

$$P_T = \frac{E_g \, \alpha \, \Delta T}{(rn/t) + 1 - \nu} \tag{40}$$

where:

P_T = thermally-induced lateral pressure (Pa)
E_g = elastic modulus for the stored material (Pa)
α = coefficient of thermal expansion (1/°C)
ΔT = ambient temperature change (°C)
r = bin radius (m)
n = ratio of stored material to bin wall elastic modulus (unitless)
t = wall thickness (m)
ν = bin wall material Poisson's ratio (unitless)

An empirical model (Eq. 41) was proposed by Reimbert and Reimbert (1976) for predicting thermally-induced loads. These authors recommended that the hoop stress in a bin wall induced by the stored material under rapidly-declining ambient-temperatures be equal to 1/3 the hoop stress of an incompressible fill-material. Equation 41 gives thermal pressures which are 30–45% of the maximum static pressure for commercial storage bins. Thermal pressures must be considered in addition to the maximum static pressure.

$$\sigma = \frac{E\,\alpha\,\Delta T}{3} \tag{41}$$

4.4.2. Measurement of Thermally-Induced Pressures

Britton (1973) and Myers (1974) were among the first to measure thermally-induced loads on grain-bin walls. Britton observed a 25% increase in strains in model bin walls caused by an 11°C temperature decline. Myers' field study also documented an increase in bin wall pressures during periods of rapidly-declining ambient temperatures. Blight (1985, 1988) and Blight and Garstang (1987) have reported field test results that confirm the significance of thermally-induced loads.

A comprehensive program was initiated in at the Pennsylvania State University (PSU) the early 1980s to systematically: 1) study and quantify thermally-induced loads using experimental data, and 2) develop a numerical model for predicting thermally-induced loads in containers storing bulk solids. Thermally-induced loads were measured using a cylindrical model bin 0.9 m (diameter) × 1.2 m (high) made of 0.8-mm thick aluminum flat sheet (Muzzelo 1983, Manbeck and Muzzelo 1985). The flat-bottom cylindrical bin, (Fig. 14), was equipped with strain guages and thermocouples and housed in an environmental chamber that accurately controlled the ambient temperature. The bulk materials tested included wheat, shelled corn, grain sorghum, and sand. The ambient temperature profiles studied have been: ambient temperatures declining linearly at rates of 10 and 20°C/h, and cyclic ambient temperatures linearly-increasing and linearly-decreasing at rates of 10°C/h (Muzzelo 1983; Zhang et al. 1986a, 1987; Li et al. 1990a). Typical measured values with wheat as the stored material are shown in Figures 15–17 for 10°C/h, 20°C/h, and cyclic variation of ambient temperature, respectively. These data show that thermally-induced loads are significant and add to static pressures with magnitudes near 15% of static pressures per 20°C temperature change.

Regression analysis of the test data showed that the variation of thermally-induced pressure can be related to the temperature change. Therefore, the rate of change of thermally-induced pressures with temperature was represented by thermal pressure coefficient (C_p) defined in Equation 42. The measured C_p values ranged from 0.150 kPa/°C to 0.525 kPa/°C. The data showed that thermally-induced loads are independent of the grain depth. The change in thermal pressure coefficient C_p showed an increase of 63.6% from the first to the second cycle during the temperature decreasing mode (Fig. 18).

$$C_p = \frac{P_T}{\Delta T} \tag{42}$$

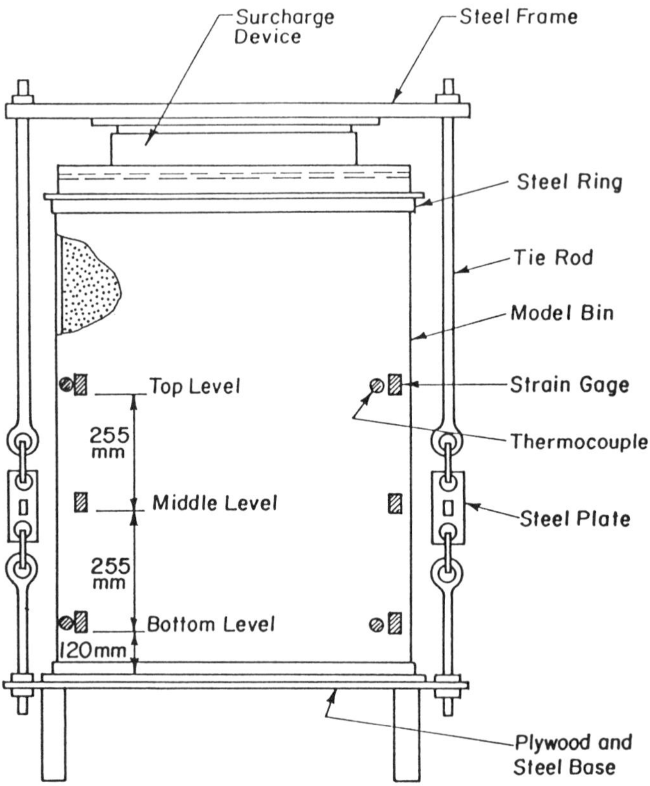

Figure 14 Schematic of the model bin wtih strain gages and thermocouples. *Source*: Li et al. (1990a).

4.4.3. *Model for Thermally-Induced Loads*

Analytical models or closed form solutions for thermally-induced loads on grain bin walls are not possible because of the complex and nonlinear interactions between the stored material and the bin wall. Therefore the PSU group embarked on the development of a numerical model using the finite element method. In this approach the grain bin system is assumed to comprise two interacting continua, namely, the stored material and the bin wall. The first generation finite element model (FEM), developed in 1986 (Zhang et al. 1986b), included rectangular elements to describe the axisymmetric bin. The stored material response was modeled using the elastoplastic theory developed for cohesionless materials (Lade 1977). The frictional interaction between the bin wall and the stored

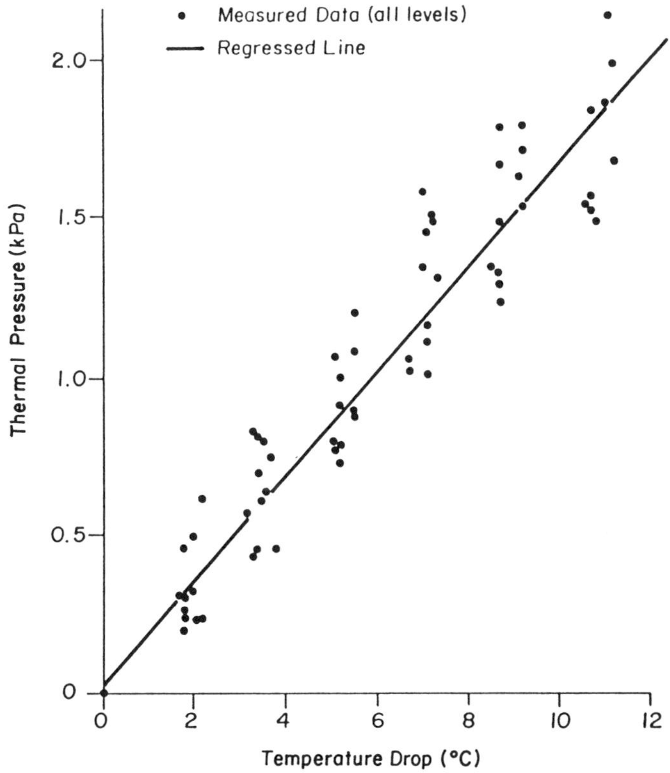

Figure 15 Measured thermally-induced bin wall pressures for 10°C/h temperature decline. *Source*: Manbeck and Muzzelo (1985).

material was assumed to be negligible (i.e., frictionless wall surface). The model predictions qualitatively agreed with the measured values. Subsequently, the first generation model was enhanced by including wall-grain interaction using the interface element (Zhang et al. 1987). The constitutive model parameters for the stored bulk material were estimated using the triaxial tester. The finite element simulation of the nonlinear response of the grain bin system compared favorably with the measured values. The validated FEM was then vectorized and run on a supercomputer to model the response of three full-size grain bins with diameter and height dimensions of: 3 m × 9 m, 6.4 m × 9 m, and 9 m × 9 m. The results showed that: 1) the bin diameter did not influence thermal pressure coefficient in regions away from the top and bottom boundaries, and 2) thermally-induced loads were proportional to temperature decline (Zhang et al. 1989).

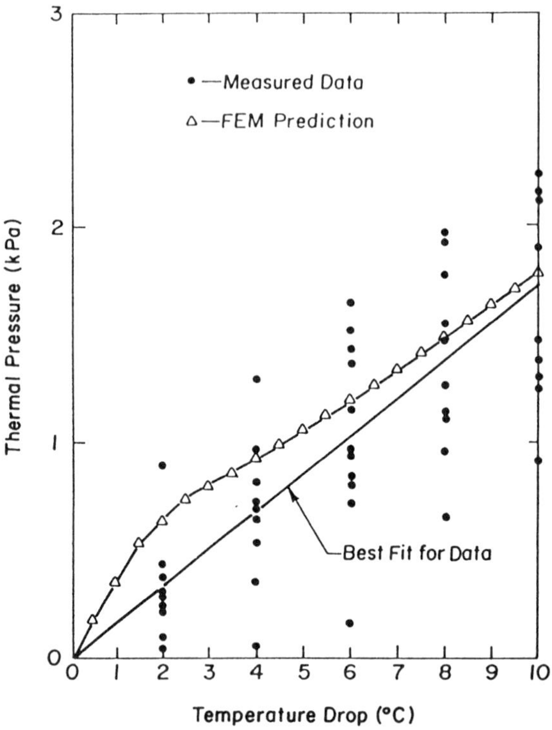

Figure 16 Measured and FEM-predicted thermally-induced bin wall pressures for 20°C/h temeprature decline. *Source*: Zhang et al. (1987).

The second generation FEM was later enhanced to model cyclic thermally-induced loads in bulk-solids storage containers. Lade's (1977) elastoplastic model could not be used because the Lade model is not recommended for cyclic loads. Therefore, an elastic viscoplastic (EVP) model based on the mixture theory as proposed by Youngs (1982) was used (Li et al. 1990b). The EVP model parameters were determined using a modified triaxial tester. These parameters were incorporated into the FEM. The mean measured and the FEM-predicted results are shown in Figure 18 (Li et al. 1991). The FEM-predicted and measured thermally-induced pressures for the first two cycles of loading were not significantly different ($P < 0.05$). Efforts are underway: 1) to further improve and enhance the capability of the FEM, 2) to develop and refine existing constitutive models to more accurately describe the stress-strain behavior of the stored material, and 3) to describe the flow properties and behavior of agricultural and food bulk solids.

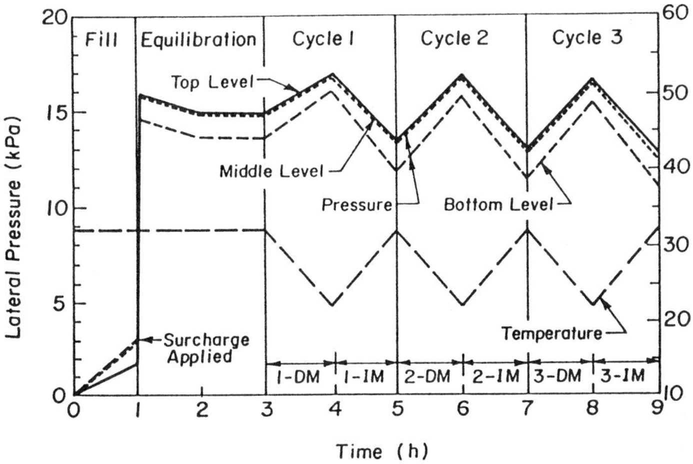

Figure 17 Measured thermally-induced bin wall pressures for cyclic ambient temperature variation. *Source*: Li et al. (1990a).

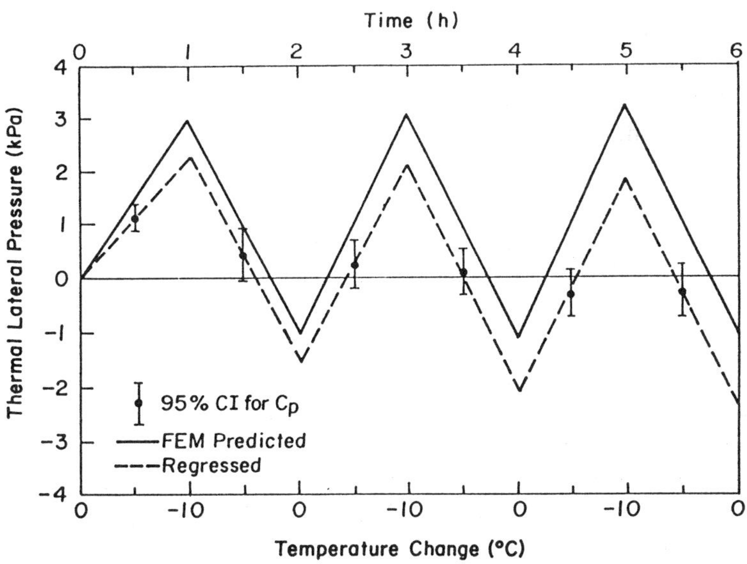

Figure 18 Mean measured and FEM-predicted thermally-induced pressures with cyclic ambient temperature variation. *Source*: Li et al. (1991).

4.5. Loads Induced by Changes in Moisture Content

Only a few articles in the scientific literature address the issue of swelling pressures in grain bins. Numerous articles document the swelling of the individual kernels during moisture absorption, but there are few data on the swelling of bulk granular media.

Dale and Robinson (1954), using pressure cells in the sidewall and floor of a 460-mm diameter by 1520-mm high by 1.9-mm thick steel bin, measured the swelling pressures induced by shelled corn with initial moisture content (mc) of 12.5–13.2% wet basis (wb). The average final moisture content of the corn was approximately 15% in two tests and 22.4% in a third test. The results (summarized graphically in Figs. 19 and 20) suggest a four- to six-fold increase in the bin wall pressures as the grain moisture content increased from 13 to 15%. For the same moisture content increase, the ratio of swelling to static pressure increased from 3.8 to 6.5 as static pressure increased from 1 to 2 kPa. This is mild evidence that swelling pressures are dependent upon initial static pressure level. The results in Figure 20 further suggest an eleven fold increase in the bin wall pressure as moisture content increased from 12.5% to 22.4%. Dale and Robinson's work, while pioneering, does not provide conclusive evidence of the dependency of the swelling pressure on static pressure level nor does it allow one to draw conclusions about the relationship between moisture content increase and the level of swelling pressures. Their data, based upon only three tests, are too limited to identify anything other than very general trends in swelling pressures.

Risch and Herum (1982) presented results of research using a 610-mm diameter by 1220-mm high by 0.5-mm thick steel bin with a spot welded vertical seam. They measured the swelling pressures exerted by shelled corn which was aerated with high humidity air (> 95% relative humidity) at temperatures ranging from 10 to 25°C. Bin-wall hoop-strains during 100 h of aeration were measured with a single vertical-row of strain gages. The initial grain moisture contents for the three tests were 16.0, 15.7, and 13.6% (wb), respectively. Final moisture contents, which were not reported in the paper, were assumed to be equal to the equilibrium moisture content of shelled corn at 95% relative humidity (rh), or 22% (wb). The results (summarized in Figs. 21 and 22) indicate that hoop stresses and swelling pressures increase at all grain depths during aeration, swelling pressures increase linearly with time from the beginning of aeration, and swelling pressures increase dramatically for small changes in initial moisture content. The maximum observed swelling pressures were 30 and 11 kPa, respectively, for initial moisture contents of 13.6 and 16.0% (wb). The corresponding ratios of swelling to calculated static pressures (static pressure levels were not measured) were 10.0 and 3.7. Their observation of smaller swelling pressures near the bin floor than at the mid-height was probably due to frictional boundary effects between the grain and bin floor.

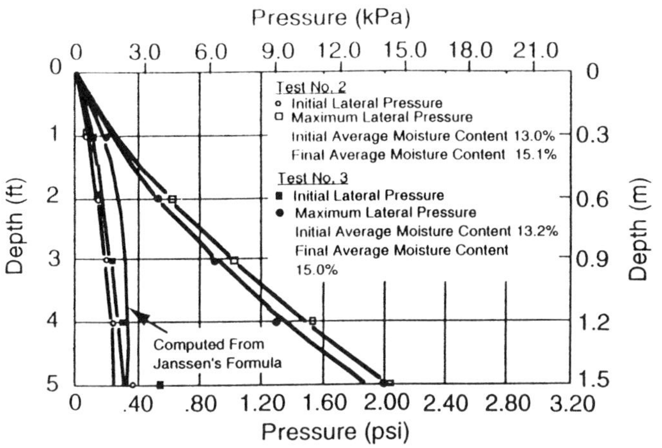

Figure 19 Pressure increase with increase in grain moisture content. *Source*: Dale and Robinson (1954).

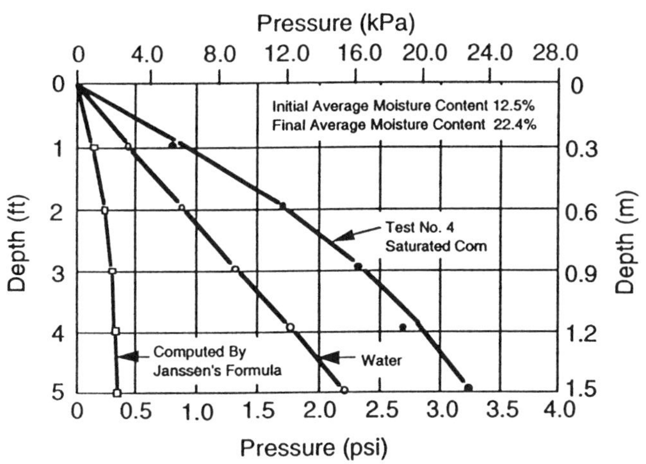

Figure 20 Relative pressures developed by dry grain, water, and saturated shelled corn. *Source*: Dale and Robinson (1954).

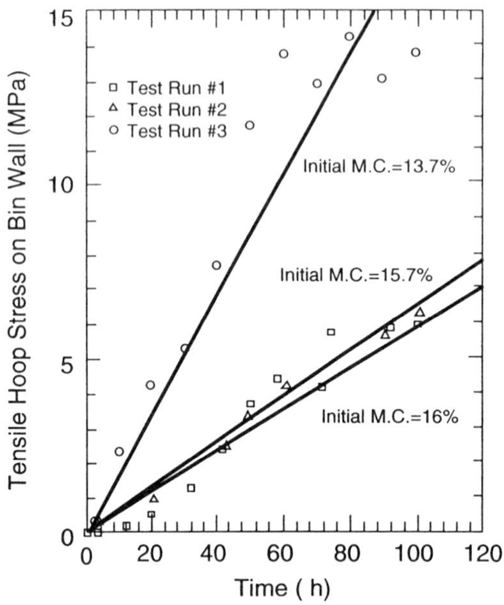

Figure 21 Stress at level 5 (bottom). *Source*: Risch and Herum (1982).

Risch and Herum (1982) also reported on a simulation procedure to predict swelling pressures. The simulation program first predicts the rate of moisture content increase of the grain by solving the simultaneous energy and mass balance equations for the grain control volume. Swelling pressures are computed by assuming the coefficient of moisture expansion for the bulk to be equal to the coefficient of moisture expansion of individual kernels and by assuming the load deformation behavior of the granular material to be linearly elastic. Within these constraints the simulation suggests that swelling pressures are significantly influenced by aeration air velocity, temperature, and relative humidity, but are not significantly influenced by initial grain temperature or moisture content. A weakness of this simulation is that it was not validated with experimental data. There is no evidence given to indicate how well the simulation procedures predicted the swelling pressures measured in the model study.

Blight (1986) measured swelling pressures for eight grains using the K_o oedometer ring (Fig. 23). He simulated a typical level of vertical pressure by applying a surcharge to the top of the grain surface. Swelling pressures were highest for grain which has a hard seed coat and a spherical shaped kernel. He ranked several grains in descending order of swelling pressure: grain sorghum, maize (corn), millet, wheat, sunflower, rice, and oats.

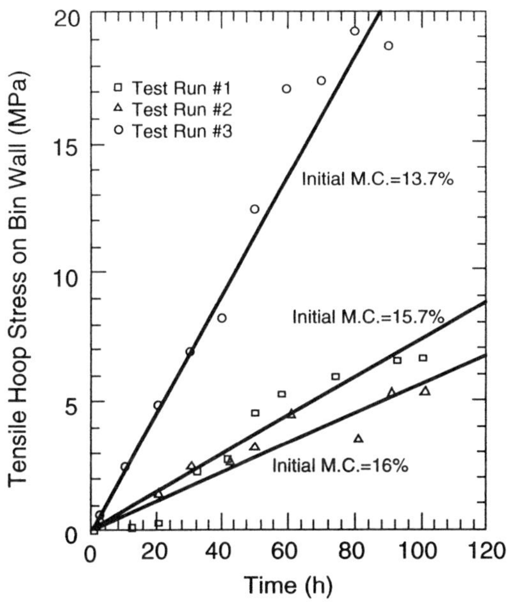

Figure 22 Stress at level 5 (mid-height). *Source*: Risch and Herum (1982).

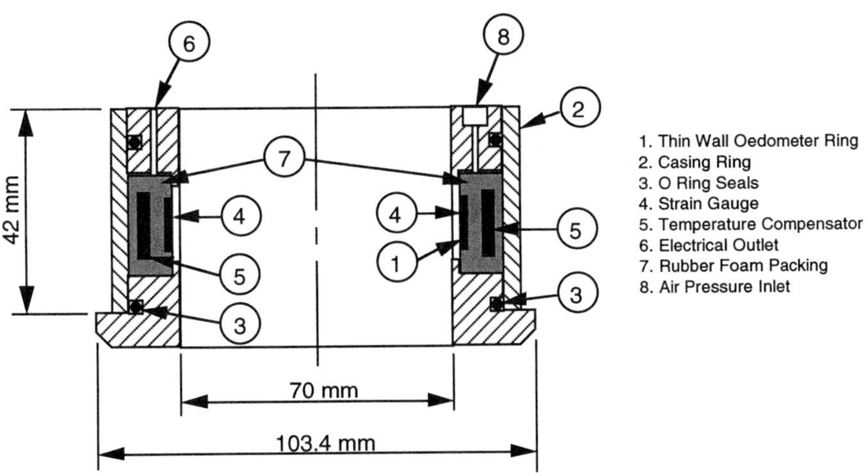

Figure 23 Section showing principal components of K_O oedometer ring. *Source*: Blight (1986).

(a)

(b)

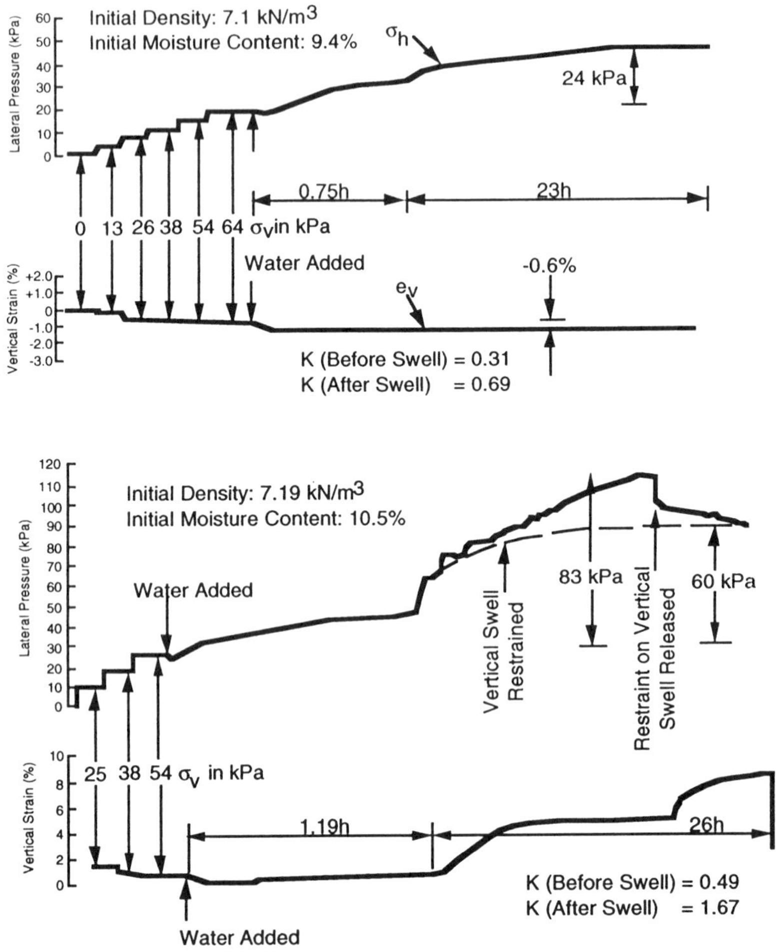

Figure 24 Swelling curves for wheat and maize (corn): (a) swelling of wheat; (b) swelling of normally consolidated maize showing effect of preventing vertical strain. *Source*: Blight (1986).

Typical swelling pressure curves for wheat and maize (corn), as observed by Blight (1986), are shown in Figure 24. The wheat sample, after being preconsolidated stepwise in 13 kPa increments to 64 kPa, was saturated by immersion in water and lateral pressures were monitored for 14 h; for wheat, the measured static and swelling pressures were 20 and 24 kPa, respectively. The initial moisture content of the wheat was 9.4% (wb). The final moisture content was not reported but presumably approached an equilibrium moisture content between 22 and 25% (wb).

Blight suggests that swelling pressures in a typical grain bin should be taken as twice the static pressures. Although he measured swelling pressures in excess of twice the static levels, he discounts the possibility of the entire bin contents being rewetted at the same time. The validity of Blight's suggestion must be considered based on the limitations of his work. These limitations include: 1) swelling pressures are given for only one level of static pressure; 2) the test sample size was extremely small, thereby raising serious questions about the influence of frictional and surcharge effects upon his results; 3) swelling pressures are given for only one initial moisture content; 4) swelling pressures cannot be correlated directly to grain moisture content and thus the results cannot be used to predict swelling pressures as a moisture front moves through grain during aeration; 5) there is no measure of the time dependency of the swelling pressures; and 6) moisture was added by immersion of the grain in water, rather than by aeration with humid air (there is little basis to support the assumption that the swelling of grains is similar when immersed and when aerated with humid air).

The swelling pressures of grains, as measured by the three investigators, are summarized in Table 1. Comparison of the three investigator's data for corn strongly suggests that swelling pressures are dependent upon the level of static pressure, the initial moisture content, and the change in grain moisture content. However, it is difficult to draw conclusions about the magnitudes of the swelling pressures for a given moisture content increase because: 1) the initial and final moisture contents were not carefully documented in two of the studies; 2) the initial static pressure was either held constant at one level through the entire study or the variation in initial static pressure was smaller (0–3 kPa) than the potential experimental errors; and 3) swelling pressures were reported only for the overall change in moisture content. The final condition was only assumed to be near saturation for any given test in two studies. The third study suggests a linear increase in swelling pressure with time of aeration but still does not relate the pressure to intermediate changes in moisture content.

It is clear that swelling pressures are potentially very large in grain storage bins and may approach the full passive pressure state. It is equally clear there are not enough data to develop a load specification. However, the data show the importance of proper aeration and management of stored grains to preclude all but minimal (1–2%) increases in moisture content of the stored grain.

4.6. Vibrational Loads

There is only anecdotal evidence that vibrational loads due to such sources as railroad sidings are of any significance. The authors are aware of only one

Table 1 Measured Swelling Pressures for Grains

Moisture content (wb)		Static pressure (kPa)	Swelling presure (kPa)	Swelling to static pressure ratio	Source
Initial (%)	Final (%*)				
Corn					
13.6	22.0	3	30	10.0	Blight (1986)
16.0	22.0	3	11	3.7	
10.5	22.0	25	60	2.4	Risch and Herum (1982)
12.5	22.5	2	22	11.0	Dale and Robinson (1954)
12.5	22.5	1.4	15.9	11.3	
12.5	22.5	1	11.7	11.7	
13.0	15.1	2	13	6.5	Dale and Robinson (1954)
13.0	15.1	1.4	5.2	3.7	
13.0	15.1	1	3.8	3.8	
Wheat					
9.4	22	20.0	24.0	1.2	Blight (1986)
Grain sorghum					
17.7	22	28	87	3.1	Blight (1986)
Sunflower seed					
7.8	22	23	23	1.0	Blight (1986)

* All final moisture contents are assumed except those for Dale and Robinson's data.

research report about vibrational loads in grain bins (Sakai et al. 1985). They tested two cylindrical model bins, 1/30 and 1/8 scale, by setting them on a horizontal shaking table. The bins were filled with coal and were constructed, respectively, of plastic and concrete. The bins were subjected to monotonically increasing amplitudes of vibration. Their key observations were: 1) 75–80% of the base shear generated by the stored material is carried by the wall; 2) the damping effect of the stored material is so large that 40% can be considered as equivalent viscous damping; and 3) the higher the input acceleration, the smaller the base shear per unit input acceleration shared by the wall. None of the grain load codes or specifications include provisions for any vibrational loads for grain storage bins. However, ACI (ACI 1991) does mention that earthquake loads may be important and refers the designer to earthquake load provisions of the 1988 Uniform Building Code (ICBO 1988).

4.7. Induced Loads by Temperature Cable

Cables, onto which are fastened temperature sensors, are often suspended from the roof of the grain bin so that grain temperatures can be monitored during storage. The frictional load between the grain and cables during emptying and flow are loads induced by the temperature cable. Wickstrom (1980) reported these loads to have caused localized failures in some grain bins. Magnitudes of the cable forces, as measured in model bins, have been reported by Thompson (1987) and Schwab et al. (1991). Schwab et al. (1992) reported that cable loads were largest on cables nearest the bin wall, cable surface material significantly influenced cable frictional loads (vinyl-coated cables experienced the largest loads, and polyethylene-coated cables experienced the lowest loads) and emptying flow rate does not significantly influence cable loads. Schwab et al. (1992) also developed the following equation to convert cable loads from the earlier model studies to cable loads in prototypes.

$$\lambda_p = 20.3\,\lambda_m \qquad (43)$$

where:

λ_p = cable friction load in prototype (N)
λ_m = cable friction load in model as reported by Thompson (1987) and Schwab et al. (1991) (N)

Maximum vertical cable loads in the prototype bin on cables located near the sidewall ranged from 1000 N to 1550 N. Maximum vertical cable loads in the prototype bin on cables located at least 1/4 diameter from the wall ranged from 500 N to 750 N.

5. CODES FOR GRAIN BIN LOADS

There are numerous codes or engineering practices available to designers of grain bins and silos. For example, by country of origin, the codes are Australia, Austria, Canada, Germany, United Kingdom, Japan, Poland, Russia, and the United States. Eleven of the codes are summarized and evaluated in the paper by Wilms (1990). A more narrowly focused engineering practice written specifically for thin-walled bins for free flowing noncohesive materials has recently been adopted by the American Society of Agricultural Engineers (ASAE 1992b). The German DIN 1055 (DIN 1987), the ACI 313, (ACI 1991), the Canadian Farm Building (CFBC 1983, 1990), and the ASAE EP433 (ASAE 1992b) are the grain-load codes discussed herein.

5.1. DIN 1055

5.1.1. Scope and Limitations

The German design (DIN 1055) code for calculating loads in bins was developed for the bulk solids industry (DIN 1987). A brief description of the scope and limitations of the DIN 1055 code and of the procedure to estimate loads during filling and discharge is provided. The relevant design equations are also included. The variables used in the description of the DIN 1055 design code are explained in the nomenclature section. Finally, a cylindrical flat-bottom bin made of steel (10-m diameter × 15-m height) with a centric discharge is used as an example to estimate loads during filling and emptying of the grain bin. A schematic of the grain bin with key dimensions and notations pertinent to DIN 1055 is shown in Figure 25.

The German standard is generally recommended for bins with height to diameter ratios (H/D) greater than 0.8. Some of the key features of DIN 1055 are: 1) estimation of filling loads, 2) estimation of discharge loads during mass flow and funnel flow conditions, and 3) estimation of loads in hoppers. The design code DIN 1055 does not cover: 1) loads induced by dust explosions, 2) loads due to the storage and handling of potentially explosive bulk materials, 3) bins with annular cross-sections, 4) loads due to internal fittings and discharge aids, 5) highly cohesive materials, and 6) bulk solids that are liable to swell. Other specific limitations are provided in the design code DIN 1055 (DIN 1987).

5.1.2. Filling Pressures

The filling pressures are computed using Janssen's method discussed in Section 3.3. The horizontal grain pressure during filling at depth y from the free surface ($P_{hf}(y)$) is predicted using Janssen's equation (Eq. 17). The vertical pressure in

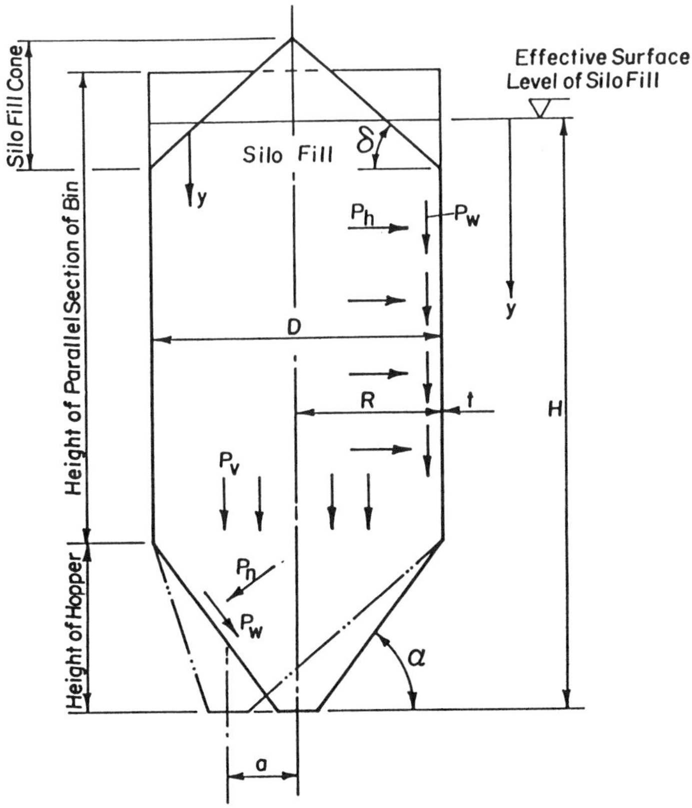

Figure 25 Schematic and definition sketch of grainbin for DIN 1055. *Source*: (DIN 1987).

the grain during filling ($P_{vf}(y)$) is estimated using Equation (16). The vertical wall pressure during filling ($P_{wf}(y)$) is calculated using Equation 18.

5.1.3. Discharge Pressures

The discharge or emptying pressure calculations are generally based on two different flow conditions (i.e., flow that induces uniform pressures and flow that induces nonuniform pressures). Typical examples of the uniform and nonuniform pressure situations are: a mass flow hopper bin with centric discharge and a funnel flow hopper bin with eccentric discharge. For uniform pressures, the loads are estimated using these relationships:

Table 2 Bulk Properties and Design Coefficients for Several Agricultural Materials

Bulk solid	Density (kg/m^3)	Pressure ratio k	Wall friction* μ	Overpressure coefficient, e_h
Wheat	917	0.60	0.25	1.4
Shelled corn	815	0.60	0.25	1.6
Barley	815	0.65	0.25	1.4
Flour	714	0.40	0.25	1.4
Sugar	968	0.60	0.45	1.2

*For smooth walls
Source: DIN (1987).

1. for horizontal pressure $P_{he} = e_h P_{hf}$
2. for vertical pressure $P_{ve} < P_{vf}$

The overpressure coefficient (e_h) given in Table 2 should be used whenever $H/D > 5.0\ \mu$, otherwise a linear interpolation should be used up to H/D of 2.5μ. A constant value is recommended for H/D below $2.5\ \mu$. The nonuniform pressure loads are calculated by using either of two methods in DIN 1055. The two methods described in this design code are based on the desired level of factor of safety. In the first method, which provides the usual level of factor of safety, the partial pressure concept is used. Herein the uniform loads, mentioned earlier, are increased using the partial pressure factor. In the second method, which provides for higher factors of safety, the uniform horizontal pressure factor concept is used. In this loading situation, also, the bin wall pressures are increased by the estimated uniform horizontal pressure factor.

5.1.4. *Bin Bottom Pressures*

For $H/D > 1.5$, the bin bottom pressure (P_b) is estimated using $P_b = 1.8P_{vf}$. This pressure can be assumed to be uniform over the bin bottom. However, when $H/D < 1.5$, the bin bottom pressure is nonuniform. For these bins, the calculations are based on the fill height for the outer radius pressures; whereas, the calculations are based on the fill height of $1.5D$ for bin center pressures.

5.1.5. *Hopper Loads*

Pressures exerted by the bulk material on the hopper are estimated in DIN 1055 using empirical equations. For a level-filled hopper or a conical mass that will level-fill the hopper, the hopper pressures are estimated using Equations 44 and

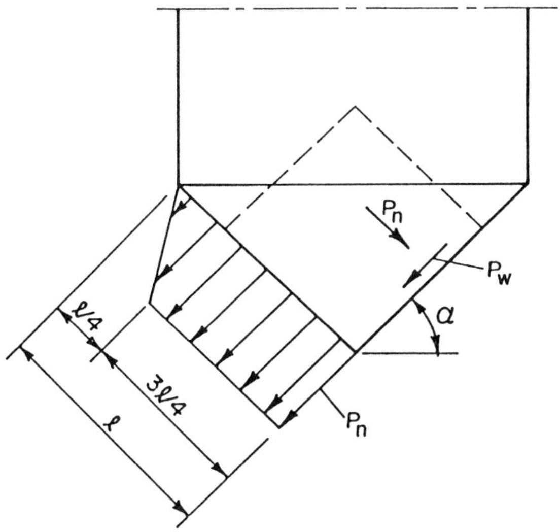

Figure 26 Normal and wall friction pressure values for a level-fill hopper. *Source*: (DIN 1987).

45, respectively, for calculating the normal and wall friction pressures as shown in Figure 26.

$$P_n = 2.4 \frac{A}{C} \rho\, gk \frac{\sin^2\alpha}{\sqrt{\mu}} \tag{44}$$

$$P_w = \frac{P_n}{2} \tag{45}$$

When the bulk material completely fills the hopper and at least some or all of the vertical portion of the bin, the outer (P_v), upper (P_{no}), and wall friction pressures are estimated using Equations 46, 47, and 48, respectively.

$$P_{no} = (P_{vf} C_b \cos^2 \alpha + P_{hf} \sin^2 \alpha)\left(1 + \frac{\sin (2\alpha)}{4\,\mu}\right) \tag{46}$$

$$P_v = P_{vf} C_b \cos^2 \alpha \tag{47}$$

$$P_w = \frac{P_n}{2} \tag{48}$$

The terms in Equations 46–48 are defined in Figure 27. In Equations 46 and 47, the coefficient C_b is equal to 1.5 (1.8 for granular materials, such as shelled

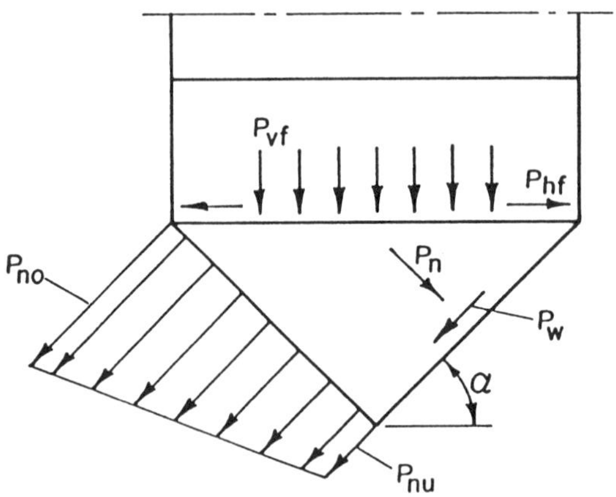

Figure 27 Normal and wall friction pressures for a filled bin. *Source*: (DIN 1987).

corn). The hopper load equations are applicable only when $\alpha > 20°$. During mass flow conditions additional pressures induced by the switch load must be considered. The switch pressure can be several times larger than the static load and its value is calculated using:

$$P_s = \rho\, gy$$

or

$$P_s = \rho\, gD$$

For design purposes, the lower of the two values is recommended.

5.1.6. *Other Loads*

The German standard also discusses methods for estimating pressures caused by the injection of compressed air and the pressures induced by silage. It includes methods for estimating the loads during buckling and bulging of the structure.

5.1.7. *Example Analysis*

The example analysis presented in this section is for a smooth-steel flat-bottom bin with a port for center discharge. The bin diameter and height are 10 m and 15 m, respectively, and the stored material has the following properties: bulk density 800 kg/m^3 and wall friction coefficient 0.20. The pressure ratio (k) used in this example problem was 0.60 (for wheat) from Table 2.

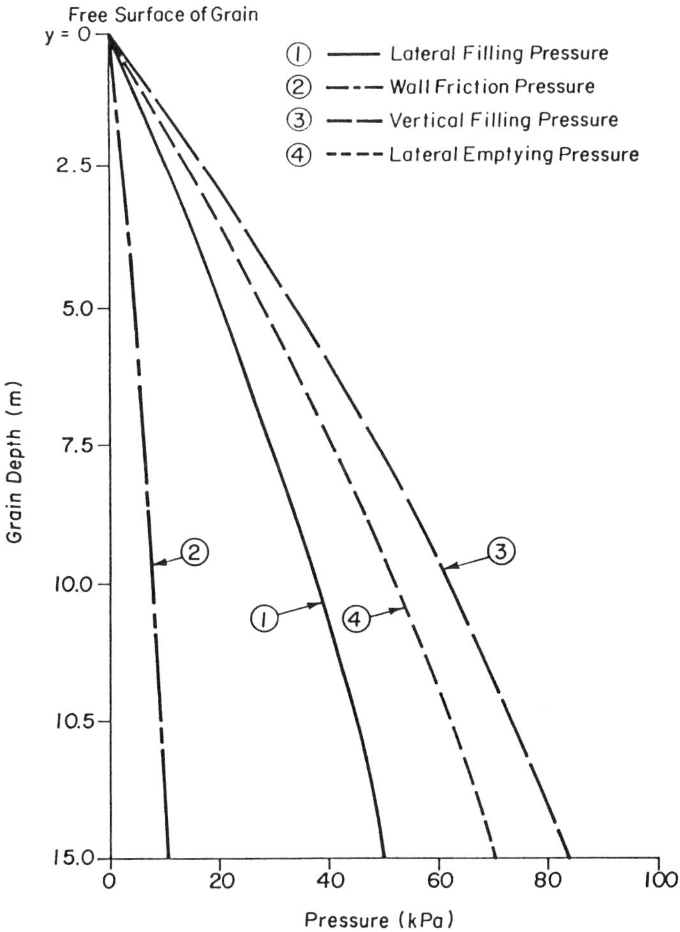

Figure 28 Lateral and vertical wall pressures for the 10 m (D) × 15 m (H) bin. *Source*: DIN 1987).

Filling pressures: Substituting the property values into Equations 16–18 leads to Equations 49–51, respectively. The units for pressure in Equations 49–51 are kPa. The results are shown in Figure 28.

$$P_{hf}(y) = 98.10\,(1 - e^{-0.048y}) \tag{49}$$

$$P_{vf}(y) = 163.51\,(1 - e^{-0.048y}) \tag{50}$$

$$P_{wf}(y) = 19.62\,(1 - e^{-0.048y}) \tag{51}$$

Emptying or discharge pressures: The emptying or discharge pressures can be calculated once the overpressure coefficient (e_h) is known. For the example analysis, $H/D = 1.5$ which is greater than 5.0 μ. Therefore, the overpressure coefficient was assumed to be 1.4 (Table 2).

5.2. ACI 313-91

5.2.1. *Scope and Limitations*

The American Concrete Institute's standard for design and construction of concrete silos and stacking tubes for storing granular materials is described in ACI 313-91 (ACI 1991). The scope and limitations of ACI 313-91 and design equations for estimating static and discharge loads are discussed in the following sections. The terms silos, bins, and bunkers have the same meaning in this section. A schematic of the grain bin with the associated notations is given in Figure 29.

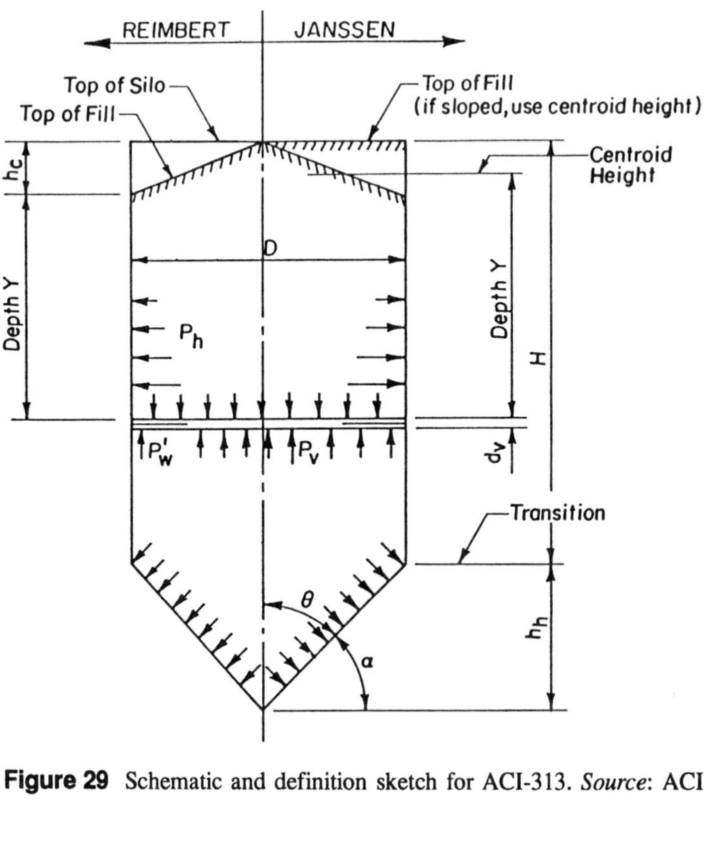

Figure 29 Schematic and definition sketch for ACI-313. *Source*: ACI (1991).

The equations for calculating bin wall loads in the ACI 313-91 standard are based on the strength design method. The primary focus of this standard is the design and construction of concrete silos and stacking tubes for storing bulk solids. The recommended design equations can also be used for metal bins. The design equations for static and emptying conditions for flat and hopper bottom bins are provided. Design methods for industrial stave silos of precast concrete are included in this standard. Silos storing silage and similar wet nonflowing materials are not covered in the ACI 313-91.

5.2.2. *Static Bin Wall and Hopper Loads*

In the ACI 313-91, calculations of static pressures exerted by the stored granular material can either be performed using Janssen's or Reimbert's method.

Janssen's Method: Equations 16 and 17 are used for estimating the vertical grain pressure (P_v) and lateral grain pressure (P_h) using Equation 14 to estimate the pressure ratio (k). The wall friction pressure (P_w) is calculated using Equation 52, which is very similar to Equation 18. The grain depth (y) in these design equations is measured from the free surface as shown in Figure 29.

$$P_w = \frac{\rho gD}{4}\left(1 - 0.8\, e^{-4\mu ky/D}\right) \tag{52}$$

Reimbert's Method: The bin wall loads at grain depth y and with surcharge height of h_c are calculated using Equations 53–55.

$$P_v = \rho\, g\left[y\left(\frac{y}{A_1} + 1\right)^{-1} + \frac{h_c}{3}\right] \tag{53}$$

$$P_h = P_{max}\left[1 - \left(\frac{y}{A_1} + 1\right)^{-2}\right] \tag{54}$$

$$P_w = \frac{\rho gD}{4}\left[1 - \left(\frac{y}{A_1} + 1\right)^{-2}\right] \tag{55}$$

For circular grain bins, the constants P_{max} and A_1 are given in Equations 56 and 57. For other cross-sections, the appropriate equations are provided in ACI 313-91.

$$P_{max} = \frac{\rho gD}{4\,\mu} \tag{56}$$

$$A_1 = \frac{D}{4\,k\,\mu} - \frac{h_c}{3} \tag{57}$$

The static pressures normal to the hopper inclined at an angle α are estimated using Equation 58, which is valid for both the Janssen's and the Reimbert's methods.

$$P_n = P_h \sin^2 \alpha + P_v \cos^2 \alpha \tag{58}$$

5.2.3. *Dynamic Bin Wall and Hopper Loads*

The design loads during emptying or discharge are sensitive to the profile of the moving and stationary mass of the stored material. Equations for estimating the design loads for concentric flows are provided in the ACI 313-91 standard. Bins discharging relatively-fine particulate material may require air to homogenize the flow of material. In these cases the design equations are different and should be factored into the design load calculations. Nonuniform flow resulting from asymmetric discharge can cause a circumferential bending moment which must also be considered in the design. Load estimation equations during nonuniform discharge are not provided.

The design loads for uniform flow are estimated by multiplying the static P_v, P_h, and P_n either by the overpressure factor (e_h) or the impact factor (e_i) whichever gives greater total pressures. The recommended values of e_h for Janssen's and Reimbert's design equations are listed in Table 3. These values are for $H/D \leq 2$ and $H/D = 3$; values for other H/D ratios are given in the ACI 313-91 standard. The recommended values of e_i are summarized in Table 4. Physical properties of selected agricultural materials are given in Table 5. Design pressures for homogenizing silos are estimated using either Equations 16–18 or Equations 53–55, or by using $P_h = P_v = 0.6\rho\, gy$, whichever is larger.

5.2.4. *Other Loads*

Other loads that can become critical and should be considered when appropriate are: dead load, snow load, thermal loads, seismic loads, and wind loads. Calculations of these loads should be performed using the methods in the ACI 318 standard (ACI 1989).

5.2.5. *Example Analysis*

The example discussed in Section 5.2.7 is analyzed here for grain with an angle of internal friction of 20°.

Static Loads: Janssen's Method. The design load equations with the above properties are given in Equations 59–61. These equations are for vertical wall pressure, horizontal (lateral) wall pressures, and wall friction pressure, respectively. The variation of bin wall pressures with grain depth is shown in Figure 30.

Table 3 Recommended Minimum Overpressure-Factors During Emptying of Grain Bins

	Overpressure factor, e_h			
	$H/D \leq 2$		$H/D = 3$	
Material level	Janssen	Reimbert	Janssen	Reimbert
$0 \leq y \leq 0.25$H	1.45	1.20	1.55	1.30
0.25H $\leq y \leq$ 0.50H	1.55	1.45	1.65	1.55
0.50H $\leq y \leq$ 0.75H	1.65	1.65	1.75	1.75
0.75H $\leq y \leq$ H	1.65	1.65	1.75	1.75
Steel bottom	1.50	1.75	1.50	1.75

Source: ACI (1991).

Table 4 Recommended Minimum Values of Impact Factor (e_i)

	Ratio of volume dumped in one load to total bin capacity				
Type of bin bottom	1:2	1:3	1:4	1:5	1:6 and less
Concrete	1.4	1.3	1.2	1.1	1.0
Steel	1.75	1.60	1.5	1.35	1.25

Source: ACI (1991).

Table 5 Physical Properties of Selected Agricultural Materials

Stored material	Bulk density ρ (kg/m^3)	Angle of repose φ (degrees)	Wall friction μ
Grains: Wheat, shelled corn, oats, rice, rye	736–990	23–37	0.26–0.42
Soybeans, peas	800–960	23	0.20
Flour	610	40	0.30
Granular sugar	1000	35	*

* Not available
Source: ACI (1991).

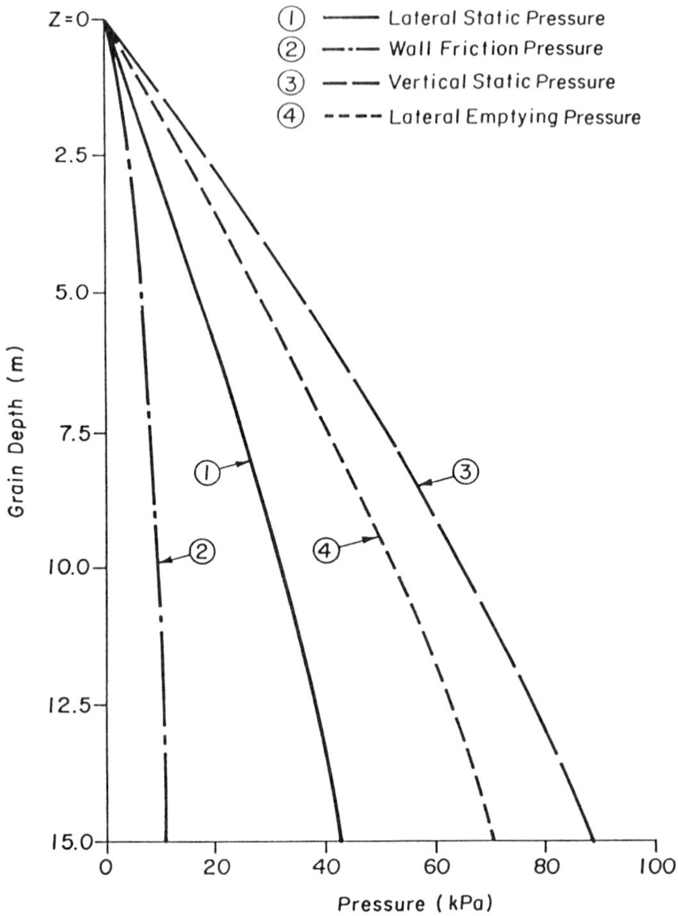

Figure 30 Calculated bin wall pressures using ACI-313 and Janssen's method.

$$P_v = 200.0\,(1 - e^{-0.039y}) \tag{59}$$

$$P_h = 98.1\,(1 - e^{-0.039y}) \tag{60}$$

$$P_w = 19.62\,(1 - 0.8\,e^{-0.039y}) \tag{61}$$

Static Loads: Reimbert's Method. The design bin wall load Equations 53–55 lead to Equations 62–64. The variation of the bin wall loads with grain depth is shown in Figure 31.

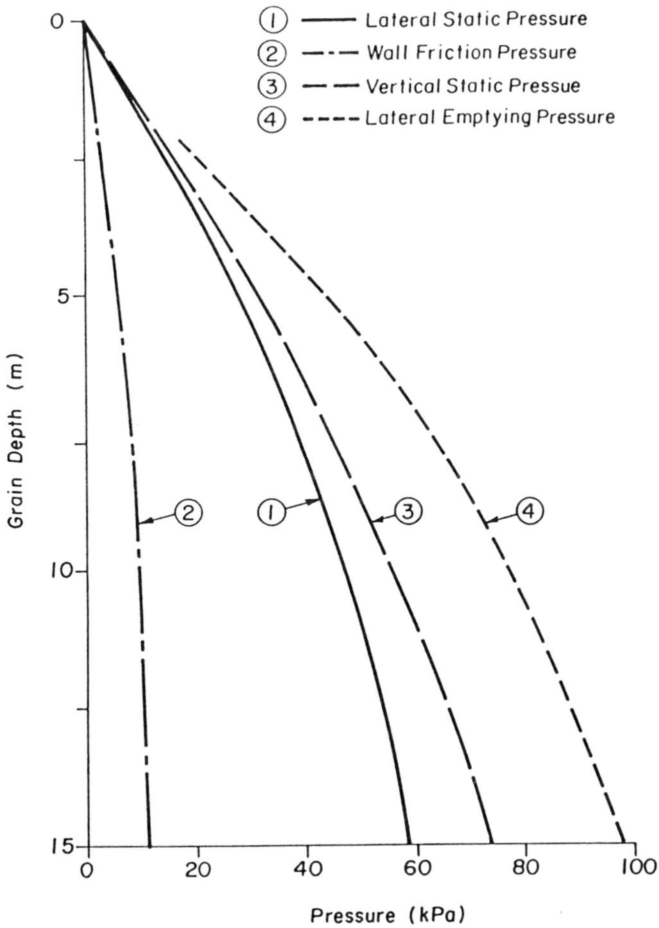

Figure 31 Calculated bin wall pressures using ACI-313 and Reimbert's method.

$$P_v = 7.85\,y\left(\frac{y}{25.51} + 1\right)^{-1} \tag{62}$$

$$P_h = 98.10\left[1 - \left(\frac{y}{25.51} + 1\right)^{-2}\right] \tag{63}$$

$$P_w = 19.62\left[1 - \left(\frac{y}{25.51} + 1\right)^{-2}\right] \tag{64}$$

Discharge Loads. For estimating the emptying or discharge bin wall loads, the overpressure factor (e_h) was assumed to be the overriding factor compared with the impact factor (e_i). Since the H/D ratio was 1.5 in the example problem, the values in the first two columns of Table 3 were used in the calculations. The lateral loads on the bin wall during emptying using the methods of Janssen and Reimbert are plotted in Figure 30 and 31, respectively.

5.3. Canadian Farm Building Code, 1990

5.3.1. Static Loads

The Canadian Farm Building Code (CFBC 1990) recommends the use of Janssen's equation (Eq. 17) to predict lateral loads on grain bin walls (CFBC 1990). However, ρg is defined (for commercial purposes) such that it incorporates a 6% increase in the grain bulk density. An extensive table of recommended bulk densities is included in an appendix to the Code.

The hydraulic radius (R) is defined as $D/4$ for a circular bin, where D is the diameter of the cross-section. For rectangular bins, the hydraulic radius is defined by Equations 65 and 66. For the long side:

$$R = \frac{2\,ab - a^2}{4\,b} \tag{65}$$

and for the short side:

$$R = \frac{a}{4} \tag{66}$$

where:

a = length of the short side of the bin cross-section (m)
b = length of the long side (m)

Coefficients of friction between the grain and the bin wall (μ) are dependent on the wall material and the type of grain and are defined in a table (CFBC 1990). Values range from a low of 0.1 for 11% mc wheat and barley on smooth steel, to a high of 0.6 for 16% mc shelled corn on concrete.

Acceptable values for the ratio of horizontal to vertical pressure (k) are also presented in a table. Wall smoothness as well as grain type are considered in selecting the appropriate value. For smooth walls, k ranges from 0.40 to 0.55; for rough walls, the values range from 0.4 to 0.8.

Vertical friction load per unit circumference of the bin to a depth y below the grain surface is predicted using Equation 67:

$$F_v = \rho\, gR\left(y - \frac{R}{k\,\mu} + \frac{R}{k\,\mu}\, e^{-k\mu y/R}\right) \tag{67}$$

Vertical pressures within the grain bulk are predicted using the horizontal to vertical pressure ratio (k). It is noted in the Appendix to the code that this value assumes a maximum friction between the wall and the grain. After a few hours, during which the friction coefficient decreases in storage, the vertical wall load increases by as much as 25%. It should be noted, however, that this is information from the Code Appendix and is not a Code requirement.

For deep bins with floors which slope at an angle between 0° and 20°, the vertical pressure is recommended as the floor design load. Normal loads on hopper bottoms with angles between 20° and 60° are predicted using equations which have been derived from a resolution of horizontal and vertical pressures.

Some special considerations are given to shallow bin conditions, notwithstanding the general application of Janssen's equation. A bin is defined as shallow when the bin depth is less than or equal to 0.75 times the bin diameter. For shallow bins in which walls slope at angles between 50° and 120° from the horizontal, pressures normal to the wall are predicted by modifying the horizontal pressures as predicted using Janssen's equation with the Reimbert coefficient (C; Eq. 68).

$$C = (\beta - \varphi)\,(90^\circ - \varphi)^{-2} \tag{68}$$

Floor pressures in shallow bins are predicted using Equation 69.

$$P_f = \rho\, gH \tag{69}$$

5.3.2. *Dynamic Loads*

Dynamic loads are determined by applying an overpressure factor to the static horizontal pressure as determined using Janssen's equation. In general, dynamic loads are not considered to occur in bins where the $H/4R$ ratio is less than 2.5 μ and it is considered to reach a maximum when $H/4R$ becomes greater than 5 μ. The magnitude of the overpressure factors is shown in Table 6, which has been reproduced from the code (CFBC 1990).

For bins which are emptied from the center and which have $H/4R$ values between the stated limits, the applicable overpressure factor is obtained by linear interpolation. For bins with discharge ports which are eccentric by at least $R/6$, the dynamic horizontal pressure is predicted by doubling the static horizontal pressure and applying this load to a vertical strip of wall having a width R and extending from the opening to the grain surface. It is unclear what load is to be applied to the remaining perimeter of the bin. It is interesting to note that there is no consideration of a dynamic effect on the vertical friction load on the wall of the bin during emptying.

Table 6 Overpressure Factors for Stored Grain

Grain stored	Overpressure factor	
	$H/4R < 2.5\mu$	$H/4R > 5\mu$
Cereal grains, shelled corn, soybeans, and canola	1.0	1.4
Flaxseed and canary seed	1.0	1.6

Source: CFBC (1990).

Pressure calculations for the 15-m high by 10-m diameter flat-bottom bin for storing material with $\rho = 800$ kg/m^3, $\varphi = 20°$, and $\mu = 0.20$ for static, lateral, and vertical wall pressures are identical to those for Janssen's method. Emptying lateral pressures are 1.4 times static pressures at y = 3, 8, and 13 m. There are no provisions for thermal loads.

5.4. ASAE EP 433: International

Engineering Practice 433 (EP 433), "Loads Exerted by Free-Flowing Grain on Bins" was developed and published by the American Society of Agricultural Engineers (ASAE 1992b). The scope of the document is restricted to "... estimating the grain pressures in centrally loaded and unloaded bins used to store free-flowing agricultural whole grain."

All static loads, both within the grain mass and on the bin wall and for both deep and shallow bins, are calculated using Janssen's equation (Eqs. 16–18). Conical surcharges at the grain surface are accommodated by increasing the grain depth by one-third the altitude of the surcharge cone. Maximum grain density (ρ) of 834 kg/m^3 is suggested for general design. If standard densities are used for other stored grains, use of a compaction factor of 1.08 is recommended.

Table 7 Material Properties and Overpressure Factors

Wall material	μ	k	F
Steel	0.30	0.5	1.4
Concrete	0.40	0.5	1.4
Corrugated steel	0.37	0.5	1.4

Source: ASAE (1992b).

Engineering Practice 433 defines any bin with a height to diameter ratio less than 2 as a funnel-flow bin. The height is taken as the distance between the discharge port (flat-bottom or hopper-bottom) and one third of the depth of any conical grain surcharge. Dynamic emptying pressures for the funnel flow bins are the same as the static pressures.

Bins with height to diameter ratios greater than 2.0 are classified as plug flow bins. Dynamic lateral wall-pressures are estimated by multiplying the static pressure (obtained using the appropriate values for μ and k in Janssen's equations) by the overpressure factor (F) (Table 7). However, in flat-bottom bins, F may be linearly reduced from 1.4 at $D/4$ from the bin bottom to 1.0 at the bin bottom.

Procedures for determining static and dynamic emptying pressures in hoppers of funnel flow bins are given in EP 433. Mass flow hoppers are not within the scope of the document. Static pressures normal to the hopper surface are estimated using Equation 70 and the hydraulic radius at the hopper-bin interface.

$$P_n = P_v \cos^2 \alpha + P_h \sin^2 \alpha \tag{70}$$

Dynamic hopper pressures at the hopper-bin interface are obtained using F (Table 7). The overpressure factor is reduced linearly with depth to 1.0 at the hopper bottom.

Friction pressures on the hopper wall are estimated using Equation 71.

$$P_t = \mu P_n \tag{71}$$

where P_t = hopper friction pressure (Pa).

Engineering Practice 433 also includes provisions for estimating grain loads on antidynamic tubes and flumes as well as provisions for estimating thermally-induced pressures in circular steel bins. Thermally-induced, lateral pressures on walls are estimated as 8% and 15% of static pressures for ambient air temperature decline rates of 10°C/h and 20°C/h, respectively.

Moisture-induced (i.e., swelling) pressures are identified as being potentially very large compared with static pressures for moisture content increases of only 3–4%. However, no design pressures are recommended. Instead, EP 433 recommends that precautions be taken to assure that the bin contents do not increase in moisture content during storage. There are no provisions for vibration-induced loads. Insufficient data are available to warrant a specification other than earthquake loading at this time.

6. GRAIN LOAD COMPARISONS

The static lateral wall-pressure, the static vertical wall-pressure, the dynamic lateral wall-pressure and thermally-induced pressures at 3, 8, and 13 m below the grain surface of a 15-m deep, flat-bottom bin having 10-m diameter, and made of smooth steel are summarized for the several theories and codes in Table 8. The stored grain has the properties: ρ = 800 kg/m^3, φ = 20°, and μ = 0.2. The notation N/A (not applicable) is used to indicate that theory or code has no provisions for calculating the design pressure.

7. FUTURE RESEARCH

Static loads are generally well defined by existing theories. However, there are no generally-accepted, rationally-based theories to adequately predict dynamic grain pressures in bins. There have been some significant efforts to develop rational numerical solutions to thermal pressure problem (Zhang et al. 1986a,b, 1989; Li et al. 1991). Clear definition of comprehensive rational solutions to predict dynamic, or quasistatic, grain bin loads is essential if the understanding of grain load phenomena is to improve and expand. Likewise, such developments in dynamic load theories are needed if grain bin design is to become more efficient and economical.

For efficient, economical, and reliable grain bin design, accurate characterization of physical properties of the stored grain is very important. Some of the properties that need to be more accurately measured for existing and future grain and seed varieties include: size and shape as a function of moisture content and bulk density for different filling conditions and moisture contents, angle of internal friction, and grain-wall coefficient of friction for different normal pressures and moisture contents. The time-effect (i.e., the effect of time-in-storage) on the above properties should also be quantified.

More research is also needed in: 1) development of improved and refined comprehensive elastic-viscoplastic and viscoelastic constitutive models for granular materials; 2) development of testers specifically designed for measurement of load-response of grains over stress paths commonly observed in grain bins; 3) determination of the magnitude of the grain mechanical properties required in the comprehensive constitutive models; 4) development of comprehensive finite element or boundary element numerical methods to compute the states of stress in grain bins; and 5) development of appropriate simplified design equations or design aides by conducting sensitivity and parametric analyses with the comprehensive numerical models.

For validation of numerical models, long-term testing of not only model bins but also full-size bins should be conducted. Such tests should not be confined to filling and emptying conditions only but also include quasi-dynamic phenomena (e.g., increased loads due to grain moisture gain and rapid changes in ambient temperature) and dynamic phenomena (e.g., vibration-induced loads and wind loads). Scale-up of results from model to full-size bins needs further attention. With the advent of parallel computation, discrete and hybrid discrete-continuum modeling of dynamic loads is also becoming feasible.

8. NOMENCLATURE

A cross-sectional area perpendicular to the longitudinal direction (m^2)
A_1 constant in Reimbert's equation as defined in Equations 39 and 57
C internal perimeter of bin (m)
C_b coefficient appearing in Equations 46 and 47
C_p thermal pressure coefficient (Pa/°C)
D diameter of the largest inscribed circle (m)
e_h overpressure factor (unitless)
e_i impact, factor (unitless)
E elastic modulus for the bin wall material (Pa)
E_g elastic modulus for the stored material (Pa)
F overpressure factor (unitless)
F_H resultant horizontal load on bin wall (N/m)
Fv resultant vertical load on bin wall (N/m)
Fy load on bin wall at depth y below top grain surface (N/m)
g mass to force conversion (9.8 N/kg)
h maximum depth of fill (m)
h_c surcharge depth (m)
H overall bin height from eave to hopper-bin interface (m)
k pressure ratio (unitless)
n ratio of stored material to bin wall elastic modulus (unitless)
P_h lateral or horizontal bin wall pressure (Pa)
P_{he} horizontal emptying pressure (Pa)
P_{hf} horizontal filling pressure (Pa)
P_{max} maximum pressure as defined in Equation 56 (Pa)
P_n pressure normal to the hopper walls (Pa)
P_v vertical grain pressure (Pa)
P_{ve} vertical emptying pressure (Pa)
P_{vf} vertical filling pressure (Pa)
P_w wall friction pressure (Pa)
P_{wf} wall friction pressure during filling (Pa)
P internal perimeter (m)
P_T thermally-induced lateral pressure (Pa)
R hydraulic radius of circular bin = A/C (m)
r radius of the largest inscribed circle; bin radius (m)
t wall thickness (m)

Table 8 Comparison of Selected Design Grain Pressures by Several Theories and Design Codes for a 10 m × 15 m (D × H) Flat-Bottom Steel Bin with Smooth Walls (ρ = 800 kg/m^3, φ = 20°, μ = 0.2)

Theory or code	Depth below grain surface (m)	Lateral wall pressure: Static (kPa)	Lateral wall pressure: Emptying (kPa)	Lateral wall pressure: Thermal[1] (kPa)	Vertical wall pressure (kPa)
Rankine (Active)	3	11.5	N/A	N/A	2.3
	8	30.7	N/A	N/A	6.1
	13	49.9	N/A	N/A	10.0
Coulomb	3	10.4	N/A	N/A	2.0
	8	27.8	N/A	N/A	5.5
	13	45.1	N/A	N/A	9.0
Janssen	3	11.1	N/A	N/A	2.2
	8	26.8	N/A	N/A	5.4
	13	39.7	N/A	N/A	8.0
Reimbert	3	10.9	N/A	5.3**	5.7
	8	26.4	N/A	5.3	6.3
	13	39.2	N/A	5.3	9.1

DIN 1055*	3	13.2	18.4	N/A	2.6
	8	31.3	43.8	N/A	6.3
	13	45.5	63.8	N/A	9.1
ACI 313-91 (Reimbert)	3	10.9	15.8	N/A	5.7
	8	26.4	43.6	N/A	8.2
	13	39.2	64.7	N/A	10.2
ACI 313-91 (Janssen)	3	19.6	23.5	N/A	3.9
	8	41.3	68.1	N/A	8.3
	13	55.1	90.9	N/A	11.0
Canadian standard	3	11.1	15.5	N/A	2.2
	8	26.8	37.5	N/A	5.4
	13	39.7	55.6	N/A	8.0
ASAE EP433	3	11.1	15.5	0.9	2.2
	8	26.8	37.5	2.1	5.4
	13	39.7	52.4	3.1	8.0

[1]Thermal pressures listed for a 10°C/h rate of temperature decline and for $\Delta T = 10°C$.
* Assuming $k = 0.6$
** $E = 200$ GPa, $\alpha = 1.2 \times 10^{-5}$ 1/°C, $\Delta T = 10°C$, wall thickness = 5 mm.

W_w vertical frictional force per unit length of wall perimeter $N \cdot m \cdot m^{-2}$
y depth from effective surface level to point of interest (m)
α angle of inclination of hopper wall (degrees); coefficient of thermal expansion (1/°C)
β slope of wall measured from the horizontal (degrees)
δ surcharge angle (degrees)
ΔT ambient temperature change (°C)
μ wall friction coefficient (unitless); bin wall material Poisson's ratio (unitless)
φ angle of internal friction (degrees)
ρ bulk density of the stored material (kg/m^3)

REFERENCES

ACI. 1989. Building code requirements for reinforced concrete (ACI 318-89). Am. Concrete Institute, Detroit, MI. 353 pp.

ACI. 1991. Standard and practice for design and construction of concrete silos and stacking tubes for storing granular materials (ACI 313-91). Am. Concrete Institute, Detroit, MI. 27 pp.

Andersen, P. 1966. Temperature stresses in steel grain storage tanks. Civil Engineering 36(1):74.

ASAE. 1992a. Agricultural building snow and wind loads (EP 288.4). pp. 453-461 in ASAE Standards, 1992. Am. Soc. of Agric. Eng., St. Joseph, MI.

ASAE. 1992b. Loads exerted by free-flowing grain on bins (EP 433). pp. 525-528 in: ASAE Standards, 1992. Am. Soc. of Agric. Eng., St. Joseph, MI.

ASCE. 1990. Minimum design loads for buildings and other structures (ASCE 7-88). Am. Soc. Civil. Eng., New York. 94 pp.

Bernstein, M. S. 1947. "Rastchot Konstruktsii s Odnostoronnimi Sviaziami" (Design of structures with one-sided connections). Moscow, Russia. 85 p. (cited by Turitzen 1963)

Blight, G. E. 1985. Temperature changes affect pressures in steel bins. Int. J. Bulk Solids Storage Silos 1(3):1-7.

Blight, G. E. 1986. Swelling pressure of wetted grain. Bulk Solids Handling 6(6):1135-1140.

Blight, G. E. 1988. Strains measured in 7500t sugar silo. Part II. Bulk Solids Handling 8(4):413-419.

Blight, G. E. and A. Garstang. 1987. Strains measured in 7500t sugar silo. Part I. Bulk Solids Handling 7(4):573-582.

BOCA. 1990. The BOCA National Building Code 1990. Building Officials & Code Administrators International Inc., Country Club Hills, IL. 552 pp.

Britton, M. G. 1973. Strain in deep bin walls due to ambient temperature decrease. Unpublished Ph.D. Thesis. Texas A & M University, College Station, TX. 114 pp.

Caughey, R. A., C. W. Tooles, and A. C. Scheer. 1951. Lateral and vertical pressure of granular material in deep bins. Iowa Eng. Exp. Station Bull. 172. Ames, IA. 32 pp.

CFBC. 1983. Canadian Farm Building Code. NRCC No. 21312. National Research Council of Canada, Ottawa, ON. 27 pp.

CFBC. 1990. Canadian Farm Building Code. NRCC No. 30627. National Research Council of Canada, Ottawa, ON. 28 pp.

Clower, R. E., I. J. Ross, and G. M. White. 1973. Properties of compressible granular materials related to forces in bulk storage structures. Trans. ASAE (Am. Soc. Agric. Eng.) 6(2):98-101, 103.

Dale, A. C. and R. N. Robinson. 1954. Pressures in deep grain storage structures. Agric. Eng. 35(8):570-573.

Deutsch, G. P. and L. C. Schmidt. 1969. Pressures on silo walls. Trans. ASME 91:450-459.

DIN. 1965. Design loads for buildings—loads in silos (DIN 1055, Blatt 6). German Standard Nov. 1964; also in Beton and Stahlbetonbau 1965 5:126-128.

DIN. 1987. Design loads for buildings—Loads in silo bins (DIN 1055). Deutsche Normen, Berlin. 6 pp.

Gaylord, E. H. and C. N. Gaylord. 1984. Design of steel bins for storage of bulk solids. Prentice-Hall, Englewood Cliffs, NJ. 359 pp.

Hough, B.H. 1957. Basic soils engineering. Ronald Press Co., New York. 513 pp.

ICBO. 1991. Uniform building code. Int. Conf. Building Officials. Whittier, CA. 1050 pp.

Isaacson, J. D. and J. S. Boyd. 1965. Mathematical analysis of lateral pressures in flat-bottomed deep grain bins. Trans. ASAE (Am. Soc. Agric. Eng.) 8(3):358-360, 364.

Janssen, H. A. 1896. On the pressure of grain in silos. Proc. Instit. Civil Eng. 124:553-555.

Jenike, A. W. and J. R. Johanson. 1969. On theory of bin loads. J. Eng. Industry, Trans. ASME. 91(2):339-344.

Jenike, A. W., J. R. Johanson, and J. W. Carson. 1973a. Bin loads-Part 2: Concept. J. Eng. Industry, Trans. ASME. 95(1):1-5.

Jenike, A. W., J. R. Johanson, and J. W. Carson. 1973b. Bin loads-Part 3: Mass flow bins. J. Eng. Industry, Trans. ASME. 95(1):6-12.

Jenike, A. W., J. R. Johanson, and J. W. Carson. 1973c. Bin loads-Part 4: Funnel flow bins. J. Eng. Industry, Trans. ASME. 95(1):13-16.

Ketchum, M. S. 1919. The design of walls, bins and grain elevators, Third Edition, McGraw-Hill, New York. 351 pp.

Kim, V. S. 1959. Grain pressure and the improvement of the construction of silos and grain elevators. Khleboissdat, Moscow. (in Russian).

Lade, P. V. 1977. Elasto-plastic stress-strain theory for cohesionless soil with curved yield surfaces. Int. J. Solids Struc. 13:1019-1035.

Li, Y., V. M. Puri, and H. B. Manbeck. 1990a. Loads in a scaled model bin subjected to cyclic ambient temperature. Trans. ASAE (Am. Soc. Agric. Eng.) 33(2):651-656.

Li, Y., V. M. Puri, and H. B. Manbeck. 1990b. Elastic-viscoplastic cyclic constitutive model parameter determination and evaluation for wheat en masse. Trans. ASAE (Am. Soc. Agric. Eng.) 33(6):1984-1995.

Li, Y., V. M. Puri, and H. B. Manbeck. 1991. Finite element model prediction of cyclic thermally induced loads in a scaled bin using elastic-viscoplastic constitutive equation. Trans. ASAE (Am. Soc. Agric. Eng.) 34(5):2207-2215.

Loewer, O. J. Jr., I. J. Ross, D. D. Kratzer, and J. N. Walker. 1977. Properties of ground shelled corn as related to forces in bulk storage structures. Trans. ASAE (Am. Soc. Agric. Eng.) 20(1):155-156.

Manbeck, H. B. 1984. Predicting thermally induced pressure in grain bins. Trans. ASAE (Am. Soc. Agric. Eng.) 27(2):482-486.

Manbeck, H. B. and L. M. Muzzelo. 1985. Measurement of thermally induced pressures in a model bin. Trans. ASAE (Am. Soc. Agric. Eng.) 28(4):1253-1258.

Manbeck, H. B., M. G. Goyal, G. L. Nelson, and M. G. Singh. 1977. Dynamic overpressures in model bins during emptying. Paper No. 77-4505. Am. Soc. Agric. Eng., St. Joseph, MI. 25 pp.

McCabe, R. P. 1974. Flow patterns in granular material in circular silos. Geotech. 24(1):45-62.

Mohsenin, N.N. 1986. Physical properties of plant and animal materials. 2nd ed. Gordon and Breach Science Publ., New York. 891 pp.

Moysey, E. B. and S. Hiltz. 1985. Friction characteristics of some grains and fertilizers. pp. 721-732 in: Proc. 10th Annual Powder and Bulk Solids Handling and Processing Conf. Rosemont, IL.

Muzzelo, L. M. 1983. Measurement of thermally induced pressures in a model grain bin. Unpublished M.S. Thesis. The Pennsylvania State University, University Park, PA. 108 pp.

Myers, K. F. 1974. Overpressures in bolted steel grain bins due to ambient temperature fluctuations. Unpublished M.S. Thesis. Ohio State University, Columbus, OH. 48 pp.

Nguyen, T. V, C. E. Brennen, and R. H. Sabersky. 1980. Funnel flow in hoppers. Trans. ASME 102:729-735.

Pieper, K. 1969. Investigation of silo loads in measuring models. J. Eng. Industry, Trans. ASME 91:365-372.

Platonov, P. N. and A. P. Kovtun. 1959. "Davlenie Zerna na Stenki Silosov Elevatorov" (The pressure of grain on silo walls). Muhkomol'no Elevatornaia Promyshlennost" 25(12):22-24. (cited by Turitzen 1963)

Puri, V. M. and H. B. Manbeck. 1991. Potential of finite element method in modeling load response of particulate materials. Paper No. 91-4076. Am. Soc. Agric. Eng., St. Joseph, MI. 31 pp.

Rankine, W. J. M. 1857. On the stability of loose earth. Phil. Trans. Royal Soc. London. 147 pp.

Reimbert, M. and A. Reimbert. 1959. Silos, traite theorique et pratique, 2nd ed., (translated by A.B. Phillips). Eyrolles, Paris. 563 pp.

Reimbert, M. and A. Reimbert. 1963. Excess pressure phenomena due to the emptying of silos: practical repair of damaged silos by means of metal tubes. Steel 9:377-382.

Reimbert, M. and A. Reimbert. 1976. Silos: theory and practice. Trans. Tech. Pub. Clausthal, Germany. 250 pp.

Reimbert, M. and A. Reimbert. 1987 Silos: Theory and practice, vertical silos, horizontal silos (retaining walls). Lavoisier Publishing Co., New York. 563 pp.

Risch, E. and F. L. Herum. 1982. Bin wall stresses due to aeration of stored shelled corn. Paper No. 82-4072. Am. Soc. Agric. Eng., St. Joseph, MI. 22 pp.

Ross, I. J., T. C. Bridges, O. J. Loewer, and J. N. Walker. 1979. Grain bin loads as affected by grain moisture content and vertical pressure. Trans. ASAE (Am. Soc. Agric. Eng.) 22(3):592-597.

Sakai, H., H. Matsumura, M. Sasaki, N. Nakamura, M. Kabayashi, and Y. Kitagusa. 1985. Study on the dynamic behavior of coal silos against earthquakes. J. Bulk Solids Handling 5(5):1021-1033.

Schwab, C. V. 1989. Grain pressure phenomena in a circular flat-bottom bin. Unpublished Ph.D. Thesis. University of Kentucky, Lexington, KY. 265 pp.

Schwab, C. V., R. A. Curtis, S. A. Thompson, and I. J. Ross. 1991. Vertical loading of temperature cables. Trans. ASAE (Am. Soc. Agric. Eng.) 34(1):269-274.

Schwab, C. V., S. A. Thompson, A. Williams, and I. J. Ross. 1992. Temperature cable load comparison between model and full-scale grain bins. Trans. ASAE (Am. Soc. Agric. Eng.) 35(1):297-302.

Smid, J. 1975. Pressure of granular material on wall of model silo. Coll. Czech. Chem. Commun. 40:2424-2436.

Soviet Code. 1965. Ukanzia po proectirovaniu silosov dlia siputshich materialov. (Instructions for design of silos for granular materials). Sovity Code CH-302-65 Gosstroy, Moscow, USSR.

Theimer, O. F. 1969. Failures of reinforced concrete grain silos. Trans. ASME 91:460-477.

Thompson, S. A. 1987. Vertical loads on cables in a model grain bin. Trans. ASAE (Am. Soc. Agric. Eng.) 30(2):485-591.

Thompson, S. A. and I. J. Ross. 1981. Compressibility and frictional coefficients of wheat. Paper No. 81-3520. Am. Soc. Agric. Eng., St. Joseph, MI. 23 pp.

Thompson, S. A. and I. J. Ross. 1984. Thermal stresses in steel grain bins using the tangent modulus. Trans. ASAE (Am. Soc. Agric. Eng.) 27(1):165-168.

Thompson, S. A., J. L. Usry, and J. A. Legg. 1986. Loads in a model grain bin as affected by various unloading techniques. Trans. ASAE (Am. Soc. Agric. Eng.) 29(2):556-561.

Turitzin, A. M. 1963. Dynamic pressure of granular material in deep bins. J. Struc. Div. Am. Soc. Civil Eng., 89(ST2):49-73.

Tyson, A. W. and H. B. Manbeck. 1980. Interaction between bin wall stiffness and emptying pressures. Paper No. 80-4502. Am. Soc. Agric. Eng., St. Joseph, MI. 24 pp.

Usry, J. L. and S. A. Thompson. 1986. Loads on vertical stiffeners on corrugated grain bins. Trans. ASAE (Am. Soc. Agric. Eng.) 29(5):1355-1363.

Walker, D. M. 1966. An approximation theory for pressures and arching in hoppers. Chem. Eng. Sci. 21:975-997.

Walker, D. M. and M. H. Blanchard. 1967. Pressures in coal hoppers. Chem. Eng. Sci. 22:1713-1745.

Weiland, W. F. 1964. Structural failures of grain storage facilities. Trans. ASAE (Am. Soc. Agric. Eng.) 7(2):130-133.

Wickstrom, L. L. 1980. Loads applied to grain bin roofs. Paper No. 80-4505. Am. Soc. Agric. Eng., St. Joseph, MI. 11 pp.

Wilms, H. 1990. Criteria for evaluation of silo design codes. Paper No. 1342, presented at the 10th International Congress of Chemical Engineering, Chemical Equipment Design and Automation. Praha, Czechoslovakia. August 26-31. 10 pp.

Yamahara, N. and P. Takahashi. 1967. Studies on pressure distribution in silos. Japanese Ingenieur-Aeitschrift. Bd. 5. September. (In Japanese).

Youngs, R. R. 1982. A three dimensional effective stress model for cyclically loaded granular soils. Unpublished Ph.D. Thesis. University of California, Berkeley, CA. 322 pp.

Zhang, Q., M. G. Britton, and R. Jaremek. 1991. Discharge loads in smooth and corrugated walled model bins for wheat, barley, and canola. Paper No. 91-4063. Am. Soc. Agric. Eng., St. Joseph, MI. 12 pp.

Zhang, Q., V. M. Puri, and H. B. Manbeck. 1986a. Finite element modeling of thermally induced pressures in grain bins. Trans. ASAE (Am. Soc. Agric. Eng.) 29(1):248-256.

Zhang, Q., V. M. Puri, and H. B. Manbeck. 1986b. Determination of elastoplastic constitutive parameters for wheat en masse. Trans. ASAE (Am. Soc. Agric. Eng.) 29(6):1739-1746.

Zhang, Q., V. M. Puri, and H. B. Manbeck. 1987. A finite element model for predicting static and thermally induced bin wall pressures. Trans. ASAE (Am. Soc. Agric. Eng.) 29(1):248-256.

Zhang, Q., V. M. Puri, and H. B. Manbeck. 1989. Static and thermal loads in full size grain bin predicted by vectorized FEM on supercomputer. Trans. ASAE (Am. Soc. Agric. Eng.) 32(2):685-690.

14

Mathematical Modeling of Heat, Moisture, and Gas Transfer in Stored-Grain Ecosystems

Digvir S. Jayas

University of Manitoba, Winnipeg, Manitoba, Canada

SUMMARY

Mathematical models of heat, moisture, and gas transfer in stored-grain ecosystems are reviewed; a philosophy of the development and use of mathematical models is presented, along with a discussion on the solution methods, boundary conditions, initial conditions, and simulation results.

In stored-grain ecosystems heat, moisture, and gas transfer occur simultaneously. The interactions between the abiotic factors (temperature, moisture content, and gas concentration) and biotic factors (grain, insects, mites, fungi, and bacteria) make the mathematical modeling of stored-grain ecosystems a challenging problem. Stored-grain ecosystems are more complex than the mathematical models used to describe them.

The differential equations that govern heat transfer in stored-grain ecosystems have been solved using analytical and numerical methods for one-, two-, and three-dimensional configurations. For a majority of the models, predicted temperatures compare well with temperatures measured under a range of real conditions. The diurnal temperature variations only affect 150 mm of commodity near the wall and the grain surface. The effect of seasonal temperature variations on changes in grain temperatures decreases with an increase in the distance from the bin wall.

There are few models of pure diffusion of moisture and gas transfer through intergranular spaces of stored grain. Mathematical models of moisture transfer due to temperature gradients in stored grain have been reported in the published literature but with only limited experimental validation.

Future research efforts should be directed towards development and validation of comprehensive models of moisture migration in stored grain; incorporation of free convection into CO_2 transfer models, including diffusion coefficients of CO_2 and other chemical fumigants as affected by kernel orientation, moisture content, foreign material in the grain, compaction, and gas concentration gradient; sorption and desorption of CO_2 and fumigants by grain; integration of biomodels of stored grain with physical models; and stochastic modeling of the transfer processes in stored-grain ecosystems. Experimental validation of the model predictions should be carried out and reported to make the models trustworthy and useful.

---

1. INTRODUCTION

A stored-grain bulk is a man-made ecological system in which deterioration of the stored product results from interactions among physical, chemical, and biological factors. The important factors are temperature, moisture, carbon dioxide (CO_2), oxygen (O_2), grain characteristics, microorganisms, insects, mites, rodents, birds, geographical location, and granary structure. Heat, moisture, and CO_2 are produced during the deterioration of grain; therefore, increases in grain temperature, moisture content, and CO_2 concentration in the intergranular air can be used as indicators of incipient grain spoilage. The generated heat, moisture, and CO_2 move within and out of the grain mass. This movement is dependent on the characteristics of the grain, storage structure, and weather. Mathematical models can be developed to seek to predict temperatures, moisture contents, and CO_2 concentrations at various locations in the stored grain. The predictions of such models can be used to decide the resolution required and location of the sensors for detection of incipient spoilage in stored grain.

The survival and reproduction of biological agents in grain are dependent to a great extent on the temperature and moisture levels. Stored-product insects can live at temperatures of 8–41°C and intergranular relative humidities (rh) in equilibrium with grain of 1–99% (Sinha and Watters 1985). Usually development and multiplication are optimum near 30°C and 50–70% rh (Howe 1965, Sinha 1973, Sinha and Watters 1985). Mites can live at temperatures of 3–41°C and relative humidities of 42–99% with the optimum for development and multiplication near 25°C and 70–90% rh. Fungi can develop at temperatures of –2–55°C and relative humidities of 70–90% with the optimum temperature near 30°C, and relative humidity around 80%. There is a considerable variation in optimum conditions for different species. Localized regions may occur in stored-grain ecosystems for optimum development and multiplication of insects, mites, and fungi even when the average conditions of the bulk would prevent pest infestation. To design and evaluate methods of reducing temperature and moisture gradients in stored grain, and to predict probability of development and locations within the bin where spoilage may occur and rates of deterioration, mathematical models of heat and moisture transfer can play a major role.

Modified atmospheres can be used for controlling populations of insects, mites, and fungi in stored grain. A modified atmosphere can provide an environment which is lethal to stored-grain pests. This is achieved by altering the proportions of CO_2, nitrogen (N_2), and O_2 in the intergranular atmosphere. Controlled atmospheres containing elevated CO_2 (> 60%), depleted O_2 (< 10%), and low relative humidities (< 50%) at high temperatures (> 27°C) are lethal within a few days of exposure to most insect and mite species that are responsible for losses in stored grain (Jay et al. 1970, White et al. 1990, Jayas et al. 1991, White and Jayas 1991, Rameshbabu et al. 1991) (see Chapter 11). Exposure time and gas concentrations are the critical factors at given environmental conditions and these vary with pest species. To create modified atmospheres, CO_2 or N_2 is usually introduced into the grain mass at single or multiple points. The uniform distribution of introduced gases and maintenance of the lethal concentrations for the required duration are critical for effectively reducing pest populations. Mathematical models of gaseous movement can be used to predict the distribution of gases within the stored mass and the loss of gases from the grain in nonairtight structures. The gas-loss models can be used to determine the amount, frequency, and location of gas introduction needed for maintenance of the lethal atmospheres. Similarly, mathematical models of the movement of other chemical fumigants that are used for disinfesting grain can be used for the prediction of chemical concentrations at various points in a bulk of stored grain.

Experimental studies can be conducted to measure changes in temperature, moisture content, and gas concentration in stored grain. But the large number of biotic and abiotic factors and their interrelationships in stored-grain ecosystems makes full-scale experimental studies expensive and time consuming, and the generalizations needed to predict the variables accurately are difficult to make. Mathematical models, if based on the principles of physical and biological sciences and if properly validated, can be used to study the effect of various biotic and abiotic parameters on the distribution of temperature, moisture, and gases (e.g., Jayas et al. 1992). The major advantages of validated models are their ability to answer "what if?" questions and their transferability to different climatic regions of the world, thus making them globally applicable rather than locally applicable as is expected of experimental studies.

In this chapter, the interrelationships of influencing factors (both internal and external to the stored grain) and mathematical models (both physical and biological) are discussed to illustrate the complexities of stored-grain ecosystems. The methods used to solve the mathematical models of heat, moisture, and gas transfer are briefly outlined along with their advantages and limitations. The results of the published models of heat, moisture, and gas transfer are presented and discussed. Possible directions for future research in the area of mathematical modeling of physical processes in stored-grain ecosystems and the potential of integration of physical models with biological models (e.g., population dynamics models) are discussed.

2. INTERACTIONS AMONG ECOSYSTEM SUBMODELS

2.1. An Overview

The transfer of heat, moisture, and gases in stored-grain ecosystems is governed by partial differential equations with associated boundary conditions. Because the transfer processes in stored grain are time-dependent, the solution of these equations also requires the specification of the initial conditions.

In general terms, boundary conditions are of three types:

1. variables to be predicted (temperature, moisture content, and gas concentration) may be specified along portions of the boundary;
2. a portion of the boundary may be impermeable to the flow of the moving quantity (heat, moisture, and gas); and
3. a portion of the boundary may be permeable to the moving quantity.

All three types of boundary conditions may or may not be present in any given problem.

The specific boundary conditions used by various researchers will be discussed in later sections where results of the published simulation studies are summarized. Values of the variables to be predicted in the stored grain at the start of the solution constitute the initial conditions.

The solution of the partial differential equations results in the prediction of the temperatures, moisture contents, and gas concentrations at various locations in the stored grain as a function of time. Temperatures, moisture contents, and gas concentrations interact with grain and the biological inhabitants in the grain (insects, mites, fungi, and bacteria).

2.2. External Factors

In stored-grain ecosystems, the movements of heat, moisture, and gases are affected by many external and internal parameters (Fig. 1). The geographical location determines the weather data which affect the stored-grain ecosystems. The weather parameters of importance are: wind velocity and direction, ambient air temperature, solar radiation, and atmospheric pressure. Relative humidity of the atmospheric air is a more significant parameter in mechanically ventilated grain than in nonventilated bulks. The relative humidity, however, affects the moisture transfer indirectly through exchanges between intergranular and outside

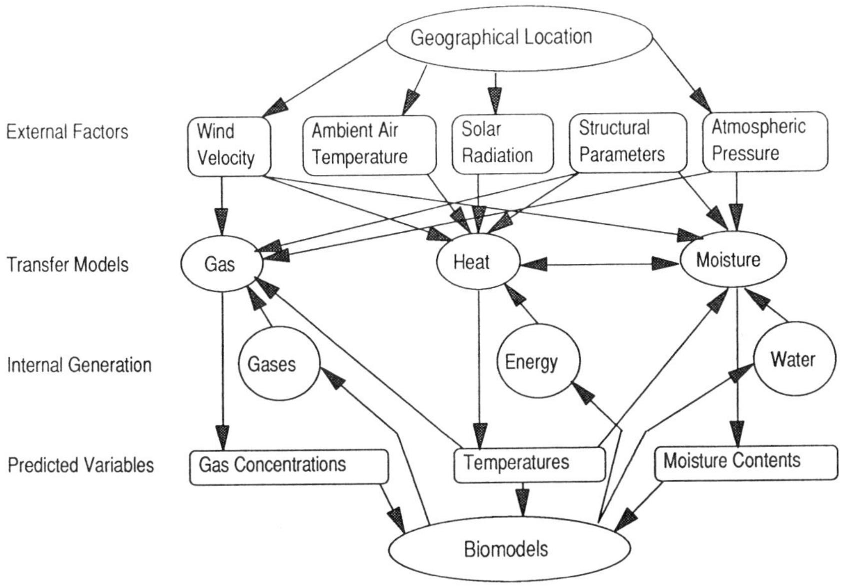

Figure 1 Interrelationships among various submodels and influencing parameters in a comprehensive stored-grain ecosystem model. The structural parameters include thermal and physical properties of the structural material; shape, size, orientation, and location of the structure; and small openings in the structure. Biomodels are models of respiration rate and population dynamics of insects, mites, fungi, and bacteria in stored grain.

air masses. The air exchanges could be caused by pressure and temperature differences between the inside and outside of the storage structure, which might have many small openings.

2.3. Internal Factors

The production of heat, moisture, and CO_2 by respiration of grain, insects, mites, fungi, and bacteria, and thermal (specific heat and thermal conductivity) and physical (bulk density and porosity) properties of the bulk are the main internal factors that affect the movement of heat, moisture, and gases in stored grain. Biomodels of the respiration of grain and associated biological agents can be used to estimate the internal generation of heat, moisture, and CO_2. The respiration of grain and pests increases with increases in the temperature and moisture content of grain but decreases with an increase in the CO_2 concentration which is toxic to most biological agents. Models of the population dynamics of bioagents can be used for the prediction of the infestation levels within the conditions of temperature and moisture content of the grain (see Chapter 6). Temperature, moisture content, and CO_2 concentration affect the internal generation of heat, moisture, and CO_2, and the thermal and physical properties which affect the transfer processes resulting in a complex system of many interacting submodels.

2.4. Factors Affecting Heat Transfer

Wind velocity, ambient air temperature, solar radiation, and structural parameters affect the transfer of heat within the stored grain (Fig. 1). The structural parameters include size and shape of the bin, type of construction material, type of foundation, and number and size of openings in the structure. Heat transfer between the storage structure and the surroundings is by convection, radiation, and conduction to the ground or to other attached bins. The convective heat transfer coefficient is affected by the wind velocity, the shape and size of the bin, and the temperature difference between the bin surface and the ambient air temperature.

The convective heat transfer coefficient increases with increasing wind velocity and temperature difference but decreases with increasing bin size. The heat flow into or out of the grain increases proportionally with an increase in the convective heat transfer coefficient. The dependence of the convective heat transfer coefficient on the bin surface temperature converts the heat transfer problem into a nonlinear problem requiring an iterative solution. The predicted surface temperature at the end of every time step is used to estimate a new convective heat transfer coefficient for the next time step.

The amount of convective heat flow into or out of the grain mass is directly proportional to the temperature difference between the bin surface and the surrounding air. The shape of the storage structure affects the convective heat transfer coefficient and thus the heat transfer from the structure to the surroundings. The size of the storage structure determines the distance that heat has to travel. When exposed to cool outside temperatures warm grain at the center of a large bin will take longer to cool than in a small bin because of the low Fourier number which depends on the low thermal diffusivity of grain. The thermal diffusivity of wheat at 14.5% moisture content (mc) wet basis (wb) is 0.105×10^{-6} m^2/s (Yaciuk et al. 1975a) compared with the thermal diffusivity of glass wool insulation (24 kg/m^3 density) of 2.26×10^{-6} m^2/s, and of copper of 112.3×10^{-6} m^2/s (Singh and Heldman 1984). When exposed to low outside temperatures, a rectangular bin with a small width-to-length ratio will cool faster than a square bin with the same perimeter because of the short travel path for heat across the width of the structure.

Depending on the thermal absorptivity and emissivity of the structural material and the surface temperature, solar radiation may result in a net heat gain or loss from the stored grain. Structural materials with high shortwave absorptivities and low longwave emissivities may result in a net heat gain by radiation. The amount of solar radiation striking the storage structure is dependent on the annual number of cloud-free hours and the latitude and altitude of the location. The orientation of rectangular buildings will have an effect on the heat gain from solar radiation. In temperate climates of high latitudes, south walls in the Northern Hemisphere and north walls in the Southern Hemisphere intercept most of the solar radiation; therefore, rectangular bins should be placed with their longer axes running north to south to help keep the grain cool. In areas of low latitudes, small overhangs on north and south walls can provide shading for much of the day. In low latitude areas, east- and west-facing walls intercept more solar radiation than the north- and south-facing walls; therefore, rectangular bins should be placed with their longer axes running east to west. The location of the storage bin with respect to other structures and trees on a farmstead should be such that the wall expected to receive the most solar radiation is shaded.

Heat exchange through the foundation will depend on the soil temperature under the bin and thermal properties of the soil and the foundation materials for flat-bottom bins on the ground. For hopper-bottom bins that are supported above ground, soil temperatures under the foundation will have no effect on heat transfer to the grain whereas for partially or fully underground bins soil temperature will affect the heat exchange through the bottom and the submerged walls. The soil conditions of the areas in the vicinity of the bins will affect heat flow by reflection of the solar radiation from the surrounding areas to the walls. Therefore, the reflectivity of the surrounding areas contributes to heat transfer in stored-grain ecosystems.

The internal factors that affect heat transfer are the physical and thermal properties of a grain bulk, respiration of the grain, and respiration and population dynamics of the fauna and microflora in the grain bulk. The CO_2 concentration and moisture content affect the internal generation of heat and thus the heat transfer process and prediction of temperature. The dependence of thermal and physical properties of stored grain on moisture content results in further influences of moisture transfer models on the heat transfer models. The effect of gas concentrations on the physical and thermal properties of stored grain is expected to be small. Because temperature also influences respiration, pest population, and thermal properties of grain the mathematical model of heat transfer in stored-grain ecosystems is nonlinear.

2.5. Factors Affecting Moisture Transfer

The external factors that directly affect the moisture transfer process are the wind velocity, atmospheric pressure, and structural parameters; their influence, however, is small. The wind velocity causes lower pressure on the leeward side compared with the windward side of a structure. If there are openings in the structure, air will be exchanged between the inside and outside of the bin as air pressures are equilibrated. Depending on the difference in water vapor concentrations between the inside and outside air masses, moisture may be transferred from or to the grain. Openings at the junctions of the wall and roof can provide direct access of surface grain to the outside air, resulting in some moisture transfer between the grain at the surface and the air (Aldis and Foster 1977). Openings in the roof may result in the entrance of wind blown rain or snow. This external water may cause localized areas of high moisture in grain and thus create considerable damage to the grain.

Most significant of the internal factors that affect the moisture transfer in stored grain are the temperature gradients. Temperature gradients are of significance normally only in large grain masses. Air density depends on the air temperature; cool air being denser than warm air. As an example, consider a 6-m diameter bin filled with wheat in the autumn (late August–early September) on the Canadian prairies. The temperature of wheat at storage could be up to 35°C. As winter approaches the grain starts cooling, although very slowly at the bin center. In the winter months (December–January), when mean outside temperatures are around –20°C, grain near the bin wall will approach the ambient temperature but in the center may still be at its initial temperature. The cool air near the wall is pulled downward by the force of gravity and pushes upward the warm less dense air in the bin center. As air moves through the grain it exchanges moisture with the grain. Usually warm air moving up in the center will release moisture when it comes in contact with the cooler grain near the top of the grain mass. The reverse may happen in the summer months and an increase in moisture may occur near the bottom of the grain bulk.

The other internal factor that influences the moisture transfer in the stored grain is the generation of moisture by the grain and its living-inhabitants. The temperatures and CO_2 concentrations influence the internal generation of water. Increasing grain temperature up to the thermal death points of organisms (e.g., insects, mites, fungi) will increase the production of moisture due to increased respiration rate of the organisms. The respiration rate of organisms usually doubles with a 10°C increase in temperature when the temperature increase is within their survival and developmental range. High CO_2 concentrations usually decrease the amount of moisture produced by grain and organisms.

2.6. Factors Affecting Gas Transfer

Wind velocity, atmospheric pressure, and structural parameters affect the movement of gases in stored grain in a similar way as they affect moisture movement. The diffusion coefficients of gases through bulk grain depend on both the temperature and moisture content of grain (Oxley 1948, Singh (Jayas) et al. 1984). The internal generation of CO_2 also depends on both the moisture content

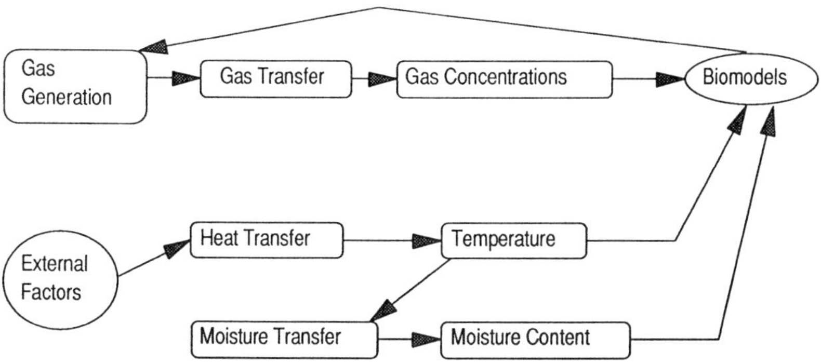

Figure 2 An example of the interaction paths for a gas transfer model in a stored-grain ecosystem.

and temperature of grain. Thus the influence of ambient air temperature, solar radiation, and thermal properties of the bin wall and foundation on CO_2 transfer is through the heat transfer model (Fig. 2). The temperature gradients can result in convection currents that further affect the movement of CO_2 in stored grain.

Other factors that can influence the transfer processes are the amount and type of foreign material in grain. The presence of foreign material (weed seeds, chaff, broken kernels, dust) affect the relevant properties of grain bulks such as diffusion coefficients of CO_2 and moisture, bulk density, porosity, specific heat, and thermal conductivity of the grain bulk. The presence of foreign material also increases the rate of respiration of the bulk, generally because of its higher moisture content than undamaged kernels. Broken kernels or damaged surfaces of kernels are more susceptible to fungi and insects than sound kernels. Therefore, their presence may increase the respiration rate of the bulk.

For short-term storage of grain (1–3 mo) such as at grain handling facilities, a comprehensive model of transfer processes may not be needed because the gradients in temperature, moisture, and gas concentrations are eliminated due to frequent movement of the grain. When treating infested grain at the handling facilities by introducing CO_2, N_2, phosphine, or other chemical fumigants, a model of gas transfer may be used for determining the uniformity of distribution of the chemical fumigants in the grain mass and for designing systems that will provide uniform distribution. For long-term storage of grain, only a comprehensive model of transfer processes can predict conditions of the grain and the physical parameters that influence the grain. Before using such a model for management of grain, proper validation is necessary. Without validation, the usefulness of models is dependent on the closeness of the behavior of the stored-grain ecosystem to the assumptions made during the derivation and solution of the models.

3. METHODS OF SOLUTION

3.1. Analytical Methods

The transfer processes in stored-grain ecosystems are described by partial differential equations because the variables to be predicted are interrelated and are dependent on position in the stored grain and time. The solution of these partial differential equations with associated boundary and initial conditions results in the prediction of physical variables (temperature, moisture content, and gas concentration) which interact with grain and the pest organisms in grain. Interrelationships among variables and transfer models require the models to be nonlinear and require iterative solutions.

Ideally, the mathematical models should be solved analytically because that will result in the most accurate predictions. The present mathematical models of stored-grain ecosystems are complex and may have inherent errors because of an incomplete understanding of the ecosystem, the stochastic or random nature of many processes and variables, and uncertainties in experimental measurements. Many simplifying assumptions are made to solve the mathematical models analytically, resulting in a model which deals with only the most pertinent features of the process and neglects many important secondary effects. The analytical solutions of such simplified models, even though exact may contain inherent errors due to simplifying assumptions which necessarily may not be valid in real systems. Also, analytical solutions are impossible to derive for irregularly shaped storages and multidimensional systems. For example, the boundary of a bag filled with grain is for all practical purposes impossible to model analytically unless it is approximated as a parallelopiped. A flat-bottom cylindrical bin filled with grain to a leveled depth has easy-to-model geometry but mathematical models for such a bin may be three-dimensional because of the random variations along the boundary due to external factors and unpredictable locations of spoilage in the stored grain. To summarize, analytical solutions cannot be obtained for most "real world" problems because they usually involve nonhomogeneous materials, irregular geometries, and difficult boundary conditions.

In stored-grain ecosystems, analytical solutions can be used only when simplifying assumptions can transfer the mathematical model usually to a one-dimensional situation with constant or predictable boundary conditions. Solutions to many partial differential equations for simple geometrical shapes (e.g., cylinder, parallelopiped) with constant physical and thermal properties and constant or sinusoidally varying boundary conditions are given in many books (e.g., Crank 1956, Carslaw and Jaeger 1959, Bird et al. 1960, Beek and Muttzall 1975, Gutfinger 1975, Cunningham and Williams 1980, Geankoplis 1983, Brodkey and Hershey 1988, Smith and Jensen 1989, Incropera and DeWitt 1990, Mills 1992).

3.2. Numerical Methods

3.2.1. *Finite Difference Method*

For realistic models of stored-grain ecosystems, approximate solutions can be obtained using numerical methods. The most commonly used numerical methods for solution of partial differential equations are the finite difference and the finite element methods. The limitation of numerical methods is that the solutions are approximate but this can be overcome by proper validation of the models. The advantage of numerical methods is their ability to handle three-dimensional problems with complex boundary conditions and varying physical and thermal properties of grain and structural materials.

In the finite difference method, the system of partial differential equations which describe the transfer processes in stored grain is transformed into a system of algebraic equations by applying a difference operator. The region of interest enclosed within the boundary is called the domain. The domain is divided into small subregions connected at nodes. The size of subregions can vary within the domain with small subregions where the expected gradients in the predicted variables are high. The variables to be predicted have distinct values at the nodes. Using the nodal values, the continuous derivatives are transformed into ratios of finite change in the variable to be predicted to the increment between the nodes or time. Using the energy and mass balances for subregions in the domain, a set of algebraic equations is obtained. After modification of the system of equations based on the boundary conditions, the system of equations is solved for values at nodes. The details on the finite difference method are given in many books (e.g., van der Houwen 1968, Smith 1985).

The finite difference method can be used readily for solving one-dimensional problems and two- or three-dimensional problems when the boundaries are parallel to the coordinate axes. For curved and irregular boundaries, the finite difference method becomes more cumbersome. Singh and Thorpe (1993) have devised a relatively simple procedure for modeling heat, moisture, and momentum transfer in grain stores of any shape.

3.2.2. *Finite Element Method*

Detailed descriptions of the finite element method are given in Segerlind (1976, 1984), Zienkiewicz (1977), Desai and Abel (1972), and Bathe (1982). As in the finite difference method, the domain is discretized into small subregions, called elements, connected at the nodes. The variables to be predicted are interpolated within the elements from nodal values. A system of algebraic equations is derived either by minimization of a functional based on variational calculus or by minimization of weighted residuals. The system of equations is modified based on boundary conditions and solved for unknown nodal values in the region. The major advantages of the finite element method are:

1. a general computer program can be written to solve similar partial differential equations in various fields and for various geometries;
2. domains with irregular boundaries can be handled easily;
3. the material property can vary within the domain;
4. various size elements can be used; and
5. mixed boundary conditions can be handled easily.

The major disadvantage of the finite element method is the preparation of input data based on the discretized domain.

4. RESULTS AND DISCUSSION OF THE PUBLISHED MODELS

4.1. Heat Transfer Models

4.1.1. Published Models

Many investigators have developed mathematical models of heat transfer for prediction of temperature in bulk grain (Muir 1970; Converse et al. 1973; Lo et al. 1975; Yaciuk et al. 1975a, 1975b; Muir et al. 1980; Metzger and Muir 1983; Gough 1985; Longstaff and Banks 1987; White 1988; O'Dowd et al. 1988; Alagusundaram et al. 1990a, 1990b; Sarker and Kunze 1991). Heat transfer problems in stored-grain ecosystems generally are three-dimensional because of variability due to solar radiation along the walls and the unpredictability of spoilage locations in the stored grain. With certain assumptions, the problems can be solved for one- or two-dimensional configurations. The heat transfer models have been solved for one- (Muir 1970, Converse et al. 1973, Lo et al. 1975, Yaciuk et al. 1975a, Longstaff and Banks 1987, White 1988, Sarker and Kunze 1991), two- (Muir et al. 1980, Metzger and Muir 1983), and three-dimensional configurations (Gough 1985, Alagusundaram et al. 1990a, 1990b). Heat flow in the radial direction only in cylindrical bins (e.g., Yaciuk et al. 1975a) or heat flow from the head space to the grain in the vertical direction only (e.g., Longstaff and Banks 1987) are modeled as one-dimensional problems. Heat flow in radial and axial directions only in cylindrical bins or heat flow in one of the horizontal directions and vertical direction in parallelopiped-shaped structures are modeled as two-dimensional problems. When heat flow is in axial and radial directions only in cylindrical bins, the problems are also known as axisymmetric. Heat flow problems in storage structures of various shapes (e.g., cylindrical, parallelopiped, W-shaped bins in Brazil, cone and pitcher shaped bins in many developing countries) are modeled as three-dimensional problems. O'Dowd et al. (1988) modeled the temperature changes of air inside warehouses caused by changes in the surroundings. His model can be considered three-dimensional. The heat transfer models have been solved using analytical (e.g., Converse et al. 1973), finite difference (e.g., Muir et al. 1980), and finite element (e.g., Alagusundaram et al. 1990b) methods. Many investigators (e.g., Converse et al. 1973, Gough 1985) also reported experimental data used for validation of their models. The predictions of the published heat transfer models were within 3°C in most cases.

4.1.2. Assumptions

Because of their complexity, the equations of comprehensive models cannot be conveniently solved. The complexity can be reduced by making assumptions about the storage systems. For example, the assumption that there is no heat flow in the circumferential direction in a cylindrical bin converts a problem to a two-dimensional heat flow problem and a further assumption that there are no temperature gradients in the axial direction converts the problem to a one-dimensional heat flow problem. With these assumptions, variations in solar radiation, convective heat transfer, shading along the walls, physical and thermal properties, and internal generation of heat within the grain mass due to temperature and moisture contents cannot be handled. For an analytical solution of heat transfer problems in stored grain, it is necessary to make these assumptions. For numerical solutions, these assumptions are not necessary but may be justified for middle portions of the bins filled with dry grain that is free of insects and mites. The top and bottom 1 m of grain have temperature gradients in the vertical direction due to heat transfer from the headspace and soil under the bin, respectively.

All of the published models have assumed that internal heat generation is negligible in the stored grain and that the thermal and physical properties are constant. These assumptions are valid for dry clean grain of uniform moisture content and having no insect or mite infestation (Steele et al. 1969, White et al. 1982a, 1982b). These assumptions are critical for an analytical solution of a heat transfer model of stored grain. For numerical solutions, the physical and thermal properties and internal generation can vary from point to point in the stored bulk. The variations in thermal and physical properties and internal heat generation may be due to nonuniform moisture contents at the time of filling or nonuniform moisture contents caused by translocation of moisture in the bin. Nonuniformity of amount and type of foreign material in the grain mass also can cause variations in these parameters.

Most published heat transfer models were solved for a cylindrical shape, mainly because it represents the most common storage structure in the countries where models were developed. The regular geometry (such as a cylinder or a parallelopiped) of the storage structure, filled to a level surface, is required when models are solved analytically. For irregular-shaped structures (e.g., W-shaped-bottom bins in Brazil), hopper-bottom bins, or bins with a nonleveled grain surface, numerical methods, especially finite element, should be used for solving heat transfer models.

4.1.3. Boundary Conditions

For solving partial differential equations it is necessary to describe boundary conditions that are to be satisfied during the solution process. The storage structure acts as an interface between the stored grain and the surroundings. Heat moves to this interface mainly by conduction. The boundary conditions define the process of heat exchange between the interface and the surroundings. For grain bins, boundary conditions must be defined at the floor, roof, and walls. For an analytical solution of only radial heat flow problems in cylindrical bins or for one-dimensional heat flow towards a wall in rectangular bins, the boundary condition at the wall is assumed to be heat transfer by convection. The outside air temperature is taken as constant or as a harmonic function of time. If historical or real-time air temperatures are to be used, the problem should be solved numerically. In one-dimensional problems, the effect of solar radiation is usually not considered although Yaciuk (1973) did include the effect of solar radiation. The conditions at the floor and roof do not play a role in one-dimensional heat flow problems.

For two- and three-dimensional heat flow problems, heat flow through the foundation is considered to be zero or heat flow is calculated using properties of the soil and foundation material and the soil temperatures under the foundation. The assumption of no heat flow through the floor could result in significant errors in the bottom portion of the bin. Changes in the air temperature near the walls and roof could be a harmonic function of time or could be historical or real-time measured data. The heat exchange to the surroundings at the walls and roof is by convection and radiation. The assumption of constant temperature at the walls and roof does not represent "real-world" problems but this assumption is valid when a model bin or a bag of grain is placed in a temperature-controlled room. Heat gain by absorption of shortwave solar radiation, reflected and emitted radiation from the surroundings, and longwave radiation emitted by the walls have to be assumed to be uniform along the walls and roof to model two-dimensional heat transfer in cylindrical or rectangular bins. For most storage bins located outside, solar radiation effects vary along the walls and roof. Only a three-dimensional model can incorporate the variable solar effects along the walls and roof.

4.1.4. *Initial Conditions*

The problem of heat flow in stored grain is time-dependent, therefore, the grain temperature at the beginning of the simulation must be known. For an analytical solution of simple problems, a uniform temperature is assumed. Although there are variations in the grain temperature as the bin is filled at harvest, the grain may be mixed while being transferred to another bin and uniform temperatures can be achieved. For numerical solutions, the initial grain temperature can be specified for each node.

The initial grain temperature can either be measured in the bin after filling or can be estimated from ambient air temperatures at the time of harvest. The estimation must consider that the grain temperature could be 7–8°C higher than the ambient temperature (Prasad et al. 1978) on a sunny day in Manitoba. As the bin is filled, initial grain temperatures for individual layers can be used for initial conditions.

4.1.5. *Simulation Results*

The results of the published models and measured temperatures at various locations in grain bins show that the diurnal variations in temperature affect only the grain within 150 mm from the wall (e.g., Muir 1970, Converse et al. 1973).

The effect of seasonal temperature changes on grain temperatures is dependent on the distance from the bin wall. Converse et al. (1973) reported measured and predicted grain temperatures at various locations in a 5.5-m diameter by 33.5-m high concrete bin. The grain temperature at the wall was assumed to vary sinusoidally with a mean of 14.2°C and an amplitude of 15.3°C, properties of the grain were assumed constant, and the initial grain temperature was uniform. The difference between the amplitudes of ambient air and wheat temperatures increased as the distance from the wall increased (Fig. 3). A plot of Yaciuk's (1973) simulation results showed that the fluctuations of temperature at the center of a bin decreased as the bin diameter increased (Fig. 4). For assessing the effect of temperature fluctuations of outside air on grain temperature, the increase in distance from the bin wall in a constant-diameter bin analyzed by Converse et al. (1973) is similar to studying the center temperature of bins of increasing diameter (e.g., Yaciuk 1973). The time lag between the seasonal changes in outside temperatures and the changes in grain temperature increases as the distance from the bin wall increases (Fig. 5). Farmers in grain-growing areas of Western Canada and the United States are purchasing large diameter bins because of cheaper storage per tonne than in smaller diameter bins. When grain storage and handling is centralized; as in, for example, Australia, Brazil, and India; bins of large size are usually used. In Winnipeg,

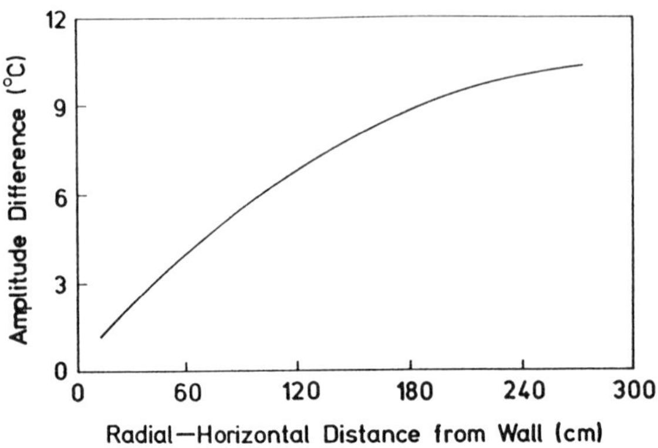

Figure 3 Difference in temperature amplitude between the outside air and the wheat as a function of distance from the vertical exposed bin wall at Cheney, Kansas. *Source*: Converse et al. (1973).

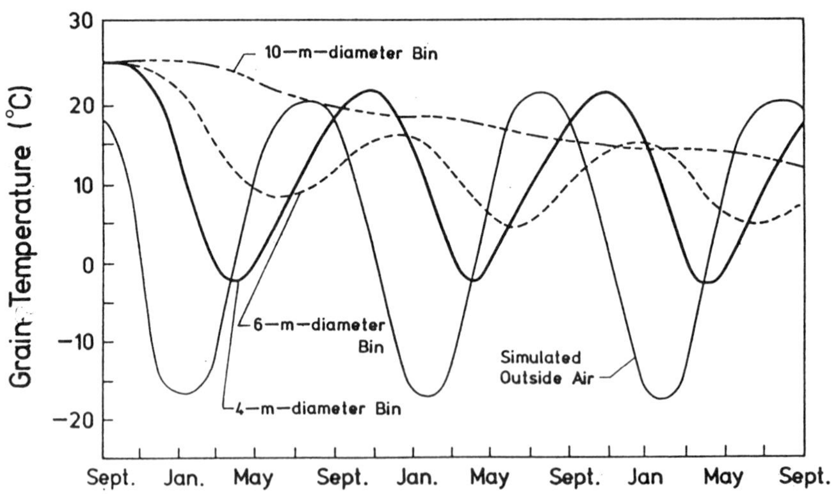

Figure 4 Effect of bin diameter on the predicted temperatures at the center of galvanized steel bins containing wheat at Winnipeg, Canada. *Source*: Based on Yaciuk's (1973) simulation results.

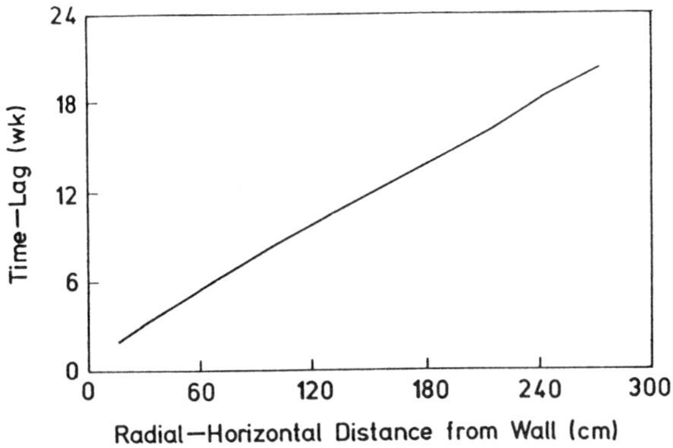

Figure 5 Time lag between outside seasonal temperature variations and wheat temperature variations as a function of distance from the exposed bin wall at Cheney, Kansas. *Source*: Converse et al. (1973).

Canada it takes only 49 d to cool below 20°C the center of a 2-m diameter galvanized steel bin filled with wheat at 35°C, but it takes 652 d for the center of a 10-m diameter bin (Table 1). Under the subtropical conditions of central Brazil, temperatures at the center of a 1.5-m diameter bin rose to near 45°C while in a 5.0-m diameter bin the peak temperature was 10°C lower (Fig. 6) (Ferreira and Muir 1981). Bins filled with grain at temperatures higher than the ambient temperature take longer to cool as the bin diameter increases. The initial grain temperature has a considerable effect on the future temperature of stored grain (Fig. 7). For an 8-m diameter steel bin filled with wheat, it takes 225 d to cool the center of the bin below 20°C when the initial temperature is 25°C and 323 d when the initial temperature is 35°C (Table 1).

To take advantage of the cold winter temperatures of the Canadian prairies for controlling pest populations in grain and for increasing grain storage life, small diameter bins (< 4 m) should be used when there is no mechanical ventilation of the grain. In subtropical and tropical climates, cooled grain should be stored in large diameter bins (> 8 m). In subtropical and tropical climates, because of the high initial temperature of the grain and high ambient air temperatures, it is impossible to cool grain for safe storage in nonventilated bins.

Yaciuk et al. (1975a) reported the effect of bin wall material on the grain temperatures in Winnipeg, Canada (Table 1). With regards to grain temperature, insulated galvanized steel is the worst material and galvanized steel is the second worst material to use for bin walls. White-painted steel or wood, and concrete are equally effective for storage of grain. In the subtropical region of Brazil, galvanized steel is also the worst material (Fig. 8) (Ferreira and Muir 1981).

Table 1 Number of Days It Takes the Center of a Bin in Winnipeg, Canada to Cool Below 20°C

Diameter (m)	Initial temperature (°C)	Steel wall	Steel wall painted white	Insulated steel wall	Insulated steel wall painted white	Wood wall painted white	Wood wall painted red	Concrete wall	Black rubber wall
2	25	45	26	—	—	26	29	25	31
	35	49	35	—	—	35	38	32	39
4	25	90	69	99	78	70	73	68	74
	35	112	95	125	109	95	98	90	99
6	25	150	129	161	140	129	132	126	134
	35	192	175	211	192	175	178	168	178
8	25	225	203	241	217	203	206	198	207
	35	323	283	523	314	284	289	273	290
10	25	329	295	356	316	296	300	288	301
	35	652	498	691	561	501	529	491	542
12	25	553	422	618	756	423	432	414	435
	35	978	719	1046	768	720	740	715	748
16	25	1026	797	1094	840	798	820	798	830
	35	1726	1283	*	1363	1285	1333	1292	1355
20	25	1637	1299	1712	1305	1251	1289	1260	1306
	35	*	*	*	*	*	*	*	*

— Indicates a bin system that was not simulated
* Indicates that the bin center does not drop below 20°C during the first 5 yr of storage.
Source: Yaciuk et al. (1975a).

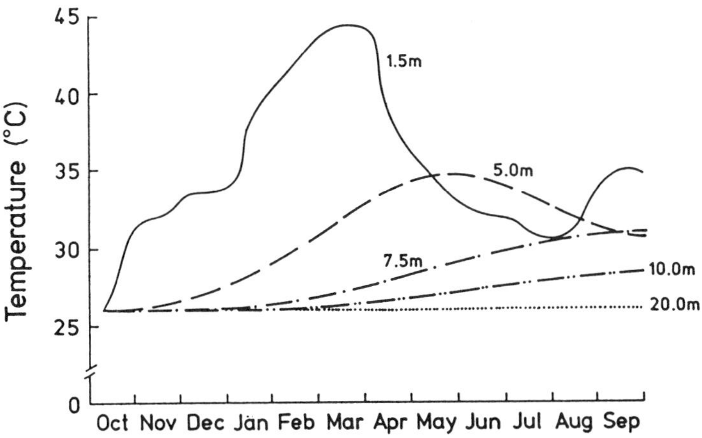

Figure 6 Effect of diameter on simulated temperatures at the center of metal silos of maize in Sao Paulo, Brazil. *Source*: Ferreira and Muir (1981).

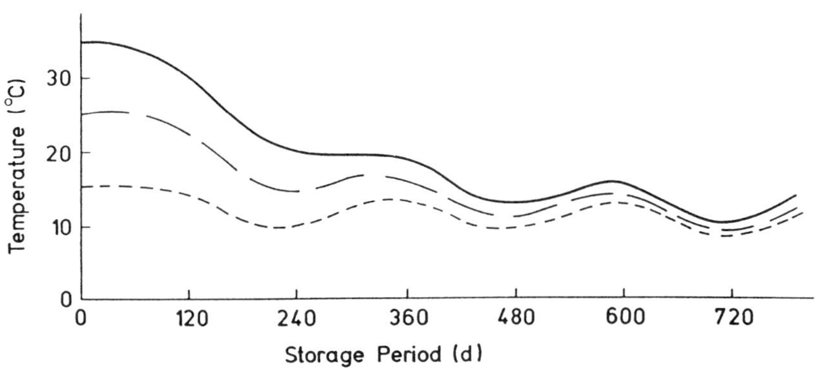

Figure 7 Effect of initial grain temperatures on the simulated temperatures at the center of an 8-m diameter steel bin at Winnipeg, Canada. *Source*: Yaciuk (1973).

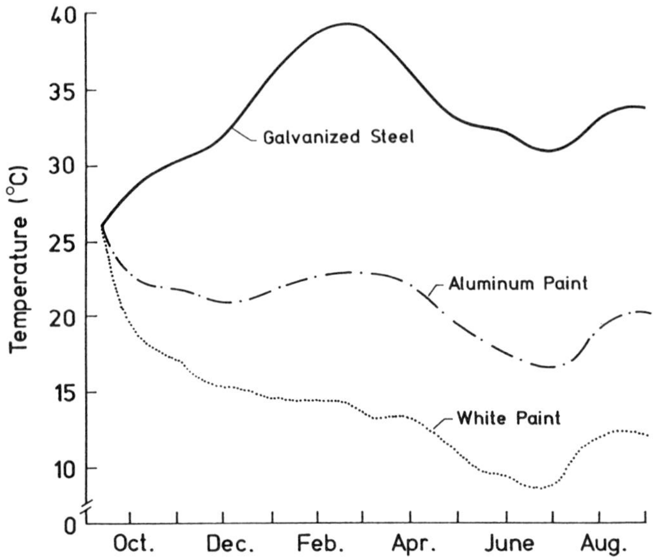

Figure 8 Effect of bin wall surface on the predicted temperatures 0.5 m from the wall of a 5-m diameter metal silo in Sao Paulo, Brazil. *Source*: Ferreira and Muir (1981).

Galvanized steel has a high absorptivity for solar radiation (about 0.89) and a low emissivity at bin wall temperatures (about 0.23) whereas for white paint these properties are about 0.18 and 0.95, respectively, thus making it a good covering material for bin walls. Because a black wall with both absorptivity and emissivity of 1.0 is as good as a white-painted wall (Table 1), it appears that the emissivity of bin wall materials at atmospheric temperatures is the predominant factor. The black wall, however, will have the highest wall temperature when solar radiation is striking it and will go through the greatest temperature change between night and day.

Solution of heat transfer problems in stored grain for one- (e.g., Converse et al. 1973) or two-dimensional (Muir et al. 1980) configurations does not consider the variable heating of the bin wall and inclined roof by solar radiation. A three-dimensional solution (Alagusundaram et al. 1990b) of heat transfer in a rapeseed bin shows that the temperature difference between points one-half radius from the center towards the south and north can be up to 15°C (Fig. 9). The difference decreases as the bin diameter increases. For a 6-m diameter bin filled with rapeseed to a depth of 4 m, the difference is 8°C, which may provide a better environment for arthropod and microorganism activity on the south side of a grain mass than on the north side in the Northern Hemisphere. Buscher-

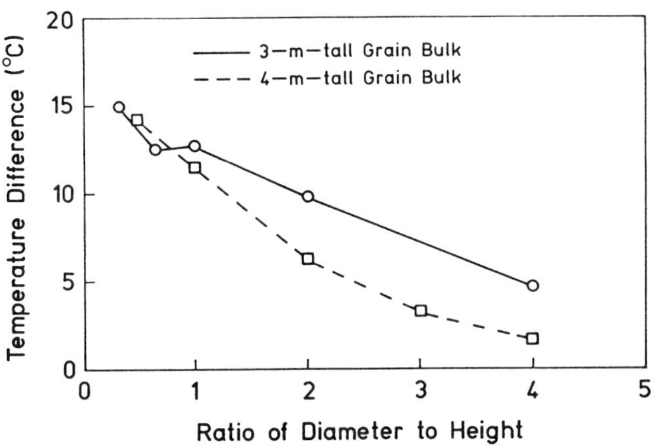

Figure 9 Average absolute temperature differences between points one-half radius from the center towards the south and north in a bin filled with rapeseed at Winnipeg, Canada. The temperatures were predicted for rapeseed initially at 20°C for a 12-mo period beginning January 1, 1974 using a 3-D finite element model. *Source*: Alagusundaram et al. (1990b).

mohle et al. (1988) measured a temperature difference of 7°C between north and south portions of corn stored in a 6.4-m diameter bin at Clemson, South Carolina. The results of these studies suggest that in northern temperate climates, because of rapid biological activity due to high temperature, grain should be monitored on the south side of the mass rather than on the north side.

For long-term maintenance of grain reserves, dry grain must be cooled and maintained at temperatures below 0°C. Such reserves can be used when famine occurs throughout the world due to a shortage of food. The published simulation models can be used for prediction of grain temperatures at various locations in the world to assess the suitability, from the temperature viewpoint, of the locations for grain reserves. Yaciuk et al. (1975b) simulated the temperature of grain stored in large bins at four locations in Canada (Fig. 10). As expected, Churchill, Manitoba located at a latitude near 60° maintains the lowest grain temperatures.

Temperature gradients in grain can cause convection currents and thus influence the movement of heat in the grain mass. The effects of convection currents were modeled by Muir et al. (1980) for heat transfer in radial and axial directions in cylindrical bins. They concluded that the inclusion of a model of convection currents does not result in more accurate predictions of temperature but computer time increases by a factor of about 25 over a conduction model

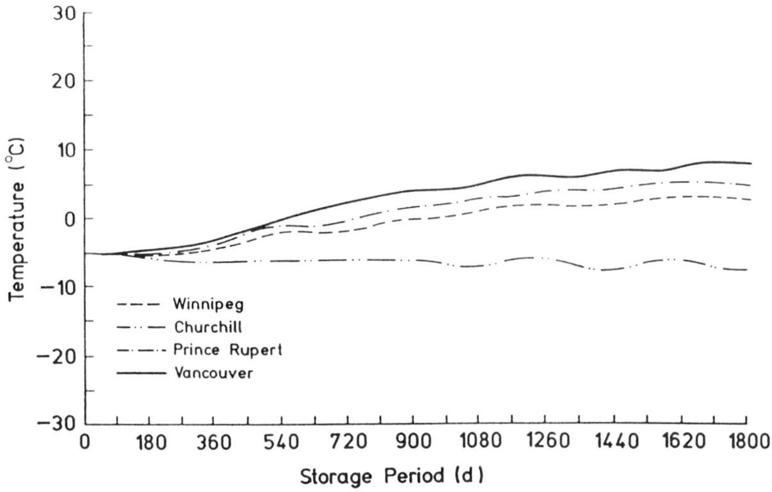

Figure 10 Predicted temperatures at the centers of 12-m diameter concrete bins at four locations in Canada. *Source*: Yaciuk et al. (1975b).

only. Smith and Sokhansanj (1990a) showed that for convection to affect the temperature of a porous medium the Rayleigh number must be large. They concluded that for small cereals, such as wheat, the Rayleigh number is not large enough to influence temperatures due to convection and thus confirmed the findings of Muir et al. (1980). The heat transfer in stored grain, therefore, can be modeled as a conduction problem with appropriate boundary conditions.

4.2. Moisture Transfer Models

4.2.1. Mechanisms of Moisture Movement

The movement of moisture in stored grain can occur by:

1. diffusion of moisture through interparticle contact,
2. leakage of water or snow through openings in the structure's roof and wall,
3. exchange of water vapor with the atmospheric air at the grain surface,
4. diffusion of moisture due to vapor gradients in the bulk, and
5. translocation of moisture due to convection currents.

The moisture transfer through interparticle contact is very slow. If the storage structure is well designed and properly constructed and maintained the leakage of moisture through openings can be considered negligible and then moisture movement due to the first three factors can be omitted. Under isothermal conditions, vapor diffusion through the intergranular spaces of the grain bulk also is a slow process. Under nonisothermal conditions, the convection currents caused by temperature gradients dominate the moisture movement in grain. Most of the published literature on moisture transfer in grain deals with nonisothermal storage of grain.

4.2.2. Moisture Movement under Isothermal Conditions

Only limited studies on the movement of moisture through bulk grain under isothermal conditions have been reported in the literature (Pixton and Griffiths 1971; Thorpe 1981; Thorpe et al. 1991a, 1991b). Considerable research has been done on the transfer of moisture from individual kernels to the surrounding air for prediction of drying and wetting of grain. These studies are considered beyond the scope of this chapter (e.g., Luikov 1975, Kumar et al. 1982, Parry 1985, Gencturk et al. 1986, Syarief et al. 1987, Walton et al. 1988, Irudayaraj et al. 1988, Parti and Dugmanics 1990).

A mathematical model of moisture movement in a bulk of grain can be used to seek to predict the time required for equilibration of moisture content in a bin filled with many loads of grain at different moisture contents. If the grain temperature is uniform, then moisture movement will be due to vapor diffusion through intergranular spaces of the bulk. Pockets of high or low temperatures in the absence of convection currents can also contribute to vapor diffusion in a bulk. The relative humidity of the surrounding air at a point in the bulk is in equilibrium with the grain and can be calculated using an equilibrium moisture isotherm for the grain. Using the temperature and equilibrium relative humidity in psychrometric relationships, the moisture content of the air can be calculated at various points in the grain and expressed as the mass of water per unit volume. The gradient in the moisture content of air or vapor pressure is the driving force for movement of moisture through the pore spaces of the granular material.

In an experiment, Pixton and Griffiths (1971) embedded small parcels (120 g) of wet wheat at 22.1% mc (wb) in 500-g lots of dry wheat at 13.7% mc at constant temperatures of 5 and 22.5°C. The rate of moisture change was slower at the lower temperature. At 22.5°C the moisture change after 65 d had stopped, but at 5°C the rate of moisture change was still proceeding slowly at up to 140 d when the experiment was discontinued, suggesting the slowness of the moisture diffusion in bulk grain. Because of hysteresis, complete equality of moisture content of the "wet" and "dry" grain was not established.

In another experiment, Pixton and Griffiths (1971) immersed 8.7% mc wheat placed in 3-cm diameter and 19-cm long tubes, open at the top only, in a large mass of wheat of uniform moisture content (17.5% wb), which provided atmospheres of constant relative humidities of 65 and 69.5% at 5 and 22.5°C, respectively, and measured the adsorption of moisture by the wheat in the tubes. To measure desorption of moisture, wheat in the tubes was at 18.1% and relative humidity was controlled by wheat at 8.7% mc. By solving the diffusion equation analytically for a one-dimensional case using the procedure given by Crank (1956), they calculated the diffusion coefficient for moisture. The diffusion coefficient was approximately 2.5×10^{-4} mm^2/s at 5°C and 8.0×10^{-4} mm^2/s at 22.5°C compared with the diffusion coefficient of water vapor in air of 23.9 mm^2/s at 8°C (Weast 1970). The transfer of water vapor through air is approximately 50 000 times faster than through intergranular space. Thorne (1981) and Thorpe et al. (1991a, 1991b) gave theoretical derivations for estimation of diffusion coefficients of water vapor through hygroscopic materials such as grain and explained the diffusion processes of Pixton and Griffiths (1971). Because moisture diffusion through the intergranular space of bulk grain is negligible compared with other transfer mechanisms, only a few mathematical models have been developed for moisture transfer by diffusion only.

4.2.3. Combined Heat and Moisture Transfer Models

Considerable literature exists on mathematical models of natural convection in inert porous media (e.g., Carberry and Bretton 1958, Holst and Aziz 1972, Combarnous and Bories 1975, Ribando and Torrance 1976, Hardee and Nilson 1977, Eckert and Faghri 1980, Chan and Banerjee 1981, Beukema et al. 1983, Close 1983, Close and Peck 1986, Prat 1986, Davidson 1986). These studies provide a theoretical understanding of simultaneous heat and moisture transfer in porous media but the results are not directly related to stored grain.

Many mathematical models have been developed for the prediction of moisture migration in stored grain due to temperature gradients (Williams 1973; Matouk 1976; Scott and Shipp 1978; Beukema 1980; Thorpe 1982; Tanaka and Yoshida 1983; Nguyen 1986a; Thorpe and Nguyen 1986; Casada and Young 1987; Dona and Stewart 1988; Smith and Sokhansanj 1990a, 1990b; Khankari et al. 1990; Uiso et al. 1991, Freer et al. 1990). Limited validation of the predicted moisture migration has been reported in the literature. Many experimental studies have been conducted for quantification of moisture migration in grain bulks (Muir et al. 1973; Stewart 1975; Gough 1985; Gough et al. 1987a, 1987b, 1990, 1991; Reed 1992). Most of the researchers have attempted to explain their experimental observations using simple calculations based on the fundamentals of physical sciences and fluid mechanics.

Thorpe (1982) and Khankari et al. (1990) have simulated moisture redistribution within the grain initially at a uniform moisture content and held between plates at constant temperatures. Thorpe (1982) and Khankari et al. (1990) compared their simulated results with the experimental data of Griffiths (1964) for

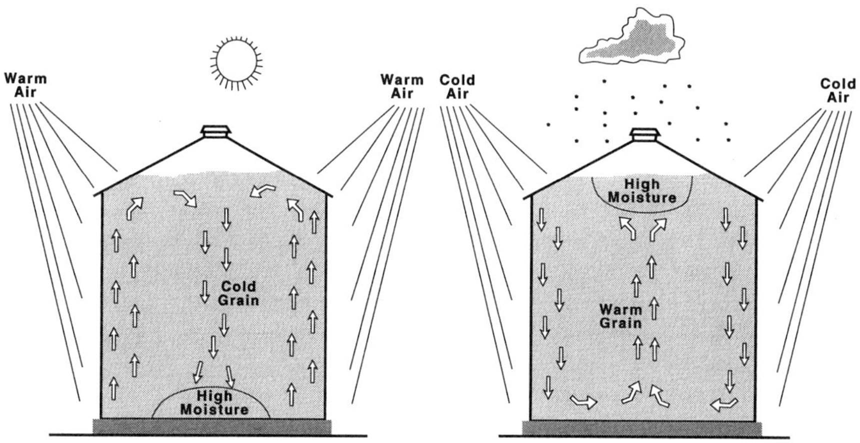

Figure 11 Moisture migration in stored grain: (left) when outdoor temperatures are greater than the grain temperatures; (right) when outdoor temperatures are lower than the grain temperature.

wheat initially at 11.9% mc (wb) between two plates at 25 and 35°C, separated by 0.203 m. The simulated and measured moisture contents near the cold plate were 14.4% (wb) and near the hot plate were 10.1% (wb) after 7 mo of storage. Stewart (1975) studied moisture movement by creating a temperature gradient in a glass jar (305-mm diameter) filled to a depth of 356 mm with whole, ground, or rolled corn at 30% mc (wb), and showed that both convection and diffusion are the driving mechanisms for moisture migration. Convection, however, is the major mechanism for moisture migration in both low and high moisture grains.

Many books (e.g., Brooker et al. 1974, Sauer 1992) and extension publications (e.g., Friesen and Huminicki 1987) have used a figure similar to Figure 11 for the explanation of the moisture migration in stored grain. Most of the mathematical models have attempted to simulate this pattern of moisture migration and have predicted moisture increases of up to 2 percentage points. Figure 11, although not based on experimental data, explains the moisture accumulation near the top surface when outdoor temperatures are lower than grain temperatures, as observed by Muir et al. (1973), Gough (1985), and Gough et al. (1987b). Gough et al. (1990) indicated that the shape of the convection current does not have to be toroidal. Air movement should be downward near the wall and upward in the central column without forming a closed loop. The air rising in the central column would disperse in the headspace by diffusion after releasing its moisture to the grain. The same air may not necessarily return near the wall.

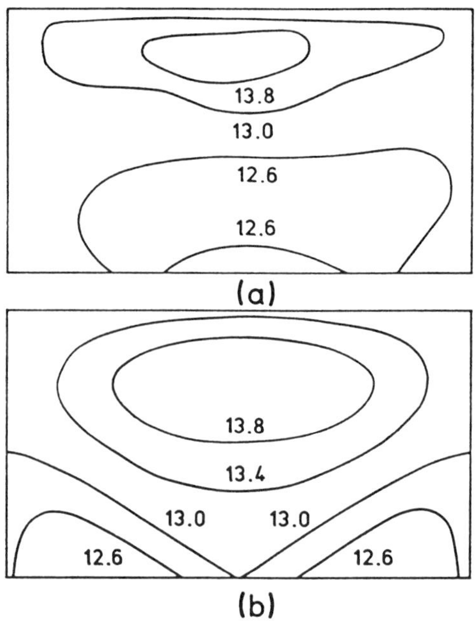

Figure 12 The distribution of moisture content (wet basis) in a cylindrical bin at Hutchinson, Kansas after 30 mo of storage: (a) experimental results; (b) simulation results. *Source*: Smith and Sokhansanj (1990b).

Similarly for moisture migration when outdoor temperatures are higher than the grain temperatures, air currents may not form closed loops in the grain or headspace. Although the isotherms based on measured temperatures in a bin were nearly symmetrical about the vertical axis in the north-south vertical plane, Gough et al. (1987b) measured up to 5 percentage points increase in moisture content on the north side of the bin in Seoul, Korea filled with wheat at 13–15% mc. Such an increase on one side of the bin cannot be explained by convection or diffusion models. Reed (1992) measured moisture movement during the autumn and winter in three round metal bins containing about 270 t of wheat per bin located on farms in Kansas. His results indicate that moisture accumulated more or less uniformly at the surface of naturally cooling grain masses but not at the center surface as depicted in Figure 11. Gough et al. (1987a, 1990) experimentally measured the air velocities using a laboratory unit simulating the process in Figure 11 and a 250-t bin, 7.4-m diameter by 7.6-m high to the eaves. The measured and predicted air velocities compared well in both situations and the magnitude ranged between 2×10^{-4} to 6×10^{-4} m/s (0.72 to 2.16 m/h) depending on the temperature gradients. In a full-size bin located in Njombe, Tanzania, the average measured increase in the top 100-mm layer of grain was 1.3 percentage points after 3 mo storage (Gough et al. 1990).

Smith and Sokhansanj (1990b) compared their simulated results with the measured results of Schmidt (1955). The simulation was continued for 30 mo for a bin, 4.3 m in diameter, filled with wheat at 13% mc (wb) to a depth of 2.4 m. Both measured and simulated results indicated an increase in moisture content near the top center of the grain mass (Fig. 12).

Nguyen (1986a, 1987) and Thorpe and Nguyen (1986) have reported simulation results for many rectangular and bunker type storages. They, however, have reported no experimental validation. For a rectangular bin (Fig. 13) and a bunker type storage (Fig. 14), the streamlines do not cross the top surface, indicating that the grain surface is covered by an impermeable membrane at the surface. For air rising up in the central column of grain with an open top surface, it should be easier to go straight up into the headspace rather than bend sideways to create a toroidal current. In the rectangular bin, two percentage points increase in moisture content is predicted in the top layer of the grain and about one percentage point decrease in moisture content in the bottom layer of the bin (Fig. 13). In the bunker silo, moisture is transferred to the region under the ridge from the two lower regions along each side (Fig. 14). The accumulation of moisture in the top region is consistent with observations in bunker storages where caked and molded masses of grain have been found along the ridge (Nguyen 1986a).

In summary, there is a need to conduct well-planned experiments to measure and explain moisture migration in stored grain and to develop mathematical models of the observed behavior in stored grain.

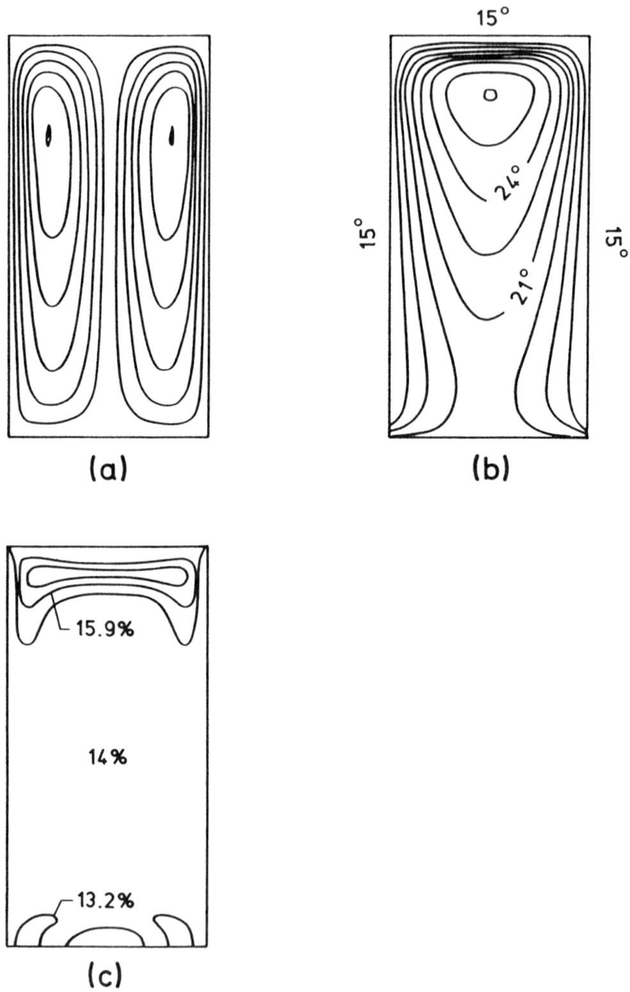

Figure 13 Simulated moisture and heat movement in a rectangular bin 6-m high by 3-m wide filled with corn initially at 30°C and 14% mc and exposed to an environment at 15°C for 75 d: (a) streamlines; (b) isotherms; and (c) moisture isosteres. *Source*: Thorpe and Nguyen (1986).

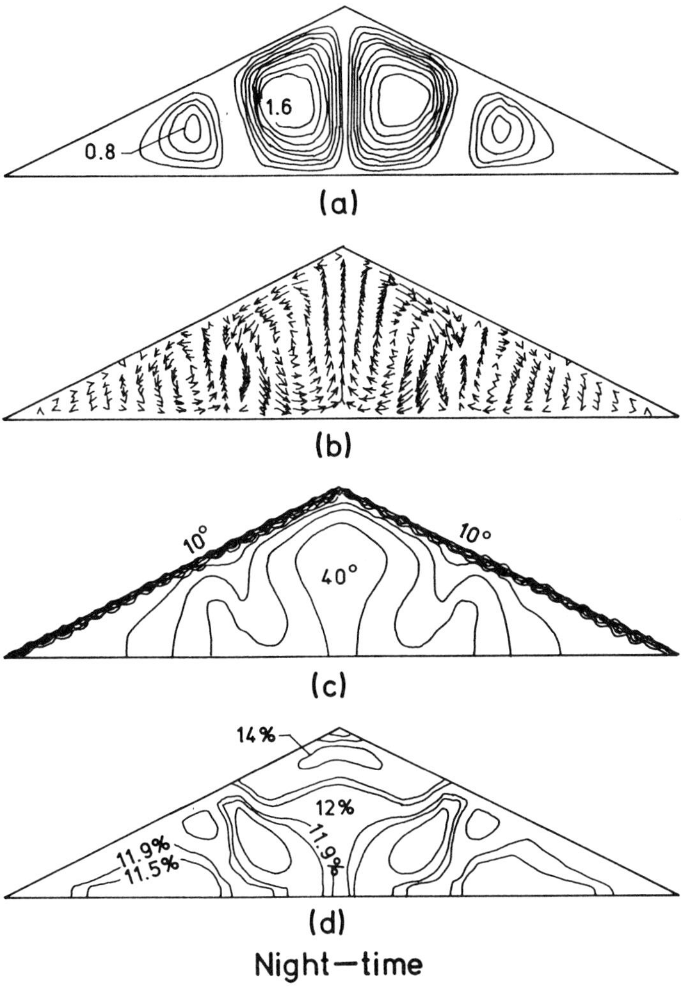

Figure 14 Simulated moisture and heat movement in a bunker silo filled with grain initially at 40°C and 12% mc and exposed to an environment at 10°C for 12 mo: (a) streamlines; (b) vector plot; (c) isotherms; and (d) moisture isosteres. *Source*: Nguyen (1986a).

4.3. Gas Transfer Models

4.3.1. Need for Gas Transfer Models

Modified atmospheres with elevated CO_2 or N_2 and depleted O_2, or synthetic fumigants can be used to disinfest stored grain of insects and mites. The success of modified atmospheres and fumigants in killing stored-product pests depends on the uniform distribution of the gas in the stored grain and maintenance of a lethal level of the gas for the required exposure period. The gas is usually introduced in the grain mass at one or more points. Mathematical models can be used to decide the number and location of the introduction points. A mathematical model of CO_2 movement from spoiling grain can be used for selection of the best location and resolution of instruments required for incipient spoilage detection. Although these needs exist, only a limited effort has been made to develop mathematical models of gas movement in stored grain.

4.3.2. Possible Assumptions in Gas Transfer Models

To model gas movement in cylindrical grain bulks, the mathematical model can be solved as an axisymmetric problem by assuming no CO_2 concentration gradient in the circumferential direction. This assumption would not be valid if spoilage was occurring at a point located nonsymmetrically from the central axis. For such situations, a three-dimensional solution would be required.

The amount of CO_2 generated by uninfested dry grain can be assumed to be zero. This assumption would not be valid if the grain was at high moisture content and nonuniform temperatures. In such situations a numerical solution could accommodate the varying amounts of CO_2 generated as a function of moisture content and temperature at various locations in the bin. Similarly, the amount of CO_2 absorbed by grain (also dependent on temperature and moisture content) can be handled in the model.

4.3.3. Potential Boundary Conditions for Gas Transfer Models

The top surface of grain is open to atmospheric air under a ventilated roof. The CO_2 leaving the grain surface will diffuse rapidly into the headspace air, CO_2 concentration at the top surface can be taken as constant at the atmospheric level (0.6 g/m^3). If the bin does not have significant ventilation because of no wind and no openings between roof and wall, the measured CO_2 concentration either as a constant or as a function of time could be used at the top surface. The top surface could be covered using an impermeable sheet resulting in zero flow across the boundary.

The walls usually have small openings in a bin. Gas would be lost to the surroundings through these openings. Two extreme boundary conditions are: 1) the wall is impermeable to the flow of the gas, and 2) the wall is fully permeable to the gas. Most storage bins probably would be between the two extremes tending towards the impermeable bin-wall condition. The permeable wall would provide a conservative boundary condition because it would predict greater loss to the surroundings. The movement of gas through the bottom can be assumed as zero in most bins because bin floors are usually made of concrete or steel; for bins with a concrete bottom, some adsorption may occur.

Small bins in developing countries are usually made of mud, reinforced with straw or bamboo. For such bins, boundary conditions at the wall and floor may be difficult to predict but can be simulated as discussed earlier. For welded steel bins, walls can be assumed impermeable to the flow of gases.

4.3.4. Potential Initial Conditions for Gas Transfer Models

The initial concentration of CO_2 in most cases can be assumed to be equal to atmospheric concentrations. For wet grain stored at high temperatures, measured CO_2 concentrations at various locations in the bin can be used as initial conditions.

4.3.5. Simulation Results

Singh (Jayas) et al. (1983) modeled the diffusion of CO_2 from three separate pockets of deteriorating grain in a circular steel bin, 6.0 m in diameter filled with wheat to a depth of 4.6 m (Fig. 15). The pocket of spoiling grain was assumed to be a disk 0.3 m deep and 1 m in diameter and producing CO_2 at a rate of 0.006 $g \cdot h^{-1} \cdot kg^{-1}$ of wheat (White et al. 1982a). Top and bottom locations of the spoiling grain represent the high moisture areas of grain that can occur in the bin due to moisture migration (Fig. 11) and thus act as the source of CO_2.

Singh (Jayas) et al. (1983) assumed that the spoiling pocket was centrally located in the bin, the sorption and generation of CO_2 was zero outside the spoiling disk, the grain mass was at uniform temperature outside the spoiling pocket, CO_2 concentration was set at atmospheric level on the top surface, the bin floor was impermeable to the flow of CO_2, and the bin wall was either permeable or impermeable to the flow of CO_2. Using simulation results they concluded that in the absence of prior knowledge of the location of potential deterioration, a sensor capable of detecting CO_2 levels of 2 g/m^3 should be located near the center of the bin. This finding was later confirmed by Muir et al. (1985) using experimental data from farm bins in Manitoba, Saskatchewan, and Minnesota.

Jayas et al. (1988) solved the CO_2 diffusion model for an axisymmetric configuration and predicted the CO_2 concentrations in a 5.8-m diameter bin filled with wheat to a depth of 4.9 m. The CO_2 was injected at a point along

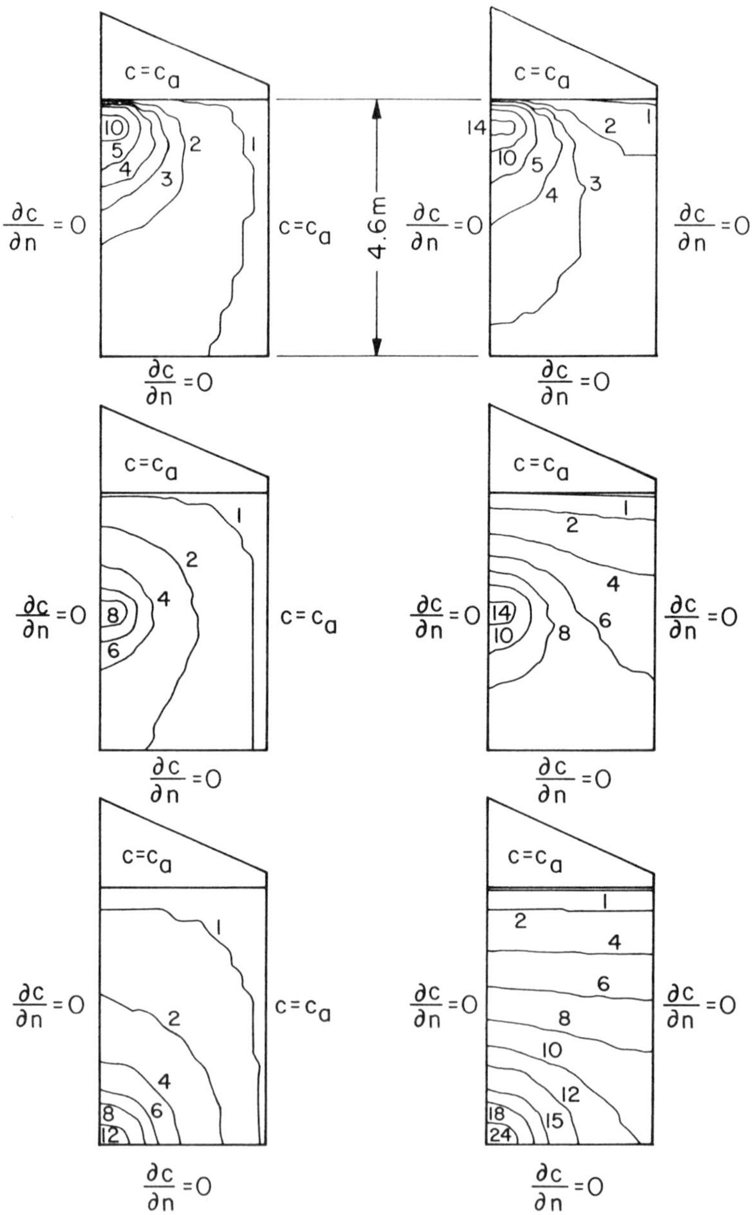

Figure 15 Lines of equi-concentration of CO_2 (g/m^3) in a 6-m diameter bin predicted to occur for three locations of the spoilage pocket (0.3 m, 2.0 m, and 4.3 m from the top surface of grain 4.6 m deep). Left-side figures are for the bin wall permeable to CO_2; right-side figures are for the bin wall impermeable to CO_2. *Source*: Singh (Jayas) et al. (1983).

the central axis of the bin at 1 m below the grain surface at a flow rate of 77.4 L/min for 1 h. The predicted and measured CO_2 concentrations were compared at various locations in the bin 23 h after injection. In the bottom two-thirds of the bin, CO_2 concentrations predicted by the model (with an impermeable wall) were higher than the measured data. They concluded that the model predictions could be improved by simulating some leakage through the bin wall.

Alagusundaram et al. (1991) solved the diffusion equation for a three-dimensional configuration under isothermal conditions. The predicted CO_2 concentrations were compared with experimental data on the movement of CO_2 through wheat contained in 1.45-m diameter model bins equipped with three different floor openings for the introduction of CO_2. The CO_2 was introduced as dry ice under the floor openings. The predicted CO_2 concentrations were lower than the measured data. The initial expansion of the CO_2 and displacement of the intergranular air caused the deviations.

Nguyen (1986b) developed an idealized mathematical model of buoyancy-driven circulation of air due to temperature gradients in grain stores. He predicted the circulation patterns in five situations of sealed storage structures: two parallelopiped-shaped structures, two horizontal structures and one bunker storage (Fig. 16). The grain was assumed to be initially at a uniform temperature of 30°C. Solar heating on the indicated wall was simulated by assuming a specified temperature of 42°C for 10 h/day. The boundary condition for the rest of the day and type of grain were not specified in the paper. Based on the simulated results he concluded that chemical fumigants should be introduced along the hot wall of a single-cell structure and under both roofs for a double-cell structure. The north-south bunker was the most effective for distribution of the chemical fumigants.

Hunter (1986) presented a method of calculating the time required for a fumigant to reach certain locations in a bin, when fumigants are distributed by forced air introduced through various duct configurations. The method can also be used when air is recirculated (e.g., Navarro et al. 1986).

No other mathematical models of movement of gases in bulk grain have been reported in the literature. Many experimental investigations (e.g., Bond et al. 1977, Gilby 1983, Banks and Annis 1984, Boland 1984, Zettler et al. 1984, Navarro et al. 1986, Banks 1990) have been conducted to study the distribution of gases and chemical fumigants in a stored bulk of grain.

5. FUTURE RESEARCH

In a comprehensive model of stored-grain ecosystems, equations of heat, moisture, and gas (CO_2) transfer should be solved simultaneously for three-dimensional configurations. The comprehensive model could be used for the prediction

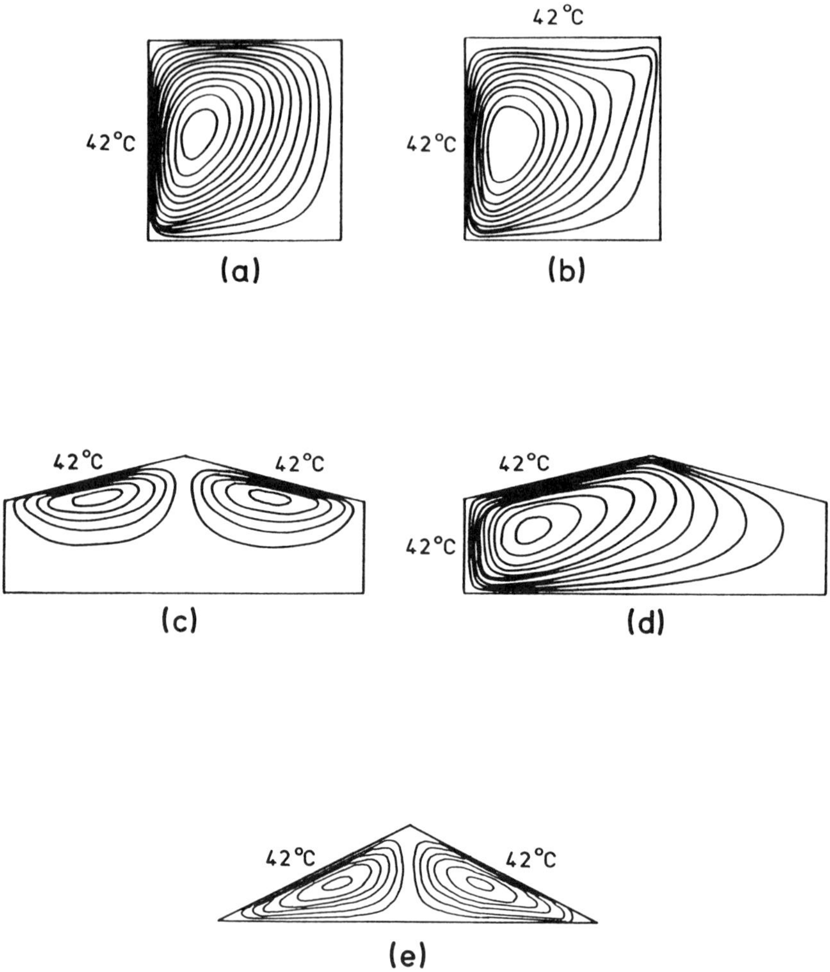

Figure 16 Circulation patterns, in various stores filled with grain initially at 30°C, due to heating by solar radiation represented by 42°C specified temperature for 10 h per day along the indicated boundaries: (a) store heated from a side; (b) store heated from side and top; (c) north-south oriented horizontal store; (d) east-west oriented horizontal store; and (e) north-south oriented bunker store. *Source*: Nguyen (1986b).

of favorable conditions for insect infestation outbreaks, potential for hot spot development, and evaluation of various design and management alternatives for safe storage of grain. There is a need for the development and validation of a more elaborate model of moisture migration in stored grain. The CO_2 transfer model must incorporate the effects of convection currents due to temperature gradients. Predictions of temperature, moisture, and gas concentration must be integrated with the biomodels of stored grain. The biomodels of interest are: moisture, heat, and CO_2 production models for grain, microorganisms, insects, and mites; CO_2 sorption model for grain; and the population dynamics models for insects, mites, and fungi under the range of varying conditions of temperature, moisture, and gas composition that are expected to occur in the bin. Experimental studies to measure the diffusion coefficient of CO_2 and other fumigants as affected by kernel orientation, moisture content, foreign material in grain, compaction, and gas concentration gradient are needed.

All environmental factors that affect the stored-grain ecosystem are random in nature. All the published models are deterministic. A challenge exists to model the transfer processes in the stored-grain ecosystem as stochastic processes, which should result in more realistic results. The developed model, however, may become too complex and lose its practical utility.

Banks (1986) reviewed the literature related to the models of sorption and desorption of fumigants by grain. He concluded that the quantitative data on the rate and magnitude of sorption-desorption on grains are fragmentary and there is a need for further research to provide full and consistent information for use in the mathematical models of fumigant transport within the grain and the loss of fumigant to the atmosphere (Banks et al. 1986a, 1986b). Similar well-planned research needs to be conducted for atmospheric gases such as CO_2.

6. REFERENCES

Alagusundaram, K., D.S. Jayas, N.D.G. White, and W.E. Muir. 1990a. Finite difference model of three-dimensional heat transfer in grain bins. Can. Agric. Eng. 32:315-321.

Alagusundaram, K., D.S. Jayas, N.D.G. White, and W.E. Muir. 1990b. Three-dimensional, finite element, heat transfer model of temperature distribution in grain storage bins. Trans. ASAE (Am. Soc. Agric. Eng.) 33:577-584.

Alagusundaram, K., D.S. Jayas, W.E. Muir, N.D.G. White and R.N. Sinha. 1991. A finite element model of three-dimensional CO_2 diffusion in grain bins. Paper 91-6558. Am. Soc. Agric. Eng., St. Joseph, MI. 13 pp.

Aldis, D.F. and G.H. Foster. 1977. Moisture changes in grain from exposure to ambient air. Paper 77-3524. Am. Soc. Agric. Eng., St. Joseph, MI. 29 pp.

Banks, H.J. 1986. Sorption and desorption of fumigants on grains: mathematical description. pp. 179-193 in: Pesticides and humid tropical grain storage systems, B.R. Champ and E. Highley (eds.). ACIAR, Proc. No. 14, Aust. Centre Int. Agric. Res., Canberra, Australia.

Banks, H.J. 1990. Behavior of gases in grain storages. pp. 96-107 in: Fumigation and controlled atmosphere storage of grain, B.R. Champ, E. Highley, and H.J. Banks (eds.). ACIAR, Proc. No. 25, Aust. Centre Int. Agric. Res., Canberra, Australia.

Banks, H.J. and P.C. Annis. 1984. Importance of processes of natural ventilation to fumigation and controlled atmosphere storage. pp. 299-323 in: Controlled atmosphere and fumigation in grain storages, B.E. Ripp (ed.). Elsevier Sci. Pub., New York.

Banks, H.J., A.K. Sharp, and A.R. Irving. 1986a. Influence of transport on gas loss from freight containers under fumigation. Part I: Experimental Investigation. Pestic. Sci. 17:207-220.

Banks, H.J., A.K. Sharp, and A.R. Irving. 1986b. Influence of transport on gas loss from freight containers under fumigation. Part II: Descriptive model. Pestic. Sci. 17:221-229.

Bathe, K.J. 1982. Finite element procedures in engineering analysis. Prentice Hall Inc., Englewood Cliffs, NJ. 735 pp.

Beek, W.J. and K.M.K. Muttzall. 1975. Transport phenomena. John Wiley & Sons, New York, NY. 298 pp.

Beukema, K.J. 1980. Heat and mass transfer during cooling and storage of agricultural products as influenced by natural convection. Centre Agric. Pub. and Doc., Wageningen, The Netherlands. 159 pp.

Beukema, K.J., S. Bruin, and J. Schenk. 1983. Three-dimensional natural convection in a confined porous medium with internal heat generation. Int. J. Heat Mass Transfer 26:451-458.

Bird, R.B., W.E. Stewart, and E.N. Lightfoot. 1960. Transport phenomena. John Wiley & Sons, New York. 780 pp.

Boland, F.B. 1984. Phosphine fumigations in silo bins. pp. 425-430 in: Controlled atmosphere and fumigation in grain storages, B.E. Ripp (ed.). Elsevier Sci. Pub., New York.

Bond, E.J., R.A. Sellen, and T. Dumas. 1977. Control of insects with phosphine in open-ended bin spouts. J. Econ. Entomol. 70:22-25.

Brodkey, R.S. and H.C. Hershey. 1988. Transport phenomena: a unified approach. McGraw-Hill, New York. 847 pp.

Brooker, D.B, F.W. Bakker-Arkema, and C.W. Hall. 1974. Drying cereal grains. AVI, Westport, CT. 265 pp.

Buschermohle, M.J., J.M. Bunn, and R.A. Spray. 1988. Moisture migration in stored grain. Paper 88-6508. Am. Soc. Agric. Eng., St. Joseph, MI. 11 pp.

Casada, M.E. and J.H. Young. 1987. Moisture migration during shipment of shelled peanuts. Paper 87-6512. Am. Soc. Agric. Eng., St. Joseph, MI. 15 pp.

Carberry, J.J. and R.H. Bretton. 1958. Axial dispersion of mass in flow through fixed beds. A.I.Ch.E. Journal 4:367-375.

Carslaw, H.S. and J.C. Jaeger. 1959. Conduction of heat in solids. Oxford Univ. Press, London. 510 pp.

Chan, Y.T. and S. Banerjee. 1981. Analysis of transient three-dimensional natural convection in porous media. ASME J. Heat Transfer 103:242-248.

Close, D.J. 1983. Natural convection with coupled mass transfer in porous media. Int. Commun. Heat Mass Transfer 10:465-476.

Close, D.J. and M.K. Peck. 1986. Experimental determination of the behavior of wet porous beds in which natural convection occurs. Int. J. Heat Mass Transfer 29:1531-1541.

Combarnous, M.A. and S.A. Bories. 1975. Hydrothermal convection in saturated porous media. Adv. Hydrosc. 10:231-307.

Converse, H.H., A.H. Graves, and D.S. Chung. 1973. Transient heat transfer within wheat stored in a cylindrical bin. Trans. ASAE (Am. Soc. Agric. Eng.) 16:129-133.

Crank, J. 1956. The mathematics of diffusion. Oxford Univ. Press, London. 347 pp.

Cunningham, R.E. and R.J.J. Williams. 1980. Diffusion in gases and porous media. Plenum Press, New York. 275 pp.

Davidson, M.R. 1986. Natural convection of gas/vapor mixtures in a porous medium. Int. J. Heat Mass Transfer 29:1371-1381.

Desai, C.S. and J.F. Abel. 1972. Introduction to the finite element method. Van Nostrand Reinhold Co., New York. 477 pp.

Dona, C.L.G. and W.E. Stewart Jr. 1988. Numerical analysis of natural convection heat transfer in stored high moisture corn. J. Agric. Eng. Res. 40:275-284.

Eckert, E.R.G. and M. Faghri. 1980. A general analysis of moisture migration caused by temperature differences in an unsaturated porous medium. Int. J. Heat Mass Transfer 23:1613-1623.

Ferreira, F.W. and W.E. Muir. 1981. Efeito de diametro e dos materias da parede de silo metalico na tepperatura de graos armazenados. (Effect of bin diameter and wall materials on the temperature of stored grain in metallic bins.) Rev. Bras. de Armaz. 6(2):11-18.

Freer, M.W., T.J. Siebenmorgen, R.J. Couvillion, and O.J. Loewer. 1990. Modeling temperature and moisture content changes in bunker-stored rice. Trans. ASAE (Am. Soc. Agric. Eng.) 33:211-220.

Friesen, O.H. and D.N. Huminicki. 1987. Grain aeration and unheated air drying. Agdex 732-1. Manitoba Dept. Agric., Winnipeg, MB. 30 pp.

Geankoplis, C.J. 1983. Transport processes and unit operations. Allyn and Bacon, Inc., Boston. 862 pp.

Gencturk, M.B., A.S. Bakshi, Y.C. Hong, and T.P. Labuza. 1986. Moisture transfer properties of wild rice. J. Food Process Eng. 8:243-261.

Gilby, A.R. 1983. Movement of halogenated fumigants through wheat. J. Stored Prod. Res. 19:199-202.

Gough, M.C. 1985. Physical changes in large scale hermetic grain storage. J. Agric. Eng. Res. 31:55-65.

Gough, M.C., C.B.S. Uiso, and C.J. Stigter. 1987a. Convection current in bulk grain. Trop. Sci. 27:29-37.

Gough, M.C., H.S. Cheigh, S.K. Kim, and T.W. Kwon. 1987b. Physical changes in stored bulk rice. J. Agric. Eng. Res. 37:59-71.

Gough, M.C., C.B.S. Uiso, and C.J. Stigter. 1990. Air convection current in metal silos storing maize grain. Trop. Sci. 30:217-222.

Gough, M.C., A.J.K. Bisbrown, and A.C. Brueton. 1991. The physical effects of tropical humid seasons on bag stacks of grain. Abs. No. 2780. Agric. Eng. Abs. 16(7):288.

Griffiths, H.J. 1964. Bulk storage of grain: A summary of factors governing control of deterioration. CSIRO Div. Mech. Eng. Report ED8, Melbourne, Australia. 49 pp.

Gutfinger, C. (ed.). 1975. Topics in transport phenomena. Hemisphere Pub. Corp., Washington, DC. 622 pp.

Hardee, H.C. and R.H. Nilson. 1977. Natural convection in porous media with heat generation. Nuclear Sci. Eng. 63:119-132.

Holst, P.H. and K. Aziz. 1972. Transient three-dimensional natural convection in confined porous media. Int. J. Heat Mass Transfer 15:73-90.

Howe, R.W. 1965. A summary of estimates of optimal and minimal conditions for population increase of some stored products insects. J. Stored Prod. Res. 1:177-184.

Hunter, A.J. 1986. Traverse times for gas flow through porous media for some particular geometries. J. Agric. Eng. Res. 35:11-23.

Incropera, F.P. and D.P. DeWitt. 1990. Fundamentals of heat and mass transfer. John Wiley & Sons, New York. 919 pp.

Irudayaraj, J., K. Haghighi, and R.L. Stroshine. 1988. Finite element simulation of grain temperature and moisture profiles. Paper 88-6514. Am. Soc. Agric. Eng., St. Joseph, MI. 11 pp.

Jay, E.G., L.M. Redlinger, and H. Laudani. 1970. The application and distribution of carbon dioxide in a peanut (groundnut) silo for insect control. J. Stored Prod. Res. 6:247-254.

Jayas, D.S., W.E. Muir and N.D.G. White. 1988. Modeling the diffusion of carbon dioxide in stored grain. Paper 88-6013. Am. Soc. Agric. Eng., St. Joseph, MI. 12 pp.

Jayas, D.S., B. Khangura, and N.D.G. White. 1991. Controlled atmosphere storage of grains. Postharvest News and Inf. 2:423-427.

Jayas, D.S., K. Alagusundaram, G. Shunmugam, W.E. Muir, and N.D.G. White. 1992. Simulation of temperatures in stored bulks of wheat using a three-dimensional finite element model. Paper 92-6527. Am. Soc. Agric. Eng., St. Joseph, MI. 21 pp.

Khankari, K.K., R.V. Morey, and S.V. Patankar. 1990. Moisture diffusion in stored grain due to temperature gradients. Paper 90-6581. Am. Soc. Agric. Eng., St. Joseph, MI. 17 pp.

Kumar, A., J.L. Blaisdell, and F.L. Herum. 1982. Generalized analytical model for moisture diffusion in a composite cylindrical body. Trans. ASAE (Am. Soc. Agric. Eng.) 25:752-758.

Lo, K.M., C.S. Chen., J.T. Clayton, and D.D. Adrian. 1975. Simulation of temperature and moisture changes in wheat storage due to weather variability. J. Agric. Eng. Res. 20:47-53.

Longstaff, R.A. and H.J. Banks. 1987. Simulation of temperature fluctuations near the surface of grain bulks. J. Stored Prod. Res. 23:21-30.

Luikov, A.V. 1975. Systems of differential equations of heat and mass transfer in capillary-porous bodies (review). Int. J. Heat Mass Transfer 18:1-14.

Matouk, A. M. 1976. Heat and moisture movements during low temperature drying and storage of maize grain. Unpublished Ph. D. Thesis, Dept. of Agric. Eng., University of Newcastle upon Tyne, United Kingdom. 217 pp.

Metzger, J.F. and W.E. Muir. 1983. Computer model of two-dimensional conduction and forced convection in stored grain. Can. Agric. Eng. 25:119-125.

Mills, A.F. 1992. Heat transfer. Irwin, Homewood, IL. 888 pp.

Muir, W.E. 1970. Temperatures in grain bins. Can. Agric. Eng. 12:21-24.

Muir, W.E., R.N. Sinha, and H.A.H. Wallace. 1973. Abiotic and biotic characteristics of grain stored in temporary farm bins. Can. Agric. Eng. 15:35-42.

Muir, W.E., B.M. Fraser, and R.N. Sinha. 1980. Simulation model of two dimensional heat transfer in controlled atmosphere grain bins. pp. 385-397 in: Controlled atmosphere storage of grains, J. Shejbal (ed.). Elsevier Sci. Pub. Co., Amsterdam.

Muir, W.E., D. Waterer, and R.N. Sinha. 1985. Carbon dioxide as an early indicator of stored cereal and oilseed spoilage. Trans. ASAE (Am. Soc. Agric. Eng.) 28:1673-1675.

Navarro, S., E.G., Jay, and J.G. Leesch. 1986. Recirculation rate requirements for adequate distribution of carbon dioxide in grain bins. Trans. ASAE (Am. Soc. Agric. Eng.) 29:1348-1354.

Nguyen, T.V. 1986a. Modeling temperature and moisture changes resulting from natural convection in grain stores. pp. 81-88 in: Preserving grain quality by aeration and in-store drying, B.R. Champ and E. Highley (eds.). ACIAR, Proc. No. 15, Aust. Centre Int. Agric. Res., Canberra, Australia.

Nguyen, T.V. 1986b. Movement of fumigants in bulk grain. pp. 195-201 in: Pesticides and humid tropical grain storage systems, B.R. Champ and E. Highley (eds.). ACIAR, Proc. No. 14, Aust. Centre Int. Agric. Res., Canberra, Australia.

Nguyen, T.V. 1987. Natural convection effects in stored grains—a simulation study. Drying Technol. 5:541-560.

O'Dowd, E.T., S.M. Kenneford, A.J.K. Bisbrown, and J. Farrington. 1988. A mathematical model, with cost implications, for predicting temperatures in seed stores. Bulletin No. 16, Overseas Development Natural Resources Institute, Chatham, Kent, United Kingdom. 30 pp.

Oxley, T.A. 1948. The scientific principles of grain storage. The Northern Publishing Co. Ltd., Liverpool, United Kingdom. 103 pp.

Parry, J.L. 1985. Mathematical modeling and computer simulation of heat and mass transfer in agricultural grain drying: a review. J. Agric. Eng. Res. 32:1-29.

Parti, M. and I. Dugmanics. 1990. Diffusion coefficient for corn drying. Trans. ASAE (Am. Soc. Agric. Eng.) 33:1652-1656.

Pixton, S.W. and H.J. Griffiths. 1971. Diffusion of moisture through grain. J. Stored Prod. Res. 7:133-152.

Prasad, D.C., W.E. Muir, and H.A.H. Wallace. 1978. Characteristics of freshly harvested wheat and rapeseed. Trans. ASAE (Am. Soc. Agric. Eng.) 21:782-784.

Prat, M. 1986. Analysis of experiments of moisture migration caused by temperature differences in unsaturated porous medium by means of two-dimensional numerical simulation. Int. J. Heat Mass Transfer 29:1033-1039.

Rameshbabu, M., D.S. Jayas, and N.D.G. White. 1991. Mortality of *Cryptolestes ferrugineus* (Stephens) adults and eggs in elevated carbon dioxide and depleted oxygen atmospheres. J. Stored Prod. Res. 27:163-170.

Reed, C. 1992. Grain cooling, moisture translocation, and theories of convection current. pp. 14-15 in: Extended abstracts: Int. Symp. Stored Grain Ecosystems, D. S. Jayas, N.D.G. White, W.E. Muir, and R.N. Sinha (eds.), Dept. Agric. Eng., Univ. Manitoba, Winnipeg, MB.

Ribando, R.J. and K.E. Torrance. 1976. Natural convection in a porous medium: effects of confinement, variable permeability, and thermal boundary conditions. ASME J. Heat Transfer 98:42-48.

Sarker, N.N. and O.R. Kunze. 1991. Finite element prediction of grain temperature changes in storage bins. Paper 91-6560. Am. Soc. Agric. Eng., St. Joseph, MI. 15 pp.

Sauer, D.B. (ed.). 1992. Storage of cereal grains and their products. Am. Assoc. Cereal Chemists, Inc., St. Paul, MN. 615 pp.

Schmidt, J.L. 1955. Wheat storage research at Hutchinson, Kansas and Jamestown, North Dakota. USDA Tech. Bull. 1113. Washington, DC. 98 pp.

Scott, C.J. and P.H. Shipp. 1978. Temperature and moisture variations in stored grains. pp. 851-869 in: Numerical methods in laminar and turbulent flow, C. Taylor, K. Morgan, and C.A. Brebbia (eds.). John Wiley & Sons, New York.

Segerlind, L.J. 1976. Applied finite element analysis. John Wiley & Sons, Inc., New York. 422 pp.

Segerlind, L.J. 1984. Applied finite element analysis (2nd Ed.). John Wiley & Sons, Inc., New York, NY. 427 pp.

Singh, R.P. and D.R. Heldman. 1984. Introduction to food engineering. Academic Press, Inc., Toronto, ON. 306 pp.

Singh (Jayas), D., W.E. Muir, and R.N. Sinha. 1983. Finite element modeling of carbon dioxide diffusion in stored wheat. Can. Agric. Eng. 25:149-152.

Singh (Jayas), D., W.E. Muir, and R.N. Sinha. 1984. Apparent coefficient of diffusion of carbon dioxide through samples of cereals and rapeseed. J. Stored Prod. Res. 20:169-175.

Singh, A.K. and G.R. Thorpe. 1993. A solution procedure for remodeling three-dimensional free convection flow in peaked bulks of grain. J. Stored Prod. Res. 29:221-235.

Sinha, R.N. 1973. Interrelations of physical, chemical, and biological variables in the deterioration of stored grains. pp. 15-47 in: Grain Storage: Part of a system, R.N. Sinha and W.E. Muir (eds.). AVI Pub. Co., Inc., Westport, CT.

Sinha, R.N. and F.L. Watters. 1985. Insect pests of flour mills, grain elevators, and feed mills and their control. Publication 1776. Research Branch, Agriculture Canada, Ottawa, ON. 290 pp.

Smith, G.D. 1985. Numerical solution of partial differential equations: finite difference method (3rd ed.). Oxford Univ. Press, Oxford. 337 pp.

Smith, H. and H.H. Jensen. 1989. Transport phenomena. Oxford Univ. Press, Oxford. 431 pp.

Smith, E.A. and S. Sokhansanj. 1990a. Natural convection and temperature of stored produce—a theoretical analysis. Can. Agric. Eng. 32:91-97.

Smith, E.A. and S. Sokhansanj. 1990b. Moisture transport caused by natural convection in grain stores. J. Agric. Eng. Res. 47:23-34.

Steele, J.L., R.A. Saul, and W.V. Hukill. 1969. Deterioration of shelled corn as measured by carbon dioxide production. Trans. ASAE (Am. Soc. Agric. Eng.) 12:685-689.

Stewart, J.A. 1975. Moisture migration during storage of preserved, high moisture grains. Trans. ASAE (Am. Soc. Agric. Eng.) 18:387-393, 400.

Syarief, A.M., R.J. Gustafson, and R.V. Morey. 1987. Moisture diffusion coefficients for yellow-dent corn components. Trans. ASAE (Am. Soc. Agric. Eng.) 30:522-528.

Tanaka, H. and K. Yoshida. 1983. Heat and mass transfer mechanisms in a grain storage silo. Proc. 3rd Int. Symp. on Eng. and Food, Dublin, Ireland. 8 pp.

Thorpe, G.R. 1981. Moisture diffusion through bulk grain. J. Stored Prod. Res. 17:39-42.

Thorpe, G.R. 1982. Moisture diffusion through bulk grain subjected to a temperature gradient. J. Stored Prod. Res. 18:9-12.

Thorpe, G.R. and T.V. Nguyen. 1986. The response of nonventilated grain to environmental temperature. pp. 230-242 in: Proc. 4th Int. Work. Conf. Stored-Product Prot., E. Donahaye and S. Navarro (eds.), Tel Aviv, Israel.

Thorpe, G.R., J.A.O. Tapia, and S. Whitaker. 1991a. The diffusion of moisture in food grains—I. The development of a mass transport equation. J. Stored Prod. Res. 27:1-9.

Thorpe, G.R., J.A.O. Tapia, and S. Whitaker. 1991b. The diffusion of moisture in food grains—II. Estimation of the effective diffusivity. J. Stored Prod. Res. 27:11-30.

Uiso, C.B.S., M.C. Gough, and C.J. Stigter. 1991. Detection, causes, and consequences of convection current in bulk stored grain. Abs. No. 4041. Agric. Eng. Abs. 16(10):416.

van der Houwen, P.J. 1968. Finite difference method for solving partial differential equations. Mathematisch Centrum, Amsterdam. 123 pp.

Walton, L.R., F.A. Payne, and I.J. Ross. 1988. Diffusion of moisture as a function of Fourier and Biot numbers. Trans. ASAE (Am. Soc. Agric. Eng.) 31:603-607.

Weast, R.C. 1970. Handbook of chemistry and physics. The Chemical Rubber Co., Cleveland, OH. pp. F-47.

White, G.G. 1988. Temperature changes in bulk stored wheat in subtropical Australia. J. Stored Prod. Res. 24:5-11.

White, N.D.G. and D.S. Jayas. 1991. Control of insects and mites with carbon dioxide in wheat stored at cool temperatures and in nonairtight bins. J. Econ. Entomol. 84:1933-1942.

White, N.D.G., R.N. Sinha, and W.E. Muir. 1982a. Intergranular carbon dioxide as an indicator of biological activity associated with the spoilage of stored wheat. Can. Agric. Eng. 24:35-42.

White, N.D.G., R.N. Sinha, and W.E. Muir. 1982b. Intergranular carbon dioxide as an indicator of deterioration in stored rapeseed. Can. Agric. Eng. 24:43-49.

White, N.D.G., D.S. Jayas, and R.N. Sinha. 1990. Carbon dioxide as a control agent for the rusty grain beetle (Coleoptera: Cucujidae) in stored wheat. J. Econ. Entomol. 83:277-288.

Williams, Jr., E.H. 1973. Moisture migration simulation in high moisture shelled corn. Unpublished M.Sc. thesis. Univ. Illinois, Urbana, IL. 53 pp.

Yaciuk, G. 1973. Temperatures in grain storage systems. Unpublished Ph.D. Thesis. Department of Plant Science, Univ. Manitoba, Winnipeg, MB. 171 pp.

Yaciuk, G., W.E. Muir, and R.N. Sinha. 1975a. A simulation model of temperatures in stored grain. J. Agric. Eng. Res. 20:245-258.

Yaciuk, G., W.E. Muir, and R.N. Sinha. 1975b. Selection of locations for long term storage of wheat. Paper 75-4057. Am. Soc. Agric. Eng., St. Joseph, MI. 20 pp.

Zettler, J.L., H.B. Gillenwater, L.M. Redlinger, J.G. Leesch, R. Davis, L.L. Mcdonald, and J.M. Zehner. 1984. Efficacy of perforated tubing in assisting phosphine distribution for in-transit fumigation of export corn. J. Econ. Entomol. 77:675-679.

Zienkiewicz, O.C. 1977. The finite element method. McGraw Hill, New York, NY. 787 pp.

15

Forced Air Distribution for In-Bin Drying and Aeration

Edward A. Smith

University of Paisley, Paisley, United Kingdom

SUMMARY

This chapter reviews the way in which airflow through grain beds has been analyzed using computer modeling. The basic equations which describe the flow are discussed, pointing out the assumptions used to develop the models and parameters which have to be determined experimentally for each type of grain and condition. Also described are the numerical and analytic methods which have been used to solve the airflow equations. Consideration is given to the extent to which the numerical solutions have been validated by comparing the results with experimental observations and analytically correct solutions. The comparison of measured and calculated values of airflow indicate that the resistance to flow is anisotropic with the air flowing horizontally more easily than vertically. The major areas in grain drying and storage where air distribution is important are also discussed. These topics include the models used to study the drying of grain beds where the airflow is nonuniform. Also reviewed is the work undertaken to optimize the operation of grain stores as well as the importance of maintaining grain quality in these calculations.

1. AIRFLOW MODELING

1.1. Introduction

This chapter reviews the use of computer modeling in the study of airflow distribution in beds of grain. The purpose of such modeling is to design systems which optimize the flow pattern so that the maximum amount of grain is dried or cooled as quickly as possible and with as low an airflow as possible.

The two equations used to model the flow of air through grain beds are:

$$\nabla p = -R\,\boldsymbol{u} \tag{1}$$

and

$$\nabla \cdot \boldsymbol{u} = 0 \tag{2}$$

where $\boldsymbol{u}$ is the filtration or superficial velocity (the volume of air flowing through a unit area in a unit time).

Equation 1 is the momentum equation in which ∇p is the pressure gradient required to move the fluid at a velocity ($\boldsymbol{u}$) against the resistance (R) caused by the grain bed. The second equation is based on the assumption that the fluid is incompressible.

In most of the agricultural engineering literature, Equation 2 has been accepted without discussion and the general form of Equation 1 is taken to be correct. Most interest has concentrated on deriving the best description of the resistance term (R).

It is necessary to use Equation 1 with the resistance (R) dependent on the air velocity. If the grain bed is isotropic then the resistance (R) is the same for flow in all directions. But there is evidence (Kumar and Muir 1986, Jayas et al. 1987) that grain beds are anisotropic with different resistances in the horizontal (x) and vertical (y) directions so that

$$\frac{\partial p}{\partial x} = -R_x\,u_x$$

$$\frac{\partial p}{\partial y} = -R_y\,u_y \tag{3}$$

Early work analyzed the airflow patterns produced by various duct arrangements (Hukill and Shedd 1955, Collins 1953, Hall 1955, Brooker 1958). In this work, the pressure throughout a bin was measured and isobars and pressure gradients were calculated. Then the air velocity and streamlines were calculated by assuming that the streamlines were perpendicular to the isobars. The velocities were not measured directly because of the difficulty in measuring the very low velocities in grain beds; air pressures are much easier to measure accurately. Hukill and Shedd (1955) showed that the calculated streamlines had a similar shape to the paths taken by smoke injected into the airstream. They also introduced the concept of traverse time, which is the time, at a point in the grain bed, taken by the air to travel from the air inlet up to that point. Points with the same traverse time are joined to form isotraverse-time lines. Hukill and

Shedd (1955) calculated the position of isotraverse-time lines in a grain bed using measured pressure values and Shedd's (1953) resistance data. They found that the measured shape of the drying front, as it moved through the grain bed, was similar to the shape of the isotraverse-time lines.

In addition to numerical calculation, several methods have been used to study the airflow pattern from ducts or perforated floors. Cromarty (1968), Lamond and Smith (1982) and Harrison (1988) used gas tracer techniques to directly measure the air traverse time. Cromarty (1968) analyzed the chilling pattern in grain bins by measuring the position of lines of constant cooling time on which the grain had cooled to the same temperature in the same time. Cromarty showed that the lines of constant cooling time had the same shape as the lines of constant traverse time; thus, the airflow pattern can be analyzed by measuring the temperature throughout the grain bed while warm or cold air is blown in. The advantage of this method is that temperature is much easier to measure than velocity. Friesen (1989) used this method to determine the airflow pattern in grain bins with partially perforated floors and to determine the time for air to pass through a grain bed with a peaked surface relative to the time taken if the surface was level.

Another method of analyzing the airflow pattern is to measure the escape velocity of the air through the top surface of the grain bed. A method of measuring these low velocities was developed by Burrell (1974) which was calibrated to measure velocities in the range 0.003–0.15 m/s. In practice, the method is used to compare values of velocity in different parts of the grain surface to find regions of low flow rate. This is a reasonably good method of fault finding but it does not give a good guide to the airflow pattern within the grain bed (Cromarty 1968).

1.2. Flow Equations

The bulk resistance to flow of a fluid through porous material was first measured by Darcy (1856). In his experiment, the porous medium was composed of uniform size particles that were randomly and loosely packed. The macroscopic flow was steady and one dimensional and the filtration velocity (u) was related to the pressure gradient by the equation:

$$\frac{dp}{dx} = -\frac{\mu}{K} u \tag{4}$$

where the permeability (K) is a constant. The Darcy model has been studied extensively and is reasonably accurate for gas flows except at very low and high velocities. The equation has some of the characteristics of the Stoke's flow equation, where viscous forces dominate inertial forces so it is not valid for high velocities. Also at low gas pressures the mean free path of the gas molecules may be of the order of the pore sizes and therefore velocity slip occurs; this Knudsen effect results in larger permeabilities. This effect and similar problems with theoretical basis of the Darcy model are discussed by Scheidegger (1974). Experimental work with grain shows that Equation 4 is not accurate when K is a constant, except at very low air velocities (ASAE 1987).

There is no general theory which can be used to model the permeability of grain beds but a variety of models have been developed from a mixture of first principles and empirical data. The simplest approach is to model the pores as tubes and use the Navier-Stokes equation to calculate the velocity-pressure drop relationship. If a parameter is introduced to account for the twisting of the tubes then we get the Carman-Kozeny equation (Dullien 1979) for the permeability:

$$K = \frac{\varepsilon^3}{36\,k\,(1-\varepsilon)}\,d^2 \tag{5}$$

In this equation, k is the "Kozeny constant" and is approximately equal to 5.0 but this "constant" is usually a function of the porosity as well as air velocity. For example, Happel and Brenner (1986) use detailed numerical and analytical methods to show that the "Kozeny constant" is not constant for flow along and perpendicular to an assembly of cylinders. In practice, a variety of porosity functions are used to model the permeability in the Carman-Kozeny equation (Dullien 1979) and there is no universally applicable relationship between permeability and porosity. Another problem occurs if the fluid velocity is too large. Kaviany (1991) reports work where the Navier-Stokes equations were solved numerically for various arrangements of cylinders which shows how the "Kozeny constant" varies with the Reynolds Number.

Ergun (1952) extended the basic Carman-Kozeny equation to deal with higher velocities and he established that (MacDonald et al. 1979):

$$-\frac{dp}{dx} = \frac{\rho\,u^2\,(1-\varepsilon)}{\varepsilon^3}\left[\frac{180\,(1-\varepsilon)}{R_e} + 1.8\right] \tag{6}$$

where the Reynolds Number (R_e)

$$R_e = (\rho\,u\,d/\mu)$$

Equation 6 is called Ergun's equation. Using the Carman-Kozeny equation for the permeability (K) (Eq. 5) we get:

$$-\frac{dp}{dx} = \frac{\mu}{K}\,u + 1.8\,\frac{1-\varepsilon}{\varepsilon^3}\,\frac{\rho}{d}\,u^2 \tag{7}$$

Thus Ergun's equation has the form:

$$\frac{dp}{dx} = au + bu^2 \tag{8}$$

where the parameters a and b depend on the geometry of the grain bed and are influenced by, among other things, the grain type and method of loading the bed. From this equation, we see that theoretical form of the resistance R is:

$$R = -(a + bu) \tag{9}$$

It is not clear how Ergun's equation should be generalized to three-dimensional flow but one possibility is:

$$\nabla p = a\boldsymbol{u} + b\,|u|\,\boldsymbol{u} \tag{10}$$

Stanek and Szekely (1974) discuss more general forms of the three-dimensional version of Ergun's equation. From Equation 7 or 8, it can be seen that if the velocity (u) is small enough then Ergun's equation reduces to Darcy's Law (Eq. 4).

Experimental work by Dybbs and Edwards (1984) show that for Darcy's Law to be accurate the velocity should be small enough so that

$$u < \frac{\mu}{\rho\, d} \tag{11}$$

The basic form of the momentum equation has been studied in detail using the "Local Volume-Averaging Method" by Bear (1988), Slattery (1981), and Whitaker (1969). This approach shows that Darcy's Law is valid for low velocities, where viscous forces dominate the flow.

1.3. Application to Grain

A large amount of experimental work has concentrated on the relationship between air velocity and pressure gradient through beds of seeds. Most of the work has been where the air velocity is one dimensional plug flow and the main results are published in the ASAE Standards (ASAE 1987). But there are a number of cases where flows at right angles are used to measure anisotropic permeability.

One of the main pieces of work was done by Shedd (1951, 1953) who presented graphs relating airflow rate to pressure drop for 22 types of grain. The airflow range was 0.005–0.3 m/s. Figure 1 shows one of these results. These data were summarized by the equation:

$$u = A\,(\Delta p)^B \tag{12}$$

where Δp is the negative of the pressure gradient $-(dp/dx)$ and the constants A and B are product dependent. They can also vary with the pressure gradient

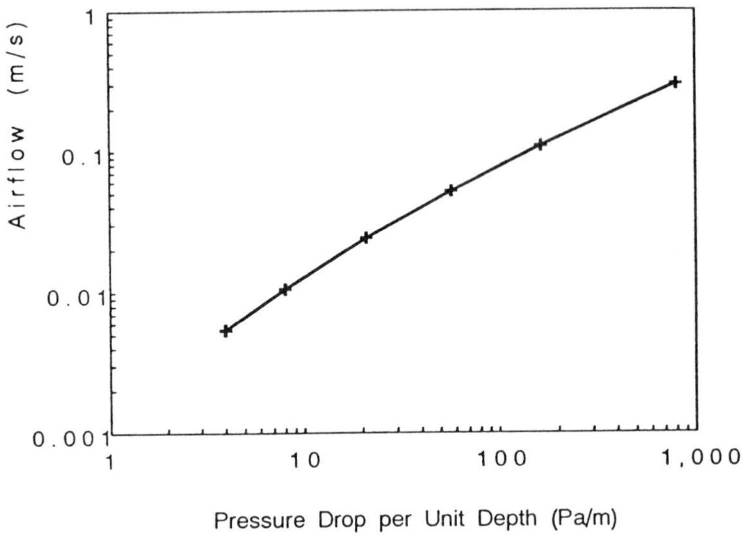

Figure 1 Experimental values of resistance of shelled corn to airflow. *Source*: Shedd (1951).

and Table 1 shows the values for shelled corn. The data are valid for loose fill of clean dry grain and corrections are suggested (ASAE 1987) for grain with a high moisture and for packed grain.

Rewriting Equation 12 in the form:

$$\frac{dp}{dx} = a\,u^b \tag{13}$$

and comparing this equation with Ergun's equation (Eq. 8) it can be seen that the constant b, lying between 1 and 2, is a compromise between the velocity (u) and the (u^2) terms in Ergun's equation.

An alternative equation was used by Hukill and Ives (1955) to summarize their data:

$$\Delta p = \frac{b_1\,u^2}{\ln(1 + b_2 u)} \tag{14}$$

This equation usually fits the data more closely than Shedd's equations over the whole velocity range but it has no theoretical basis. Also, the Hukill and Ives equation is more costly computationally to evaluate, so that most numerical methods use Shedd's equation. To increase the accuracy of Shedd's

Table 1 Values of A and B for shelled corn in $u = A\,(\Delta P)^B$. The velocity u is measured in m/min.

Range of pressure gradient (ΔP) (kPa/m)	A	B
0–0.008	61.7	0.950
0.009–0.021	41.0	0.867
0.022–0.057	27.8	0.768
0.058–0.164	23.1	0.702
>0.2	20.4	0.628

Source: Brooker (1961).

equation, the constants A and B are determined for several short ranges of velocity (Brooker 1969, Jindal and Thompson 1972) as shown in Table 1.

Bakker-Arkema et al. (1969) used Ergun's equation (Eq. 6) to model the pressure drop-velocity relationship in beds of grains. They found that the equation was accurate but that it required an extra parameter which depended on the variety of the seeds. Patterson et al. (1971) successfully used Ergun's equation, as modified by Bakker-Arkema et al. (1969), to calculate the airflow through beds of corn and beans.

Hunter (1983) used Equation 8 based on Ergun's equation:

$$\Delta p = au + bu^2 \tag{15}$$

to model the flow of air through a variety of seed beds. The constants a and b were obtained by regression analysis; unlike the work by Bakker-Arkema et al. (1969) no attempt was made to relate the constants a and b to the geometry of the seeds or the seed beds.

Matthies and Petersen (1974) studied a slightly different form of Ergun's equation, based on Equation 6. They found that the equation accurately modeled their airflow-pressure drop data.

It is advantageous to use Ergun's equation—or the forms developed by Bakker-Arkema et al. (1969), Matthies and Petersen (1974), or Hunter (1983)—in calculating the velocity and pressure in grain beds because it is simpler to deal with analytically than either Shedd's equation or Hukill and Ives' equation. Also it is easier to evaluate numerically than Hukill and Ives' equation.

From Equation 15 used by Hunter it can be seen that:

$$\Delta p = au \tag{16}$$

which is Darcy's Law, when the velocity is so low that $u \ll (a/b)$.

Thus, Hunter's equation reduces to Darcy's Law if the velocity is small enough. For wheat, experiments show that the ratio $(a/b) = 0.3$ so if $u \ll 0.3$ m/s then Darcy's Law is accurate. This suggests that the velocities have to be smaller than (0.3)/10 (say about 0.01 m/s or less), which is too low for drying but reasonable for aeration. This condition is supported by theory which says in Equation 11 that Darcy's Law is accurate if $u < (\mu/\rho d)$ which becomes $u <$ 0.01 m/s for wheat-sized particles.

The importance of this condition is that if the velocity is small enough then the resistance

$$R = -(a + bu)$$

becomes $R = -a$ so that the basic airflow Equations 1 and 2 become:

$$\nabla p = -a\boldsymbol{u} \tag{17}$$

$$\nabla \cdot \boldsymbol{u} = 0 \quad \text{or} \quad \nabla^2 p = 0 \tag{18}$$

This equation, called Laplace's equation, can be solved very accurately for the pressure and the velocity. For larger velocities, the full Ergun equation (Eq. 10) is required and the resulting equation to be solved for pressure is far more complicated and the solutions are less accurate.

The resistance to airflow is influenced by a number of factors such as the bulk density of the grain, its moisture content, and the presence of foreign material. In principle, it should be possible to relate these effects to the constants in the basic Ergun's equation or its variations. The work that has been done uses the equation:

$$\nabla p = au + bu^2 + c \tag{19}$$

and determines the constants a, b, and c by regression analysis. Theoretically, the constant c should be zero but Bern and Charity (1975) found that nonzero values gave a better fit to their data. The constants a and b were related to the kernel density and the bulk density of the grain. Haque et al. (1982) took $c =$ 0 while a was related to the grain moisture content. The influence of foreign material on grain resistance to airflow was studied by Haque et al. (1978), Jayas and Sokhansanj (1989), and Sokhansanj et al. (1988). The equation of Hukill and Ives (1955) was modified by a factor which depended on the mass fraction of the fines. Grama et al. (1984) studied the effect of the amount and size distribution of foreign material on the resistance to airflow.

There are indications that grain beds have an anisotropic resistance to airflow. An experiment by Lamond and Smith (1982) suggested that the resistance to airflow depended on the direction of the airflow. Kumar and Muir

(1986) measured the resistance of wheat and barley and found that the resistance to airflow in the vertical direction was larger than the value in the horizontal direction, up to as much as two times larger. Jayas et al. (1987) found that for canola, horizontal resistance was about 0.6 times the vertical resistance. Sokhansanj et al. (1988) found that for lentils, the horizontal resistance was about 0.52 times the vertical resistance.

2. SOLVING THE AIRFLOW EQUATIONS

2.1. Analytic Solution

The pressure within a grain bed is calculated from the equation:

$$\nabla \cdot \left(\frac{1}{R} \nabla p \right) = 0 \tag{20}$$

where the resistance to airflow (R) depends on the air superficial velocity. Analytic solutions to this equation can be found for some very restricted cases. The value of these analytic solutions is that they can be used to test the accuracy of the numerical solutions. They also supply insight into the effect of the resistance on the flow pattern.

When the flow is purely radial, the equation has an analytic solution which has been used by Marchant (1976), Smith (1982a), and Miketinac and Sokhansanj (1985) to test the accuracy of their numerical work. For radial flow, the momentum equation is:

$$\frac{1}{R}\frac{dp}{dr} = -u \tag{21}$$

Also, the incompressibility condition

$$\nabla \cdot \boldsymbol{u} = 0$$

implies that:

$$u = \frac{\text{constant}}{\text{r}} \tag{22}$$

If we take the resistance from Ergun's equation, then:

$$R = a + b\,|u| \tag{23}$$

with $a = 3131$ and $b = 10756$ for wheat (Hunter 1983). Other forms of resistance can be taken, for example that based on Shedd's resistance data (Miketinac and Sokhansanj 1985) or Hukill and Ives equation (Smith 1982a). If Equation 21 is solved for pressure (p), with the boundary condition that $p = 100$ at $r = 0.1$ and $p = 0$ at $r = 1$. The solution is:

$$p = -37.4233 \ln(r) + 1.5366\left(\frac{1}{r} - 1\right) \tag{24}$$

Using this in Equation 21 to calculate the velocity u gives:

$$u = \frac{0.012}{r} \tag{25}$$

This solution will be used later to test the accuracy of the finite element method.

Another type of analytic solution can be obtained when the velocity u is so low that $b\ u << a$ in Equation 23 for the resistance (R). In this case, the pressure-velocity equation reduces to:

$$-\nabla p = a\boldsymbol{u} \tag{26}$$

So with

$$\nabla \cdot \boldsymbol{u} = 0$$

we get:

$$\nabla^2 p = 0 \tag{27}$$

which is Laplace's equation. This equation has analytic solutions for a wide variety of shapes of grain beds and types of boundary conditions. This is the approach taken by Spencer (1969) and Hunter (1983).

For any analytic function $F(z)$ of the complex variable $z\ (= x + iy)$:

$$F(z) = p(x,y) + i\ q(x,y) \tag{28}$$

the real part (p) and imaginary part (q) of $F(z)$ are solutions of Laplace's equation. Using the technique of conformal mapping, these solutions can be placed on realistically shaped beds of grain. Another useful result is that the curves $p(x,y)$ = constant represent the isobars while the curves $q(x,y)$ = constant are the streamlines.

By placing a single source of air at the point $x = 0$, $y = 0$ and using a sine transform to map the solution of

$$\nabla^2 p = 0$$

onto the bin shown in Figure 2, the complex potential is:

$$F(z) = \frac{-aQ}{\pi \ell} \ln\left(\sin \frac{\pi z}{w}\right) \tag{29}$$

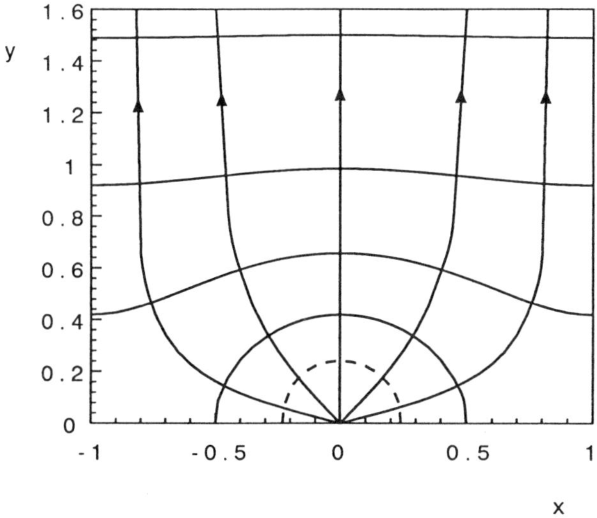

Figure 2 Isobars and streamlines from a source at the origin. The air duct is placed on an isobar shown by (- - - - - - - -).

The real part is:

$$p(x,y) = \frac{-aQ}{\pi\ell} \ln\left[\left(\sin\frac{\pi x}{w}\cosh\frac{\pi y}{w}\right)^2 + \left(\cos\frac{\pi x}{w}\sinh\frac{\pi y}{w}\right)^2\right]^{1/2} \tag{30}$$

and the imaginary part is:

$$q(x,y) = \frac{-aQ}{\pi\ell} \tan^{-1}\left(\frac{\tanh \pi y/w}{\tan \pi x/w}\right) \tag{31}$$

Thus the isobars, given by $p(x,y)$ = constant, are:

$$y = \frac{w}{2\pi}\cosh^{-1}\left(k_1 + \cos\frac{2\pi x}{w}\right) \tag{32}$$

where the different values of the constant k_1 give different isobars.

Similarly the streamlines, given by $q(x,y)$ = constant, are:

$$y = \frac{w}{\pi}\tanh^{-1}\left(k_2 \tan\frac{\pi x}{w}\right) \tag{33}$$

and different values of the constant k_2 give different streamlines. Figure 2 shows a range of isobars and streamlines. Spencer (1969) used this method to estimate the pressure drop required in on-floor-duct systems.

Hunter (1983) notes that this approach can be extended to the case where:

$$-\nabla p = (a + b\,|u|)\,\boldsymbol{u} \tag{34}$$

with the nonlinear term $b|u|$ included. He suggests that this nonlinear term influences the value of the pressure gradient but does not significantly alter the airflow problem. He says that there is little difference between the streamlines for the nonlinear case and the streamlines calculated using Laplace's equation. This is probably true in most cases but there are examples where there are large pressure gradients or highly curved streamlines where there is a noticeable difference. An example of this was examined by Brooker (1969) and Miketinac and Sokhansanj (1985); this problem is discussed in more detail in Section 2.2.

It is clear from Equation 34 that if the nonlinear term, $b|u|$, is present then the resistance to flow is increased. The regions with a large velocity will increase in resistance more than the regions of low velocity. The result is that the streamlines will move away from regions of high velocity. This process was examined analytically by de Ville and Smith (1993) who considered the case where $b|u|$ was small compared with a. The method of nonlinear perturbation expansions was used to obtain an analytic solution to the problem where the airflow is predominantly in one direction. This showed how the streamlines were changed as the parameter b was increased relative to a and the resultant effect this had on the air traverse time.

2.2. Numerical Solution

2.2.1. *Finite Difference Method*

The first numerical method which was used to calculate the pressure distribution through grain beds was the finite difference method (Brooker 1961). Starting with the equation:

$$u = A\,(\Delta p)^B \tag{35}$$

based on Shedd's resistance data. Brooker (1961) generalized this to

$$u_x = A\left[\left(\frac{\partial p}{\partial x}\right)^2 + \left(\frac{\partial p}{\partial y}\right)^2\right]^m \frac{\partial p}{\partial x} \tag{36}$$

$$u_y = A\left[\left(\frac{\partial p}{\partial x}\right)^2 + \left(\frac{\partial p}{\partial y}\right)^2\right]^m \frac{\partial p}{\partial y} \tag{37}$$

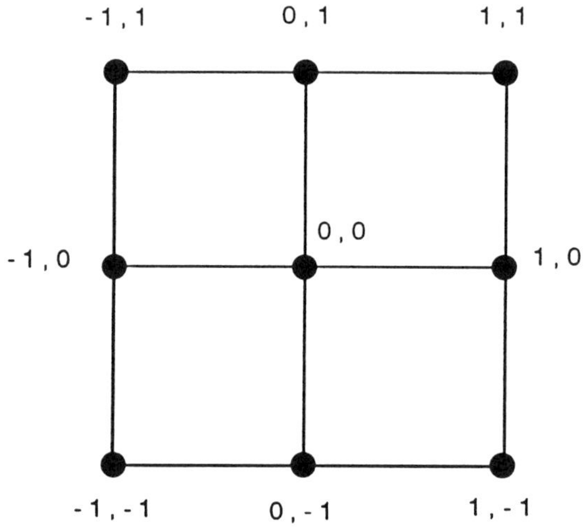

Figure 3 The computational molecule showing the node 0, 0 at which the pressure is calculated.

where

$$m = \frac{B-1}{2}$$

The incompressibility condition $\nabla \cdot \boldsymbol{u} = 0$ was then used to form the equation:

$$\left[\left(\frac{\partial p}{\partial x}\right)^2 + \left(\frac{\partial p}{\partial y}\right)^2\right]\left(\frac{\partial^2 p}{\partial x^2} + \frac{\partial^2 p}{\partial y^2}\right) + 2m\left[\left(\frac{\partial p}{\partial x}\right)^2 \frac{\partial^2 p}{\partial x^2} + 2\frac{\partial p}{\partial x}\frac{\partial p}{\partial y}\frac{\partial^2 p}{\partial x \partial y} + \left(\frac{\partial p}{\partial y}\right)^2 \frac{\partial^2 p}{\partial y^2}\right] = 0 \tag{38}$$

Using the finite difference method to solve this equation, the cross-section of the grain bin was covered by a grid with nodes spaced a distance h apart. To calculate the pressure $p_{0,0}$ at a node 0,0 the group of nine nodes surrounding 0,0 are used (Fig. 3). Then the derivatives of pressure at 0,0 can be approximated by, for example:

$$\frac{\partial p}{\partial x} = \frac{1}{2h}(p_{1,0} - p_{-1,0}) \tag{39}$$

and

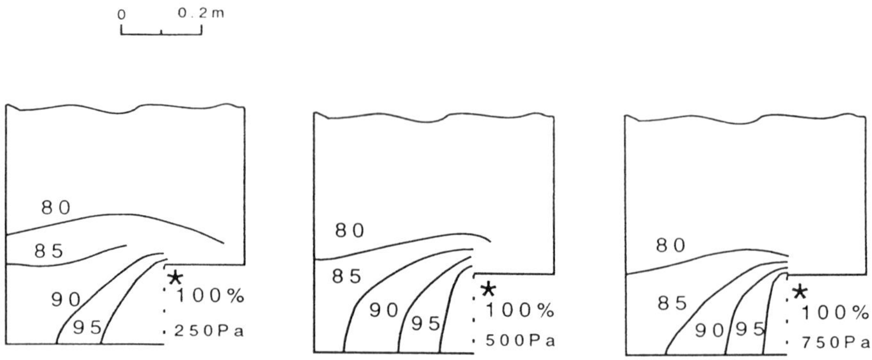

Figure 4 Brooker's experimental results showing isobars at 80–95% of the duct pressure for the three duct pressures. The re-entrant corner is marked by *. *Source*: Brooker (1969).

$$\frac{\partial^2 p}{\partial x \partial y} = \frac{1}{4h^2}(p_{1,1} + p_{-1,-1} - p_{1,-1} - p_{-1,1}) \tag{40}$$

using the standard finite difference method (Ames 1977).

This process converts the partial differential equation into a system of algebraic equations which can be solved for the nodal values. The difficulty is that the equations are nonlinear but a system of nonlinear over-relaxation (Ames 1977) can be used to iterate towards a solution.

To test the accuracy of the numerical results, Brooker (1969) carried out a series of experiments which modeled the situation where an on-floor duct was used to aerate a grain bed (Fig. 4). The duct contains a re-entrant corner at the point marked * in the figure. At this point, the air has to turn the corner very quickly and the equation used to model airflow through grain beds cannot model the flow round this sharp corner. This leads to problems in obtaining an accurate numerical simulation of this problem.

A significant feature of this experiment can be seen in Figure 4 where the 85% isobar line changed shape as the duct pressure was increased from 250 Pa to 500 Pa. If the pressure values were modeled by Laplace's equation (if $m = 0$ in Equation 38) then there would be no change in the shape of the isobars. Thus, these results suggest that Laplace's equation is not able to accurately model the isobars; this is unusual, as noted by Hunter (1983).

To simulate this experiment, Brooker (1961) chose the values of the parameters A and B in the resistance equation (Eq. 35) as shown in Table 1.

The calculated position of the 85% isobar is shown in Figure 5 for all three values of the duct pressure. Thus, the position of this isobar was not calculated

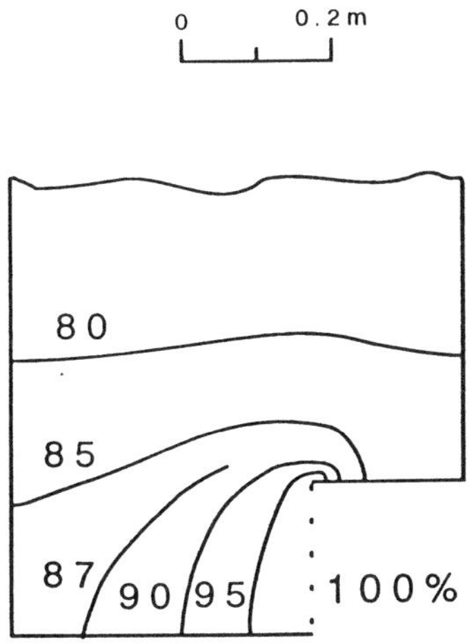

Figure 5 The isobars calculated by Brooker using his model. The same results were obtained for all three values of the duct air pressure (250, 500, and 750 Pa). *Source*: Brooker (1969).

very accurately in this corner of the grain bed but the values of pressure were calculated reasonably well in the rest of the bed. It has been suggested (Miketinac and Sokhansanj 1985) that the difficulty is caused by the re-entrant corner. The numerical solution has to be calculated carefully in this region by, for example, choosing very small grid sizes near this corner. Also it has been noted by Segerlind (1982) and Miketinac and Sokhansanj (1985) that it is difficult to deal with values of B which change with velocity, as used by Brooker (1961) in his finite difference method. However, Brooker's method can be justified (Miketinac and Sokhansanj 1985).

Jindal and Thompson (1972) used Brooker's method to calculate the pressure and airflow pattern in a triangular shaped bed of sorghum. The experimentally determined values of pressure were in close agreement with the calculated values. But in this experiment there were no large pressure gradients or re-entrant corners which would have more seriously tested the numerical method.

2.2.2. *Finite Element Method*

The finite element method is the numerical technique which is most commonly used to calculate the pressure and velocity distribution in grain beds (Marchant 1976, Segerlind 1982; Smith 1982a; Miketinac and Sokhansanj 1985; Jayas et al. 1990). The main advantages of this method are that it deals easily with curved boundaries and there are many computer packages available to implement the method which, in view of the complexity of the technique, is an important factor.

One of the first applications of the finite element method to the calculation of airflow through grain beds was by Marchant (1976), who calculated the isobars in the experiments reported by Brooker (1958) and Barrowman and Boyce (1966). Marchant reported good agreement with these experimental results; even for the case (Brooker 1958, 1961, 1969) where the re-entrant corner caused problems for others (Segerlind 1982; Miketinac and Sokhansanj 1985).

The basic equation for calculating the pressure is:

$$\nabla \cdot \left(\frac{1}{R} \nabla p \right) = 0 \tag{41}$$

where the resistance (R) is a function of velocity, for example Equation 23 based on Ergun's equation. This equation is solved for pressure (p) with the boundary condition that $p = 0$ on the air outlet surface, p is specified on the air inlet surface and finally the pressure gradient is zero through the solid walls of the grain bin. Once the pressure distribution has been calculated, the velocity is given by:

$$\boldsymbol{u} = \frac{-1}{R} \nabla p \tag{42}$$

To solve Equation 41, the grain bed is divided into elements as shown in Figure 6 where a quarter of a circular region is divided into rectangular elements. The elements could be of other shapes, such as triangles or they may have curved sides. But errors occur if the elements are too distorted.

Within each element, the pressure p is approximated by a polynomial in x and y; usually a quadratic polynomial but other types are also used. The polynomial is defined in terms of the p value at each of the nodes on the edge of each element. If a linear polynomial is used then there is one node at each corner of the element as shown in Figure 6; more nodes are required for higher-order polynomials. The finite element method calculates the pressure at each node, the more there are nodes the more accurate usually is the solution. More nodes are introduced into a problem by adding more elements, or by using a higher order polynomial, or by using a combination of both.

It is usually found that linear or quadratic polynomials in a triangular or rectangular element give accurate results. Elements should be concentrated in regions where there is a high pressure gradient (Miketinac and Sokhansanj 1985). Computer packages greatly help in deciding on the best type of elements and how many there should be because they allow easy selection of element types as well as rapid experimentation.

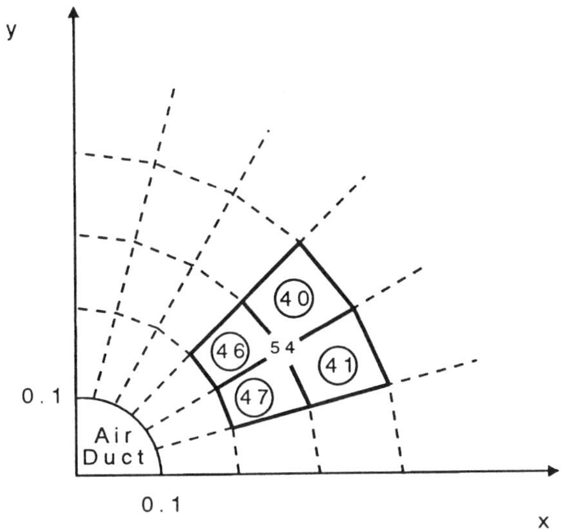

Figure 6 Some of the elements and nodes used in the finite element method solution of the radial flow problem. The elements numbered around node 54 are referred to in Table 2 and the text.

The nodal values of pressure p are found by minimizing the discretized version of the functional:

$$f(p) = \int\int \frac{1}{R}\left[\left(\frac{\partial p}{\partial x}\right)^2 + \left(\frac{\partial p}{\partial y}\right)^2\right] dx\, dy \tag{43}$$

which is mathematically equivalent to solving Equation 41. This is the approach described by Zienkiewicz and Taylor (1989) and widely used in modeling the flow of air through grain. Minimizing Equation 43 results in a nonlinear system of equations:

$$[H]\ \{p\} + \{B\} \tag{44}$$

In this equation, $\{p\}$ represents the vector of all the nodal values of pressure, $[H]$ is the stiffness matrix, and $\{B\}$ is a vector which depends on the boundary conditions. In this system of equations, the matrix $[H]$ depends on the resistance (R) of the grain bed so it will vary with velocity and hence with the nodal values of pressure. This makes the equation nonlinear and a special solution method is required (Winslow 1967).

To test their solution method, Miketinac and Sokhansanj (1985) compared their calculated values of pressure (Fig. 7) with the experimental values obtained

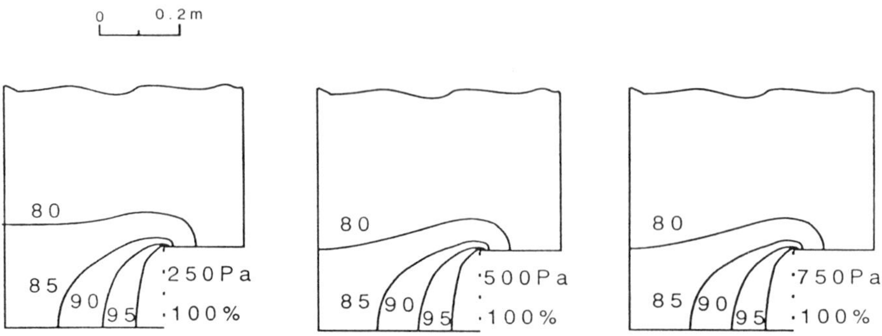

Figure 7 Finite element simulation of Brooker's experiment. *Source*: Miketinac and Sokhansanj (1985). Reprinted by permission of John Wiley & Sons, Ltd.

by Brooker (1969) which are shown in Figure 4. The results are quite close to each other. The main difference is that the experimental 85% line changes shape significantly as the duct pressure increases but the calculated 85% line does not change shape by much. Miketinac and Sokhansanj (1985) suggest that some of the differences could be due to experimental error in measuring the isobars and in obtaining the values of A and B in Equation 35.

In another test of accuracy of the method Marchant (1976), Smith (1982a), and Miketinac and Sokhansanj (1985) compared their calculated values with the analytic solution to the radial flow problem shown in Figure 6. This problem has already been solved analytically in Section 2.1 and was solved numerically using the finite element method with linear, rectangular elements with one node at each corner. The analytic and numerical results are shown in Figure 8 for values along the streamline which goes vertically up along the y axis. These results show that the pressure was calculated accurately.

With these pressure values the velocity u was calculated using the pressure gradient in the equation:

$$u = -\frac{1}{R}\frac{dp}{dr} \tag{45}$$

Table 2 shows the values of velocity at node 54 and this illustrates a typical problem in this method of solution; the calculated value of velocity at node 54 is different in each element. These values of velocity should all be the same and the differences are caused by the method of solution. The table shows that the differences are large, with the values of u_y ranging from 0.043 to 0.070 m/s.

This problem often occurs when the finite element method is used to calculate the pressure and velocity distribution through a bed of grain; the

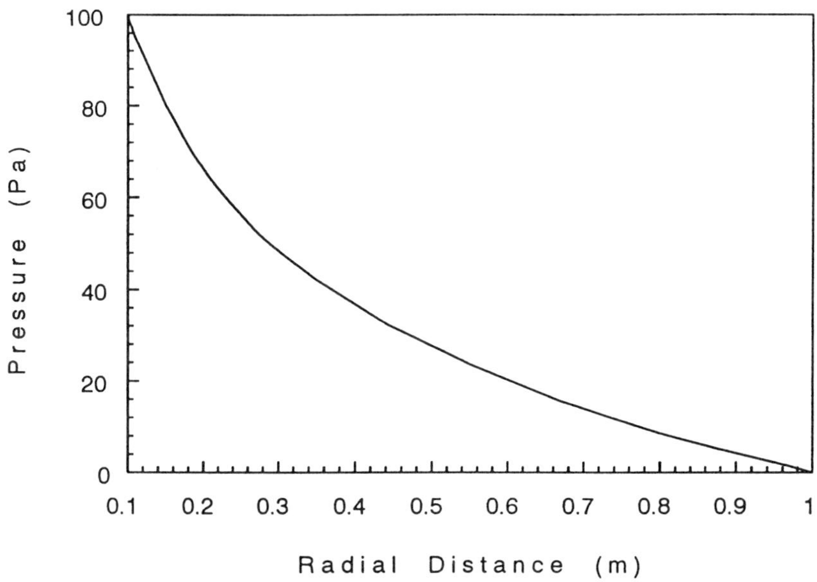

Figure 8 Pressure values in the radial flow problem. The analytic and numerical curves are identical.

Table 2 Values of Velocity (m/s) at Node 54 in Figure 6 from Each of the Four Elements Around the Node.

Velocity (m/s)	Element no.				Average for four elements
	40	41	46	47	
u_x	0.0886	0.1032	0.0919	0.1070	0.0977
u_y	0.0680	0.0428	0.0705	0.0443	0.0564
u	0.1117	0.1117	0.1158	0.1158	0.1138

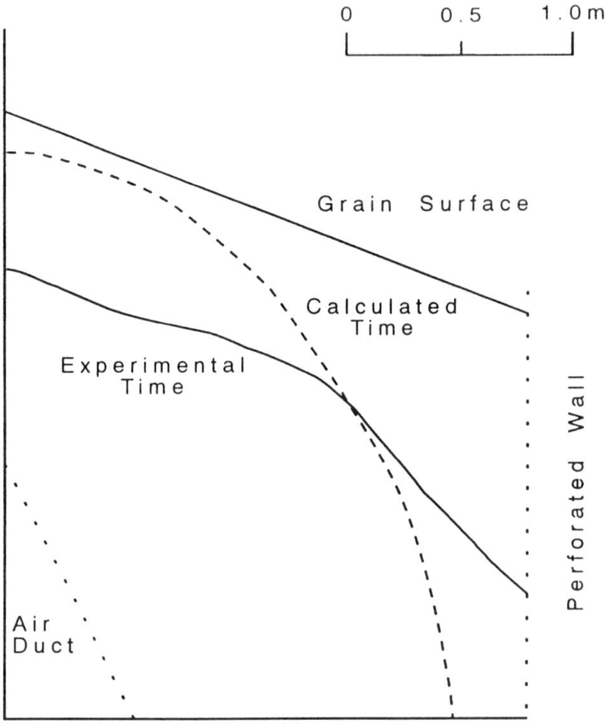

Figure 9 Calculated and measured values of the 10 s line of constant air-traverse time in a bed of barley.

pressures are calculated accurately but the velocities are not. The problem arises because in the method of solution, the pressure is continuous in x and y across the element boundaries but the velocity is not continuous. To avoid this problem, the average value of nodal values of velocity is normally used. But in what circumstances will the average value of velocity not be sufficiently accurate? There are methods (Zienkiewicz and Taylor 1989) for smoothing the velocities across the boundaries of the elements and it remains to be seen if the average value of velocity is as accurate as the smoothed values.

It is important that the velocities are calculated accurately because it is the velocity values which control the accuracy of the heat and mass transfer simulation. The calculation of velocity is relevant to the calculation of the air traverse time which is used to predict the drying pattern of the grain bed (Smith 1982b, Chapman et al. 1989, Smith et al. 1992b).

Very little work has been done to experimentally test the accuracy of the finite element method at calculating the velocity of the air through the grain bed. One result reported by Lamond and Smith (1982) shows that there could be a problem. Figure 9 shows the cross-section of a grain bed in which the airflow is two dimensional. The vertical wall on the right of the bed is perforated so the air flows freely through that wall. A tracer gas was used to measure the air traverse time throughout the grain bed. By measuring the resistance to airflow, of this variety of grain, it was determined that:

$$R = \frac{b_1 u}{\ln(1 + b_2 u)} \tag{46}$$

with $b_1 = 68178.0$ and $b_2 = 16.34$.

Using the finite element method to solve this problem, the air traverse time was calculated and Figure 9 shows the 10 s isotraverse line. The calculated isotraverse line is very different from the measured line and it suggests that the resistance of the grain bed to airflow is anisotropic, with the air flowing horizontally more easily than vertically. This led to experiments which measured the anistropic resistance of grain beds (Kumar and Muir 1986, Jayas et al. 1987).

3. AIR DISTRIBUTION FROM DUCTS

Many experiments and much analytical work has been used to study the flow of air through grain beds. The work by Brooker (1965), Barrowman and Boyce (1966), and others considered the shape of the duct and air escape area in connection with the airflow pattern through the grain bed. But this is only part of the problem because in their work the air pressure in the duct is known but this is not a realistic situation. In most systems used on farms, this pressure is not known and has to be calculated from the fan pressure.

The variation in pressure along ducts was studied by Purser and Greig (1967) but they used ducts which were not covered with grain. Several mathematical models have been developed to calculate the air pressure and velocity along ducts and then through the grain bed (Shove and Hukill 1963, Steele and Shove 1969, Marchant and Nellist 1977, Marchant 1976, Burrell et al. 1982, Smith 1982b, Hunter 1986, Fick et al. 1990). These models were developed from the study of perforated ducts used in ventilating buildings and tunnels by Haerter (1963) who derived an equation relating duct pressure to duct air velocity for a system where air moves in a constant diameter duct.

In grain ventilation, the air encounters a resistance to flow both along and through the wall of the duct and through the grain bed. The equation for flow along the duct has to be solved together with the equation for flow through grain. These equations are simplified so that they can be solved and this means that the results are only approximately correct. Probably the most accurate method is that used by Marchant and Nellist (1977) and Burrell et al. (1982) where the finite element method was used to model the flow of air through the grain bed. Hunter (1986) uses his more approximate but more rapid method for calculating flow through the grain bed, in which it is assumed that at any point in the duct the air pressure is proportional to the average velocity of the air through the duct wall.

When comparing the numerical results with experimental values, Burrell et al. (1982) found that the calculated values could be very accurate but that the numerical results were very sensitive to the regain coefficient in the duct as well as the grain resistance. They suggest that these parameters have to be known very accurately before the mathematical models can be used to predict the distribution of air from ventilation ducts under grain. Burrell et al. (1982) showed that the region coefficient is important when they found that the velocity distribution was much better if the air was blown into the duct rather than sucked out. The distribution was further improved if the diameter of the duct was reduced in stages along the store. Also, simulations by Fick et al. (1990) showed that airflow uniformity was significantly higher if the air was blown into the duct and the pressure had to be much larger if the air was sucked from the ducts.

Much of the advice on the design and operation of ducted, grain ventilation systems comes as empirical rules on duct spacing and air velocities in the ducts (Holman 1960, Burrell 1974, Foster and Tuite 1982). Formulae have been developed to calculate the extra airflow required for ducted bins compared with fully perforated floors (Ives et al. 1959).

It is recommended that the perforations in the floor or ducts should be at least 7–10% of the surface area of the duct or floor area but many ducts have at least twice that area. The surface area of the duct should be large enough that the velocity, of the air through the grain next to the duct, is not more than 0.15–0.25 m/s (Holman 1960). If the surface area of the duct is too small then the velocity will be too large and too much energy will be lost in the high velocity region resulting in a reduced airflow in the rest of the grain bed.

Often, more than one ventilating duct is required and there are several ways to lay out a duct system. The aim is to keep the air paths through the grain as nearly equal as possible. If there is a path which is significantly shorter than the rest then most of the air will flow in that direction. It has been shown experimentally that the longest path should be less than 1.5 times the length of the shortest path although larger values are possible in small stores of dry grain (Burrell, 1974). Where multiple ducts are used, they should not be further apart than the depth of the grain bed. The distance from the wall to the nearest duct should not exceed half of the grain depth.

4. AIRFLOW AND GRAIN DRYING

4.1. Introduction

Computer programs used to simulate grain drying are normally designed for heated air drying or near-ambient (low temperature) drying. The work on heated-air drying is discussed in Chapter 16. This section is concerned with near-ambient drying. In this type of drying, the ambient air is heated slightly as it passes over the fan or it may be heated using supplemental heaters, or both.

Grain-drying models were developed from a simple consideration of the thin-layer drying of grain and the laws of heat and mass balance (Henderson and Henderson 1968). This development continued with the work of O'Callaghan et al. (1971), Bakker-Arkema et al. (1974), and Brooker et al. (1974) where the basic description of the grain drying process was given. Subsequent work has not significantly altered this basic concept although many improvements have been made.

The airflow distribution has a strong influence on the drying process but in most of the work on simulating the near-ambient drying of grain, the simplest airflow pattern (plug flow) has been used (Bowden et al. 1983). Plug flow is the ideal pattern because the uniform velocity ensures that the whole grain bed dries as rapidly as possible. This airflow pattern is only achieved by using a fully perforated floor and level grain surface but in many drying and storage systems the use of on-floor-ducts gives a nonuniform pattern. Although much work has been done to model the airflow distribution in these situations, very little has been done to model the drying process.

This section discusses the models which have been used to simulate near-ambient temperature drying and combines models of airflow and drying with a nonuniform airflow distribution.

4.2. Drying Equations

The grain drying equations are based on the following principles of heat and mass transport:

1. *Moisture balance between air and grain.*
2. *Energy balance for the air and for the grain.* This includes the convective heat transfer between the air and the grain as well as the energy lost (gained) by the grain through evaporation (condensation).

In addition to these principles, there is the experimentally determined information on the drying characteristics of individual grains.

3. *Thin-layer drying rate.* The change in the moisture content of a small quantity of grain is related to the temperature and humidity of the surround-

ing air by the thin-layer drying equation. This equation includes data on the equilibrium moisture content of the grain and the rate of moisture flow through the grain and between the air and grain. Throughout near-ambient drying, the grain is close to equilibrium with the air, thus the accuracy of the drying rate equation greatly affects the accuracy of the simulation.

The above information is used to form four equations which can be solved for the four unknowns: air temperature, air humidity, grain temperature, and grain moisture content. The main transport processes are forced convection and heat and moisture transfer between the air and grain. The other processes such as thermal conduction, moisture diffusion or turbulent dispersion are much smaller and are usually ignored (Smith et al. 1992b).

A common problem encountered when solving these equations is that the calculated relative humidity can exceed 100%. This occurs as a result of errors in the numerical method when the relative humidity is close to 100%, which can happen in near-ambient drying. A method of dealing with this problem was developed by Ingram (1976).

4.3. Uniform Velocity

In most of the work on solving the drying equations it has been assumed that the air velocity is uniform throughout the grain bed. Thus, the velocity is easily calculated during the drying and shrinkage of the grain bed, so the main problem is modeling the processes of heat and mass transport.

Early use of numerical methods to solve these equations was made by O'Callaghan et al. (1971) and by Brooker et al. (1974). With the computer hardware available, the simulations were very slow and it was impractical to use these simulation methods to model the low temperature drying of grain. Approximations to the drying equations were developed which used the fact that for most of the drying process the grain and air are close to equilibrium.

The most extreme approximation is to assume that the air temperature is equal to the grain temperature while the grain moisture content is related to the temperature and humidity of the air by the equilibrium moisture content equation. The result of this approximation is that the number of unknown quantities is reduced to two, the air temperature and humidity. The two equations can be solved very accurately using the numerical method of characteristics (Sutherland et al. 1971, Ingram 1979). In their work it is assumed that the exchange of heat and moisture between the air and grain is so rapid that as the air flows through the grain bed it is always in equilibrium with the grain. This is reasonably accurate if the airflow rate is very slow.

This equilibrium analysis shows that the drying process is characterized by two fronts, a rapidly moving temperature front in which there is very little change in moisture content and a more slowly moving moisture front in which most of the drying is carried out. The temperature front closely follows a line of constant grain moisture content on the psychrometric chart while the moisture front follows the constant wet bulb temperature line. An interesting result of this analysis is that the temperature front becomes wider as it moves through the grain bed. With changing ambient air conditions, new fronts are constantly forming and moving through the bed and these interact with each other. This interaction results in an evening out of the grain temperature and moisture content profiles (Sutherland et al. 1983). This implies that the extremes of ambient temperature and humidity are not passed on to the grain, which tends to remain at a more uniform temperature and moisture content during drying except near the air inlet.

These results can also be obtained by the more complete drying models which do not assume equilibrium throughout the grain bed (Smith and Bailey 1983). The advantage of the equilbrium models is that the numerical methods are accurate and fast and there is the possibility of analytic solutions. These solutions could be used to analyze the general behavior of the drying process, particularly the factors which influence the velocity of the fronts.

Another approximation which has been made is to assume that the air temperature is equal to the grain temperature throughout the grain bed (Bakker-Arkema et al. 1976). This assumption is reasonable when the heat transfer process is more rapid than the moisture transfer process. Also, the grain moisture content is calculated, not from the equilibrium moisture content equation, but from the thin-layer drying equation. Bowden et al. (1983) developed a near-equilibrium model based on this assumption. When compared with experimental results, the simulation was accurate but in their experiments the inlet air temperature was held constant and near to the initial grain temperature. It may be that with variable inlet air-temperature and humidity this approximation will not be sufficiently accurate, especially if supplementary heating is used.

The most common type of near-equilibrium model is one where both the temperature and moisture content of the grain are moving towards equilibrium. Thompson (1972) developed a model in which the air and grain come into equilibrium over one time step of several hours. It was based on the work of Bloome and Shove (1971) and was intended for near-ambient drying with low air flow rates. The Thompson model also included the effect of respiration on the temperature of the grain, which can be significant at low air flow rates of 0.001 m/s or less.

The Thompson model was tested by Pfost et al. (1977) and Morey et al. (1979b). This led to a number of modifications, the main one being that the drying rate was modeled on the thin-layer drying equation rather than assuming that the moisture content of the grain reached equilibrium within one time step of the calculation.

Models based on that of Morey et al. (1979b) have been tested by Van Ee and Kline (1979), Bowden et al. (1983) and Sharp (1984b). The model was found to be accurate and often performed better than the models designed for high-temperature drying. Mittal and Otten (1980) made further improvements to account for bed shrinkage and they used more accurate thin-layer drying and rewetting equations.

At present, the most accurate model for studying near-ambient drying is the type developed by Morey et al. (1979b); in the future, there is likely to be increasing use of one drying model which is valid for both near-ambient and heated-air drying. Current high-temperature models are intrinsically more accurate despite their poor showing compared with the near-equilibrium model in the past (Bowden et al. 1983). The reason for using a separate model for near-ambient drying was that the high-temperature model took too long to run, but this is no longer the case with modern computer hardware. Also, there is a need to accurately calculate the temperature and moisture content history of the grain to predict accurately the change in quality during drying and storage. The simulations are more likely to be accurate if the high temperature modeling approach is used.

Relatively little consideration has been given to grain shrinkage during drying. Shrinkage has little effect in situations where the airflow is uniform but the uniform shrinkage of the grain bed does have an effect on the air flow rate. Most models of this process are empirical (Mittal and Otten 1980, Bowden et al. 1983). Some results (Bowden and Lamond 1981) suggest that the average dry-bulk density of the grain bed is linearly related to the average moisture content of the bed. But the process is likely to be more complex than that. During the initial period of drying (Bowden and Lamond 1981) the bed depth changes rapidly because the flow of air through the bed causes the grains to move and settle more closely together. Also the grains tend to stick together and then move suddenly; if there is sufficient mold growth then the grain bed can become solid and form a bridge.

4.4. Nonuniform Velocity

The interaction between the airflow pattern and the process of heat and mass transfer becomes more important when the airflow is not uniform. In this case, the shrinking grains will alter the shape of the bed and the resulting change in the airflow distribution will alter the drying pattern. No model has been developed to describe this process because of the difficulty in predicting how the bed changes shape during drying.

With uniform flow, the heat and moisture is only transported in the direction of the flow and no transport process acts across the streamlines. But with nonuniform flow (Hukill and Shedd 1955, Ives et al. 1959, Williams 1965), experiments show that the lines of constant moisture content are not at right angles to the streamlines, similar results have been found for temperature (Cromarty 1968, Friesen 1989). Thus, there are moisture and temperature gradients which drive energy and moisture across the streamlines. The main transport processes involved are thermal conduction and the turbulent dispersion of moisture. Order of magnitude calculations (Smith et al. 1992b) suggest that these processes are so small, in comparison with transport by forced convection along the streamlines, that they can be ignored. But in regions of low velocity (less than 0.005 m/s), these transport processes across the streamlines will be important.

By ignoring thermal conduction and turbulent dispersion, the problem reduces to modeling the transport of heat and moisture along the streamlines. For this, the standard drying models can be used with the modification that the velocity of the air can now change along the streamlines (Smith 1982b). This method gives the temperature and moisture history along one streamline, the simulation can be repeated along several streamlines to cover the whole bed. However, a simpler procedure is to use lines of constant traverse time which Hukill and Shedd (1955) suggested have the same shape as the drying front as it moves through the grain bed. Experiments (Hukill and Shedd 1955, Ives et al. 1959, Williams 1965) support this view, although account has to be taken of the shrinkage of the grain bed. The concept of isotraverse-time lines has proved to be very useful; Chapman et al. (1989) have suggested that these lines give a more accurate representation of the drying pattern than isobars. This is certainly the case if the grain has an anisotropic resistance to airflow because in this case the streamlines will not be perpendicular to the isobars, and the isotraverse-time lines will be different from the isobars.

The lines of constant-traverse time have exactly the same shape as the drying front when the equilibrium model of Sutherland et al. (1971) and Ingram (1979) are valid. This is because in these models the velocity of the drying front is proportional to the air velocity. These equilibrium models are only valid when the grain and air are in equilibrium throughout the drying process. This assumption gives only an approximately correct description of the drying behavior of the grain bed. The near-equilibrium model of Morey et al. (1979b) is required to give an accurate description. This is probably also true of the lines of constant-traverse time, they probably only give an approximately correct description of the shape of the drying front but not such detailed information as the shape of the lines of constant moisture content.

Smith et al. (1992a, b) simulated the drying of grain in bins with partially perforated floors and compared their results with their experimental results (Fig. 10). They found that the lines of constant-traverse time outlined the position of the drying front as it moved through the grain bed. The problem of grain shrinkage and the changing shape of the bed was dealt with empirically. The grain bed was assumed to remain static for a period of time and then to

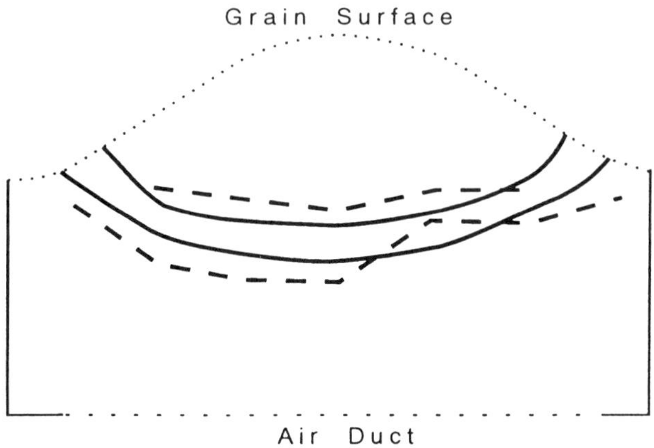

Figure 10 Cross-section of cylindrical bin showing the leading (20% db) and trailing (15% db) edges of the drying front. The experimental (- - -) and computed (——) values are shown. *Source*: Smith et al. (1992b).

change suddenly to the next measured position. The airflow and drying was simulated for a static bed over each time interval. A model of the process of bed deformation is required to simulate the full interaction between the drying process and the changing shape of the bed.

There are equations which are used to study the shrinkage of porous material (Zienkiewicz and Shiomi 1984) and these have been used to study the deformation of soil. These models have not been applied to the shrinkage of grain beds because, unlike particles of soil, cereal grains can be deformed individually and this greatly complicates the analysis.

5. OPERATION OF GRAIN STORES

5.1. Introduction

Simulation models of airflow and near-ambient drying have been used to investigate the complex problem of the management and control of in-store drying and preservation (Sharp 1984a, Ryniecki and Nellist, 1991). Once the computer program has been validated by comparison with full-scale experiments, it can be used to analyze the control problem far more cheaply than using experiments to determine the best control strategy.

The aim of this type of study is to determine the best control strategy for ambient air driers, which could be used eventually, in an automatic system of control. Ambient air driers are difficult to operate and a great deal of experience is required to dry the grain safely and economically to the required moisture content. Thus, any successful automation of this process would be very useful.

The best control strategy is usually the one which reduces the grain to a predetermined moisture content, using as little energy as possible while the deterioration in quality is kept to an acceptable level.

Most studies of the control process aim to achieve a low energy consumption because a feature of near-ambient drying is that it uses the drying potential of the ambient air and there is an interest in maximizing this potential. But from studies of the whole system of harvesting and drying, Muir et al. (1983) found that the cost of drying and storing the grain was very much less than the overall cost of harvesting; the main cost is in buying and maintaining the drying equipment and storage structures. Fraser and Muir (1980) found that electricity gave the lowest contribution to the overall cost of harvesting, while the highest contribution came from the drying equipment. Despite this, most studies of the control process aim to achieve the minimum use of energy. This is a reasonable approach so long as it does not interfere with the main aim of drying: safely reducing the grain to a predetermined moisture content.

With the main cost of harvesting and storing grain being the cost of the drying equipment and storage structures there tends to be little economic advantage in using extra equipment to speed up the drying process. For example, the use of solar energy in grain drying tends to be uneconomic (Morey et al. 1979b). Also, extra equipment to stir the grain bed can be uneconomic. Such equipment, however, is useful if it helps to dry the grain safely in conditions of wet weather or wet grain where drying is very difficult. Stirring the grain bed speeds up the drying process and helps to overcome the problem of overdrying near the air inlet. Also, stirring loosens the grain bed and temporarily increases the airflow (Morey et al. 1979a).

A significant factor in the economic analysis of the harvesting and storage of grain is the effect of government policy on the sale of grain. One important consideration is that in some countries, grain must be sold with its moisture content below a set level; in Canada, wheat must be below 14.5% mc and corn must be below 15.5% mc. Also the grain is sold by its wet mass (dry matter plus water) so there is an economic advantage in selling the grain as wet as possible to maximize the amount of water that is sold. This is purely a matter of government regulation since wheat could be dried to a much lower level than 14.5% without loss of quality. The result of this policy is the concept of "overdrying". Fraser and Muir (1980) found that the penalty for overdrying, below the level required by regulation was nearly twice the cost of electricity.

5.2. Grain Quality

The optimum control strategy is determined by the quality of the grain; there would be no need to dry the grain if its quality was not important. Quality is measured in a number of ways: degree of mold infection, loss of germination capacity, insect and mite activity, amount of foreign material, and loss of dry matter.

Data reported by Kreyger (1972) have been used to develop models to predict the occurrence of visible mold (Sharp 1984a, Bowden et al. 1983, Fraser and Muir 1981). The occurrence of visible mold is a rather subjective measure of quality because it depends on how carefully you look. An alternative measure is the loss of dry matter and the production of carbon dioxide which results from respiration of grain, microflora, and insects; these are quantities which can be objectively determined. This is the approach developed by Steele et al. (1969) and used to study the ambient air drying of maize in the United States and Canada (Colliver et al. 1983, Morey et al. 1979a, Mittal and Otten 1982, Pierce and Thompson 1979). This work was adapted for use in drying wheat by Morey et al. (1981). For drying maize, the quality becomes unacceptable when the dry matter loss reaches a level of 0.5%, this is equivalent to approximately 7.33 g CO_2/kg dry matter. At this level it is expected that mold growth will be visible. The carbon dioxide production for wheat was measured by White et al. (1982); they note that a 0.1% dry matter loss is not acceptable when seed germination is important.

Spoilage indices have been formed to model grain quality; one is based on mold growth and the other is based on dry matter loss or CO_2 production. These methods of modeling grain spoilage have been examined by Brook (1987a). He concluded that they provide a reasonable estimate of the allowable storage time for cereal grains and can be used during the simulation of ambient air drying. He also found that the spoilage index based on mold growth was safer than the index based on dry matter loss. This was because the mold index tended to underestimate the allowable storage time and this provided a safety margin.

Wallace et al. (1983) tested the accuracy of a deterioration model that was based on measurements of germination, CO_2 production, and visible mold. They found that the model could calculate a usable storability index with a high factor of safety.

Methods have been developed to model the decrease in germination during drying and storage. These methods are based on the work by Roberts (1972) on the decrease in germination under constant storage conditions. This work was modified to cope with the variable conditions which occur during drying (Nellist 1981, Muir et al. 1983, Bowden et al. 1983, Smith and Bailey 1983). The work by Bowden et al. (1983) showed that although the method could be useful it was not sufficiently accurate, at that time, to be used in analyzing the control of ambient air driers. Later work on high-temperature drying shows that the method can be accurate (Nellist and Bruce 1987).

5.3. Control Strategies

The commonly used methods of controlling the drying process incorporate a fan control and heater control. It is important that the fan is selected to supply sufficient air to dry the grain safely. Appropriate airflow rates will depend on the maximum initial moisture content of the grain as well as the typical weather conditions (Fraser and Muir 1981, Sharp 1984a, Smith 1984). Normally, airflow is not a control parameter during drying but is fixed by the choice of fan and the depth of the grain bed. The normal process of fan control is to switch the fan off when the relative humidity of the air exceeds a set value. In addition to fan control, a heater control can be considered. Normally the heat is added in stages depending on the relative humidity of the air.

Using fan control alone, a typical operating point in the United Kingdom is to switch the fan off if the relative humidity of the air is above 85% when drying grain which is initially at or below 18% mc (wb) (ADAS 1982); however, the fan should be left on for wetter grain. Work in the United States and Canada shows that intermittent fan operation often results in an increase in dry matter loss (Mittal and Otten 1982, Morey et al. 1979a). There are indications that this loss increases as the relative humidity set point is decreased (Smith and Bailey 1983). The most common advice is to run the fan continuously until the drying front has reached the surface of the grain bed and then use fan control to reduce the grain moisture content to the required level. Foster (1953) found that the best results were obtained by running the fan continuously when the average grain moisture content is above 15.5% (wb) and then using humidistat control when the moisture content is below this level. Similar results have been found by others (Morey et al. 1979a, Pierce and Thompson 1979). Morey et al. (1979a) found that for corn drying in Minnesota, continuous ventilation is preferable to humidistat control. Mittal and Otten (1982) recommend starting the drying process with continuous fan operation and no supplemental heat. Humidistat control is then only used when the moisture content of the top surface is less than 17%.

Supplemental heat can be used during wet weather when it is not possible to safely dry the grain using only a fan. The temperature rise from the heater, however, should be minimized because of the increased loss in grain quality with higher temperatures. Simulations by Smith and Bailey (1983) suggest that increasing the energy input from air heaters results in an increase in the dry matter loss. Pierce and Thompson (1979) found that supplemental heat had no effect on the minimum airflow rate, except in humid climates. They also found (Pierce and Thompson 1980) that with continuous fan operation, supplemental heat simply increased the overdrying problem. Stirring the grain, however, improved the situation. Van Ee and Kline (1979) suggest that increasing the fan capacity is more effective than adding heat.

5.4. Optimization of the Control Process

There is interest in determining the best control strategy; one which, over a prolonged period of time of variable weather conditions, will safely dry grain to a predetermined level with least energy cost. Colliver et al. (1983) considered the use of predicted weather conditions to determine the optimum control strategy for fan and heater operation. However, they pointed out that an accurate, reliable, and inexpensive relative humidity transducer is required before such control methods can be implemented. This scheme has been implemented in the United States (Simonton et al. 1981).

Peart et al. (1985) point out that, although the relative humidity of the inlet air is important, other factors can be used in the control strategy. They suggest, among other quantities, the temperature profile through the grain bed and the grain moisture content at the top and bottom of the bed.

Brook (1987b) used a simplified version of the drying equations to determine the optimum temperature rise for each time step of the computer simulation. Ryniecki and Nellist (1991) have developed a method for optimizing the controlling parameters of a control strategy based on the standard model of near-ambient drying. This work has been applied to the United Kingdom as well as Canadian weather conditions (Ryniecki et al. 1991).

6. FUTURE RESEARCH

The computer models which are used to calculate the airflow distribution in grain beds have been found to give reasonably accurate results. Most work has concentrated on calculating the pressure distribution in grain beds and comparing these results with experimental values. These results are reasonably accurate for most situations but there are problems, particularly where there are sharp corners in the air ducts and where there are large pressure gradients. Relatively little work has been undertaken to directly measure the airflow distribution and the work that has been done suggests that the resistance to airflow is anisotropic; however, more measurements of the airflow pattern are required. An accurate, indirect way of doing this would be to measure the temperature distribution when cool (warm) air is blown into a grain bin. The lines of constant cooling (warming) time are equivalent to the lines of constant traverse time for the air.

In modeling the resistance to airflow, Shedd's equation has been found accurate and easy to use in numerical calculation, particularly when the values of the constants A and B are determined for short ranges of velocity. Equations based on Ergun's equation are also easy to use and are equally accurate if the constants are again defined for short ranges of velocity. Ergun's equation has the advantage that it has a strong theoretical basis and reduces to Darcy's law when the velocity is low enough. These resistance equations are defined for one-dimensional flow and it is not clear how they should be extended to three-dimensional flow but a possible form of the equation is given in this chapter.

The computer models which are used to study the near-ambient drying of grain have been shown to give reasonably accurate values of moisture content and temperature. Quality indices have been developed which can be used to determine when the quality of the grain will fall to an unacceptable level. These indices are safe in that they underestimate the decay time.

Work has been undertaken to model the combination of nonuniform airflow and drying. These models give the correct qualitative behavior but they have yet to be validated in detail. The work indicates that the lines of constant air traverse time give a good indication of the shape of the drying front as it travels through the grain bed. The existing drying models give a reasonable estimate of the time taken by the drying front to move through the grain bed. The main work which remains to be done is in modeling the shape of the bed as it shrinks during drying; little work has been done to model this complex process.

7. NOMENCLATURE

A, B	constants used in Shedd's equation (Eq. 35)
a, b	coefficients in Ergun equation (Eq. 8)
b_1, b_2	constants in Hukill and Ives equation (Eq. 14)
d	mean particle diameter (m)
K	permeability (m^2)
ℓ	length of air duct (m)
p	pressure (Pa)
$p_{i,j}$	pressure at node number i, j in the finite difference method (Pa)
Q	airflow rate along a duct ($m^3 \cdot m^{-2} \cdot s^{-1}$)
q	streamfunction
R	resistance to airflow ($Pa \cdot m^{-2} \cdot s$)
Re	Reynolds number
r	radial distance (m)
u	superficial velocity ($m^3 \cdot m^{-2} \cdot s^{-1}$)
$\mathbf{u}$	velocity vector (m/s)
w	width of bin (m)
Δp	pressure drop (Pa/m)

ε porosity
μ viscosity (Pa · s)
ρ air density (kg/m^3)
∇ gradient vector

REFERENCES

ADAS. 1982. Bulk grain driers. Ministry of Agriculture, Fisheries and Food. Booklet 2416. Alnwick, United Kingdom. 29 pp.

Ames, W. F. 1977. Numerical methods for partial differential equations. Nelson, New York. 365 pp.

ASAE. 1987. Standards—1987. Am. Soc. Agric. Eng., St. Joseph, MI. 561 pp.

Bakker-Arkema, F. W., R. J. Patterson, and W. G. Bickert. 1969. Static pressure-airflow relationships in packed beds of granular biological materials such as cherry pits. Trans. ASAE (Am. Soc. Agric. Eng.) 12:134-136, 140.

Bakker-Arkema, F. W., L. E. Lerew, S. F. de Boer, and M. G. Roth. 1974. Grain dryer simulation. Mich. State Univ., Agric. Exp. Station Res. Rep. 224, East Lansing, MI. 80 pp.

Bakker-Arkema, F. W., J. Haight, M. G. Roth, and D. B. Brooker. 1976. Required weather data for simulation of solar grain drying. Paper 76-3020. Am. Soc. Agric. Eng., St. Joseph, MI. 12 pp.

Barrowman, R. and D. S. Boyce. 1966. Air distribution from lateral ducts in barley. J. Agric. Eng. Res. 11:243-347.

Bear, J. 1988. Dynamics of Fluids in Porous Media. Dover Publications Inc., New York. 764 pp.

Bern, C. J. and L. F. Charity. 1975. Airflow resistance characteristics of corn as influenced by bulk density. Paper 75-3510. Am. Soc. Agric. Eng., St. Joseph, MI. 8 pp.

Bloome, P. D. and G. C. Shove. 1971. Near equilibrium simulation of shelled corn drying. Trans. ASAE (Am. Soc. Agric. Eng.) 14:709-712.

Bowden, P. J. and W. J. Lamond. 1981. Drying two-meter depths of Golden Promise Barley with near ambient air, harvest 1980. Dept. Note SIN/327, Scottish Inst. Agric. Eng., Penicuik (unpubl). 22 pp.

Bowden, P. J., W. J. Lamond, and E. A. Smith. 1983. Simulation of near-ambient grain drying. J. Agric. Eng. Res. 28: 279-300.

Brook, R. C. 1987a. Modeling grain spoilage during near-ambient grain drying. Div. Note DN 1388, AFRC Inst. Eng. Res., Silsoe, the United Kingdom. 20 pp.

Brook, R. C. 1987b. Control system development for near-ambient grain drying. Div. Note DN 1404, AFRC Inst. Eng. Res., Silsoe, the United Kingdom. 60 pp.

Brooker, D. B. 1958. Lateral duct airflow patterns in grain drying bins. Agric. Eng. 39:348-351.

Brooker, D. B. 1961. Pressure patterns in grain-drying systems established by numerical methods. Trans. ASAE (Am. Soc. Agric. Eng.) 4:72-77.

Brooker, D. B. 1965. Nonlinear airflow patterns in grain-drying systems. Univ. Missouri, Agric. Res. Bull. 892, Columbia, MO. 40 pp.

Brooker, D. B. 1969. Computing air pressure and velocity distribution when air flows through a porous medium and nonlinear velocity pressure relationships exist. Trans. ASAE (Am. Soc. Agric. Eng.) 12:118-120.

Brooker, D. B., F. W. Bakker-Arkema, and C. W. Hall. 1974. Drying cereal grains. AVI, Westport, CT. 265 pp.

Burrell, N. J. 1974. Aeration. pp. 407-441 in: Storage of Cereal Grains and Their Products. 2nd Ed. C. M. Christensen (ed.). Am. Assoc. Cereal Chem., St. Paul, MN.

Burrell, N. J., E. A. Smith, and D. M. Armitage. 1982. Air distribution from ventilation ducts under grain. J. Agric. Eng. Res. 27:337-354.

Chapman, J. E., R. V. Morey, H. A. Cloud, and J. L. Nieber. 1989. Airflow patterns in flat storage aeration systems. Trans. ASAE (Am. Soc. Agric. Eng.) 32:1368-1376.

Collins, T. 1953. Flow patterns of air through grain during drying. Agric. Eng. 34:11-15.

Colliver, D. G., R. M. Peart, R. C. Brook, and J. R. Barrett. 1983. Energy usage for low temperature grain drying with optimized management. Trans. ASAE (Am. Soc. Agric. Eng.) 26:594-600.

Cromarty, A. S. 1968. A gas tracer technique for predicting chilling patterns in stored barley. J. Agric. Eng. Res. 13:1-11.

Darcy, H. 1856. Les Fontaines Publiques de la ville de Dijon. Dalmont, Paris. 647 pp.

de Ville, A., and E. A. Smith. 1995. Airflow through beds of cereal grains. Appl. Math. Model. (submitted for publication).

Dullien, F. A L. 1979. Porous Media: fluid transport and pore structure. Academic Press, New York. 396 pp.

Dybbs, A. and R. V. Edwards. 1984. A new look at porous media fluid mechanics—Darcy to turbulent: pp. 199-256 in: Fundamentals of Transport Phenomena in Porous Media. J. Bear and M. Y. Corapcioglu (eds.). Martinus Nijhoff Ltd., The Hague, Netherlands.

Ergun, S. 1952. Fluid flow through packed column. Chem. Eng. Prog. 48:89-94.

Fick, R. J., R. V. Morey, and H. A. Cloud. 1990. Airflow distribution from corrugated aeration ducts. Trans. ASAE (Am. Soc. Agric. Eng.) 33:1319-1326.

Foster, G. H. and J. Tuite. 1982. Aeration and stored grain management. pp. 117-143 in: Storage of Cereal Grains and their Products. C. M. Christensen (ed.). Am. Assoc. Cereal Chem., St. Paul, MN.

Foster, G. W. 1953. Minimum airflow requirements for drying grain with unheated air. Agric. Eng. 34:681-684.

Fraser, B. M. and W. E. Muir. 1980. Cost predictions for drying grain with ambient and solar heated air in Canada. Can. Agric. Eng. 22:55-59.

Fraser, B. M. and W. E. Muir. 1981. Airflow requirements predicted for drying grain with ambient and solar-heated air in Canada. Trans. ASAE (Am. Soc. Agric. Eng.) 24:208-210.

Friesen, O. H. 1989. Airflow patterns in grain bins with partially perforated floors. Paper 89-6040. Am. Soc. Agric. Eng., St. Joseph, MI. 25 pp.

Grama, S. N., C. J. Bern, and C. R. Hurburgh. 1984. Airflow resistance of mixtures of shelled corn and fines. Trans. ASAE (Am. Soc. Agric. Eng.) 27:268-272.

Haerter, A. 1963. Flow distribution and pressure change along slotted or branched ducts. Am. Soc. Heat.Ref. Air Cond. Eng. J. 5:47-59.

Hall, C. W. 1955. Analysis of air flow in grain drying. Agric. Eng. 36:4-7.

Happel, J. and H. Brenner. 1986. Low Reynolds number hydrodynamics. Martinus Nijhoff Ltd., The Hague, Netherlands. 534 pp.

Haque, E., G. H. Foster, D. S. Chung, and F. S. Lai. 1978. Static pressure drop across a bed of corn mixed with fines. Trans. ASAE (Am. Soc. Agric. Eng.) 21:997-1000.

Haque, E., Y. N. Ahmed, and C. W. Doyoe. 1982. Static pressure drop in fixed beds of grain as affected by grain moisture content. Trans. ASAE (Am. Soc. Agric. Eng.) 25:1095-1098.

Harrison, P. 1988. Grain aeration with perforated floor and duct inlets. Paper 88-404. Can. Soc. Agric. Eng., Saskatoon, SK. 12 pp.

Henderson, J. M. and S. M. Henderson. 1968. A computational procedure for deep bed drying. J. Agric. Eng. Res. 13:87-95.

Holman, L. E. 1960. Aeration of grain in commercial storages. Marketing Res. Rep. No. 178. US Dept. Agric., Washington. 45 pp.

Hukill, W. V. and N. C. Ives. 1955. Radial airflow resistance of grain. Agric. Eng. 36:332-335.

Hukill, W. V. and C. K. Shedd. 1955. Nonlinear air flow in grain drying. Agric. Eng. 36:462-466.

Hunter, A. J. 1983. Pressure difference across aerated seed bulk for some common duct and store cross-sections. J. Agric. Eng. Res. 28:437-450.

Hunter, A. J. 1986. Approximate analysis of variation of static pressure along seed bulk aeration ducts. J. Agric. Eng. Res. 31:17-39.

Ingram, G. W. 1976. Deep bed drier simulation with intra-particle moisture diffusion. J. Agric. Eng. Res. 21:263-272.

Ingram, G. W. 1979. Solution of grain cooling and drying problems by the method of characteristics in comparison with finite difference solutions. J. Agric. Eng. Res. 24:219-232.

Ives, N. C., W. V. Hukill, and R. A. Saul. 1959. Grain ventilation and drying patterns. Trans. ASAE (Am. Soc. Agric. Eng.) 2:95-101.

Jayas, D. S. and S. Sokhansanj. 1989. Design data on resistance of airflow through canola (rapeseed). Trans. ASAE (Am. Soc. Agric. Eng.) 32:295-296.

Jayas, D. S., S. Sokhansanj, E. B. Moysey, and E. M. Barber. 1987. The effect of airflow direction on the resistance of canola (rapeseed) to airflow. Can. Agric. Eng. 29:189-192.

Jayas, D. S., S. Sokhansanj, E. B. Moysey, and E. M. Barber. 1990. Predicting airflow pressure patterns in canola bins. Can. Agric. Eng. 32:249-254.

Jindal, V. K. and T. L. Thompson. 1972. Air pressure patterns and flow paths in two-dimensional triangular shaped piles of sorghum using forced convection. Trans. ASAE (Am. Soc. Agric. Eng.) 15:737-741.

Kaviany, M. 1991. Principles of Heat Transfer in Porous Media. Springer-Verlag, New York. 626 pp.

Kreyger, J. 1972. Drying and storing grains, seeds and pulses in temperate climates. Pub. 205, Inst. Storage and Processing (IBVL), Wageningen, Netherlands. 333 pp.

Kumar, A. and W. E. Muir. 1986. Airflow resistance of wheat and barley affected by airflow direction, filling method and dockage. Trans. ASAE (Am. Soc. Agric. Eng.) 29:1423-1426.

Lamond, W. J. and E. A. Smith. 1982. Modeling low temperature drying of grain in anisotropic beds. Third Int. Drying Symp., Birmingham, United Kingdom, 12 pp.

MacDonald, I. F., M. S. El-Sayed, K. Mow, and F. A. L. Dullien. 1979. Flow through porous media—Ergun equation revisited. Ind. Eng. Chem. Fund. 18:199-208.

Marchant, J. A. 1976. The prediction of airflows in crop drying systems by the finite element method. J. Agric. Eng. Res. 21:417-429.

Marchant, J. A. and M. E. Nellist. 1977. Air pressure and flow in ducted crop systems. 1. Single ducts. J. Agric. Eng. Res. 22:303-310.

Matthies, H. J. and H. Petersen. 1974. New data for calculating the resistance to airflow of stored granular material. Trans. ASAE (Am. Soc. Agric. Eng.) 17:1144-1149.

Miketinac, M. J. and S. Sokhansanj. 1985. Velocity-pressure distribution in grain bins—Brooker's model. Int. J. Num. Meth. Eng. 21:1067-1075.

Mittal, G. S. and L. Otten. 1980. Simulation of low temperature drying of corn for Ontario conditions. Paper 80-3519. Am. Soc. Agric. Eng., St. Joseph, MI. 38 pp.

Mittal, G. S. and L. Otten. 1982. Simulation of low temperature corn drying. Can. Agric. Eng. 24:111-118.

Morey, R. V., H. A. Cloud, R. J. Gustafson, and D. W. Peterson. 1979a. Management of ambient air drying systems. Trans. ASAE (Am. Soc. Agric. Eng.) 22:1418-1425.

Morey, R. V., H. A. Cloud, R. J. Gustafson, and D. W. Peterson. 1979b. Evaluation of the feasibility of solar energy grain drying. Trans. ASAE (Am. Soc. Agric. Eng.) 22:409-417.

Morey, R. V., H. A. Cloud, and D. J. Hansen. 1981. Ambient air wheat drying. Trans. ASAE (Am. Soc. Agric. Eng.) 24:1312-1316.

Muir, W. E., W. J. Lamond, G. W. Ingram, and P. H. Bailey. 1983. Optimum systems for harvesting with in-bin drying in northern Britain. J. Agric. Eng. Res. 28:359-371.

Nellist, M. E. 1981. Predicting the viability of seeds dried with heated air. Seed Sci. and Technol. 9:439-455.

Nellist, M. E. and D. M. Bruce. 1987. Drying and cereal quality. Aspects Appl. Biol. 15:439-456.

O'Callaghan, J. R., D. J. Menzies, and P. H. Bailey. 1971. Digital simulation of agricultural drier performance. J. Agric. Eng. Res. 16:223-244.

Patterson, R. J., F. W. Bakker-Arkema, and W. G. Bickert. 1971. Static pressure-airflow relationship in packed beds of granular biological materials such as grain. Trans. ASAE (Am. Soc. Agric. Eng.) 14:172-174, 178.

Peart, R. M., Y. C. Li, and J. R. Barrett. 1985. Simulation of a computerized grain drying decision support system. Paper 85-3014. Am. Soc. Agric. Eng., St. Joseph, MI. 5 pp.

Pfost, H. B., S. G. Maurer, L. E. Grosh, and G. Foster. 1977. Fan management systems for natural air driers. Paper 77-3526. Am. Soc. Agric. Eng., St. Joseph, MI. 16 pp.

Pierce, R. O. and T. L. Thompson. 1979. Solar grain drying in the North Central Region—simulation results. Trans. ASAE (Am. Soc. Agric. Eng.) 22:178-187.

Pierce, R. O. and T. L. Thompson. 1980. Management of solar and low temperature grain drying systems. Trans. ASAE (Am. Soc. Agric. Eng.). 23:1020-1023.

Purser, R. N. and D. J. Greig. 1967. Pressure variations in lateral ducts. J. Agric. Eng. Res. 3:184-193.

Roberts, E. H. 1972. Viability of seeds. Chapman and Hall, London. 448 pp.

Ryniecki, A. and M. E. Nellist. 1991. Optimization of control systems for near-ambient grain drying. J. Agric. Eng. Res. 48:1-17.

Ryniecki, A., W. E. Muir, and D. S. Jayas. 1991. Optimization of control systems for near-ambient wheat drying under Maritime and Continental climates. Paper 91-6551. Am. Soc. Agric. Eng., St. Joseph, MI. 17 pp.

Scheidegger, A. E. 1974. The Physics of flow through porous media. University of Toronto Press, Toronto. 353 pp.

Segerlind, L. J. 1982. Solving the nonlinear airflow equation. Paper 82-3017. Am. Soc. Agric. Eng., St. Joseph, MI. 18 pp.

Sharp, J. R. 1984a. The design and management of low temperature grain driers. J. Agric. Eng. Res. 29:123-131.

Sharp, J. R. 1984b. Experimental seed bed low temperature drying of wheat to validate simulation models. Div Note DN 1227, Nat. Inst. Agric. Eng., Silsoe (unpubl). 33 pp.

Shedd, C. K. 1951. Some new data on resistance of grains to airflow. Agric. Eng. 32:493-495,520.

Shedd, C. K. 1953. Resistance of grains and seeds to airflow. Agric. Eng. 34:616-619.

Shove, C. G. and V. W. Hukill. 1963. Predicting pressure patterns in perforated grain ventilation ducts. Trans. ASAE (Am. Soc. Agric. Eng.) 2:115-122.

Simonton, H. R., J. M. Allison, and B. D. McLenden. 1981. A microprocessor based grain drying control system. Paper 81-5027. Am. Soc. Agric. Eng., St. Joseph, MI. 28 pp.

Slattery, J. C. 1981. Momentum, energy and mass transfer in continua. R. F. Krieger Pub. Co. Inc., Huntington, NY. 682 pp.

Smith, E. A. 1982a. 3-Dimensional analysis of air velocity and pressure in beds of grain and hay. J. Agric. Eng. Res. 27:101-117.

Smith, E. A. 1982b. Simulation of grain-drying when air-flow is nonparallel. J. Agric. Eng. Res. 27:21-33.

Smith, E. A. 1984. Minimum energy to dry barley without mould growth in near ambient driers. J. Agric. Eng. Res. 30:289-303.

Smith, E. A. and P. H. Bailey. 1983. Simulation of near ambient grain driers: II Control Strategies. J. Agric. Eng. Res. 28:301-317.

Smith, E. A., D. S. Jayas, W. E. Muir, K. Alagusundaram, and V. H. Kalbande. 1992a. Simulation of grain drying in bins with partially perforated floors. Part I: isotraverse lines. Trans. ASAE (Am. Soc. Agric. Eng.) 35:909-915.

Smith, E. A., D. S. Jayas, W. E. Muir, K. Alagusundaram, and V. H. Kalbande. 1992b. Simulation of grain drying in bins with partially perforated floors. Part II: calculation of moisture content. Trans. ASAE (Am. Soc. Agric. Eng.) 35:917-922.

Spencer, H. B. 1969. Pressure drop in on-floor-duct drying systems. J. Agric. Eng. Res. 14:165-172.

Sokhansanj, S., D. S. Jayas, and A. A. Falachinski. 1988. Resistance of bulk lentils to airflow. Paper 88-6535. Am. Soc. Agric. Eng., St. Joseph, MI. 23 pp.

Stanek, V. and J. Szekely. 1974. Three-dimensional flow of fluid through nonuniform packed beds. AIChE 20:974-980.

Steele, J. L. and G. C. Shove. 1969. Design charts for flow and pressure distribution in perforated ducts. Trans. ASAE (Am. Soc. Agric. Eng.) 3:220-224.

Steele, J. L., R. A. Saul, and W. V. Hukill. 1969. Deterioration of shelled corn as measured by carbon dioxide production. Trans. ASAE (Am. Soc. Agric. Eng.) 12:685-689.

Sutherland, J. W., P. J. Banks, and H. J. Griffiths. 1971. Equilibrium heat and moisture transfer in air flow through grain. J. Agric. Eng. Res. 16:368-386.

Sutherland, J. W., P. J. Banks, and W. B. Elder. 1983. Interaction between successive temperature or moisture fronts during aeration of deep grain beds. J. Agric. Eng. Res. 28:1-19.

Thompson, T. L. 1972. Temporary storage of high moisture shelled corn using continuous aeration. Trans. ASAE (Am. Soc. Agric. Eng.) 15:333-337.

Van Ee, G. R. and G. L. Kline. 1979. Management strategies for corn production and low temperature drying. Paper 79-3032. Am. Soc. Agric. Eng., St. Joseph, MI. 50 pp.

Wallace, H. A. H., P. L. Sholberg, R. N. Sinha, and W. E. Muir. 1983. Biological, physical and chemical changes in stored wheat. Mycopath. 82:65-76.

Whitaker, S. 1969. Advances in theory of fluid motion in porous media. Ind. Eng. Chem. 61:14-28.

White, N. D. G., R. N. Sinha, and W. E. Muir. 1982. Intergranular carbon dioxide as an indicator of biological activity associated with the spoilage of stored wheat. Can. Agric. Eng. 24:35-42.

Williams, W. F. 1965. Pressure losses and drying rates in grain ventilated with various on-floor duct systems. J. Agric. Eng. Res. 10:271-276.

Winslow, A. M. 1967. Numerical solution of the quasi-linear Poisson's equation in a nonuniform triangle mesh. J. Comp. Phys. 1:149-172.

Zienkiewicz, O. C. and T. Shiomi. 1984. Dynamic behaviour of saturated porous media, the generalized Biot formulation and its numerical solution. Int. J. Num. Anal. Meth. Geomech. 8:71-96.

Zienkiewicz, O. C. and R. L. Taylor. 1989. The Finite Element Method. Fourth edition. McGraw-Hill Book Company, New York. 648 pp.

16

Heated-Air Grain Drying

Martin E. Nellist and David M. Bruce

Silsoe Research Institute, Silsoe, Bedford, United Kingdom

SUMMARY

Heated-air dryers are machines which normally both dry and cool the grain in a few hours at most. They have large power consumptions, mostly as heat to raise the air temperature and this heat is normally supplied to the air by combustion of fossil fuels either directly or through a heat exchanger and then to the grain. The development of mathematical models of the heat and moisture transfer and their incorporation into computer simulations of individual dryer types has been crucial to furthering the understanding, the design, and the operation of these machines.

The basis of the mathematical models is described and sources of data on the most important constituent properties of the grain are reviewed. In addition to physical properties, the biological properties determining the fitness of the grain for various end uses are necessary to define the limits, particularly of air temperature, within which a dryer must work. Models for loss of seed viability and of baking quality are described. Milling quality can be reduced by stress-cracking which may be a consequence of rapid drying, even though the cracking may not occur until the grain is in storage. The techniques of dryeration and combination drying are discussed; both these systems reduce cracking because rapid drying is not continued down to storage moisture levels. Surveys of grain contamination have not shown that there is a danger to human health from direct-firing with oil or natural gas.

The use of computer simulation in studies of the design and use of real dryers is illustrated in relation to performance at steady state and for correction and interpretation of dryer test results. Simulation of dryer behavior at unsteady state has been a powerful tool in the development of improved dryer control algorithms. The challenge for the future is to move towards optimal control of both cost and grain quality.

1. INTRODUCTION

In the context of this chapter, heated-air grain drying is defined as drying by the through-circulation of air which has been raised to temperatures in the range of 40–120°C but in some cases to as high as 275°C. Under these conditions, the drying of most cereal grains is completed within 0.5–5 h and, therefore, is carried out in dryers of the batch or continuous-flow type. More than one pass through the dryer may be necessary for certain crops (e.g., rice or grain with high moisture) and normally there will be a final cooling stage. Some drying occurs during cooling, particularly in the "dryeration process" discussed in Section 3.2.3.

Heated-air dryers have a high power requirement; a dryer capable of drying wheat at 10 t/h from a moisture content (mc) of 20–15% wet basis (wb) requires around 750 kW. In most drying installations, it would not be feasible (or economic) to use electrical power, and because most (between 95 and 97%) of the energy is used as heat, it is normally supplied from the combustion of fossil fuels. In direct-fired dryers, the products of combustion remain in the hot gas which, after dilution with additional air, is passed directly through the grain. Whether or not this practice constitutes a danger to human health is discussed in Section 3.2.5. In indirectly-fired dryers, the heat is transferred from the combusted gas to the drying air through a heat exchanger. This is less efficient but the loss of heat can be reduced by the use of a condensing heat-exchanger.

The latent heat of vaporization of water is of the order of 2.4 MJ/kg of water evaporated but the specific energy consumption (SEC) of heated-air dryers lies in the range 3–16 MJ/kg of water evaporated depending upon dryer design, type of grain, and moisture removed, particularly the final moisture level. Isaacs and Mühlbauer (1975) showed that even in an ideal dryer it would not be possible to reduce the SEC below about 3 MJ/kg. This is because drying is essentially an adiabatic process; although some heat is transferred from the drying air to the grain, most of the sensible heat is transferred into latent heat of vaporization so that the air temperature reduces but its enthalpy remains substantially constant. Figure 1 (Isaacs and Mühlbauer 1975) was drawn by dividing the heat required to raise air, at 10°C and 80% relative

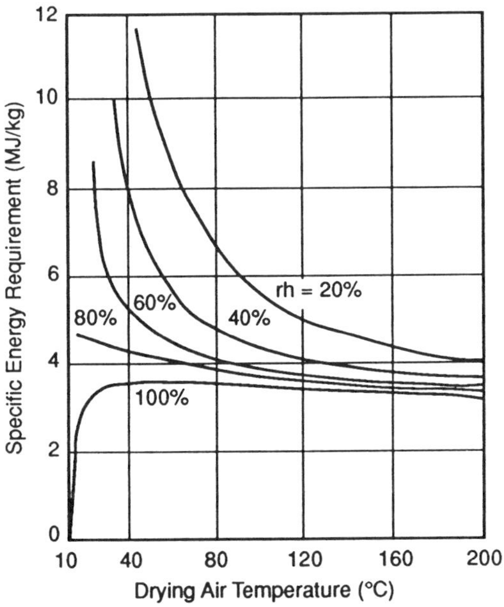

Figure 1 Effect on the specific heat energy consumption, MJ/kg water evaporated, for an "ideal" adiabatic dryer using air heated from 10°C and 80% rh to temperatures in the range of 40–200°C and exhausting at relative humidities of 20–100%. Source: Isaacs and Mühlbauer (1975).

humidity (rh) to temperatures in the range 40–200°C by the amount of water evaporated for varying percent of adiabatic saturation. Even at 100% saturation, the SEC does not reduce below 3 MJ/kg and, in practice, 100% saturation is only realized, if at all, during the short period in which the grain is still extremely wet. For heated-air drying the importance of Figure 1 is that as the drying air temperature increases the penalty for partial saturation reduces rapidly and at 200°C, the specific heat consumption is low (4 MJ/kg) even for air exhausting at 20% rh.

Another reason for increasing the drying air temperature is the reduction in the mass of air that has to be pumped per unit mass of grain. Figure 2 (Nellist 1982) plots, for a range of typical specific energy consumptions, the effect of a rise in air temperature on the mass of air required per unit mass of dried grain for a 20–15% moisture removal. Clearly if it was possible to move from temperature rises of 20–60°C to rises of 100–200°C, then the reduction in total air flow would be considerable. In theory, this would not only reduce fan power

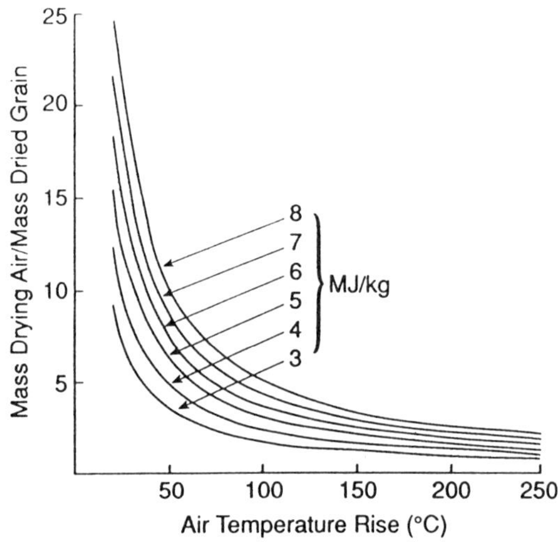

Figure 2 Effect of air temperature rise and specific heat energy consumption, MJ/kg water evaporated, on the mass ratio of drying air to dried grain for a moisture removal of 5% wb (20–15% wb). *Source*: Nellist (1982).

but would also make a contribution to the reduction in pollution from emissions, including noise. Thus, once the commitment is made to the use of heated air, the goal should be to heat as little of it as possible to as high a temperature as possible without causing unacceptable damage to the grain. These conflicting requirements of dryer efficiency and grain quality have lain at the heart of developments in heated-air drying.

A major constraint on the design of equipment for grain drying is the shortness of the harvest season and hence limited annual use. A dryer which works for only 4 wk of the year has to be considerably cheaper than those used continuously throughout the year, as in a food-processing plant, for example. Grain dryers, therefore, have to be relatively simple in a mechanical, if not a thermodynamic sense. Most of them, both batch and continuous types, rely on gravity to move a "packed-bed" of grain through the dryer at a rate determined by a simple discharge mechanism. Normally the depth of the grain bed relative to the direction of the air flow is kept sufficiently shallow (typically 0.1–0.4 m) for the fan pressures (and hence power) required for pumping the air to be relatively low (less than 1.96 kPa) and within the range of inexpensive agricultural fans of both centrifugal and axial types. Cory (1991a,b,c) has reviewed the design and performance of fans for crop drying purposes.

2. TYPES OF DRYER

2.1. A Dryer Classification

There are many detailed differences in mechanical design among dryers but the most important basis for the classification of dryer types is the relative direction of the flow of the grain and air. The flow directions are crucial because of their effect on the temperatures which may have to be endured by the grain. The air may cross-the path of the grain, it may flow along in parallel with it, and it may flow along in a parallel against it. Thus, the three basic types are: 1) cross flow, 2) concurrent-flow, and 3) counter-flow. Dryers may consist of one of these types or be a combination of two or all three. They may also incorporate partial recirculation or re-use of the drying and cooling air before it is exhausted to atmosphere.

2.2. Batch and Cross-Flow Dryers

In the simplest of batch dryers, the grain is placed inside a large shallow bin that has a perforated base and rests upon a hot-air plenum from which the heated air is able to move up through the bed of grain and discharge to atmosphere. In the direction of the air flow, the depth of the grain bed has to be such that the gradient in grain moisture content and temperature established across it is not too great (i.e., grain is neither excessively over- nor under-dried). For the air flows and pressures generated by typical fans, the depths will be of the order of 150–500 mm.

To turn a horizontal, flat-bed, batch dryer into a continuous-flow machine in which wet grain enters at one end and dry grain exits at the other, it is possible to devise a reciprocating mechanism (NIAE 1962) or moving beds (NIAE 1964). An inclined tray can make use of gravity to move the grain. Even greater use can be made of gravity if the tray is placed vertically with the grain now constrained between two perforated or louvred walls. In another arrangement, the air enters and leaves the grain through a series of layers of horizontal ducts of inverted V-shape and of which alternate layers are used for air inlet and exit, respectively. Such an arrangement can still be used as a batch dryer with the grain bed remaining static throughout drying and cooling. If the grain is moved down such a column of either type then it may simply be recirculated until dry (recirculating batch dryer) or the grain flow rate may be adjusted to give the residence time equivalent to the batch drying time and the grain discharged to store. The dryer is then in continuous-mode. In this mode, part of the grain chamber will normally be reserved for cooling with unheated air. This means that, unlike in a batch dryer, cooling time in a continuous-flow dryer varies with drying time and in some circumstances may not be adequate to cool the grain completely.

A simple cross-flow dryer is one in which the air flows across the path of the grain, as in a dryer with perforated walls. Here, the main problem of design is to reduce the gradient of moisture and temperature which develops across the bed. This can be done by splitting the drying column into an inner and outer part which can be moved at different rates, as in the design proposed by Bakker-Arkema et al. (1982) and Bakker-Arkema and Schisler (1984), or which can be interchanged or inverted part of the way down the dryer (Otten et al. 1980).

Where the grain column is moving through a dryer with layers of horizontal ducts, the relative flow of air and grain becomes more complex and combines elements of cross-, concurrent- and counter-flow. In Europe, this type of dryer is termed "mixed-flow." In North America it is often referred to as a rack-type dryer or an LSU dryer after Louisiana State University where they seem to have been first used in the United States. The most thorough and comprehensive single text dealing with mixed-flow drying is that by Olesen (1987).

2.3. Counter-Flow Dryers

In counter-flow dryers, the hot-air enters the driest part of the bed and leaves from the wettest. Since the air will exhaust close to saturation, counter-flow dryers have a potential for high efficiency. However, the grain exits the dryer at or near the air inlet temperature so this has to be low enough not to cause damage to the grain. The commonest and most successful type is the in-bin counter-flow dryer which is built in a round bin (Michigan State University 1981). Wet grain is spread uniformly across the top of a bed which may vary in depth from 0.5 to 3 m as dried grain is removed from the bottom of the bin by an auger rotating across the surface of the perforated base. In this case, the grain is cooled in a separate bin. In another version (Mathews 1980), both drying and cooling are arranged to occur in the same bin.

2.4. Concurrent-Flow Dryers

In concurrent-flow dryers, the air moves in the same direction as the grain. Such dryers are able to make use of the latent heat transfer to cool the drying air so that, even though the air temperature may be high, maximum grain temperatures do not exceed safe levels.

In the late 1970s and 1980s, several concurrent-flow dryers were developed in the United States (Bakker-Arkema et al. 1984) but they have not been popular in Europe where they pose both technical and management problems. While the high temperatures do give high drying rates, they also lead to large moisture gradients in the grain kernels and it is necessary to dry in short bursts with resting periods in between for moisture equilibration. Also there are mechanical difficulties in arranging the concurrent-flow and it is expedient to make use of the nonventilated resting bed as a barrier to flow. The net result is a very tall multistage dryer having a very long residence time even if actual drying times are short. Dryers with residence times of several hours require continual running night and day and require a continuity of grain supply. They are not compatible with European farms which, in general, are smaller than those of North America. Concurrent-flow dryers are more suitable for centralized storages and food-processing plants than for farms.

3. DRYER PERFORMANCE

3.1. Calculation

3.1.1. Equations of Heat and Moisture Transfer

The progress of drying in a heated-air grain dryer is affected by the interactions of several variables, some of which are subject to considerable and uncontrollable variation. Ambient conditions, for example, are changing continuously and contribute in part to fluctuations in the temperature of the heated-air. More importantly, the initial moisture content of the inlet grain is frequently changing. Hence dryers seldom, if ever, run at steady state and attempts to understand their behavior based upon experiments alone have never been very successful. The breakthrough in understanding came in the mid-1960s with the work of Bakker-Arkema et al. (1967a, 1967b) and Thompson (1970) in the United States and first Boyce (1965, 1966) and later O'Callaghan et al. (1971) in the United Kingdom. These workers showed that differential equations describing heat and moisture change in a bed of grain similar to those proposed by Van Arsdel (1955) could be solved by numerical means using digital computers. The development of the four partial differential equations commonly used to describe convective heat and moisture transfer in ventilated granular beds can be found in many texts (e.g., Brooker et al. 1992) but the following is a summary of that given by Sharp et al. (1982). The four equations are:

Moisture (mass) balance

$$G \frac{\partial H}{\partial x} = -\rho_g \frac{\partial M}{\partial t} - \varepsilon \rho_a \frac{\partial H}{\partial t} \tag{1}$$

Heat balance

$$G (c_a + c_v H) \frac{\partial T}{\partial x} = \rho_g c_v (T - \theta) \frac{\partial M}{\partial t} - h_a (T - \theta) - \rho_a \varepsilon (c_a + c_v H) \frac{\partial T}{\partial t} \tag{2}$$

Heat transfer

$$\rho_g (c_g - c_w M) \frac{\partial \theta}{\partial t} = h_a (T - \theta) + L_b \frac{\partial M}{\partial t} \rho_g \tag{3}$$

Moisture (mass) transfer

$$\frac{dM}{dt} = - k (M - M_c) \tag{4}$$

where:

c_a = specific heat of dry air ($J \cdot kg^{-1} \cdot K^{-1}$)
c_g = specific heat of dry grain ($J \cdot kg^{-1} \cdot K^{-1}$)
c_v = specific heat of water vapor ($J \cdot kg^{-1} \cdot K^{-1}$)
c_w = specific heat of water ($J \cdot kg^{-1} \cdot K^{-1}$)
G = mass flow rate of air ($\mathrm{kg \cdot m^{-2} \cdot s^{-1}}$)
H = absolute humidity of air (kg/kg dry air)
h_a = grain bed heat transfer coefficient ($J \cdot kg^{-1} \cdot K^{-1}$)
k = drying rate constant (s^{-1})
L_b = latent heat of vaporization of bound water (J/kg)
M = moisture content of grain (kg/kg dry matter)
M_e = equilibrium moisture content of grain (kg/kg dry matter)
T = air temperature (°C)
t = time (s)
x = depth in bed from air inlet (m)
ε = void ratio (decimal fraction)
ρ_a = density of air ($\mathrm{kg/m^3}$)
ρ_g = bulk density of grain ($\mathrm{kg/m^3}$)
θ = grain temperature (°C)

It has been shown (Spencer 1972, Bakker-Arkema et al. 1967b) that the terms involving the partial differentials $\partial T/\partial t$ and $\partial H/\partial t$ are small and can be neglected so that Equations 1 and 2 become Equations 5 and 6:

$$G \frac{\partial H}{\partial x} = - \rho_g \frac{\partial M}{\partial t} \tag{5}$$

$$G (c_a + c_v H) \frac{\partial T}{\partial x} = \rho_g c_v (T - \theta) \frac{\partial M}{\partial t} - h_a (T - \theta) \tag{6}$$

These or similar sets of equations can be solved by a variety of numerical methods (see, for example, Bakker-Arkema 1970). The papers by Sharp (1982) and Parry (1985) constitute a comprehensive survey of the development of the various formulations of the equations and their methods of solution. Although the method of solution is important, particularly in extreme situations, experience has been that there is very little difference between results from different models provided that the psychrometic and material properties used in the models are the same.

3.1.2. Mass Transfer Rate

Of the grain properties required, the most important is that represented by Equation 4, the rate of moisture loss. It has long been known that the loss of moisture from a wet, porous, body into a drier atmosphere is analogous to the loss of heat from a hot body and follows a law of decay which is approximately exponential. In the simple case when all the resistance to drying resides at the grain surface, then by analogy with heat transfer, the drying rate of individual grains can be expressed by Newton's equation, as in Equation 4.

Although it is relatively easy to show experimentally that this is not an ideal description of the drying rate of real grains, it is unfortunately not easy to substitute improved equations which have general application and acceptance. Thus in practice, Equation 4 is still widely used, while, in the meantime, research into what is commonly called "the thin-layer drying rate" of grain has continued more or less unabated over 40 yr.

The integral of Equation 4 is:

$$M = (M_o - M_e) \exp(-k\,t) + M_e \tag{7}$$

where: M_o = initial moisture content of grain (kg/kg dry matter). Equation 7 can be rewritten as:

$$MR = \exp(-k\,t) \tag{8}$$

where:

$$MR = \frac{M - M_e}{M_o - M_e}$$

and is termed the moisture ratio.

Equation 7 is a particular solution of Fick's law of diffusion for a body in which all the resistance to diffusion is concentrated at the surface. In more realistic solutions, which take account of the internal diffusion of moisture through the grain solid, the single exponential term is replaced by an exponential series (Crank 1975, Parry 1985). These solutions account for the initially higher drying rate obtained while a moisture gradient is established across the grain but do not allow for the reduction in moisture diffusivity within the grain (and hence a reduction in the value of k) with decreasing moisture content.

Several empirical modifications of Equation 8 have been suggested. Of these the equation of Page (1949), is both effective and relatively simple:

$$MR = \exp(-kt^n) \tag{9}$$

where n is an exponent determined by experiment. For corn the equation of Thompson et al. (1968) is sometimes used:

$$t = a \ln(MR) + b \ln(MR)^2 \tag{10}$$

where a and b are coefficients determined by experiment.

One way to account for the reduction in diffusivity with decreasing moisture content is to solve the diffusion equation numerically by considering the grain as a number of concentric shells each of which has a moisture-dependent diffusivity (Bruce 1985). Continuing increases in computing power have offset the disadvantages of greater computing time required by such models and there are situations, such as modeling dryers with tempering (resting) sections, where their use is essential. Again, the reader is referred to Parry (1985) for a review of "thin-layer equations." Sokhansanj and Cenkowski (1988), Jayas et al. (1988), and Cenkowski (1989) have reviewed thin-layer drying equations and summarized values of the coefficients for most of the important grains.

The dependence of the rate coefficient, k, upon air temperature is often expressed by means of the Arrhenius equation for the rate of a chemical reaction:

$$k = A \exp[-B/(T + 273)] \tag{11}$$

where A and B are experimentally determined coefficients. While having a basis in theory, this equation does not allow for any dependency upon other variables, particularly relative humidity. Note that most exposed- or thin-layer experiments are carried out using an air flow well in excess of that necessary to remove surface moisture and therefore, the value k so determined is independent of air flow. One drawback of Equation 11 is that, because typical drying temperatures have a small range when expressed in K, minor variations in the values of k can cause large variations in the values of A and B derived from very similar data. Better correlations are often obtained by expressing k as a function of the Celsius temperature, sometimes with the inclusion of a separate term for relative humidity. Since the driving force for drying may be regarded as the difference in vapor pressure between the ventilating air and the grain surface, k is usually found to be a near-linear function of either the vapor pressure or the vapor pressure deficit (Nellist 1976a).

Many authors have used multiple linear regression to express the variation in k as a function of several variables, often including additional terms representing interactions (products) or transformed values (e.g., logarithms) of the variables. Although these can be relied upon to fit the experimental data well, they have no real basis in theory and the danger is that they also describe the "noise" specific to the particular experiments.

There is a need to bring together the many studies on thin-layer drying rate to derive a single expression for each crop species. The difficulty is, however, that such a task would need to begin with the original data for the drying curve so that a uniform approach could be taken to fitting the curve and expressing the dependency of the coefficients. Unfortunately, because of the large number of data, the published papers rarely provide the data necessary to carry out such a harmonization. It might also be argued that the better route is to further the understanding of the intraparticle drying process so that a fundamental knowledge of diffusivity within grain solids would allow the drying rate of any grains to be computed from a knowledge of shape and structure. This approach was taken by Fortes and Okos (1981) and Fortes et al. (1981), to some extent by Bruce (1985), Sokhansanj and Bruce (1986, 1987) and later by Haghighi et al. (1990).

Another area of debate is whether it is legitimate to rewet the grain used for exposed-layer drying studies (and full-scale drying tests). Apart from the fact that few studies would have been possible unless they had been carried out with artificially rewetted grain, the evidence (Hustrulid 1962, 1963; Sokhansanj et al. 1983, 1984) suggests that for mature grain, properly equilibrated, the drying rates of naturally-wet and artificially-wet grain are virtually the same, certainly for practical purposes. It is, of course, necessary to draw a distinction between mature grains—in which the conversion of soluble sugars to starch is complete—and immature grains containing liquid components, which once desiccated cannot be reconstituted. The results of Dietrich (1957) illustrate this problem. It is also necessary to ensure that rewetted grain has time to equilibrate before use. Fortes et al. (1981) developed a model of the intrakernel drying of wheat based on nonequilibrium thermodynamics and using the concept of liquid and vapor conductivities rather than liquid diffusion. They showed that using material properties derived from drying curves the model accurately predicted rewetting curves. They also concluded that the upper limit of moisture content to which the model applied and beyond which the wheat was immature was 28.5% wb (40% db).

Much research has been devoted to describing the drying rate as a function of the drying conditions but less attention has been given to the rate of heat transfer (represented in Eqs. 2, 3, and 6 by the term h_a). The reason is that heat transfer is far more rapid than moisture transfer and many workers follow Pabis and Henderson (1962) in assuming uniform kernel temperature and using an average value of the heat transfer coefficient. In exposed-layer experiments with wheat, barley, and rapeseed, Bruce (1986) measured both the mass loss and the surface temperature. Using these data, Sokhansanj and Bruce (1987) developed a new model with which to calculate the temperature gradient within the kernel.

Like the moisture diffusion model of Bruce (1985), this model represents the kernel as a series of concentric spheres and uses a moisture-dependent thermal diffusivity. Haghighi et al. (1990) model moisture and temperature change in a single kernel using the finite element method.

3.1.3. Equilibrium Moisture Content

The equilibrium moisture content (M_e) is the other important grain property included in Equation 4. It represents the moisture content which the grain will attain if dried for an infinite time at a particular relative humidity and temperature. Whereas the mass transfer rate is strongly dependent upon temperature (or vapor pressure), the equilibrium moisture content is more strongly dependent upon relative humidity. Many equations have been proposed to describe the equilibrium relationships between grain moisture content, air temperature, and humidity; Chen and Morey (1989) reported that van den Berg and Bruin (1981) compiled a list of 77 theoretical, semi-theoretical, and purely empirical equations.

Following an initial investigation of five different equations for corn, Pfost et al. (1976) concluded that equations with three coefficients described the data as well and were easier to use than those with four coefficients. On this basis they selected the Modified-Henderson equation:

$$M_e = \left[- \frac{\ln (1 - \varphi)}{a_1 (T + a_2)} \right]^{1/a_3} \tag{12}$$

where:

φ = relative humidity (decimal)
a_1, a_2, and a_3 are coefficients

and the Chung-Pfost equation:

$$M_e = - \frac{1}{b_3} \ln \left[\frac{- \ln \varphi \, (T + b_2)}{b_1} \right] \tag{13}$$

where b_1, b_2, and b_3 are coefficients that are fitted to data from 20 food grains.

Nellist and Dumont (1979) fitted data for wheat from 13 sources to Equations 12 and 13, and also to the equation of Smith (1947) modified by the inclusion of a temperature term:

$$M_e = c_1 - c_2 \ln (1 - \varphi) - c_3 \ln (T) \tag{14}$$

where c_1, c_2, and c_3 are coefficients.

Equation 14 gave slightly better fits than either Equation 12 or Equation 13 and has been used in heated-air dryer models (Bruce 1984, Nellist 1987) based on the drying model of Nellist (1974). Chen and Morey (1989), fitted Equations 12 and 13 and two other equations, the Modified-Halsey:

$$M_e = \left[\frac{\ln \varphi}{\exp\,(d_1 - d_2 T)}\right]^{-1/d_3} \tag{15}$$

where d_1, d_2, and d_3 are coefficients, and the Modified-Oswin:

$$M_e = (e_1 + e_2 T) \cdot \left(\frac{\varphi}{1 - \varphi}\right)^{-1/e_3} \tag{16}$$

where e_1, e_2, and e_3 are coefficients, to 36 data sets for 18 crops. They found that the Modified-Henderson (Eq. 12) and Chung-Pfost (Eq. 13) equations gives the best fits to data for starch grains and fibrous materials but that for seeds high in oil and protein, the Modified-Halsey equation (Eq. 15), is preferred.

Data for fitting to these equations are normally derived from "static" experiments of two types. In the first type, the grain is brought into equilibrium with a known relative humidity and in the second, the relative humidity generated by a sample of known moisture content is determined. Mainly because of its speed, the second method appears to be more popular but the first more closely represents the drying (desorption) or wetting (adsorption) process and is equivalent to continuing an exposed-layer drying experiment for a long time. In practice, exposed-layer drying experiments (particularly at temperatures used for heated-air drying) are seldom continued to equilibrium since the moisture value is well below that of economic interest. But when the data are fitted to a thin-layer equation the value of the equilibrium moisture content (M_e) may be estimated by the curve-fitting procedure. Because of the inability of most thin-layer equations to describe adequately the drying curve along all of its length, these values of M_e are often higher than the real long-term static values. The use of these so-called "dynamic" equilibrium moisture contents in a simulation model, particularly one including cooling, may prevent the target moisture value being achieved. Therefore, whenever possible, the value of M_e should be determined in separate static experiments but with the same material used for the thin-layer experiments.

3.1.4. *Net Heat of Desorption*

One reason why the results obtained from computer models based on different formulations of the heat and mass transfer equations are so similar is that the process is very much dominated by the latent heat of vaporization. It is, therefore, important to quantify the amount by which the latent heat of vaporization of water held in "bound" form (L_b) exceeds that of "free" water (L_f). The

Table 1 Values of the Coefficients in Equation 17.

Crop	Coefficients a	b	Moisture range (db)
Barley	1.0	−19.9	$0.09 < M < 0.22$
Bean	0.5	−16.0	$0.09 < M < 0.28$
Maize (corn)	2.1	−17.0	$0.10 < M < 0.24$
Rice	3.2	−21.7	$0.10 < M < 0.14$
Sorghum	1.2	−19.6	$0.10 < M < 0.24$
Soybean	0.4	−13.9	$0.06 < M < 0.26$
Wheat, durum	0.8	−18.1	$0.10 < M < 0.26$
Wheat, hard	1.7	−17.6	$0.10 < M < 0.26$
Wheat, soft	3.9	−23.6	$0.10 < M < 0.20$

Source: Cenkowski et al. (1990a)

difference $(L_b - L_f)$ is termed the net heat of desorption, but the two quantities are often expressed as the ratio L_b/L_f or $[(L_b/L_f) - 1]$. Othmer (1940) showed that the ratio of L_b/L_f is the slope of the straight line obtained by plotting (for a range of temperatures) the logarithms of the partial pressures of water vapor generated within a seed sample at a single moisture content against the equivalent saturated vapor pressures of water. The variation with temperature of the partial vapor pressure in the grain can be calculated from the relationship between relative humidity and equilibrium moisture content M_e, given by equations of sorption isotherms. Using this method Gallaher (1951) expressed the dependence of L_b/L_f on moisture content by the equation:

$$L_b = L_f\,[1 + a \exp(bM)] \tag{17}$$

where:

L_b = latent heat of vaporization of bound water (J/kg)
L_f = latent heat of vaporization of free water (J/kg)
a, b = empirical coefficients found to be 23 and −40, respectively, for wheat

Cenkowski et al. (1990a) used the same method to derive values of the coefficients a and b for a range of grain crops (Table 1). Partial vapor pressures were calculated from the Chung equation and coefficients for that equation published in ASAE Standards (ASAE 1987). A comparative plot of $[(L_b/L_f) - 1]$ against moisture content (Fig. 3) indicates that there are differences between

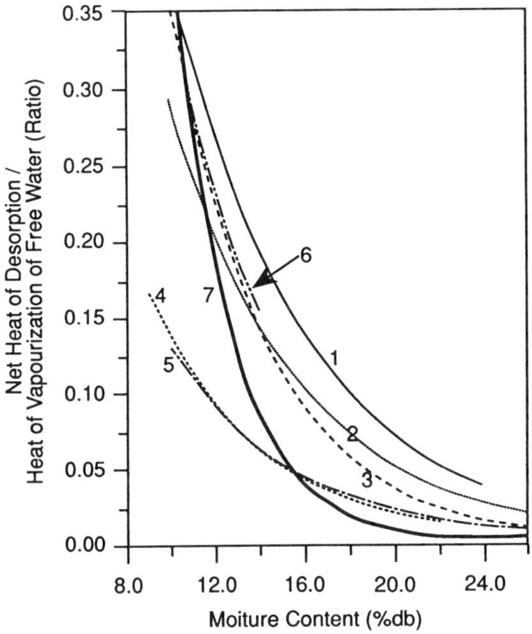

Figure 3 Effect of grain moisture content upon the ratio of net heat of desorption to the latent heat of vaporization of free water for: 1) maize, 2) hard wheat, 3) soft wheat, 4) barley, 5) durum wheat, 6) rice; plotted from Equation 17 using the coefficients of Cenkowski et al. (1990) (Table 1). Curve 7 is that given for wheat by Gallaher (1951).

grains but that in every case the net heat of desorption increases rapidly as moisture reduces below about 15% db (13% wb).

Hunter (1987) used sorption isotherm data from a number of sources to derive values of the coefficients in an alternative expression for the ratio of the latent heats:

$$\frac{L_b}{L_f} - 1 = \frac{a \ln b\,M - (m/M_z)^n\, c \ln d\,M}{1 - (M/M_z)^n} \tag{18}$$

where M_z is a transitional moisture content given by:

$$M_z = \left(\frac{b^a}{d^c}\right)^{1/(c-a)} \tag{19}$$

Table 2 Values of the Coefficients in Equation 18

Crop	n	m	a	b	c	d	P_o	M_z
Barley	8.478	1.779	−0.2625	4.610	−0.0091	1.249	6.918e+005	0.207
Maize	11.476	3.788	−0.5016	5.632	−0.09832	3.104	1.158e+005	0.154
Peanuts	11.295	4.446	−0.3040	7.490	−0.0712	3.997	1.212e+005	0.110
Rapeseed	7.336	2.647	−0.2298	10.589	−0.01706	2.446	6.518e+006	0.084
Rice	12.056	4.967	−0.2991	5.143	−0.14485	3.811	9.748e+005	0.147
Sorghum	56.381	9.715	−0.1766	5.187	−0.06118	4.282	7.403e+007	0.174
Soybeans	10.358	5.665	−0.2230	8.166	−0.04524	2.896	8.538e+006	0.094
Sunflower	6.072	3.045	−0.2780	12.673	−0.04931	4.479	5.053e+006	0.063
Wheat	17.671	4.968	−0.2113	4.740	−0.03483	2.621	4.457e+006	0.188

P_o is the intercept pressure (Pa) from which relative humidity can be calculated from Hunter's Equations 1 and 5.
M_z has been calculated from Equation 19.
Source: Hunter (1987).

and *a*, *b*, *c*, and *d* are coefficients (Table 2). Hunter (1989) followed a slightly different approach in developing an expression for $[(L_b/L_f) - 1]$ for Australian paddy rice based on the isotherm equation used by Putranon et al. (1979, 1980). The approach would work with other isotherm equations provided they can be differentiated with respect to temperature at constant moisture content. An equation for $[(L_b/L_f) - 1]$ can then be written directly in terms of the coefficients of the isotherm equations.

3.1.5. Other Physical Properties

Most of the other necessary physical properties such as specific heat capacity and bulk density can be found in standard works of reference and are briefly discussed in Chapter 12. A property which is not often discussed in the context of heated-air dryers is the resistance of the grains to air flow. As was explained in the introduction, in most heated-air dryers the air flow paths are relatively short, the fan pressures required are generally low and energy consumption is small (often < 3%) in relation to heat energy consumption. Many dryers are either not fitted with or have only crude control over air flow. In such dryers, differences in resistance between crops (e.g., wheat and rapeseed) can cause significant differences in flow and hence evaporative capacity. Most if not all of the available data (see Alagusundaram and Jayas 1990 for a review) have been measured on static beds whereas in a continuous-flow dryer, the grain has to expand to move and as a result may be expected to have a lower resistance. Furthermore, in dryers such as the mixed-flow type where air flow paths are complex, it will be important to describe these flows (and those of the grain) if grain temperatures are to be predicted with sufficient accuracy to allow the computation of grain damage. Cenkowski et al. (1990b) have used the finite element method to model the flow of air in a mixed-flow dryer. They did not model the movement of the grain but found some evidence that packing density varied with position and that the pressure resistance was anisotropic. It may be expected that the design of heated-air dryers will eventually benefit from the modeling of three-dimensional air flows which is now an important topic in research into bulk storage drying (e.g., Alagusundaram et al. 1989).

3.2. Grain Quality

3.2.1. Grain Viability

Grain dryers are machines designed to preserve the quality of grain for a variety of end uses. They do this by rapidly removing moisture which would otherwise cause deterioration by microbial growth. The problem with heated-air dryers is that the heat and speed of drying may cause damage. Arguably, the most sensitive measure of quality which can be affected by heated-air drying is the grain viability as expressed by the germination capacity. Germination capacity is important because a proportion of all grain crops must be retained for seed and a reduction in germination capacity reflects damage to biochemical systems within the grain and may be an indication of a reduction in related measures of quality for other end uses (e.g., baking quality of wheat). Even where another quality parameter (e.g., oil quality) is not immediately affected, it is possible that its subsequent deterioration may be faster in dead seeds than in live seeds.

Efforts to understand the mechanisms of damage to viability began when Groves (1917) carried out a series of treatments in which grain was heated but not dried. Similar experiments have been comprehensively reviewed by Nellist (1978). The virtue of this type of experiment is that, because grain moisture content and temperature are held virtually constant, the only variable factor is exposure time. This reduced number of variables gives results which are more repeatable and more easily interpreted than those from drying tests where all three factors are varying simultaneously. Initially, it was possible to derive equations describing a critical temperature above which an unacceptable level of damage occurs. For situations where grain temperature and moisture content remain more or less constant (e.g., storage), the critical temperature is useful and capable of direct application. Under drying conditions, however, both the moisture content and temperature are continuously changing and no single combination of the variables can represent the conditions experienced by all the grain in the batch. In an attempt to relate the published results of germination loss in laboratory drying experiments to the "sealed heating" results, Nellist (1978) showed that damage usually occured in those drying treatments for which the final grain temperature had exceeded the critical grain temperature (calculated from the drying time and final moisture content) but the comparison gave no indication of the extent of the damage.

A method suggested by Voroshilov and Kuperschmidt (1973) of integrating the "sealed" heating equation is similar in concept to the "spoilage index" used for low temperature drying by Frazer and Muir (1981). The sealed heating equations are used to calculate, at each time step and for each layer, a maximum exposure time. This time, divided into the length of time step, gives the fraction of the maximum used up in the time step. When successive fractions of safe exposure time have accumulated to unity, that layer is deemed to have exceeded the threshold value. The problem remains that, although damaged layers may be identified, the overall depression in germination cannot be assessed.

A method by which the damage can be accumulated in units which indicate the actual depression in germination was developed by Nellist (1981) from a study (Nellist 1978) of the work on accelerated aging by Roberts (1960) and colleagues. Roberts realised that, although the time-scale over which viability is lost varies with temperature and moisture content, under constant conditions the pattern of seed deaths is always the same. That is, at first the rate of death is so low that changes in viability are difficult to detect. Gradually, more seeds begin to die and the death rate accelerates to a maximum at which time half of the seeds have died. Thereafter it necessarily declines to a very low level. Thus, seed deaths tend to be normally distributed in time and plots of the accumulated percentage of deaths (or more conveniently the percentage of seeds remaining viable; this is the germination percentage) against time produce a sigmoid shaped curve typical of the normal probability integral. Replacing probability (P), by the decimal value of the germination (G_s) can be written as:

$$G_s = \frac{1}{\sqrt{2\pi}} \int_{-\infty}^{x=(\bar{t}-t)/\sigma} \exp\left(-\frac{1}{2}X^2\right) dX \tag{20}$$

where:

- t = exposure time (s),
- $\bar{t}$ = the exposure time at which $G_s = 0.5$ (i.e., 50% germination, the half life of the seed sample (s)
- σ = the standard deviation of the distribution.

The quantity X is a value termed the standard normal deviate and is symmetrical about zero at the point $G_s = 0.5$ (i.e., $\bar{t} = t$). X depends on $\bar{t}$, t, and σ but for any value of X there is only one value of G_s; also, X varies linearly with t. Thus, if X_o is the value of X at $t = 0$ then:

$$X = X_o - \frac{t}{\sigma} \tag{21}$$

A branch of statistical analysis called probit analysis (Finney 1971) is based upon Equations 20 and 21, and replacement of the probability scale with its transformation to X is the basis of the probability axis in graphical statistics. Ellis and Roberts (1980) have coined the term "probit viability" to describe the values of X. The standard deviation (σ) is a measure of the dispersion of the deaths about the mean exposure time (or half life) and can be shown to be that time in which the germination reduces from 84 to 50%.

For a particular moisture content and temperature there will be a single death rate, $1/\sigma$, which applies at all points on the transformed probit curve. Thus for a given exposure at a given condition, the final value of X depends upon the initial value (X_o). This explains why grain with low initial germination is more easily damaged than that of good quality. For example, one unit (or standard deviation) of probit damage will reduce an initial germination of 99% ($X_o = 2.33$) and of 90% ($X_o = 1.28$) by 8% and 29%, respectively.

A most important finding by Ellis and Roberts (1981) was that the value of σ was a function of seed moisture and temperature only and was not affected

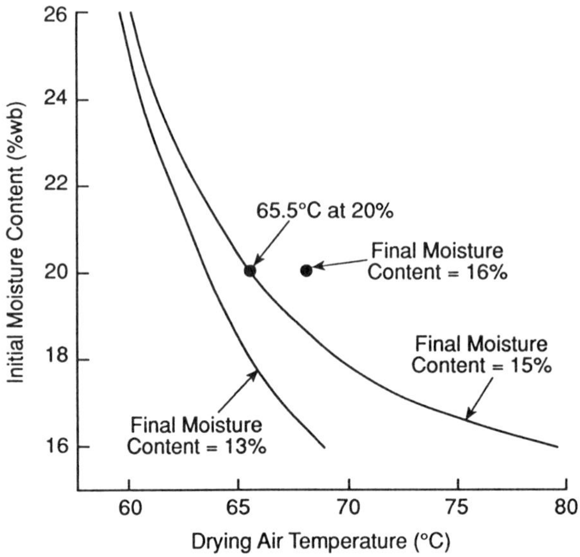

Figure 4 Combinations of drying air temperature, initial moisture content, and final moisture at which the computed probit loss of viability for a cross-flow dryer is 0.028 units, an undetectable amount of damage for good quality grain. *Source*: Nellist and Bruce (1987).

by the initial quality of the seed. As a consequence, it is possible to add reduction in probit from successive environments to estimate the total probit loss. The final probit viability is calculated by subtracting this loss from the initial probit value; the final viability can be found by converting probit back to germination. The procedure is similar to using logarithms.

The computer simulations of batch and continuous-flow drying developed and used at Silsoe Research Institute (Bruce 1984, Nellist 1987) incorporate this procedure (Nellist 1981) to calculate probit loss. The routines work well provided the prediction of grain temperature and moisture content is good (which is true in the cross-flow and batch models, for example, but not in the mixed-flow model where the complex pattern of grain and air flow is not yet modeled in detail). Nellist and Bruce (1987) have shown that to avoid the need to assume an initial value of germination, the amount of damage caused by a grain dryer at given operating conditions can be characterized in terms of probit loss rather than germination loss. For example, evidence from a series of dryer tests suggested that 65.5°C was a safe drying air temperature for cross-flow dryers drying wheat from 20 to an average of 15% mc (Nellist 1982). For the dryer described by Nellist (1987), having a drying-to-cooling ratio of 3:1, the predicted probit loss at this operating condition is 0.028 units. Other conditions of drying air temperature, and initial and final moisture contents which give the same loss, were used to draw the "isoprobit" lines of Figure 4 (Nellist and Bruce 1987). These curves show that the safe drying temperature reduces with increasing initial moisture content and also with decreasing final moisture content. This is because of the increase in residence time necessary to remove additional water. Although not verified by physical experiment, circumstantial evidence suggests that this is a realistic envelope of safe temperatures for this type of dryer. Further work is necessary to determine the death rates for other crops and to improve the prediction of grain temperatures in dryers of complex design such as the mixed-flow dryer. Similar information on the concurrent-flow dryer has been developed by Lescano and Tyrrell (1987) for the drying of maize seed.

The probit approach is also applicable to the death of insects and has been used in the development of methods for disinfesting wheat (Evans et al. 1983, Thorpe 1987) and dried grass (Sokhansanj et al. 1990, Wood and Sokhansanj 1990). In moist climates (e.g., the United Kingdom), there may be potential for using heated-air dryers for thermal disinfestation of stored cereal grains (see Chapter 11).

3.2.2. *Baking Quality*

The concept of using sealed heating tests to reduce the number of variables can be applied to the loss of baking quality. Becker and Sallans (1956) found that reduction in the relative volume of loaves baked from heat-treated grain followed an exponential law of decay with time. The quantities of grain required for baking loaves made it impractical to carry out large numbers of sealed heating treatments; a more practical approach is to proceed directly to thin layer-drying tests.

Wassermann, L., et al. (1983) and Bruce (1986) used thin-layer drying tests to investigate the effect of grain temperature, moisture, and exposure time on germination and baking quality. Their results (Fig. 5) (Nellist and Bruce 1987) are in good agreement. Both graphs illustrate the rapid increase in damage rate with grain temperature between 70–80°C and a less rapid increase with moisture between 18–22% (wb). Schreiber et al. (1981) fitted a model to data of Wassermann, L., et al. (1983) which predicts a constant rate of loss of specific loaf volume at a given grain moisture and temperature:

$$-\frac{dQ}{dt} = k_o \exp\ (a\theta + bW) \tag{22}$$

where:

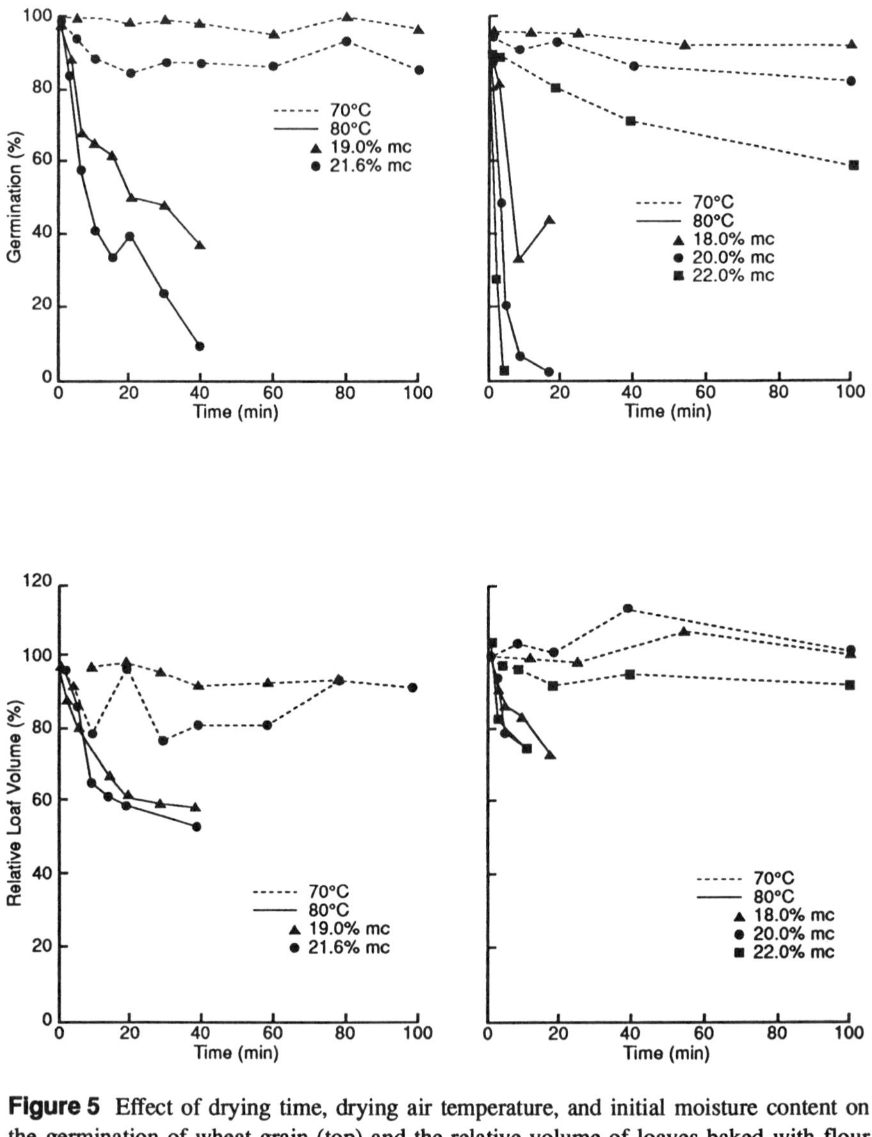

Figure 5 Effect of drying time, drying air temperature, and initial moisture content on the germination of wheat grain (top) and the relative volume of loaves baked with flour from the wheat grain (bottom) dried in thin layers by Wassermann, L. et al. (1983) (left) and Bruce (1986) (right). *Source*: Nellist and Bruce (1987).

Q = specific loaf volume (cm^3/g)
W = grain moisture content (% wb)
k_o, a, and b are coefficients determined by experiment

Using a diffusion model consisting of ten concentric spherical shells to represent the drying behavior of the individual grains (Bruce 1985), Bruce (1992) computed the rate of loss of relative loaf volume for each shell using Equation 23:

$$-\frac{dV_r}{dt} = V_r \exp\{130 + 41.4M_s - 427000/[(R(\theta_s + 273)]\} \tag{23}$$

where:

V_r = contribution to relative loaf volume (V/V_c) of a shell of the 10-shell grain model
V = loaf volume (cm^3)
V_c = loaf volume of control (unheated) sample (cm^3)
M_s = moisture content of a shell of the 10 shell grain model (% db)
θ_s = temperature of a shell of the 10 shell grain model (°C)
R = gas constant = 8.3143 $J \cdot mol^{-1} \cdot K^{-1}$

The mean relative loaf volume for the 10-shell model was calculated for each of the times when the grain in the experimental tests had been sampled and a test loaf baked. The coefficients of Equation 23 were derived from 112 thin-layer drying tests by nonlinear optimization and validated (Bruce 1992) by comparison with data of Wassermann, L., et al. (1983). Figure 6 (Bruce 1992) shows the agreement between relative loaf volume predicted over time by Equation 23 and the experimental data from 29 drying tests with wheat initially at 22% (wb).

3.2.3. *Milling Quality*

For maize and rice destined for dry and wet milling, it is important to avoid creating stress cracks which cause the starch granules to fracture during milling. At one time it was thought that rapid removal of moisture caused stress cracks; it was then realized that they were more likely to be caused by rapid cooling (Foster 1973).

A possible explanation (Nellist 1984) for this is as follows. When a grain kernel enters a dryer, the moisture is more or less uniformly distributed throughout the endosperm. The first moisture to be removed by the drying air will be that at or close to the surface; for further drying to take place, moisture must migrate from the inner to the outer layers. Therefore, a moisture gradient is established from the outside to the inside of the grain and, because of the tendency of the dry material to shrink around the wet inner core, severe stresses

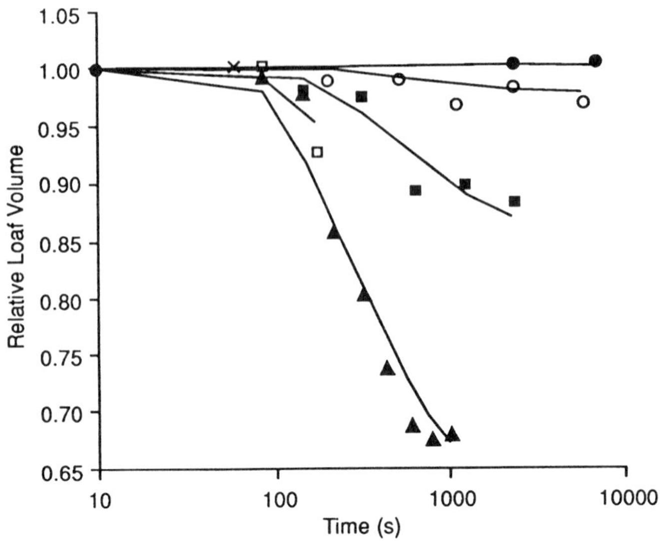

Figure 6 Relative volumes of loaves predicted by Equation 23 compared with values observed for wheat samples dried in thin layers at temperatures of 65 (●), 70 (○), 75 (■), 80 (□), and 85°C (×) for the cultivar Broom, and 80°C (▲) for the cultivar Avalon. *Source*: Bruce (1992).

can be induced, particularly in large grains such as maize. Because thermal diffusivity is much greater than the moisture diffusivity, the temperature gradients across the grain are small, so small that they can normally be neglected. The temperature of the grain, however, does affect the moisture diffusivity and the speed with which the moisture equalizes if drying is stopped. Thus, rapid cooling after heated-air drying hinders moisture equalization, increases the stresses and causes the grains to tear apart. On the other hand, if the grain is kept hot, the moisture equalizes quite quickly, the stresses are relieved and cracking during subsequent cooling is greatly reduced. This is a reasonable hypothesis if the average moisture content of the whole kernel (and hence the moisture content of the core) is relatively high; (> 17% wb). Kunze (1984), reviewing the causes of cracking in rice, points out that the moisture equalization, while allowing the outer layers to expand, may cause the inner core (which is losing moisture) to contract and come under increased stress. Thus, cracks or fissures develop slowly during subsequent storage of rapidly dried rice or in rice which is so dry that it reabsorbs moisture from the atmosphere.

In the dryeration technique (Foster 1964, 1973; Thompson and Foster 1967; McKenzie 1969; Sabbah et al. 1972), partially-dried grain is transferred from

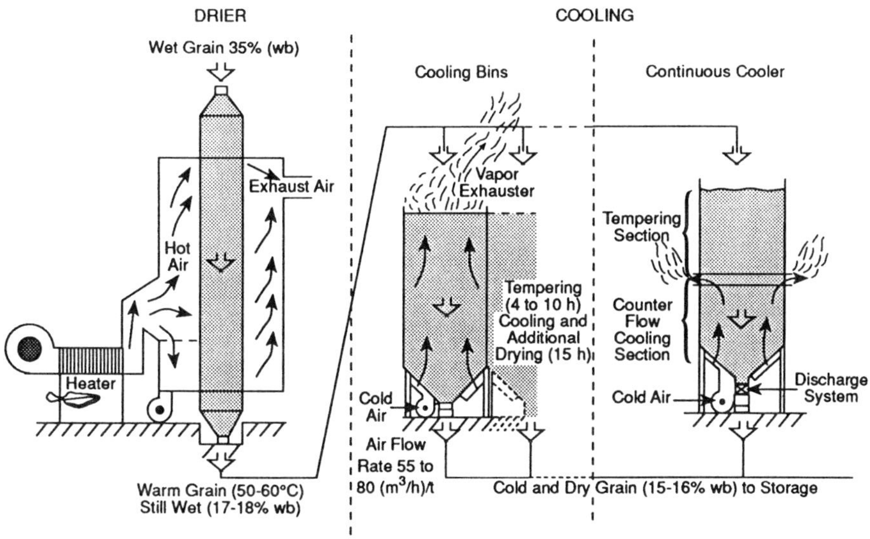

Figure 7 Steps in the dryeration process for maize showing cooling in either batch or continuous mode. *Source*: Lasseran (1977).

the dryer (without cooling and at about two percentage points above the normal storage moisture content) into a "steeping" bin where it is allowed to "sweat" for several hours before final cooling. After sweating, the surface layers of the kernel are both warm and relatively wet and most of the remaining moisture is removed by evaporative cooling. This means that the dryeration process not only prevents cracking but is more thermally efficient than conventional drying and rapid cooling.

The technique has found favor in France, where corn is dried from moisture contents as high as 40% wb. For corn (Lasseran 1982), the grain is first dried to 17–19% (wb) and transferred at 50–60°C to a sweating bin for 8–15 h. It is then cooled for 12–14 h with an air flow of 40–60 $(m^3/h)/m^3$ of grain. Cooling is normally done by a batch process in bins. For corn tempering, it is necessary to have at least four bins since each batch may require up to 32 h consisting of fill (8 h), temper (4 h), cool (15 h), and empty (5 h). An alternative approach is to use a special continuous-flow cooling bin (Fig. 7), in which only the bottom half is ventilated in counter-flow while grain descending through the top half tempers in still air.

Although stress cracking is not a problem for the small grains (e.g., wheat, barley, oats, and rapeseed), the introduction of the in-bin counterflow dryer (Michigan State University 1981) has introduced dryeration to grain farms not growing corn. Nellist (1984) used computer simulation to examine the application of the technique to the drying of wheat from 20 to 15% mc (wb). The potential thermal efficiency of the combined system and increased capacity of the dryer were confirmed. Tempering times and minimum grain temperatures were based on limiting germination damage to less than 0.2 units of probit viability.

Gustafson et al. (1983) commented that the dryeration process as such is not a practical system for large-scale facilities and so they explored the possibilities of reducing tempering times. They found that delaying cooling of corn for as little as 30 min after drying decreases breakage by 96% and increases moisture removal during subsequent cooling by 70%. A system in which partial drying by heated air is immediately followed by further drying with air at near-ambient temperatures is termed "combination drying." Otten and Brown (1982) showed that this system gives much of the energy saving and reduced breakage of dryeration. Sokhansanj and Lischynski (1991) showed that combination drying of wheat saves energy compared with conventional heated-air drying but the effect on stress-cracking was not studied. It seems probable that this technique works because the switch from rapid to slow drying is made at moisture contents above those which impose excessive stress.

In fundamental studies, Litchfield and Okos (1988) and Haghighi and Segerlind (1988a,b) have attempted to model stress cracking in individual kernels. Such work will lead to much better understanding. The topic is complex but important and will continue to be the subject of further research.

3.2.4. Stock Feed Quality

The "least demanding" use for grain is as feed for stock and, in the case of ruminants and provided the grain is not charred, it may be both more palatable and more easily digested in the rumen if it has been slightly toasted (Sullivan et al. 1974). In rations for nonruminant stock, the grain component may be an important source of protein; therefore, protein denaturation during drying should be limited. Mühlbauer and Christ (1974) found reduction in the availability of the protein lysine a useful measure of damage and suggested an upper acceptable limit of 10% (Fig. 8). Combinations of temperature and time of exposure which Mühlbauer and Christ found to give such a reduction are plotted in Figure 8 as a dotted line. In comparing results from a number of workers, Nellist (1978) found this line separates those experimental treatments which have damaged grain from those which have not. Thus, it seems that an air temperature of 120°C can be tolerated for periods up to 1 h and 100°C can be tolerated for up to 3 h. In practice, damage to feed grain is not a problem since most dryers are simply not capable of raising the grain to these temperatures. Also, the maximum operating temperature is limited by the difficulty of cooling hot grain.

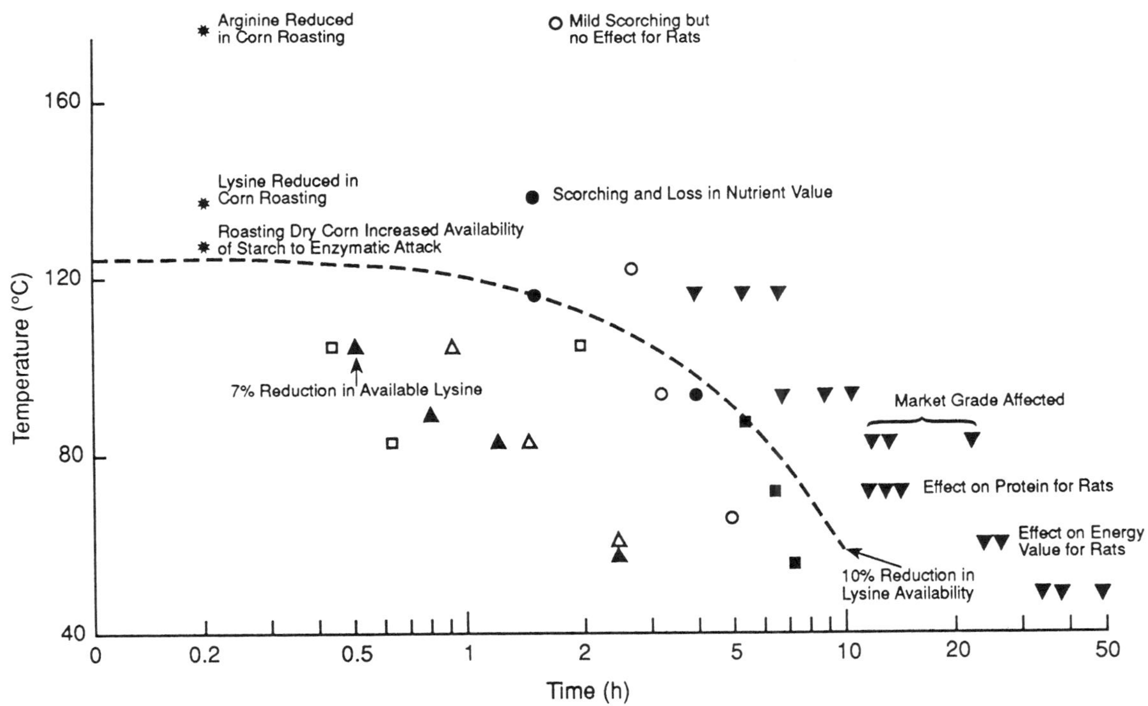

Figure 8 Combinations of air temperature and exposure time found by Mühlbauer and Christ (1974) to reduce lysine availability by 10% (dotted line) compared with results from eight other workers and showing that adverse treatments generally fall outside the line. *Source*: Nellist (1978).

3.2.5. Grain Contamination

An objection to the use of direct-fired furnaces in heated-air drying is that the products of combustion pass through the grain and cause contamination which may be injurious to human health. Of the several substances present in combustion gas, the two groups which cause most concern are the polycyclic aromatic hydrocarbons (PAH) and the nitrous oxides (NO_x). The PAH are considered to be carcinogenic and one of these Benzo(a)pyrene (BaP) seems to be either the most common or the most dangerous. Bolling (1964) found evidence of an increase in BaP in grain dried in oil-fired dryers but levels were less than those caused by growing the grain close to industrial centers. Fisher et al. (1971) failed to find any increase in BaP in wheat dried in three direct oil-fired dryers. Hutt and Winkler (1978), (reported in English by Hutt et al. 1978) found increases in PAH in grain dried by direct oil-firing but the levels gave some cause for concern only when the burners were poorly adjusted. The increase from burning industrial gas was less than that for oil. Larsson et al. (1991) failed to find significant increases in PAH when drying with natural gas. Wilkins et al. (1991) point out that natural gas is a very pure fuel and produces little PAH.

However, natural gas can produce NO_x which are perceived as a problem because with some foods they combine to form nitrosamines. Wet malt (i.e., germinated barley) is a good example and for direct-firing with natural gas in the malting industry, low NO_x burners have been developed. However, NO_x does not appear to increase the level of nitrosamines in ungerminated grain (Wilkins et al. 1991).

The conclusion seems to be that there is no danger of grain contamination from direct firing when using fuel-oil in a well-adjusted burner or when using natural gas. As analytical methods improve and food standards are made more strict, the use of direct firing will be questioned again.

3.3. Dryer Simulation

Section 2 briefly describes the types of dryer, and Sections 3.1 and 3.2 have detailed the basis of the equations and the properties necessary to calculate the drying and cooling processes. Once these calculations are embodied into a computer program it becomes possible to adapt the programs to mimic or simulate the behavior of real grain dryers and to explore their performance in a way which would be impossible by experiment alone. Computer simulations of grain drying have been described by Boyce (1965, 1966), Bakker-Arkema et al. (1967a,b), Henderson and Henderson (1968), Thompson et al. (1968), Spencer (1969, 1972), O'Callaghan et al. (1971), Husain et al. (1972), Nellist (1974, 1987), Sutherland (1975), Ingram (1976), Parry (1983), and Bruce (1984) and this list is not exhaustive! Some authors (e.g., Bakker-Arkema et al. 1974, Nellist 1974, Parry 1983) include listings or reference to listings of their computer

Table 3 A Simple Example of the Use of Computer Simulation. Computed Performance of a Crossflow Dryer Drying Corn from 40 to 14% (wb) at a Drying Air Temperature of 80°C and Air Flow of 1500 $m^3m^{-2}h^{-1}$, Showing the Advantage of Using Three Passes

	Three pass			
Parameter	1st	2nd	3rd	Single pass
Throughput (t/h)	2.5	2.5	2.5	0.3
Residence time (h)	0.28	0.28	0.28	2.27
M_o (% wb)	40	32.4	23.6	40.0
M_f (% wb)	32.4	23.6	14.0	14.0
Moisture extraction rate (t/h)	0.32	0.33	0.30	0.13
Maximum grain temperature (°C)	64	74	78	80

Source: Extracted from Meiering and Hofkes (1977).

programs. Thus, it is possible to gain access to programs although as far as the authors are aware, there is no commercial version providing software support.

Initially, the models were used to simulate experimental results from recorded input variables and their use greatly increased understanding of deep-bed drying. The next stage was to use them to answer "what if" questions about design and operation. To give a simple example, Table 3—which is extracted from results computed for a cross-flow dryer (Meiering and Hofkes 1977)—shows the advantage in terms of increased throughput and lower maximum grain temperatures of drying very wet corn (40%) in three passes rather than in one. The following authors have published studies of continuous-flow and batch dryers: Thompson (1970), Bakker-Arkema et al. (1971, 1973), O'Callaghan et al. (1971), Farmer et al. (1972), Lerew et al. (1972), Paulsen and Thompson (1973), Morey and Cloud (1973), Young and Dickens (1975), Ingram (1976), Stevens and Thompson (1976), Sutherland (1975, 1977, 1984), Sabbah et al. (1979), Meiering et al. (1977), Meiering and Hofkes (1977), Bakker-Arkema et al. (1977), Brook and Bakker-Arkema (1977, 1978a,b, 1980), Pierce and Thompson (1981), Walker and Bakker-Arkema (1981), Miller and Whitfield (1984), Nellist (1987), Patil and Lechoczky (1985), and Kulshreshthha et al. (1988).

One of the most interesting of these studies was that of Pierce and Thompson (1981) who used simulation to define the performance of a cross-flow dryer over a wide range of operating conditions when drying corn. They also used a graphical representation (Fig. 9) which represents four variables in a two-dimensional plot. Figure 9 shows the advantage of increasing drying air temperature to increase grain flow rate and to reduce energy use, and the advantage

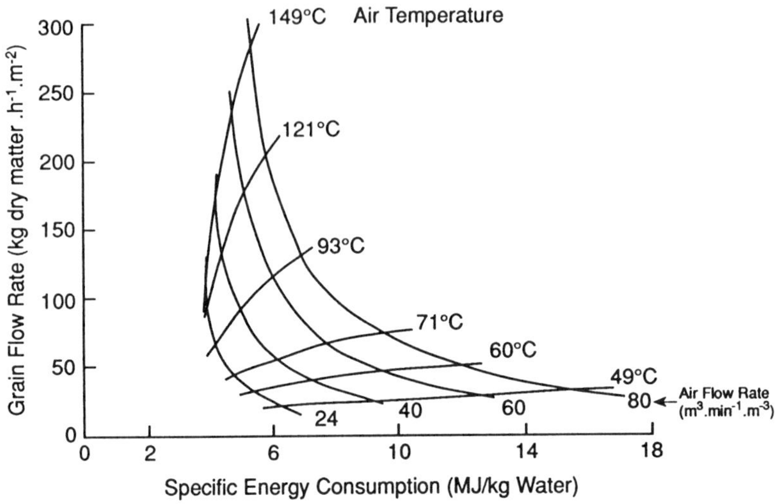

Figure 9 Computed effect of varying drying air temperature and air flow rate on the grain flow rate and specific energy consumption for a cross-flow dryer drying corn from 25 to 15% wb. *Source*: Pierce and Thompson (1981).

of reducing air flow, particularly at low temperatures, to reduce energy use. There is another reason why these results are of interest. The model used by Pierce and Thompson (1981) is based on that developed originally by Thompson et al. (1968) and is different in a number of respects from that used by Nellist (1987) for modeling the cross-flow drying of wheat. However, when the grain properties used by Pierce and Thompson (1981) are inserted into the Nellist model, the predicted results are in close agreement and provide support for the view that grain properties are more important than model formulation. The other interesting result of that unpublished exercise is that when the performance of the dryer is recalculated using the properties of wheat, then the range in energy consumption is much narrower for the same operating conditions. This is a consequence of the higher drying rate of wheat, the smaller of the two grains. A similar chart for the performance of the cross-flow dryer modeled by Nellist (1987) is presented in Figure 10 (Nellist 1989). In this case, the initial moisture content has replaced air flow as one of the variables. At any temperature, increasing the initial moisture content reduces the specific energy consumption and the grain flow rate. Once again, increasing the air temperature increases the grain flow rate and generally reduces the specific energy consumption. At the 16% level, however, when the moisture removal is only 1%, there is a

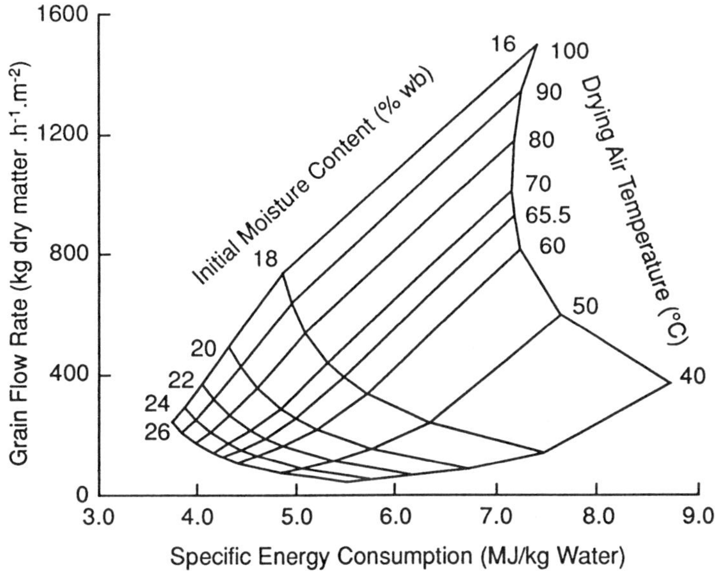

Figure 10 Computed effect of varying drying air temperature and initial moisture content on the grain flow rate and specific energy consumption for a cross-flow dryer (Nellist 1987) drying wheat to a final moisture content of 15%. *Source*: Nellist (1989).

temperature (≃ 70°C) at which the specific energy consumption is a minimum and above which it increases. This interaction between initial moisture content and temperature occurs because, for such a small moisture removal, the grain flow rate is high; the drying air is poorly used and exhausts at a high temperature; and a disproportionate amount of heat is required to remove the moisture from the grain. Virtually no drying occurs in the cooling section and cooling is ineffective because of the short residence time. Figure 11, for the same dryer as Figure 10, shows the value of increasing drying air temperature to reduce the energy consumption (and hence drying cost) but now also shows the penalty incurred by drying to very low moisture contents. A prime motive for improving dryer controls (see Section 4) has been to reduce the amount of overdrying that frequently occurs when a dryer is controlled manually. These charts indicate the advantage of increasing drying air temperature but the extent to which this can be done is bounded by the limits set by grain quality considerations. The safe combinations of initial and final moisture contents and drying air temperatures plotted in Figure 4 (Section 3.2.1) are for the same dryer as in Figures 10 and 11.

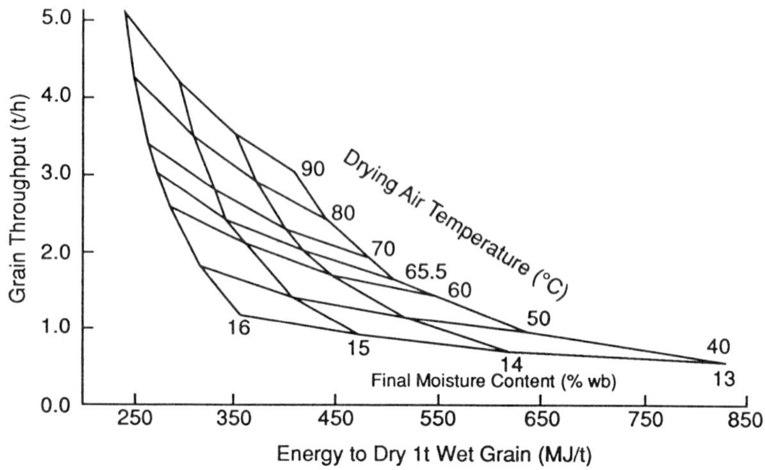

Figure 11 Computed effect of varying drying air temperature and final moisture content on the throughput of wet grain and the energy consumption per tonne of wet grain for a cross-flow dryer (Nellist 1987) drying wheat from an initial moisture content of 20% wb. *Source*: Nellist (1989).

3.4. Dryer Testing

Physical testing, whether formal or informal, has always been a necessary part of the development and marketing of heated air dryers. In the United Kingdom, experience of testing dryers (Bailey 1959) for thermodynamic performance at the former National Institute of Agricultural Engineering was distilled into a British Standard, BS 3986:1966 (British Standards Institute 1966). This standard prescribed procedures for measuring the performance of batch dryers and continuous-flow dryers when operating at steady state. It was written specifically for drying rewetted wheat from 21%, and used interpolation of results from several tests to find the evaporation rate at output moisture contents in the range of 17–14%. In the United States, Bakker-Arkema et al. (1978) proposed some standard conditions for testing maize dryers and introduced the concept of a "hybrid drying factor" to account for the differences in drying rate between maize hybrids. They also suggested a "dryer efficiency factor" to be obtained by comparing experimental and simulated results. This ignored the possibility of experimental error in the test results and of systematic error in the computed predictions. Keener and Glenn (1978) also suggested prescriptions for moisture removal and discussed the effect of differences in grain drying rate and air flow resistance on the results. Hall (1981) addressed the problem of adjusting dryer test results to standard conditions and presented a series of charts of correction factors derived from the results of a number of workers. A grain dryer evaluation program began in Canada in 1979 and a testing facility was built (Wassermann, J. D., et al. 1983). The evaluation procedures are based upon the earlier standards and proposals from Europe and America, and contain prescriptions for testing dryers on wheat, barley, rapeseed, and corn. Tests with corn were carried out with naturally-moist grain but wheat, barley, and rapeseed were conditioned to required moisture contents (Sokhansanj et al. 1980). Wassermann, J. D., et al. (1983) achieved partial corrections of dryer capacity to standard conditions by log-log regression against initial moisture content and of fuel consumption by regression against ambient temperature. Specific fuel and specific energy consumptions were based upon these corrected quantities.

In subsequent tests at the Silsoe Research Institute, computer simulation was used both to interpret test results and to extrapolate performance to a wider range of conditions than can be explored by physical testing alone. Out of this work has come a completely revised British Standard, BS 3986:1991 (British Standards Institution 1991) which differs from the 1966 version in several respects. The two most important differences are the introduction of a new procedure for correcting evaporation rates to standard or defined conditions and of procedures for estimating the uncertainty in the estimation of the derived measures of performance. The procedure for correcting the evaporation rate is to multiply the observed value by the ratio of the rate predicted for the specified conditions by a computer model to the value predicted for the observed conditions by the same model. If a computer model of the particular dryer is not available then the standard gives approximate formulae from which a correction factor can be calculated. These formulae (Nellist and Bruce 1992) were derived from an analysis of the performance of the cross-flow dryer described by Nellist (1987).

The procedure for estimating the uncertainty in the derived measures of performance involves combining the standard errors of the mean (SEM) and the degrees of freedom of the measured constituent-variables, such as moisture contents and fuel consumption, to provide the estimates of the SEM and the effective degrees of freedom of the derived variable. The 95% confidence limits can then be calculated. Nellist and Bruce (1992) used a spreadsheet approach to illustrate the application of BS 3986:1991 to the initial calculation of results, the correction of those results to standard conditions and the calculation of confidence limits. They also illustrated the importance of accurate measurement of input and output moisture contents and the sharp increase in uncertainty as moisture removals fall below 5% wb.

Dryer design and trade are international and there is a need, therefore, to standardize procedures on a world scale. The first part of an International Standard should meet this need as far as the basic testing procedures and calculations are concerned. A second part is planned to deal with the problems specific to particular crops (e.g., the feasibility of rewetting, the assessment of drying damage, and the effects of differences in drying rate). These test procedures relate to operation at steady state (or at least quasi-steady state). In reality, few dryers are fed grain of constant moisture content and operate at steady state. Unsteady-state operation implies dryer control which is discussed in the next section.

4. DRYER CONTROL

4.1. The Control Problem

Grain which is wet is at risk of molding in storage, so it must be dried to an appropriate moisture content. Further drying will increase its safe storage life but this incurs further costs; it uses more fuel, reduces the rate at which the dryer can process the harvested grain, and reduces the mass of grain available for sale. Unless the grain is sold on a dry matter basis, grain which is dried below the appropriate level is worth less. Because of these costs of under- and overdrying, the best course of action is to dry all the grain to the target moisture content (TMC). The inlet (or input) moisture content (IMC) is rarely steady; when drying wheat in the United Kingdom, changes in the IMC may alter the drying load by a factor of four over a short period. Therefore, the dryer must be adjusted to reduce the effect of such changes in the moisture content of the discharged grain.

To achieve this control, it is possible, in principle, to alter the grain discharge rate, the temperature of the drying air, the air flow rate, or a combination of these. Other strategies, such as air or grain recirculation or zonal temperature control, are possible but are too complex and expensive for agricultural dryers (which must have low capital cost). Altering the grain discharge rate is simple to achieve and can give the required range of adjustment whereas, by altering the air flow, it is only possible to have a limited influence on the drying before the interaction of the fan characteristics and crop aerodynamics become limiting. Changing the drying air temperature is an effective moisture control action but too high a temperature may result in thermal damage to the grain. Unlike industrial dryers, grain dryers do not normally have to keep pace with other equipment, so altering the discharge rate of grain is the most suitable action.

The control problem was analyzed by Marchant (1985) who showed that both instability and slow response can result from feedback control of discharge rate based on the error at the dryer output, namely: the output moisture content (OMC) less TMC. Depending on the input IMC, the sensitivity (i.e., change in OMC divided by the change in error) can be eight times larger for an upward step of 5% mc than for a downward step of the same magnitude. By using the inverse discharge rate, which is proportional to residence time, Marchant (1985) formulated a model of the dryer from which most of the nonlinearity is removed, enabling a control system with consistent response to be designed.

Space only permits a brief review of the field of dryer control. The reader is referred to reviews by Gui et al. (1988) and Brooker et al. (1992).

4.2. Sensing

Sensing the moisture content of a hygroscopic material like grain is not straightforward; instruments developed for industrial use are either unsuitable because they fail to allow for the moisture gradient in dried grains (e.g., infrared reflectance) or are too complex and expensive (electrical methods incorporating density compensation by gamma ray or neutron absorption, or by weighing a controlled volume). Simpler, capacitance-based instruments are becoming available (Bruce and McFarlane 1990) which, with calibration adjustment to an off-line meter calibrated to oven standards, can give a cost-effective, on-line moisture signal. An alternative method of sensing moisture content is to use the cooling of the air caused by evaporation of moisture from the grain.

Provided that more than 2% mc is being removed, an approximately linear relationship exists between the exhaust air temperature and the grain moisture content (Nellist 1976b), so this temperature can be used to form an error signal for control. However, the method cannot be used if the drying air temperature is changing, or if too little moisture is being removed, or during start-up. Because of the low cost and simplicity of the sensor itself, this sensing system may remain popular on smaller dryers.

4.3. Feedback Control

Feedback control uses an error signal (i.e., the difference between the measured value of an output of the drying process and a target value for that output) to alter an input to the process so as to reduce the error. Feedback has the potential to bring the dryer output (either grain moisture content or exhaust air temperature) to the target value if input conditions are steady but the result of a disturbance to the IMC are only detected (and correction can therefore only start) once the output starts to change. Hence, there will be a mass of grain, equal to or greater than the dryer holding capacity, discharged before the target can be regained.

Fábián and Samu (1980) used a capacitive moisture sensor in a feedback controller on a maize dryer and reported a significant reduction in overdrying. A feedback controller for a mixed-flow dryer, based on Marchant's (1985) linearization, was described by Whitfield (1988a,b). The moisture content was measured directly by a capacitive sensor in the cooling bed and a proportional plus integral algorithm was developed to adjust the inverse discharge rate of grain. Whitfield (1986) used a dynamic simulation model to derive a describing function for the dryer which enables the correct values of the controller parameters to be selected. This controller is now marketed in the United Kingdom by Carier Bulk Materials Handling Ltd. Nybrant and Regnér (1985) and Nybrant (1988) developed a feedback controller, using the exhaust air temperature as the measured output, to control a cross-flow dryer. Nybrant (1989) used feedback to control a concurrent-flow dryer, where the exhaust air temperature and the approximate maximum grain temperature were fed back and controlled accurately, but the OMC control was less good.

4.4. Feedforward

If the disturbances to the process are measurable, as in the case of IMC changes to a grain dryer, then action can be taken to reduce the severity of an output error resulting from an input disturbance, before that output error occurs. This is termed feedforward control. However, the output disturbance cannot be removed. In a conventional grain dryer, it is not possible to move some part of the grain at a different speed to the rest. Therefore, a control action affects all the grain in the dryer and there is no action which removes the effect on the OMC of a step change in IMC. Two problems arise; what discharge rate is required to achieve the TMC with the new IMC and how should the change to that rate be made? The discharge rate to bring the dryer to the target must be predetermined either by simulation of the dryer or by experiment. Because moisture content sensing by the exhaust air temperature method is not feasible near the dryer inlet, direct sensing of moisture content is required in feedforward control.

Olesen (1987) assumed the moisture loss to be proportional to residence time. He used a shift register approach, and took all the predicted action when the change in IMC reached the middle of the drying bed. Stone (1984) showed how the delay, before the point when all the predicted action was taken, could be optimized to minimize the integrated sum of squares of output moisture error. Forbes et al. (1984) used a decay-type dryer model for feedforward control in which the time constant of the model was estimated on line. They employed a simulation of density change during drying to ensure that the IMC and OMC values of each batch of grain were correctly matched for use in the parameter estimation algorithm. This type of feedforward controller, again with feedback via parameter estimation, was implemented on several cross-flow dryers by Eltigani et al. (1986) and Bakker-Arkema et al. (1989).

4.5. Feedback Plus Feedforward Control

Feedforward control has the limitation that, unless the predicted discharge rate is correct, the output will always be in error. As models are always less than a perfect representation of the real process, the error may be too large to be acceptable. The predictive benefits of feedforward can be combined with feedback to remove the residual error, at the expense of requiring two sensors one each for the input and the output grain.

Nybrant (1986) showed that the addition of feedforward constitutes a major improvement over a feedback-only system. McFarlane and Bruce (1991) added a feedforward term to Whitfield's (1988a, b) feedback controller, after linearizing it to synchronize the discrete sensing and control actions with grain position in the dryer rather than fixed in time. Their feedforward action is a first-order lag change to the discharge rate, which is 90% completed in one residence time. Feedforward gain, equal to half the feedback gain, is scheduled as a linear function of drying air temperature. The reaction by the feedback term to the changes initiated by the feedforward term requires the gains of the two terms to be carefully matched.

4.6. Model-Based Control

4.6.1. Definition

Although feedforward control requires a model of the relationship between input and output at steady state, model-based control requires a more complex model which represents the dynamics of the process. The model can be used in two ways, termed optimal and adaptive control.

4.6.2. Optimal Control

The future inputs of the dryer, either measured or assumed, together with its present internal states, are used in the model to predict the outputs of the dryer for some period ahead. Possible control actions can be evaluated in their effect on the outputs of the dryer and the optimal course of action selected and implemented. An objective function (calculated from the outputs of the dryer) is formulated with which to select the optimal course of action. Holtman and Zachariah (1969a,b) used this approach and found that a linear model of drying gave as good control as a more accurate but cumbersome simulation. The controller was not implemented or tested in a practical situation. Nikolic and Mojsovski (1987) used this approach together with linear programming to implement optimal moisture control of a rice dryer but no performance data were reported.

4.6.3. Adaptive Control

Recognizing that the process model is imperfect, adaptive control uses measurements from the dryer and outputs of the model to estimate one or more parameters of the model itself. In this way the model is continually adapted so its output matches that of the dryer.

Besides allowing for imperfections in the model, adaptivity can compensate for changes in parameters such as the drying air temperature, the ambient conditions and the drying characteristics of the grain. However, the more parameters there are to be estimated, the more measurements and therefore the more time is needed for the model to adapt. A parameter such as the drying air temperature can be measured directly and accounted for explicitly in an algorithm, or its effect on the controller parameters can be estimated. The choice of which method is more appropriate will have to take account of the greater complexity and slower parameter estimation of an adaptive algorithm versus the cost of direct measurement of air temperature and accuracy of prediction of its effect. If, however, the drying characteristics of the crop vary and the control performance is sensitive to them, then the problem is ideally suited to adaptive methods because on-line measurement of drying properties is not feasible. Adaptive control is also a means of avoiding the need to remove nonlinearities from the control problem because the process parameters adapt to the most suitable values. However, if the disturbances in nonmodeled variables occur faster than the estimation process can converge, the adaptive approach will be ineffective.

Nybrant and Regnér's (1985) use of adaptive control on a laboratory dryer was advantageous because the sensing method, exhaust air temperature, has nonlinear characteristics. They reported a period of two residence times for estimates of their three parameters to stabilize after a step change of IMC and a further one residence time for accurate control to be achieved.

Moreira and Bakker-Arkema (1990) used two linear models, one with a large number of parameters to be estimated and the second with two parameters. They reported good control of two commercial cross-flow dryers by both controllers. Though the parameter estimation time was not given or commented on, the two parameter model was preferred because of its simplicity. Model-based adaptive controllers were also described by Eltigani et al. (1986) and by Bakker-Arkema et al. (1989).

4.7. Future of Dryer Control

Developments in artificial intelligence such as fuzzy logic and expert systems will make possible further advances in control of dryers but without good sensing methods they are limited in their usefulness. Improvements in sensing means improved accuracy and perhaps reduced cost for adequate performance and long-term reliability in an arduous environment. The performance of bench moisture meters for grain is a long way from ideal, so the performance of a device located perhaps 10 m above ground in a grain duct where it is inconvenient to check is not likely to be better. It is possible that models of heat and mass transfer in grain could be used to estimate the moisture content of grain from the air temperature and air and grain flow rates. If it could be implemented on-line, the oscillating cell method of weighing, used in at least one bench meter, would make possible density compensation for electrical capacitance methods of moisture sensing.

For the controllers described in this chapter, the drying air temperature is assumed to be a measurable or nonmeasurable disturbance because it is selected by the operator. If the controller had authority to alter this temperature, it could ensure grain is not overdried when the dryer's maximum output rate is reached. Zhang and Litchfield (1991) developed a fuzzy logic controller which alters both the drying air temperature and the discharge rate of grain. It reduced the corn breakage in a laboratory dryer compared with manual control. Rice and Kaya (1986) presented a controller which alters the setpoint of the drying air temperature to achieve maximum drying rate without scorching; it also alters the grain flow rate once the maximum air temperature has been reached. If the model of the dryer were to contain a calculation of the effects of drying on the quality loss, then the controller could maximize the drying air temperature, and hence the discharge rate and energy efficiency, within a constraint of acceptably small quality loss.

The cost of controller implementation could perhaps be reduced if the algorithm, sensor function, and output action were programmed on the programmable logic controller (PLC) which is becoming more widely used for controlling the sequencing of the dryer and handling system. This would reduce the need for communication between controller and dryer management system if control over drying air temperature were implemented.

With the continuing increase in microcomputer power per unit cost, more powerful control techniques will become feasible to implement but sensors remain a key element of the technology and their performance to price ratio will not fall alongside microcomputers.

5. CONCLUSIONS

Because of their limited annual use, heated-air dryers employed at farms and elevators have to be relatively simple machines of low capital cost. Their capacity and efficiency are improved by the use of high air temperatures but limits to these are set by constraints on thermal damage set by the end-use for the grain.

The development of mathematical models and computer simulations of dryers has been vital to the better understanding of function and to the improvement in design and operation. The models may be formulated in different ways and solved at varying levels of numerical complexity but, given the same constituent properties of the grain, generally yield similar results. The general basis of the models has been presented. Empirical and semi-empirical equations are commonly used to describe the average drying rate of the grain but improvements in three-dimensional modeling and in computer power will lead eventually to their replacement by numerical solutions of the diffusion equation for single kernels. While equilibrium moisture contents should be compatible with the drying rate coefficient with which they are used, they should also be a realistic representation of the long term static value; net heats of desorption can be derived from equations describing the equilibrium moisture content.

More precise application of quality parameters as constraints in the computer simulation will depend upon a better understanding and quantification of the complex flows of air and grain, particularly in the ubiquitous mixed-flow dryer.

The biological basis of grain quality has been reviewed. For seed grain, the loss of viability can be described by the probit hypothesis of accelerated seed death and has provided a means of decoupling the interactions of dryer conditions and initial grain quality. The loss of baking quality of wheat, as expressed by relative loss of loaf volume can also be modeled with good accuracy for thin-layers. It has yet to be applied within computer simulations of dryer types. For grains such as maize and rice, their quality for milling is affected by internal fissuring resulting from stresses set up by differential expansion and shrinkage of the endosperm. Stress cracking is a complex subject for which there is a massive body of literature. Understanding and modeling of stress cracking in individual kernels has lagged behind the development of techniques of dryeration and combination drying which have been found to be effective in reducing it. The important feature of these systems seems to be that the grain enters a slow-drying regime while still two or more percentage points above the target final moisture content.

Contamination of the grain by direct-fired dryers has not been shown to be a danger to human health. For firing by oil the concern has been with the uptake of PAH which have a carcinogenic effect. Natural gas does not produce PAH but can produce NO_x. With some foods, NO_x react to form nitrosamines but this does not appear to happen in normal ungerminated grains. Nevertheless, direct-firing is banned in some countries, and as long as analytical chemistry yields ever more sensitive tests for carcinogenic compounds, the debate will continue.

The first application of computer simulation of dryers was to investigate design and operation at nominal steady state. Studies of this type have been referred to but not discussed in detail. As an example the performance of a cross-flow dryer was used to show that increasing air temperature does not always increase efficiency. For small moisture removals, too much heat may be lost to the exhaust air. Large increases in energy consumption can also result from drying to moisture contents, one or two percentage points below the safe storage level.

Developments in the understanding of dryer behavior have led to developments in methods of physical testing and revision of testing standards. Key features of the revised British Standard are the use of computer simulation (or formulae derived from simulation) to adjust test results to standard conditions and the introduction of procedures to estimate confidence limits for the derived results.

The models referred to and the physical testing relate to operation at steady-state. In real life, this seldom, if ever, occurs so that dryer control becomes important. Unsteady-state versions of dryer models have provided the data necessary to develop control algorithms for a new generation of controllers superior to those based upon simple control of exhaust air temperature. The challenge is to develop means of optimal control which will operate the dryer so as to minimize an objective function combining running costs with the quality constraints and the end-use value of the grain.

Engineers and scientists have made great progress with limited resources. No-one is more aware than the scientists themselves that of the limitations of many of their experiments, both physical and computed, and that more could be done to improve the confidence with which dryers can be designed and operated for high quality and efficiency.

6. NOMENCLATURE

a and b = coefficients in Equations 10, 17, 19, 22
a_1, a_2 and a_3 = coefficients in the Modified-Henderson equation, Equation 12
b_1, b_2 and b_3 = coefficients in the Chung-Pfost equation, Equation 13
c and d = coefficients in Equation 19
c_1, c_2 and c_3 = coefficients in the modified Smith equation, Equation 14
c_a = specific heat of dry air ($J \cdot kg^{-1} \cdot K^{-1}$)
c_w = specific heat of dry grain ($J \cdot kg^{-1} \cdot K^{-1}$)
c_v = specific heat of water vapor ($J \cdot kg^{-1} \cdot K^{-1}$)
c_w = specific heat of water ($J \cdot kg^{-1} \cdot K^{-1}$)
d_1, d_2 and d_3 = coefficients in the Modified-Halsey equation, Equation 15
e_1, e_2 and e_3 = coefficients in the Modified-Oswin equation, Equation 16
h_a = grain bed heat transfer coefficient ($J \cdot m^{-3} \cdot K^{-1}$)
k = drying rate constant (s^{-1})
k_o = coefficient in Equation 22
n = exponent of time in Equation 9
t = time (s)
$\bar{t}$ = the exposure time at which $G_s = 0.5$
x = depth in bed from air inlet (m)
A and B = coefficients in Equation 11
G = mass flow rate of air ($J \cdot m^{-2} \cdot K^{-1}$)
G_s = germination capacity as a decimal fraction (i.e., 0.5 = 50%)
H = absolute humidity of air (kg/kg dry air)

L_b = latent heat of vaporization of bound water (J/kg)
L_f = latent heat of vaporization of free water (J/kg)
M = moisture content of grain (kg/kg dry matter)
M_e = equilibrium moisture content of grain (kg/kg dry matter)
M_o = initial moisture content of grain (kg/kg dry matter)
MR = $(M - M_e)/(M_o - M_e)$ and is termed the moisture ratio
M_s = moisture content of a shell of the 10 shell grain model (kg/kg dry matter)
M_z = a transitional moisture content (decimal db) defined by Equation 18
Q = specific loaf volume (cm^3/g)
R = gas constant = 8.3143 $J \cdot mol^{-1} \cdot K^{-1}$
T = air temperature (°C)
V = loaf volume (cm^3)
V_c = loaf volume of control (unheated) sample (cm^3)
V_r = contribution to relative loaf volume, V/V_c, of a shell of the 10-shell grain model
W = grain moisture content (% wb)
X = standard normal deviate = $(T - \bar{t})/\sigma$
X_o = value of X at $t = 0$
ε = void ratio (decimal fraction)
θ = grain temperature (°C)
θ_s = temperature of a shell of the 10-shell grain model (°C)
ρ_a = density of air (kg/m^3)
ρ_g = bulk density of grain (kg/m^3)
σ = the standard deviation of the distribution
φ = relative humidity (decimal)

REFERENCES

Alagusundaram, K. and D. S. Jayas. 1990. Air flow resistance of grains and oilseeds. Postharvest News Info. 1(4): 279-283.

Alagusundaram, K., D. S. Jayas, O. H. Friesen, and N. D. G. White. 1989. A model to predict three-dimensional air pressure patterns in grain bins. Paper 89-4545. Am. Soc. Agric. Eng., St. Joseph, MI. 14 pp.

ASAE. 1987. Moisture relationships of grains. ASAE Standard D245. 4, Am. Soc. Agric. Eng., St. Joseph, MI.

Bailey, P. H. 1959. The measurement of crop drier performance. J. Inst. British Agric. Eng. 15: 59-70.

Bakker-Arkema, F. W. 1970. Numerical techniques in cooling and drying simulations (I) Solution of ordinary differential equations. (II) Solution of partial differential equations. pp. 05a:1-11. In: Proc. Inst. Simul. Cool. Drying Agric. Prod., Mich. State Univ., East Lansing, MI.

Bakker-Arkema, F. W., A. Ahmadnia-Sokhansanj, and R. Green. 1977. High temperature wheat drying. Paper 77-3527. Am. Soc. Agric. Eng., St. Joseph, MI. 19 pp.

Bakker-Arkema, F. W., W. G. Bickert, and R. V. Morey. 1967a. Gekoppelter Wärme-und Stoffaustausch während des Trochnungsvergangs in einem Behälter mit Getreide. (Simultaneous heat and mass transfer during grain drying.). Landtech. Forsch. 17: 175-180.

Bakker-Arkema, F. W., W. G. Bickert, and R. J. Patterson. 1967b. Simultaneous heat and mass transfer during the cooling of a deep bed of biological products under varying inlet conditions. J. Agric. Eng. Res. 12: 297-307.

Bakker-Arkema, F. W., T. W. Evans, and D. M. Farmer. 1971. Simulation of multiple-zone grain drying. Trans. ASAE (Am. Soc. Agric. Eng.) 14: 935-938.

Bakker-Arkema, F. W., D. M. Farmer, and L. E. Lerew. 1973. Optimum grain dryer design through simulation. Ann. Technol. Agric. 22: 275-290.

Bakker-Arkema, F. W., C. Fontana, R. C. Brook, and C. M. Westelaken. 1984. Concurrent flow rice drying. Drying Technol. 1: 171-191.

Bakker-Arkema, F. W., L. E. Lerew, S. F. De Boer, and M. G. Roth. 1974. Grain dryer simulation. Res. Rep. No. 224, Mich. State Univ., East Lansing, MI. 80 pp.

Bakker-Arkema, F. W., L. E. Lerew, R. C. Brook, and D. B. Brooker. 1978. Energy and capacity performance evaluation of grain dryers. Paper 78-3523. Am. Soc. Agric. Eng., St. Joseph, MI. 14 pp.

Bakker-Arkema, F. W., R. G. Moreira, and A. Y. Eltigani. 1989. Automatic control of crossflow grain dryers. pp. 2199-2265. In: Proc. 11th Int. Cong. Agric. Eng. (CIGR), Dublin, Ireland, 4-8th September, 1989.

Bakker-Arkema, F. W., J. C. Rodriquez, I. P. Schisler, B. Stevenson, and C. H. Westlaken. 1982. A new commercial cross-flow dryer: The differential grain-speed dryer. Paper 82-3007. Am. Soc. Agric. Eng., St. Joseph, MI. 18pp.

Bakker-Arkema, F. W. and I. P. Schisler. 1984. Differential grain-speed crossflow grain drier. Paper 84-3522. Am. Soc. Agric. Eng., St. Joseph, MI. 11 pp.

Becker, H. A. and H. R. Sallans. 1956. A study of the relation between time, temperature, moisture content and loaf volume by the bromate formula in the heat treatment of wheat and flour. Cereal Chem. 33: 254-265.

Bolling, H. 1964. The influence of exhaust gases on cereal drying. Die Muhle 101: 759-762.

Boyce, D. S. 1965. Grain moisture and temperature changes with position and time during through drying. J. Agric. Eng. Res. 10: 333-341.

Boyce, D. S. 1966. Heat and moisture transfer in ventilated grain. J. Agric. Eng. Res. 11: 255-265.

British Standards Institution. 1966. Methods of test for agricultural grain driers. British Standard 3986: 1966. BSI, London.

British Standards Institution. 1991. Methods of test for drying performance of agricultural grain dryers. British Standard, BS 3986:1991, BSI, London.

Brooker, D. B., F. W. Bakker-Arkema, and C. W. Hall. 1992. Drying and storage of cereal grains and oilseeds. Van Nostrand Reinhold, New York. 450 pp.

Brook, R. C. and F. W. Bakker-Arkema. 1977. Design of commercial concurrent-flow grain driers. Paper 77-3017. Am. Soc. Agric. Eng., St. Joseph, MI. 12 pp.

Brook, R. C. and F. W. Bakker-Arkema. 1978a. Dynamic programming for process optimisation 1. An algorithm for design of multi-stage grain dryers. J. Food Process Eng. 2: 199-211.

Brook, R. C. and F. W. Bakker-Arkema. 1978b. Simulation for design of commercial concurrent flow grain dryers. Trans. ASAE (Am. Soc. Agric. Eng.) 21: 978-981.

Brook, R. C. and F. W. Bakker-Arkema. 1980. Design of multi-stage corn dryers using computer optimisation. Trans. ASAE (Am. Soc. Agric. Eng.) 23: 200-203.

Bruce, D. M. 1984. Simulation of multiple-bed concurrent-, counter-, and mixed-flow grain driers. J. Agric. Eng. Res. 30: 361-372.

Bruce, D. M. 1985. Exposed-layer barley drying: three models fitted to new data up to 150°C. J. Agric. Eng. Res. 32: 337-348.

Bruce, D. M. 1986. The NIAE exposed-layer drier—catalogue of data and of programs for data analysis. Divisional Note DN1311, National Inst. Agric. Eng., Silsoe, United Kingdom. 37 pp.

Bruce, D. M. 1992. A model of the effect of heated-air drying on the bread baking quality of wheat. J. Agric. Eng. Res. 52: 53-76.

Bruce, D. M. and N. J. B. McFarlane. 1990. Automatic control of dryers. pp. 85-95 in Energy Innovation and the Agro-Food Industry. Proc. Int. Sympo. Corté, P., Fabry, C. and Ferrero, G. L. (eds). Commission of European Communities, Luxembourg.

Cenkowski, S. 1989. A comparison of grain thin layer mathematical models. Zeszyty Naukowe Akademii Rolniczej we Wroclawin, Rolnictwa 50(178): 39-47.

Cenkowski, S., D. S. Jayas, and D. Hao. 1990a. Latent heat of vaporization for main crops and selected foods. Paper No. 90-408, Can. Soc. Agric. Eng., Saskatoon, SK., 9p.

Cenkowski, S., M. J. Miketinac, and A. W. Kelm. 1990b. Air flow patterns in a mixed-flow dryer. Can. Agric. Eng. 32: 85-90.

Chen, C. C. and R. V. Morey. 1989. Comparison of four EMC/ERH equations. Trans. ASAE (Am. Soc. Agric. Eng.) 32: 983-990.

Cory, W. T. W. 1991a. Fans for today's agriculture. Agric. Eng. 46: 2-8.

Cory, W. T. W. 1991b. Fans for today's agriculture—Part 2. Centrifugal fans and mixed-flow fans. Agric. Eng. 46: 34-38.

Cory, W. T. W. 1991c. Fans for today's agriculture. Part 3: The problem of noise. Agric. Eng. 46: 66-70.

Crank, J. 1975. The mathematics of diffusion. Oxford University Press, Oxford. 414 pp.

Dietrich, N. 1957. Drying of naturally moist and artificially-moistened wheat grains. Landtech. Forsch. Munch. 7: 140-145.

Ellis, R. H. and E. H. Roberts. 1980. Improved equations for the prediction of seed longevity. Ann. Bot. 45: 13-30.

Ellis, R. H. and E. H. Roberts. 1981. The quantification of ageing and survival in orthodox seeds. Seed Sci. Technol. 9: 373-409.

Eltigani, A. Y., J. C. Anderson, F. W. Bakker-Arkema, and S. Shivvers. 1986. Automatic control of a commercial crossflow grain drier. Paper 86-6501. Am. Soc. Agric. Eng., St. Joseph, MI. 15 pp.

Evans, D. E., G. R. Thorpe, and T. Dermott. 1983. The disinfestation of wheat in a continuous-flow fluidised bed. J. Stored Prod. Res. 19: 125-137.

Fábián, Z. and M. Samu. 1980. Automation control of the moisture content in grain driers. Grundl. Landtechnik 30: 174-178.

Farmer, D. M., F. W. Bakker-Arkema, and S. F. Deboer. 1972. Simulation and optimal design of a commercial concurrent flow grain dryer—the Anderson model. Paper 72-847. Am. Soc. Agric. Eng., St. Joseph, MI. 12 pp.

Finney, D. J. 1971. Probit Analysis. Third Ed. Cambridge University Press, Cambridge. 333 pp.

Fisher, N., R. M. Cooper, and B. M. Bell. 1971. Processing and the possible contamination of wheat and baked products with benzo(a)Pyrene. pp. 7-10 in Paper 20/71, 78th Health Cong. Royal Soc. Health, Eastbourne, the United Kingdom.

Forbes, J. F., B. A. Jacobson, E. Rhodes, and G. R. Sullivan. 1984. Model based control strategies for commercial grain drying systems. Can. J. Chem. Eng. 62: 773-779.

Fortes, M. and M. R. Okos. 1981. Nonequilibrium thermodynamics approach to heat and mass transfer in corn kernels. Trans. ASAE (Am. Soc. Agric. Eng.) 24: 761-769.

Fortes, M., M. R. Okos, and J. R. Barrett. 1981. Heat and mass transfer analysis of intra-kernel wheat drying and rewetting. J. Agric. Eng. Res. 26: 109-125.

Foster, G. H. 1964. Dryeration—a corn drying process. USDA Agric. Marketing Ser. Bull. 532. U. S. Dept. Agric., Washington.

Foster, G. H. 1973. Dryeration: heated air drying and corn quality. Ann. Technol. Agric. 22: 233-244.

Frazer, B. M. and W. E. Muir. 1981. Air flow requirements predicted for drying grain with ambient and solar-heated air in Canada. Trans. ASAE (Am. Soc. Agric. Eng.) 24: 208-210.

Gallaher, G. L. 1951. A method of determining the latent heat of agricultural crops. Agric. Eng. 32: 34.

Groves, J. F. 1917. Temperature and life duration of seeds. Bot. Gaz. 63: 169-189.

Gui, X. Q., J. Bentsman, and J. B. Litchfield. 1988. Control of grain drying: State of the art and open problems. Paper 88-3502. Am. Soc. Agric. Eng., St. Joseph, MI. 34 pp.

Gustafson, R. J., Y. Mahmoud, and G. E. Hall. 1983. Breakage susceptibility reduction by short-term tempering of corn. Trans. ASAE (Am. Soc. Agric. Eng.) 26: 918-922.

Haghighi, K. and L. J. Segerlind. 1988a. Failure of biomaterials subjected to temperature and moisture gradients using the finite element method. 1. Thermo-hydro viscoelasticity. Trans. ASAE (Am. Soc. Agric. Eng.) 31: 930-938.

Haghighi, K. and L. J. Segerlind. 1988b. Failure of biomaterials subjected to temperature and moisture gradients. 2. Stress analysis of an isotropic sphere during drying. Trans. ASAE (Am. Soc. Agric. Eng.) 31: 938-945.

Haghighi, K., J. Irudayaraj, R. L. Stroshine, and S. Sokhansanj. 1990. Grain kernel drying simulation using the finite element method. Trans. ASAE (Am. Soc. Agric. Eng.) 33: 1957-1965.

Hall, G. E. 1981. Drier test data standardisation (SOC method). Paper 81-3010. Am. Soc. Agric. Eng., St. Joseph, MI. 18 pp.

Henderson, J. M. and S. M. Henderson. 1968. A computational procedure for deep-bed drying analysis. J. Agric. Eng. Res. 13: 87-95.

Holtman, J. B. and G. L. Zachariah. 1969a. Continuous crossflow modeling for optimal control. Trans. ASAE (Am. Soc. Agric. Eng.) 12: 430-432.

Holtman, J. B. and G. L. Zachariah. 1969b. Computer controls for grain driers. Trans. ASAE (Am. Soc. Agric. Eng.) 12: 433-437.

Hunter, A. J. 1987. An isotere equation for some common seeds. J. Agric. Eng. Res. 37: 93-105.

Hunter, A. J. 1989. On the heat of sorption of Australian paddy rice. J. Agric. Eng. Res. 44: 237-239.

Husain, A., C. S. Chen, J. T. Clayton, and L. F. Whitney. 1972. Mathematical simulation of mass and heat transfer in high moisture foods. Trans. ASAE (Am. Soc. Agric. Eng.) 15: 732-736.

Hustrulid, A. 1962. Comparative drying rates of naturally moist, remoistened and frozen shelled corn. Trans. ASAE (Am. Soc. Agric. Eng.) 5: 64-67.

Hustrulid, A. 1963. Comparative drying rates of naturally moist, remoistened and frozen wheat. Trans. ASAE (Am. Soc. Agric. Eng.) 6: 304-308.

Hutt, W., A. Meiering, W. Oelschlager, and E. Winkler. 1978. Grain contamination in drying with direct heating. Can. Agric. Eng. 20: 103-107.

Hutt, W. and E. Winkler. 1978. Influence of direct heating on the contents of polycyclic aromatic hydrocarbons in grain. Grundl. Landtechnik 28: 240-245.

Ingram, G. W. 1976. Deep bed drier simulation with intra-particle moisture diffusion. J. Agric. Eng. Res. 21: 263-272.

Isaacs, G. W. and W. Mühlbauer. 1975. Possibilities for and limits to energy saving in maize drying. Landtechnik 30: 397-401.

Jayas, D. S., S. Cenkowski, and W. E. Muir. 1988. A discussion of the thin layer drying equation. Paper 88-6557. Am. Soc. Agric. Eng., St. Joseph, MI. 7p.

Keener, H. M. and T. L. Glenn. 1978. Measuring performance of grain drying systems. Paper 78-3521. Am. Soc. Agric. Eng., St. Joseph, MI. 15 pp.

Kulshreshthha, M, R. D. Singh, and S. K. Srivastava. 1988. Grain drying with intermittently heated air—a simulation approach. pp. 65-67 in: Sixth Int. Drying Sympo. (IDS'88) Volume 2, Versailles, France.

Kunze, O. 1984. Physical properties of rice related to drying the grain. Drying Technol. 2: 369-387.

Larsson, B. K., S. Regnér, and P. Baeling. 1991. Polycyclic aromatic hydrocarbon and volatile N-nitrosamine content of wheat grain before and after direct gas-fired drying. J. Sci. Food Agric. 56: 373-384.

Lasseran, J. C. 1982. New developments in energy preservation for maize drying. Maize: Recent Progress. pp. 53-76 in: Chemistry and Technology. Academic Press, New York.

Lerew, L. E., F. W. Bakker-Arkema, and R. C. Brook. 1972. Simulation of a commercial cross-flow drier: the Hart-carter model. Paper 72-829. Am. Soc. Agric. Eng., St. Joseph, MI. 11 pp.

Lescano, C. L. and D. E. Tyrrell. 1987. Change in viability of maize during high-temperature drying. Drying Technol. 5: 497-510.

Litchfield, J. B. and M. R. Okos. 1988. Prediction of corn kernel stress and breakage induced by drying, tempering and cooling. Trans. ASAE (Am. Soc. Agric. Eng.) 31: 585-594.

Marchant, J. A. 1985. Control of high temperature continuous flow grain driers. Agric. Eng. 40: 145-149.

Mathews, B. C. 1980. M-C drybine. Continuous in-bin grain dryer, dry and slow cool one-pass system. Paper 80-3516. Am. Soc. Agric. Eng., St. Joseph, MI. 6 pp.

McFarlane, N. J. B. and D. M. Bruce. 1991. Control of mixed-flow grain driers: development of a feedback- plus-feedforward algorithm. J. Agric. Eng. Res. 49: 243-258.

McKenzie, B. A. 1969. Dryeration, a faster way to handle wet grain. World Farming 11: 10-13.

Meiering, A. G., T. B. Daynard, R. Brown, and L. Otten. 1977. Drier performance and energy use in corn drying. Can. Agric. Eng. 19: 49-54.

Meiering, A. G. and H. J. Hofkes. 1977. The determination of drying results and energy requirements of grain drying with the help of computer simulation. Grundl. Landtechnik 27:1-8.

Michigan State University. 1981. Shivvers in-bin counterflow dryer. Test Report 3. 1. Agric. Eng. Dept., Mich. State Univ., East Lansing, MI. 13 pp.

Miller, P. C. H. and R. D. Whitfield. 1984. The predicted performance of a mixed-flow grain drier. J. Agric. Eng. Res. 30: 373-380.

Moreira, R. G. and F. W. Bakker-Arkema. 1990. A feedforward/feedback adaptive controller for commercial cross-flow grain driers. J. Agric. Eng. Res. 45: 107-116.

Morey, R. V. and H. A. Cloud. 1973. Simulation and evaluation of multiple column crossflow grain drier. Trans. ASAE (Am. Soc. Agric. Eng.) 16: 984-987.

Mühlbauer, W. and W. Christ. 1974. The admissable drying time of maize kernels for animal nutrition with different grain temperatures. Grundl. Landtechnik 24: 161-164.

Nellist, M. E. 1974. The drying of ryegrass seeds in deep layers. Unpublished Ph. D. Thesis. Univ. Newcastle upon Tyne, the United Kingdom. 617 pp.

Nellist, M. E. 1976a. Exposed layer drying of ryegrass seeds. J. Agric. Eng. Res. 21: 49-66.

Nellist, M. E. 1976b. Drying of hay and grain. Agric. Eng. 31: 55-60.

Nellist, M. E. 1978. Safe temperatures for drying grain. Report No. 29, National Inst. Agric. Eng., Silsoe, the United Kingdom. 90 pp.

Nellist, M. E. 1981. Predicting the viability of seeds dried with heated air. Seed Sci. Technol. 9: 439-455.

Nellist, M. E. 1982. Developments in continuous-flow grain driers. Agric. Eng. 37: 74-80.

Nellist, M. E. 1984. Some developments in heated-air grain drying. Farm Bldg. Eng. 1: 25-31.

Nellist, M. E. 1987. Modeling the performance of a cross-flow grain drier. J. Agric. Eng. Res. 37: 43-57.

Nellist, M. E. 1989. Application of computer simulation techniques to the design and use of grain dryers. Paper to Workshop on Grain Drying and Bulk Handling and Storage Systems in ASEAN. Pitsanuloke, Thailand. 21p.

Nellist, M. E. and D. M. Bruce. 1987. Drying and cereal quality. Aspects Appl. Biol. 15: 439-456.

Nellist, M. E. and D. M. Bruce. 1992. Grain dryer testing—procedure, calculation and adjustment of results, and effect of errors. J. Agric. Eng. Res. 52:35-51.

Nellist, M. E. and S. Dumont. 1979. Desorption isotherms for wheat. Dev. Note DN/CDV/983/06010, National Inst. Agric. Eng., Silsoe, the United Kingdom. 24 pp.

Nikolic, D. and A. Mojsovski. 1987. On-line control of cereals drying process. pp. 484-489 in: Intelligent Autonomous Systems. An Int. Conf., Amsterdam.

NIAE. 1962. Report on test of Gascoigne 3/70 grain drier. Report No. 314, National Inst. Agric. Eng., Silsoe, United Kingdom. 8 pp.

NIAE. 1964. Report on test of Allmet 50 cwt grain drier d 400/63 series. Test Report For Users, No. 414, National Inst. Agric. Eng., Silsoe, United Kingdom. 11 pp.

Nybrant, T. 1986. Modeling and control of agricultural driers. Thesis for Doctorate of Technology, University of Uppsala, Sweden. 134 pp.

Nybrant, T. G. 1988. Modeling and adaptive control of continuous grain driers. J. Agric. Eng. Res., 40: 165-173.

Nybrant, T. G. 1989. Modeling and adaptive control of concurrent-flow driers. Comput. Electro. Agric. 3: 243-253.

Nybrant, T. G. and P. J. S. Regnér. 1985. Adaptive control of continuous grain driers. Paper 85-3011. Am. Soc. Agric. Eng., St. Joseph, MI. 19 pp.

O'Callaghan, J. R., D. J. Menzies, and P. H. Bailey. 1971. Digital simulation of agricultural drier performance. J. Agric. Eng. Res. 16: 223-244.

Olesen, H. T. 1987. Grain drying. Publ. by Innovation Development Engineering ApS, Thisted, Denmark. 280 pp.

Othmer, D. F. 1940. Correlating vapor pressure and latent heat data. Ind. Eng. Chem. 32: 841.

Otten, L and R. B. Brown. 1982. Low temperature and combination corn drying in Ontario. Can. Agric. Eng. 24: 51-55.

Otten, L., R. Brown, and K. Anderson. 1980. A study of a commercial crossflow grain drier. Can. Agric. Eng. 22: 163-170.

Pabis, S. and S. M. Henderson. 1962. Grain drying theory. III. The air/grain temperature relationship. J. Agric. Eng. Res. 7: 21-26.

Page, G. 1949. Factors influencing the maximum rates of air drying of shelled corn in thin layers. Unpublished M. Sc. Thesis, Purdue Univ., IN.

Parry, J. L. 1983. Mathematical modeling and computer simulation of heat and mass transfer in agricultural grain drying. Unpublished Ph. D. Thesis, Cranfield Inst. Technol, Cranfield, United Kingdom. 177 pp.

Parry, J. L. 1985. Mathematical modeling and computer simulation of heat and mass transfer in agricultural grain drying: A review. J. Agric. Eng. Res. 32: 1-29.

Patil, N and L. Lechoczky. 1985. Drying analysis of recirculating crossflow dryer design using computer simulation. Int. Agrophy. 1: 269-288.

Paulsen, M. R. and T. L. Thompson. 1973. Effects of reversing air flow in a crossflow grain drier. Trans. ASAE (Am. Soc. Agric. Eng.) 16: 541-543.

Pfost, H. B., S. G. Maures, D. S. Chung, and G. A. Milliken. 1976. Summarising and reporting equilibrium moisture data for grains. Paper 76-3520. Am. Soc. Agric. Eng., St. Joseph, MI. 11 pp.

Pierce, R. O. and T. L. Thompson. 1981. Energy use and performance related to crossflow dryer design. Trans. ASAE (Am. Soc. Agric. Eng.) 24: 216-220.

Putranon, R., R. G. Bowrey, and J. Eccelston. 1979. Sorption isotherms for two cultivars of paddy rice grown in Australia. Food Technol. Austral. 31: 510-515.

Putranon, R., R. G. Bowrey, and R. T. Fowler. 1980. The effect of moisture content on the heat of sorption and specific heat of Australian paddy rice. Food Technol. Austral. 32: 56-59.

Rice, L. S. and A. Kaya. 1986. A distributed control application for continuous drying. pp. 1662-1668 in: Proc. 1986 Am. Control Conf. Vol III, IEEE, New York.

Roberts, E. H. 1960. The viability of cereal seed in storage in relation to temperature and moisture. Ann. Bot. 24(93): 12-31.

Sabbah, M. A., G. H. Foster, G. C. Haugh, and R. M. Peart. 1972. Effect of tempering after drying on cooling of shelled corn. Trans. ASAE (Am. Soc. Agric. Eng.) 15: 763-765.

Sabbah, M. A., G. E. Meyer, H. M. Keener, and W. L. Roller. 1979. Simulation studies of reversed-direction air-flow drying method for soybean seed in a fixed bed. Trans. ASAE (Am. Soc. Agric. Eng.) 22: 1162-1166.

Schreiber, H., W. Mühlbauer, L. Wassermann, and H. Kuppinger. 1981. Influence of drying on the quality of wheat—study of the reaction kinetics. Z. Lebensm Unters. Forsch. 173: 169-175 (NIAE Translation No. 517).

Sharp, J. R. 1982. A review of low temperature drying simulation models. J. Agric. Eng. Res. 27: 169-190.

Sharp, J. R., P. J. Bowden, and W. J. Lamond. 1982. Simulation of grain driers using near ambient air. pp. 79-89 in: Proc. Third Int. Drying Sympo. Vol 2. J. C. Ashworth (ed.). Drying Research Ltd, Wolverhampton, the United Kingdom.

Smith, S. E. 1947. The sorption of water vapor by high polymers. J. Am. Chem. Soc. 69: 646-651.

Sokhansanj, S. and D. M. Bruce. 1986. Heat transfer coefficient in drying granular materials. pp. 962-866 in: Drying '86, A. S. Mujumdar (ed.), Hemisphere, New York.

Sokhansanj, S. and D. M. Bruce. 1987. A conduction model to predict grain temperatures in grain drying simulation. Trans. ASAE (Am. Soc. Agric. Eng.) 30: 1181-1184.

Sokhansanj, S. and S. Cenkowski. 1988. Equipment and methods of thin-layer drying—a review. pp. 159-170 in Sixth Int. Drying Sympo. (IDS'88). Versailles, France.

Sokhansanj, S., W. P. Lampman, and J. MacAuley. 1983. Investigation of grain tempering on drying tests. Trans. ASAE (Am. Soc. Agric. Eng.) 26: 293-296.

Sokhansanj, S. and D. E. Lischynski. 1991. Combination drying of wheat. Can. Agric. Eng. 33: 273-277.

Sokhansanj, S., E. O. Nyborg, J. D. Macaulay, and T. Ullyott. 1980. Grain sample preparation for the performance testing of commercial grain dryers. Paper 80-3514. Am. Soc. Agric. Eng., St. Joseph, MI. 21 pp.

Sokhansanj, S., D. Singh (Jayas), and J. D. Wasserman. 1984. Drying characteristics of wheat, barley, and canola subjected to repetitive wetting and drying cycles. Trans. ASAE (Am. Soc. Agric. Eng.) 27: 903-906, 914.

Sokhansanj, S., H. C. Wood, and V. S. Venkatesan. 1990. Simulation of thermal disinfestation of hay in rotary drum dryers. Trans. ASAE (Am. Soc. Agric. Eng.) 33: 1647-1651.

Spencer, H. B. 1969. A mathematical simulation of grain drying. J. Agric. Eng. Res. 14: 226-235.

Spencer, H. B. 1972. A revised model of wheat drying process. J. Agric. Eng. Res. 17: 189-194.

Stevens, G. R. and T. L. Thompson. 1976. Improving crossflow grain dryer design using simulation. Trans. ASAE (Am. Soc. Agric. Eng.) 19: 778-781.

Stone, M. L. 1984. Potential for feedforward control in driers. pp. 672-681 in: Agricultural Electronics—1983 and Beyond. Volume 2. Am. Soc. Agric. Eng., St. Joseph, MI.

Sullivan, J. E., P. M. Costa, F. N. Owens, A. H. Jensen, K. E. Wikoff, and E. E. Hatfield. 1974. The effect of heat on the nutritional value of corn. pp. 45-63 in corn quality in world markets. L. D. Hill (ed.). Interstate Printers Publishers Inc., Danville, IL.

Sutherland, J. W. 1975. Batch grain drier design and performance prediction. J. Agric. Eng. Res. 20: 423-432.

Sutherland, J. W. 1977. Batch grain drier design and performance prediction. CSIRO Tech. Rep. Tr 16. Highett, Victoria, Australia. 45 pp.

Sutherland, J. W. 1984. The design and prediction of performance of batch driers for paddy rice. Int. J. Devel. Technol., 2: 119-129.

Thompson, R. A. and G. H. Foster. 1967. Dryeration—high speed drying with delayed aeration cooling. Paper 67-843. Am. Soc. Agric. Eng., St. Joseph, MI. 16 pp.

Thompson, T. L. 1970. Simulation for optimal grain drier design. Trans. ASAE (Am. Soc. Agric. Eng.) 13: 844-848.

Thompson, T. L., R. M. Peart, and G. H. Foster. 1968. Mathematical simulation of corn drying—A new model. Trans. ASAE (Am. Soc. Agric. Eng.) 11: 582-586.

Thorpe, G. R. 1987. Thermodynamic performance of a continuous-flow fluidized bed grain disinfestor and drier. J. Agric. Eng. Res. 37: 27-41.

van den Berg, C. and S. Bruin. 1981. Water activity and its estimation in food systems: Theoretical aspects. pp. 1-61 in: Water activity influences on Food Quality, L. B. Rockland and G. F. Steward (eds.). Academic Press, New York.

Van Arsdel, W. B. 1955. Simultaneous heat—and mass-transfer in a nonisothermal system, through-flow drying in the low-moisture range. Chem. Eng. Prog. Sympo. Series 16: 47-58.

Voroshilov, A. I. and B. A. Kuperschmidt. 1973. Maximum permissible time for heat treatment of seeds under nonstandard temperature conditions. Dokl. Vses. Akad. S. Kh. Nauk. 11: 42-44.

Walker, L. P. and F. W. Bakker-Arkema. 1981. Energy efficiency in concurrent-flow rice drying. Trans. ASAE (Am. Soc. Agric. Eng.) 24: 1352-1356.

Wassermann, J. D., G. E. Frehlich, and R. M. Bartel. 1983. Performance testing of grain driers. Paper 83-3509. Am. Soc. Agric. Eng., St. Joseph, MI. 18 pp.

Wassermann, L., W. Mühlbauer, and H. Schreiber. 1983. Influence of drying on wheat quality 3. Effect of grain moisture content on drying behavior and baking properties. Getreide Mehl und Brot. 9: 268-274.

Whitfield, R. D. 1986. An unsteady-state simulation to study the control of concurrent and counter-flow grain driers. J. Agric. Eng. Res. 33: 171-178.

Whitfield, R. D. 1988a. Control of a mixed-flow drier. Part 1. Design of the control algorithm. J. Agric. Eng. Res. 41: 275-287.

Whitfield, R. D. 1988b. Control of a mixed-flow drier. Part 2. Test of the control algorithm. J. Agric. Eng. Res. 41: 289-299.

Wilkins, K., K. P. Poulsen, and E. Green. 1991. Direct drying of foods with natural gas as fuel. Trans. Inst. Chem. Eng. 69(C): 182-188.

Wood, H. C. and S. Sokhansanj. 1990. Heat treatment of chopped alfalfa in rotary drum dryers. Drying Technol. 8: 533-541.

Young, J. H. and J. W. Dickens. 1975. Evaluation of costs for drying grain in batch or cross-flow systems. Trans. ASAE (Am. Soc. Agric. Eng.) 18: 734-739.

Zhang, Q and J. B. Litchfield. 1991. Fuzzy logic control for a continuous crossflow grain dryer. Paper 91-6559. Am. Soc. Agric. Eng., St. Joseph, MI, 19 pp.

17

Technology Transfer in Management of Stored-Grain Pests

Phillip K. Harein

Stored Product Pest Management, Inc., Crosslake, Minnesota

Steven A. Clarke

University of Minnesota, St. Paul, Minnesota

SUMMARY

Technology transfer carries information from researchers to users but unfortunately research and extension efforts are often widely separated. It is important to involve producers in research that has potential application. Storage of grain is viewed differently by producers, handlers, consumers, laborers, and administrators, and these views affect how grain will be managed. They affect the degree of resources put into research and extension on postharvest crop preservation.

In the United States, several successful National Stored-Grain Pest Management "Train-the-Trainer Programs" were conducted by the University of Minnesota from 1985 to 1989. These examples offer a useful guide for future extension efforts and include the use of improved technology such as satellite teleconferencing.

An example of a successful postharvest technology transfer program in a developing country is given for the African country of Rwanda. The governments of Rwanda and the United States financed the research on quality and quantity loss in beans and sorghum and the transfer of information obtained. Technical assistance was provided by the University of Minnesota. Technology transfer consisted of producer visits to the research sites, user feedback, graduate student training, development of a training course, and visits to storage sites by specialists to give advice. The research laboratory also provided grain sample analysis and advice, a regional conference was held, and several radio broadcasts presented.

In all situations, the keys to successful postharvest technology transfer are starting research with an extension outlook, involving clients from the outset, solving problems of concern to users, building communication links and respect, creating demand for services and information, and utilizing the available human resource pool.

---

1. INTRODUCTION

Agricultural researchers generally develop new information on products or techniques which then are formally or informally forwarded to extension personnel for distribution. This top-down system has been most successful in areas of high-value crops and well-to-do producers. However, the separation between research and extension has been a definite barrier in the transfer of technology to small and resource-poor producers, especially in developing countries. In addition, extension personnel, when compared with researchers, may have relatively little education, low status in the community, and are often poorly paid. Consequently, they are frequently classified as inadequate sources of appropriate answers to local problems and needs (Collinson 1985).

One method to improve the training of producers is to encourage their involvement in both the practical research and the extension effort (Andrew and McDermott 1985). However, on-farm research should not substitute for an effective extension system which addresses all the problems of technology transfer. To be most effective, farmers must be fully involved in planning, implementing and concluding the research in which they are expected to participate. Otherwise, whatever technology is developed may not be acceptable (Stoop 1988). The demonstration procedure, conducted by producers in coordination with extension workers, is one of the best methods of initiating changes that improve crop production or preservation (Ballantyne 1987). Various methods have been devised to determine farmer needs and knowledge, including surveys (formal and informal), on-farm trials, demonstrations, meetings, and field days (Ewell 1988, Biggs 1989). Each method should be considered.

An important means of disseminating technology is via students, both at the undergraduate and graduate level. Students provide a constant source of technology to the rest of society, whether to industry, government, or educational institutions. This type of technology transfer will continue only as long as students are exposed to stimulating and high-quality research and extension programs. Increased emphasis should be placed on a more direct form of technology transfer in which the discoveries, innovative ideas, and specialized expertise are made available to those who can put them to practical use. Most land-grant universities in the United States have a tradition in this area.

The new focus on technology transfer becomes more urgent with the increasingly rapid pace of technological change and the international environment in which it occurs. No longer can any country afford to take decades to test practical applications of new discoveries. To remain competitive in both the academic and the agricultural communities requires a constant awareness of and almost instant access to the very latest knowledge and technology in a given field. There is never enough technology transfer. Technology transfer is one of the most important ingredients of the worldwide resource-sharing process. The agricultural Extension Service's primary objective—the transfer of knowledge—is becoming more important every day. However, only the most productive and acceptable technology transfer systems must be selected and employed if they are to have meaningful impact.

Of equal importance is developing a specific group of personnel to facilitate technology transfer. Such a group must be in complete agreement on their goals and objectives, develop advanced methods to accomplish these targets, and set a specific level of expectation. These people must be well-trained and those who are trained must in turn continue the process of training others. There must be a commitment to alter their training process, if justified, and they must continue to upgrade their own expertise. Stored-grain managers are becoming increasingly well-educated and more sophisticated. Consequently, it is absolutely necessary to constantly update the subject matter as well as the methodology and means of communication.

Technology transfer really includes the results of research, development, and the diffusion of new knowledge including both hard products and soft information. However, extension activities are often implemented via information and marketing services in agribusiness, which tend to be biased towards their own products.

There always seems to be disagreement among extension specialists and agents as to what segment of the population should receive the most attention in terms of personnel, time, or funding. The tendency to provide major resources to the established "commercial" clientele has merit in a train-the-farmer philosophy. Needless to say, however, the informational requirements of different clientele can vary considerably. Subject matter packages and methods of distribution need to be developed for each distinct group of clientele.

As stated earlier, transfer of technology is the final step in a cycle including innovation, technical and commercial development, and distribution. It certainly includes equipment and the practical information required for its use. However, too often the "use" aspect of knowledge remains with the inventor. Technology transfer must be a multilevel process of communication involving a variety of senders and receivers of ideas and materials. It encompasses both public and private enterprise. The challenge centers on developing indigenous capacity to generate and adapt agricultural technology to local conditions. This is crucial in developing countries where the gap between new technology and local conditions is often large. The focus of the transfer is also important; for instance, most of the farm work in Africa is done by women but information packages are oriented towards the male population.

2. STORED-GRAIN LOSSES

The magnitude of stored-grain losses varies from year to year and from country to country. Losses resulting directly or indirectly from pests in excess of 30–50% have been reported in some developing countries where food is often in short supply (Hall 1970). Although economically important, in most of the developed countries the percentage loss of stored grain is much lower, and often unnecessary, because the means to prevent losses are already known.

Providing an adequate food bank is possible in several ways, although the major method is production. However, we need not rely solely on production, especially with grain and cereal products. Protecting harvested crops in storage from pests can significantly increase available food resources. Field production losses can be recovered by replanting or by the partial regeneration of damaged plants but no such avenues exist following damage to harvested grain. Losses in storage should also be included in the cost of producing the grain, as are the cost for field preparation, seed, pest management (weeds, diseases, and insects), and harvesting.

Stored-grain pests have little concern for man-made boundaries unless the bounded areas provide environmental limits (temperature, moisture, food, and space) that restrict pest development. No country, whatever its geographical situation or developmental status, can afford to dispense with efforts to manage stored-grain pests. Unfortunately, however, the presence of pests or pest-generated filth in stored grain and cereal products is too often accepted as a normal condition, even in developed countries. Unbelievably, some agricultural producers and stored-grain managers still think that harvested grain somehow "spontaneously generates" insect life and that grain is satisfactorily and desirably "matured" only when infested.

The management of stored-grain to prevent its deterioration from all factors, including insects, is not an easy task, especially with ever-increasing sanitation requirements in the processing of food and environmental standards. Under present technological conditions, a certain level of pest infestation and pesticide residue in our food must be accepted if meeting our supply requirements is to be both economical and feasible.

Considerable information has been generated on stored-grain pest management by both private and public agencies. Unfortunately, relatively little of these data have been utilized by stored-grain managers. There are many reasons for this lack of progress and the reasons vary widely within and between countries.

3. ENVIRONMENTAL QUALITY

The environmental quality of stored grain is an important concept of pest management and since humans are an important factor of environmental quality, we should identify some of our concepts of stored grain.

Concepts about stored grain will depend, in part, on our relationship to it, whether we are producers, handlers, consumers, laborers, or administrators. In addition, these concepts change as one gains experience and knowledge. There have been at least four basic concepts identified in the past. You might believe in any one of them, any combination, all, or none.

First, stored grain may be considered as just another commodity like lumber, cement, or gravel. All such commodities must go through marketing channels, including the tasks of loading, shipping, unloading, bagging, stocking, and storing. Generally, with this concept, relatively little attention is given to the commodities' condition. Consequently, when stored grain is also regarded strictly as a commodity, very little attention is paid to its quality. After the grain is stored, the responsibility is considered complete until the grain must be moved again. Regular inspection and quality control are rarely a part of the operation.

Second, stored grain is often referred to as simply a measure of wealth or as a source of wealth, or both. For people with this concept, grain in storage is just like money in the bank. Interest in the maintenance of its quality may be temporary, if at all. When the grain begins to deteriorate, it is sold or moved. Subsequent owners or holders are often liable for any losses or increased potential for loss from deterioration.

Third, grain may represent a living organism to some stored-grain managers. Actually, each viable seed is living. It respires, has a potential for growth and development, but eventually dies after prolonged storage. Seeds must receive essentially the same care required by other living organisms if they are to remain healthy. However, their condition is influenced by many environmental and biotic factors, both external and internal. The more important of these factors include temperature, humidity, oxygen, carbon dioxide, microorganisms, insects, and rodents. Regular checks of storage conditions becomes a necessity.

Fourth, stored grain is recognized by some managers as food. When first provided by nature grain is relatively free of contaminants. However, grain that is neglected or improperly handled following harvest may soon become polluted and befouled by insects, rodents, and microorganisms. Grain once contaminated can never be adequately treated to become entirely clean or pure again. Regular inspection and quality control become part of the manager's responsibility.

4. RESEARCH AND EXTENSION

Compared with the relative amount of resources available for crop production, the amount assigned to research and extension efforts devoted to stored-grain management is minimal. To make matters worse, researchers' efforts have often been designed as basic research rather than designed to provide information that can be utilized immediately by stored-grain managers. Consequently, extension stored-product specialists have often been forced to conduct their own applied research to solve and answer immediate problems.

5. STORED-GRAIN TECHNOLOGY-TRANSFER IN DEVELOPED COUNTRIES

5.1. General Considerations for a Program in the United States

In developed countries, the avenues available for technology transfer are unlimited compared with those available in developing countries. However, the principles for successful technology transfer, as reviewed earlier, apply to all situations. Basically, successful technology transfer requires leadership, dedication, and cooperation from many segments involved in stored-grain management. This audience must include students, stored-grain managers, educators, and pertinent personnel from government and industry.

Many of the methods, problems, solutions, organizations, etc., alluded to earlier were relied upon at the University of Minnesota. This provided an appropriate program in the transfer of information to an individual stored-grain manager or to a national audience, who in their turn became disseminators of current technology of stored-grain management in their individual states. It is not necessary to elaborate on the program details required for the successful transfer of information to all levels, be it individual, county, state, regional, national, or international. Consequently, the following example will hinge primarily on the program conducted by the University of Minnesota to provide learning sessions for personnel interested in proper stored-grain management methods.

5.2. U.S. National Stored Grain Conference

5.2.1. Situation

The purpose of the transfer of information program was to develop and conduct the National Stored-Grain Pest Management "Train-the-Trainer Programs." It reviewed the recommended chemical and nonchemical methods to manage insects and molds that reduce the quality and quantity of stored grain.

Insects and molds cause significant losses to stored grain each year. These losses will continue unless stored-grain management systems are upgraded by farmers and commercial stored-grain managers. More use of nonchemical control methods must be incorporated into the stored-grain pest management program to prevent and halt infestations. Improved pest-monitoring methods and materials must also be considered for early detection of pest species.

Pesticides still remain essential to keep grain storage losses under control. The increased restrictions placed on the application of certain insecticides because of insect resistance and excessive residues (creating potential cancer-producing products), however, requires that we use the remaining insecticides even more efficiently and safely. The changing technology developed for the new chemicals that prevent pest infestations (especially in certain integrated pest management practices), requires constant information transfer to those involved with the proper use of pesticides and pest management programs.

5.2.2. *Objectives*

The results of a questionnaires to each state extension specialist involved in stored-grain management strongly supported the need for the "Train-the-Trainer" program. The Minnesota Extension Service, University of Minnesota, agreed to:

- provide leadership in developing and arranging a program to present current information on stored-grain pest management to a pertinent agricultural extension specialist and a regulatory official from each state interested in stored-grain management;
- identify training materials (publications, slide sets, video tapes, stored-grain management recommendations, etc.) that could be used in stored-grain management programs within individual states;
- coordinate and conduct on-site training, including hands-on demonstrations of proper stored-grain management;
- provide copies of materials developed as a result of this program for distribution to all state pesticide education co-ordinators and state lead agencies.

Thus, the "Train-the-Trainer" program included conprehensive information on:

- stored-grain losses
- federal and state regulations
- pest identification and biology
- insect pest management
- safety considerations
- stored-grain economics—expert systems

Table 1 Evaluation of a Stored-Grain Management National Teleconference (N = 63)

	Very good	Good	Average	Poor
Overall evaluation				
Program format	25	34	4	0
Program content	22	34	6	1
Speaker expertise	26	30	7	0
Graphics and tables on video	16	31	14	1
Subjects presented	24	32	6	1
Total length of program	16	40	7	1
Specific topics evaluation				
Insect management	24	32	7	1
Stored grain molds	26	28	9	0
Engineering: aeration, drying	29	26	7	0
Economics	11	21	24	5
Interactions of the above areas	11	33	11	1

- initiating training programs
- pest-management hands-on demonstrations

In addition to current pamphlets and training materials from various universities and industry, the participants were provided a current manual on stored-grain management from the University of Minnesota (Cink and Harein 1989). Demonstrations by industry at grain storage sites were appreciated by the participants as they were able to witness many pest-management efforts applied under actual field conditions.

5.3. Satellite Teleconference

The "Train-the-Trainer" was updated and reinforced by a subsequent satellite teleconference on stored-grain management. It was apparent that the effort was successful and it served as a major link to transmit current technology to personnel. A summary of the teleconference evaluations on stored-grain management was based on sixty-three evaluations of the "Train-the-Trainer" program. A numerical rating of the format and topics presented are shown in Table 1. In addition, each state extension specialist received a video tape copy of the entire teleconference to use in future stored-grain meeting within their respective states. This type of program can also be used on an international basis.

6. STORED-GRAIN TECHNOLOGY TRANSFER IN DEVELOPING COUNTRIES

6.1. General Considerations

Research and extension activities focused on improving postharvest technology have also received considerably fewer resources in developing countries than preharvest, production-oriented aspects of grain and other food supply systems. Despite the apparent universality of the fundamental principles of stored-grain management, their widespread application, though desirable, may not always be cost-effective for those involved in storing grain or be considered sufficiently important to warrant substantial government support in many developing countries.

Relatively large amounts of precious and limited resources are often needed to accurately quantify the true magnitude and nature of the storage losses. Given the complexities of both the biological as well as environmental factors in tropical areas (where most developing nations are located), the very nature of subsistence farming and the myriad types of losses which occur, it is extremely difficult to measure losses, especially when attempting to aggregate them at the provincial, regional, or country level. These factors are also responsible for the high degree of variability in losses from season-to-season, village-to-village, and country-to-country. For this reason, intensive in-depth surveys of stored-grain losses, if very carefully designed and conducted, may give "good" results but only for one point in time. They are difficult to repeat because of their cost; thus it is almost impossible to prove if they are representative of actual situations. Surveys do have value, however, in enhancing awareness of grain storage systems, in providing training to scientists and staff involved, and in developing contacts with clientele and potential users or recipients of improved technology.

It is important to form and develop a pool of local scientists and professional staff dedicated to understanding and improving stored-grain technology. Though principles are universal, their application is not. This pool (or working group) is multidisciplinary by necessity and can:

- serve as the collectors and centralized repository for indigenous knowledge and traditional storage practices for their country;
- develop science- and research-based approaches to understanding and assessing traditional methods;
- formulate and test new, promising technologies;
- provide advisory and consulting services for those in need;
- operate grain quality laboratories for conducting necessary analyses in support of their applied research and outreach activities;
- provide training (workshops, short courses, demonstrations) to groups of extension agents, producers, grain traders, and managers of stored grain at all levels in the food supply system; and

- take the lead in promoting the need for attaching importance to postharvest problems, and in educating national policy-makers on this critical phase of food production.

Any research conducted should be applied to specific problems. This requires that storage specialists be well linked and networked into the international scientific community to take advantage of new knowledge and technology that can be adapted to local conditions. Activities should be client-oriented and demand-driven. Some services, once established and valued, should be fee-based and revenue-generating to make the overall program more self-supporting. Outreach and extension activities should be started as quickly as possible and integrated with on-going research work whenever the opportunity arises. Although this is frequently a glaring weakness in developing countries, there has to be a tight link between the research and extension components if the program is to be successful.

6.2. Rwanda

6.2.1. Problem Definition

Rwanda is a small, land-locked country located just south of the equator in the highlands of east central Africa. It is called the "land of a thousand hills" because of its steeply rolling topography. Temperatures are quite moderate, averaging 16–24°C depending on elevation, and uniform throughout the year. The rainfall pattern is bimodal with two distinct rainy seasons (February–May and October–December) with an annual precipitation of 800–2000 mm.

Rwanda has a population of 7 million and is one of the most densely populated countries in Africa. The population is largely rural (95%), most of whom live in small, scattered farmsteads (average size: 1 ha) rather than villages. Almost all Rwandans are engaged in agriculture which is predominantly one of subsistence. The main food crops are beans, bananas, sorghum, sweet potatoes, corn (maize), peas, cassava, and Irish potatoes. The main cash crops are coffee, tea, and pyrethrum, which are grown for export and constitute Rwanda's primary source of foreign exchange earnings.

The Rwandan Institute of Agricultural Sciences (ISAR) has primary responsibility for agricultural research conducted in Rwanda. ISAR has pursued research on food and industrial crops, cropping and farming systems, and forestry and animal husbandry, but to date has placed little emphasis on stored products. In 1984, research on food-storage problems was initiated at the National Office for the Development and Marketing of Food Crops and Animal Products (OPROVIA) through the GRENARWA II-Research Project. This project was set up to conduct research on the storage of beans and sorghum.

6.2.2. *The GRENARWA II-Research Project*

The GRENARWA II-Research Project (OPROVIA 1988) was a 5-yr project (1984-1988) financed jointly by the Rwandan and United States Government and based at OPROVIA in the capital city of Kigali. The University of Minnesota provided technical assistance under contract to AID. Support for an improved system for the storage and marketing of grains and pulses has been a major thrust of the total U.S. Agency for International Development (USAID) presence in Rwanda since the mid-1970s.

The major problem to be addressed by the project was the loss in both quantity and quality of beans and sorghum during storage. The principal objectives of the research were to:

- determine the extent of losses of beans and sorghum during storage at different levels in the postharvest system, and the causes of these losses; and
- develop and test appropriate technologies to reduce these losses, especially for storage periods exceeding 6 to 8 mo.

The research program was divided into six separate but interdependent components:

- *Grain storage survey*, whose objectives were to inventory current methods and conditions of storage of beans and sorghum at three levels (farm, cooperative, and OPROVIA), and to estimate the magnitude and causes of losses incurred.
- *Bean variety survey*, whose objectives were to inventory the bean mixtures and varieties grown in Rwanda (including development of a catalogue and reference collection) and to determine varietal preferences of producers and consumers.
- *Bean cookability and sensory quality studies*, whose objectives were to determine the effects of different storage conditions (e.g., moisture content, temperature, and time) on cookability of beans as well as sensory preferences of consumers, and to study the possibilities of alternative preservation methods to avoid or minimize the problem of bean hardness.
- *Insect resistance study*, whose objective was to test the resistance to stored-grain insects of the principal bean varieties grown in Rwanda.
- *Alternative storage method evaluation*, whose objectives were to develop and test potential improvement in storage structures and management techniques to reduce losses occurring during storage.
- *Development of grain quality standards*, whose objectives were to develop standards and testing procedures for beans and sorghum to facilitate marketing and to improve storability of these crops.

In addition, the project had a certain number of more general, long-term objectives including:

- The creation of a complete Food Quality Laboratory at OPROVIA for testing and control of quality (both physical and sensory) of stored grains and other foodstuffs; and
- The development of a Center of Postharvest Technology at OPROVIA to ensure the continuity of research, dissemination of research results and information, and provision of training and advisory services.

6.2.3. Approach to Technology Transfer

Besides the six original research components, a seventh component was established by the project team in Rwanda to provide a framework for a program to disseminate results generated by the research program. By making this activity an explicit and integral part of the project, it was also intended to focus the attention of the staff on the ultimate goal and the true measure of success of their research.

This seventh component did not exist in the contract of the University of Minnesota nor in the initial project documents. As a result, no specific personnel were assigned nor funds reserved by the funding agency for this component. It was added, however, by project management to ensure the application and practical utilization of information collected within the research program. All staff and all six research components were expected to contribute to the effort. As a result, the project furnished various technical, advisory, and extension services to third parties. To illustrate, a number of examples are given below.

- Consultative visits were requested by the Farming Systems Project at Rewerere for producers in the communes covered by the storage project with the aim of studying existing storage problems and recommending actions and solutions to be undertaken. Several farms were visited and meetings were held with groups of producers. Findings were explained, general storage principles described with the help of large posters, and their views were solicited on the major storage problems. Several demonstrations of alternative insect control methods were established.
- An appraisal was made at the National Center for Small Animal Husbandry at Kabuye of the management of large-scale storage of grain and other raw products used in the manufacture of animal feed. Problems were identified by a staff team and recommendations for improvement were given both verbally and in a written report to the Center's management.
- The project collaborated in the conduct of theses research for two Masters-level students from the National University of Rwanda. The subjects of the

research were the evaluation of insecticidal properties of local plants and the varietal resistance of beans to bruchids (storage insects).

- OPROVIA and the Cooperative Training and Research Center jointly implemented a separate project entitled "Postharvest Management in Rwandan Cooperatives," which provided training in storage and marketing of agricultural commodities. The GRENARWA II-Research Project was responsible for the organization, monitoring, and evaluation of the training in grain-storage management of 120 managers and Board members of some thirty producer cooperatives.
- An agreement was signed between OPROVIA and the Global Development Project of Butare which provides for the GRENARWA II-Research Project to make appraisals of storage problems prevailing in the zone covered by the Project and to organize training on storage practices for managers of cooperatives and commune-level agricultural extension agents in the Butare prefecture. The training included on-site follow-up visits to inspect facilities and advise managers on corrective measures for their specific problems.
- The project provided services for the benefit of OPROVIA, private merchants, cooperatives, and other agricultural projects on various commodities (beans, sorghum, rice, and food aid products among others). On numerous occasions, the project laboratory conducted quality analyses of samples submitted by these groups, explained results, and provided advice on corrective measures when appropriate.
- The project presented two papers, one on the results of the component "Study on Sensory Qualities and Cookability of Beans" and the other on the results of the component "Inventory of Bean Varieties Grown in Rwanda."
- Project team members presented three papers on different aspects of the research being conducted and its application to reducing storage losses of food.
- Several radio broadcasts on RADIO RWANDA were written and presented by project staff. Topics included various aspects of the research program underway and project results of benefit to Rwandan farmers. Radio is the prime mass media in Rwanda and an important means of communicating with its largely rural population.

6.2.4. Impact

It is extremely difficult to measure impact of the technology transfer activities in quantitative terms such as tonnes or percent of food saved or losses reduced. Without any intervention at all, losses are highly variable and depend on a multitude of factors. Farmer knowledge and actual use of sound stored-grain principles as well as extension efforts to spread and enhance that knowledge base are only one set of those factors. Measurement of activities alone can be indicative or suggestive of impact, but it does not constitute evidence that the "message" was received and understood by users. Demand for information or services may be as good an indicator of usefulness and impact that we have. Demand in this sense means unsolicited requests for assistance from those involved in storing and marketing grain. Over the course of this project, there was a steady increase in these requests, especially from grain merchants, large millers, managers of grain cooperatives, and managers of other large-scale storage units. There was little demand directly from producers except through a few government-operated development projects and the farm-level survey conducted by project staff. The fact that the Rwandan government has continued to provide resources and kept the group intact following the termination of USAID funding, and for the first time began relying on staff for laboratory analyses and advice about quality and storability of grain imports prior to purchase, are also clear signs that the project is meeting national or public needs.

It is perhaps premature to judge the success of this project or its technology transfer activities. After all, the project was only a 5-yr effort and thus relatively young. It constitutes the country's first concerted attempt to focus on postharvest problems and it was designed primarily not as an extension but as a research project. This is logical since the first requirement is always a solid base of data and knowledge on which to build appropriate extension programs.

7. CONCLUSIONS AND FUTURE DIRECTION

It is our opinion that the keys to future success for postharvest technology transfer in either developed or developing countries are the process and approach used: starting the research phase with an extension mentality and outlook, involving from the outset potential clients and users of the technology in the process, solving the problems potential users perceive as important; building mutual respect and understanding as well as communication links, and creating demand for information and services available. Another key is the human resource pool: the expertise, ingenuity, and teamwork of the staff, especially the national staff, on whom future progress ultimately depends.

To have technology transfer, we must be good listeners in order to fully comprehend the problems voiced by producers including stored-grain managers or to fully understand the solution as developed by an array of sources such as researchers, extension specialists, consumers, etc.

One of an unsuccessful listener's major problems is to interrupt the source before completing all the required information. Consequently, interrupters risk missing important points and can destroy the logical progression of the discussion. Technology transfer requires that we as listeners give 100% of our attention to the discussion even if the subject might be secondary to our current interest. Sometimes it may help all concerned to provide a "designated" listener. This could be an assistant or an associate who takes notes but whose verbal contributions should be kept to a minimum. Perhaps a final effort to aid technology transfer by improving our ability to listen is to pattern our behavior from a person who seems to be one of the best listeners for information we attempt to distribute. Every effort to improve listening properly will enhance all efforts in technology transfer.

REFERENCES

Andrew, C.O. and J.K. McDermott. 1985. Farmers linkages to agricultural extension and research: A question of cost-effectiveness. Farming Systems Support Project. Univ. Florida, Gainsville, FL. 10 pp.

Ballantyne, A.O. 1987. Extension work in the small farm sector. Agric. Admin. Ext. 24:141-147.

Biggs, S.D. 1989. Resource-poor farmer participation in research: A synthesis of experience from nine national agricultural research systems. OFCOR Cooperative Study No. 3. The Hague: ISNAR. 37 pp.

Cink, J. and P. Harein. 1989. Stored grain pest management. Burgess Printing Co., Minneapolis, MN. 282 pp.

Collinson, M.E. 1985. FSR: Diagnosing the problems. Research, Extension, Farmer: A two-way continuum for agricultural development. M.M. Cernea, J.K. Coulter, and J.F.A. Russel, (eds.). World Bank, Washington. 171 pp.

Ewell, P.T. 1988. Organization and management of field activities in research: A review of experience in nine countries. OFCOR Comparative Study No. 2. The Hague: ISNAR. 55p.

Hall, D.W. 1970. Handling and storage of food grains in tropical and subtropical areas. UN Food and Agriculture Organization Agric. Dev. Paper 90. Rome. 350 pp.

OPROVIA. 1988. Synthesis of research, recommendations, and prospects for the future. GRENARWA II-research project. Ministry of Agriculture, Republic of Rwanda, Kigali, Rwanda. 74 pp.

Stoop, W. 1988. NARS linkages in technology generation and transfer. Working Paper No. 11. The Hague: ISNAR. 31 pp.

18

Ecological Aspects of Feed-Mill Operation

J. T. Mills

Agriculture and Agri-Food Canada, Winnipeg, Manitoba, Canada

SUMMARY

A feed mill is a human-made, unstable ecosystem with a continually replenished energy supply. Most of the food energy is transferred to farm animals through product sales. Some of the energy, in dust and other residues, remains in the mill and provides sustenance for feed-mill pests. This review describes the feed-mill ecosystem in relation to pest ecology and to factors affecting feed production and storage in mills and on farms. Emphasis is placed on management decision-making strategies particularly in relation to quality control. Feed-mill management is approached from the viewpoints of site, staff, raw materials, production, finished products and distribution. Regulatory and product-value factors are discussed as well as future research needs. Most research and development in feed mills will likely occur in the problem areas of cost reduction, quality control, regulatory requirements and health and safety. Examples of feed-mill operations and attendant problems are given for Indonesia and Peru. Relatively few ecological studies have been conducted in feed mills in developing countries.

1. INTRODUCTION

The ecosystem concept has been applied to stored-grain research in Canada (Sinha 1973a,b, White and Sinha 1980), the United States (Arbogast and Muller 1988), and the United Kingdom (Coombs and Woodroffe 1963). These studies were conducted from a holistic viewpoint with multidisciplinary inputs in an attempt to unravel the complex interrelationships occurring among biotic and abiotic factors operating within the ecosystem. Most of the studies were conducted with binned grain, loose grain in piles, or bagged grain. Relatively few ecological studies, however, have been done on processed feeds or feed ingredients in commercial mills or farms. Much information exists in the literature on the management of stored feeds and on related storage problems; however, it is often fragmented and needs to be examined from a comprehensive viewpoint.

This chapter reviews the feed-mill ecosystem, feed-mill problems and their effects, and factors affecting the production, storage, and distribution of feeds and their ingredients in feed mills and recipient premises. Emphasis is placed on management decision-making strategies particularly in relation to quality control. Regulatory and product-value factors are discussed as well as the future need for multidisciplinary research and feed-mill handling and storage problems in developing countries.

2. THE FEED-MILL ECOSYSTEM

2.1. Overview

To understand the ecology of feed mills, consider a feed mill as an ecological system in which the abiotic physical-chemical environment, raw and finished feed products and associated residues, and the biotic assemblage of flora and fauna interact with one another (Fig. 1). A feed mill is a human-made, unstable ecosystem (Margalef 1963) of a relatively simple structure with a continually replenished food energy supply.

Energy is the most important item in an animal's diet, and feeding standards and ration formulations in mills are largely based on its measurement (Church 1984). Energy in feeds originates mainly from cereals, oilseeds, pulse seeds, animal products (including tallow), mineral and vegetable oils, and molasses. Energy may also be derived from additions such as seed screenings or, in developing countries, other inexpensive materials such as rice polishings or ground shrimp skins. This energy is utilized by farm animals and also by feed-mill flora and fauna. Most of the energy in the feed-mill ecosystem is transferred to farm animals through product sales but is continually replenished from fresh materials. Some of the energy and nutrients that remain in the mill as residues constitute an important food source for flora and fauna which may infest and contaminate raw and finished feed products (Fig. 1). Because of frequent interference from the physical environment and human beings, the feed-mill system, like the stored-grain ecosystem (Sinha 1973a), is considered unstable.

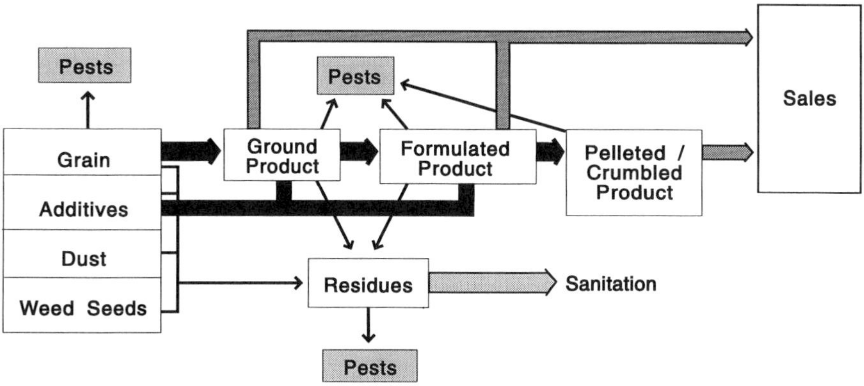

Figure 1 Schema of a feed-mill ecosystem showing the movement of food nutrients, energy and waste products. ➡ transfer of food nutrients and energy within the mill; ➡ transfer of food nutrients and energy to farm animals; ➡ transfer of residues outside mill; ⟶ flow of nutrients and energy to pests.

The feed-mill ecosystem—unlike flour mill, grain elevator, or warehouse ecosystems—has certain features which are conducive to pest problems, including a warm, partially or nonenclosed production area, diverse operations involving production of many different rations, and complex sanitation requirements (Table 1). Because of these and other peculiarities, the feed-mill staff have to be continually alert to prevent, detect, and control pest problems within their work environment.

A feed-mill ecosystem is composed of two parts; mill premises (Fig. 2A) and premises for storing or using manufactured products (Fig. 2B). The latter may be located nearby or as far away as another country. The main functional areas of a generalized feed mill with associated flow of materials are shown in Figure 2A (areas 1–5). The main types of premises storing and marketing feed-mill products are shown in Figure 2B (areas 6–10). A more detailed examination of the management requirements, and quality control concerns of each of areas 1–10 will be dealt with later.

The operating machinery of a feed mill do not form a largely enclosed system as in a flour mill. Consequently, there are much more dust and residues in the internal working space. The internal mill environment consists of central warm (or heated) milling and pelleting areas connected by pipes, ducts, conveyors, or doors to adjacent cool or unheated areas for storage of raw materials and bulk or bagged products. During mill operation there is a continuous two-way flow of contaminants into and out of the internal mill environment (Fig. 3).

Table 1 Salient Features of Feed Mill, Flour Mill, and Grain Elevator Ecosystems

	Feed mill	Flour mill	Grain elevator
Environment	Warm year-round in production area	Warm year-round in production area	Unheated in temperate climates; warm year-round in tropical climates
System	Partially or nonenclosed	Largely enclosed	Partially or nonenclosed
Sanitation	Difficult because of plant and equipment design and perceived need for less care	Relatively easy because of pneumatic systems and perception of need for sanitation because of human consumption	Moderately difficult
Materials	Numerous, domestic and imported oilseed, cereal, pulse, and animal products; nutrients; drugs	Few, mainly domestic cereals and flour	Few, mainly domestic cereals, oilseeds, and legumes
Storage	Plant and animal raw products in bins and tanks; supplements, micro and macromixes in bags in warehouses; custom and complete feeds as meal or pellets in bins	Cereal grains in bins, flour in bags or silos, supplements, bleach	Cereal grains, oilseeds, and legumes in bins, bags
Changes required during operations	Many frequent changes between batches of different feeds, in formulation, and pelleting	Some changes	Handling of different crops depending on harvest season, separate handling of varieties and types of grains for special end use, including for feed

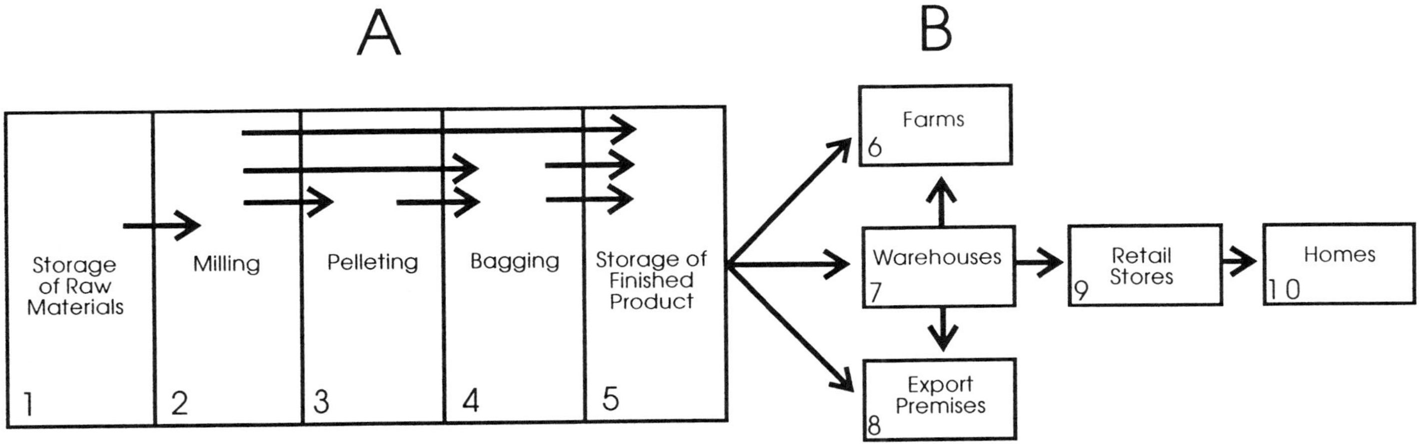

Figure 2 Schema showing movement of materials in: A), feed mills (areas 1–5); and B), premises storing and marketing domestic or export feed products (6–10). ⟶ flow of materials.

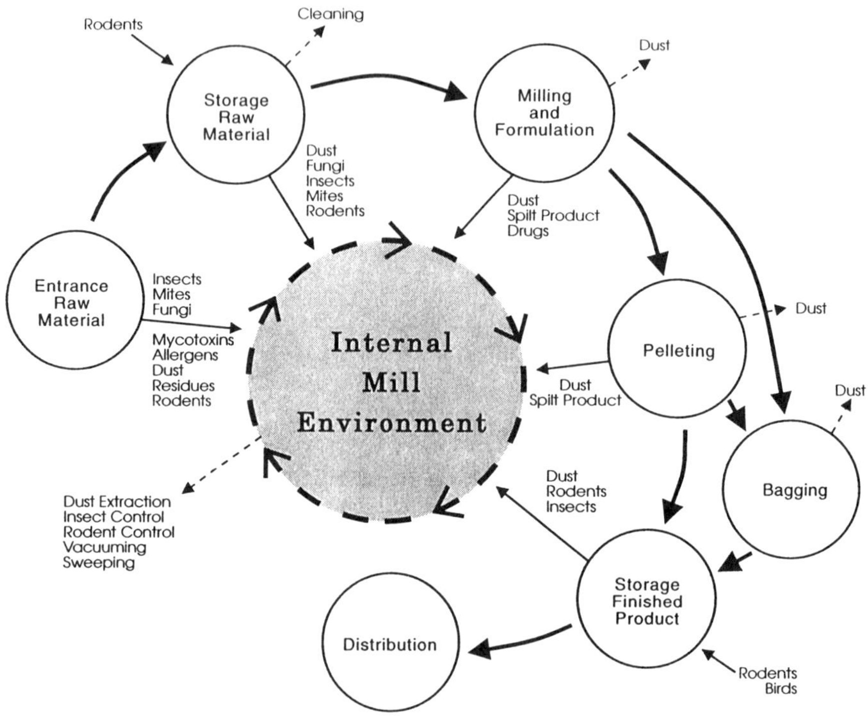

Figure 3 Schema showing feed-mill processes and associated flow of materials (outer ring) with their contributions to contamination of the internal mill environment (inner ring). ➡ flow of materials; ⟶ contaminants entering; - - - - -➝ residues and contaminants leaving; ■■■■➧ recycling of residual contaminants.

These include grain dust and residues present on incoming grains or generated during handling, fungal spores and mycelium, possibly mycotoxins, insects (alive and dead, entire and fragmented), mites, rodent hairs and feces, spilt product, toxic additives (e.g., coccidiostats used in poultry feed are toxic to horses and dogs), medicaments, and allergenic substances. The level of contaminants in areas outside the internal mill environment is reduced through rejection of raw materials, dust extraction from milled and pelleted products, and sales (Fig. 3). Some, but not all, of the contaminants in the internal mill environment are removed through dust extraction, vacuuming, sweeping, and rodent and insect control. The residual contaminants that are constantly recycled inside the mill constitute an ongoing operational problem. They are a source of mold, insect, and mite infestations, rodent contamination, and possible health, fire, and explosion hazards.

2.2. Molds

Mold growth in field crops or stored grain reduces the starch and lipid content of animal feeds, with consequent increases in fiber, and an overall reduction in digestible energy; palatability is also often adversely affected (Blaney and Williams 1991). Growth of spores and hyphal fragments of molds in stored materials and plant and animal debris are determined largely by substrate water content (or more precisely the water activity), temperature, gas composition, and the presence of other fungi (Magan and Lacey 1988). The fungi colonizing grain have been broadly classified into two groups known as field and storage fungi (Christensen and Kaufmann 1969); some fungi in each group produce mycotoxins.

Stored feeds can gain moisture through moisture migration as a result of nonuniform temperatures. Temperature changes in a silo can occur because of fluctuating daytime and nighttime temperatures or because of the effect of the sun shining on one wall in conjunction with lower temperatures on the shaded side. Moisture migration leads to pockets of moisture accumulation, mold development, and eventually spoilage problems. Löwe and Friedrich (1983) in Germany simulated the effect of the sun shining on one side of a large silo filled with pig pellets. Condensed water accumulated at the surface and at the outer walls, and spoilage occurred after 1 wk. The molds clumped the pellets, impeding their flow and later their transportation. Even before any visible molds were evident, the flow characteristics declined because the pellets picked up moisture and became swollen and softened. A decline in flowability of pellets is indicative of moisture changes and possibly also of mold and mycotoxin development (Mills 1989).

The safe moisture content for processed feeds is usually taken as that in equilibrium with a maximum of 70% relative humidity (rh), the level at which molds begin to grow (Table 2). Most animal feed pellets however, are produced

Table 2 Moisture Content and Equilibrium Relative Humidity of Pig Feed Pellets at 25°C

Moisture content	Equilibrium relative humidity%
9.4	40
10.6	50
12.2	60
14.0	70
17.0	80
22.7	90

Source: Henderson (1985).

at moisture contents equilibrating with relative humidities below 70% rh. To prevent mold spoilage problems, pellet moisture contents and storage temperatures should be monitored. Mold inhibitors may also be added to the finished product.

In nature, storage fungi decompose plant or animal substrates reducing their volume and recycling substrate energy and nutrients for other forms of life. In feed-mill materials, storage fungi perform these same functions but the effects of their activities (e.g., production of mycotoxins) may adversely influence product quality (Blaney and Williams 1991), human and animal health, and safety.

Extensive growth by storage fungi can bind grains or other materials together forming crusts which prevent air circulation, promote heating, provide refuges for insects, and create hazards for persons entering the bins, as there can be voids under the crust. Such molds may produce mycotoxins and allergens. They may either provide a source of nutrients for insects (Pellitteri and Boush 1983) or be harmful to their development (Sinha 1966). Clumps of molded material may clog pipes, block-off parts of bins, and in extreme instances, may cause partial or complete structural collapse (Ravenet 1978). Bins need to be properly designed and grain properly stored to prevent resultant mold problems and blockage of the handling system by raw materials.

Few ecological studies of storage fungi occurring in situ in feed and flour mills have been made. Several studies have investigated the concentration of spores in the atmosphere of flour mills and other premises and their effects on worker health (Gupta and Singh 1983, Lacey and Crook 1988). Some laboratory studies have been made on field and storage fungi and on the mycotoxins produced by them in and on commercial feeds and their ingredients (Buckle 1986, Ylimaki 1981). One comprehensive laboratory study of corn entering the premises of 82 feed manufacturers in the midwestern United States during 1988–1989 found that the predominant molds were *Fusarium* spp. and 19.5% of samples contained mycotoxins (Russell et al. 1991). The natural occurrence and significance of mycotoxins in the principal commodities used in animal feeds, together with the methods used to detect them, are reviewed by Buckle and Scudamore (1990). Lists of filamentous fungi occurring in food and feed have been compiled by Pitt and Hocking (1985) and Samson et al. (1986).

2.3. Insects

Insects occur in stored materials and plant and animal debris as eggs, larvae, pupae, and adults. Their development depends on substrate temperature and moisture, the presence of suitable niches, an ongoing supply of suitable food, and other factors. Insects develop to pest levels when the temperature of binned stocks increase above 15–20°C, when residual material is not removed from bins, and when there is a low turnover of stocks.

Some stored-product insects develop and feed inside kernels, consuming the entire seed, whereas others feed mainly on the germ. Many of the insects associated with stored grains are detritus feeders and damage broken, cracked, or processed grains. Some insects also feed on mold mycelium. The detritus and fungivorous insects are also pests, as their feces and body parts contaminate grains and their metabolic activity may permit storage molds to grow. In addition, the feeding action of insects facilitates the entrance of molds and mites into seeds (Mills 1986).

Several ecological studies on stored-product insects in feed and flour mills and warehouses have been carried out, notably those on the colonization and infestation of a new provender (feed) mill in the United Kingdom (Ministry of Agriculture, Fisheries, and Food 1981a,b). Other studies are by: Rilett and Weigel (1956) on the Coleoptera present in feed and flour mills at Buffalo, NY; Buchelos (1981) on Coleoptera populations in flour mills at Piraeus, Greece; Dyte (1965) on insect infestations in the machinery of flour mills in the United Kingdom; Howe (1950) on the distribution and abundance of the beetle *Ptinus tectus* Boield in a warehouse in the United Kingdom; and Loschiavo and Okumura (1979) on the occurrence of stored-product insects in flour and feed processing premises in Hawaii.

In feed mills and related premises, the main effect of insect activity is on product quality, directly influencing customer satisfaction and the commercial reputation of the manufacturer. Additional costs are incurred by the need for fumigation of insect-infested shipments or by rejection of feed because of off-odors.

2.4. Mites

A few species of mites feed on whole grain while many species can feed on broken grain, weed seeds, and molds associated with the grain and can thrive in moist grains and grain products at 15–17% moisture content. The mites spread mold spores present on and in their bodies through the grain mass, and encourage development of spoilage molds (Mills 1989). The occurrence of many mites lowers seed and product quality through visible pest presence and damage, and development of unpleasant odors and allergies in animals and humans. The distribution of stored-product mites in bulks of wheat, barley, and rapeseed at a range of humidities was studied by Armitage (1984).

Several ecological studies on stored-product mites associated with feeds have been conducted in feed and flour mills and farms in the United Kingdom. The mite fauna of a Scottish flour and feed mill was examined by Salmond (1956). Following reports of mites adversely affecting farm animal performance, up to 500,000 mites/kg were detected in stored cattle-feed concentrates on farms; the likely infestation sources were residues in farm feed lofts (Ministry of Agriculture, Fisheries and Food 1978, 1981c, Wilkin and Thind 1984). Examin-

Table 3 Frequency of Occurrence and Rank of the 10 Most Common Insects in Feed Mills in Canada, 1969–1981 [a]

			Incidence in food material				Incidence in each location		
Rank	Insect[b]	Occurrence in reports[c] (%)	Wheat n=11[d] (%)[e]	Soybean n=25[d] (%)	Dust, debris n=83[d] (%)	Equipment, boots n=81[d] (%)	Wall, floor n=260[d] (%)	Empty bins n=16[d] (%)	Several locations n=52[d] (%)
1	*Tribolium castaneum* (Herbst)	14	0	8	15	10	17	0	21
2	*Tribolium confusum* (Jacquelin du Val)	13	0	8	8	22	10	13	27
3	*Attagenus* spp.	12	27	0	11	5	17	6	10
4	*Nemapogon granella* (L.)	6	9	24	6	6	3	19	8
5	*Cryptolestes ferrugineus* (Stephens)	6	9	24	6	5	2	6	2
6	*Tenebrio molitor* (L.)	6	0	4	6	6	5	13	4
7	*Oryzaephilus surinamensis* (L.)	5	0	0	2	12	4	0	4
8	*Plodia interpunctella* (Hübner)	4	18	0	6	4	3	0	0
9	*Tribolium* spp.	4	0	0	4	3	4	0	10
10	*Tenebroides mauritanicus* (L.)	3	9	0	6	4	2	6	0

[a] *Source*: Sinha and Watters (1985).
[b] Total species reported = 94.
[c] Total reports were 413; 55% indicated infestation.
[d] n = no. of reports.
[e] Percent incidence of *Attagenus* sp. was 27 % of n = 11 reports.

ation of eight feed mills showed some infestation of raw materials but freshly prepared cattle feed was virtually mite-free.

2.5. Interrelations Among Molds, Insects, and Mites

Molds, insects, and mites occur in storages, within equipment, in handling and processing debris, and in corners and crevices within feed mills. Complex ecological relationships are known to occur among these pests in stored-grain ecosystems on farms (Sinha 1973b). Although no detailed studies have been undertaken, it is considered likely that similar phenomena occur among these biotic components within the storage and working areas of feed mills. Within these areas, populations of molds, insects, and mites are frequently disturbed by the rapid turnover of materials and frequent in-house sanitation procedures. Nevertheless, enough pest organisms survive in debris in overlooked areas or inaccessible niches so that a threat to product quality remains. Certain insect pest species are associated with particular materials and locations within feed mills (Table 3). Such knowledge is of considerable importance when planning pest management strategies.

3. STORAGE AND MANAGEMENT RELATED PROBLEMS IN FEED MILLS

3.1. Overview

A feed-mill ecosystem is influenced by abiotic factors (e.g., temperature and moisture) and biotic factors (e.g., insects, mites, and molds). Abiotic and biotic factors, poor building and equipment design, and/or poor management can influence the raw materials constituting the product, production environment and condition both before and after sale; the resulting problems in the stored product or in its environment are often difficult to unravel as more than one interrelated factor may be involved. Because of these interrelationships, a knowledge of the complete feed-mill ecosystem is required before storage and management-related problems can be comprehended, and before counteracting strategies are developed.

A feed mill is made up of a series of storage and processing units, as shown in simplified form in Figure 4. The following account describes the influence of production, storage, distribution, and other factors on management decision-making in a typical feed mill (Fig. 5, Areas 1-6). Emphasis is placed on management and the decision-making process with inputs from various areas and sources, both external and internal. It is a generalized account taking in all situations since feed mills vary considerably in their management. Additional reading on feed-mill planning and management are given by Coleman (1981) and Littleton (1981).

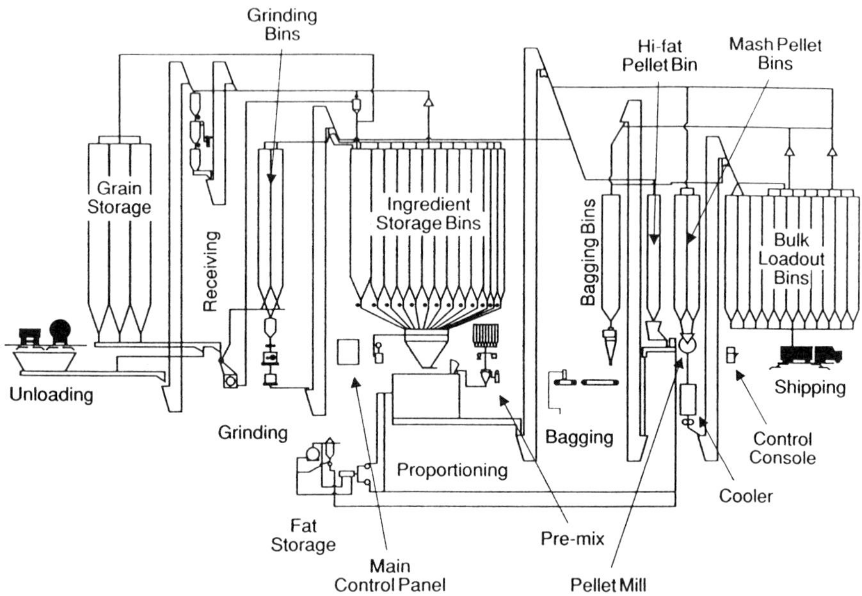

Figure 4 Simplified diagram of a feed mill.

3.2. Site

Before construction of a feed mill, the advice of a registered professional engineer should be sought on the most suitable types of milling and storage structures for the climatic region and for their intended use. Once selected, the structures should be located on a well-drained site (Fig. 5, Area 1) on properly designed foundations thus avoiding cracked floors, ingress of drainage water, and fungal spoilage problems (Mills 1989).

The mills should be carefully designed to eliminate lodging of residues and insect harbourages and for ease of cleaning. Each construction stage should be carefully monitored to discover and immediately rectify mistakes in assembly, especially those providing insect harbourages, moisture accumulation, or easy access by rodents or birds.

The exterior surroundings of the mill and storage structures should be kept clear of vegetation, plant residues, wooden pallets, unused equipment and debris as these harbour pests (Mills 1989). Similarly, residues should be periodically removed from gutters, roofs, near exhaust vents, the vicinity of garbage containers, under dock levellers, railroad sidings, and all other possible areas of contamination (Gentry 1984).

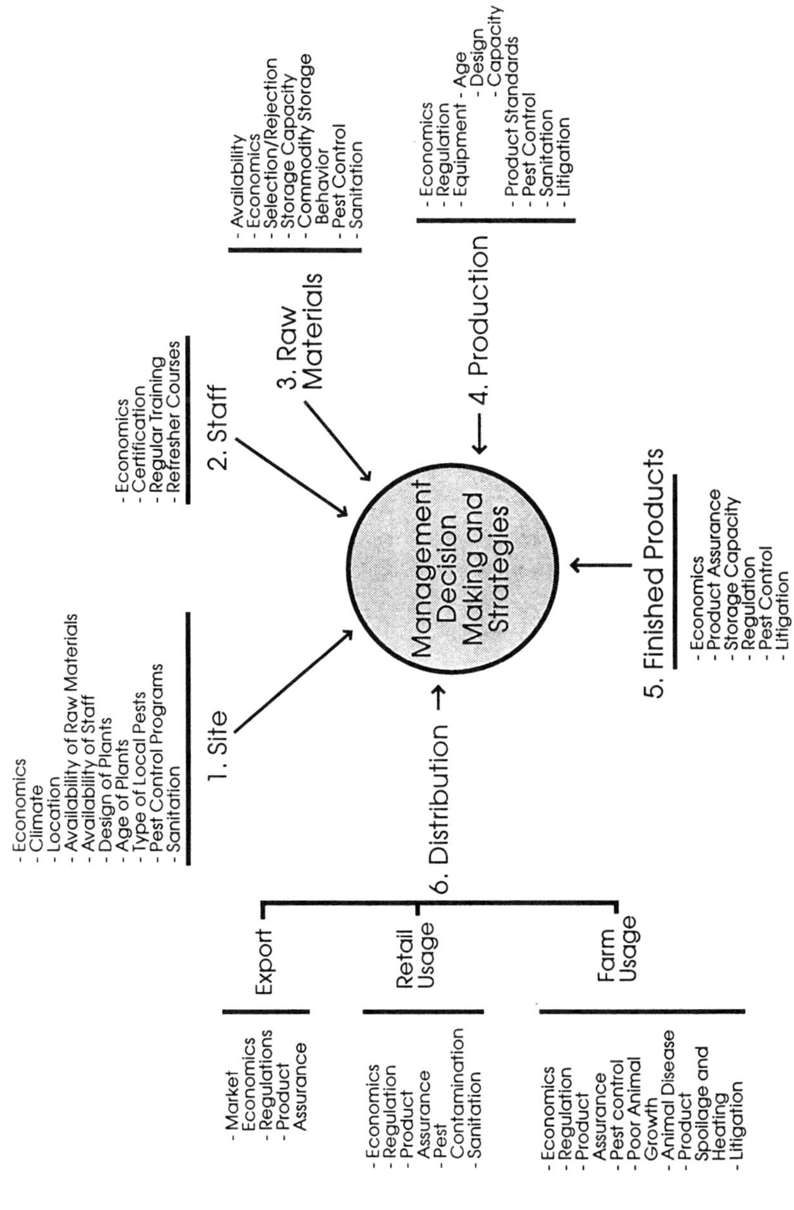

Figure 5 Influence of production, storage, and distribution factors on management-decision making in feed mills.

3.3. Staff

Factors of importance in staffing (Fig. 5, Area 2) include salary economics, availability of regular and refresher training courses, professional certification, and staff motivation and turnover. Staff transport may be a factor in developing countries.

At Farmland Industries, Kansas City, MO, in-plant and off-location classroom training are provided to mill personnel. All local cooperative feed-mill managers are required to attend a clinic every two years for update training in feed manufacturing practices and the exchange of ideas. Personnel training has established the basis for a successful quality control program (Unruh 1984).

Skilled staff are essential for efficient operations. In particular, the batching panel operator, responsible for inventory control, weighing and drug additions, and the pellet mill operator, responsible for pellet production, need to be responsible for their inputs and to be aware of their importance (Charles 1975).

3.4. Raw Materials

Managers receiving loads of grain and other materials (Fig. 5 Area 3) for processing in a facility have the right to refuse loads that have doubtful keeping quality or that are in poor condition, thus preventing later handling, storage, and quality problems (Ferket and Jones 1989, Mills 1989). The quality of materials should be defined in the supply contracts. Purchasing materials from unknown suppliers is risky; their reputation and facilities, if possible, should first be checked.

Loads may be contaminated with chemicals, pests and pest products, other grains, and excessive debris or physical objects (e.g., stones). Plant materials (Passmore 1990) and rendered and other animal products may be contaminated with *Salmonella* bacteria (Bentley 1985). Ingredients should be inspected on receipt for evidence of water damage as well as contamination by such *Salmonella* carriers as rodents, birds, and insects (Nape 1968). Cereal grains may be frost-damaged, shrivelled, spoiled, or heated which are all conditions that reduce the nutritive value of grains. For economic reasons, a manager may decide to incorporate materials with minor contamination or of low quality with good quality, contamination-free materials in the product mix. This decision should be carefully considered in view of potential complaints of animal sicknesses or poor performance, possible legal suits by purchasers, and real affects on the company's reputation.

In ideal circumstances, raw materials are stored for a maximum of 2 wk with sufficient stocks kept on hand only for short-term plant operations. In certain situations, however, raw materials may be stored for up to 3–4 mo. For example, winter blizzards in eastern Canada often restrict movement of incoming supplies and freezing of canals in the Netherlands may interrupt barge traffic bringing in raw materials.

Generally, the short-term nature of raw material storage precludes ongoing storage problems in mills. Because storage time is greater on farms, problems are more prevalent there (see Section 4).

3.5. Production

Mixed or pelleted feeds usually contain ground grains; sometimes rolled or cracked grains are used. It is most important that a feed mill be equipped with efficient and well-maintained processing equipment (Fig. 5, Area 4). It is nearly impossible to produce high-quality feed from equipment that leaks material, blows out dust, or has a high potential for cross-contamination. The batch scale must weigh accurately; the feed mixer must mix completely and efficiently; and protective equipment such as cleaners and magnets must operate properly (Unruh 1984).

3.6. Finished Products

In Canada, 70–80% of the feed sold is handled in bulk form; the remainder is bagged (Fig. 5, Area 5). It is traditional to maintain enough bin capacity to handle one day's production. Bulk bins must be designed carefully to minimize segregation and eliminate hang-ups. Steep hopper slopes in the range of 70° to the horizontal, a three to one ratio of height to bin diameter, and a large discharge opening generally give the best results and keep segregation to a minimum (Tremere 1982).

To assure that bags of feed in a warehouse are kept as fresh as possible, a "first in–first out" storage system is required. Feeds that are 90-d old should be removed from the feed inventory and worked back into similar feeds (Unruh 1984). When pallets of bagged feeds are overlooked for long periods within warehouses they can become infested with insects; this causes serious product and plant contamination problems.

3.7. Distribution

The finished product is transported from the feed mill to warehouses, farms, or export premises; from the warehouses it is further distributed to retail stores and homes (Fig. 2B, Areas 6-10; Fig. 5, Area 6). Although the product has left the mill, there is an ongoing responsibility by the feed company for the quality of its manufactured product, particularly from the viewpoint of formulation and other agreed-upon contractual components. The feed mill is however not responsible for inappropriate usage, handling, or storage of their product by others resulting in animal health problems or loss of product quality.

3.8. Quality Control

Quality control is important in every phase of feed-mill operation. Ideally, a quality control program should be developed before the mill is constructed. In designing a new mill, attention should be given to ease of sanitation, machinery clean-out, dust control, and location of sampling ports for incoming ingredients and finished feeds. In existing mills, changes that can be made include: replacing spouting that is too small; eliminating bottlenecks in spouting and bins where feed can bridge; incorporating features which prevent ingredients and other feeds from becoming cross-contaminated; identifying controls with a number and a name plate stating contents or purposes; and installing sampling ports and developing a sampling procedure for ingredients and finished feed (Charles 1975).

Controlling the quality of the finished feed products is most important. Quality control procedures should include dating of code tags, correctly labeling the product, checking bag weights, taking a total bag count, checking bulk shipping weights of each feed run, and making a critical visual inspection before shipment. A representative sample from each product run, bulk, or bag is collected and retained for 90 d, and at least one sample from the first run of a feed product each month should be forwarded to the laboratory for nutrient analysis (Unruh 1984). With broiler chicken feed, samples only need to be kept for 60 d since the broiler chicken cycle is 40 d. Samples of materials should be taken in a manner consistent with generally recognized good sampling techniques or practices (Garfield 1989).

There is nothing that can cause more damage to a feed-mill's reputation for quality than a customer who is dissatisfied with the condition of the product received. Excessive breakage and generation of fine particulates increases feed wastage. The durability test provides a useful measure of pellet integrity in the mill. Necessary adjustments to the pellet system can be made to produce a high quality, durable, and marketable pelleted product (Unruh 1984).

The process of feed manufacturing has the potential for carryover of drug residues. Mixers, conveyors, elevating legs, dust systems, and delivery trucks may harbour sizeable amounts of residues which can be incorporated into the next batch of feed. The problem of sulfonamide residues in edible swine tissues in the United States during the 1980s forced feed manufacturers (and farmers) to examine their systems carefully to discover the source of carryover drug residues. Recommended solutions included regular flushing of mixers and conveyors after every run of sulfonamide feed; consolidating sulfonamide feed manufacture to 1–2 d a week; setting up a separate mixing, conveying, storage, and delivery system for such feeds; and cleaning or flushing of vehicles after every delivery of sulfonamide feeds (Wilcox 1982).

The feed manufacturing process does not destroy *Salmonella* bacteria in contaminated feedstuffs. Studies of contaminated materials in an Ontario feed mill showed that some of the bacteria survived the pelleting process (Hacking et al. 1978). Recovery studies of *Salmonella* from material in five midwestern US feed mills demonstrated that meat meal was the primary contamination source. There are two lines of defense against the *Salmonella* organism. The first is a purchasing program set up in association with suppliers having an effective *Salmonella* control program in their plants; the second is good housekeeping, good manufacturing, and sanitation practices carried out within the mill (Nape 1968).

More information on quality control in feed mills is given in Charles (1975), Friend (1978), Holland (1982), and McEllhiney (1979), and on quality control guidelines for feed-mill employees in Wilcox (1987). Charles (1975) describes the main responsibilities of milling, sanitation, mechanical maintenance, management, technical, and clerical personnel. Sutton (1983) provides comprehensive quality control checklists in the areas of personnel, maintenance of mill and equipment, control of incoming ingredients, and quality of finished feed.

4. STORAGE PROBLEMS ON FARMS

4.1. Overview

On-farm storage facilities include bins, silos, or hoppers for raw ingredients and for finished products. Most raw material storage problems can be prevented through a knowledge of safe storage guidelines, preparation for storage, and management of stocks (detailed in the following sections). Finished product storage problems, however, are potentially more severe because, as the product is reconstituted, it breaks down through handling, adheres to bin walls, and becomes an excellent substrate for mold and mycotoxin development. Moreover, farmers often order too much of a particular type of feed which remains in the bin permitting mold development; moldy feed may subsequently cause animal health problems. Therefore, finished product bins need to be regularly inspected and cleaned, and stocks ordered according to usage patterns (Charles 1985).

4.2. Safe Storage Guidelines

The relative risk levels of spoilage and heating in a range of stored commodities likely to be used for animal feeds are listed in Table 4; risk levels were determined according to the known history of storage problems, the total oil content, the presence of residual oil, and other factors. Safe storage guidelines

Table 4 Relative Risk of Spoilage and Heating in Stored Commodities Used for Animal Feeds

	Risk level				
	Very high (Class 1)	High (Class 2)	Moderate (Class 3)	Moderate-low (Class 4)	Low (Class 5)
Type of product	Oilseeds with certain oils, grass products	Meals with high oil content, grass products	Oilseeds	Cereals with high oil content	Cereals, pulses
Problem frequency	Very common	Common	Common	Occasional	Occasional
Inert gas requirement	Essential to preferred	None	None	None	None
Examples	Alfalfa pellets	Corn meal Fishmeal Hay Rice bran	Canola/rapeseed Cottonseed Flaxseed Safflower seed Soybean Sunflower seed	Animal feeds Corn/maize Canola/rapeseed meal	Barley Domestic buckwheat Fababeans Field beans Oats Peanuts Peas Rice Screenings Sorghum Wheat Wheat bran, shorts, middlings

Source: Mills (1989).

and storage characteristics of 35 grains, oilseeds, pulses, and their processed products are described by Mills (1989).

4.3. Preparation for Storage

Vertical silos used for storing large amounts of animal feeds prone to heating should be located in the open, well clear of buildings, and away from any other combustible material (Fire Protection Association 1968). Silos or bins should be emptied completely at regular intervals and examined within to detect and remove adherent material that can provide refuge for insects and mites. After cleaning, the floors and walls should be thoroughly sprayed with an appropriate contact insecticide to kill any remaining insects which might infest the new feed. Old grain and debris near doors, or under aeration or drier floors should be removed and then burned or buried. Structures should be kept in a good state of repair and weather-tight to keep out wind-driven rain or snow which provides moisture for mold development. Cracks in concrete walls or floors, gaps in the metal wall, and gaps between the metal wall and concrete base need to be repaired and sealed to keep out moisture and to reduce the number of hiding places for insects (Mills 1989).

4.4. Management of Stocks

As the material is stored in the bin, it should be sampled periodically and its range of moisture content determined. A high moisture content is the greatest concern, so the binned material is often ventilated to eliminate or minimize any moisture pockets and temperature differences. Once binned, the material requires frequent inspection to avoid mold spoilage and other storage problems. Permanent metal ladders on both the outside and inside of storage structures facilitate such inspections. The material is probed regularly for signs of spoilage and heating, particularly in the top of the bin, or remote sensors that monitor excessive carbon dioxide and temperature are installed as warning devices. Pitfall traps may be used to detect insect infestation. For more information on the prevention and detection of spoilage and heating problems in stored stocks, see Mills (1989) and Mills et al. (1990).

4.5. Quality Control

A good on-farm, quality control program should include the following elements: a knowledge of the feeds used and their frequency of usage, inventory identification, stocks management, pest sanitation, and a good relationship with feed-mill management to ensure finished feed requirements are met and problems promptly handled.

5. THE FEED INDUSTRY IN CANADA

The animal feed manufacturing industry in Canada totalled approximately Cdn $2.6 billion in sales in 1986. More than Cdn $2.0 billion of materials were purchased, some of which were byproducts from other food industries (Industry, Science and Technology Canada 1988). Feed grains—mainly barley, corn, wheat, and oats (Statistics Canada 1989)—make up about 60% of most complete feed rations (by volume). Swine and poultry feeds account for about 70% of complete feed rations sold (Statistics Canada 1990).

The feed industry is primarily oriented towards serving the local domestic market. On a geographical basis, 70% of feed production is in eastern and central Canada. Most manufactured feeds are sold within a 100-km radius of a plant but higher-valued, specialty feeds and ingredients are traded over a much wider area. The regional distribution of feed mills parallels that of livestock and poultry across the country. The industry is located primarily in Ontario, Quebec, and Alberta, although there are feed plants in every province.

Some integration of feed manufacturing associated with livestock and poultry businesses does exist, although this may not be as advanced in Canada as it is in the United States or elsewhere. As a result, Canadian feed mills tend to produce a wider range of feed products than some of their more specialized foreign counterparts (Industry, Science and Technology Canada 1988).

6. FEED-MILL HANDLING AND STORAGE PROBLEMS IN DEVELOPING COUNTRIES

6.1. Overview

Temperatures in developing countries are often warm or hot and conducive to year-round insect, mold, and rodent development. Such pests develop in growing crops, storages, farms, and feed mills. Raw materials entering mills may be contaminated with pesticides because of numerous treatments applied in the field and in farm storage, and need to be screened out before entry. Stored-product insects fly year-round into storages and processing plants; many are exceptionally destructive (e.g., moths, weevils, and grain borers). Insecticides are frequently applied to mill structures to reduce insect populations, and also to raw materials if storage is prolonged. Frequent application of insecticides leads to accumulation of residues and the development of insecticide resistance by pest species.

High atmospheric relative humidities, coupled with high temperatures, favor development of molds and sometimes mycotoxins, some of which are carcinogens; hence, raw material entering mills needs to be evaluated and, if necessary, screened out. Within the mill, moisture condensation and leakage through roofs may occur during wet seasons, favoring mold growth, increased spoilage, and material handling problems.

Relatively few ecological studies have been conducted in feed and flour mills in developing countries. Salama and Salem (1973) studied at monthly intervals changes in insect fauna and in micro- and macroclimatic conditions within a flour mill in Egypt. Hodges et al. (1985) evaluated various insect traps as aids to pest management in bagged rice stores in Indonesia. Quality control in a Mexican feed mill was described by Maltos (1973). Bauduret (1990) described the occurrence of *Aspergillus flavus* Link, aflatoxins, and salmonellae in mixed poultry feeds and raw materials on farms in Reunion Island, Indian Ocean. The following two accounts illustrate feed-mill operations and attendant problems in Indonesia and Peru. In addition, Hertrampf (1985) describes the feed industries of 24 countries, primarily within Africa, the Middle East, and Asia. The information is formatted to assist those engaged in compound feed manufacturing and livestock and poultry rearing in developing countries.

6.2. Indonesia

Feed-mill plants are concentrated in the major cities of Jakarta, Semarang, and Surabaya (Java), and Medan and Tanjung Karang (Sumatra). Most large feed mills, producing 60,000–250,000 t/yr, have integrated broiler, layer, breeder, and aquaculture (shrimp) operations. Indonesia produced 3.4 Mt of compound feeds for monogastric, ruminant, and aquatic animals in 67 mills in 1990.

Corn is the major energy-derived feedstuff used; others include rice bran, sorghum, wheat bran, and cassava root chips. Soybean meal and fish meal are the major protein-derived feedstuffs with meat and bone meal, hydrolyzed chicken feather meal, groundnut cake, rapeseed meal, sesame seed meal, and copra cake used in lesser amounts.

Indonesia produces corn, rice, sorghum, bone meal, and fish meals but the quantity is currently insufficient to meet domestic requirements for compound feed. Corn was imported from Thailand, China, and Vietnam in 1991 because of an acute drought in Indonesia; fish meals have been imported from Chile, Peru, and Thailand; meat and bone meals from Australia, New Zealand, Europe, and Thailand; and chicken feather meal from Australia, the United States, and Thailand. Nonconventional commercial feed ingredients include Leucaena leaf meals, dried cassava starch residues, crude palm oil, and dried brewers' grains; nutritive feed additives include molasses, fats and oils, synthetic amino acids, vitamins, and minerals which are purchased through local agencies or directly from overseas.

Compound feeds are formulated using least-cost computer programs. Most feedmills have laboratory facilities (with varying analytical capabilities) that perform quality control assessments on raw materials and finished products. Analyses of feeds for amino acids, trace mineral elements, vitamins, drugs, pesticides, mycotoxins, pathogenic microorganisms, and other specific compounds are usually carried out by overseas laboratories.

Some feed mills have vertical bins of concrete or corrugated steel which are used for storage of feed ingredients and finished products. When bins are not cleaned, a build-up of adhesive or sticky materials may occur, thereby creating ideal conditions for mold and insect development. Postharvest losses caused by molds, insects, and rodents are aggravated in Indonesia by high temperatures, humid conditions, and a lack of drying, storing, and handling facilities.

Feedstuffs are stored in warehouses (horizontal storages) for 1–2 wk and they are often transported a few days after production. Feedstuffs are generally packed in 50-kg bags and delivered to farmers by truck, ship, or both. For interisland shipments, bags are lined with additional layers to withstand rough handling at harbors and to reduce exposure to wet conditions during shipment. There is no bulk delivery of finished products because farms are not equipped with silo and automatic feeding facilities and because of poor roads leading to most livestock farms. Broiler feeds are generally marketed as pellets or crumbles and layer feeds as mash or meals. In our Metro feed mills, the annual production averages 75,000 t. Of this amount, broiler feeds account for nearly 50%, layer feeds 30%, aquatic feeds 10%, and the remainder for pigs and ruminant concentrates (R.I. Hutagalung, personal communication 1991).

6.3. Peru

Feed-mill operations in Peru are concentrated along the coast of the Pacific Ocean, a region which is desert-like but with high atmospheric relative humidity. Most of the industry is vertically integrated with poultry production. The integrators, initially involved only with poultry, are now also involved with swine, beef cattle fattening in feed lots, and shrimp cultivation.

Most feeds are composed of the following ingredients: corn (about 60% of the balanced feeds); fish meal (the most important source of protein); cotton seed meal (the main protein source for dairy cattle); soybean meal (an important ingredient in poultry rations); and wheat bran (widely used in dairy cattle and swine rations).

Rice polishings are extensively used in the northern and northeastern regions of the country, where rice cultivation is important. Corn supplies, from the coastal valleys and from the northeastern tropical area of Peru, meet about 50% of the demand; the rest is imported, mainly from the United States. Corn imported from the United States shows a relatively high proportion of broken grain and a high proportion of fine particulates. Corn from both the coastal region and the northwestern tropical area show high contamination with stones, cobs, and chaff. Because of environmental conditions and a lack of grain drying facilities, grain from the northeastern region arrives with moisture levels above 15%, which creates problems related with heating and mold development.

Fish meal presents problems of variable protein content, gizzerosine, and *Salmonella* contamination. Cottonseed meal and rice polishings are frequently adulterated with the addition of hulls. Rice polishings may be rancid and cottonseed meal may have a high gossypol content.

Few feed mills store feedstuffs in bins. Relatively favorable environmental conditions (little rain) permits storage of grains and other feedstuffs in open spaces on concrete floors. A plastic sheet is sometimes used to cover the pile of grain during the winter, when a fine drizzle creates problems. Some leaks through the plastic cover provide moisture for mold development. Bins are mostly used to store finished products but are not frequently cleaned, creating conditions for insect and mold development in adherent material.

Feeds are frequently sampled to check for composition and efficiency of mixing is measured by determination of drug concentrations in the feed.

Because of the integrated operation with poultry production and the fears of *Salmonella* cross-contamination, strict measures of control are exerted at the entrance of feed mills. Visitors are not always welcome.

Most of the feed produced by integrators is handled in bulk form. Polypropylene bags are mostly used for bagged feed. Bagged feeds are labeled; labels indicate the class of the animals for which the feed has been formulated, including the main ingredients, levels of protein, weight, and date of preparation. The feed produced by integrators stays a short time in warehouses, and is usually shipped 2 or 3 d after production. Because of the high cost of the equipment, mechanical pelleting is not a common practice and those who produce pelleted feeds have problems with pellet quality (J. Kalinowski, personal communication 1991).

7. REGULATION

The manufacture, sale, and importation of feeds is regulated by federal, provincial, or state, and sometimes local statutes. In Canada, for example, the manufacturing of livestock feeds as well as their importation and sale is regulated by the Feeds Act (1960 with later amendments) which provides penalties for improper actions. The Act is administered by Agriculture and Agri-Food Canada and also by Health and Welfare Canada because of the medications in feeds and possible contamination by field crop pesticides. The constituents of complete feeds, supplements, macropremixes, micropremixes, and minerals are regulated in Canada by Section 22 of the Feeds Regulations (1983 with later amendments). The Feeds Act (1960) and Feeds Regulations (1983) have specific labeling requirements, which include feed-additive drug labeling information set out in Medicating Ingredient Brochures. In addition, feed mills are inspected by occupational health and safety inspectors of Labour Canada to ensure the well-being of employees. In the United States, the Model Feed Bill (developed by the Association of American Feed Control and the American Feed Manufacturer's Association) covers appropriate definitions, registration of brand feed names, labeling and other regulations (Church 1984).

The limits of undesirable contaminants in feeds are regulated in many countries. The maximum tolerated levels for aflatoxins in animal feed stuffs in 33 countries has been summarized (up to 1987) by van Egmond (1991). In Canada, the maximum allowable content of aflatoxin in a single or mixed feed is 20 ppb (Feeds Regulations 1983, with later amendments). In the United States, the Food and Drug Administration regulatory level for aflatoxins in feedstuffs and feed stuff ingredients is 20 ppb (Food and Drug Administration 1982, compliance policy guide 7126:33). In the United Kingdom, the maximum level for aflatoxin B_1 is 20 ppb for complete feedstuffs for pigs and poultry (except piglets and chicks, for which it is 10 ppb) (Feeding Stuffs Regulations 1982, No. 1143) (Buckle 1986).

8. ECONOMIC CONSIDERATIONS

During feed manufacture, the value of the product is augmented. Feed products, however, are bulky and of relatively low value, and companies cannot afford long-distance transportation costs; exceptions to this include some high-valued or specialty products like vitamin and mineral micromixes, milk replacers, and fish feeds. In Canada and the United States, feed ingredients represent about 80% of feed production costs and the availability and prices of local grains and proteins affect feed production costs in a given area. Feed mills in ingredient-deficient regions (e.g., the New England States, the Atlantic Provinces and British Columbia) can have feed costs 25% higher than those in the surplus regions (e.g., the midwestern United States, Ontario, and the Canadian prairies). An increase in on-farm mixing has led to a shift in demand towards premixes and feed concentrates from which farmers can prepare complete feeds using their own grains. On-farm mixing is predominant in areas of good grain production but in grain-deficient areas the demand for commercially prepared rations is stronger.

Because the purchase of ingredients is the single most important factor affecting production costs, firms must pay close attention to price movements on commodity markets. With low profit margins, firms must also closely watch their inventory and purchasing practices. Cost structures and profitability vary significantly from plant-to-plant and firm-to-firm, depending on supplier and end-user industries.

Some firms have closed their smaller, older facilities to centralize their operations into fewer, larger facilities at the best possible locations for access both to raw materials and to markets for finished products. The major strategic consideration behind local feed plant investment is the desire to maintain or increase market share (Industry, Science and Technology, Canada 1988).

These and other economic considerations affect directly or indirectly the incidence of storage and management problems in feed mills. For example, the closure of older, smaller plants will finally remove intractable pest problems originating from refuges within poorly designed or maintained premises. Also, when new plants are built, they are designed to ensure production of feed material of optimal quality. The purchase of poor quality, low-cost ingredients may have severe consequences to the feed-mill operator because of attendant pest problems, animal unthriftiness or sickness, leading to declining commercial reputation and possible loss of market share.

9. FUTURE TRENDS

Future problems in feed mills are anticipated to occur in the areas of cost reduction, quality control, regulatory requirements and health and safety (Table 5). Because of their complexity, specific problems within these areas will likely be tackled using both commercial and scientific inputs. For example, in order to determine the effects of nonantibiotic feed additives on animals, humans, microflora, and microfauna (Table 5), a team of mill managers, veterinarians, medical doctors, mycologists, parasitologists might be constituted. There are differences in the commercial and scientific approaches to problem solving (Table 5); generally, however, inputs will be necessary from both viewpoints for optimal problem solution. Some of the approaches used for problem solving will probably utilize biotechnology. Biotechnological inputs might include increased usage of nonantibiotic feed additives (Lyons 1987), usage of DNA probes for monitoring control by food preservatives of *Salmonella* contamination in meals (Pumfrey and Nelson 1991), development and usage of monoclonal antibodies for mycotoxin detection (Morgan 1989), and use of pheromones for monitoring insect pests (Burkholder 1985). Research and development efforts will continue on the production and usage of aquacultural feeds, particularly in Asian Pacific Rim countries (Woodroofe 1990), and nonconventional feedstuffs (e.g., poultry waste) following earlier work in the United States (Hamblin 1980) and Europe (Bod'a 1990). Improved staff education and training will play a vital role in achieving reduced costs, improved quality control, better compliance with regulatory requirements, and, most importantly, an improved awareness of human and animal health. Studies have shown that both irritant and allergic contact dermatitis occurred in workers exposed to drug and other additives in

Table 5 Feed Mill Problem Areas with Some Commercial and Scientific Approaches for Their Solution

	Solution approach	
Problem area	Commercial	Scientific
Reduced costs	Increased usage of nonantibiotic additives (e.g., pH additives).	Effect of nonantibiotic additives on animals, humans, microfauna, microflora.
	Increased production and usage of aquacultural feeds.	Effect of inhibitors in aquacultural feeds on fish.
	Increased production and usage of nontraditional feeds (e.g., poultry manure).	Effect of nontraditional feeds on animals, humans, microflora, microfauna.
	Audit of all energy used in the mill.	Energy audit of materials entering and leaving mills to identify energy wastage.
Raw material quality	Detection and rejection of substandard raw product entry.	New approaches to feed ingredient microscopy.
Insect infestation	Harborage elimination through design improvements.	Engineering studies; detection of harbourages.
	Reduction of pests through determination of acceptable levels, identification of key species, improved control and sanitation methods, staff education.	Determination of pest levels in relation to ecological changes (e.g., after manufacture of different types of feed) leading to maximal pest control with minimal use of pesticides.
	Monitoring of pests using pheromones.	Characterization of new pheromones and their usage in feed mills through pitfall traps.
Feed contamination by mycotoxins	Identification and rejection of mycotoxin-contaminated materials before entry.	Use of monoclonal antibodies to detect specific mycotoxins.
Regulatory requirements	Improved staff dialog and knowledge; better procedures.	Production of user-friendly manual written from an ecological viewpoint.
Health and safety	More emphasis on long-term health and safety.	Determination of animal medication levels in debris within mill.
		Determination of long-term health exposure risks/mortalities of mill employees.

feed mills in northern Italy (Mancuso et al. 1990). Also, severe lead intoxication of cattle caused by contaminated feed occurred in the northern Netherlands in 1989 (Baars et al. 1990). Results from multidisciplinary research and development should permit plant workers and managers to view their workplace as a special human-managed ecosystem with its own nutrient and energy flow, special niches for pests, and seasonal changes. Most importantly, through this perception, such personnel will be better able to deal with the complex multidisciplinary problems arising in the feed mills of the future.

The occurrence of insects, fungi, mites, drugs, and other additives within the working spaces of feed mills has been described but their long-term impact on human and animal health is presently unknown. To provide answers to these questions, a detailed study was undertaken in a Manitoba feed mill (Mills and White, in press). The study monitored the ecological changes occurring in the mill mainly from a pest viewpoint. Emphasis was placed on the use of rapid and convenient methods of examination, incorporating the latest Canadian and foreign technological advances, to categorize the ecological situation with time.

ACKNOWLEDGEMENTS

The author thanks G. Elias, Dr. D. R. Hickling, A. Humphries, Dr. M. Carvajal, Dr. H. Li Pun, W. Steeds, J. van Loon, G. Vis, and Dr. V. F. Wright for their critical inputs and useful suggestions. Thanks are due to the director, Canadian International Grains Institute, Winnipeg for copyright permission to reprint Figure 4 from *Grains and Oilseeds: Handling, Marketing, Processing*, 4th Edition. Special thanks are extended to Dr. Rudy I. Hutagalung, nutritionist, P. T. Metro Inti Sejahtera animal feed mill, Bekasi Barat, Indonesia and to Dr. Juan Kalinowski, Universidad National Agraria La Molina, La Molina, Lima, Peru for summarizing ecological aspects of feed-mill operation in their respective countries.

REFERENCES

Arbogast, R. T. and M. A. Muller. 1988. Insect succession in a stored grain ecosystem in southeast Georgia, USA. Ann. Entomol. Soc. Am. 81: 899-912.

Armitage, D. M. 1984. The vertical distribution of mites in bulks of stored produce. pp. 1006-1013 in: Acarology VI. Volume 2. D. A. Griffiths and C. E. Bowman (eds.). Ellis Horwood, Chichester, United Kingdom.

Baars, A. J., H. van Beek, I. J. R. Visser, G. Vos, W. van Delft, G. Fennema, G. W. Lieben, K. Lautenbag, J. H. M. Niewenhuijs, P. A. de Lezenne Coulander, F. H.

Pluimers, G. van de Haar, Tj. Jorna, L. G. M. Th. Tuinstra, P. Zanstra, and B. Bruins. 1990. Loodintoxicatie bij rundvee in Noord-Nederland Oktober 1989—Januari 1990 (Lead intoxication in cattle in the northern part of the Netherlands October 1989—January 1990). Tijdschr. Diergeneeskd. 115: 882-890.

Bauduret, P. 1990. A mycological and bacteriological survey on feed ingredients and mixed poultry feeds in Reunion Island. Mycopathologia 109: 157-164.

Bentley, A. H. 1985. Salmonella control in Canada with particular reference to the feed industry. Proc. Am. Assoc. Feed Microsc. 33: 47-49.

Blaney, B. J. and K. C. Williams. 1991. Effective use in livestock feeds of mouldy and weather-damaged grain containing mycotoxins—case histories and economic assessments pertaining to pig and poultry industries of Queensland. Aust. J. Agric. Res. 42: 993-1012.

Bod'a, K. (ed.). 1990. Nonconventional feedstuffs in the nutrition of farm animals. Dev. Anim. Vet. Sci. 23: 13-33.

Buchelos, C. T. 1981. Coleoptera populations at flour mills and related areas. Ann. Inst. Phytopathol. Benaki (N. S.) 13: 6-29.

Buckle, A. E. 1986. Moulds and mycotoxins in cereal and other components of compound animal feeds. pp. 55-60 in: Spoilage and Mycotoxins of Cereals and other Stored Products. Flannigan, B. (ed.). C. A. B. International, Slough, the United Kingdom.

Buckle, A. E. and K. A. Scudamore. 1990. The occurrence, detection and significance of mycotoxins in animal feeding stuffs. pp. 411-428 in: Feedstuff Evaluation. Wiseman, J. and D. J. A. Cole (eds.). Butterworths, London.

Burkholder, W. E. 1985. Pheromones for monitoring and control of stored product insects. Annu. Rev. Entomol. 30: 257-272.

Charles, O. W. 1975. Quality control assurance in the feed mill—it doesn't cost it pays. University of Georgia Cooperative Extension Service, Athens, GA. Leaflet 215. 8 pp.

Charles, O. W. 1985. Feed bin maintenance. Poultry tips. University of Georgia College of Agriculture, Athens, GA. 2 pp.

Christensen, C. M. and H. H. Kaufmann. 1969. Grain Storage: The Role of Fungi in Quality Loss. University of Minnesota Press, Minneapolis, MN. 153 pp.

Church, D. C. 1984. Livestock Feeds and Feeding, 2nd Ed., O&B Books Inc., Corvallis, OR. 549 pp.

Coleman, W. 1981. Coping with reality. Feed Management 32: 12-13, 16, 18, 20, 22, 26.

Coombs, C. W. and G. E. Woodroffe. 1963. An experimental demonstration of ecological succession in an insect population breeding in stored wheat. J. Anim. Ecol. 32: 271-279.

Dyte, C. E. 1965. Studies on insect infestation in the machinery of three English flour mills in relation to seasonal temperature changes. J. Stored Prod. Res. 1: 129-144.

Feeds Act. 1960. Royal Statutes of Canada, Chapter F-9. Supply and Services Canada, Ottawa. 7 pp.

Feeds Regulations and Amendments. 1983. Canada Gazette, Part II, 27 July 1983. 117: 2813-2846.

Ferket, P. R. and F. T. Jones. 1989. North Carolina research leads to ingredient quality findings. Feedstuffs (September 25, 1989), p. 32, 39-40, 57-59, 70.

Fire Protection Association. 1968. Fire safety with silos. Fire Prot. Assoc. J. 78: 37-39.

Food and Drug Administration. 1982. Compliance Policy Guide 7126.33. Washington.

Friend, C. L. 1978. Feed industry quality control guidelines. pp. 229-234 in: Canadian Feed Manufacturing Technology. Can. Feed Ind. Assoc., Ottawa.

Garfield, F. M. 1989. Sampling in the analytical scheme. J. Assoc. Off. Anal. Chem. 72: 405-411.

Gentry, J. W. 1984. Inspection techniques. pp. 34-42 in: Insect Management for Food Storage and Processing. F. J. Baur (ed.). Am. Assoc. Cereal Chem., St. Paul, MN.

Gupta, B. N. and K. P. Singh. 1983. Mycological studies in small flour mills. Biol. Mem. 8: 95-102.

Hacking, W. C., W. R. Mitchell, and H. C. Carlson. 1978. *Salmonella* investigation in an Ontario feed mill. Can. J. Comp. Med. 42: 400-406.

Hamblin, D. C. 1980. Commercially processing and selling poultry waste as a feed ingredient. J. Animal Sci. 50: 342-344.

Henderson, S. 1985. The relationship between moisture content and equilibrium relative humidity of pig feed. J. Stored Prod. Res. 21:131-134.

Hertrampf, J. W. 1985. The Feed Industry in Oversea Countries. Verlag Moritz Schäfer. Detmold, Germany. 230 pp.

Hodges, R. J., J. Halid, R. P. Rees, J. Meik, and J. Sarjono. 1985. Insect traps tested as an aid to pest management in milled rice stores. J. Stored Prod. Res. 21: 215-229.

Holland, R. 1982. Management and quality control of feed mills. pp. 1-3 in: Proc. Maryland Nut. Conf. Feed Manufact. College Park, MD.

Howe, R. W. 1950. Studies on beetles of the family Ptinidae III. A two year study of the distribution and abundance of *Ptinus tectus* Boield. in a warehouse. Bull. Entomol. Res. 41: 371-394.

Industry, Science and Technology Canada. 1988. Industry Profile. Animal Feeds 1988. Ottawa, 7 pp.

Lacey, J. and B. Crook. 1988. Fungal and actinomycete spores as pollutants of the workplace and occupational allergens. Ann. Occup. Hyg. 32: 515-533.

Littleton, I. F. 1981. Planning a new feed mill or modifying an existing plant? Poultry Notes (April 1981): 5-18.

Loschiavo, S. R. and G. T. Okumura. 1979. A Survey of stored product insects in Hawaii. Proc. Hawaiian Entomol. Soc. 13: 95-118.

Löwe, R. and W. Friedrich. 1983. The storage behavior of pelletized feed in silo bins. Kraftfutter 66: 94, 96, 98.

Lyons, T. P. 1987. The role of biological tools in the feed industry. pp. 1-49 in: Biotechnology in the Feed Industry. T. P. Lyons (ed.). Alltech Technical Publications, Nicholasville, KY.

Magan, N. and J. Lacey. 1988. Ecological determinants of mold growth in stored grain. Intern. J. Food Micro. 7: 245-256.

Maltos, H. R. 1973. Quality control in a Mexican feed mill. Proc. Am. Assoc. Feed Microsc. 21: 67-74.

Mancuso, G., M. Staffa, A. Errani, R. M. Berdondini, and P. Fabbri. 1990. Occupational dermatitis in animal feed mill workers. Contact Dermatitis 22: 37-41.

Margalef, R. 1963. On certain unifying principles in ecology. Am. Nat. 97: 357-374.

McEllhiney, R. R. 1979. Quality control in the feed mill. Feed Management 30: 28-30.

Mills, J. T. 1986. Postharvest insect-fungus associations affecting seed deterioration. pp. 39-51 in: Physiological-Pathological Interactions Affecting Seed Deterioration. Crop Sci. Soc. Am., Madison, WI. Special Publ. 12.

Mills, J. T. 1989. Spoilage and heating of stored agricultural products. Prevention, detection and control. Agric. Can. Publ. 1823E, Ottawa. 101 pp.

Mills, J. T. (ed.). 1990. Protection of farm-stored grains and oilseeds from insects, mites, and molds. Agric. Can. Publ. 1851E, Ottawa. 45 pp.

Mills, J. T. and N. D. G. White. 1994. Seasonal occurrence of insects and mites in a Manitoba feed mill. Proc. Entomol. Soc. Man. (in press).

Ministry of Agriculture, Fisheries and Food. 1978. Dairy rations. Pest. Infest. Cont. Lab. Rpt 1974-76. Her Majesty's Stationery Office, London. 43 pp.

Ministry of Agriculture, Fisheries and Food. 1981a. Storage Pests 1979. pp. 25-26 in: Agricultural Science Service Research and Development Report. Reference Book 251. Her Majesty's Stationery Office, London.

Ministry of Agriculture, Fisheries and Food. 1981b. Storage pests 1980. pp. 33-34 in: Agricultural Science Service Research and Development report. Book 251(80). Her Majesty's Stationery Office, London.

Ministry of Agriculture, Fisheries and Food. 1981c. Dairy Rations. pp. 30-31 in: Pest infest. Cont. Lab. Rpt. 1977-79. Her Majesty's Stationery Office, London.

Morgan, M. R. A. 1989. Mycotoxin immunoassays with special reference to ELISAS Tetrahedron 45: 2237-2250.

Nape, W. F. 1968. Recovery of *Salmonella* from material in feed mills. U.S. Livestock Sanit. Ass. Proc. 72: 144-156.

Passmore, S. M. 1990. The occurrence of *Salmonella* spp. in imported animal feedings stuffs of vegetable origin. J. Appl. Bacteriol. 69(6): XVII.

Pellitteri, P. and G. M. Boush. 1983. Stored-product insect pests in feed mills in southern Wisconsin. Wisconsin Acad. Sci. Arts and Letters 71: 103-112.

Pitt, J. I. and A. D. Hocking. 1985. Fungi and Food Spoilage. Academic Press, Sydney, Australia. 413 pp.

Pumfrey, L. and C. E. Nelson. 1991. Use of a most probable number method modified with a deoxyribonucleic acid probe to monitor control by food preservatives of natural *Salmonella* contamination in animal meat meals. Poultry Sci. 70: 780-784.

Ravenet, J. 1978. Silos II. Deformations—failures—explosions—accident prevention. Editores Técnicos Associados S. A., Barcelona, Spain. 364 pp.

Rilett, R. O. and R. D. Weigel. 1956. A winter survey of Coleoptera in feed and flour mills. J. Econ. Entomol. 49: 154-156.

Russell, L., D. F. Cox, G. Larsen, K. Bodwell, and C. E. Nelson. 1991. Incidence of molds and mycotoxins in commercial animal feed mills in seven midwestern states, 1988-89. J. Animal Sci. 69: 5-12.

Salama, H. S. and S. A. Salem. 1973. Seasonal fluctuations in the population of insect fauna in flour mills. Z. Angew. Entomol. 74: 160-169.

Salmond, K. F. 1956. The insect and mite fauna of a Scottish flour mill. Bull. Entomol. Res. 47: 621-630.

Samson, R. A., J. C. Frisvad and O. Filtenborg. 1986. Taxonomic schemes for food-borne fungi. pp. 247-270 in: Methods for the Mycological Examination of Food. A. D. King, J. I. Pitt, L. R Beuchat and J. E. L. Corry (eds.). Plenum Press, New York.

Sinha, R. N. 1966. Development and mortality of *Tribolium castaneum* and *T. confusum* (Coleoptera: Tenebrionidae) on seed-borne fungi. Ann. Entomol. Soc. Am. 59: 192-201.

Sinha, R. N. 1973a. Ecology of storage. Ann. Technol. Agric. 22: 351-369.

Sinha, R. N. 1973b. Interrelations of physical, chemical and biological variables in the deterioration of stored grains. pp. 15-47 in: Grain Storage: Part of a System. R. N. Sinha and W. E. Muir (eds.). AVI Publishing Co., Westport, CT.

Sinha, R. N. and F. L. Watters. 1985. Insect pests of flour mills, grain elevators and feed mills and their control. Agric. Can. Publ. 1776, Ottawa. 290 pp.

Statistics Canada. 1989. Food Industries. Feed Industry. 1986 Census of Manufactures. Catalogue 32-250. Ottawa. February 1989.

Statistics Canada. 1990. Shipments of animal and poultry feeds semi-annual catalog 32-004, Ottawa. Vol. 11(2) March 1990.

Sutton, C. D. 1983. Feed mill quality controll: it costs to do without it. Poultry Digest. (March 1983). p. 116-118, 120.

Tremere, A. W. 1982. Feed grains—utilization and processing. pp. 882-905 in: Grains and Oilseeds, Handling, Marketing and Processing. 3rd Ed. Can. Int. Grains Inst., Winnipeg, MB.

Unruh, D. L. 1984. Practical quality control of feedstuffs. pp. 27-29 in: Official Proceedings 32nd Annual Meeting Am. Assoc. Feed Microsc., Manhattan, KS. J. B. Windsor (ed.), American Association of Feed Microscopists, Mechanicsburg, PA.

van Egmond, H. P. 1991. Limits and regulations for mycotoxins in raw materials and animal feeds. pp. 424-436 in: Mycotoxins and Animal Foods. J. E. Smith and R. S. Henderson (eds.). CRC Press, Boca Raton, FL.

White, N. D. G. and R. N. Sinha. 1980. Principal component analysis of interrelations in stored wheat ecosystems infested with multiple species of insects. Res. Popul. Ecol. 22: 33-50.

Wilcox, R. A. 1982. Aspects of quality control in feed manufacturing. p. 44 in: Official Proceedings 30th Annual Meeting Am. Assoc. Feed Microsc. Ames, IA. J. B. Windsor (ed.), American Association of Feed Microscopists, Mechanicsburg, PA.

Wilcox, R. A. 1987. Quality control guidelines for feed mill employees. Kansas State University Cooperative Extension Service, Manhattan, KS. MF 880. 6 pp.

Wilkin, D. R. and B. B. Thind. 1984. Stored product mites detection and loss assessment in animal feed. pp. 608-620 in: Proc. 3rd Int. Working Conf. Stored Prod. Entomol., Manhattan, KS.

Woodroofe, J. M. 1990. The aquacultural feed boom. World Grain 8: 27-31.

Ylimaki, A. 1981. The mycoflora of cereal seeds and some feedstuffs. Ann. Agric. Fenn. 20: 74-88.

19

Insect Pest Management in Stored-Rice Ecosystems

Anand Prakash and Jagadiswari Rao

Central Rice Research Institute, Cuttack, Orissa, India

SUMMARY

Rice is an important staple food throughout much of the world and is damaged by several species of insects during storage. Insects cause considerable quantitative, qualitative, and viability losses in stored rice every year. The quantitative or mass losses caused by insects are 5–25% per year. The species of insects infesting stored rice, their pest status, sources and mode of infestations, and losses caused by them are discussed with emphasis on rice storage ecosystems in tropical countries. Seventeen species of insects are reported to infest stored rice, with *Sitrotoga cerealella* the main pest. Insect pest management in stored rice using holistic ecosystem approaches involves alterations of abiotic components (e.g., grain moisture, temperature inside the storage, condition, type, and inner atmosphere of storage structures) in addition to the use of synthetic and botanical grain protectants. Synthetic insecticides are not encouraged as protectants on rice for consumption but are recommended to disinfest storage structures and containers as wells as to protect seeds. Plant products are suggested as protectants against storage insects when the rice is meant for human consumption. Several nonconventional means of insect pest management (e.g., use of radiant energy, insect sterilants, pheromones, repellents, antifeedants, growth regulators, chitin synthesis inhibitors, and grain resistance) are discussed.

1. INTRODUCTION

Rice is the most important food crop for more than one-half of the world's population, so losses during rice storage pose a threat to the existence of an enormous number of people. Even though the increase in rice production in most tropical countries has not kept pace with increased demand, increasing quantities of rice are being stored (Prakash et al. 1987).

Storage of rice to minimize spoilage is more problematic in the developing and under-developed countries than in the more developed countries. During storage, several organisms—insects, rodents, fungi, mites, and birds—cause considerable losses every year; insects are the key pests (Prakash and Kauraw 1982c). Reliable estimates of overall losses in rice at national and international levels are not available because of varied and imperfect loss estimation procedures; however, the losses are much greater than generally appreciated (Prakash and Rao 1985b). The harvesting periods of rice grown in wet and dry seasons generally coincide with rain and high humidity. Thus, harvested as well as processed rice grains retain high moisture, which predisposes them for considerable damage by insects and molds (Prakash 1983).

Insects cause quantitative, qualitative, and viability losses in stored rice. Insects contaminate the grain with their dead bodies, excreta, and scales. Microorganisms and insect contaminants can be health hazards (Prakash and Kauraw 1982b, Prakash and Rao 1985b). Therefore, it is of prime importance to avoid these losses not only to save valuable human food but also to maintain seed quality and the economic value of the crop. Losses caused by insects in rice storages can be minimized by using synthetic insecticides or botanical products as grain protectants, by toxifying the storage atmosphere either by fumigants or by manipulating its gaseous concentrations, by killing infestations using radiant energy or physical methods, or by using nonconventional methods like growth regulators and their analogues, sex pheromones, sterilants, repellents, and antifeedants (Prakash and Rao 1985a, 1986a; Prakash et al. 1987).

In this chapter, we discuss insect pest management of stored rice with special emphasis on the application of the ecosystem approach under tropical conditions. The ecosystem approach includes alterations of the abiotic components of storage such as grain moisture content, temperature, condition, type, and inner atmosphere of storage structures; additional components include the use of synthetic and botanical grain protectants and other nonconventional methods. Insect pests are one of the important biotic components of rice storage ecosystems and are considered with special reference to their species composition, pest status, sources and modes of infestation, and losses caused by them.

2. SPECIES AND PEST STATUS OF INSECTS INFESTING STORED RICE

Rice stored as paddy (in-the-hull rough rice), brown, and milled or polished rice, in both raw and parboiled forms, and in rural as well as in commercial storages has been damaged by 17 species of insects, namely: Angoumois grain moth, *Sitotroga cerealella* (Olivier); lesser grain borer, *Rhyzopertha dominica* (Fabricius); rice weevil, *Sitophilus oryzae* (L.); rice moth, *Corcyra cephalonica* (Stainton); saw-toothed grain beetle, *Oryzaephilus surinamensis* (L.); red flour beetle, *Tribolium castaneum* (Herbst); confused flour beetle, *T. confusum* (Jacquelin du Val); almond moth, *Ephestia cautella* (Walker); corn weevil, *Sitophlus zeamais* (Motsch.); long headed flour beetle, *Latheticus oryzae* (Waterhouse); flat grain beetle, *Cryptolestes pusillus* (Schonherr); yellow mealworm, *Tenebrio molitor* (L.); cadelle, *Tenebroides mauritanicus* (L.); corn sap beetle, *Carpophilus dimidiatus* (Fabricius); Indian meal moth, *Plodia interpunctella* (Hubner); khapra beetle, *Trogoderma granarium* (Everts); and cereal psocid, *Liposcelis bostrychophilus* (Badonel.) (Hall and McFarlane 1961; Soegiarto et al. 1981; McFarlane 1982; Pande and Das 1984; Prakash et al. 1984a, 1987). *Sitotroga cerealella*, *R. dominica*, *S. oryzae*, *C. cephalonica*, and *O. surinamensis* were considered the major pests based on their regular occurrence and extent of damage. The other 12 species were of minor importance. Nevertheless, their pest status as major or minor pests, varied with types of storage and forms of rice (Prakash et al. 1984a, Prakash and Rao 1987a). Incidence of *S. zeamais* in milled rice was recorded occasionally with infestations of *S. oryzae* (Singh et al. 1978). Similarly, cereal psocids were considered of minor importance (McFarlane 1982).

Sitotroga cerealella was a dominating pest throughout the storage period only in paddy and infestation by this moth was higher in bags than in bulk storages (Prakash et al. 1984a). In paddy stored in bags, *S. cerealella* appeared to be the major pest but in all other types of containers it was second in importance (Douglas 1941; Prakash et al. 1981b,d, 1984a). *Rhyzopertha dominica* in bulk-stored paddy was given a pest status similar to *S. cerealella* (Cogburn 1977).

Rhyzopertha dominica and *S. oryzae* infested both husked and milled grains. Their infestation commenced after about 3 mo of storage in paddy, usually after infestation by *S. cerealella*, whereas in milled rice these species appeared after 1.5 mo of storage. Unlike *S. cerealella*, both species were found more in bulk storages than in bags. On the basis of percent grain damage, *R. dominica* was the foremost pest in bulk storage of paddy but was second in importance in milled rice, both in bulk and bag storages (Prakash et al. 1984a). *Sitophilus oryzae* was third in importance on paddy in both types of storage and also on milled rice stored in bulk, though it was found to be the second most important pest of milled rice in bags (Prakash et al. 1987). *Sitophilus oryzae*, however, has been ranked in a similar status to *R. dominica* (Rouse et al. 1958).

Pest status of nonboring insects were studied based on their larval and adult populations. *Tribolium castaneum* infested both husked and milled rice. Infestation by grain borers in husked grains was found to be conducive for the attack of *T. castaneum* (Prakash et al. 1981b, 1984a, 1987). In milled rice, this beetle also appeared after infestation by *R. dominica* and *S. oryzae* and the population of *T. castaneum* was the most abundant of the minor pests in most of the grain samples (Prakash et al. 1984a).

Oryzaephilus surinamensis was found in milled rice samples only after 6 mo of storage in both bags and bulk containers and it ranked as the fifth most common insect pest (Prakash et al. 1984a). In Orissa (Berhampur), an unusual incidence of this beetle was recorded in bags and also in underground storage during 1984 and it was the most destructive storage pest in milled rice (Prakash and Rao 1987a). *Carpophilus cephalonica* was the foremost pest in milled rice samples collected from storage bins and rice mills but did not invade husked rice (Rahman and John 1979, Prakash et al. 1984a, 1987). Other pests (e.g., *L. oryzae* and *C. pusillus*) were also found in milled rice in both bags and bulk storage along with *O. surinamensis*. Infestations by *E. cautella, T. molitor*, and *T. mauritanicus* were restricted to a few milled rice samples collected from rice mills and farm godowns after 4–5 mo of storage. In general, these insects were considered to be minor pests; however, their status varied in both bulk and bag storage (Prakash et al. 1987).

3. LOSSES DUE TO STORAGE INSECTS

Insects present during storage cause mass losses by directly feeding on grain kernels. Estimation of losses in mass vary from 5 to 25% in a storage period of 3–12 mo (Agrawal and Pingale 1965; Calderwood 1967; Mookherjee 1969; Indian Grain Storage Institute 1979; Central Institute of Agricultural Engineering 1979; Gujar 1980; Huq 1980; Prakash and Kauraw 1982a; Prakash et al. 1984a, 1987; Samajapati et al. 1984). In general, reliable figures for losses cannot be estimated because adequate methodology for assessment of storage losses is not available (Prakash and Rao 1985b). Losses vary depending on type of storage structure, storage period, grades and forms of grain, local climatic conditions, and several other factors (Prakash and Rao 1985b, Prakash et al. 1987). Normally, rural storages are small, scattered, and are made up of locally available materials, which complicates reliable loss assessments (Prakash et al. 1987).

Storage insects feed on grain kernels and destroy the germ; this not only reduces its value directly but also causes a loss in quality of the grain. Grain containing dockage due to insect feeding or half-eaten kernels are not acceptable to most consumers (Sinha and Campbell 1975, Prakash and Rao 1985b). The total number of amino acids was found to be reduced by insect infestation in raw and parboiled rice (Sudhakar and Pande 1983).

Contamination of grains with insect excretions, scales, body fragments, and dead bodies is a common phenomenon when insects multiply in stored grains (Prakash and Rao 1985b). Insect fragments and excrement contamination were closely related and dependent upon the feeding species, and caused loss of palatability and bad odors (Prakash and Rao 1985b, Sinha and Watters 1985). The odor from excrement was observed to be musty in the case of *R. dominica* and unpleasant in the case of *T. castaneum* (Khare 1972). Paddy grains highly contaminated with *S. cerealella* appear golden-yellow due to the presence of scales on the grains (Prakash et al. 1987). Body parts of insects ingested with food grains cause antibody formation and occasionally allergic reactions in people (Perlman 1977). Navarro and Calderon (1979) reported positive skin reactions are caused by *T. castaneum*, *S. oryzae*, and *R. dominica*. These insects also caused allergic reactions when exposure to allergens was through contact, injection, or inhalation (Gorham 1975).

Insects infesting stored grains act as carriers of several pathogenic and toxicogenic fungi and provide ecological conditions for better fungal development and accentuate biochemical degradation (Mishra et al. 1961, Hussain et al. 1968, Sinha and Muir 1973, Srinath et al. 1976). Either through their body parts or excreta, insects contaminate the grains with several storage fungi by their vertical and horizontal movements in rice storages. In turn, these fungi damage the germs of seeds and enhance the multiplication of the insects by providing an alternate food source and by modifying the temperature and moisture environment. The fungi can cause health hazards by producing mycotoxins (Kauraw and Prakash 1980, Gangopadhyay and Chakrabarti 1981, Prakash et al. 1982a). Storage insects also harbor in their gut potentially harmful bacteria such as: *Salmonella* spp., a common cause of food poisoning; *Streptococcus* spp.; *Escherichia coli*; and viruses capable of causing health hazards (Christensen and Kauffmann 1969).

In paddy storage, seed viability was decreased by insects and fungi, whose development was related to grain moisture and temperature (Prakash 1983). Insect infestation of stored seed directly affected its germination, and grains bored by storage insects failed to germinate (Prakash et al. 1987, Howe 1973); however, it was difficult to estimate the viability losses due to insects alone where visible damage symptoms were not clear. For example, young tiny larvae of *S. cerealella* can die soon after feeding on the seed germ (the softest part of the grain); as a result, seed viability is lost although seed mass remained almost unchanged (Prakash et al. 1987). Such viability loss was not detectable using methods recommended in seed certification programs. Loss of viability due to insects in stored paddy was 6.9% after 6 mo of storage (Mookharjee 1969).

Khare (1972) reported paddy seed germination of only 66% when the infestation of insects inside the seed was 16.6%. Prakash and Kauraw (1982a) reported losses in paddy seed germination of 3.7–41.0% with infestations of 1.4–14.1% during 3–9 mo of storage.

Rhyzopertha dominica, a notorious pest, also destroys wooden bases of storage structures and polyethylene-lined containers by making small holes, which facilitate the entry of other pests like *Tribolium* spp. and *S. oryzae* (Pruthi and Singh 1950, Prakash et al. 1987). Damage to wooden framing in cargo ships was an excellent example of destruction caused by this grain borer (Hall and McFarlane 1961).

4. SOURCES OF INFESTATIONS

Infestations of insects in stored rice are classified into four groups based on the source of the insects (Prakash et al. 1987). Horizontal infestations occur when insects like *S. cerealella* and *S. oryzae* directly infest the grains of mature standing crop in paddy fields by laying eggs on the grains; the infestation migrates from fields to storage during postharvest operations (Fernando 1959, Bisara 1968, Singh et al. 1978, Ganeshlingham and Krishnarajah 1979). Vertical infestations in bulk storages of rice usually commence from top to bottom (or vice versa) through gradual migration (Prakash et al. 1987). Latent infestations are common in old storage containers (e.g., jute bags, used or old undisinfested containers, walls, roofs and floors of godowns, and grain transporting carriers) because tiny eggs and adults of storage insects can hid in these sources and then infest newly stored rice (Prakash and Rao 1986b). Cross-infestations commonly occur in rice storages because the wide feeding range of the insects that allows them to spread from one cereal to another in the same store house (Prakash et al. 1987).

5. MANAGEMENT OF RICE STORAGE INSECTS USING ECOSYSTEM APPROACHES

5.1. Grain Moisture

Insect infestation in rice storage can be minimized by modifying abiotic components of the storage ecosystem (e.g., grain moisture, temperature), toxification or manipulation of the gaseous concentrations in the storage atmosphere, using synthetic and botanical grain protectants, and other nonconventional methods.

Grain moisture is one of the important factors which regulates biological activity of insects in storages (Prakash et al. 1981a,b,c,d). Atmospheric humidity and grain moisture are closely related factors. Even though stored grains are dried to a safe moisture level before storage, the moisture content in due course comes into equilibrium with the humidity in the air.

At temperature of 26–30°C, insect infestations gradually increase with an increase of grain moisture from 14 to 21% and their multiplication rate is greatest at an optimum range of grain moisture of 15.5–18.0%. The optimum grain moisture is maintained at 70–85% relative humidity (rh) in paddy (Prakash 1983). Paddy at 13.5–14.0% moisture content (mc) do not allow insects to multiply in storage (Prakash and Kauraw 1982a). Calderwood (1967), reported that paddy rice at 13% mc could be safely stored for 6 mo and even longer at lower moisture contents. Milled rice at 12% mc and adequately polished can be stored safely for more than 12 mo (Pillaiyer 1979). Thus, if stored rice is maintained at or below 14% mc, insects cannot multiply to damage the grains even though they remain in the storage. It is possible to maintain this safe moisture level either by storing the appropriately dried grain in fairly airtight structures, wherein moisture from the outside atmosphere cannot migrate into the storage, or by installing devices to reduce humidity in the storage atmosphere.

It is usually necessary to dry the grain to bring down its moisture content to a safe level to minimize or avoid losses caused by insects. Paddy grains are dried in two stages. First, the harvested crop is dried loose or in bundles in the paddy fields or on the threshing floor to reduce the grain moisture from 28–24% to 18–16%. After threshing, the grain is dried to reduce the moisture from 18–16% to 14–13% (Prakash and Kauraw 1982b,c) using sun drying or mechanical dryers. Since an economic, mechanical dryer for paddy is not available in developing and under-developed countries, only sun drying is commonly used in these countries. Cracking of grains and splitting of their husk during sun drying should be prevented because this grain damage facilitates infestation by insects as well as infection by microbes (Cramford 1956, Prakash et al. 1980a,b). Sun drying requires attention by the farmers to control the drying and to prevent wetting during unexpected short periods of rain. Even now, "controlled" drying is achieved by changing the thickness of the layers of paddy on drying floors, bamboo mats, or polyethylene sheets.

Sun drying or solar heat treatment of paddy can stop the initial horizontal infestation of the grain before storage (Nawab et al. 1980). Whenever possible, staggering sun drying of threshed paddy with short exposures to the sun spread over several days (9:00 AM to 11:00 AM for 8 days) reduces horizontal, vertical, and cross-infestations by insects. Grain drying under high sunlight intensity, especially from 11:00 AM to 2:00 PM should be avoided because it causes splitting of the husks or cracking of the grains (Indian Grain Storage Institute 1979, Central Institute of Agricultural Engineering 1979, Prakash et al. 1980a,b). Similarly, crop drying on moist soil in paddy fields or in threshing floors causes more grain and husk defects due to the sharp differentiation of grain temperature as compared with drying of the loose crop on polyethylene sheets or in small erected crop bundles with the panicles upward (Prakash et al. 1987).

5.2. Temperature

Temperature is one of the important factors that governs the rate of metabolism, reproduction, growth, development, behavior, and distribution of storage insects (Prakash 1983). An optimum temperature range for rice storage insects is 28 to 33°C (Simwat and Chahal 1970, Prakash et al. 1981b,d). But in rural storages, which are usually not airtight, insects as well as the stored grain are continuously subjected to variations of temperature because storage temperature fluctuates with the outside environment. Fluctuations of temperature above and below the optimum temperature range are common in areas like northern India, where temperature decreases to as low as 5°C during winter and increases to as high as 45°C during summer (Prakash 1983). In coastal areas, temperature fluctuations are smaller, often remaining near the optimum temperature range. Thus, storage losses are greater and more constant in coastal areas than in northern India (Prakash and Rao 1983).

Insect development is inhibited at temperatures below the threshold of development and still lower temperatures are fatal. In certain conditions, insects develop cold hardiness and adapt to the rigor of very low and even freezing temperatures with considerable variation. Generally, rice storage insects are unable to withstand low temperature for a long period. Even more hardy insects are not able to tolerate temperatures below −17°C for more than a day. Salt (1936), however, reported a supercooling point of −27°C in the case of *S. cerealella* and *E. cautella* larvae. No hatching was recorded in eggs of *S. cerealella* at 16.7°C (Alder 1960) and 95% egg mortality was reported when eggs of *O. surinamensis, T. castaneum*, and *E. cautella* were exposed to −10°C for 7, 8, and 5 h, respectively (Mullen and Arbogast 1979). Similarly, Obretenchev (1983) found that adults of *O. surinamensis* died at 0°C after 6 d, at −5°C after 60 h, at −10°C after 4 h, and at −15°C after 47 min.

Storing grains at low temperatures is one of the safest means to control the multiplication of storage insects and microbes; this is the concept of all the commercial cold storages. Seed viability also remains constant at low temperatures. Each 5°C reduction in seed temperature doubles seed life (Harrington 1972). In most of the rice growing countries, the practice of storing rice at low temperatures (0–5°C) is limited due to the high cost of installation and maintenance. However, cooperative cold storages for vegetables are common. Thus, if the cost of cold storage can be lowered or the value of high quality rice is increased, the use of low temperature can be an effective approach for insect pest management in rice storages.

5.3. Storage Structures and Practices

A storage structure is very important and is a basic component of the storage ecosystem. Structures generally used for rice storage in most of the tropical rice growing countries are of traditional types, which not only allow moisture to migrate from the outside atmosphere into the stored grain but cracks and crevices are frequently present in these structures that limit effective cleaning and disinfestation. These structures become repositories for dust and moldy grain and harbor latent insect infestations (Prakash et al. 1980a,b, Prakash and Rao 1983). Therefore, it is necessary to improve these structures to make them moisture proof for safe storage. If improved structures are made available at economically feasible prices to farmers, it will be of great significance to insect pest management. Otherwise, prior to rice storage, traditional structures should be thoroughly cleaned of all debris, plastered properly wherever necessary, and disinfested with any of the contact insecticides or fumigants listed in Table 1. Old gunny bags should also be disinfested with insecticides in the same manner to avoid latent and cross-infestations of insects (Prakash and Rao 1983).

Impregnation of bags with dust formulations of selected insecticides provide a protective barrier against cross-infestation of insects (Prakash et al. 1981d). Polyethylene-lined gunny bags or polyethylene bags of 300–400 gage thickness also protect paddy from cross-infestation of insects for up to 2 yr as well as stopping moisture migration to the paddy stored in these bags (Rao et al. 1972, McFarlane 1979, Kumaravel 1984). However, certain storage insects (e.g., *R.*

Table 1 Doses of Insecticides Recommended for Treatment of Storage Structures and Paddy Seeds

Insecticide or fumigant	Treatment of storage structures (a.i. @ 2 L/25 m^2)	Treatment of grains
Malathion 50 EC	0.05%	0.05% a.i. @ 2 L/100 kg
5% dust	–	1% w/w
Fenitrothion 50 EC	0.04%	0.04% a.i. @ 2 L/100 kg
Lindane (gamma BHC) 50 EC	0.04%	–
Etrimfos 50 EC	0.001%	0.001% a.i. @ 2 L/100 kg
2% dust	–	1% w/w
Nuvan 100 EC	–	0.001% a.i. @ 2 L/100 kg
Deltamethrin 2.5% (WP)	0.001%	0.08% a.i. + 500 g ICP/100 kg seed

ICP = inert chalk powder
a.i. = active ingredient

dominica and *S. oryzae*) can penetrate polyethylene linings (Prakash and Rao 1985b). The loss of seed viability was also greater in polyethylene-lined bags than in jute bags without polyethylene (Huq 1980).

Storing unthreshed paddy on open threshing floors or courtyards is a common storage practice in eastern and southern Indian states and also in countries neighboring India. Paddy thus stored is exposed to insect infestations as well as to a number of other biotic (rodents and birds) and abiotic agents. If rain falls on the paddy, fungi and bacteria can cause considerable quantitative and qualitative losses (Prakash and Kauraw 1982a,c, Prakash et al. 1987).

Storage of milled rice for 1–2 yr in underground pits lined with burnt earth bricks or mortar is another common practice in coastal states of India; these pit structures are lined and insulated by ropes made up of paddy straw. The rice usually remains free from insect attack due to a lack of oxygen in these structures. Frequently, however, moisture migration from the soil to the storage leads to grain spoilage by fungi and bacteria in these hermetic structures, especially in the areas where the ground-water table is high. These structures can be improved by making them moisture proof (Prakash and Rao 1983, Prakash et al. 1987).

5.4. Use of Synthetic Insecticides and Fumigants as Grain Protectants

Rice grain itself is a vital biotic component of the rice storage ecosystem. Admixing insecticidal dust with the grain and treating grain with emulsifiable formulations of insecticides as paddy protectants are important means to minimize viability and mass losses by insects (Prakash and Rao 1983). Treatment of rice with synthetic insecticides is not recommended if the rice is meant for human consumption because of direct or indirect health hazards (Prakash et al. 1980a,b, Prakash and Kauraw 1982c; 1987b). Insecticides, however, show no adverse effects on viability and are recommended for seed treatment (Prakash and Kauraw 1982b,c). Several synthetic insecticides have been tested against rice storage insects (McGaughey 1970; Rai and Croal 1973; Cogburn 1976; Chellapa and Chelliah 1976a; Rai 1977; Prakash and Pasalu 1981; Prakash et al. 1983b, 1984b, 1987) but only a few of them are economical and are recommended (Table 1).

To stop horizontal or field infestations by storage insects insecticides were tested by spraying the mature crop in the paddy fields before harvest (Kumar et al. 1982); however, such treatments did not prevent storage losses (Central Rice Research Institute 1987).

Fumigants are synthetic volatile chemicals which are used to toxify the storage atmosphere for insect control. A number of fumigants (e.g., carbon disulphide, chloropicrin, methyl bromide, 3:1 mixture of ethylene dichloride and carbon tetrachloride (EDCT), ethylene dibromide, and phosphine) have been evaluated for insect control in rice (Tilton and Cogburn 1965, Bell 1976, Bond et al. 1977, Quinlan and McGaughey 1983, Prakash and Rao 1985a). Because of their cost and health risks only two fumigants—ethylene dibromide and phosphine—are recommended for effective and economical use in rice storage (Prakash and Rao 1985a).

In addition to its contact poisoning property, dichlorvos is also considered to be an effective grain protectant because of its vapor action; however, it does not penetrate bulk grain and kills only moth populations in the empty headspace in godowns (Girish 1969). Dichlorvos alone was effective against *R. dominica*, *S. oryzae*, *T. castaneum*, and *S. cerealella* in rough rice and also in combination with gardona (McGaughey 1970, 1973a,b). Dichlorvos strips suspended in the headspace in galvanized-metal bins containing rough rice reduced the infestation of *S. cerealella* by 99% within 6 wk and eliminated the population within 12 wk (Cogburn et al. 1983). It showed, however, insignificant protection of stored rice from insects when tested against natural infestations in local storage containers (Prakash and Rao 1985a).

Ethylene dibromide is an effective fumigant against *S. oryzae* and *T. castaneum* (Khare 1972) and also against *E. cautella* (Dhaliwal 1974). It is recommended to be used only on grains meant for consumption because it has an adverse effect on seed viability (Yadav 1983, Prakash and Rao 1985a). Methyl bromide and methyl dibromide in large-scale fumigations of milled rice in bag storages controlled *R. dominica*, *S. oryzae* (Halliday 1967), moth pests (Bell 1976), and *O. surinamensis* (Kazem-Zahedi 1978).

Phosphine gas liberated from aluminum phosphide is the most commonly used fumigant in warehouses and local bulk containers (Prakash and Kauraw, 1982a,c). Effective doses vary greatly depending upon the type of storage and storage temperature (Bagent 1969). In bulk storages of rice, phosphine tablets producing 1–3 g/m^3 effectively control insects (Tyler et al. 1983). Phosphine possesses no residual toxicity in stored rice according to Gillenwater et al. (1981) but residues toxic to humans were reported in rice stocks in commercial warehouses after fumigation (Kumar et al. 1981).

6. BOTANICAL GRAIN PROTECTANTS

Plant materials which are found effective, economic, and available locally (Table 2) may be integrated into insect pest management of stored rice (Prakash et al. 1981a,c; 1986a; 1990). Products from Neem, *Azardirachta indica*, (e.g., leaf, kernel powder, oil, and crude extract) possess repellent, antifeedant, and feeding deterrent properties against storage insects (Savitri and Rao 1976, Devi and Mohandas 1982, Prakash et al. 1982b). Root powder and oil of sweet flag, *Acorus calamus*, protect rice from storage insects (Abraham et al. 1972b). Products like

Table 2 Plant Products Recommended for Use in Rice Storages

Products	Botanical name	Doses	Protection period
Begonia leaves	*Vitex negundo*	2% w/w	270
Bel leaves	*Aegle marmelos*	2% w/w	180
Wild sage leaves	*Lippia geminata*	2% w/w	180
Garlic extract	*Alium sativum*	0.02% @2L/100 kg	180
Pyrethrum 2 EC	*Chrysanthemum cineraraefolium*	0.01% @2L/100 kg	120
Neem oil	*Azadirachta indica*	1% w/w	120

leaves of begonia, *Vitex negundo* (Prakash et al. 1981a 1982b, 1989a,b), leaves of bel, *Aegle marmelos* (Prakash et al. 1983a), leaves of wild sage, *Lippia geminata* (Prakash and Rao 1984), leaves of pudina, *Mentha spicata* and bhang, *Cannabis sativa* (Prakash et al. 1982b), have also been evaluated as protectants against storage insects for paddy. Extract of garlic, *Alium sativum*, repels *S. cerealella* and *R. dominica* in paddy stored under controlled conditions and also provided significant protection under natural storage conditions (Prakash et al. 1982b). Extract of mochrand, *Eclipta alba*, which showed ovicidal activity against *S. cerealella* in a bioassay test (Prakash et al. 1979a) did not show promising grain protection under natural storage conditions (Prakash et al. 1982b).

The cake of undi, *Callophyllum inophyllum*, showed promising grain protection but adversely affected seed viability (Prakash et al. 1982b); however, leaves of kharanja, *Pongamia glabra*, protected stored rice from storage insects in rural Orissa (Govindaswami and Patnaik 1958). Several other plant products were also reported to protect rice from storage insects, including: dried leaves of the malabar nut tree, *Justicia adhatoda* (Abraham et al. 1972b), extract of sikkakai, *Acacia concinna*, ajuba, *Adhatoda vasica* (Chellapa and Chelliah 1976a), and root bark powders of *Celastrus angulatus* and *Tripterygium welfordi* (Zhang and Zhao 1983).

7. MANIPULATION OF STORAGE ATMOSPHERE

In a storage ecosystem manipulation of the concentrations of the atmospheric gases (e.g., carbon dioxide, oxygen, and nitrogen) kill insect populations (Bailey and Banks 1980). Reduced atmospheric pressure (0.27 Pa) in storages caused complete mortality of the insects without any significant loss in germination of paddy seeds (Khan et al. 1976).

Storage atmospheres with oxygen concentrations less than 1.7% (Ali-Niazee 1973, Navarro 1978, Navarro and Calderon 1979) or less than 1% (Storey 1977) are toxic to all stages of the flour beetle. Atmospheric oxygen concentrations less than 2% are toxic to *S. oryzae, R. dominica*, and *O. surinamensis* (Navarro 1978).

In a storage atmosphere, a 60% concentration of carbon dioxide maintained for 4 d kills all stages of *R. dominica* and *S. oryzae* at 23–28°C (Jay 1971), whereas 20% CO_2 caused 95% mortality to *T. castaneum* and an increase in temperature from 15.6 to 27.7°C increased the poisoning susceptibility (Ali-Niazee 1973). Brown and polished rice can be stored safely for 2 yr in nylon-polythene bags after replacing the air with CO_2 (Kuotsong-Tehi and Chow 1978). Bags of stored milled rice enclosed in a polyvinyl chloride membrane and then dosed with dry ice provided an atmosphere of 70% CO_2 for 22 d and provided absolute protection from the attack of *R. dominica, S. oryzae*, and *T. castaneum* (Annis et al. 1984).

Nitrogen concentration was also manipulated in storage atmospheres for control of insects, especially when it replaced oxygen (Pearson and Sorrenson 1970). The effects of nitrogen and carbon dioxide were tested on the eggs of *E. cautella* and CO_2 was more effective than nitrogen using 100% of each gas in the storage atmosphere; hatching was delayed in both gases (Bell et al. 1980). Absolute mortality of adult *S. oryzae* and *R. dominica* was achieved in a controlled atmosphere containing 2 kg of nitrogen in a 111 L cylinder filled with grain and exposed for 7 d (Kumar et al. 1980).

The manipulation of gases in a storage atmosphere offers a good line of approach to pest control in storage. The cost of operation, however, is prohibitive and the viability of seeds can be affected adversely (Roberts and Adballa 1968); however, in a controlled atmosphere of 1% oxygen replaced by nitrogen or in 70% carbon dioxide, seed viability was retained (Banks 1981).

8. USE OF NONCONVENTIONAL METHODS

8.1. Radiant Energy

Radiant energy of various wave lengths (i.e., gamma-rays, infrared rays, visible and ultraviolet light) and energy from various atomic particles (e.g., neutrons, alpha particles, and electrons) has been tested for use either for direct killing of *S. cerealella* by heating or for paddy drying (Prakash et al. 1987). Use of gamma-rays was in practice for several years and proved to be an excellent means of insect control, particularly for large volumes of stored products including rice (Prakash et al. 1981d). The rays are usually passed through the grain to check latent infestation by either directly killing the insect stages or causing sterility (Prakash and Rao 1983).

Eggs of *S. cerealella* treated with 140 Gy of gamma radiation failed to hatch (Quereshi et al. 1970). Exposure of immature and adult stages of *R. dominica* to 132 Gy of radiation (five times at an hourly interval) caused no mortality to adults (Tilton et al. 1966) but at 200 Gy its pupae showed 100% sterility in the emerged adults (Abdul-Martin and Hooper 1974). *Sitophilus oryzae* adults treated with 20 Gy radiation for five successive generations showed no resistance to radiation sensitivity (Brower et al. 1973) and an exposure of 400 Gy was required to completely inhibit the hatching of their eggs and to cause absolute sterility to adults (Abdul-Martin and Hooper 1974). A treatment of 16 kGy to the adults of O. surinamensis caused 100% sterility (Jafferies 1966), whereas its fecundity and fertility were completely eliminated at 25.8 kGy exposure (Chawla et al. 1973). The development of the eggs and larvae of *E. cautella* was arrested at 300 and 200 Gy treatments, respectively (Cogburn et al. 1973); at 400 Gy exposure, treated males mated with untreated females for five successive matings enhanced its fecundity from 5–25% (Gonen 1977) but adults showed 99% sterility when treated with 500–600 Gy radiation (Brower 1979, Younes 1980). Complete sterility was achieved at 450 Gy treatment to the newly emerged adults of *C. cephalonica* (Chand and Sehgal 1979).

Infrared rays are used for killing storage insects inside or among grain kernels by heating but these rays are readily absorbed by most stored products; therefore, the penetration of the rays is not deep in the grains (Prakash and Rao 1983). The possibilities of using infrared rays (Schroeder 1960, 1961) were evaluated for the control of insects such as *R. dominica* (Tilton and Schroeder 1963, Cogburn et al. 1971, Tilton et al. 1966), *S. oryzae* (Tilton and Schroeder 1963) and *T. castaneum* (Kirkpatrick and Cagle 1978). Similarly, ultraviolet rays from a UV wet light trap disinfest and reduce insect populations in rice warehouses before filling with grain (Sukardi 1981).

Microwaves were also tested for the control of *R. dominica* (Kirkpatrick et al. 1973). Comparative effectiveness of radio waves of 39 and 2450 MHz frequencies were studied for the control of rice weevil. Complete mortality was obtained at 39 MHz, which raised the grain temperature to 46°C, and at 2450 MHz, which raised grain temperature to 80°C (Nelson and Stetson 1974).

Use of radiant-energy, however, involves a high cost of installing the initial facilities and transporting the grain to and from the irradiators (Prakash and Rao 1988). In addition, gamma-ray treatment of rice having a high moisture content just after harvest produced marked changes in chemical composition (Hamid et al. 1966).

Subsonic pressure waves of 18 MHz at two intensity levels (90 and 120 dB) affected the size of progeny in *Tribolium* spp. (Mullen 1975). Treatment of stored grains at 0.25 pdB and coupled with partial vacuum of 4.7 kPa for 10 min eliminated infestations of *S. cerealella*, *R. dominica*, and *S. oryzae* (Tilton and Vardell 1982).

8.2. Sterilants

Sterilants include radiant energy (e.g., gamma rays) and certain chemicals (e.g., apholate, thiotepa, tepa, and metapa, etc.) which cause sterility in both male and female insects; male sterility is the most practical for control of storage insects (Prakash and Rao 1983). Gamma irradiation was used to sterilize pupae and adults of *R. dominica* at 140 and 115 Gy (Abdel-Martin and Hooper 1974) and adults of *E. cautella* (Brower 1979).

Several chemicals were evaluated as chemosterilants against these insects, such as: furadene and nitrofurazone against *Sitophilus* spp. and *S. cerealella* (Thorpe and George 1963); apholate against *E. cautella* (Gangrade and Pant 1973); and tepa, metapa, thiotepa, phosphinic acid, and bis (1-aziridinyl) ethyl ester against the pupae and adults of *S. cerealella* (Pande and Teotia 1978). Of several methods available for insect sterilization, such as chemosterilization, radio-sterilization, cytoplasmic incompatibility, hybrid sterility, etc., use of chemosterilants is the most effective and can safely be used in combination with conventional insecticidal sprays and attractants (Prakash et al. 1987).

8.3. Repellents

The repellent and antifeedant properties of plant products have already been discussed as botanical grain protectants (see Section 5.4). Certain synthetic chemicals are effective repellents, including: piperonal butoxide for *Tribolium* spp. (Guy et al. 1970); diethyltoluamide (Deet) for *O. surinamensis*, *R. dominica*, and *T. castaneum* (Khan and Wohlegemuth 1980); 2-pentynyl and 2-monynyl mandelate for *Tribolium* spp. (Gillenwater et al. 1981); ethyl mandelate, dibutyl adepate, and benzoic acid for *T. castaneum* (Khan et al. 1976); and pirimiphos-methyl for larvae of *T. castaneum* (Mullen and Arbogast 1979).

8.4. Antifeedants

Synthetic chemicals show antifeedant activity when impregnated on packing materials and gunny bags, including: AC-24055 (4,3,3-Dimethyltriagenon acetanilide) for *T. castaneum* (Loschiavo 1969); brestan, brestanol, plictran, and AC-24055 for *S. cerealella* (Rajan and Dale 1971); and mandelic acid and rotenone for *R. dominica* and *T. castaneum* (Noak and Schmidt 1981).

8.5. Pheromones

Sex pheromones are chemicals isolated from female insects and utilized to attract males of the same species. Aggregation pheromones are also produced by many insects and attract both sexes to a suitable food source. Since these chemicals are natural substances that regulate essential behavior of the insect species, insects are less likely to become resistant (Vick et al. 1978). The synthetic sex pheromone, (7 A, 11 E)-7-11-hexadecadienyl acetate, affects mating of *S. cerealella* in the closed environment of farm godowns; the pheromone also traps males (Su and Mohny 1974, Vick et al. 1978, 1979). The synthetic sex pheromone, *cis*-9, *trans*-12-tetra decadein-1-ol-acetate, successfully disturbs the mating of *E. cautella* in laboratory conditions (McGregor and Kramer 1976).

8.6. Use of Growth Regulators

Growth regulators and their analogues can be used to upset the normal development of insects. These chemicals are highly active, very selective in action, and relatively nontoxic to other animals and are referred to as "third generation pesticides" (Novak 1975).

A number of chemicals are effective molting hormones, including: altozar (methoprene) against *R. dominica* and *S. oryzae* (Strong and John 1973); dimilin (TH-6040) against coleopteran pests (McGregor and Kramer 1976); expoxyphenyl ether against *R. dominica, T. castaneum, Ephestia* sp., and *Sitophilus* sp. (Hoppe 1976); methoprene and hydroprene against *R. dominica, S. oryzae, E. cautella*, and *O.surinamensis* (Gonen 1975, Amos and William 1977, Mian and Mulla 1982); TH-6041 and TH-6042 against beetle pests (Kramer and McGregor 1978a); and diflubenzuron and triflubenzuron against *R. dominica, S. oryzae*, and *O. surinamensis* (Mian and Mulla 1982).

Juvenile hormones and their analogues are effective for the control of numerous insects, including: altosid (hydroforene) against *S. cerealella* (Babu and Panwar 1976), *E. cautella* (Piserov and Shaaya 1979), and *C. cephalonica* (Brower 1979); JH-25(7-epoxy-1-(phenyl phenoxy-3,7-dimethyl-2-octane) against *Tribolium* spp. (Rahman and John 1979); JHa, methyl (3,7,11-trimethyl-7,11-dichloro-2)-2 doclecenoate against *R. dominica* (Bhatnagar-Thomas 1973); pyridyl against both coleopteran and lepidopteran pests (Kramer and McGregor 1978b); Ro 20-3600[(E)-5-[(5-(3-ethyl-3-methyl-oxiranyl)-3-methyl-2-pentenyl]oxy]-1,3-benzodioxole)] against *C. cephalonica* (Fajardo and Morallo-Rajesus 1980); Mobil-9089 (1-CL, 1-dimethl ethyl)-4-(1-(4-epoxy phenyl)-2 nitrobutyl benzene) against *R. dominica, S. cerealella, O.surinamensis*, and *T. castaneum* (Kramer and McGregor 1979a); and AI_3-63392(2-fluoro-N-[[4-(phenylmethoxy) phenyl]-methylene] benzenamine) against *T. castaneum* (Vir 1981).

8.7. Chitin Synthesis Inhibitors

Certain chemicals which inhibit the synthesis of chitin during development of insects, are called chitin synthesis inhibitors (CSI) and have been given status of a "new class of insecticides" (Marx 1977, Majumder 1978). Of several such chemicals tested, only a few are effective in suppressing the development of insects and lead to their death; these include: N(4-(4-nitrophenoxy-3,5-dichlorophenyl amino carbonyl)-2-chlorobenzamide against *S. oryzae, R. dominica, O.*

surinamensis, E.cautella, and *Tribolium* sp. (Kramer and McGregor 1979b); LY-127063, N-1(5-(4-Bromophenyl-2-pyrazinyl amino) carbonyl, and 11-2-chlorobenzamide against *S. cerealella, R. dominica, S. oryzae, E. cautella, O. surinamensis*, and *Tribolium* sp. (Kramer and McGregor, 1980); and AI_3-63233[2,6-difloro-N-[[(4-trifluromethyl)phenyl]amino]carbonyl] benzamide against *T. castaneum* (Saxena and Mathur 1981).

8.8. Grain Resistance

Research on grain resistance to insect pests of stored rice was initiated in the early 1970s (Cogburn 1975). Use of grain resistance in breeding programs to produce varieties resistant against storage insects has distinct possibilities provided that the characters and factors responsible for such resistance are identified and intensively studied (Prakash et al. 1982c). Responses of the grains of the rice cultivars to oviposition, development, adult emergence of the storage insects, and grain damage caused by them were well studied by several workers (Abraham et al. 1972a; Cogburn 1973, 1974; McGaughey 1974; Chellapa and Chelliah 1976b; Russell 1976; Sardar 1976; Nigam et al. 1977; Prakash et al. 1979b, 1982a,b 1983c, 1986b). According to them, grain characteristics (i.e., grain grade, scentiness of grains, silica content, thickness and texture of husk, relative hardness of grain kernels, protein and amylose contents, etc.) were associated with differential responses. The resistant or least susceptible rice cultivars to storage insects are listed in Table 3. Based on available information of the correlation of morpho-chemical grain characters and parameters to evaluate grain resistance to storage insects more than one character is associated with such varietal preferences in a complex way.

9. CONCLUSION

In rice storage, grain moisture at or below a safe level and storage temperature below the threshold of pest development are the most important abiotic factors that inhibit insect multiplication. They must be integrated into an effective insect pest management program in rice storage ecosystems. Grain moisture content should be brought down to the safe level through appropriate drying of the crop and processed grains. If economically feasible, low temperature storage is one of the most effective approaches to avoid storage losses. Similarly, manipulation of storage atmospheric gaseous concentrations, if economically feasible, is also a viable means to control insects in storage.

Use of traditional storage practices—storing of unthreshed paddy at open places (open storage) and in underground storages (hermetic storage)—should be improved to prevent moisture migration, latent and cross-infestations of

Table 3 Rice Cultivars Resistant to Storage Insects

Cultivar	Insect species	References
Badshabhog*	*S. cerealella*	Chatterjee et al. 1977
Cauvery	*S. cerealella*	Chellapa and Chelliah 1976b
CR 404-56	*R. dominica*	Prakash et al. 1983c
Dawn	All storage insects	McGaughey 1973c, 1974
IR 8	*R. dominica*	Chellapa and Chelliah 1976b
	S. cerealella	Nigam et al. 1977
Madhukar	*S. oryzae*	Awasthi et al. 1983
Peta	*S. cerealella*	Halliday 1967, Suzuki and Juliano 1975
Pusa 2-21	*R. dominica*	Prakash 1982a,b
	S. oryzae	
Ptb.7	*S. cerealella*	Abraham et al. 1972a
Rambhog	*S. oryzae*	Sudhakar and Pandey 1983
Shakti	*S. cerealella*	Prakash 1982a,b; Prakash et al. 1983c
	R. dominica	
	S. oryzae	
T-9	*S. oryzae*	Sudhakar and Pandey 1983
T-412*Vani	*S. cerealella*	Chatterjee et al. 1977
Vani	*R. dominica*	Prakash 1982a,b
	S. oryzae	
Vijaya	*R. dominica*	Rout et al. 1976, Prakash 1982a,b
	S. oryzae	

* Reported to be moderately resistant or least susceptible.

storage insects, and problems related to other pests like rodents and birds. Before storage, grain should be dried to a safe moisture level. Traditional structures should be made moisture proof. It is advisable to check the field infestation of storage insects prior to storage. Before storing grain, storage structures (godowns) should be properly disinfested either by using insecticides or fumigants to avoid latent insect infestation. If grains are found to be infested they should be treated or fumigated. Use of plant products as grain protectants is advantageous for rice to be used as seed and for consumption. Use of synthetic insecticides, however, is advised for only seed.

Use of nonconventional methods should be encouraged for commercial storage, integrated with conventional methods depending on the availability of the product and its feasibility. In addition, emphasis should be given to explore the nonconventional approaches to make them acceptable. Estimation of the cost of insect pest management practices is necessary as these need to be added to the grain price. Because precise loss assessments are difficult, economic justification of control methods can be almost impossible.

REFERENCES

Abdel-Martin, A. S. M. and G. H. S. Hooper. 1974. Susceptibility of *Rhzopertha dominica* to ionising raditions. J. Stored Prod. Res. 10:199-207.

Abraham, C. C., B. Thomas., K. Karunakaran, and R. Gopalkrishnan. 1972a. Relative susceptibility of different varieties of paddy to infestation by Angoumois grain moth, *Sitotroga cerealella* Oliv. (Gelechiidae: Lepidoptera) as influenced by amylose content of the endosperm. Bull. Grain Technol. 10:263-266.

Abraham, C. C., B. Thomas., K. Karunakaran, and R. Gopalkrishnan. 1972b. Relative susceptibility of different varieties of paddy to infestation by Angoumois grain moth, *Sitotroga cerealella* Oliv. infesting stored paddy in Kerala. Agric. Res. J. Kerala 10:56-59

Agrawal, N. S., and S. V. Pingale. 1965. Losses suffered by paddy, rice and wheat storage in India. Seminar on postharvest handling of food grains in Bangalore 10-13 July, 1965, 13 pp.

Alder, V. E. 1960. Effect of low temperature on the eggs of the Angoumois grain moth, Indian meal moth and confused flour beetle. J. Econ. Entomol. 53:973-974.

Ali-Niazee, M. T. 1973. Oxygen toxicity in *Tribolium* spp. Environ. Entomol. 2:293-295.

Amos, T. G. and S. P. William. 1977. Insect growth regulators. Some effects of methoprene and hydroprene on productivity of several stored grain insects. Aust. J. Zool. 25: 201-206.

Annis, P. C., H. J. Banks, and T. Sukardi. 1984. Insect control in stacks of bagged rice using carbon dioxide treatment and an experimental PVC enclousure. CSIRO Bull., Canberra, Australia. 38 pp.

Awasthi, B. K., V. Pandey, and P. M. Nigam. 1983. Relative resistance of different rice varieties to *Sitophilus oryzae* Linn. Bull. Entomol. 24:59-60.

Babu, T. S. and M. S. Panwar. 1976. Effect of altosid a juvenile hormone analogue on the embryonic development of Angoumois grain moth *Sitotroga cerealella* Oliv. Indian J. Entomol. 38:395-396.

Bagent, J. L. 1969. Fumigate stored rice to reduce losses of insects. Rice J. 72:15-16.

Bailey, S. W. A. and R. W. Banks. 1980. A review of the recent studies on the effect of controlled atmosphere on stored product pests. pp. 101-118 in: Controlled Atmosphere Storage of Grains, J. Shejbal (ed.). Elsevier Scientific Pub. Co., Amsterdam.

Banks, H. J. 1981. Effects of controlled atmosphere storage on grain quality. Food Tech. Australia 33:335-340.

Bell, C. H. 1976. The tolerance of immature stages of four stored product moth to methyl bromide. J. Stored Prod. Res. 12:1-10.

Bell, C. H., E. C. Sparatt, and D. J. Mitchell. 1980. The effect of nitrogen and carbon dioxide on eggs of *Ephestia cautella* walk, and *E. kuehniella* Zell. (Lepidoptera: Pyralidae). Bull. Entomol. Res. 70: 293-298.

Bhatnagar-Thomas, P. L. 1973. Control of insect pests of stored grains using juvenile hormone-analogues. J. Econ. Entomol. 66:277-278.

Bisara, S. I. 1968. Comparison of some aspects of flight activity of *Sitophilus zeamais* Mots. and *S. oryzae* Linn. Bull. Soc. Entomol. Egypt 52:477-500.

Bond, E. J, R. A. Sellen, and T. Dumas. 1977. Control of insects with phosphine in open ended bin spouts. J. Econ. Entomol. 70:22-25.

Brower, J. H. 1979. Suppression of laboratory populations of *Ephestia cautella* Walk. (Lepidoptera: Pyralidae) by release of males with cytoplasmic incompatibility. J. Stored Prod. Res. 15:1-4.

Brower, J. D., M. M. Hassain, and W. T. Elivin. 1973. Radiation sensitivity of successively irradiated generation of *Tribolium castaneum* and *S. oryzae*. J. Stored Prod. Res 9:43-44.

Calderwood, D. L. 1967. Bulk storage of rice under climatic conditions prevailing in southern US. Rice J. 70 36-39.

Central Institute of Agricultural Engineering, Bhopal. 1979b. Improvement of rice storage structures and practices to minimise losses. Proc. VII Ann. Work. Postharv. Technol. (ICAR) CPCRI, Kasargod (India). 76 pp.

Central Rice Research Institute. 1987. Final Project Report (RPF-III) of "Studies on stored grain pests of rice and methods of minimising losses caused by them (1978-87). Central Rice Research Institute, Cuttack, India. 49 pp.

Chand, A. T. and S. S. Sehgal. 1979. Sexual competitiveness of gamma irradiated adult rice moth, *Corcyra cephalonica* Staint. Entomon. 4:111-125.

Chatterjee, S. M., R. C. Dani, and G. W. Govindaswami. 1977. Evaluation of rice varieties for resistance of *Sitotroga cerealella*. J. Entomol. Res. 1:74-77.

Chawala, S. S., J. M. Perron, L. Heist, and G. W. Corrivault. 1973. Sensitivity of *Oryzaephilus surinamensis* to Cesium 137. II. Irradiation of larvae. J. Econ. Entomol. 66:605-608.

Chellapa, K. and S. Chelliah. 1976a. Studies on the efficacy of malathion and certain plant products in control of *Sitotroga cerealella* Oliv. and *Rhyzopertha dominica* Fabr. Madras Agric. J. 63: 190-192.

Chellapa, K. and S. Chelliah. 1976b. Varietal susceptibility of rice to infestation by *S. cerealella* Oliv. and *R. dominica* Fabr. Madras Agric. J. 63 (2):190-192

Christensen, C. M. and H. H. Kauffmann. 1969. Grain storage: the role of fungi in quality loss. Univ. of Minnesota Press, Minneapolis, MN. 153 pp.

Cogburn, R. R. 1973. Insect resistance of stored rice. Rice J. 76:72-73.

Cogburn, R. R. 1974. Domestic rice varieties apparent resistance to rice weevil, lesser grain borer and Angoumois grain moth. Environ. Entomol. 3:681-685.

Cogburn, R. R. 1975. Stored rice insect research. Rice J. 7:78.

Cogburn, R. R. 1976. Pirimiphos methyl as a grain protectant for stored rough rice: small bin test. J. Econ. Entomol. 69:369-373.

Cogburn, R. R. 1977. Resistance to Angoumois grain moth in some varieties of rough rice from USDA-World collection. J. Econ. Entomol. 70:753-754.

Cogburn, R. R, H. B. John, and E. W. Tilton. 1971. Combination of gamma and infra-red radiation for control of Angoumois grain moth in wheat. J. Econ. Entomol. 64:923-925.

Cogburn, R. R, E. W. Tilton, and J. H. Brower. 1973. Almond moth-gamma radiation effect on the life stages. J. Econ. Entomol. 66:745-751

Cogburn, R. R., D. L. Calderwood, B. D. Webb, and M. A. Marchetti. 1983. Protecting rough rice stored in metal farm bins from insect attack. J. Econ. Entomol. 76: 1377-1383.

Cramford, R. Q. 1956. Breakage of rice during milling. Tropical Stored Prod. Inf. 3:64-66.

Devi, D. A. and N. Mohandas. 1982. Relative efficacy of some antifeedants and deterrents against insect pests of stored paddy. Entomon. 7:261-264

Dhaliwal, G. S. 1974. On the degree of susceptibility of two species of storage insects to different fumigants. Bull. Grain Technol. 12:64-65.

Douglas, W. A. 1941. Field infestation of insects that injure rice in storage. Circ. USDP No. 602, 8 pp.

Fajardo, V. and B. Morallo-Rajesus. 1980. Effect of ten insect growth regulators on *Plodia interpunctella* Hubn. and *Corcyra cephalonica* Staint. Philippine Entomol. 4:169-175.

Fernando, H. E. 1959. Storage loss of paddy due to *Sitotroga cerealella*. Rice Newsl. 7:20-25.

Ganesalingham, V. K. and S. R. Krishnarajah. 1979. Infestation of *Sitotroga cerealella* Oliv. under field and storage in northen Sri Lanka. Ceylon J. Sci. (Bio-Sci.) 13:159-165.

Gangopadhyay, S. and N. K. Chakrabarti. 1981. Mycotoxin in stored rice. Curr. Sci. 50:272-275.

Gangrade, G. A and N. C. Pant. 1973. Inducting sterility with apholate in *Ephestia cautella* Walk. (Lepidoptera: Phycitidae) by oral, topical and injection methods. Indian J. Entomol. 35:85-94.

Gillenwater, H. B., T. P. McGrovern, and L. L. McDonald. 1981. Repellents for adult *Tribolium confusum*, alkynyl mandelates. J. Georgia Entomol. Soc. 16:106-112

Girish, G. K. 1969. An insecticide/fumigant for protection of stores and food processing plants. Bull. Grain Technol. 7:92-109.

Gonen, M. 1975. Effect of gamma radiation on *Ephestia cautella* Walk. (Lepidoptera: Phycitidae). IV. Sensitivity of maturing adult to sterilising dose. J. Stored Prod. Res. 11:97-101.

Gonen, M. 1977. Susceptibility of *Sitophilus granarius* and *S. oryzae* (Coleoptera: Curculionidae) to high temperature after exposure to supra-optimal temperature. Entomol. Exp. Appli. 23:243-248.

Gorham, J. R. 1975. Filth in foods: Implication for health. J. Milk Food Technol. 38:409-418.

Govindaswami, S. and N. K. C. Patnaik. 1958. Storage methods in Jeypore tract. Rice Newsletter (India) 6:5-7.

Guy, R. H., H. A. Highland, and C. E. Metts. 1970. Repellency of selected compounds to *Tribolium castaneum*. J. Econ. Entomol. 63:1847-1850.

Gujar, G. T. 1980. A note on survey of storage loss in paddy in Gujrat state. Bull. Grain Technol. 18:236-237.

Hall, D. W. and J. A. McFarlane. 1961. Postharvest problems with paddy and rice in British Overseas Territories. p. 28. in: Proc. IX Meeting FAOUS, IRC Working Party on Production and Protection. Ibadan, Nigeria.

Halliday, D. 1967. Field studies of fumigation of produce from nothern Algeria with phosphine. pp. 35-59 in: Rep. Nigeria Stored Prod. Res. Inst. Ibadan, Nigeria.

Hamid, M. A., M. T. Nadeem, W. A., Farooqi, and A. S. Bhatt. 1966. Immediate effect of gamma-radiation on the physical conditions, chemical composition and cooking quality of stored rice. Int. Rice Comm. Newsl. 15:26-29.

Harrington, J. F. 1972. Problems of seed storage. pp. 251-263 in: Seed Ecology, W. Heydecker (ed.). Butterworths Publ., London.

Hoppe, T. 1976. Micro-plot trial with an epoxyphenyl ether (insect growth regulator) against several pests of stored wheat. J. Stored Prod. Res. 12:205-209.

Howe, R. W. 1973. Loss of viability in seed in storage attributed to infestation of insects and mites. Seed Sci. Tech. 1: 563-586.

Huq, A. K. F. 1980. Some estimates on farm level storage of rice in Bangaladesh. Trop. Stored Prod. Info. 39:15-18.

Hussain, S. S, S. S. Quereshi, S. A. Husan, and M. Anwarullah. 1968. Stored grain insects as carrier of fungi. Pakistan J. Sci. Indust. Res. 11:297-298.

Indian Grain Storage Institute. 1979. Report of All India Co-ordinated Project on Storage Losses for the year 1976-1979. Indian Grain Storage Institute, Hapur, India. 137 pp.

Jafferies, D. J. 1966. Susceptibility of saw toothed grain beetle *Oryzaephilus surinamensis* L. to gamma radiation. Entomol. Rad. Disinf. Grain 119-130.

Jay, E. G., 1971. Suggested conditions and procedures for using carbon dioxide to control insects in grain storage facilities. USDA ARS Bull. No. 51/56. 6 pp.

Kazem-Zahedi. 1978. The effect of methyl bromide on *Oryzaephilus surinamensis* L. Proc. 30th Int. Symp. Crop Prot. Medeleligen van de Faultsit Landbouweten-schappen Rijksuniversiteit Gen. 43:909-910.

Kauraw, L. P and A. Prakash. 1980. The influence of storage fungi on the development of lesser grain borer in paddy cultivar CR 1014. Seed Res. 8:67-70.

Khan, M. A. and R. Wohlegemuth. 1980. Diethyl-tolumide as a repellent against stored product pests. Anzeiger fur Schadlings Kunde Pflanzenschutz Unweltischutz 53:126-127.

Khan, R. A., B. S. Prabhakar, and R. K. Mookhrajee. 1976. Vacuum storage of paddy seed. Bull. Grain Technol. 14:154-155.

Khare, B. P. 1972. Insect pests of stored grains and their control in Uttar Pradesh. G. B. Pant Univ. Agric. Technol., Pantnagar Res. Bull. No. 5. 152 pp.

Kirkpatrick, R. L. and A. Cagle. 1978. Controlling insects in bulk wheat with infrared radiation. J. Kansas Entomol. Soc. 51:386-393.

Kirkpatrick, R. L., J. Brown, and E. W. Tilton. 1973. Gamma, infrared, and microwave radiation combination for control of *Rhyzopertha dominica*. J. Stored Prod. Res. 9:19-23.

Kramer, K. R. and H. E. McGregor. 1978a. Activity of TH-6041 and TH-6042 against stored product insects. J. Econ. Entomol. 71:825-826.

Kramer K. J. and H. E. McGregor. 1978b. Activity of pyridyl and pheryl ether analogues of juvenile hormone against Coleoptera and Lepidoptera in stored grain. J. Econ. Entomol. 71:132-134.

Kramer K. J. and H. E. McGregor. 1979a. Reproductive inhibition of stored product insects by Mobil-9089. J. Kansas Entomol. Soc. 52:331-333.

Kramer K. J. and H. E. McGregor. 1979b. Activity of seven chitin synthesis inhibitors against the development of stored product insects. Environ. Entomol. 8:274-276.

Kramer, K. J. and H. E. McGregor. 1980. Susceptibility of stored product insects to chitin inhibitors. J. Kansas Entomol. Soc. 53:627-630.

Kumar, A., R. B. Doharey, G. P. Pandey, and B. K. Verma. 1980. Efficacy of nitrogen gas against stored grain insect pests. Bull. Grain Technol. 18:192-194.

Kumar, A., J. L. Srivastava, G. P. Pandey, and B. K. Verma. 1981. Phosphine residue in multifumigated rice stocks stored in commercial warehouses. Bull. Grain Technol. 19:81-85.

Kumar, A., G. P. Pandey, R. B. Doharey, and B. K. Verma. 1982. Field trial with some newer organophosphatic insecticides against pests of stored grain. Pesticides 16:7-10.

Kumarvel S. 1984. Selection of a container for safe storage of rice germ. Agric. Sci. Digest 4:65-66.

Kuotsong-Tehi and T. Y. Chow. 1978. Storage of rice grain in plastic bag packages and phostoxin. Kio Hsua-h Fa Cham Yuen Kman 6:639-650.

Loschiavo, S. R. 1969. The effect of antifeeding compound AC-24055 on survival development and reproduction of some stored product insects. J. Econ. Entomol. 62:102-107.

Majumder, S. K. 1978. The effects of different types of storage structures on quality and condition of grain storage. Indian Food Packer 32:48-57.

Marx, J. L. 1977. Chitin synthesis inhibitors:a new class of insecticides. Science 197: 1170-1172.

McFarlane, J. A. 1979. The effects and significance of insect infestation on milled rice. pp. 455-461 in: South-East Asia Cooperative PostHarvest Research and Development Program Workshop, Bangkok.

McFarlane, J. A. 1982. Damage to milled rice by psocids. Trop. Stored Prod. Info. 44:3-10.

McGaughey, W. H. 1970. Evaluation of dichlorvos for insect control in stored rough rice. J. Econ. Entomol. 63:1863-1870.

McGaughey, W. H. 1973a. Angoumois grain moth control in rough rice with a gardona-dichlorvos mixture. J. Econ. Entomol 66:1353.

McGaughey, W. H. 1973b. Dichlorvos vapour for insects control in rice mill. J. Econ. Entomol. 66:1147-1151.

McGaughey, W. H. 1973c. Resistance ot the lesser grain borer in Dawn and Labelle varieties of rice. J. Econ. Entomol. 66:1005.

McGaughey, W. H. 1974. Insect development in milled rice, effect of varieties, degree of milling, parboiling and broken kernels. J. Stored Prod. Res. 10:81-86.

McGregor, H. E. and J. J. Kramer. 1976. Activity of dimilin (TH-6040) against coleopteran pests of stored wheat and corn. J. Econ. Entomol. 69:479-480.

Mian, L. S and M. S Mulla. 1982. Biological activities of insect growth regulators against four stored product coleopterans. J. Econ. Entomol. 75:80-85.

Mishra, C. P., C. M. Christensen, and A. C. Hodson. 1961. The Angoumois grain moth *Sitotroga cerealella* and storage fungi. J. Econ. Entomol. 54:1032-1033.

Mookherjee, P. B. 1969. Integrated control of storage pests. pp. 97-98 in: Proc. Int. Seminar Integ. Pest Control (ESI-IARI), Indian Agricultural Research Institute, New Delhi, India.

Mullen, M. A. 1975. Infra sound retards in development of *Tribolium castaneum* and *T. confusum*. J. Stored Prod. Res. 11:111-113.

Mullen, M. A. and R. T. Arbogast. 1979. Time-temperature-mortality relationship of various stored product insect eggs and chilling times for selected commodities. J. Econ. Entomol. 72:476-478.

Navarro, S. 1978. The effect of low oxygen tensions on three stored product insect pests. Phytoparasitica 6:51-58.

Navarro, S. and M. Calderon. 1979. Exposure of *Ephestia cautella* Walk. (Lepidoptera: Phycitidae) to low pressure: effect on adults. Experimentis 32: 620-621.

Nawab, A., S. M. Illyas, and A. Alam. 1980. Crop dryers use of solar energy and agricultural wastes. Tech. Bull. CIAE, Bhopal (India) No. CIAB/80-12. 67 pp.

Nelson, S. O. and L. E. Stetson. 1974. Compartive effectiveness of 39 and 2450 MHz electric fields for control of rice weevils in wheat. J. Econ. Entomol. 67:592-595.

Nigam, P. M. B. K. Awasthi, and V. Pandey. 1977. Studies on comparative resistance of paddy varieties to *Rhyzopertha dominica* (Bostrichidae: Coleoptera). Bull. Grain Technol. 15:121-122.

Noak, S and H. U. Schmidt. 1981. Studies on the effect of repellents and insecticides used for impregnation of packing materials for protection against stored product pests. Zeitschrift fur Angewandte Entomologie 92:202-212.

Novak, V. J. A. 1975. Insect Hormones, 2nd Edition. Chapman and Hall. London. 600 pp.

Obretenchev, D. 1983. The effect of low and high lethal temperatures in individual stages of sawtoothed grain beetle, *Oryzaephilus surinamensis* L. Rasteniey dnj Nauki 20:45-50.

Pande, Y. D. and K. Das. 1984. Relative abundance and extent of losses caused by insects to stored grain in Tripura. Bull. Grain Technol. 22:209-214.

Pande, G. P. and T. P. S. Teotia. 1978. Chemosterilization of *Sitotroga cerealella* Oliv. Effect of promising chemosterilants by oral administration and residual deposits to moths. Bull. Grain Technol. 16:193-198.

Pearson, N. K. and J. W. Sorrenson. 1970. Use of gaseous nitrogen for controlling stored product insects in cereal grains. Cereal Chem. 47:679-680.

Perlman, F. (ed.). 1977. Food allergens in immunological aspects of food. AVI Publ. Co., Westport, CT. 153 pp.

Pillaiyer, P. 1979. Requirements for long term storage of rice. Bull. Grain Technol. 17:228-232.

Piserov, V. and E. Shaaya. 1979. Effect of altosid on development of *Ephestia cautella*. pp. 33-40 in: Special Publ. Div. Sci. Publ. Bet Dagan, Israel, No. 140.

Prakash, A. 1982a. Varietal resistance of stored rice grains to *Rhyzopettha dominica* F. (Bostrichiidae: Coleoptera). Uttar Pradesh J. Zool. 2:89-92.

Prakash, A. 1982b. Varietal resistance of stored rice grains to *Sitophilus oryzae* L. (Curculionidae: Coleoptera). Bull. Grain Technol. 20:8-12.

Prakash, A. 1983. Factors affecting losses due to insects and their management in rice storage ecosystems. pp. 50-54 in: Proceedings of the National Symposium on Insect Ecology and Resource Management, U. P. Zool. Soc., Muzaffarnagar, India.

Prakash, A. and L. P. Kauraw. 1982a. Studies on insects, mites, and fungal infestatin in rural and farm stored rice in Orissa. J. Res. Assam Agric. Univ. 3:190-196.

Prakash, A. and L. P. Kauraw. 1982b. A package of practices for pest management in rice storage. Farmers and Parliament 17:15-16.

Prakash, A. and L. P. Kauraw. 1982c. Problems and quality control of paddy seed. Seeds Farms 8:13-16.

Prakash, A. and I. C. Pasalu. 1981. Evaluation of some synthetic insecticides as seed grain protectant against *Sitotroga cerealella* Oliv. (Gelechiidae: Lepidoptera). Oryza 18:130-131.

Prakash, A and J. Rao. 1983. Insect pests and their management in rice storage. Page 23 in: National Symp. Maximising and Stabilising Yields in Rainfed Rice Production System, Association of Rice Research Workers, Cuttack, India.

Prakash, A. and J. Rao. 1984. Wild sage leaves: a paddy grain protectant in storage. Oryza 212:209-213.

Prakash, A. and J. Rao. 1985a. Evaluation of certain fumigants as grain protectants in stored rice. Bull. Grain Technol. 23:111-117.

Prakash, A. and J. Rao. 1985b. Losses due to insects in stored rice. Bull. Grain Technol. 23:77-82.

Prakash, A. and J. Rao. 1986a. Evaluation of plant products as antifeedants against rice storage insects. pp. 201-205 in: Proceedings National Symp. Pesticide Residues and Environmental Pollution, Uttar Pradesh Zoological Society, Muzaffarnagar, India.

Prakash, A. and J. Rao. 1986b. Angoumois grain moth—a most serious pest of stored paddy. Bull. Grain Technol. 24:240-247.

Prakash, A. and J. Rao. 1987a. Damage due to *Oryzaephilus surinamensis* in stored milled rice. Bull. Grain Technol. 25:271-274.

Prakash, A. and J. Rao. 1987b. Use of chemicals as grain protectants in storage ecosystem and its consequence. Bull. Grain Technol. 25:65-69.

Prakash, A. and J. Rao. 1988. Use of radiant energy in postharvest entomology of rice. Bull. Grain Technol. 26: 238-243.

Prakash, A., I. C. Pasalu, and K. C. Mathur. 1979a. Ovicidal activity of *Eclipta alba* Hassk. (Compositae). Curr. Sci. 48:1090.

Prakash, A., I. C. Pasalu, and K. C. Mathur. 1979b. Development of *Sitotroga cerealella* Oliv. in relation to some morphological characters. J. Entomol. Res. 3:226-228.

Prakash, A., I. C. Pasalu, and K. C. Mathur. 1980a. Save stored paddy from insects. Indian Farming 23:45.

Prakash, A., I. C. Pasalu, and K. C. Mathur. 1980b. Evaluation of some plant products as grain protectants against insect pests of stored paddy. Bull. Grain Technol. 18:25-28.

Prakash, A., I. C. Pasalu, and K. C. Mathur. 1981a. Begonia leaves: a grain protectant in paddy storage. Bull. Grain Technol. 19:59-61.

Prakash, A., I. C. Pasalu, and K. C. Mathur. 1981b. Population studies on insect pests of stored rice in natural and controlled conditions. Oryza 18:90-92.

Prakash, A., I. C. Pasalu, and K. C. Mathur. 1981c. Plant products in insect pest management of stored grains (a review). Bull. Grain Technol. 19:213-219.

Prakash, A., I. C. Pasalu, and K. C. Mathur. 1981d. Problems of rice storage and insect pest management in rural India. pp. 14-19 in: Proc. of the Strategies of Pest Management, Entomol. Soc. India, New Delhi, India.

Prakash, A., I. C. Pasalu, and K. C. Mathur. 1982a. Development of *Sitotroga cerealella* Oliv. in relation to some stored grain fungi. Indian J. Entomol 44:298-300.

Prakash, A., I. C. Pasalu, and K. C. Mathur. 1982b. Evaluation of some plant products as grain protectants in paddy storage. Int. J. Entomol. 1:75-77.

Prakash, A., I. C. Pasalu, and K. C. Mathur. 1982c. Grain resistance to storage insects of rice (a review). Bull. Grain Technol. 20:124-133.

Prakash, A., I. C. Pasalu, and K. C. Mathur. 1983a. Use of bel (*Aegle marmelos*) leaf powder as a protectant against storage pests of paddy. Kerala J. Agric. Res. 21:79-80.

Prakash, A., I. C. Pasalu, and K. C. Mathur. 1983b. Satisfar: a paddy grain protectant in storage. Pesticides 17:24-25.

Prakash, A., I. C. Pasalu, and K. C. Mathur. 1983c. Varietal resistance of stored paddy grains to *Sitotroga cerealella* Oliv. Oryza 20:135-137.

Prakash, A, I. C. Pasalu, and J. Rao. 1984a. Insect infestations and their pest status in stored rice in Orissa. Trop. Stored Prod. Info. 47:15-20.

Prakash, A., I. C. Pasalu, and K. C. Mathur. 1984b. Allitin: a grain protectant in paddy storage. Indian J. Entomol. 46:376-378.

Prakash, A., J. Rao, and J. P. Kulshreshtha. 1986a. Evaluation of plant products for management of rice storage insects. pp. 23-28 in: Res. Plan. Work. Bot. Pest Control of Rice Based Croppintg Systems, International Rice Research Institute, Manila, Philippines.

Prakash, A., I. C. Pasalu, and K. C. Mathur. 1986b. Ovipositional and developmental response of lesser grain borer and rice weevil in relation to certain morphological grain characters of rice grains. Indian J. Agric. Res. 20:159-162.

Prakash, A., J. Rao, I. C. Pasalu, and K. C. Mathur. 1987. Rice storage and insect pest management. B. R. Publishing House, New Delhi, India. 337 pp.

Prakash, A., J. Rao, and S. N. Tewari. 1989a. Exploitation of natural plant products against pests and diseases in rice ecosystem. pp. 33-36 in: Proc. Symp. Growth, Dev. and Res. Cons., Nature Conservators, Muzaffarnagar, India.

Prakash, A., J. Rao, S. P. Gupta, and T. C. Binh. 1989b. Evaluation of certain plant products as paddy grain protectants against Angoumois grain moth. *Sitotroga cerealella* Oliv. J. Nature Conserv. 1:7-13.

Prakash, A., J. Rao, S. P. Gupta, and J. Behera. 1990. Evaluation of botanical pesticides as grain protectants against rice weevil. *Sitophilus oryzae* L. pp. 13-14 in: National Symp. Problems and Prospects of Botanical Pesticides in Integrated Pest Management, Indian Society of Tobacco Science, Rajahmundry, India.

Pruthi, H. S. and M. Singh. 1950. Pests of stored grains and their control. Indian J. Agric. Sci. (Special Number) 18:1-52.

Quinlan, J. K. and W. H. McGaughey. 1983. Fumigation of empty grain drying bins with chloropicrin, phosphine and liquid fumigant mixtures. J. Econ. Entomol. 76: 184-187.

Quereshi, Z. A., D. A. Wilbur, and R. B. Mills. 1970. Irradiation of early instars of Angoumois grain moth. J. Econ. Entomol. 63:1241-1247.

Rahman, M. and M. S. John. 1979. Effect of temperature on the development of *Corcyra cephalonica* Staint. (Lepidoptera: Pyralidae). Bangaladesh J. Zool. 7:95-99.

Rai, B. K. 1977. Control of storage pests of paddy by admixing with primiphos methyl. Bull. Grain Technol. 15:176-181.

Rai, B. K. and K. Croal. 1973. Chemical control of *Sitotroga cerealella* infesting stored paddy in bags. Indian J. Entomol. 35:57-62.

Rajan, A. P. A. and D. Dale. 1971. Studies on the use of antifeedants for protecting stored paddy from Angoumois grain moth, *Sitotroga cerealella* Oliv. Bull. Grain Technol. 15:123-125.

Rao, K., S. A. Jacob, and M. S. Mohan. 1972. Resistance of flexible packing materials to some important pests of stored products. Indian J. Entomol. 34:94-101.

Roberts, E. H. and F. M. Adballa. 1968. Effect of nitrogen on viability of seed. Ann. Bot. 32:97-127.

Rouse, P., L. M. Rolston, and C. Lincoln. 1958. Insects in farm stored rice. Bull. Arkansas Agric. Exp. Sta. No. 600. 25 pp.

Rout, G., B. Senapati, and T. Ahmed. 1976. Studies on the relative susceptibility of some high yielding varieties to rice weevil, *Sitophilus oryzae* (Curculioniidae: Coleoptera). Bull. Grain Technol. 20:137-139.

Russell, M. P. 1976. Resistance of commercial rice varieties to *Sitotroga cerealella* Oliv. (Lepidoptera: Gelechiidae). J. Stored Prod. Res. 12:105-109.

Salt, R. W. 1936. Studies on the freezing process in insects: Tech. Bull. Minn. Expt. Stat. No. 116. p 46.

Samajapati, J. N., A. Md. Rashid, and S. A. Md. Sheikh. 1984. Quantitative and qualitative losses in paddy and wheat in Bangladesh. Agric. Mech. Asia, Africa, Latin Am. 15:71-75.

Sardar, M. A. 1976. Susceptibility of certain varieties of stored rice to moth *Sitotroga cerealella* Oliv. Indian J. Entomol. 38:388-390.

Savitri, P. and S. C. Rao. 1976. Studies on the admixture of neem kernel powder with paddy in control of important storage pests of paddy. Andhra Agric. J. 23 137-143.

Saxena, S. C. and G. Mathur. 1981. Suppression of adult emergence on treated eggs of *Tribolium castaneum* (Herbst) by new synthesised disubstituted benzoyl-phenyl urea compounds. Curr. Sci. 50:336.

Schroeder, H. W. 1960. Infra-red drying of rough rice. II. Short grain type, Calrose and Caloro rice. Rice J. 63:6-8.

Schroeder, H. W. 1961. Infra-red drying of rough rice. III. Medium grain type Nato and Mangolia. Rice J. 64:11-12.

Simwat, G. S. and B. S. Chahal. 1970. Effect of temperature and relative humidity on the longevity of adults of *Tribolium castaneum* (Herbst) (Tenebrionidae: Coleoptera). Bull. Grain Technol. 8:107-110.

Singh, K., R. K. Agrawal, and P. K. Srivastava. 1978. Infestation of grain moth, *Sitotroga cerealella* Oliv. and maize weevil. *Sitophilus zeamais* Mots. on standing crops in the fields. Bull. Grain Technol. 16:125-127.

Sinha, R. N. and A. Campbell. 1975. Energy loss in stored grain by pest infestation. Canada Agric. 20:15-17.

Sinha, R. N. and W. E. Muir. 1973. Grain storage—part of a system. Avi Publ. Co., Westport, CT. 471 pp.

Sinha, R. N. and F. L. Watters. 1985. Insect pests of flour mills, grain elavators and feed mills and their control. Agric. Canada. Pub. No. 1776E, Ottawa. 290 pp.

Soegiarto, C. J., R. Sinaga, M. Hudaya, A. Nasroh, Harsono, Y. A. A. G. Bagas, M. Sidik, and H. Sukardi. 1981. Development of insect pest population in certain rice godowns in Java and hypothetical appraisal on some ecological influences. pp. 127-128 in: Proc. BIOTROP Symp. Pests of Stored Products. Bogor, Indonesia.

Srinath, D., A. N. Raghunathan, and S. K. Majumdar. 1976. Stored product insects as carrier of toxigenic fungi. Indian J. Entomol. 38:189-191.

Storey, C. L. 1977. Effect of low oxygen atmospheres on mortality of red and confused flour beetles. J. Econ. Entomol. 70:253-255.

Strong, R. H. and D. John. 1973. Comparative effectiveness of 15 insect growth regulators against several pests of stored products. J. Econ. Entomol. 66:1167-1173.

Su, C. F. H. and P. G. Mohny. 1974. Alkaliviscorram and other properties of starch of tropical rice. Agric. Biol. Chem. 39:811-817.

Sudharkar, T. R. and N. D. Pandey. 1983. Chemical factors in resistance of wheat, raw and parboiled rice to *Sitophilus oryzae* Linn. I. amino acid. Indian J. Entomol. 43:364-368.

Sukardi, T. 1981. The role of light traps in controlling pest of stored products. pp. 189-196 in: Proc. BIOTROP Symp. Pest of Stored Products. Bogor, Indonesia.

Suzuki, H. and B. O. Juliano. 1975. Alkaliviscogram and other properties of starch of tropical rice-resistance factor. Agric. Biol. Chem. 3:811-817.

Thorpe, D. R. and W. W. George. 1963. Nitrofurane as chemosterilant of stored grain insect. J. Econ. Entomol. 56:404-407.

Tilton, E. W. and R. R. Cogburn. 1965. Phosphine fumigation of rough rice in upright bins. Rice J. 68:8-9.

Tilton, E. W. and H. W. Schroeder. 1963. The effect of infrared radiation on immature insects in kernels of rough rice. J. Econ. Entomol. 56:227-230.

Tilton, E. W. and H. H. Vardell. 1982. Combination of microwave and partial vacuum for control of four stored product insects in stored grains. J. Georgia Entomol. Soc. 17:106-112.

Tilton, E. W., E. B. Wendell, and R. R. Cogburn. 1966. Effect of gamma radiation on *Rhyzopertha dominica, Sitophilus oryzae, Tribolium castaneum* and *Lasioderma serricorne*. J. Econ. Entomol. 59:1363-1368.

Tyler, P. S, R. W. Taylor, and D. P. Prees. 1983. Insect resistance to phosphine fumigation in food warehouses in Bangladesh. Int. Pest Control 25:10-13.

Vick, K. W., J. A. Coffelt, and M. A. Sullivan. 1978. Disruption of pheromone communications in Angoumois grain moth with synthetic sex pheromone. Environ. Entomol. 7:528-532.

Vick, K. W., J. Kvenberge, J. A Coffelt, and C. Steward. 1979. Investigations of sex phermone traps from simultaneous detection of Indian meal moth and Angoumois grain moth. J. Econ. Entomol. 72:245-249.

Vir, S. 1981. AI 3-63392 A highly effective JHa for control of *Tribolum castaneum* (Herbst) (Coleoptera: Tenebrionidae). Curr. Sci. 50:502-503.

Yadav, T. D. 1983. Seed fumigation as an aspect to seed storage technology. Seed Research 11:240-247.

Younes, M. W. F. 1980. Sterilization studies on the adult stage of almond moth, *Cadra cautella* Walk. (Lepidoptera: Phycitidae). Res. Bull. Faculty Agric. Ain Shams Univ., Egypt, No. 1393. 14 pp.

Zhang, X. and S. H. Zhao. 1983. Relative efficacy of some seed protectants of plant origin against lesser grain borer. J. Grain Storage (China) 1:1-8.

20

Future Systems for Grain Storage and Transportation

Gerald W. Moore

Economic Development Board of Cabinet, Government of Manitoba, Winnipeg, Manitoba, Canada

SUMMARY

Canada's system for storing and transporting grain was first developed on the basis of the necessarily-short distances that grain could be hauled by horses and wagons. The cross-continental railway system was constructed to move grain from many local delivery points to the eastern and western seaports for the export market. Beginning in 1970, the storage and transport system rapidly evolved to fewer, high-throughput delivery points along a reduced railway system. As the storage capacity of the commercial system in Canada rapidly reduced, the responsibility for the storing and quality maintenance of the crop shifted to the farmers.

1. INTRODUCTION

Nearly 60% of the protein consumed by human beings in the world's developing countries comes from grains. In the developed nations, the diet has higher levels of protein from meats, fruits, and vegetables; in these countries, grains are vitally important to the production of meat, milk, and eggs. It is estimated that over 50% of the grains grown in the world are fed to animals. The dependence of the world's population on grains for food, makes it clear that grain handling and storage technology must maintain the quality of grain products and ensure safe storage of the product from point of production to ultimate consumption.

The costs incurred to plant and harvest grain crops in any country of the world are significant. In some countries, the costs are in terms of intensive labor, while in others, the costs are in terms of high capital investment. When we invest labor and capital resources into the production of grain, it follows that effective and efficient storing and handling systems are required to maintain the quality of the grain and grain products. The system must also be capable of moving grain by a number of modes, including trucks, railways, ships, and (in some developing parts of the world) in bags moved by animal drawn wagons.

When the requirements of the world's grain marketing and handling system are studied from the structural, logistical, economic, social, and political points of view, the summary of findings shows a challenging and complex set of issues. For example, some countries do not have the capital resources to build the type of system required to store grain and maintain its quality. If the product is not consumed within a short period of time, it is likely to be lost to rodents, insect infestations, fungal sprouting spoilage or, equally as bad, left in the field to rot. Other countries spend billions of dollars on subsidies to support the incomes of their grain farmers, while at the same time moving product into the world market at extremely depressed prices.

The reason for mentioning the economic and political aspects of the business is to ensure we appreciate the significant impact that government policy at home and abroad can have on the ultimate physical design of a grain handling system.

The future design of grain handling and transportation systems, of which storage is a key component, needs study in all parts of the world. The Canadian system has been studied, studied, and restudied, and offices and libraries are filled with the literature of this on-going work. It is not an easy industry to address, for it involves a great many issues, numerous participants, and the views of all players do not always coincide.

It would take a book in itself, or perhaps more accurately many books to cover the world grain handling and transporting systems. Therefore, I have elected to emphasize the Canadian system, its benefits and problems, and in some instances, to compare it to other countries' systems where a comparison provides an opportunity for rethinking directions for the future. Such comparisons are not necessarily appropriate where climatic conditions, geography, population density, and other factors may be considerably different.

2. THE HISTORY OF WESTERN CANADA'S COUNTRY GRAIN SYSTEM

Canada's prairie grain handling system has grown out of the fiber of the people, organizations, and governments that have participated with and in its development. The story begins in 1884 when the first prairie wheat was shipped from Thunder Bay, Ontario to eastern Canada. This movement was made possible by the completion (in 1883) of the Canadian Pacific Railway from the head of Lake Superior to Winnipeg.

At the turn of the century, about 0.6 M ha of western Canada's virgin prairie land had been turned into fields of grain. Today, there are about 35 M ha under cultivation. The expansion of cultivated land and the opening up of western Canada went hand in hand with the development of the railways. This history remains with us today, for the railways are a key component of our grain handling system and they will remain so in the future. Little communities and their country grain handling systems sprang up at 10–15 km intervals along the railway tracks. Why so close together? The main reason was undoubtedly the limited capability of the farmers of the day to haul grain from the farm with their teams of horses and wagon-boxes.

The first commercial grain handling and storage facilities to be constructed were warehouses designed to accept farmers' bagged-grain. These facilities were constructed adjacent to the railway lines. This location accommodated effective loading of railcars when market demand required the product to be moved forward. By 1890, the country elevators were being designed and constructed along the railways to handle grain in bulk. These elevators were fundamentally simple in design. The grain is elevated by bucket elevator and distributed to a number of high bins. The grain flows by gravity from these bins into a weighing scale and then to the same or to a shipping bucket-elevator from which it flows by gravity to adjacent railcars. Since wood is abundant in Canada and historically is the cheapest construction option, it was the primary material used in the construction of these facilities. Today that has changed and concrete and steel have replaced wood as the main construction material.

3. THE CHANGING COUNTRY ELEVATOR SYSTEM

As the prairie grain production base grew in western Canada, there were over 295 companies and individuals operating country elevators by 1916. In the 1950s, there were some 5000 elevator buildings in operation on the prairies but, in the late 1950s and early 1960s, the grain handling system started to consolidate. Today, there are only eight companies left that are of a significant size in the business. The bottom line is that all companies must eliminate uneconomic units to remain in business.

The change appears dramatic when we review it in 10 year periods since 1960 (Table 1). Capacity of the country elevator system has reduced by over 50% within the last two decades. From about 1930 to 1960, however, the entire system changed very little.

Table 1 The Changes in Country Elevator Numbers and Storage Capacity

Year	Licensed elevators	Elevator capacity (Mt)
1960	5,296	12.9
1970	4,971	14.2
1980	3,324	8.7
1991	1,539	7.2

4. WESTERN CANADA'S CURRENT GRAIN HANDLING SYSTEM

The western Canadian grain handling and storage system in 1991 was comprised of 1539 elevators with a storage capacity of 7.2 Mt. These facilities were managed by about 1500 managers located at 990 different stations on the prairies. The system provides the following service:

1. Annual grain production on the prairies is 45–50 Mt. (Production is 56% wheat, 27% barley, 9% canola, and 8% flax, rye, and oats).
2. Annual movement of grains and oilseeds is 30–34 Mt. (This is grain moved by the elevator system to both domestic and export markets).
3. Annual exports out of the country are 25–30 Mt.
4. Average amount of grain handled per elevator is 22,000 t.
5. Ratio of average amount handled per year to storage capacity (referred to as turn-over ratio) is 4.5.

The Canadian system today is relatively small on a plant by plant basis. For the bulk-grain handling system to become more efficient, considerable changes will be needed in the future. There are two main reasons for increasing the size and capability of each plant in the system: 1) the cost per unit of storage and throughput, and 2) the ability to load long strings of railcars within a short period of time. Because the costs of handling and transportation are paid by farmers (except for a transportation subsidy paid by the Canadian Government), the service industries are concerned about their costs getting out of line. Farmers have the option to load their own railcars and bypass the elevator system.

5. COUNTRY ELEVATOR CONSTRUCTION COSTS

Country elevator costs have escalated dramatically over the past decade. This is one of the reasons that has caused the decline in the number of companies and elevators. Companies are concerned that the costs may become too high for the service provided. Should this become the case, farmers will look for other more efficient methods of handling their grain. While the elevators being built today are larger and have greater handling capability, the costs of these plants would be 6 or 7 times previous costs, even if we were to design them along the lines of those of 20 years ago.

To maintain service fees for handling grain at reasonable levels, it is essential that the annual throughput of the elevators increase significantly. Increased volume is necessary to cover higher operating costs as well as increased capital costs. Elevators being built today are primarily constructed of slip-form concrete. Storage capacities are in the 10,000–12,000 t range. These plants are being designed with approximately 30 separate bins with sizes of 100–1500 t.

The cost of these facilities when the rail trackage service and land development are included, run about Cdn $360–400/t of storage. Therefore a plant with about 10,000 t of storage will cost a minimum of Cdn $3.6 million. To justify this type of investment on a sound economic basis the facility should handle 100,000 t of grain annually. These figures are based on a business operation that should maintain the service fees at reasonable levels because these fees are paid by farmers. The railways will play a major role in the success of this type of elevator operation. Long strings of railcars (i.e., about 50) are required to allow the elevator to function in an effective and efficient manner. The move towards this consolidation is needed to keep in balance the costs for handling, storage, and transportation from the farm to the ultimate market.

6. THE RAILWAY NETWORK

Canada's prairie grain system is serviced by approximately 29,000 km of rail line; there are 28,000 railcars in the fleet, including 26,000 covered hopper cars. The remaining 2000 or so cars are restored boxcars that are used on light service rail lines. Over half of the export grain moves from the prairies to the ports of Vancouver and Prince Rupert on the west coast of Canada. The major portion of the remaining balance moves east to the head of the Great Lakes (Thunder Bay) and then by lake boat to the St. Lawrence Seaway. In recent years, this route is becoming less used because of the added costs of transfers between the different modes of transportation (i.e., railway to terminal, to lake boat, to transfer elevator, to ocean going vessel). A small amount of grain moves through the port of Churchill located on Hudson Bay. This is Canada's arctic port and it only has open water 3 months of the year.

The western Canadian grain system moves about 7 railcars per week from an average country elevator. This kind of performance does not auger well for efficiencies in the rail system. A few years ago, the Grain Transportation Agency, which is responsible for administering the Western Grain Transportation Act, pointed out that the average turn around time on the hopper car fleet was about 20 days for all lines and all ports.

The hopper car fleet loses a number of cars every year through normal wear and tear. If efficiencies are not improved dramatically, then as many as 9900 new covered hopper cars may be required in order to maintain grain movement volume. This would cost of about 800 million dollars Cdn. However, a more likely alternative is that the railcar fleet will be used more efficiently. The Canadian National Railway, one of the two main rail carriers, have stated that if one day can be taken off the cycle time of railcars in the grain fleet, it would be the same as adding 700 cars to the fleet. A one day improvement would be a saving of nearly 60 million dollars. The rail fleet moving coal in Canada, for example, has a cycle time of about 14 days. This is a bench mark against which the future grain movement system can measure itself, albeit that the conditions are not completely comparable.

When new grain markets began to open up in the 1970s, the Canadian rail system was not in a position to accommodate the market. The government acted by purchasing hopper cars for grain service; today, about 70% of the grain car fleet are government-purchased hopper cars.

Railway development in western Canada was closely governed by national transportation legislation. The Crows Nest Pass Rates, which were legislated in 1897, set the rate at which the railways could charge for hauling grain and prevented the railways from increasing their tariff for hauling grain until new legislation was implemented in 1984. The rail freight rates charged under the "Crow Rate" legislation varied only by distance and were approximately 0.3 cents per tonne-km. This kept the shipping costs to the farmers at very reasonable levels, with the average shipping tariff being Cdn $4.85/t. While this appeared good to the farmers, the downside was the deterioration of the railway fleet. In the 1950s, the railways began to lose money hauling grain; this prompted an immediate lack of interest by the railways to haul grain. As a result, new capital was not put into the grain service and the grain fleet and the grain-dependent branch lines began to deteriorate. In turn, grain elevator companies viewed this action very carefully when making their investment decisions, because to invest in upgraded or new elevator equipment on a rail line in poor condition is a very questionable investment. In such cases, it is likely that the railway company wants to abandon the line completely but requires clearance under legislation to allow this to happen.

In summary, the railways are a key service cost in the Canadian grain system. On average, the cost of moving a tonne of grain by rail from the prairies to export terminals is about Cdn $34/t; on the other hand, elevator tariffs are about Cdn $7–8/t. The railways prime responsibility in the system is to get the grain to ports for export or to domestic end users. To provide efficiencies in grain handling and transportation in the future, it is essential that the grain companies and the railways work together in designing a system that is efficient and economic. This means that the facilities at the point of origination and the facilities at the point of destination must accommodate fast loading and unloading of large volumes of grain.

7. THE TERMINAL ELEVATOR SYSTEM

Terminal elevators are located in Vancouver and Prince Rupert on the west coast and in Thunder Bay at the head of the Great Lakes for grain going east. There are also large transfer elevators located in other locations on the Great Lakes and on the lower St. Lawrence River. The main function of the transfer elevators on the lower St. Lawrence is to transfer grain from lake ships to ocean-going vessels. This system has generally served the Canadian industry very well. The throughput capability of these plants has seldom, if ever, been tested on a sustained basis considering all ports; however, as volumes moving west continue to increase, some increased capacity will likely be required on the west coast.

The terminal elevators at Prince Rupert, Vancouver, and Thunder Bay are the prime facilities for cleaning all exported grain. In these plants, grains and oilseeds are cleaned to export standards to ensure Canada retains the highest quality levels in purity of sample for our overseas customers. Grain cleaning technology has improved significantly, large banks of cylinder machines used in tandem with large rotary-screen machines have allowed terminal operators to design systems that have a one-pass cleaning and reclaim operation. In other words, as grain is unloaded from a railcar, it can be elevated, cleaned, and weighed for loading into a ship. These plants are also equipped with grain dryers. The dryers are not used every year because in most years prairie grain is harvested in a dry condition.

The west coast throughput capacity is about 20 Mt, 14–15 Mt through Vancouver and 5 Mt through Prince Rupert on an annual basis. In the east, the terminals in Thunder Bay have the capability to handle over 20 Mt/yr and Churchill has a capacity of approximately 800,000 t for the 3 mo which it operates. If Canada's long-term goal is to export 35 Mt/yr, or a more optimistic 40 Mt/yr by the year 2000, the terminal handling capacities are sufficient.

Improvements in Canada's grains and oilseed export capabilities are related directly to efficient railcar movements, supported by efficient country loadings coupled with the terminals' handling and cleaning capacity and the timely arrival of vessels. What then should a system of the future look like?

8. FUTURE SYSTEMS FOR GRAIN STORAGE AND TRANSPORTATION

8.1. Storage

Canadian farmers have adequate storage on the farm to put an average year's entire crop under shelter. There are some exceptions, such as an extremely large crop may cause farmers to store grain in open piles on the ground. This is normally a very small percentage of the crop, likely less than 1–2% in years when it does occur. Thus, we have on-farm storage in the range of 50 Mt and another 14.5 Mt of storage in primary elevators, terminals, transfer houses, and processing facilities.

During the Second World War, extensive temporary storage facilities were built along side country elevators. These were normally low-cost large annexes to allow grain to be moved off the farm. Today, the general view is that on-farm storage can keep the costs of storing grain below that of any commercial enterprise close to terminals or country elevators, even though some have tried.

The volume of Canadian grain exports are twice as big now (and could be 2.5 times as big) as the storage capacity of the off-farm system through which they move to market. Commercial grain-storage capacity in Canada has been decreasing, mainly at the country elevators and it is likely this trend will continue into the immediate future. Farm storage is cheaper to build than the kind of storage that goes with high-volume handling plants. This means that farmers must be responsive to market opportunities and move their grain from the farm into the system when it is required to fill market sales. The trucking service in Canada is more than adequate to haul grain from the farms to country elevators.

Condominium storage is being tried by some farmers when replacement of older on-farm facilities is needed. This is not a new idea and it has been tried in areas of the United States. The concept is that a group of individuals form a company or business structure to share a storage facility. The plant may be constructed by a private investor and rented out, or it may be constructed by a group of individuals with each shareholder owning a defined portion of the space. In Canada, such facilities are being placed adjacent to railways to avoid an extra truck movement. An additional improvement is that the new farm storage bins are of steel construction, frequently with hopper bottoms to allow for easy unloading, and are often fit with aeration or drying equipment to enhance the safe keeping of the grain.

8.2. Other Countries' Storage

In the United States, there is more than adequate room in their commercial facilities to accommodate annual grain export requirements. The statistics show that the US system can store 2.5 times the average amount of grain exported every year. The US system is also used to store government stocks of grain. These are stocks bought by the government from farmers at the loan rate, which is the system used by the United States to support farm cash flow needs when markets are slow and prices depressed.

Australia also has a much larger commercial storage system than Canada relative to their annual exports. The Australian commercial system can store virtually all of the country's annual production. Climatic conditions in that country are far more conducive to damage by insects than in Canada. The warmer weather in Australia does not give grain handlers the same opportunity to kill pests by moving grain, as do the cold winters in Canada. The extended harvest period in Australia compared with Canada, does not pressure their farmers to get the crop off and under cover like the coming of a prairie winter does for Canadian farmers. Thus, Australian farmers have very little on farm storage and they move their grain directly into the commercial storage system. This is not a practical or economic approach for the future Canadian system mainly because of the capital investment in the existing farm storage.

Another major world exporter of grains, Argentina, also has more than adequate commercial storage to accept deliveries of their farmers' entire annual production. Maintenance of quality and the need to provide adequate service to their customers has pressured Argentina to increase their commercial storage capability over the past decade.

European grain storage varies from country to country but, for the European Community as a whole, there is more than enough commercial storage to cover average annual grain exports.

Relative to other major world exporters, Canada has the unique position of holding the largest percentage of its harvested crop in storage on farms. Looking ahead, I can see the trend continuing toward even less storage in the commercial system, particularly in country elevators. The farm storage system will provide the main base from which grain will be withdrawn to meet market demand. This would also support more domestic processing of our field crops should the community decide to embark on more value-added products for the market. This system will place a continued responsibility on engineers and scientists to ensure farm operators have the equipment and pest control support to retain the quality of the crop in store on the farm.

8.3. Handling

Over the last 30 years, the western Canadian grain industry has lost 3718 primary country elevators at 1074 locations to which grain can be delivered and well over 3 Mt of storage capacity adjacent to rail lines. The word "lost" is not the appropriate word because the consolidation was mostly planned. Consolidation was absolutely essential to ensure the country elevator system could meet its responsibility to handle and store grains and oilseed, at costs considered reasonable by farmers. This change did not take place because of grain companies' whims or desires to close elevators. It came (and is still coming about) because the cost of newer facilities and machinery requires a large production area to supply the grain necessary to pay for these large new facilities. To do other than consolidate and re-organize the system would place the grain companies in a position where tariff levels would far exceed what is deemed reasonable for the service provided. Even today some farmers by-pass the country elevator system by directly loading railcars to avoid paying the country elevators' service charges. This activity suggests that more on track storage or service facilities are not the answer to the Canadian grain handling system of the future. Today's farm trucks and handling equipment allow farmers more flexibility. Other alternative grain handling systems are considered in the following subsections.

8.4. High-Throughput Elevators

Consider a system of elevators designed to handle 100,000 t of grain each, on an annual basis, with a storage capacity of 10,000 t. These are the plants, or are close in design to some of the plants, that are currently being built to serve the Canadian industry of the future. If export and domestic movement targets for the future are 35 Mt, then about 350 facilities of this type would be required. This would mean a reduction in plants of about 1200 from what we currently have and a reduction in on-track storage from 7.2 Mt to 3.5 Mt.

Such a system would require an entirely new information network. A new onus would be placed on farmers to deliver when called upon. For example, if the on-track elevator system would operate efficiently at 70% capacity, then there would be less than 3 wk visible supply of grain in the network. Currently, when planning the logistics of transportation, there is a 3 wk lead time before the actual movement takes place. Some dramatic changes would be essential for improved information flow between various participants in the system to accommodate the reduction in grain stored in facilities on-track.

For those industry players responsible for the maintenance of grain quality while in store, the emphasis would be placed at the farm to ensure that when grain was delivered into the commercial system the quality was satisfactory. The rate of grain movement through this type of system would limit the opportunity for grain to be infested or go out of condition within the plant itself. The equipment and materials used in construction would provide an excellent protective environment for grain while briefly in store on-track. Grain cleaning and drying equipment is also likely to play a role in the services of these plants.

8.5. Elevators on Wheels

There is an opportunity when rethinking a storage and grain transportation system to look at ideas that may not be conventional but may effectively and efficiently get the job done. One such concept—called an "elevator on wheels" —has been discussed at some length but to date a prototype has not been tried.

Instead of building elevators, a grain company would purchase a string of railcars. For example, if it costs $3.6 million to construct a 10,000 t modern high-throughput prairie-elevator and if a hopper car costs $85,000, then the money could be used to purchase 42 covered hopper railcars. This would provide approximately 4000 t of mobile storage. The major advantage is that such equipment is not limited to one location but can be moved to any station where storage is required; hence the name "elevator on wheels."

For this concept to work, some changes would be required in the railcar allocation system. The Canadian Wheat Board is responsible for the exports of all wheat and most of the barley out of Canada, and for all domestic wheat used for human consumption within Canada. In an average year, approximately 80% of the grain handled by the system is controlled and marketed by the Board. A company purchasing railcars (instead of building an elevator) would require changes in the allocation system that would allow it to retain its cars outside of the entire railway car pooling program. If it could not do this, it would lose to competitors the advantage of the purchased cars, by giving up its share of the system cars because of the percentage allocation. In other words, under the present system the privately owned railcars would be considered part of the total fleet, and the shipping percent allocation would remove any advantage of having purchased these cars.

To make this approach possible, it would also require that earnings could be made on loading and storing grain placed in these cars. A loading device that could transfer grain from trucks to long strings of railcars would be needed. The device should be powered to move along the string of cars to take best advantage of minimum track requirements and it should be capable of weighing the grain accurately.

This concept makes a lot of economic sense but provides a great challenge to policy makers and further challenges to the engineering community. The key for this type of system to be successful relies on farmers having adequate grain storage on the farm. The question then arises: is a transfer house needed between farm storage and the railcar? The "elevator on wheels" is just an idea to compare with other alternatives for future development of the industry. The logistics may be too difficult to implement.

8.6. Product-Specific Plants

Specialty plants can be constructed to handle only one product. For example, canola crushers build plants to receive the seed required for their processing operations. Such facilities require less segregation of the incoming product. The result is that bins can be far fewer, much bigger, much cheaper, and the overall design is much more cost effective. This idea would have different plants designed for wheat, barley, flax, and so on. However, this may limit service opportunities too much to be a viable competitive option.

8.7. Inland Terminals

In addition to country elevators and the new 10,000 t concrete plants described earlier, there are several facilities on the prairies that can handle large volumes of grain, perhaps in the neighborhood of 10 or more times that of our newly-designed country plants. Many of these plants were constructed by the Federal Government in the early decades of this century. The plants are made of slip-form concrete and are in good physical repair today. These plants have storage capacities of 50,000–100,000 t and have cleaning equipment installed that can clean grain to export standards.

In the last few years, these plants have been purchased by grain companies and are used daily in providing service to Canadian farmers. To build one of these plants today equipped with cleaning and drying capacity would probably cost 75–100 million dollars Cdn. This estimate is based on an approximation of the differences between these plants and the latest terminal built at Prince Rupert, BC. I don't see any more plants of this size being constructed on the Prairies.

8.8. Producer Cars

Farmers have the opportunity to load railcars directly from the farm and there has been considerable pressure from farmers (and some of their organizations) to allow a far greater percentage of the grain to be moved in this manner. The challenge with this approach is somewhat comparable to the "elevator on wheels" concept mentioned earlier. To ensure that efficiencies are created in the rail segment of the system, farmers would have to be able to load long strings of cars within a short period of time. For example, it might make sense to be able to load 50 cars in one string within a 24 h period. This would require considerable organization on the part of farmers and the industry. The advantage of this concept is clear; far less capital is required in equipment but a reorganization of operational and quality control procedures would be required.

8.9. Summary of the Canadian System

The Canadian system will evolve with something old, something current, and something new. The trends toward improved on-farm storage, consolidation of the country elevator system, the development of higher-throughput slip-form concrete plants will continue; no more transfer elevators will be constructed, the terminal plants will be refurbished, and it is likely that some type of new terminal facility will be constructed on the west coast.

The design and function of a countries' grain system is dependent upon the environment within which it has to work. By environment, I mean climate, distance to market, available modes of transportation, requirements for domestic consumption, requirements for export markets, national policy, international policy, quality control, size of the production area and production yields; overriding all these points (and many others) is a sense of economic efficiency to find a way to prevent people on this globe from going to bed hungry at night.

9. BUILDING A NEW SYSTEM FOR SAUDI ARABIA

A few years ago, I had the opportunity with some colleagues to visit Saudi Arabia. Most of our time in the country was spent with people involved in wheat production, grain handling, flour milling, and the feed milling industry.

In the early to mid-1970s the government of the country implemented a program to become selfsufficient in the production of wheat. What has happened since then is an almost unbelievable story. Vast tracks of the open desert were irrigated and the red tan color of the sand was converted to the healthy rich green of growing wheat. Yields in excess of 6 t/h were obtained. The new production capability required a system to handle, store, and maintain the quality of the inventory. The plants designed are slip-form concrete with many bins and large storage capacities. The facilities were designed to accept farm deliveries by semi-trailer trucks directly from the farms.

There were a lot of forward-looking ideas used in the planning of the Saudi system that deserve mention. They constructed the grain handling and storage systems next to the flour mills and the feed mills. The mills are in different buildings on site but are reached easily by conveyor. The mills also have some storage to service the processing within each plant but the bulk of the commercial storage is in the grain handling and storage facilities. These "integrated" plants were constructed near large cities (e.g., Riyadh, Buraydah, and Dammam) which facilitates easier transportation and distribution of the processed products to a major portion of the country's population. Wheat delivered to the plants was based on 9% moisture and discounts for moisture in excess of that level were applied. Any wheat harvested above 12% was not allowed into the storage system. In such a warm country these regulatory guidelines are sensible.

The plants themselves are well built, very secure buildings that will service the Saudi industry for years to come. Certainly, the Saudi system is a considerable contrast to what we have in Canada but their goals and objectives are different than Canada's. They have had an opportunity to build a new system without having to consider an older system and have taken good advantage of that opportunity.

10. NEW QUALITY CONTROL TECHNOLOGY

10.1. Automatic Grading

Researchers at the University of Manitoba in Winnipeg and in other parts of the world have been working on a new grain-grading technology that will allow millers and handlers to measure the quality of wheat so as to reduce the need for milling and baking tests. The technology will provide the information in minutes if their research is successful. The capability of grading grain in minutes will impact on the types of facilities built and how they are operated. New technology can guide us and we must be ever mindful of it.

10.2. Electron Beam Irradiation

Researchers in many countries of the world are working on disinfestation of cereal grains and their ground products by nuclear radiation and other physical methods (see Chapter 11). Many of these technologies would be suitable for grain in transit before it goes into storage.

The kind of work being done in these emerging areas is essential to the well being of the world grains industry. If the world population is going to reach 10 billion in another half century, then we have a lot of work to do. The key to our success will be based on how well we support new research efforts throughout the world and how well we apply the findings of that research in our commercial systems.

11. CONCLUSION

The world grain industry's greatest challenge still lies ahead. The challenge is for farmers to produce crops necessary to feed the forecasted world population growth. The challenge is for governments to find an economic solution for sharing the food with all peoples around the world, one that includes not letting the production industry decline in the process. The challenge is to ensure we can deliver quality grain products for human and animal consumption in an effective and efficient manner.

Index